REVIEWS in ~~~~~~~~~~ GY
and GEOCHEMISTRY

Volume 40   2000

— SULFATE MINERALS —

Crystallography, Geochemistry, and Environmental Significance

Editors:

Charles N. Alpers   *U.S. Geological Survey, Sacramento, California*

John L. Jambor   *Leslie Research and Consulting and Department of Earth and Ocean Sciences, University of British Columbia, Vancouver*

D. Kirk Nordstrom   *U.S. Geological Survey, Boulder, Colorado*

FRONT-COVER PHOTOGRAPH: Melanterite [(Fe^{2+},Zn,Cu)SO$_4$·7H$_2$O] stalagmites (blue and blue-green) from the Mattie deposit, Richmond tunnel, Iron Mountain, California. Yellow and brown minerals are products of melanterite oxidation and dehydration. Field of view is 10 × 7 cm.

BACK-COVER PHOTOGRAPHS: a . Melanterite stalactite (blue-green, with orange, Fe^{3+}-rich water inside) from the Mattie deposit, Richmond tunnel, Iron Mountain, California. Photo copyright: Bud Eagle 1992. Field of view: 5 × 8 cm. **b.** Copiapite stalagmite (yellow) with halotrichite needles (white) from the Richmond mine, Iron Mountain, California. Field of view: 25 × 40 cm. **c.** Voltaite from the Richmond Mine, Iron Mountain, California. Canadian Museum of Nature specimen CMNOC 3525. Photo courtesy of George Robinson. Width of specimen: 8 mm. **d.** Coquimbite (pale purple) with copiapite, and voltaite from the Richmond mine, Iron Mountain, California. Canadian Museum of Nature specimen CMNOC 3526. Photo courtesy of George Robinson. Width of specimen as viewed is 2 cm.

Series Editor: **Paul H. Ribbe**
Virginia Polytechnic Institute and State University
Blacksburg, Virginia

MINERALOGICAL SOCIETY of AMERICA

Washington, DC

REVIEWS IN MINERALOGY
AND GEOCHEMISTRY

(Formerly: REVIEWS IN MINERALOGY)

ISSN 1529-6466

Volume 40

Sulfate Minerals: Crystallography, Geochemistry, and Environmental Significance

ISBN 0-939950-52-9

** This volume is the second of a series of review volumes published jointly under the banner of the Mineralogical Society of America and the Geochemical Society. The newly titled *Reviews in Mineralogy and Geochemistry* has been numbered contiguously with the previous series, *Reviews in Mineralogy*.

Additional copies of this volume as well as others in this series may be obtained at moderate cost from:

THE MINERALOGICAL SOCIETY OF AMERICA
1015 EIGHTEENTH STREET, NW, SUITE 601
WASHINGTON, DC 20036 U.S.A.

— SULFATE MINERALS —

Crystallography, Geochemistry, and Environmental Significance

40   *Reviews in Mineralogy and Geochemistry*   40

FOREWORD

The review chapters in this volume were the basis for a short course on sulfate minerals sponsored by the Mineralogical Society of America (MSA) November 11-12, 2000 in Tahoe City, California, prior to the Annual Meeting of MSA, the Geological Society of America, and other associated societies in nearby Reno, Nevada. The conveners of the course (and editors of this volume of *Reviews in Mineralogy and Geochemistry*), Charles Alpers, John Jambor, and Kirk Nordstrom, also organized related topical sessions at the GSA meeting on sulfate minerals in both hydrothermal and low-temperature environments. A special issue of a journal yet to be identified is being planned for the publication of research articles based on several of these presentations. Taken together, the MSA short course and the related GSA sessions represent the most comprehensive grouping of technical meetings ever devoted to sufate minerals.

ERRATA (if any) may be found at the MSA website together with access to the black-and-white and some color representations of many **STRUCTURE DRAWINGS** in Chapter 1:

http://www.minsocam.org
(click on *Rev Mineral Geochem* entry)

Paul H. Ribbe, Series Editor
Virginia Tech, Blacksburg

December 9, 2000

PREFACE

Sulfate is an abundant and ubiquitous component of Earth's lithosphere and hydrosphere. Sulfate minerals represent an important component of our mineral economy, the pollution problems in our air and water, the technology for alleviating pollution, and the natural processes that affect the land we utilize. Vast quantities of gypsum are consumed in the manufacture of wallboard, and calcium sulfates are also used in sculpture in the forms of alabaster (gypsum) and papier-maché (bassanite). For centuries, Al-sulfate minerals, or "alums," have been used in the tanning and dyeing industries, and these sulfate minerals have also been a minor source of aluminum metal. Barite is used extensively in the petroleum industry as a weighting agent during drilling, and celestine (also known as "celestite") is a primary source of strontium for the ceramics, metallurgical, glass, and television face-plate industries. Jarosite is a major waste product of the hydrometallurgical processing of zinc ores and is used in agriculture to reduce alkalinity in soils. At many mining sites, the extraction and processing of coal or metal-sulfide ores (largely for gold, silver, copper, lead, and zinc) produce waste materials that generate acid-sulfate waters rich in heavy metals, commonly leading to contamination of water and sediment. Concentrated waters associated with mine wastes may precipitate a variety of metal-sulfate minerals upon evaporation, oxidation, or neutralization. Some of these sulfate minerals are soluble and store metals and acidity only temporarily, whereas others are insoluble and improve water quality by removing metals from the water column.

There is considerable scientific interest in the mineralogy and geochemistry of sulfate minerals in both high-temperature (igneous and hydrothermal) and low-temperature (weathering and evaporite) environments. The physical scale of processes affected by aqueous sulfate and associated minerals spans from submicroscopic reactions at mineral–water interfaces to global issues of oceanic cycling and mass balance, and even to extraterrestrial applications in the exploration of other planets and their satellites. In mineral exploration, minerals of the alunite-jarosite supergroup are recognized as key components of the advanced argillic (acid-sulfate) hydrothermal alteration assemblage, and supergene sulfate minerals can be useful guides to primary sulfide deposits. The role of soluble sulfate minerals formed from acid mine drainage (and its natural equivalent, acid rock drainage) in the storage and release of potentially toxic metals associated with wet–dry climatic cycles (on annual or other time scales) is increasingly appreciated in environmental studies of mineral deposits and of waste materials from mining and mineral processing.

This volume compiles and synthesizes current information on sulfate minerals from a variety of perspectives, including crystallography, geochemical properties, geological environments of formation, thermodynamic stability relations, kinetics of formation and dissolution, and environmental aspects. The first two chapters cover crystallography (Chapter 1) and spectroscopy (Chapter 2). Environments with alkali and alkaline earth sulfates are described in the next three chapters, on evaporites (Chapter 3), barite-celestine deposits (Chapter 4), and the kinetics of precipitation and dissolution of gypsum, barite, and celestine (Chapter 5). Acidic environments are the theme for the next four chapters, which cover soluble metal salts from sulfide oxidation (Chapter 6), iron and aluminum hydroxysulfates (Chapter 7), jarosites in hydrometallugy (Chapter 8), and alunite-jarosite crystallography, thermodynamics, and geochronology (Chapter 9). The next two chapters discuss thermodynamic modeling of sulfate systems from the perspectives of predicting sulfate-mineral solubilities in waters covering a wide range in composition and concentration (Chapter 10) and predicting interactions between sulfate solid solutions and aqueous solutions (Chapter 11). The concluding chapter on stable-isotope systematics (Chapter 12) discusses the utility of sulfate minerals in understanding the geological and geochemical processes in both high- and low-temperature environments, and in unraveling the past evolution of natural systems through paleoclimate studies.

We thank the authors for their comprehensive and timely efforts, and for their cooperation with our various requests regarding consistency of format and nomenclature. Special thanks are due to the numerous scientists who provided peer reviews, which substantially improved the content of the chapters. This volume would not have been possible without the usual magic touch and extreme patience of Paul H. Ribbe, Series Editor for *Reviews in Mineralogy and Geochemistry*. Finally, we thank our families for their support and understanding during the past several months.

Charles N. Alpers
U.S. Geological Survey, Sacramento

John L. Jambor
Leslie Research and Consulting &
Department of Earth and Ocean Sciences,
University of British Columbia,
Vancouver

D. Kirk Nordstrom
U.S. Geological Survey, Boulder

October 6, 2000

RiMG Volume 40.   SULFATE MINERALS:
Crystallography, Geochemistry, and Environmental Significance

Table of Contents

1   The Crystal Chemistry of Sulfate Minerals

Frank C. Hawthorne, Sergey V. Krivovichev, Peter C. Burns

2   X-ray and Vibrational Spectroscopy of Sulfate in Earth Materials

Satish C. B. Myneni

3   Sulfate Minerals in Evaporite Deposits

Ronald J. Spencer

4   Barite–Celestine Geochemistry and Environments of Formation

Jeffrey S. Hanor

5   Precipitation and Dissolution of Alkaline Earth Sulfates: Kinetics and Surface Energy

A. Hina and G. H. Nancollas

6   Metal-sulfate Salts from Sulfide Mineral Oxidation

John L. Jambor, D. Kirk Nordstrom, Charles N. Alpers

7   Iron and Aluminum Hydroxysulfates from Acid Sulfate Waters

J. M. Bigham, D. Kirk Nordstrom

8   Jarosites and Their Application in Hydrometallurgy

John E. Dutrizac, John L. Jambor

⑨   Alunite-Jarosite Crystallography, Thermodynamics, and Geochronology

R. E. Stoffregen, C. N. Alpers, J. L. Jambor

⑩   Solid-Solution Solubilities and Thermodynamics: Sulfates, Carbonates and Halides

Pierre Glynn

11   Predicting Sulfate-Mineral Solubility in Concentrated Waters

Carol Ptacek, David Blowes

12   Stable Isotope Systematics of Sulfate Minerals

Robert R. Seal, II, Charles N. Alpers, Robert O. Rye

1   The Crystal Chemistry of Sulfate Minerals

Frank C. Hawthorne

Department of Geological Sciences
University of Manitoba
Winnipeg, Manitoba, Canada R3T 2N2

Sergey V. Krivovichev and Peter C. Burns

Department of Civil Engineering and Geological Sciences
156 Fitzpatrick Hall, University of Notre Dame
Notre Dame, Indiana 46556

INTRODUCTION

Sulfur is the fifteenth most abundant element in the continental crust of the Earth (260 ppm), and the sixth most abundant element in seawater (885 ppm). Sulfur (atomic number 16) has the ground-state electronic structure $[Ne]3s^2 3p^4$, and is the first of the group VIB elements in the periodic table (S, Se, Te, Po). In minerals, sulfur can occur in the formal valence states S^{2-}, S^0, S^{4+}, and S^{6+}, corresponding to the *sulfide* minerals, *native sulfur*, the *sulfite* minerals, and the *sulfate* minerals. In the sulfide minerals, S^{2-} functions as a simple anion (e.g. $CuFeS_2$, chalcopyrite) and as a compound S_2 anion (e.g. FeS_2, pyrite). In the sulfosalts, S^{2-} functions as a component of a complex anion (e.g. AsS_3 in tennantite, $Cu_{12}As_4S_{13}$). In the sulfite minerals, S^{4+} has four valence electrons available for chemical bonding, and occurs in triangular pyramidal coordination with O. In the sulfate minerals, S^{6+} has six valence electrons available for bonding, and occurs in tetrahedral coordination with O. In addition, there are the *thiosulfate* minerals, in which S is in the hexavalent state, but is coordinated by three O^{2-} anions and one S^{2-} anion. Chemists frequently write the thiosulfate group as S_2O_3; however, we write it as SO_3S to emphasize that one of the S atoms is an anion and is involved in a tetrahedral group. Although the focus of this chapter is the sulfate minerals, we will deal also with the sulfite and thiosulfate minerals, as they occur in the same types of geochemical environments.

CHEMICAL BONDING

We adopt a pragmatic approach to matters involving chemical bonding. We use bond-valence theory (Brown 1981) and its developments (Hawthorne 1985a, 1994, 1997) to consider structure topology and hierarchical classification of structures, and we use molecular-orbital theory to consider aspects of structural energetics, stereochemistry and spectroscopy of sulfate minerals. These approaches are compatible, as bond-valence theory can be considered as a simple form of molecular-orbital theory (Burdett and Hawthorne 1993; Hawthorne 1994).

STEREOCHEMISTRY OF SULFATE TETRAHEDRA IN MINERALS

The variation of S–Ø (Ø: unspecified anion) distances and Ø–S–Ø angles is of great interest for several reasons:

(1) mean bond-length and empirical cation and anion radii play a very important role in systematizing chemical and physical properties of crystals;

(2) variations in individual bond-lengths give insight into the stereochemical behavior of structures, particularly with regard to the factors affecting structure stability;

1529-6466/00/0040-0001$10.00

(3) there is a range of stereochemical variation beyond which a specific oxyanion or cation-coordination polyhedron is not stable; it is of use to know this range, both for assessing the stability of hypothetical structures (calculated by DLS [Distance Least-Squares] refinement) and for assessing, the accuracy of experimentally determined structures.

Here, we examine the variation in S–Ø distances in minerals and review previous work on polyhedral distortions in SO_4 tetrahedra. Data for 206 (SO_4) tetrahedra were taken from 112 refined crystal structures with $R \leq 6.5\%$ (for a definition of R, see Ladd and Palmer 1994) and standard deviations of ≤ 0.005 Å on S–O bond-lengths. Ionic radii are from Shannon (1976).

The symbol $\langle \ \rangle$ signifies a mean value; thus $\langle S–O \rangle$ is the mean value of the four S–O distances in a sulfate tetrahedron. A grand $\langle \ \rangle$ value is the mean of a series of mean values; thus a grand $\langle S–O \rangle$ distance is the mean of a series of $\langle S–O \rangle$ distances.

Variation in $\langle S–Ø \rangle$ distances

The variation in $\langle S–O \rangle$ distances is shown in Figure 1a. The grand $\langle S–O \rangle$ distance is

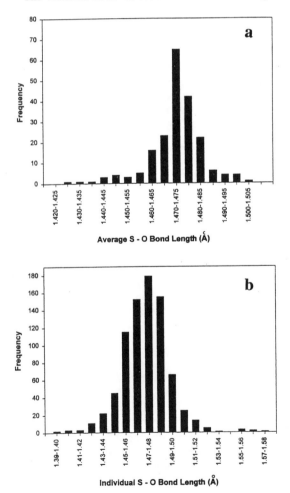

Figure 1. (a) Variation in mean S–O distance (n = 206), $\langle S–O \rangle$, in minerals containing (SO_4) tetrahedra; (b) variation in S–O distance (n = 824) in minerals containing SO_4 tetrahedra.

1.473 Å, the minimum and maximum $\langle$S–O$\rangle$ distances are 1.430 and 1.501 Å, respectively, and the range of variation is 0.071 Å. Shannon (1976) lists the radius of [4]S as 0.12 Å and the radius of [3.25]O^{2-} as 1.36 Å (assuming a mean anion-coordination number of 3.25); the sum of the constituent radii is thus $0.12 + 1.360 = 1.48$ Å, in accord with the grand $\langle$S–O$\rangle$ distance of 1.473 Å. Brown and Shannon (1973) showed that variation in $\langle$M–O$\rangle$ (M = cation) distance can correlate with bond-length distortion Δ ($= \Sigma[l(o)–l(m)]/l(m)$; l(o) = observed bond-length, l(m) = mean bond-length) when the bond-valence curve of the constituent species shows a strong curvature, and when the range of distortion is large. There is no significant correlation between $\langle$S–O$\rangle$ and Δ; this is in accord with the small degree of curvature of the bond-valence curve for S–O given by Brown (1981).

Variation in S–O distances

The variation in S–O distances is shown in Figure 1b; the grand mean S–O distance is 1.473 Å. The minimum and maximum observed S–O distances are 1.394 and 1.578 Å, respectively, and the range of variation is 0.173 Å; the distribution follows a skewed Gaussian curve. According to the bond-valence curve for S^{6+}–O (the universal curve for second-row elements) from Brown (1981), the range of variation in S–O bond-valence is 1.13–1.92 vu (valence units).

General polyhedral distortion in sulfate minerals

As noted above, there is no general correlation between $\langle$S–O$\rangle$ and bond-length distortion. However, as pointed out by Griffen and Ribbe (1979), there are two ways in which polyhedra may distort (i.e. depart from their holosymmetric geometry): (1) the central cation may displace from its central position (bond-length distortion); (2) the anions may displace from their ideal positions (edge-length distortion); Griffen and Ribbe (1979) designated these two descriptions as BLDP (Bond-Length Distortion Parameter) and ELDP (Edge-Length Distortion Parameter), respectively. Figure 2 shows the variation in both these parameters for the second-, third-, and fourth-period (non-transition) elements in tetrahedral coordination. Some general features of interest (Griffen and Ribbe 1979) are apparent from Figure 2:

(1) A BLDP value of zero only occurs for an ELDP value of zero; presuming that ELDP is a measure also of the O–T–O angle variation (T is a tetrahedrally coordinated cation), this is in accord with the idea that variation in orbital hybridization (associated with variation in O–T–O angles) must accompany variation in bond-length.

(2) Large ranges of BLDP are associated with small ranges of ELDP, and vice versa. The variation in mean ELDP correlates very strongly with the grand mean tetrahedral-edge length for each period (Fig. 3).

Griffen and Ribbe (1979) suggested that the smaller the tetrahedrally coordinated cation, the more the tetrahedron of anions resists edge-length distortion because the anions are in contact, whereas the intrinsic size of the interstice is larger than the cation which can easily vary its cation–oxygen distances by 'rattling' within the tetrahedron.

S^{6+} $\leftrightarrow$ T^{n+} substitution in minerals and its influence on (S,T)–O distances

In most sulfate structures, there is no cation substitution at the tetrahedrally coordinated S site. However, there are a few exceptions. Vergasovaite, $Cu_3O[(SO_4)((Mo,S)O_4)]$, is the only known mineral with Mo^{6+} $\leftrightarrow$ S^{6+} substitution. In this mineral, one of two tetrahedrally coordinated sites is occupied by both S and Mo, with a Mo:S ratio of about 3:1 (Berlepsch et al. 1999). Recently, Krivovichev et al. (unpublished results) studied a crystal of vergasovaite with the formula $Cu_3O[(SO_4)(MoO_4)]$, i.e. one site

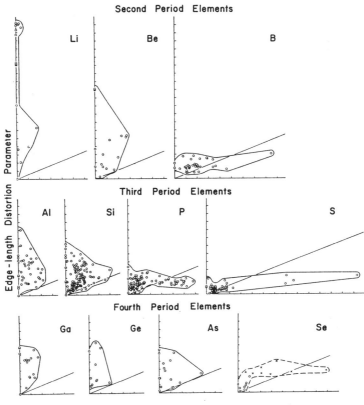

Figure 2. Variation in BLDP (Bond-Length Distortion Parameter) and ELDP (Edge-Length Distortion Parameter) for second-, third-, and fourth-period non-transition elements in tetrahedral coordination by oxygen (after Griffen and Ribbe 1979).

contained only Mo. The $\langle$(Mo,S)–O$\rangle$ distance reported by Berlepsch et al. (1999) is 1.715 Å, whereas the crystal studied by Krivovichev et al. has a $\langle$Mo–O$\rangle$ of 1.753 Å.

There are two minerals with proven $Cr^{6+} \leftrightarrow S^{6+}$ substitution. In hashemite, $Ba[(Cr,S)O_4]$ (Duesler and Foord 1986; Pasero and Davoli 1987), the Cr:S ratio is about 9:1, although a broad range of Cr^{6+}–S^{6+} substitution is known for its synthetic analogs. Carlosruizite, $K_6(Na,K)_4Na_6Mg_{10}\{(Se,S,Cr)O_4\}_{12}(IO_3)_{12}(H_2O)_{12}$ (Konnert et al. 1994), a Se analog of fuenzalidaite, is an interesting example of a structure in which Se^{6+}, S^{6+}, and Cr^{6+} occur at the same tetrahedrally coordinated site. There are twelve tetrahedrally coordinated cations in the unit formula of carlosruizite: 6.4 Se, 4.4 S, and 1.2 Cr. The mean $\langle$(Se,S,Cr)–O$\rangle$ is 1.588 Å, whereas the expected [4]Se–O distance is about 1.63 Å. Heterovalent $P^{5+} \leftrightarrow S^{6+}$ isomorphism has been observed in woodhouseite, $Ca[Al_3(OH)_6((S_{0.5}P_{0.5})O_4)_2]$ (Kato 1977) and svanbergite, $Sr[Al_3(OH)_6((S_{0.5}P_{0.5})O_4)_2]$ (Kato and Miura 1977). Cahill et al. (submitted) determined the structure of mitryaevaite, $[Al_5(PO_4)_2\{(P,S)O_3(O,OH)\}_2F_2(OH)_2(H_2O)_8](H_2O)_{6.56}$, with $P^{5+} \leftrightarrow S^{6+}$ substitution at one of the tetrahedrally coordinated sites. The charge-balance mechanism in this structure can be described by the scheme: $[P^{5+}O_3(OH)]^{2-} \leftrightarrow [SO_4]^{2-}$. Beudantite, $Pb(Fe,Al)_3[(As,S)O_4]_2(OH)_6$ (Szymanski 1988), is an example of $As^{5+} \leftrightarrow S^{6+}$ heterovalent

Table 1. S–O and S–OH bond lengths in SO₃(OH) groups in sulfate minerals.

Mineral	Formula	S–O (Å)	<S–O> (Å)	S–OH (Å)	Reference
matteuccite	Na(HSO₄)(H₂O)	1.446, 1.447, 1.458	1.450	1.578	Catti et al. (1975)
mercallite	K(HSO₄)	1.441, 1.443, 1.467	1.450	1.574	Payan and Haser (1976)
		1.438, 1.444, 1.475	1.452	1.565	Payan and Haser (1976)
letovicite	(NH₄)₃H(SO₄)₂*	1.455, 1.460, 1.461	1.459	1.529	Leclaire et al. (1985)

* in letovicite, the H atom is disordered over two positions

substitution; the tetrahedrally coordinated site contains 53.5% As and 46.5% S, with a ⟨(As,S)–O⟩ distance of 1.605 Å.

Hydrogen bonding in sulfate minerals

In general, each O atom in an (SO₄) tetrahedron receives ~1.5 vu (valence units) from the bond involving the S atom. The O–H bond of an hydroxyl group commonly has a bond valence of ~0.8 vu, and the acceptor anion receives ~0.2 vu. However, the bonds may be more symmetric, depending on local bond-valence requirements, and symmetrical hydrogen bonds, although rare, do occur: donor and acceptor anions each receive ~0.5 vu. The O atoms of (SO₄) tetrahedra commonly accept hydrogen bonds, consistent with their bond-valence requirements. Mercallite, K(HSO₄) (Payan and Haser 1976), may contain an example of a symmetrical hydrogen bond. However, the H atom involved has a site-occupancy of 0.5, and the O···H distances are 1.66 and 1.85 Å instead of 1.2–1.3 Å that is typical for symmetrical hydrogen bonds.

The sulfate minerals mercallite, letovicite, and matteuccite involve sulfate

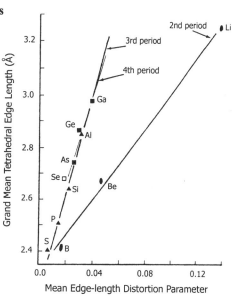

Figure 3. Variation in grand mean tetrahedral edge-length with mean ELDP for the second-, third- and fourth-period elements of the periodic table; redrawn from Griffen and Ribbe (1979).

tetrahedra that contain an hydroxyl group, i.e. acid-sulfate groups: SO₃(OH). Owing to bond-valence requirements, the S–OH bond must be significantly longer than is typical for S–O bonds. The S–OH bond should have ~1.2 vu, which corresponds to an S–OH distance of ~1.58 Å (according to the curve of Brown 1981). The remaining S–O bonds in the tetrahedron must shorten to provide sufficient bond-valence to the cation. Table 1 lists S–O and S–OH bond lengths in mercallite, letovicite, and matteuccite. The observed bond lengths are in good agreement with bond-valence considerations. Kemnitz et al. (1996a) noted that S–O bonds are ~0.1 Å shorter than S–OH bonds in alkali-metal and ammonium hydrogen sulfates. The cation polyhedra, hydrogen-bonding systems, and other crystal-chemical aspects of alkali-metal hydrogen sulfates are reviewed by Kemnitz et al. (1996a,b).

Table 2. Thiosulfate compounds.

Formula			Cation/S_2O_3 Ratio	Reference
Na$_2$		S$_2$O$_3$	2	Teng et al. (1984)
Na$_2$		S$_2$O$_3$ (H$_2$O)$_{2/3}$	2	Hesse et al. (1993)
Na$_2$		S$_2$O$_3$ (H$_2$O)$_5$	2	Uraz and Armağan (1977)
Na$_3$	Au^{1+}	(S$_2$O$_3$)$_2$ (H$_2$O)$_2$	2	Ruben et al. (1974)
[1]K$_2$		S$_3$O$_6$	1	Christidis and Rentzeperis (1985)
K$_2$	S^{2+}	(S$_2$O$_3$)$_2$ (H$_2$O)$_{3/2}$	1.5	Marøy (1971)
Mg		S$_2$O$_3$ (H$_2$O)$_6$	1	Elerman et al. (1983)
Ni^{2+}		S$_2$O$_3$ (H$_2$O)$_6$	1	Elerman et al. (1978)
Cd		S$_2$O$_3$ (H$_2$O)$_2$	1	Baggio et al. (1997)
Ba		S$_2$O$_3$ H$_2$O	1	Aka et al. (1980)
Ba	Te^{2+}	(S$_2$O$_3$)$_2$ (H$_2$O)$_3$	1	Gjerrestad and Marøy (1973)

[1] contains O$_3$S^{6+}–S^{2-}–S^{6+}O$_3$ group

STEREOCHEMISTRY OF THIOSULFATE TETRAHEDRA

The thiosulfate group consists of a central S^{6+} cation surrounded by four anions, three O^{2-} and one S^{2-}, arranged at the vertices of a tetrahedron (Fig. 4); this group is conventionally written as S$_2$O$_3$, but it is much more informative to write it as (S^{6+}O$_3$S^{2-}). As sidpietersite, Pb$^{2+}_4$(S^{6+}O$_3$S^{2-})O$_2$(OH)$_2$ (Cooper and Hawthorne 1999), is the only thiosulfate mineral for which structural data are available, here we examine stereochemical variations in some synthetic thiosulfate compounds. These are listed in Table 2; all are refined to R indices between 2 and 9%, and provide us with 23 distinct thiosulfate groups.

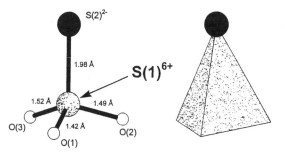

Figure 4. The thiosulfate group in sidpietersite shown as atoms (left) and as a tetrahedron (right) in which the S^{2-} anion is identified as a large black circle, S^{6+} is the random-dot-shaded circle, and O^{2-} are the unshaded circles.

Variation in ⟨S–Ø⟩ distances

The variation in ⟨S^{6+}–O⟩ distances in thiosulfate structures is shown in Figure 5a. The grand ⟨S^{6+}–O⟩ distance in thiosulfate compounds is 1.459 Å, the minimum and maximum ⟨S^{6+}–O⟩ distances are 1.429 and 1.476 Å, respectively, and the range of variation is 0.047 Å. The grand ⟨S^{6+}–S^{2-}⟩ distance is 2.038 Å, the minimum and maximum S^{6+}–S^{2-} distances are 1.965 and 2.123 Å, respectively (Fig. 5b), and the range of variation is 0.158 Å. The grand ⟨S^{6+}–O⟩ distance of 1.459 Å is fairly close to the sum of the empirical radii of Shannon (1976) for [4]S^{6+} and [3.25]O^{2-}: 0.12 + 1.360 = 1.480 Å ([3.25] is an average value for the coordination of O in sulfate minerals). However, there is no reason that the ⟨S^{6+}–O⟩ distance should be equal to the sum of the constituent radii as there is another ligand in the

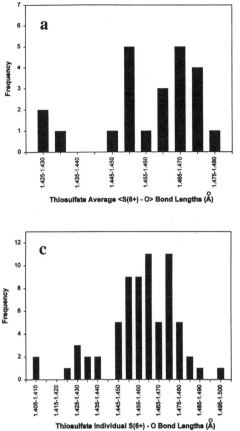

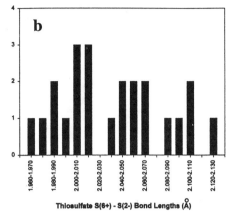

Figure 5. (a) Variation in $\langle S\text{–}O\rangle$ distances in structures containing $(S^{6+}O_3S^{2-})$ tetrahedra (n = 23); (b) variation in $S^{6+}\text{–}S^{2-}$ distances in structures containing $(S^{6+}O_3S^{2-})$ tetrahedra (n = 23); (c) variation in S–O distances in structures containing $(S^{6+}O_3S^{2-})$ tetrahedra (n = 69).

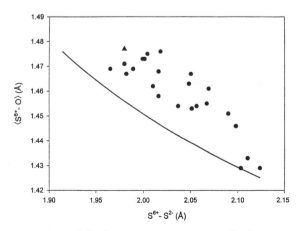

Figure 6. The $\langle S^{6+}\text{–}O\rangle$ distance as a function of $S^{6+}\text{–}S^{2-}$ distance in thiosulfate groups from the synthetic compounds listed in Table 1 and sidpietersite (black triangle). The curve shows the relation for ideal bond-valence satisfaction at the central S^{6+} cation.

coordination polyhedron: S^{2-}. According to the valence-sum rule (Brown 1981; Hawthorne 1994, 1997), there should be an inverse relation between $\langle S^{6+}-O \rangle$ and $S^{6+}-S^{2-}$ in thiosulfate groups; Figure 6 shows this to be the case. The data for sidpietersite lie on the general trend, albeit near the lower end of the range of $S^{6+}-S^{2-}$ distances. The curve in Figure 6 shows ideal agreement with the valence-sum rule calculated using the observed $\langle S-O \rangle$ distances. There is reasonable agreement between the slope of the curve and that of the data, but the curve is displaced approximately 0.02 Å below the trend of the data. The bond-valence curve for $S^{6+}-O$ can be considered as reliable. The value reported by Brese and O'Keeffe (1991) is the same as that reported by Brown (1981), and gives close-to-ideal bond-valence sums at S^{6+} in (SO_4) groups. The problem, therefore, is likely to be with the $S^{6+}-S^{2-}$ bond-valence curve of Brese and O'Keeffe (1991).

Variation in S–O distances

The variation in $S^{6+}-O$ distances is shown in Figure 5c; the grand $\langle S-O \rangle$ distance is 1.459 Å. The minimum and maximum observed S–O distances are 1.408 and 1.497 Å, respectively, and the range of variation is 0.089 Å. According to the bond-valence curve for $S^{6+}-O$ (the universal curve for second-row elements) from Brown (1981), the range of variation in S–O bond-valence is 1.41 to 1.83 vu. According to the bond-valence curve for $S^{6+}-S^{2-}$ (from Brese and O'Keeffe 1991), the variation in $S^{6+}-S^{2-}$ bond-valence is 0.87 to 1.33 vu.

The formal valences of S in the thiosulfate group

On the basis of XANES spectroscopy, Vairavamurthy et al. (1993) proposed that the valences of S in thiosulfate are 5^+ and 1^- instead of the conventionally assigned 6^+ and 2^-. This proposal may be tested in a very simple manner using structural data for sidpietersite and bond-valence theory. The bond-valence incident at the O anions of the thiosulfate group, from the rest of the structure, is calculated independent of the formal charges of S in the structure. The valence-sum rule (Brown 1981) states that the sum of the bond valences at an atom is equal to the magnitude of the formal valence of that atom. Hence the difference between the formal valence of oxygen (O^{2-}) and the sum of the bond valence incident at that oxygen (exclusive of the S–O bond) gives the bond valence of the S–O bond. Summing the bond valences thus calculated for the S–O bonds gives a value of 4.83 vu. If the formal charge on S^{6+} were actually 5^+, the bond-valence for the $S^{6+}-S^{2-}$ bond would be 0.17 vu. This result seems unlikely from several perspectives. First, the four long $Pb-S^{2-}$ bonds would be required to supply 0.83 vu, at a mean $Pb^{2+}-S^{2-}$ bond-valence of 0.21 vu. This requirement is not in accord with the incident bond-valence sums around the Pb sites, as it would require a bond-valence sum around one specific Pb site of ~2.36 vu. Second, it is unlikely that the thiosulfate group would be a prominent complex in aqueous solutions at a range of pH (and Eh) values if it were defined by such a weak $S^{6+}-S^{2-}$ bond. Third, it seems intuitively unlikely that a group involving such a weak bond would occur with such reproducible stereochemistry in a range of structures. Thus it does not seem possible that the formal valences of S in the thiosulfate group are 5^+ and 1^-.

STEREOCHEMISTRY OF FLUOROSULFATE TETRAHEDRA

The fluorosulfate group consists of a central S^{6+} cation surrounded by four anions, three O^{2-} and one F^-, arranged at the vertices of a tetrahedron: $(SO_3F)^-$. Reederite-(Y), ideally $Na_{15}Y_2(CO_3)_9(SO_3F)Cl$ (Grice et al. 1995), is the only known fluorosulfate mineral, and fortunately, structural data are available. Here, we examine stereochemical variations in 16 synthetic fluorosulfate compounds. These are listed in Table 3; all are refined to R indices between 2 and 9%, and provide us with 27 distinct fluorosulfate groups.

Table 3. Fluorosulfate compounds.

Cs	Sb^{5+}	$(SO_3F)_6$	Zhang et al. (1996)	(H_3O)	(SO_3F)	Mootz & Bartmann (1991)
Cs$_2$	Pt	$(SO_3F)_6$	Zhang et al. (1996)	H	(SO_3F)	Bartmann & Mootz (1990)
Cs	Au	$(SO_3F)_4$	Zhang et al. (1996)	I$_2$	$(SO_3F)_2$	Birchall et al. (1990)
Cs	H	$(SO_3F)_2$	Zhang et al. (1996)	Se$_{10}$	$(SO_3F)_2$	Collins et al. (1986)
Cs		(SO_3F)	Zhang et al. (1996)	(S_4N_4)	$(SO_3F)_2$	Gillespie et al. (1981a)
Ir	$(CO)_3$	$(SO_3F)_6$	Wang et al. (1996)	(S_6N_4)	$(SO_3F)_2$	Gillespie et al. (1981b)
Pd	$(CO)_2$	$(SO_3F)_2$	Wang et al. (1994)	Li	(SO_3F)	Zak & Kosicka (1978)
Sn		$(SO_3F)_2$	Adams et al. (1991)	$(XeF)_2(AsF_6)$	(SO_3F)	Gillespie et al. (1977)
Au$_2$		$(SO_3F)_6$	Willner et al. (1991)	(XeF)	(SO_3F)	Bartlett et al. (1972)
				NH$_4$	(SO_3F)	O'Sullivan et al. (1970)

Variation in S–Ø distances

The variations in S–O and S–F distances are shown in Figure 7. The $\langle S–O \rangle$ distance is 1.428 Å and the $\langle S–F \rangle$ distance is 1.540 Å. These values, and the distribution within Figure 7, are what one expects from bond-valence considerations. As the formal charge of F is 1$^-$, the S–F bond-valence must be ≤1 vu; hence the mean bond-valence of the S–O bonds must be ≥ $(6 - 1)/3$, i.e. ≥ 1.67 vu. This value translates into a maximum $\langle S–O \rangle$ distance of 1.440 Å, in accord with the observed $\langle S–O \rangle$ distance of 1.428 Å for SO$_3$F groups; this value also compares with the grand $\langle S–O \rangle$ value of 1.473 Å for SO$_4$ groups. For the bond-valence requirements at the central cation to be satisfied, one would expect an inverse relation between the S–F and $\langle S–O \rangle$ distances; however, this is not the case, as the data show only random scatter. According to the bond-valence curve for S^{6+}–O (Brown 1981), the range of S–O bond-valence is 1.33–2.56 vu. The latter value is not physically realistic; the maximum possible bond-valence of an S–O bond is 2.0 vu, which translates into a minimum possible S–O distance of 1.38 Å. Inspection of Figure 7 shows that five S–O values are less than this minimum possible distance, and hence must be in error.

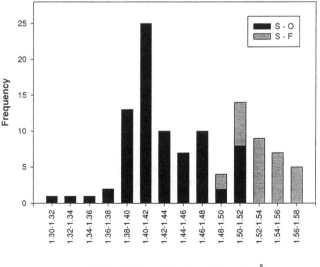

Individual S -O and S - F bond-lengths (Å)

Figure 7. Histogram of S–O and S–F distances in fluorosulfate structures.

MOLECULAR-ORBITAL STUDIES OF SO₄ POLYHEDRA

Molecular-orbital (MO) calculations have been used by theoretical chemists for many years, primarily to predict geometries, energetics, and stabilities of molecules. MO methods are based on quantum mechanics, and range from empirical and semi-empirical methods, which include an experimentally determined component, to *ab initio* methods, which include no experimentally determined parameters (for a summary of these methods, see Tossel and Vaughan 1992). MO methods have been applied with considerable success to the study of small molecules, and available computational sophistication and power have permitted application of MO methods to mineralogically relevant problems for the past two decades (Gibbs 1982).

Sulfate minerals have been the topic of a considerable number of MO calculations. To date, all MO calculations for these structures have been done using molecular clusters that are designed to be an approximation of the local environment in a structure; many long-range effects in a periodic structure are ignored by such calculations. The calculations for sulfate minerals involve various MO methods [extended Hückel (Bartell et al. 1970; Gibbs et al. 1972), Hartree-Fock self-consistent field (HF SCF) (Lindsay and Gibbs 1988; Ramondo et al. 1991; Sliznev and Solomonik 1996; Hill et al. 1997), density functional theory (DFT) (McKee 1996a,b; Wang et al. 1999)] applied to small clusters that model both individual (SO₄) polyhedra, and larger clusters that allow examination of such structural aspects as S–O–S bond angles, S–O bond lengths, and polyhedral linkages.

The role of 3d-orbitals in bonding in sulfates

Cruickshank (1961) stressed the role of 3d-orbitals in bonds between second-row elements and oxygen or nitrogen. Later, this hypothesis in its original form was criticized by Bartell et al. (1970) and finally rejected [see Cruickshank (1985) for an extensive review]. However, *ab initio* studies show that it is essential to include d-functions in the basis sets used for calculations involving third-row elements, and that the 3d-functions strengthen (and shorten) the S–O bonds significantly (Cruickshank 1985; Cruickshank and Eisenstein 1985, 1987).

Stability of the (SO₄)²⁻ tetrahedron

Calculations have shown that an isolated $(SO_4)^{2-}$ tetrahedron is not stable (Boldyrev and Simons 1994): additional ions or molecules are essential to stabilize the tetrahedron (McKee 1996a,b; Wang et al. 1999). Four (H_2O) molecules and $(SO_4)^{2-}$ form relatively stable $(SO_4(H_2O)_4)^{2-}$ anions in the gas phase (Blades and Kebarle 1994). The equilibrium geometry of the $(SO_4(H_2O)_4)^{2-}$ cluster has been calculated by McKee (1996a,b) using DFT. The S–O distances obtained were 1.521 and 1.531 Å, values that are longer than those observed in crystals.

(H₂SO₄) and (H₂S₂O₇) clusters: prediction of equilibrium geometry

Lindsay and Gibbs (1988) and Hill et al. (1997) studied the minimum-energy geometries of (H_2SO_4) and $(H_2S_2O_7)$ clusters (Fig. 8). These clusters contain three different types of oxygen atoms: O_{nbr}: non-bridging oxygen atoms bonded to S atoms only; O_{br}: bridging oxygen atoms bonded to two S atoms in the $(H_2S_2O_7)$ cluster; OH: oxygen atoms bonded to both S and H atoms. Table 4 gives a comparison of the geometric parameters calculated by Lindsay and Gibbs (1988) and Hill et al. (1997), and the parameters observed in crystal structures of H_2SO_4 (Kemnitz et al. 1996c) and $H_2S_2O_7$ (Hoenle 1991). Data in Table 4 show good agreement between calculated and observed geometries. Note that S–OH distances are ~0.1 Å longer than S–O_{nbr} distances, consistent with the findings of Kemnitz et al. (1996a) for alkali-metal hydrogen sulfates.

Figure 8. (H_2SO_4) and $(H_2S_2O_7)$ clusters studied by the HF SCF method by Lindsay and Gibbs (1988) and Hill et al. (1997).

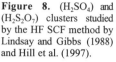
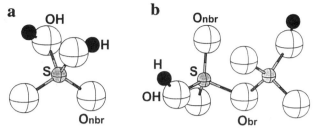

Table 4. Geometric parameters (Å and °) for H_2SO_4 and $H_2S_2O_7$ clusters: comparison of observed and calculated geometries.

Parameter	MO [6–31G**][1]	MO [6–311++G**][2]	Crystal[3]
	H_2SO_4		
S–O	1.411	1.411	1.426
S–OH	1.569	1.557	1.537
HO–S–OH	101.8	not given	105.2
HO–S–O	107.4	not given	110.7, 105.9
O–S–O	123.5	not given	117.9
	$H_2S_2O_7$		
S–O_{br}	1.623	1.609	1.618
S–O_{nbr}	1.405	1.397	1.412
S–OH	1.553	1.547	1.524
S–O_{br}–S	125.6	not given	122.1
O_{br}–S–OH	95.8	not given	102.8
O_{br}–S–O_{nbr}	108.1	not given	105.6
O_{nbr}–S–O_{nbr}	123.3	not given	122.2
O_{nbr}–S–OH	109.1	not given	109.2

[1] Lindsay and Gibbs (1988); [2] Hill et al. (1997); [3] H_2SO_4 (Kemnitz et al. 1996a,b), $H_2S_2O_7$ (Hoenle 1991) (only mean bond lengths and angles are given)

Lindsay and Gibbs (1988) calculated the deformation energies of the bridging S–O_{br}–S angle in $H_2S_2O_7$. All geometric parameters of the cluster were fixed, and single-point energy calculations were done at 20° intervals for S–O_{br}–S from 100 to 180°. Lindsay and Gibbs (1988) observed an energy minimum at $\theta = 125.6°$, with ΔE = -64.500 kJmol^{-1}. For comparison, the analogous parameters (θ_{min} and ΔE) for silicates and phosphates obtained by O'Keeffe et al. (1985) are 141.0° and -13.600 kJmol^{-1}, and 145.0° and -9.373 kJmol^{-1}, respectively. This is in good agreement with experimental data on synthetic polysulfates that show a narrower range of bridging S–O–S angles than phosphates or silicates. For this reason, Lindsay and Gibbs (1988) suggested that sulfate glasses are unlikely to adopt structures based on polymerized sulfate tetrahedra.

Bond angles in $(SO_4)^{2-}$ tetrahedra

Observed stereochemistries in (TO_4) groups indicate that there is considerable variation in O–T–O angles in crystals. Molecular orbitals can be envisaged as combinations of atomic orbitals *via* overlap, and hence the resulting bond-overlap populations are expected to be affected by variation in O–T–O angles, an idea that was confirmed for silicates by

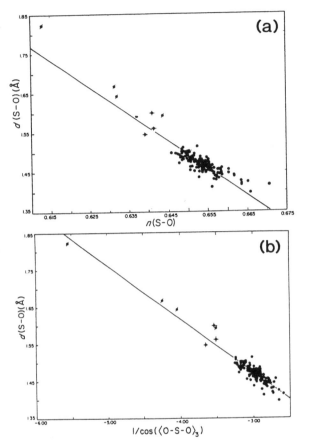

Figure 9. (a) Variation in observed S–O bond-length, d(S–O), with Mulliken bond-overlap population calculated for a constant bond-length of 1.49 Å and the observed O–S–O angles in selected sulfate structures; (b) variation in observed S–O bond-length as a function of $1/\cos\langle O\text{–}S\text{–}O_3\rangle$ in selected sulfate structures; points with a single / involve bridging anions in condensed sulfates, points with a + involve donor anions in acid-sulfate groups (after Louisnathan et al. 1977).

Louisnathan and Gibbs (1972a). Louisnathan and Gibbs (1972b) and Louisnathan et al. (1977) examined sulfate structures (primarily minerals) from this perspective, and showed that observed $\langle$S–O$\rangle$ bond-lengths are highly correlated with Mulliken bond-overlap populations (Fig. 9a). In these calculations, the S–O bond-lengths were fixed at 1.49 Å (in order to prevent inducing such a correlation through the connection of bond-overlap population to bond length) and the observed O–S–O angles were used. In this way, the effect of variations in O–S–O angles was examined while the S–O distances in the calculations were fixed. Louisnathan et al. (1977) showed that observed S–O bond-lengths are very strongly correlated with O–S–O angles (Fig. 9b) *via* the expression $1/\cos(\langle O\text{–}S\text{–}O\rangle_3)$, where $\langle O\text{–}S\text{–}O\rangle_3$ is the mean value of the three O–S–O angles involving the S–O bond under consideration. These correlations rationalize the variations in stereochemistry of the (SO$_4$) group in minerals: the final configuration is a compromise between connectivity and bonding requirements, and correlated adjustment of both S–O distances and O–S–O angles is the molecular-orbital mechanism whereby this is achieved.

Alkali metal–sulfate clusters

Wang et al. (1999) used DFT to study $M^+(SO_4)^{2-}$ and $[M^+(SO_4)^{2-}]_2$ (M = Na, K) alkali-metal sulfate-ion pairs. The results for $Na^+(SO_4)^{2-}$ and $K^+(SO_4)^{2-}$ are particularly interesting, as several K- and Na-sulfate minerals are known. In general, the calculations show that the most stable $M^+(SO_4)^{2-}$ configurations are those in which the M^+ cations are bonded to three O atoms of the (SO_4) tetrahedron (face-sharing of S and $M\mathcal{O}_n$ coordination polyhedra). The $M^+(SO_4)^{2-}$ configuration, with the M^+ cation bonded to two O atoms, has a higher energy. The difference between face- and edge-sharing configurations is 0.48 $kcal\,mol^{-1}$ for M = Na, and 3.30 $kcal\,mol^{-1}$ for M = K. This is consistent with the structures of Na- and K-sulfate minerals. Whereas face-sharing between K_n polyhedra and (SO_4) tetrahedra is common (e.g. arcanite, glaserite, etc.), in Na sulfates (e.g. thenardite), edge-sharing between Na_n polyhedra and (SO_4) tetrahedra dominates.

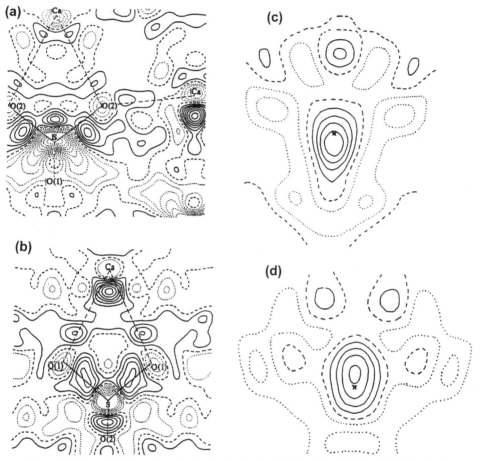

Figure 10. Dynamic deformation-density distribution in anhydrite (after Kirfel and Will 1980). (a) in the mirror plane (010) at y = 0; (b) in the mirror plane (100) at x = 1/4 ; (c) and (d) difference densities in planes halfway between S and O atoms and normal to the S–O bonds. Levels at 0.1 $e\,\text{Å}^{-3}$, zero lines broken, negative contours dotted.

Experimental studies of electron density

Kirfel and Will (1980) studied the room-temperature electron-density distribution in anhydrite, $CaSO_4$, with special attention to the charge distribution within and around the $(SO_4)^{2-}$ anion. The $(SO_4)^{2-}$ anion in anhydrite has $mm2$ point symmetry, with oxygen atoms on the mirror planes. Figure 10 shows the dynamic deformation-density distribution in the sulfate group. Kirfel and Will (1980) suggested several explanations for the features in the deformation electron-density maps. Figure 10a shows the mirror plane on (010) that contains the O(2)–S–O(2) bond angle. The peak > 0.50 e Å^{-3} between the S and O(2) atoms, evidently associated with π-bond overlap, is shifted toward a line connecting the S and Ca atoms. This may be due to Coulombic forces involving the adjacent Ca cation. The two symmetrical S–O(2) peaks are connected by a band of positive deformation density that has a maximum on the bisectrix of the O(2)–S–O(2) bond angle. Kirfel and Will (1980) associated this maximum with interaction of the $3d_z$ orbital of S with the $2p$ orbitals of the O(2) atoms according to the π-bond model for sulfate suggested by Cruickshank (1961). However, there is no maximum on the bisectrix of the O(1)–S–O(1) bond angle (Fig. 10b) where π-bond peaks are also connected by a band of positive deformation density. Figures 10c and 10d depict deformation densities in planes halfway between the S and O atoms and normal to the S–O bonds. It is evident that the π-bond charge accumulations deviate significantly from ideal cylindrical symmetry. According to Kirfel and Will (1980), this shows that the electron-density distribution within the (SO$_4$) tetrahedron is affected by external bonding conditions, i.e. by attractive Coulombic forces from Ca cations in the case of anhydrite. Similar results were also reported by Christidis et al. (1983) for monoclinic Fe$_2$(SO$_4$)$_3$.

Theoretical studies of electron densities

Hill et al. (1997) studied minimum-energy geometries and electron-density distributions for (H$_2$SO$_4$) and (H$_2$S$_2$O$_7$) clusters at the HF SCF level with a 6–311++G** basis set. Topological analysis of the electron-density distribution showed that bond critical points between S and O atoms are 0.549 and 0.545 Å away from the S atom in (H$_2$SO$_4$) and (H$_2$S$_2$O$_7$) clusters, respectively. The electron density at bond critical points for S–O bonds are 2.200 and 2.246 e·Å^{-3} for the (H$_2$SO$_4$) and (H$_2$S$_2$O$_7$) clusters, respectively. Analysis of the theoretical electron-density distribution allowed Hill et al. (1997) to characterize the S–O bonds in terms of their covalency. Applications of various theoretical criteria (see Hill et al. 1997) showed that S–O bonds in sulfates are predominantly covalent, as expected from the relative electronegativities, χ, of S and O atoms (χ$_S$ = 2.44; χ$_O$ = 3.50; Allred and Rochow 1958).

Models of chemical bonding

In this chapter, we couch several of our arguments in terms of bond-valence theory (Brown and Shannon 1973; Brown 1981; Hawthorne 1992, 1994, 1997). Louisnathan et al. (1977) examined the relation between the MO and bond-valence approaches. Brown and Shannon (1973) showed how to calculate the relative covalency of a bond using the bond-valence approach. Louisnathan et al. (1977) showed that, for the (SO$_4$) group, this covalency is strongly correlated (Fig. 11) with the electrical charge of the O atom, calculated using a fixed S–O distance of 1.49 Å, the observed O–S–O angles, and extended Hückel MO theory. This explicit connection between MO theory and bond-valence theory for the (SO$_4$) group is in accord with later arguments concerning the similarity of these two approaches (Burdett and Hawthorne 1993; Hawthorne 1994, 1997).

HIERARCHICAL ORGANIZATION OF CRYSTAL STRUCTURES

Ideally, the physical, chemical, and paragenetic characteristics of a mineral should arise as natural consequences of its crystal structure and the interaction of that structure with the environment in which it occurs. Hence, an adequate structural hierarchy of minerals should provide an epistemological basis for the interpretation of the role of minerals in Earth processes. We have not yet reached this stage for any major class of minerals, but significant advances have been made. For example, Bragg (1930) classified the major rock-forming silicate minerals according to the geometry of polymerization of

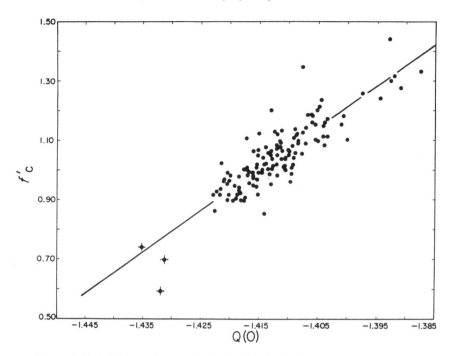

Figure 11. Variation in covalency, $f'c$, of an S–O bond, calculated according to the method of Brown and Shannon (1973), with the electric charge of the constituent O anion, Q(0), calculated for a constant bond-length of 1.49 Å and observed O–S–O angles; symbols as in Figure 9 (after Louisnathan et al. 1977).

$(Si,Al)O_4$ tetrahedra, and this scheme was extended by Zoltai (1960) and Liebau (1985); it is notable that the scheme parallels Bowen's reaction series (Bowen 1928) for silicate minerals in igneous rocks. Much additional insight can be derived from such structural hierarchies, particularly with regard to controls on bond topology (Hawthorne 1983a, 1994) and mineral paragenesis (Moore 1965, 1970, 1973; Hawthorne 1984; Hawthorne et al. 1987).

Hawthorne (1983a) proposed that structures be ordered or classified according to the polymerization of those cation coordination polyhedra with higher bond-valences. Higher bond-valence polyhedra polymerize to form *homo-* or *heteropolyhedral clusters* that constitute the *fundamental building block (FBB)* of the structure. The *FBB* is repeated, commonly polymerized, by translational symmetry operators to form the *structural unit*, a complex (typically anionic) polyhedral array (not necessarily connected), the excess charge of which is balanced by the presence of *interstitial* species (usually large, low-valence cations; Hawthorne 1985a). The possible modes of cluster polymerization are (1) unconnected polyhedra; (2) finite clusters; (3) infinite chains; (4) infinite sheets; and (5) infinite frameworks.

All structural classifications of sulfate minerals have considered sulfate-mineral structures as based on heteropolyhedral structural units. In their crystal-chemical classification, Bokii and Gorogotskaya (1969) subdivided 70 sulfate minerals into two large groups: (1) minerals with isolated (SO_4) tetrahedra, and (2) (SO_4) tetrahedra linked to other cation polyhedra. The sulfate minerals of both groups were subdivided into finite-cluster, infinite-chain, infinite-sheet, and infinite-framework structures. Sabelli and Trosti-

Ferroni (1985) classified sulfate minerals into the same four categories and provided a general scheme that included the ~200 mineral species known at that time. Rastsvetaeva and Pushcharovsky (1989) considered ~600 sulfate minerals and inorganic compounds and classified them on the basis of mixed-anion radicals (~heteropolyhedral structural units) (Sandomirsky and Belov 1984). Pushcharovsky et al. (1998) considered the main structural subdivisions and structural formulae of sulfate minerals within the framework of the structural classification suggested by Lima-de-Faria (1994) and recommended by the Subcommittee on the Nomenclature of the Inorganic Structure Types of the International Union of Crystallography Commission on the Crystallographic Nomenclature (Lima-de-Faria et al. 1990).

POLYMERIZATION OF SO_4 AND OTHER $T\emptyset_4$ TETRAHEDRA

Bond valence is a measure of the strength of a chemical bond, and, in a coordination polyhedron, can be approximated by the formal valence divided by the coordination number. Thus, in an SO_4 group, the mean bond-valence is $6/4 = 1.5$ vu. The valence-sum rule (Brown 1981) states that the sum of the bond valences incident at an atom is equal to the magnitude of the formal valence of that atom. Thus, any oxygen atom linked to the central S^{6+} cation receives ~1.50 vu from the S^{6+} cation, and hence must receive ~0.50 vu from other coordinating cations. There are several ways in which this bond-valence requirement may be satisfied:

(1) The oxygen atom is bonded to another tetrahedrally coordinated cation; this method is not common in minerals as most tetrahedrally coordinated cations have sufficiently large bond-valences that such an S–O–T linkage violates the valence-sum rule. This argument is not completely satisfactory, as even synthetic polymerized sulfates exist. For example, $K_2S_2O_7$ has been synthesized (Lynton and Truter 1960); it has an S–O distance of 1.642 Å, far longer than any S–O distance recorded in minerals (Fig. 1), and is presumably unstable under natural Earth conditions. Nevertheless, lopezite, $K_2Cr_2O_7$, is a mineral with O bridging two hexavalent cations (Cr^{6+}), a configuration similar to that in $K_2S_2O_7$, so the reason for the absence of the latter configuration is not completely clear.

Similarly, the sulfate oxyanion does not polymerize with any of the common tetra-hedral oxyanions in minerals [e.g. (PO_4), (AsO_4), (VO_4), (SiO_4), (AlO_4)] because the ideal bond-valence sums incident at the bridging anion would significantly exceed 2 vu. This is not the case for the (BeO_4) and (LiO_4) groups; polymerization of these groups with (SO_4) would give ideal bond-valence sums incident at the bridging anion of $1.50 + 0.50 = 2.00$ vu and $1.50 + 0.25 = 1.75$ vu, respectively. Nevertheless, these arrangements have not been observed in minerals.

(2) The oxygen atom is bonded to additional octahedrally coordinated cations; this mechanism is extremely common in minerals. Di- and tri-valent octahedrally coordinated cations have ideal bond-valences of 0.33 and 0.50 vu, respectively, and hence there are several different types of linkage that are compatible with the valence-sum rule, e.g. $1.50 + 0.50 = 2.00$ vu; $1.50 + (2 \times 0.33) = 2.16$ vu (small to moderate deviations from the valence-sum rule are compensated by antipathetic variations in the bond lengths).

(3) The oxygen atom bonds to additional higher-coordinated alkali and/or alkaline-earth cations; this mechanism is also common in minerals. Mono- and di-valent cations have bond valences in the ranges of 0.08-0.15 and 0.17-0.29 vu, respectively, and hence there are many types of linkage compatible with the valence-sum rule, e.g. $1.50 + (3 \times 0.15) = 1.95$ vu; $1.50 + (2 \times 0.25) = 2.00$ vu.

(4) The oxygen atom is bonded to additional octahedrally coordinated and higher-

coordinated alkali and/or alkaline-earth cations; this is also common in minerals.
(5) The oxygen atom may act as a hydrogen-bond acceptor.

Most sulfate minerals consist of (SO_4) tetrahedra polymerizing with MO_n polyhedra, and hence we will organize their structures on this basis.

A STRUCTURAL HIERARCHY FOR SULFATE MINERALS

There are ~370 sulfate-mineral species, and structural information is available for ~80% of them. About 140 sulfate-mineral structure-types are known, and these are considered here.

The utility of this approach for consideration of the minerals with single tetrahedral oxyanions (sulfates, phosphates, arsenates, vanadates, borates) has been shown by Hawthorne (1985a, 1986, 1990, 1992, 1994, 1997), Hawthorne et al. (1996), Burns et al. (1995), Grice et al. (1999), Eby and Hawthorne (1993), and Schindler et al. (2000a,b). It should be noted, however, that most of the minerals considered in the above works are based on structural units involving polymerization of ($MØ_6$) octahedra (M = divalent or trivalent cation: Mg^{2+}, Fe^{2+}, Mn^{2+}, Zn^{2+}, Al^{3+}, Fe^{3+}, etc.) and ($TØ_4$) tetrahedra (T = pentavalent or hexavalent cation: As^{5+}, P^{5+}, V^{5+}, S^{6+}, Cr^{6+}, Mo^{6+}, etc.). However, numerous sulfate minerals do not contain divalent- or trivalent-cation octahedra; hence we must introduce additional categories of structures. Thus, we subdivide the sulfate minerals into the following groups: (1) sulfates with divalent- and trivalent-metal octahedra; (2) sulfates with non-octahedral cation polyhedra and Na sulfates; (3) sulfates with anion-centered tetrahedra.

The structure diagrams presented here generally have the following shading patterns: (SO_4) tetrahedra are shaded with pale gray, colorless or narrow-spaced lines, octahedra and higher coordinations are shaded with dark gray or wide-spaced lines; H atoms are shown as small shaded circles; H bonds are shown by dotted lines.

STRUCTURES BASED ON SULFATE TETRAHEDRA AND DIVALENT AND/OR TRIVALENT CATION OCTAHEDRA

Graphical representation of octahedral–tetrahedral structures

The majority of minerals have structures that are based on complex anions built by polymerization of octahedra and tetrahedra. To analyze the connectivity of the octahedral–tetrahedral structures, Hawthorne (1983a) considered polyhedral clusters from a graph-theoretic viewpoint. Polyhedra are represented by the chromatic vertices of a labelled graph in which different colors (e.g. black and white) represent coordination polyhedra of different type. Linking of polyhedra can be denoted by the presence of an edge or edges between vertices representing linked polyhedra. The number of edges between vertices denotes the number of atoms common to both polyhedra. Thus, no edge between two vertices represents disconnected polyhedra (Fig. 12a), one edge between two vertices represents corner-sharing between two polyhedra (Fig. 12b), two edges between two vertices represents edge-sharing between two polyhedra (Fig. 12c), and three edges between two vertices represents triangular-face-sharing (Fig. 12d). The octahedra and tetrahedra are denoted as white and black vertices, respectively. Figure 12e shows the polyhedral cluster $[M_2(TØ_4)_2Ø_8]$ and its graphical representation (M = octahedrally coordinated cation; T = tetrahedrally coordinated cation; $Ø$ = unspecified ligand); round brackets and curly brackets denote a polyhedron or a group, e.g. (SO_4), (H_2O); square brackets denote linked polyhedra, e.g. $[M(TO_4)_2Ø_4]$.

Moving from polyhedral representation to graphical representation, geometrical

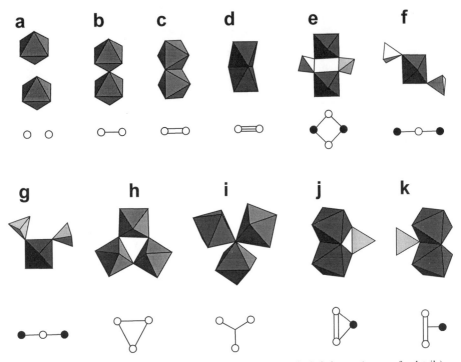

Figure 12. Examples of graphical representation of octahedral–tetrahedral clusters (see text for details).

information is lost. This is illustrated in Figures 12f and 12g, which show two different possible arrangements of the corner-linked cluster $[M(T\emptyset_4)_2\emptyset_4]$. Both of these clusters are described by identical graphs; such clusters are called *geometrical isomers*. Information on geometrical isomerism is lost in the graphical representation; examples are shown in Figures 12h to 12k. In some cases, we can distinguish between these configurations by allowing (more than two) edges in the graph to intersect without forming a vertex (e.g. Figs. 12i, 12k). Although these modifications do not follow conventional notation, they are useful in representing different structures (e.g. Hawthorne et al. 1996), and are used here.

The octahedral–tetrahedral structural units can be arranged according to the dimension of their polymerization: (a) unconnected (isolated) polyhedra; (b) finite clusters; (c) infinite chains; (d) infinite sheets; (e) infinite frameworks.

Structures with unconnected SO_4 groups

The graphs of this class of structures are shown in Figures 13a and 13b; they consist of white vertices and isolated black and white vertices. Sulfate minerals of this class are given in Tables 5 and 6, and their structures are shown in Figures 14-19. In these minerals, sulfate tetrahedra and $(M\emptyset_6)$ octahedra are linked together by hydrogen bonding and/or by large low-valence interstitial cations. In the minerals of the **hexahydrite group** (Table 5), $\{M^{2+}(H_2O)_6\}(SO_4)$ (Fig. 14a), and **retgersite**, $\{Ni(H_2O)_6\}(SO_4)$ (Fig. 14b), the (SO_4) tetrahedra and $\{M^{2+}(H_2O)_6\}$ octahedra are linked by hydrogen bonding only; this involves linkage from donor ligands of the octahedra to acceptor ligands of the tetrahedra, together with weak hydrogen bonding between ligands of different octahedra. The minerals of the **epsomite group**, $\{M^{2+}(H_2O)_6\}(SO_4)(H_2O)$, M = Mg, Zn, Ni (Fig. 14c), and the **melanterite group**, $\{M^{2+}(H_2O)_6\}(SO_4)(H_2O)$, M = Fe^{2+}, Co, Mn^{2+}, Cu^{2+}, Zn (Fig.

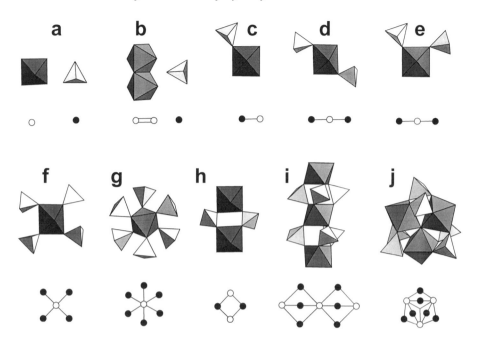

Figure 13. Finite octahedral–tetrahedral clusters in the sulfate minerals and their graphs.

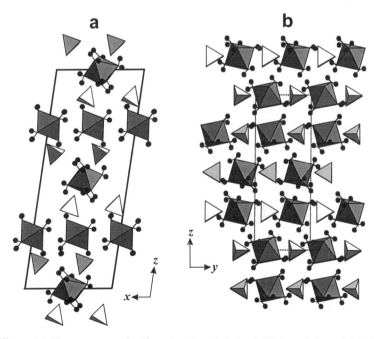

Figure 14. The structures of sulfate minerals with isolated (SO₄) tetrahedra and ($M\emptyset_6$) octahedra: (a) hexahydrite; (b) retgersite.

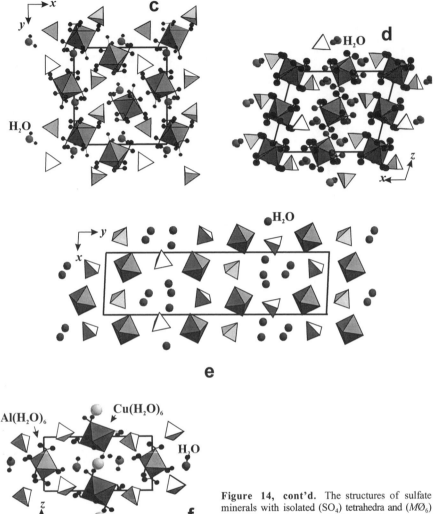

Figure 14, cont'd. The structures of sulfate minerals with isolated (SO_4) tetrahedra and $(M\emptyset_6)$ octahedra: (c) epsomite; (d) bieberite; (e) alunogen; (f) aubertite.

14d), are characterized by (H_2O) in addition to that required to coordinate the M^{2+} cations (Table 5). In the melanterite group, there is direct hydrogen bonding between octahedra and tetrahedra, and there is also hydrogen bonding through the additional interstitial (H_2O) (acting as both a hydrogen-bond donor and a hydrogen-bond acceptor) to both tetrahedral and octahedral ligands. In the epsomite group, the pattern of hydrogen bonding is slightly different; there is direct hydrogen bonding between octahedra and tetrahedra, weak hydrogen bonding between adjacent octahedra, and hydrogen bonding through interstitial (H_2O) to both octahedra and tetrahedra. In **alunogen**, $\{Al(H_2O)_6\}_2(SO_4)_3(H_2O)_{4.4}$ (Fig. 14e), there are interstitial (H_2O) groups linking tetrahedra and octahedra together through additional hydrogen bonds. In the structure of **aubertite**, $\{Cu^{2+}(H_2O)_6\}\{Al(H_2O)_6\}(SO_4)_2Cl(H_2O)_2$ (Fig. 14f), hydrogen bonding between octahedra and tetrahedra involves interstitial Cl as well as (H_2O) groups.

Table 5. Sulfate minerals with isolated (SO_4) tetrahedra and $(M\phi_6)$ octahedra with the general formula $(M\phi_6)_p(SO_4)_q(H_2O)_r$, $(r = 0-5)$. *

Name	Formula	a (Å)	b (Å)	c (Å)	α (°)	β (°)	γ (°)	S. G.	Fig.	Ref.
	$(M\phi_6)(SO_4)$									
hexahydrite	$\{Mg(H_2O)_6\}(SO_4)$	10.110	7.212	24.410	–	98.3	–	$C2/c$	14a	[1]
bianchite	$\{Zn(H_2O)_6\}(SO_4)$	10.10	7.20	24.49	–	98.3	–	$C2/c$	14a	(2)
chvaleticeite	$\{Mn^{2+}(H_2O)_6\}(SO_4)$	10.05	7.24	24.3	–	98.0	–	$C2/c$	14a	(2)
ferrohexahydrite	$\{Fe^{2+}(H_2O)_6\}(SO_4)$	10.08	7.28	24.59	–	98.4	–	$C2/c$	14a	(2)
moorhouseite	$\{Co(H_2O)_6\}(SO_4)$	10.03	7.23	24.23	–	98.4	–	$C2/c$	14a	(2)
nickelhexahydrite	$\{Ni(H_2O)_6\}(SO_4)$	9.880	7.248	24.171	–	98.5	–	$C2/c$	14a	[3]
retgersite	$\{Ni(H_2O)_6\}(SO_4)$	6.783	a	18.288	–	–	–	$P4_12_12$	14b	[4]
	$(M\phi_6)_p(SO_4)_q(H_2O)_r$									
epsomite	$\{Mg(H_2O)_6\}(SO_4)(H_2O)$	11.887	12.013	6.861	–	–	–	$P2_12_12_1$	14c	[5]
goslarite	$\{Zn(H_2O)_6\}(SO_4)(H_2O)$	11.799	12.050	6.822	–	–	–	$P2_12_12_1$	14c	(2)
morenosite	$\{Ni(H_2O)_6\}(SO_4)(H_2O)$	11.768	12.021	6.758	–	–	–	$P2_12_12_1$	14c	[6]
melanterite	$\{Fe^{2+}(H_2O)_6\}(SO_4)(H_2O)$	14.072	6.503	11.041	–	105.6	–	$P2_1/c$	14d	[7]
bieberite	$\{Co(H_2O)_6\}(SO_4)(H_2O)$	14.048	6.494	10.925	–	105.2	–	$P2_1/c$	14d	[8]
mallardite	$\{Mn^{2+}(H_2O)_6\}(SO_4)(H_2O)$	14.15	6.50	11.06	–	105.6	–	$P2_1/c$	14d	(2)
boothite	$\{Cu^{2+}(H_2O)_6\}(SO_4)(H_2O)$	13.89	6.50	10.64	–	105.6**	–	–	14d	(2)
zinc-melanterite	$\{Zn(H_2O)_6\}(SO_4)(H_2O)$	14.04	6.51	10.99	–	105.3†	–	–	14d	(2)
alunogen	$\{Al(H_2O)_6\}_2(SO_4)_3(H_2O)_{4.4}$	7.245	26.975	6.061	90.0	97.7	91.9	$P\bar{1}$	14e	[9]
aubertite	$\{Cu^{2+}(H_2O)_6\}\{Al(H_2O)_6\}(SO_4)_2Cl(H_2O)_2$	6.282	13.192	6.26	91.8	94.7	82.5	$P\bar{1}$	14f	[10]
magnesioaubertite	$\{Mg(H_2O)_6\}\{Al(H_2O)_6\}(SO_4)_2Cl(H_2O)_2$	6.31	13.20	6.29	91.7	94.5	82.6	$P\bar{1}$	14f	[2]
svyazhinite	$\{Mg(H_2O)_6\}\{Al(H_2O)_6\}(SO_4)_2F(H_2O)_2$	6.217	13.306	6.255	90.05	93.3	82.03	$P\bar{1}$	14f	(2)

* M = Cu^{2+}, Fe^{2+}, Ni, Mg, Zn, Mn^{2+}, Co, Al; ϕ = O, OH, H_2O, F, Cl. ** From morphological axial ratios assuming isomorphism with melanterite. † From re-indexing of the powder-diffraction pattern of Liu Tiegeng et al. (1995).

A reference in square brackets (e.g., [1]) indicates that a structure has been refined; a reference in round brackets [e.g., (2)] indicates a structure has not been refined. *References*: [1] Zalkin et al. (1964), [2] Gaines et al. (1997), [3] Ptasiewicz-Bak et al. (1983), [4] Calleri et al. (1984),[6] [5] Stadnicka et al. (1987), [5] Calleri et al. (1984),[6] Iskhakova et al. (1991), [7] Baur (1964), [8] Kellersohn et al. (1991), [9] Menchetti and Sabelli (1974), [10] Ginderow and Cesbron (1979)

Table 6. Sulfate minerals with isolated (SO_4) tetrahedra and $(MØ_6)$ octahedra.*

Name	Formula	a (Å)	b (Å)	c (Å)	β (°)	S. G.	Fig.	Ref.
picromerite	$K_2\{Mg(H_2O)_6\}(SO_4)_2$	9.072	12.212	6.113	104.8	$P2_1/a$	15a	[1]
boussingaultite	$(NH_4)_2\{Mg(H_2O)_6\}(SO_4)_2$	9.316	12.596	6.198	107.1	$P2_1/a$	15a	[2]
cyanochroite	$K_2\{Cu^{2+}(H_2O)_6\}(SO_4)_2$	9.066	12.130	6.149	104.4	$P2_1/a$	15a	[3]
mohrite	$(NH_4)_2\{Fe^{2+}(H_2O)_6\}(SO_4)_2$	9.170	12.419	6.297	106.7	$P2_1/a$	15a	[4]
nickel-boussingaultite	$(NH_4)_2\{Ni(H_2O)_6\}(SO_4)_2$	6.244	12.469	9.195	107.0	$P2_1/c$	15a	[5]
lonecreekite	$(NH_4)\{Fe^{3+}(H_2O)_6\}(SO_4)_2(H_2O)_6$	12.302	a	a	–	$Pa\overline{3}$	15b	(6)
potassium alum	$K\{Al(H_2O)_6\}(SO_4)_2(H_2O)_6$	12.157	a	a	–	$Pa\overline{3}$	15b	[7]
sodium alum	$Na\{Al(H_2O)_6\}(SO_4)_2(H_2O)_6$	12.213	a	a	–	$Pa\overline{3}$	15b	[8]
tschermigite	$(NH_4)\{Al(H_2O)_6\}(SO_4)_2(H_2O)_6$	12.248	a	a	–	$Pa\overline{3}$	15b	[9]
tamarugite	$Na\{Al(H_2O)_6\}(SO_4)_2$	7.353	25.225	6.097	95.2	$P2_1/a$	15c,d	[10]
amarillite	$Na\{Fe^{3+}(H_2O)_6\}(SO_4)_2$	8.419	10.841	12.472	95.5	$C2/c$	15c,d	[11]
mendozite	$Na\{Al(H_2O)_6\}(SO_4)_2(H_2O)_5$	21.75	9.11	8.300	92.5	$C2/c$	15g	[12]
kalinite	$K\{Al(H_2O)_6\}(SO_4)_2(H_2O)_5$	19.92	9.27	8.304	98.8	$C2/c$	15g	(6)
chukhrovite	$Ca_4F\{AlF_6\}_2(SO_4)(H_2O)_{12}$	16.710	a	a	–	$Fd\overline{3}$	16a-d	[18]
chukhrovite-(Ce)	$(Ca,Ce,Nd)_4F\{AlF_6\}_2(SO_4)(H_2O)_{10}$	16.800	a	a	–	$Fd\overline{3}$	16a-d	[19]
ettringite	$Ca_6\{Al(OH)_6\}_2(SO_4)_3(H_2O)_{26}$	11.260	a	21.480	–	$P31c$	16e,f,g	[13]
bentorite	$Ca_6\{Cr^{3+}(OH)_6\}_2(SO_4)_3(H_2O)_{26}$	22.35	a	21.41	–	$P6_3/mmc$	16e,f,g	(6)
charlesite	$Ca_6\{Al(OH)_6\}_2(SO_4)_2\{B(OH)_4\}(H_2O)_{26}$	11.16	a	21.21	–	$P31c$	16e,f,g	(6)
sturmanite	$Ca_6\{Fe^{3+}(OH)_6\}_2(SO_4)_2\{B(OH)_4\}(H_2O)_2$	11.16	a	21.79	–	$P31c$	16e,f,g	(6)
jouravskite	$Ca_3\{Mn^{4+}(OH)_6\}(SO_4)(CO_3)(H_2O)_{12}$	11.060	a	10.500	–	$P6_3$	16e,f,g	[14]
thaumasite	$Ca_3\{Si(OH)_6\}(SO_4)(CO_3)(H_2O)_{12}$	11.04	a	10.390	–	$P6_3$	16e,f,g	[15]
despujolsite	$Ca_3\{Mn^{4+}(OH)_6\}(SO_4)_2(H_2O)_3$	8.560	a	10.760	–	$P\overline{6}2c$	16e,f,h	[16]
fleischerite	$Pb_3\{Ge(OH)_6\}(SO_4)_2(H_2O)_3$	8.867	a	10.875	–	$P\overline{6}2c$	16e,f,h	[17]
schaurteite	$Ca_3\{Ge(OH)_6\}(SO_4)_2(H_2O)_3$	8.525	a	10.803	–	$P\overline{6}2c$	16e,f,h	(6)

* $M = Cu^{2+}$, Fe^{2+}, Ni, Mg, Zn, Al, Fe^{3+}, Cr^{3+}, Mn^{4+}, Si, Ge

References: [1] Kannan and Viswamitra (1965), [2] Maslen et al. (1988), [3] Robinson and Kennard (1972), [4] Figgis et al. (1992), [5] Tahirov et al. (1994), (6) Gaines et al. (1997), [7] Larson and Cromer (1967), [8] Cromer et al. (1967), [9] Abdeen et al. (1981), [10] Robinson and Fang (1969), [11] Li et al. (1990), [12] Fang and Robinson (1972), [13] Moore and Taylor (1970), [14] Granger (1969), [15] Zemann and Zobetz (1981), [16] Gaudefroy et al. (1968), [17] Otto (1975), [18] Mathew et al. (1981), [19] Bokii and Gorogotskaya (1965)

In the minerals of the **picromerite group** (Table 6), e.g. **cyanochroite**, $K_2\{Cu^{2+}(H_2O)_6\}(SO_4)_2$, the $\{M^{2+}(H_2O)_6\}$ and (SO_4) groups are linked by K into sheets parallel to (100) (Fig. 15a), which are linked together by hydrogen bonding from the octahedral ligands to the acceptor anions of the sulfate group. The structure of **sodium alum**, $Na\{Al(H_2O)_6\}(SO_4)_2(H_2O)_6$, consists of $\{Al(H_2O)_6\}$, $\{Na(H_2O)_6\}$, and (SO_4) groups linked by hydrogen bonding (Fig. 15b). All (H_2O) groups are coordinated to cations; any additional (H_2O) in a structure of this type would entail (H_2O) held in the structure solely by hydrogen bonding. In the structure of **tamarugite**, $Na\{Al(H_2O)_6\}(SO_4)_2$ (Fig. 15c,d), Na is octahedrally coordinated by oxygen atoms of the (SO_4) tetrahedra. The two (NaO_6) octahedra share an edge to form $[Na_2O_{10}]$ dimers that are linked through (SO_4) groups to form infinite chains along [001]. Figures 15e,f show the chain and its graph, respectively. A similar situation occurs in the structure of **mendozite**,

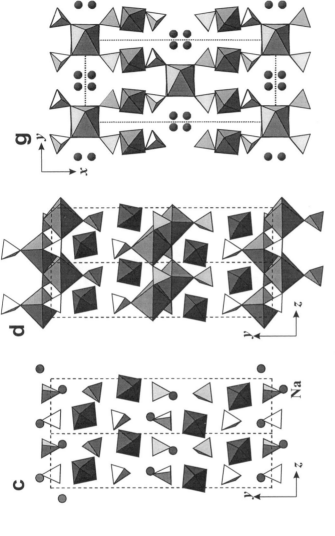

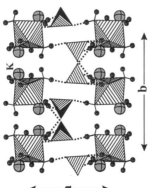

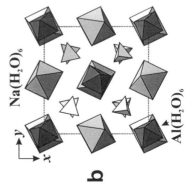

Figure 15. The structures of sulfate minerals with isolated (SO_4) tetrahedra and (MO_6) octahedra: (a) cyanochroite; (b) sodium alum; (c) tamarugite, with Na atoms shown as circles; (d) tamarugite with (NaO_6) octahedra; (g) the structure of mendozite. [Figures 15e,f,h,i on next page.]

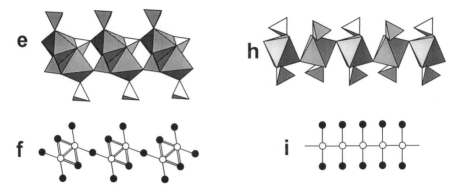

Figure 15, cont'd. The structures of sulfate minerals with isolated (SO_4) tetrahedra and ($M\emptyset_6$) octahedra: (e) the chain of ($Na\emptyset_6$) octahedra and (SO_4) tetrahedra in tamarugite; (f) the graph of the chain in tamarugite; (h) the chain of ($Na\emptyset_6$) octahedra and (SO_4) tetrahedra in mendozite; (i) the graph of the chain in mendozite.

$Na\{Al(H_2O)_6\}(SO_4)_2(H_2O)_5$ (Fig. 15g), where corner-sharing $\{NaO_4(H_2O)_2\}$ octahedra form infinite $[Na\emptyset_5]$ chains that are parallel to [001] and are decorated by (SO_4) tetrahedra (Figs. 15h,i). Note that these chains are topologically identical to the $[M\emptyset_5]_0$ chains described by Moore (1970) as 7-Å chains, although their repeat distance is longer (~8.3 Å, Table 6) because of the size of Na. In the structure of **chukhrovite**, $Ca_4F\{AlF_6\}_2(SO_4)(H_2O)_{12}$, four ($Ca\emptyset_7$) polyhedra share one F atom to form, together with four (AlF_6) octahedra, the rather complex cluster shown in Figure 16a. These clusters are linked together to form a framework with (SO_4) tetrahedra in the cavities (Fig. 16c). An alternative description is that the structure is built from (FCa_4) tetrahedra (Fig. 16b), (AlF_6) octahedra, (SO_4) tetrahedra, and (H_2O) groups (Fig. 16d).

The structures of the minerals of the **ettringite**, $Ca_6\{Al(OH)_6\}_2(SO_4)_3(H_2O)_{26}$, and **fleischerite**, $Pb^{2+}_3\{Ge(OH)_6\}(SO_4)_2(H_2O)_3$, **groups** (Table 6) are based on the column (Fig. 16e) formed by edge-sharing of $\{Mn(OH)_6\}$ octahedra with triplets of ($Ca\emptyset_8$) polyhedra (Fig. 16f) girdling the column at intervals along [001]. The columns are isolated in the minerals of the ettringite group (Fig. 16g); intercolumn linkage is provided by hydrogen bonding *via* the (SO_4) tetrahedra in the interstices between the columns. The columns are linked into a three-dimensional framework directly *via* (SO_4) tetrahedra in the minerals of the fleischerite group (Fig. 16h).

Structures with finite clusters of polyhedra

Sulfate minerals based on finite clusters of octahedra and tetrahedra are listed in Table 7; Figures 13c-j show all the cluster types and their graphs.

In **creedite**, $Ca_3[Al_2(OH)_2F_8](SO_4)(H_2O)_2$, and **jurbanite**, $[Al_2(OH)_2(H_2O)_8](SO_4)_2$ $\cdot(H_2O)_2$, there is an octahedral edge-sharing dimer of the form $[Al_2\emptyset_{10}]$. In creedite, linkage between the octahedral clusters, (SO_4) tetrahedra, and (H_2O) groups is provided by Ca cations (Figs. 16i,j), whereas linkage of structural subunits in jurbanite is provided by hydrogen bonding only (Fig. 17a). Creedite and jurbanite can be considered as transitional between the unconnected-polyhedra structures and the finite-cluster structures because of the connectivity of the octahedra.

The structures of **minasragrite**, $[V^{4+}O(SO_4)(H_2O)_4](H_2O)$, and **xitieshanite**, $[Fe^{3+}(H_2O)_4Cl(SO_4)](H_2O)_2$, are based on a simple corner-sharing cluster of a tetrahedron and an octahedron (Fig. 13c). These clusters are linked by hydrogen bonding

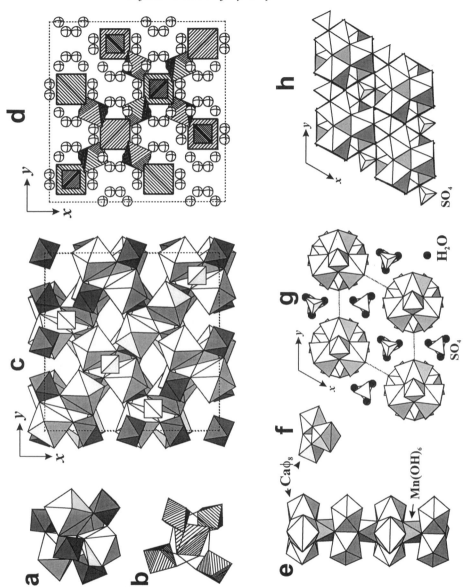

Figure 16. (a) The cluster of four (CaØ$_7$) polyhedra and (AlF$_6$) octahedra in chukhrovite; (b) the same cluster shown as an (FCa$_4$) tetrahedron surrounded by four (AlF$_6$) octahedra; (c) the linkage of clusters, shown in Figure 14a, into a framework in chukhrovite; (d) representation of the structure of chukhrovite as (FCa$_4$) tetrahedra (large tetrahedra), (AlF$_6$) octahedra, (SO$_4$) tetrahedra (small tetrahedra), and (H$_2$O) groups (circles); (e) the column of (MØ$_6$) octahedra and triplets of (CaØ$_8$) polyhedra in the minerals of the ettringite and fleischerite groups; (f) the triplet of (CaØ$_8$) polyhedra in ettringite and fleischerite; the columns are (g) isolated in the minerals of the ettringite group, and (h) linked into a framework in the minerals of the fleischerite group.

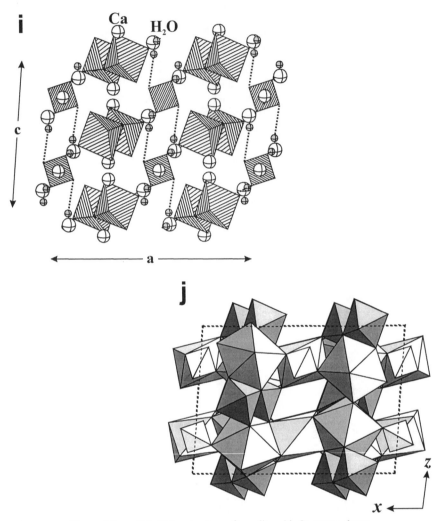

Figure 16, cont'd. (i) the structure of creedite with Ca atoms shown as circles; (j) polyhedral view of the structure of creedite.

through interstitial (H_2O) groups (Fig. 17b). The **blödite-group** minerals, $Na_2[M^{2+}(H_2O)_4(SO_4)_2]$, the **leonite-group** minerals, $K_2[M^{2+}(H_2O)_4(SO_4)_2]$, **chenite**, $(Pb^{2+}_4(OH)_2[Cu^{2+}(OH)_4(SO_4)_2]$ (Fig. 18a), and related minerals (Table 7) are based on the simple *trans* $[M(T\emptyset_4)_2\emptyset_4]$ cluster (Fig. 13d, 17c) that is also a constituent in phosphate minerals (e.g. anapaite, $Ca_2[Fe^{2+}(PO_4)_2(H_2O)_4]$ and schertelite, $(NH_4)_2[Mg(PO_3OH)_2(H_2O)_4]$. In **blödite**, these clusters are linked through edge-sharing octahedral $[Na_2\emptyset_{10}]$ dimers into sheets parallel to (001) (Fig. 17d,e) in which the clusters are arranged at the vertices of a plane centered rectangular lattice. Intersheet linkage occurs *via* the interstitial alkali cations and hydrogen bonding, as shown in Figure 17d. In the leonite-group mineral **mereiterite**, $K_2[Fe^{2+}(H_2O)_4(SO_4)_2]$, there are sheets of unconnected *trans* $[M(T\emptyset_4)\emptyset_4]$ clusters (as in blödite, Fig. 17d) parallel to (001) that are linked by interstitial K and a network of hydrogen bonds (Fig. 17f). In **chenite**, (SO_4) tetrahedra are

Table 7. Sulfate minerals with finite clusters of (SO$_4$) tetrahedra and (MØ$_6$) octahedra.*

Name	Formula	a(Å)	b(Å)	c(Å)	β(°)	S.G.	Graph	Fig.	Ref.
creedite	Ca$_3$[Al$_2$(OH)$_2$F$_8$](SO$_4$)(H$_2$O)$_2$	13.936	8.606	9.985	94 4	C2/c	13c	16i,j	[1]
jurbanite	[Al$_2$(OH)$_2$(H$_2$O)$_8$](SO$_4$)$_2$(H$_2$O)$_2$	8.396	12.479	8.155	101.9	P2$_1$/n	13c	17a	[2]
minasragrite	[V^{4+}O(SO$_4$)(H$_2$O)$_4$](H$_2$O)	6.976	9.716	12.902	110.9	P2$_1$/c	13d	–	[3]
blödite	Na$_2$[Mg(H$_2$O)$_4$(SO$_4$)$_2$]	11.03	8.140	5.490	100.7	P2$_1$/a	13e	17c,d	[5]
nickelblödite	Na$_2$[Ni(H$_2$O)$_4$(SO$_4$)$_2$]	10.87	8.07	5.46	100.7	P2$_1$/a	13e	17c,d	(6)
changoite	Na$_2$[Zn(H$_2$O)$_4$(SO$_4$)$_2$]	11.077	8.249	5.532	100.2	P2$_1$/a	13e	17c,d	(7)
leonite	K$_2$[Mg(H$_2$O)$_4$(SO$_4$)$_2$]	11.769	9.539	9.889	95.3	C2/m	13e	17f	[8]
mereiterite	K$_2$[Fe^{2+}(H$_2$O)$_4$(SO$_4$)$_2$]	11.841	9.553	9.942	94.9	C2/m	13e	17f	[9]
chenite[1]	Pb^{2+}$_4$(OH)$_2$[Cu^{2+}(OH)$_4$(SO$_4$)$_2$]	5.791	7.940	7.976	97.7	P$\bar{1}$	13e	18a,b	[10]
römerite[2]	[Fe^{3+}(SO$_4$)$_2$(H$_2$O)$_4$]$_2$[Fe^{2+}(H$_2$O)$_6$]	6.463	15.309	6.341	90.5	P$\bar{1}$	13f	18c	[11]
halotrichite	[Fe^{2+}(SO$_4$)(H$_2$O)$_5$]{Al(H$_2$O)$_6$}$_2$(SO$_4$)$_4$(H$_2$O)$_5$	6.179	24.29	20.51	101.0	P2$_1$/c	13f	18d	(6)
pickeringite	[Mg(SO$_4$)(H$_2$O)$_5$]{Al(H$_2$O)$_6$}$_2$(SO$_4$)$_4$(H$_2$O)$_5$	6.17	24.2	20.8	95.0	P2$_1$/c	13f	18d	(6)
bilinite	[Fe^{2+}(SO$_4$)(H$_2$O)$_5$]{Fe^{3+}(H$_2$O)$_6$}$_2$(SO$_4$)$_4$(H$_2$O)$_5$	6.208	24.333	21.255	100.3	P2$_1$/c	13f	18d	(6)
apjohnite	[Mn^{2+}(SO$_4$)(H$_2$O)$_5$]{Al(H$_2$O)$_6$}$_2$(SO$_4$)$_4$(H$_2$O)$_5$	6.198	24.347	21.266	100.3	P2$_1$/c	13f	18d	[12]
dietrichite	[Zn(SO$_4$)(H$_2$O)$_5$]{Al(H$_2$O)$_6$}$_2$(SO$_4$)$_4$(H$_2$O)$_5$	6.24	24.434	21.379	100.1	P2$_1$/c	13f	18d	(6)
wupatkiite	[Co(SO$_4$)(H$_2$O)$_5$]{Al(H$_2$O)$_6$}$_2$(SO$_4$)$_4$(H$_2$O)$_5$	6.189	24.234	21.204	100.3	P2$_1$/c	13f	18d	(6)
quenstedtite[3]	[Fe^{3+}(H$_2$O)$_4$(SO$_4$)$_2$][Fe^{3+}(H$_2$O)$_5$(SO$_4$)](H$_2$O)$_2$	6.184	23.600	6.539	101.7	P$\bar{1}$	13d,f	19a	[13]
polyhalite[4]	K$_2$Ca$_2$[Mg(SO$_4$)$_4$(H$_2$O)$_2$]	11.69	16.330	7.600	90.0	F$\bar{1}$	13g	19b,c	[14]
leightonite	K$_2$Ca$_2$[Cu^{2+}(SO$_4$)$_4$(H$_2$O)$_2$]	11.67	16.52	7.492	–	Fmmm	13g	–	(6)
rozenite	[Fe^{2+}(SO$_4$)(H$_2$O)$_4$]	5.97	13.640	7.980	90.4	P2$_1$/n	13i	19d	[15]
starkeyite	[Mg(SO$_4$)(H$_2$O)$_4$]	5.922	13.604	7.905	90.8	P2$_1$/n	13i	19d	[16]
ilesite	[Mn^{2+}(SO$_4$)(H$_2$O)$_4$]	5.94	13.76	8.01	90.8	P2$_1$/n	13i	19d	(6)
aplowite	[Co(SO$_4$)(H$_2$O)$_4$]	5.952	13.576	7.908	90.5	P2$_1$/n	13i	19d	[17]
boyleite	[Zn(SO$_4$)(H$_2$O)$_4$]	5.95	13.60	7.96	90.3	P2$_1$/n	13i	19d	(6)
ungemachite	K$_3$Na$_8$[Fe^{3+}(SO$_4$)$_6$](NO$_3$)$_2$(H$_2$O)$_6$	10.898	a	24.989	–	R$\bar{3}$	13h	19e,f	[18]
humberstonite	K$_3$Na$_6$(Na,Mg)$_2$[Mg(SO$_4$)$_6$](NO$_3$)$_2$(H$_2$O)$_6$	10.906	a	24.395	–	R$\bar{3}$	13h	19e,f	[19]
coquimbite	[Fe^{3+}$_2$(SO$_4$)$_6$(H$_2$O)$_6$]{Fe^{3+}(H$_2$O)$_6$}(H$_2$O)$_6$	10.922	a	17.084	–	P$\bar{3}$1c	13j	19g	[20]
paracoquimbite	[Fe^{3+}$_2$(SO$_4$)$_6$(H$_2$O)$_6$]{Fe^{3+}(H$_2$O)$_6$}(H$_2$O)$_6$	10.926	a	51.300	–	R$\bar{3}$	13j	19g	[21]
metavoltine	K$_2$Na$_6$[Fe^{3+}$_3$O(SO$_4$)$_6$(H$_2$O)$_3$]{Fe^{2+}(H$_2$O)$_6$}(H$_2$O)$_6$	9.575	a	18.170	–	P3	13k	19h	[22]

* M = Fe^{2+}, Ni, Mg, Cu^{2+}, Mn^{2+}, Zn, Al, V^{3+}, Fe^{3+} ¹ α = 112.0°; γ = 100.4°; ² α = 90.5°; γ = 85.2°; ³ α = 94.2°; γ = 96.3; ⁴ α = 91.6°; γ = 91.9

References: [1] Giuseppetti and Tadini (1983), [2] Sabelli (1985a), [3] Tachez et al. (1979), [4] Zhou et al. (1988), [5] Rumanova and Malitskaya (1959), Hawthorne (1985b), (6) Gaines et al. (1997), (7) Schlüter et al. (1999), [8] Jarosch (1985), [9] Giester and Rieck (1995), [10] Hess et al. (1988), [11] Fanfani et al. (1970), [12] Menchetti and Sabelli (1976b), [13] Thomas et al. (1974), [14] Schlatti et al. (1970), [15] Baur (1960), [16] Baur (1962), [17] Kellersohn (1992), [18] Groat and Hawthorne (1986), [19] Burns and Hawthorne (1994), [20] Fang and Robinson (1970a), [21] Robinson and Fang (1971), [22] Giacovazzo et al. (1976a)

linked to Jahn-Teller-distorted {Cu^{2+}(OH)$_4$O$_2$} octahedra *via* apical (and therefore weak) bonds. The strongest bonding in chenite is within sheets parallel to the ($\bar{1}$10) plane and is formed by Pb^{2+} and Cu^{2+} cations and (OH)$^-$ groups (Fig. 18a). The structure of the sheet (Fig. 18b) can be described as based on {(OH)Pb^{2+}$_3$} and [(OH)Cu^{2+}Pb^{2+}$_2$] triangular pyramids or {Pb^{2+}(OH)$_3$} triangular pyramids, [Pb^{2+}(OH)$_4$] tetragonal pyramids, and [Cu^{2+}(OH)$_4$] squares.

The structures of **römerite**, [Fe^{3+}(SO$_4$)$_2$(H$_2$O)$_4$]$_2${Fe^{2+}(H$_2$O)$_6$}, and the minerals of the **halotrichite group**, [M^{2+}(SO$_4$)(H$_2$O)$_5$]{M^{3+}(H$_2$O)$_6$}$_2$(SO$_4$)(H$_2$O)$_5$, are also based on the [$M(T\text{Ø}_4)_2\text{Ø}_4$] cluster, but in the *cis* rather than the *trans* arrangement (Fig. 13e). In addition to this cluster, römerite contains isolated {Fe^{2+}(H$_2$O)$_6$} octahedra (Fig. 18c), and these elements are linked by hydrogen bonding both between unconnected octahedra and

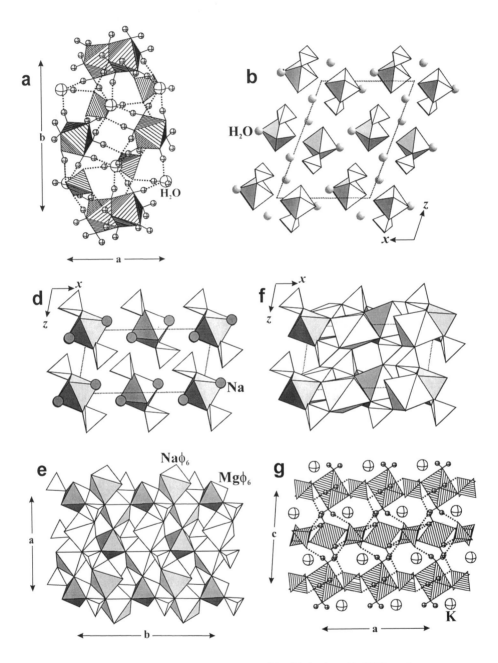

Figure 17. The structures of (a) jurbanite; (b) xitieshanite; (c) blödite, shown as [Mg(H$_2$O)$_4$(SO$_4$)$_2$] clusters and Na atoms; (d,e) blödite as a framework of (MgØ$_6$) octahedra, (NaØ$_6$) octahedra, and (SO$_4$) tetrahedra; (f) mereiterite.

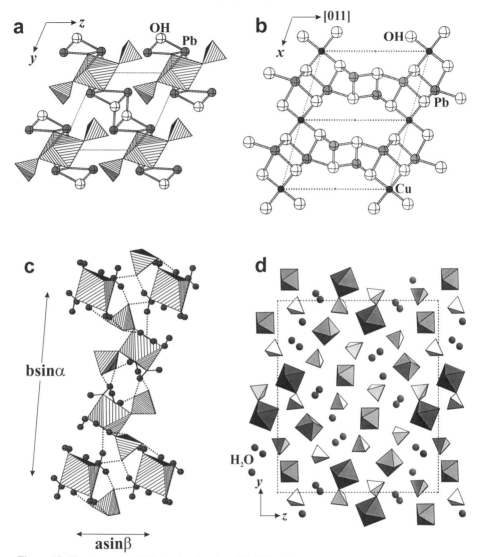

Figure 18. The structures of (a) chenite, based on [Cu(OH)₄(SO₄)₂] clusters, Pb atoms, and (OH)-anions; (b) the sheet of Pb and Cu atoms and (OH)-groups in chenite; (c) römerite; (d) apjohnite.

tetrahedra, and between octahedra and tetrahedra of the same $[M(TO_4)_2\varnothing_4]$ cluster. **Apjohnite**, a mineral of the halotrichite group, contains isolated $[Mn^{2+}(SO_4)_2(H_2O)_4]$ clusters, $\{Al(H_2O)_6\}$ octahedra, (SO_4) tetrahedra, and (H_2O) groups (Fig. 18d) linked solely by hydrogen bonding. The structure of **quenstedtite**, $[Fe^{3+}(H_2O)_4(SO_4)_2]$ $[Fe^{3+}(H_2O)_5(SO_4)](H_2O)_2$, consists of *cis* $[M(T\varnothing_4)_2\varnothing_4]$ clusters, $[M(T\varnothing_4)\varnothing_5]$ clusters, and (H_2O) groups. The clusters are arranged in layers parallel to (010). There are layers of $[M(T\varnothing_4)\varnothing_5]$ clusters either side of $y = 0$, and layers of $[M(TO_4)_2\varnothing_2]$ clusters either side of $y = 1/2$. Like layers are linked directly *via* hydrogen bonds, with donor groups in one layer and acceptor groups in the other layer. Unlike layers are linked *via* hydrogen bonding that involves intermediary (H_2O) groups (at $y \sim \pm 1/4$) that do not bond to any cations

(Fig. 19a). In **polyhalite**, $K_2Ca_2[Mg(SO_4)_4(H_2O)_2]$, the fundamental $[M(TØ_4)_4Ø_2]$ cluster is an $(MgØ_6)$ octahedron that shares four of its vertices with four (SO_4) tetrahedra (Fig. 13f). These clusters are arranged in sheets parallel to (010) (Fig. 19b) and are linked into a framework by Ca and K cations (Fig. 19c). **Leightonite**, $K_2Ca_2[Cu^{2+}(SO_4)_4(H_2O)_2]$, has the same stoichiometry as polyhalite, the cell dimensions of these two minerals are similar, and both have an F-centered unit cell. However, leightonite is reported as orthorhombic and polyhalite as triclinic with $\alpha \approx \beta \approx \gamma \approx 90°$ (Table 7). The difference in symmetry between these two structures is not understood.

The structures of the minerals of the **rozenite group**, e.g. **aplowite**, $[Co(SO_4)(H_2O)_4]$, are based on the $[M_2(TØ_4)_2Ø_8]$ cluster (Fig. 13h), linked solely by hydrogen bonds both within and between adjacent clusters (Fig. 19d). The structure of **ungemachite**, $K_3Na_8[Fe^{3+}(SO_4)_6](NO_3)_2(H_2O)_6$, is based on the $[M(TØ_4)Ø_6]$ cluster shown in Figure 13g. In his work on glaserite-related structures, Moore (1973) called such a cluster *a pinwheel*. The $[Fe(SO_4)_6]$ pinwheels in ungemachite, together with triangular-prismatic-coordinated Na cations, NO_3 groups, and K cations, form layers parallel to (001) (Fig. 19e). Each $[FeØ_6]$ octahedron shares two faces with $\{NaO_3(H_2O)_3\}$ octahedra to form a linear octahedral trimer parallel to [001] (Fig. 19f). The only interlayer linkage is provided by hydrogen bonds from (H_2O) groups of the $\{NaO_3(H_2O)_3\}$ octahedra to sulfate groups of neighboring layers. The structures of **coquimbite**, $[Fe^{3+}_3(SO_4)_6(H_2O)_6]$-$\{Fe^{3+}(H_2O)_6\}(H_2O)_6$, and its polytype, **paracoquimbite**, $[Fe^{3+}_3(SO_4)_6(H_2O)_6]$-$\{Fe^{3+}(H_2O)_6\}(H_2O)_6$, are based on the $[M_3(TO_4)_6]$ cluster shown in Fig. 13j, isolated $\{Fe^{3+}(H_2O)_6\}$ octahedra, and (H_2O) groups. The $[M_3(TO_4)_3]$ clusters are oriented with their long axis parallel to [001]. The $\{Fe^{3+}(H_2O)_6\}$ octahedra lie in the plane at $z = 0$ and 1/2 (Fig. 19g), forming layers of octahedra that are connected into sheets by hydrogen bonds; thus coquimbite and paracoquimbite are polytypic. **Metavoltine**, $K_2Na_6[Fe^{3+}_3O(SO_4)_6(H_2O)_3]_2\{Fe^{2+}(H_2O)_6\}(H_2O)_6$, is built from a complex but elegant $[M_3(TØ_4)_6Ø_4]$ cluster (Fig. 13j) that also occurs in the structures of **Maus's salts** (Scordari et al. 1994 and references therein). These clusters form layers parallel to (001) at $z = 1/4$ and 3/4 (Fig. 19h), and are linked in the plane of the layer by K. Linkage between the layers occurs *via* hydrogen bonds through layers of (H_2O) groups at $z \sim 0$ and 1/2 and involving the isolated $\{Fe^{2+}(H_2O)_6\}$ groups.

Structures with infinite chains

The graphs of chains of octahedra and tetrahedra that occur in sulfate minerals are shown in Figure 20; sulfate minerals based on these chains are listed in Table 8.

The structure of **aluminite**, $[Al_2(OH)_4(H_2O)_3](SO_4)(H_2O)_4$, is based on chains of edge-sharing octahedra (Fig. 21a) that extend along [100] and are linked to (SO_4) groups by hydrogen bonds involving (H_2O) groups. There are two types of H_2O groups in aluminite: (1) those bonded to cations; (2) those not bonded to cations (Fig. 21b). The latter are held in the structure solely by hydrogen bonds, and serve to propagate bonding from hydrogen-bond donors attached to one cation to hydrogen-bond acceptors attached to another cation out-of-range of direct hydrogen bonding. Aluminite is the only known example of an undecorated chain of octahedra in sulfate minerals. In all other minerals based on chains, $(MØ_6)$ octahedra and $(TØ_4)$ tetrahedra link together by sharing common vertices.

The structures of the **chalcanthite-group** minerals, $[M^{2+}(H_2O)_4(SO_4)](H_2O)$, are based on the $[M(TO_4)Ø_4]$ chain with alternating corner-sharing octahedra and tetrahedra (Fig. 21c). The chains extend along [110] and are cross-linked by hydrogen bonds between the octahedral (H_2O) ligands and the tetrahedral ligands. In addition, there is an (H_2O) group that is not directly bonded to a cation; this acts as both a hydrogen-bond

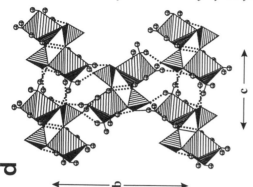

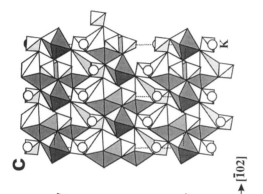

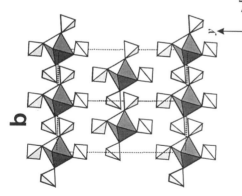

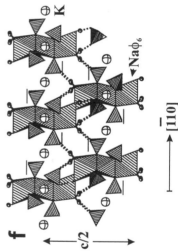

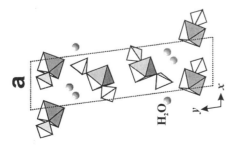

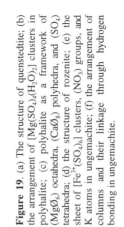

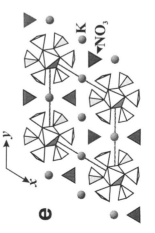

Figure 19. (a) The structure of quenstedtite; (b) the arrangement of $[Mg(SO_4)_4(H_2O_2)]$ clusters in polyhalite; (c) polyhalite as a framework of $(MgØ_6)$ octahedra, $(CaØ_8)$ polyhedra, and (SO_4) tetrahedra; (d) the structure of rozenite; (e) the sheet of $[Fe^{3+}(SO_4)_6]$ clusters, (NO_3) groups, and K atoms in ungemachite; (f) the arrangement of columns and their linkage through hydrogen bonding in ungemachite.

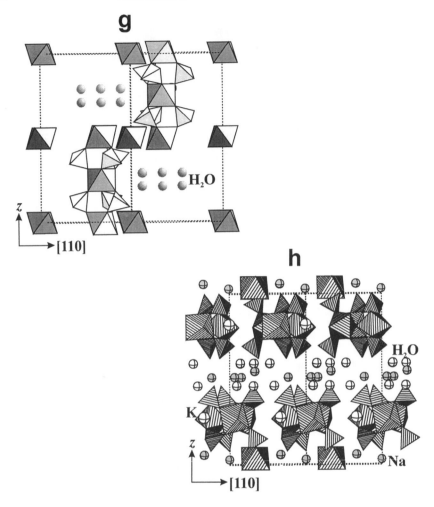

Figure 19, cont'd. (g) The structure of coquimbite, and
(h) the structure of metavoltine.

acceptor and donor, promoting linkage along and between the chains (Fig. 21d). This
$[M(TO_4)]$ chain also occurs in the structures of the arsenate minerals liroconite,
$Cu^{2+}{}_2[Al(AsO_4)(OH)_4](H_2O)_4$, and brassite, $[Mg\{AsO_3(OH)\}(H_2O)_4]$.

The $[M(T\emptyset_4)\emptyset_3]$ chain is the basis of **butlerite**, $[Fe^{3+}(OH)(H_2O)_2(SO_4)]$,
parabutlerite, $[Fe^{3+}(OH)(H_2O)_2(SO_4)]$, and **uklonskovite**, $Na[MgF(H_2O)_2(SO_4)]$. In
this chain (Figs. 20c, 21e), the tetrahedra alternate along the chain and link to *trans* vertices
of the octahedra. This is a 7-Å chain in the terminology of Moore (1970); the repeat
distance of the chains is approximately 7 Å, and this is usually apparent in the cell
dimensions of minerals containing these chains. In butlerite and parabutlerite, the
$[Fe^{3+}(SO_4)(OH)(H_2O)_2]$ chains are linked solely by hydrogen bonds, as there are no
interstitial cations (Figs. 21e to 21g). The chains are extremely similar in these dimorphous
minerals, and the principal structural difference is in the relative disposition of adjacent

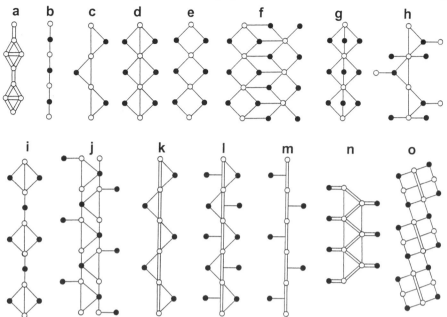

Figure 20. Graphs of octahedral–tetrahedral chains in the structures of sulfate minerals; octahedra are white, tetrahedra are black. See text for details.

chains (see Hawthorne 1990 for details). Uklonskovite consists of topologically identical $[Mg(SO_4)(OH)(H_2O)_2]$ chains, linked by [8]-coordinated Na (Fig. 21h). Hydrogen bonding provides additional linkage from an octahedral ligand in one chain to a tetrahedral ligand in the adjacent chain.

Fibroferrite, $[Fe^{3+}(OH)(H_2O)_2(SO_4)](H_2O)_3$, is a magnificent structure (Fig. 22a,b). The $[M\emptyset_5]$ vertex-sharing chain of octahedra with *cis*-corner-sharing octahedra is unusual; chains of this stoichiometry usually link through *trans* octahedral vertices (e.g. Fig. 21e). The chain in fibroferrite links *via cis* octahedral vertices, thus accounting for its helical configuration (Fig. 22a). This chain is closely related to the chain in butlerite, parabutlerite, and uklonskovite (Fig. 19c) in that they are geometrical isomers and have the same graph (Fig. 20c). Adjacent chains link both by direct hydrogen-bonding between chains, and by a hydrogen-bonding network involving (H_2O) groups not linked to any cations (Fig. 22b). The structures of **sideronatrite**, $Na_2[Fe^{3+}(OH)(SO_4)_2](H_2O)_3$, and **metasideronatrite**, $Na_4[Fe^{3+}_2(OH)_2(SO_4)_4](H_2O)_3$, are based on the chain that has the graph shown in Figure 20d; this is the type-VII 7 Å chain of Moore (1970). These minerals show considerable disorder in terms of the relations between adjacent chains in their structures (Scordari 1981c). This chain topology is also present in the phosphate mineral tancoite, $Na_2LiH[Al(PO_4)_2(OH)]$ (Hawthorne 1983b). The structures of the **botryogen group**, $[M^{2+}Fe^{3+}(OH)(H_2O)_6(SO_4)_2]$ (H_2O), are also based on the 7 Å chain of the form $[M(TO_4)_2\emptyset]$ (type-I according to Moore 1970). The chain in botryogen links to a $(Mg\emptyset_6)$ octahedron through one of its vertices (Fig. 22c). The packing of the chains in the structure is shown in Figure 22d; interchain linkage involves only hydrogen bonding.

The chain in **kröhnkite**, $Na_2[Cu^{2+}(H_2O)_2(SO_4)_2]$, can be described as based on *cis* $[M(TO_4)_2\emptyset_4]$ clusters (Fig. 13e) polymerized by corner-sharing between polyhedra (Fig. 20e, 22e). The chains are linked together by interstitial Na and hydrogen bonds from

Table 8. Sulfate minerals with chains of (SO$_4$) tetrahedra and ($M\Phi_6$) octahedra.*

Name	Formula	a (Å)	b (Å)	c (Å)	α (°)	β (°)	γ (°)	S. G.	Graph	Fig.	Ref.
aluminite	[Al$_2$(OH)$_4$(H$_2$O)$_3$](SO$_4$)(H$_2$O)$_4$	7.440	15.583	11.700	–	110.2	–	$P2_1/c$	20a	21a,b	[1]
chalcanthite	[Cu^{2+}(H$_2$O)$_4$(SO$_4$)](H$_2$O)	6.141	10.736	5.986	82.3	107.4	102.7	$P\bar{1}$	20b	21c,d	[2]
pentahydrite	[Mg(H$_2$O)$_4$(SO$_4$)](H$_2$O)	6.314	10.565	6.030	81.1	109.8	105.1	$P\bar{1}$	20b	21c,d	[3]
siderotil	[Fe^{2+}(H$_2$O)$_4$(SO$_4$)](H$_2$O)	6.26	10.63	6.06	92.1	110.2	102.9	$P\bar{1}$	20b	21c,d	(4)
jokokuite	[Mn^{2+}(H$_2$O)$_4$(SO$_4$)](H$_2$O)	6.37	10.77	6.13	98.8	110.0	102.2	$P\bar{1}$	20b	21c,d	(4)
butlerite	[Fe^{3+}(OH)(H$_2$O)$_2$(SO$_4$)]	6.500	7.370	5.840	–	108.4	–	$P2_1/m$	20c	21e,f	[5]
parabutlerite	[Fe^{3+}(OH)(H$_2$O)$_2$(SO$_4$)]	7.380	20.130	7.220	–	–	–	$Pmnb$	20c	21g	[6]
uklonskovite	Na[MgF(H$_2$O)$_2$(SO$_4$)]	7.202	7.214	5.734	–	113.2	–	$P2_1/m$	20c	21h	[7]
fibroferrite	[Fe^{3+}(OH)(H$_2$O)$_2$(SO$_4$)](H$_2$O)$_3$	24.177	a	7.656	–	–	–	$R\bar{3}$	20c	22a,b	[8]
botryogen	[MgFe^{3+}(OH)(H$_2$O)$_6$(SO$_4$)$_2$](H$_2$O)	10.526	17.872	7.136	–	100.1	–	$P2_1/n$	20h	22c,d	[9]
zincobotryogen	[ZnFe^{3+}(OH)(H$_2$O)$_6$(SO$_4$)$_2$](H$_2$O)	10.517	17.847	7.133	–	100.1	–	$P2_1/n$	20h	22c,d	[10]
kröhnkite	Na$_2$[Cu^{2+}(H$_2$O)$_2$(SO$_4$)$_2$]	5.807	12.656	5.517	–	108.3	–	$P2_1/c$	20e	22e,f	[11]
krausite	K[Fe^{3+}(H$_2$O)$_2$(SO$_4$)$_2$]	7.920	5.146	9.014	–	102.8	–	$P2_1/m$	20f	22g,h	[12]
sideronatrite	Na$_2$[Fe^{3+}(OH)(SO$_4$)$_2$](H$_2$O)$_3$	7.290	20.560	7.170	–	–	–	$Pnn2$	20d	–	[13]
metasideronatrite	Na$_4$[Fe^{3+}$_2$(OH)$_2$(SO$_4$)$_4$](H$_2$O)$_3$	7.357	16.002	7.102	–	–	–	$Pbnm$	20d?	–	(14)
ferrinatrite	Na$_3$[Fe^{3+}(SO$_4$)$_3$](H$_2$O)$_3$	15.566	a	8.690	–	–	–	$P\bar{3}$	20g	23a,b	[15]
copiapite	[Fe^{3+}(OH)(H$_2$O)$_4$(SO$_4$)$_3$]$_2${Fe^{2+}(H$_2$O)$_6$}(H$_2$O)$_6$	7.390	18.213	7.290	93.7	102.1	99.3	$P\bar{1}$	20i	23c,d	[16]
magnesiocopiapite	[Fe^{3+}$_2$(OH)(H$_2$O)$_4$(SO$_4$)$_3$]$_2${Mg(H$_2$O)$_6$}(H$_2$O)$_6$	7.370	18.890	7.240	91.3	102.4	99.0	$P\bar{1}$	20i	23c,d	[17]
cuprocopiapite	[Fe^{3+}$_2$(OH)(H$_2$O)$_4$(SO$_4$)$_3$]$_2${Cu^{2+}(H$_2$O)$_6$}(H$_2$O)$_6$	7.31	18.15	7.25	92.5	102.3	100.4	$P\bar{1}$	20i	23c,d	(4)
ferricopiapite	[Fe^{3+}$_2$(OH)(H$_2$O)$_4$(SO$_4$)$_3$]$_2${(Fe^{3+}$_{0.67}$□$_{0.33}$)(H$_2$O)$_6$}(H$_2$O)$_6$	7.390	18.213	7.290	93.7	102.2	99.3	$P\bar{1}$	20i	23c,d	[18]
calciocopiapite	[Fe^{3+}$_2$(OH)(H$_2$O)$_4$(SO$_4$)$_3$]$_2${Ca(H$_2$O)$_6$}(H$_2$O)$_6$	7.44	18.79	7.22	94.7	104.7	102.2	$P\bar{1}$	20i	23c,d	(4)
zincocopiapite	[Fe^{3+}$_2$(OH)(H$_2$O)$_4$(SO$_4$)$_3$]$_2${Zn(H$_2$O)$_6$}(H$_2$O)$_6$	7.33	18.72	7.35	91.5	102.1	98.7	$P\bar{1}$	20i	23c,d	(4)
aluminocopiapite	[Fe^{3+}$_2$(OH)(H$_2$O)$_4$(SO$_4$)$_3$]$_2${Al$_{0.67}$□$_{0.33}$}(H$_2$O)$_6$}(H$_2$O)$_6$	7.251	18.161	7.267	94.0	102.2	98.0	$P\bar{1}$	20i	23c,d	(19)
destinezite	[Fe^{3+}$_2$(OH)(H$_2$O)$_2$(PO$_4$)(SO$_4$)](H$_2$O)	9.570	9.716	7.313	98.7	107.9	63.9	$P\bar{1}$	20j	23e,f	[20]
linarite	Pb^{2+}[Cu^{2+}(OH)$_2$(SO$_4$)]	9.701	5.650	4.690	–	102.7	–	$P2_1/m$	20k	23g,h	[21]

		a	b	c	α	β	γ	Sp. Gr.			
wherryite	$Pb^{2+}_7[Cu^{2+}(OH)(SO_4)(SiO_4)]_2(SO_4)_2$	20.789	5.787	9.142	—	91.2	—	$C2/m$	20l	23i,j	[22]
tsumebite	$Pb^{2+}_2[Cu^{2+}(OH)(PO_4)(SO_4)]$	7.85	5.80	8.70	—	111.5	—	$P2_1/m$	20l	—	[23]
arsentsumebite	$Pb^{2+}_2[Cu^{2+}(OH)(AsO_4)(SO_4)]$	7.84	5.92	8.85	—	112.6	—	$P2_1/m$	20l	—	[4]
caledonite	$Pb^{2+}_5[Cu^{2+}_2(OH)_6(SO_4)_2](SO_4)(CO_3)$	20.089	7.146	6.560	—	—	—	$Pmn2_1$	20m	24a,b	[24]
chlorothionite	$K_2[Cu^{2+}Cl_2(SO_4)]$	7.732	6.078	16.292	—	—	—	$Pnma$	20n	24c,d	[25]
amarantite	$[Fe^{3+}_2O(H_2O)_4(SO_4)_2](H_2O)_3$	8.976	11.678	6.698	95.7	90.4	97.2	$P\bar{1}$	20o	24e,f	[26]
hohmannite	$[Fe^{3+}_2O(H_2O)_4(SO_4)_2](H_2O)_4$	9.148	10.922	7.183	90.3	90.8	104.4	$P\bar{1}$	20o	24e,g	[27]

* M = Cu, Fe, Ni, Mg, Zn, Mn, Co, Al; ϕ = O, OH, H_2O, F, Cl

References: [1] Sabelli and Ferroni (1978), [2] Bacon and Titterton (1975), [3] Baur and Rolin (1972), [4] Gaines et al. (1997), [5] Fanfani et al. (1971), [6] Borene (1970), [7] Sabelli (1985b), [8] Scordari (1981a), [9] Süsse (1968), [10] Hexiong and Pingqiu (1988), [11] Hawthorne and Ferguson (1975a), [12] Effenberger et al. (1986), [13] Scordari (1981b), [14] Scordari et al. (1982), [15] Scordari (1977), Mereiter (1976), [16] Fanfani et al. (1973), [17] Süsse (1970), [18] Bayliss and Atencio (1985), (19) Jolly and Foster (1967), [20] Peacor et al. (1999a), [21] Effenberger (1987), [22] Cooper and Hawthorne (1994), [23] Fanfani and Zanazzi (1967), [24] Giacovazzo et al. (1973), [25] Giacovazzo et al. (1976b), [26] Süsse (1967), Giacovazzo and Menchetti (1969), [27] Scordari (1978)

Figure 21. (a) The chain of edge-sharing (AlO_6) octahedra in aluminite; (b) the linkage of octahedral chains, (SO_4) tetrahedra, and (H_2O) groups through hydrogen bonds (shown as dashed lines) in aluminite; (c) the chain of alternating corner-sharing octahedra and tetrahedra in the minerals of the chalcanthite group; (d) interchain linkage to (SO_4) tetrahedra and (H_2O) groups through hydrogen bonds (shown as dashed lines) in chalcanthite.

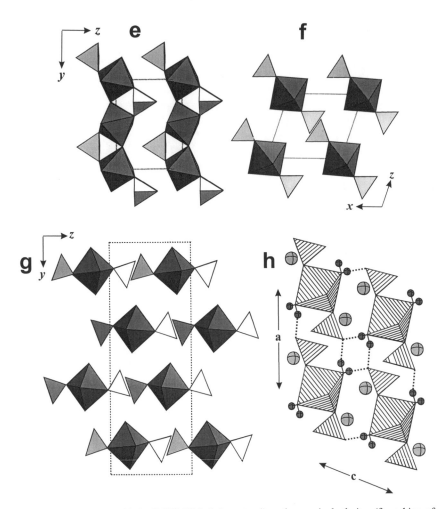

Figure 21, cont'd. (e) the $[M(TO_4)\varnothing_3]$ chains extending along y in butlerite; (f) packing of $[M(T\varnothing_4)\varnothing_3]$ chains in butlerite; (g) packing of $[M(T\varnothing_4)\varnothing_3]$ chains in parabutlerite; (h) the hydrogen bonds between adjacent $[M(T\varnothing_4)\varnothing_3]$ chains in uklonskovite (Na atoms are shown as large circles).

octahedral donor ligands to tetrahedral acceptor ligands (Fig. 22f). This type of $[M(TO_4)_2\varnothing_4]$ chain is common in arsenate and phosphate minerals, and is the basis of the minerals of the brandtite, $Ca_2[Mn^{2+}(AsO_4)_2(H_2O)_2]$, talmessite, $Ca_2[Mg(AsO_4)_2(H_2O)_2]$, and fairfieldite, $Ca_2[Mn^{2+}(PO_4)_2(H_2O)_2]$, groups (Hawthorne 1985a). **Krausite**, $K[Fe^{3+}(H_2O)_2(SO_4)_2]$, is based on a ribbon of double kröhnkite chains (Figs. 20f, 22g) linked by interstitial K (Fig. 22h). This ribbon also occurs as a fragment of the sheet that is the basis of the structure of ransomite.

Ferrinatrite, $Na_3(H_2O)_3[Fe^{3+}(SO_4)_3]$, is based on a chain consisting of octahedra linked by corner-sharing with tetrahedra such that all octahedron vertices link to tetrahedra, and half the tetrahedron vertices link to octahedra (Fig. 23a); the graph of this arrangement is shown in Figure 20g. This chain is closely related to the finite cluster in coquimbite and paracoquimbite (Fig. 13i). These columns occur at the vertices of a 3^6 plane net (Fig. 23b),

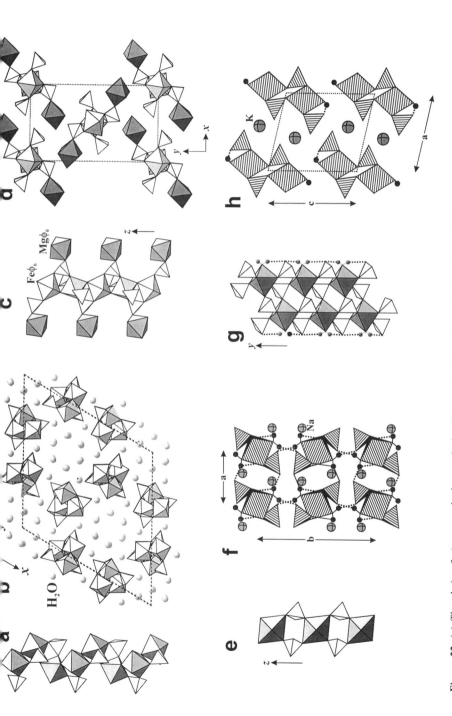

Figure 22. (a) The chain of *cis*-corner-sharing octahedra (decorated with tetrahedra) in fibroferrite; (b) the arrangement of chains and (H_2O) groups in fibroferrite; (c) the open-branched octahedral–tetrahedral chain in botryogen; (d) packing of chains in botryogen; (e) the $[M(TO_4)_2\phi_4]$ chain in krӧhnkite; (f) packing of $[M(TO_4)_2\phi_4]$ chains in krӧhnkite; (g) the double-krӧhnkite-like ribbon in krausite; (h) packing of ribbons in the structure of krausite.

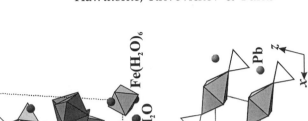

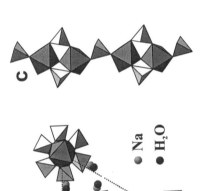

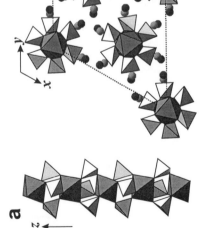

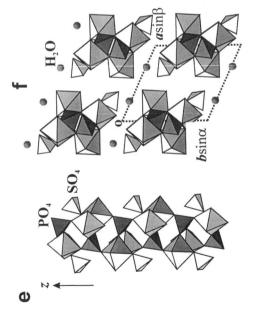

Figure 23. (a) The $[M(TO_4)_3]$ octahedral–tetrahedral chain in ferrinatrite; (b) the arrangement of $[M(TO_4)_3]$ chains in ferrinatrite; (c) the $[M_2(TO_4)_3O_3]$ chain in the minerals of the copiapite group; (d) the structure of copiapite; (e) the chain in destinezite; (f) packing of the chains in destinezite; (g) the $[M(TO_4)O_2]$ chain in linarite; (h) packing of $[M(TO_4)O_2]$ chains in linarite.

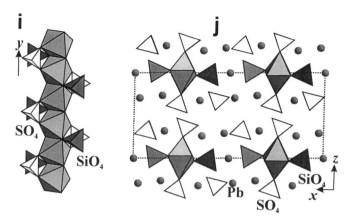

Figure 23, cont'd. (i) the $[M(TO_4)_2\varnothing]$ chain in wherryite; (j) packing of $[M(TO_4)_2\varnothing]$ chains in wherryite.

and are linked by interstitial Na and (H_2O) groups.

In the **copiapite-group** minerals, $[Fe^{3+}_2(OH)(H_2O)_4(SO_4)_3]_2\{M^{2+}(H_2O)_6\}(H_2O)_6$, corner-sharing $[M_2(TO_4)_2\varnothing_7]$ clusters link through additional tetrahedra to form infinite $[M_2(TO_4)_3\varnothing_5]$ chains (Fig. 20i) that extend along [101] (Fig. 23c,d). These chains are linked together by hydrogen bonds involving unconnected $\{Fe^{2+}(H_2O)\}$ octahedra and interstitial (H_2O) groups. There is some question as to the chemical formulae of ferricopiapite and aluminocopiapite. In Table 8, these two minerals are written with the unconnected $\{M(H_2O)_6\}$ octahedra as $M = (Fe^{3+}_{0.67}\square_{0.33})$ and $(Al_{0.67}\square_{0.33})$, as per Fanfani et al. (1973). However, these formulae seem unlikely from a bond-valence perspective. Formulae have also been written as $[Fe^{3+}_4O(OH)(H_2O)_8(SO_4)_6]\{Fe^{3+}(H_2O)_6\}(H_2O)_6$, but it is not clear that this fits with the available crystal-structure data; more structural work is needed. The $[M_2(TO_4)_2\varnothing_7]$ cluster is the basis of an extensive hierarchy of (mainly phosphate) structures (Hawthorne 1979). The structure of **destinezite**, $[Fe^{3+}_2(OH)(H_2O)_5(PO_4)(SO_4)](H_2O)$, is based on the complex chain shown in Figure 23e; the connectivity of this chain is clearly represented by its graph given in Figure 20j. The chains in destinezite and copiapite are somewhat related; this is most easily seen by comparison of their graphs (Fig. 20i,j). First, break one of the o–t linkages in graph 20i; next, fuse two of these modified graphs *via* o–t linkage: the result is the graph of Figure 20j. The chains in destinezite extend along [001] (Fig. 23e) and are linked solely by hydrogen bonding both directly and through intermediate (H_2O) groups (Fig. 23f).

The chain shown in Figure 23g is the basis of the structure of **linarite**, $Pb^{2+}[Cu^{2+}(OH)_2(SO_4)]$; octahedra share two *trans* edges to form an $[M\varnothing_4]$ chain that is decorated by flanking tetrahedra that adopt a staggered arrangement either side of the chain (Fig. 20k). In projection, the chains are arranged at the vertices of a primitive monoclinic lattice (Fig. 23h), and are linked primarily by [9]-coordinated Pb^{2+}. The $[M(TO_4)\varnothing_2]$ can be recognized as a parent chain for many other octahedral–tetrahedral structures. The structure of **wherryite**, $Pb^{2+}_7[Cu^{2+}(OH)(SO_4)(SiO_4)]_2(SO_4)_2$ (Fig. 22i,j), is based on linarite chains that are decorated by (SiO_4) tetrahedra (Figs. 20l, 23i). The $[M(TO_4)_2\varnothing]$ chains extend along the c axis and pack at the vertices of a primitive orthogonal plane lattice (Fig. 23j). There are additional (SO_4) groups in wherryite that do not link to the $(Cu^{2+}\varnothing_6)$ octahedra; the chains and the isolated (SO_4) tetrahedra are linked by [7]- and [8]-coordinated Pb^{2+} cations (Fig. 23j). The chain shown in Figure 23i is a common constituent of oxysalt minerals, occurring in brackebuschite-, fornacite- and vaquelinitegroup minerals (Hawthorne 1990). Among these groups, only **tsumebite**, $Pb^{2+}_2[Cu^{2+}(OH)(PO_4)(SO_4)]$, and **arsentsumebite**, $Pb^{2+}_2[Cu^{2+}(OH)(AsO_4)(SO_4)]$, are sulfate minerals. In the structure of tsumebite, $[Cu^{2+}(OH)(PO_4)(SO_4)]$, chains extend along the b axis, giving the typical ~5.5 Å repeat, and are cross-linked by [9]-coordinated Pb^{2+} cations. The structure of **caledonite**, $Pb^{2+}_5[Cu^{2+}_2(OH)_6(SO_4)_2](SO_4)(CO_3)$, is based on an $[M(TO_4)\varnothing_3]$ chain of edge-sharing octahedra decorated by (SO_4) tetrahedra (Fig. 24a). Note that the chain corresponds to the graph of Figure 20m. Thus both the chains shown in Figures 23g and 24a correspond to Figure 20k; these chains are geometrical isomers and hence have the same graph. As in wherryite, there are additional (SO_4) tetrahedra in the structure of caledonite and linkage is provided by [8]- and [9]-coordinated Pb^{2+} (Fig. 24b).

The $[M(TO_4)\varnothing_2]$ chain in **chlorothionite**, $K_2[Cu^{2+}Cl_2(SO_4)]$, is extremely unusual in that it involves edge-sharing between sulfate tetrahedra and divalent-metal octahedra (Fig. 24c); this type of edge-sharing seems to be much more common in Cu^{2+} minerals than in other divalent-metal oxysalts. Moreover, the stoichiometry of the chain of octahedra is $[MO_4]$, but the shared edges on an octahedron are *cis* instead of the more usual *trans* (compare Figs. 24a and 24c), which also leads to second-nearest-neighbor octahedra sharing vertices; this is more easily apparent in the graph of the $[M(TO_4)\varnothing_2]$ chain in Figure 20n. The chains are cross-linked by [6]-coordinated K, an unusually low coordination number for K (Fig. 24d).

The structures of **amarantite**, $[Fe^{3+}_2O(H_2O)_4(SO_4)_2](H_2O)_3$, and **hohmannite**, $[Fe^{3+}_2O(H_2O)_4(SO_4)_2](H_2O)_4$, are based on the chain shown in Figure 24e (its graph is given in Figure 20o). Two octahedra share an edge to form an $[M_2\varnothing_{10}]$ dimer. Two additional octahedra link to either end of the shared edges of the dimer to form an $[M_4\varnothing_{20}]$ tetramer. Two tetrahedra link between corners of adjacent octahedra, and further link adjacent decorated tetramers to form the complex chain shown in Figures 24e and 20o. The structures of amarantite and hohmannite (Figs. 24f, and 24g respectively) differ with regard to the amount of interstitial (H_2O) groups, which results in different packing of the chains.

Structures with infinite sheets

The sulfate minerals with infinite sheets can be subdivided into two groups: (a) minerals with brucite-like sheets, and (b) other minerals.

The structures belonging to the first group (Table 9) are based on sheets of octahedra closely related to the sheet of octahedra in brucite, $Mg(OH)_2$. The first examples are the minerals of the **hydrotalcite group**, ideally $[M^{2+}_6M^{3+}_2(OH)_{16}](TO_n)(H_2O)_m$, where $TO_n = (CO_3)$ in hydrotalcite itself and (SO_4) in many other minerals of this group (Table 9). A key feature of this group is the substitution of M^{3+} cations for M^{2+} cations within the brucite-like sheet, as this makes the sheet positively charged, rather than negatively charged as is the case for the structural unit in most minerals. Thus, the interstitial species of the minerals in this group require a net negative charge, which is the reason for the presence of interstitial $(SO_4)^{2-}$ and $(CO_3)^{2-}$ groups in these minerals. Many of these minerals are poorly crystalline because of extensive stacking disorder. However, this is not so for **shigaite**, $Na[AlMn^{2+}_2(OH)_6]_3 (SO_4)_2(H_2O)_{12}$, the structural unit of which is a sheet of edge-sharing $(Mn^{2+}\varnothing_6)$ and $(Al\varnothing_6)$ octahedra (Fig. 25a). These sheets are held together by hydrogen bonding involving $\{Na(H_2O)_6\}$ octahedra, (SO_4) tetrahedra, and (H_2O) groups located in the interlayer (Figs. 25b,c). This group shows a large diversity in interlayer species, and the relations of adjacent sheets cause significant variation in unit-cell parameters. Bookin and Drits (1993) and Bookin et al. (1993) examined the hydrotalcite group of minerals and showed that polytypism is mainly a function of the nature of the interlayer anion.

The decoration of the brucite-like sheet of edge-sharing octahedra by tetrahedra

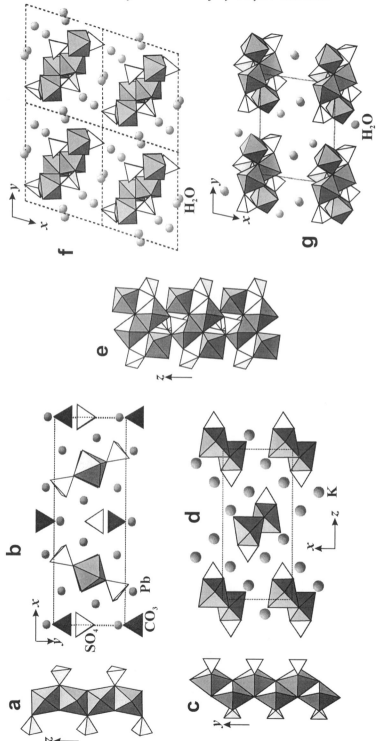

Figure 24. (a) the [$M(TØ_4)Ø_3$] chain in caledonite; (b) packing of chains and the interstitial species in caledonite; (c) the [$M(TØ_4)Ø_2$] chain in chlorothionite; (d) packing of chains in chlorothionite; (e) the complex [$M_2(TØ_4)_2Ø_3$] chain in amarantite; (f,g) the arrangement of chains in (f) amarantite and (g) hohmannite.

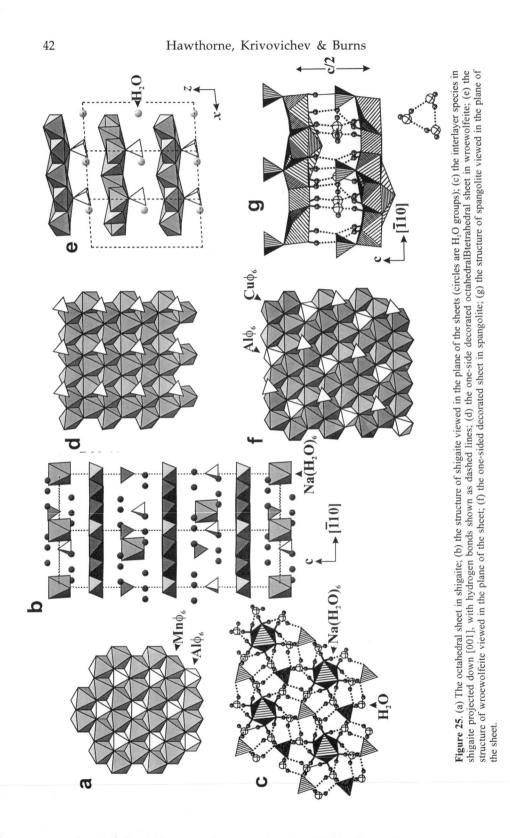

Figure 25. (a) The octahedral sheet in shigaite; (b) the structure of shigaite viewed in the plane of the sheets (circles are H_2O groups); (c) the interlayer species in shigaite projected down [001], with hydrogen bonds shown as dashed lines; (d) the one-side decorated octahedral-tetrahedral sheet in wroewolfeite; (e) the structure of wroewolfeite viewed in the plane of the sheet; (f) the one-sided decorated sheet in spangolite; (g) the structure of spangolite viewed in the plane of the sheet.

Table 9. Sulfate minerals with brucite-like octahedral ($M\emptyset_2$) sheets.

Name	Formula	a (Å)	b (Å)	c (Å)	β (°)	S. G.	Fig.	Ref.
honessite	$[Ni_6Fe^{3+}{}_2(OH)_{16}](SO_4)(H_2O)_4$	3.083	a	26.71	–	$R3m$	–	(1)
hydrohonessite	$[Ni_6Fe^{3+}{}_2(OH)_{16}](SO_4)(H_2O)_7$	3.09	a	10.80	–	?	–	(1)
glaucocerinite	$[Zn_5Al_3(OH)_{16}](SO_4)_{1.5}(H_2O)_9$	3.057	a	32.52	–	?	–	(1)
woodwardite	$[Cu^{2+}{}_4Al_2(OH)_{12}](SO_4)(H_2O)_{2-4}$	3.00	a	27.3	–	?	–	(1)
hydrowoodwardite	$[Cu^{2+}{}_{1-x}Al_x(OH)_2](SO_4)_{x/2}(H_2O)_n$, $x < 0.67$, $n >\sim 3x/2$	3.07	a	31.9	–	$R\bar{3}m$	–	(2)
carrboydite	$(Ni,Cu^{2+})_{14}Al_9(OH)_{43}(SO_4,CO_3)_6(H_2O)_7$	9.14	a	10.34	–	?	25a,b,c	(1)
motukoreaite	$Na_{0.6}[Mg_{5.6}Al_{3.4}(OH)_{18}](SO_4,CO_3)_2(H_2O)_{12}$	9.172	a	33.510	–	$R\bar{3}m$	25a,b,c	[3]
shigaite	$Na[AlMn^{2+}{}_2(OH)_6]_3(SO_4)_2(H_2O)_{12}$	9.512	a	33.074	–	$R\bar{3}$	25a,b,c	[4]
wermlandite	$[Mg_7(Al,Fe^{3+})_2(OH)_{18}](Ca,Mg)(SO_4)_2(H_2O)_{12}$	9.303	a	22.570	–	$P\bar{3}c$	25a,b,c	[5]
mountkeithite	$(Mg,Ni)_{11}(Fe^{3+},Cr^{3+})_3(OH)_{24}(SO_4,CO_3)_{3.5}(H_2O)_{11}$	10.698	a	22.545	–	?	25a,b,c	(1)
kuzelite	$[Ca_4Al_2(OH)_{12}](SO_4)(H_2O)_6$	5.759	a	26.795	–	$R\bar{3}$	25a,b,c	[6]
langite	$[Cu^{2+}{}_4(OH)_6(H_2O)(SO_4)](H_2O)$	7.118	6.034	11.209	90	Pc	–	[7]
posnjakite	$[Cu^{2+}{}_4(OH)_6(H_2O)(SO_4)](H_2O)$	10.578	6.345	7.863	118.0	Pa	–	[8]
wroewolfeite	$[Cu^{2+}{}_4(OH)_6(H_2O)(SO_4)](H_2O)$	6.045	5.646	14.337	93.4	Pc	25d,e	[9]
spangolite	$[Cu^{2+}{}_6Al(OH)_{12}Cl(SO_4)](H_2O)_3$	8.254	a	14.354	–	$P31c$	25e,f	[10]
schulenbergite	$[Cu^{2+}{}_7(OH)_{10}(SO_4)_2(H_2O)_2](H_2O)$	8.211	a	7.106	–	$P\bar{3}$	26a	[11]
devilline	$Ca(H_2O)_3[Cu^{2+}{}_4(OH)_6(SO_4)_2]$	20.870	6.135	22.191	102.7	$P2_1/c$	26b	[12]
lautenthalite	$Pb(H_2O)_3[Cu^{2+}{}_4(OH)_6(SO_4)_2]$	21.642	6.040	22.544	108.2	$P2_1/c$	26b	(13)
campigliaite	$Mn(H_2O)_4[Cu^{2+}{}_4(OH)_6(SO_4)_2]$	21.707	6.098	11.245	100.3	$C2$	26c	[14]
ktenasite	$Zn(H_2O)_6[Cu^{2+}{}_4(OH)_6(SO_4)_2]$	5.589	6.166	23.751	95.6	$P2_1/c$	26d	[15]
niedermayrite	$Cd(H_2O)_4[Cu^{2+}{}_4(OH)_6(SO_4)_2]$	5.543	21.995	6.079	92.04	$P2_1/c$	26e	[16]
chalcophyllite	$[Cu^{2+}{}_9Al(OH)_{12}(H_2O)_6(AsO_4)_2](SO_4)_{1.5}(H_2O)_{12}$	10.756	a	28.678	–	$R\bar{3}$	28a,b	[17]
mooreite	$[Mg_9Zn_4Mn^{2+}{}_2(OH)_{26}](SO_4)_2(H_2O)_8$	11.147	20.350	8.202	92.7	$P2_1/a$	27c,d	[18]
lawsonbauerite	$[(Mn^{2+},Mg)_9Zn_4(OH)_{22}](SO_4)_2(H_2O)_8$	10.50	9.64	16.41	95.2	$P2_1/c$	–	[19]
torreyite	$[(Mg,Mn^{2+})_9Zn_4(OH)_{22}](SO_4)_2(H_2O)_8$	10.619	9.292	16.486	95.4	$P2_1/c$	–	(1)
christelite[1]	$Zn(H_2O)_4[Zn_2Cu^{2+}{}_2(OH)_6(SO_4)_2]$		6.336	10.470	90.06	$P\bar{1}$	–	[20]
serpierite	$Ca(H_2O)_3[Cu^{2+}{}_4(OH)_6(SO_4)_2]$	22.186	6.25	21.853	113.4	$C2/c$	–	[21]
orthoserpierite	$Ca(H_2O)_3[Cu^{2+}{}_4(OH)_6(SO_4)_2]$	22.10	6.20	20.39	–	$Pca2_1$	–	(1)
namuwite	$[(Zn,Cu^{2+})_4(OH)_6(H_2O)(SO_4)](H_2O)_3$	8.331	a	10.54	–	$P\bar{3}$	–	[22]
gordaite	$[Zn_4(OH)_6Cl(SO_4)]\{Na(H_2O)_6\}$	8.356	a	13.025	–	$P\bar{3}$	27e,f	[23]
bechererite	$[Zn_7Cu^{2+}(OH)_{13}\{(SiO(OH)_3\}(SO_4)]$	8.319	a	7.377	–	$P3$	27g	[24]
ramsbeckite	$[(Cu^{2+},Zn)_{15}(OH)_{22}(SO_4)_4](H_2O)_6$	16.088	15.576	7.102	90.22	$P2_1/a$	27h,i	[25]

[1] $\alpha = 94.32°$, $\gamma = 90.27°$ *References*: (1) Gaines et al. (1997), (2) Witzke (1999), [3] Rius and Plana (1986), [4] Cooper and Hawthorne (1996b), [5] Rius and Allmann (1984), [6] Allmann (1977), Pollmann et al. (1997), [7] Galy et al. (1984), [8] Mellini and Merlino (1979), [9] Hawthorne and Groat (1985), [10] Hawthorne et al. (1993), Merlino et al. (1992), [11] Mumme et al. (1994), [12] Sabelli and Zanazzi (1972), (13) Medenbach and Gebert (1993), [14] Menchetti and Sabelli (1982), [15] Mellini and Merlino (1978), [16] Giester et al. (1998), [17] Sabelli (1980), [18] Hill (1980), [19] Treiman and Peacor (1982), [20] Adiwidjaja et al. (1996), [21] Sabelli and Zanazzi (1968), [22] Groat (1996), [23] Adiwidjaja et al. (1997), [24] Hoffmann et al. (1997), Giester and Rieck (1996), [25] Effenberger (1988)

produces sheets that are the basis of many Cu^{2+} and Zn sulfate minerals. A detailed consideration of the bond topology of these structures and other possible structural arrangements is given by Hawthorne and Schindler (2000). A key feature of these structures is the local bond-valence requirement that three $Cu–\emptyset_{apical}$ bonds must meet at a single anion if an (SO_4) or (H_2O) group is also to attach to that anion. The disposition of the various $Cu–\emptyset_{apical}$ bonds gives rise to different types of sheets. One-side-decorated sheets occur in the structures of the polymorphs **wroewolfeite**, $[Cu^{2+}{}_4(OH)_6(H_2O)(SO_4)](H_2O)$ (Fig. 25d,e), **langite**, $[Cu^{2+}{}_4(OH)_6(H_2O)(SO_4)](H_2O)$, and **posnjakite**, $[Cu^{2+}{}_4(OH)_6(H_2O)(SO_4)](H_2O)$, as well as **spangolite**, $[Cu^{2+}{}_6Al(OH)_{12}Cl(SO_4)](H_2O)_3$ (Fig. 25f,g). The sheets are linked through hydrogen bonds involving interlayer (H_2O) groups. In spangolite, these (H_2O) groups form the

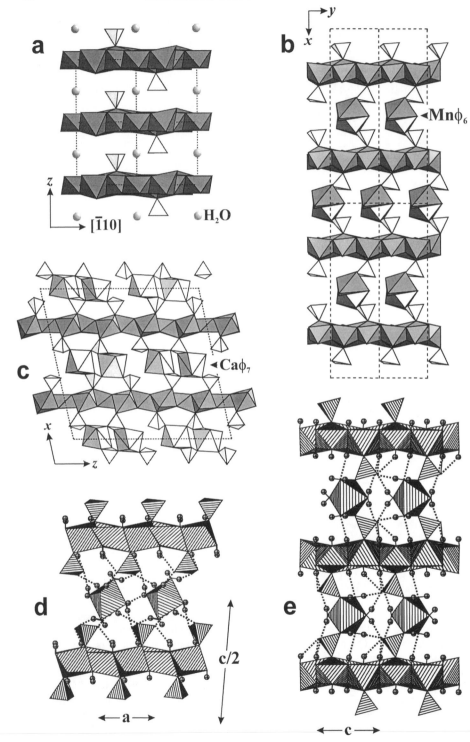

Figure 26 (opposite page). The structures of (a) schulenbergite; (b) devilline; (c) campigliaite; (d) ktenasite; (e) niedermayrite.

H-bonded cluster $(H_2O)_3$ shown in Figure 25g. Two-sided tetrahedral decoration of the brucite-like sheet occurs in the structures of **schulenbergite**, $[(Cu^{2+},Zn)_7(OH)_{10}(SO_4)_2(H_2O)_2](H_2O)$ (Fig. 26a), **devilline**, $Ca(H_2O)_3[Cu^{2+}_4(OH)_6(SO_4)_2]$ (Fig. 26b), **campigliaite**, $Mn^{2+}(H_2O)_4[Cu^{2+}_4(OH)_6(SO_4)_2]$ (Fig. 26c), **ktenasite**, $Zn(H_2O)_6-[(Cu^{2+},Zn)_4(OH)_6(SO_4)_2]$ (Fig. 26d), **niedermayrite**, $Cd(H_2O)_4[Cu^{2+}_4(OH)_6(SO_4)_2]$ (Fig. 26e), **christelite**, $Zn(H_2O)_4$ $[Zn_2Cu^{2+}_2(OH)_6(SO_4)_2]$, and **serpierite**, $Ca(H_2O)_3-[(Cu^{2+},Zn)_4(OH)_6(SO_4)_2]$. Among these minerals, only schulenbergite contains no interlayer cations; in other structures, the interlayer cations (Ca, Zn, Mn^{2+}, Cd, Pb^{2+}) provide additional linkage between the sheets (devilline, campigliaite, niedermayrite, serpierite) or form unconnected polyhedra which are held in the structure by hydrogen bonds only (ktenasite; Fig. 26d). The hydrogen bonding is essential in linking the sheets together in these minerals (Fig. 26). It is probable that numerous other minerals of this motif await discovery.

If some octahedral sites in the brucite-like sheet are vacant, the resultant holes are usually capped by tetrahedra above and/or below the plane of the sheet. Figure 27a shows the octahedral-tetrahedral sheet in **chalcophyllite**, $[Cu^{2+}_9Al(OH)_{12}(H_2O)_6(AsO_4)_2](SO_4)_{1.5}-(H_2O)_{12}$. In this mineral, the octahedral vacancies are capped by (AsO_4) tetrahedra, and (SO_4) groups occur in the interlayer, together with (H_2O) groups, providing hydrogen bonding between adjacent sheets (Fig. 27b). In the structures of **mooreite**, $[Mg_9Zn_4Mn^{2+}_2(OH)_{26}]$ $(SO_4)_2(H_2O)_8$, and **lawsonbauerite**, $[(Mn^{2+},Mg)_9Zn_4(OH)_{22}]-(SO_4)_2(H_2O)_8$, the octahedral vacancies are capped by $(ZnØ_4)$ tetrahedra above and below the plane of the sheet (Fig. 27c). The free vertices of these tetrahedra connect to interlayer $(MØ_6)$ octahedra (M = Mg, Mn^{2+}) that link the structure in the third dimension (Fig. 27d). In the structures of **namuwite**, $[(Zn,Cu^{2+})_4(OH)_6-(H_2O)(SO_4)](H_2O)_3$, **gordaite**, $[Zn_4(OH)_6Cl(SO_4)][Na(H_2O)_6]$, **bechererite**, $[Zn_7Cu^{2+}(OH)_{13}\{(SiO(OH)_3\}(SO_4)]$, and **ramsbeckite**, $[(Cu^{2+},Zn)_{15}(OH)_{22}(SO_4)_4]-(H_2O)_6$, the octahedral-tetrahedral sheets with vacant octahedral positions are decorated by attached $(TØ_4)$ tetrahedra (T = S, Si). Linkage between sheets is provided by hydrogen bonding involving interlayer (H_2O) groups (namuwite) or both interlayer (H_2O) groups and $\{Na(H_2O)_6\}$ octahedra (gordaite; Figs. 27e,f), or by corner-sharing of tetrahedra of adjacent sheets (bechererite and ramsbeckite, Figs. 27g and 27h,i, respectively). The structure of gordaite is closely related to the structure of $Ca[Zn_8(SO_4)_2(OH)_{12}Cl_2](H_2O)_9$, a new phase discovered on slag dumps at Val Varenna, Italy, in association with bechererite (Burns et al. 1998).

The structures of other sulfate minerals based on infinite sheets are listed in Tables 10 and 11. Some of their graphs are given in Figure 28. Note that some sheets cannot be represented as planar graphs.

The structures of minerals in the **alunite supergroup**, $(M^+,M^{2+})[M^{3+}_3(OH)_6(TO_4)_2]$; M^+ = K, Na, Ag^+, Tl^+, (NH_4), (H_3O); M^{2+} = Ca, Pb^{2+}, Sr, Ba; M^{3+} = Al, Fe^{3+}, Ga; T = S, As, P; (Lengauer et al. 1994; Jambor 1999; Dutrizac and Jambor, this volume) are based on octahedral-tetrahedral sheets (Fig. 28a, 29a). The crystallographic information on the minerals of this group is listed in Table 10. Octahedra occur at the vertices of a 6^3 plane net, forming six-membered rings with the octahedra linked by sharing corners. At the junction of three six-membered rings is a three-membered ring, and one set of apical vertices of those three octahedra link to a tetrahedron. The resultant sheets are held together by interstitial cations and hydrogen bonds (Fig. 29b). Ballhorn et al. (1989) and Kolitsch et al. (1999) suggested the minerals of the alunite supergroup as potential host structures for the long-term immobilization of radioactive fission-products and heavy toxic metals. This suggestion has a reasonable structural basis, as the interstitial cation is encapsulated

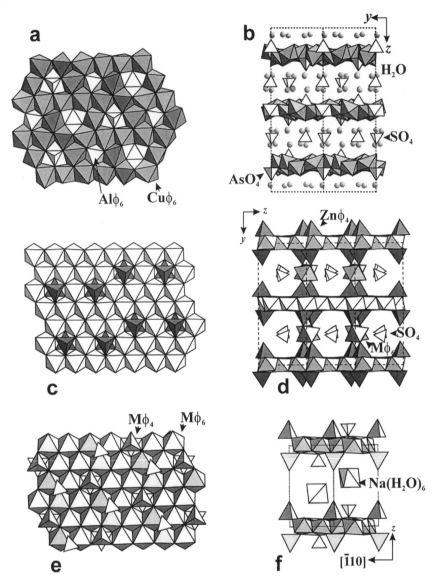

Figure 27. The octahedral–tetrahedral sheets in (a) chalcophyllite and (c) mooreite; the structures of (b) chalcophyllite and (d) mooreite viewed in the plane of the sheet; the octahedral–tetrahedral sheet in (e) gordaite; the structure of (f) gordaite.

between adjacent octahedral-tetrahedral sheets.

The structure of **felsöbányaite**, $[Al_4(OH)_{10}(H_2O)](SO_4)(H_2O)_3$, ("basaluminite") is based on the octahedral sheet shown in Figure 29c. Eight octahedra share edges to form a 2×4 cluster, and these clusters link by sharing an octahedral edge to form a zigzag ribbon that extends along [010]. These ribbons link by sharing corners with ribbons at different layers to form an interrupted sheet of edge- and corner-sharing octahedra. The resultant

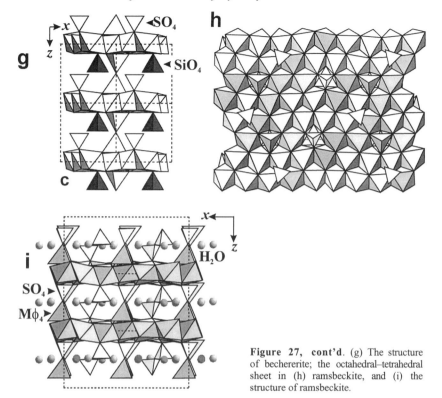

Figure 27, cont'd. (g) The structure of bechererite; the octahedral–tetrahedral sheet in (h) ramsbeckite, and (i) the structure of ramsbeckite.

sheet (Fig. 29c) is parallel to ($\bar{1}$01); the (SO$_4$) tetrahedra and (H$_2$O) groups occur in the interlayer (Fig. 29d).

The structure of **kornelite**, [Fe$^{3+}_2$(H$_2$O)$_6$(SO$_4$)$_3$](H$_2$O)$_{1.25}$, is based the sheet shown in Figure 29e (its graph is shown in Figure 28b). The sheet consists of fragments of a *trans*-linked [$M(TO_4)\emptyset_4$] chain that link to other fragments *via* a *cis* linkage. The open modulated sheets are linked by hydrogen bonds through (H$_2$O) groups in the interlayer (Fig. 29f). The sheet in **rhomboclase**, (H$_5$O$_2$)[Fe^{3+}(H$_2$O)$_2$(SO$_4$)$_2$], has the simple graph shown in Figure 28c. This sheet is based on finite polyhalite-like clusters linked in two dimensions (Fig. 29g). Rhomboclase possesses an extended hydrogen-bonding scheme involving interstitial (H$_5$O$_2$)$^+$ dimers (Fig. 29h). In **vanthoffite**, Na$_6$[Mg(SO$_4$)$_4$], glaserite-ungemachite pinwheels link to form thick slabs parallel to (100) **(Fig. 30a)**; linkage between the slabs is provided by Na in the interlayer (Fig. 30b). **Yavapaite**, K[Fe^{3+}(SO$_4$)$_2$], is based on the sheet shown in Figure 30c. The [$M(TO_4)_2$] sheet can be described as a coalescence, *via* corner-sharing, of kröhnkite-like [$M(TO_4)\emptyset_2$] chains; this [$M(TO_4)_2$] sheet also occurs in merwinite, Ca$_3$[Mg(SiO$_4$)$_2$], and brianite, Na$_2$Ca[Mg(PO$_4$)$_2$]. Interstitial [10]-coordinated K links the sheets together (Fig. 30d). The structure of **natrochalcite**, Na[Cu$^{2+}_2$(OH)(H$_2$O)(SO$_4$)$_2$], is based upon sheets consisting of linarite-like chains linked by corner-sharing of tetrahedra of the chain with the octahedra of adjacent chains (Fig. 30e). Intersheet linkage is provided by interstitial Na and by hydrogen bonds; Giester and Zemann (1987) have argued that the hydrogen-bearing interstitial species should be written as (H$_3$O$_2$)$^-$. **Goldichite**, K[Fe^{3+}(H$_2$O)$_2$(SO$_4$)$_2$](H$_2$O)$_2$, consists of sheets based on an [$M_2(TO_4)_4\emptyset_6$] cluster that links into a thick corrugated slab (Fig. 30g,h) by sharing corners between octahedra and

Table 10. Sulfate minerals with $[M_3(OH)_6(SO_4)_2]$ sheets.*

Name	Formula	a (Å)	c (Å)	S. G.	Fig.	Ref.
alunite	$K[Al_3(OH)_6(SO_4)_2]$	7.020	17.223	$R\bar{3}m$	29a,b	[1]
ammonioalunite	$(NH_4)[Al_3(OH)_6(SO_4)_2]$	7.013	17.885	$R\bar{3}m$	29a,b	(2)
natroalunite	$Na[Al_3(OH)_6(SO_4)_2]$	6.990	16.905	$R\bar{3}m$	29a,b	[3]
jarosite	$K[Fe^{3+}_3(OH)_6(SO_4)_2]$	7.315	17.224	$R\bar{3}m$	29a,b	[1]
ammoniojarosite	$(NH_4)[Fe^{3+}_3(OH)_6(SO_4)_2]$	7.325	17.374	$R\bar{3}m$	29a,b	(2)
natrojarosite	$Na[Fe^{3+}_3(OH)_6(SO_4)_2]$	7.3346	16.747	$R\bar{3}m$	29a,b	(2)
argentojarosite	$Ag[Fe^{3+}_3(OH)_6(SO_4)_2]$	7.348	16.551	$R\bar{3}m$	29a,b	(2)
hydronium jarosite	$(H_3O)[Fe^{3+}_3(OH)_6(SO_4)_2]$	7.347	16.994	$R\bar{3}m$	29a,b	(2)
beaverite	$Pb[(Fe,Cu,Al)_3(OH)_6(SO_4)_2]$	7.205	16.994	$R\bar{3}m$	29a,b	[4]
dorallcharite	$Tl[Fe^{3+}_3(OH)_6(SO_4)_2]$	7.330	17.663	$R\bar{3}m$	29a,b	[5]
osarizawaite	$Pb[(Al,Cu,Fe^{3+})_3(OH)_6(SO_4)_2]$	7.075	17.248	$R\bar{3}m$	29a,b	[6]
schlossmacherite	$(H_3O),Ca[Al_3(OH)_6(SO_4,AsO_4)_2]$	6.998	16.67	$R\bar{3}m$	29a,b	(2)
woodhouseite	$Ca[Al_3(OH)_6(PO_4)(SO_4)]$	6.993	16.386	$R\bar{3}m$	29a,b	[7]
svanbergite	$Sr[Al_3(OH)_6(PO_4)(SO_4)]$	6.992	16.567	$R\bar{3}m$	29a,b	[8]
hinsdalite	$Pb[Al_3(OH)_6(PO_4)(SO_4)]$	7.029	16.789	$R\bar{3}m$	29a,b	[9]
beudantite	$Pb[Fe^{3+}_3(OH)_6(AsO_4)(SO_4)]$	7.315	17.035	$R\bar{3}m$	29a,b	[10]
hidalgoite	$Pb[Al_3(OH)_6(AsO_4)(SO_4)]$	7.0706	16.975	$R\bar{3}m$	29a,b	(2)
kemmlitzite	$Sr[Al_3(OH)_6(AsO_4)(SO_4)]$	7.027	16.51	$R\bar{3}m$	29a,b	(2)
plumbojarosite	$Pb[Fe^{3+}_3(OH)_6(SO_4)_2]_2$	7.306	33.675	$R\bar{3}m$	–	[11]
minamiite	$(Ca_{1-x}Na_{2-x}\square_{1-x})[Al_3(OH)_6(SO_4)_2]$	6.981	33.490	$R\bar{3}m$	–	[12]
huangite	$Ca[Al_3(OH)_6(SO_4)_2]_2$	6.983	33.517	$R\bar{3}m$	–	(13)
walthierite	$Ba[Al_3(OH)_6(SO_4)_2]_2$	6.992	34.443	$R\bar{3}m$	–	[13]
corkite	$Pb^{2+}[Fe^{3+}_3(OH)_6(PO_4)(SO_4)]$	7.280	16.821	$R3m$	–	[14]
gallobeudantite	$Pb^{2+}[Ga_3(OH)_6(AsO_4)(SO_4)]$	7.225	17.03	$R3m$	–	[15]

* M = Fe, Al, Ga

References: [1] Menchetti and Sabelli (1976a), (2) Lengauer et al. (1994), [3] Okada et al. (1982), [4] Breidenstein et al. (1992), [5] Balic Zunic et al. (1994), [6] Giuseppetti and Tadini (1980), [7] Kato (1977), [8] Kato and Miura (1977), [9] Kolitsch et al. (1999), [10] Szymanski (1988), [11] Szymanski (1985), [12] Ossaka et al. (1982), [13] Li et al. (1992), [14] Giuseppetti and Tadini (1987), [15] Jambor et al. (1996)

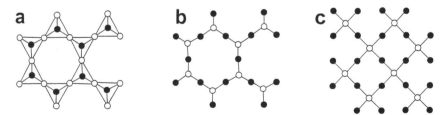

Figure 28. Graphs of the octahedral–tetrahedral sheets in the structures of sulfate minerals.

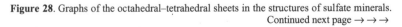

Continued next page → → →

Table 11. Sulfate minerals with sheets of (SO_4) tetrahedra and $(M\phi_6)$ octahedra.

Name	Formula	a (Å)	b (Å)	c (Å)	β (°)	S. G.	Graph	Fig.	Ref.
felsöbányaite	$[Al_4(OH)_{10}(H_2O)](SO_4)(H_2O)_3$	13.026	10.015	11.115	104.3	$P2_1$	–	29c,d	[1]
kornelite	$[Fe^{3+}_2(H_2O)_6(SO_4)_3](H_2O)_{1.25}$	14.300	20.120	5.425	96.8	$P2_1/n$	36b	29e,f	[2]
rhomboclase	$(H_5O_2)[Fe^{3+}(H_2O)_2(SO_4)_2]$	9.724	18.333	5.421	–	$Pnma$	36c	29g,h	[3]
vanthoffite	$Na_6[Mg(SO_4)_4]$	9.797	9.217	8.199	113.5	$P2_1/c$	36d	30a,b	[4]
yavapaite	$K[Fe^{3+}(SO_4)_2]$	8.152	5.153	7.877	94.9	$C2/m$	36e	30c,d	[5]
natrochalcite	$Na[Cu^{2+}_2(OH)(H_2O)(SO_4)_2]$	8.809	6.187	7.509	118.7	$C2/m$	36f	30e,f	[6]
goldichite	$K[Fe^{3+}(H_2O)_2(SO_4)_2](H_2O)_2$	10.387	10.486	9.086	101.7	$P2_1/c$	36g	30g,h	[5]
slavikite	$Na[Fe^{3+}_5(H_2O)_6(OH)_6(SO_4)_6]$ $[Mg(H_2O)_6]_2 (SO_4)(H_2O)_{15}$	12.200	a	35.130	–	$R\overline{3}$	36h	31a,b	[7]
guildite	$[Cu^{2+}Fe^{3+}(OH)(H_2O)_4(SO_4)_2]$	9.786	7.134	7.263	–	$P2_1/m$	36i	31c,d	[8]
ransomite	$[Cu^{2+}Fe^{3+}_2(SO_4)_4(H_2O)_6]$	4.811	16.217	10.403	–	$P2_1/c$	–	31e,f	[9]
poughite	$[Fe^{3+}_2(H_2O)_3(TeO_3)_2(SO_4)]$	9.660	14.200	7.86	–	$P2_1nb$	–	32a,b	[10]
fuenzalidaite	$K_6(Na,K)_4Na_6$ $[Mg_{10}(H_2O)_{12}(IO_3)_{12}(SO_4)_{12}]$	9.4643	a	27.336	–	$P\overline{3}c$	–	32c,d, e	[11]

M = Fe^{2+}, Ni, Mg, V, Zn, Al, Fe^{3+}

References: [1] Farkas and Pertlik (1997), [2] Robinson and Fang (1973), [3] Mereiter (1974), [4] Fischer and Hellner (1964), [5] Graeber and Rosenzweig (1971), [6] Giester and Zemann (1987), [7] Süsse (1973), [8] Wan et al. (1978), [9] Wood (1970), [10] Pertlik (1971), [11] Konnert et al. (1994)

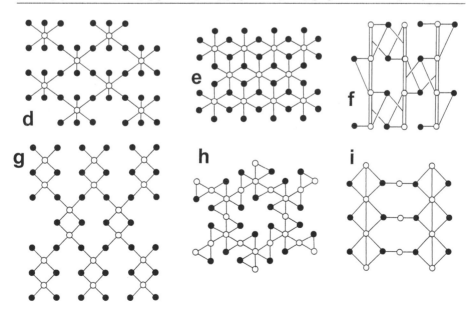

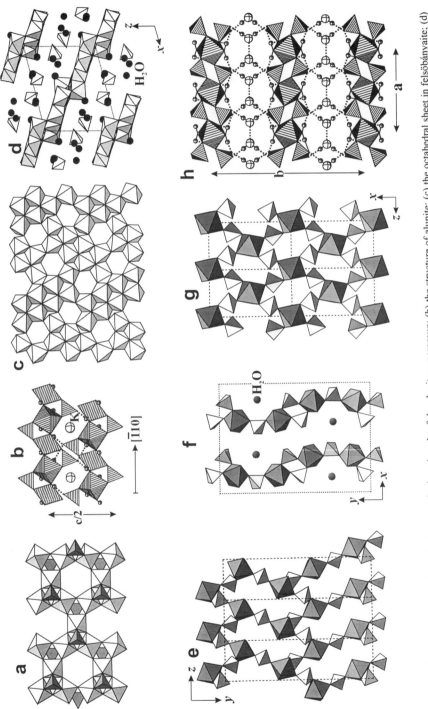

Figure 29. (a) The octahedral–tetrahedral sheet in the minerals of the alunite supergroup; (b) the structure of alunite; (c) the octahedral sheet in felsöbányaite; (d) the structure of felsöbányaite projected down [010]; (e) the $[M_2(TO_4)_3O_6]$ sheet in kornelite; (f) the structure of kornelite projected down [001]; (g) the $[M(TO_4)_2O_2]$ sheet in rhomboclase; (h) the intersheet species and hydrogen-bond network in rhomboclase.

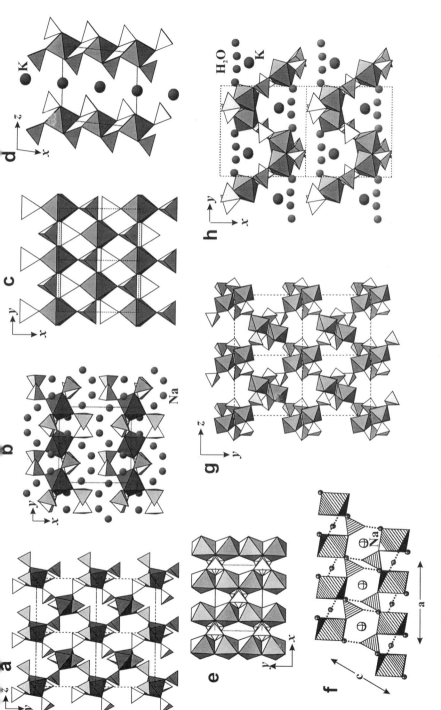

Figure 30. (a) The $[M(TØ_4)_4]$ sheet in vanthoffite; (b) fitting together of the corrugated sheets in vanthoffite; (c) the $[M(TØ_4)_2]$ sheet in yavapaite; (d) the structure of yavapaite projected down [010]; (e) the $[M(TØ_4)_2Ø_2]$ sheet in natrochalcite; (f) linkage of sheets in the structure of natrochalcite; (g) the $[M(TØ_4)_2Ø_2]$ sheet in goldichite; (h) packing of sheets in the structure of goldichite.

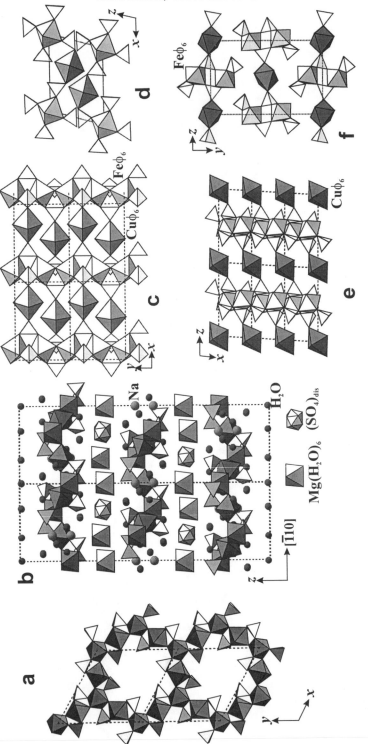

Figure 31. (a) The open sheet in slavikite; (b) the structure of slavikite projected down [110]; note that the (SO₄) tetrahedra in the interlayer are disordered; (c) the sheets in the structure of guildite; (d) the structure of guildite projected down [010]; (e) the sheet in ransomite; (f) the structure of ransomite viewed down [010].

tetrahedra. These sheets link in the [100] direction through interstitial [9]-coordinated K and a network of hydrogen bonds.

Slavikite, $Na[Fe^{3+}_5(H_2O)_6(OH)_6(SO_4)_6][Mg(H_2O)_6]_2(SO_4)(H_2O)_{15}$, is based on an unusually open sheet of corner-sharing octahedra and tetrahedra, consisting of fragments of the $[M(TO_4)\emptyset]$ butlerite-like 7 Å chain (Fig. 31a); the holes in the sheet are occupied by Na and (H_2O) groups. The interlayer consists of $\{Mg(H_2O)_6\}$ octahedra and disordered (SO_4) tetrahedra that are held in the structure only by hydrogen bonds (Fig. 31b). The structures of **guildite**, $[Cu^{2+}Fe^{3+}(OH)(H_2O)_4(SO_4)_2]$ (Fig. 31c,d), and **ransomite**, $[Cu^{2+}Fe^{3+}_2(SO_4)_4(H_2O)_6]$ (Fig. 31e,f), are based on sideronatrite-like and krausite-like chains, respectively, linked by Jahn-Teller-distorted ($Cu\emptyset_6$) octahedra and by hydrogen bonding. **Poughite**, $[Fe^{3+}_2(H_2O)_3(TeO_3)_2(SO_4)]$, and **fuenzalidaite**, $K_6(Na,K)_4Na_6$-$[Mg_{10}(H_2O)_{12}(IO_3)_{12}(SO_4)_{12}]$, are based on mixed octahedral-tetrahedral-triangular sheets. In poughite, edge-sharing octahedral dimers are linked to one (SO_4) tetrahedron each to form $[Fe^{3+}_2O_6(H_2O)_2(SO_4)_2]$ clusters that are cross-linked by (TeO_3) triangles (Fig. 32a). The sheets are held together by hydrogen bonding only (Fig. 32b). The structure of fuenzalidaite is based on the complex pinwheel sheet shown in Figures 32c,d. The pinwheels consist of ($Mg\emptyset_6$) octahedra and both (SO_4) tetrahedra and (IO_3) triangles. These complex sheets are connected through other pinwheels that involve ($Na\emptyset_6$) octahedra with K in the interstices (Fig. 32e). Fuenzalidaite was discovered in Chilean nitrate deposits and is the only known natural iodate-sulfate.

Inspection of the finite clusters, chains, and sheets in sulfate minerals and their graphs shows the advantages provided by graphical representation of octahedral–tetrahedral structures. It gives a clear and simple representation of the way in which the polyhedra link together. This shows also that Nature chooses a small number of building blocks, connects them in a very simple and elegant way, and packs the resultant units (clusters, chains, or sheets) economically to use space as efficiently as possible.

Structures with infinite frameworks

The sulfate minerals in this category are listed in Table 12. Unfortunately, the topological aspects of the framework structures cannot easily be summarized in a concise graphical fashion because of the complexity that results from polymerization in three dimensions.

The structure of **bonattite**, $[Cu^{2+}(SO_4)(H_2O)_3]$, is based on a skewed arrangement of $[M(TO_4)\emptyset_4]$ chalcanthite-like chains (Fig. 21c) polymerized by corner-sharing of polyhedra from adjacent chains (Fig. 33a). The structure of **kieserite**, $[Mg(SO_4)(H_2O)]$, is shown in Figure 33b,c. The $[M(T\emptyset_4)\emptyset]$ framework can be constructed from $[M(T\emptyset_4)\emptyset_3]$ chains of the type found in butlerite, parabutlerite, and uklonskovite (Fig. 21e). The chains pack in a C-centered array and are cross-linked by sharing corners between octahedra and tetrahedra of adjacent chains. This is a very common arrangement in a wide variety of silicate, phosphate, arsenate, vanadate, and sulfate minerals. The structures of **millosevichite**, $[Al_2(SO_4)_3]$ and **mikasaite** $[Fe^{3+}_2(SO_4)_3]$ (Fig. 33d,e), and **langbeinite**, $K_2[Mg_2(SO_4)_3]$, are based on glaserite-ungemachite-type pinwheels linked in three dimensions (Fig. 33f). The structure of **chalcocyanite**, $[Cu^{2+}(SO_4)]$, is based on linarite-type $[M(TO_4)\emptyset_2]$ chains cross-linked by the tetrahedral vertices not linked to the central octahedral chain (Fig. 33g,h). In **löweite**, $Na_{12}[Mg_7(H_2O)_{12}(SO_4)_9](SO_4)_4(H_2O)_3$, there are two types of (SO_4) groups: the first type participates in formation of the open $[Mg_7(SO_4)_9(H_2O)_{12}]$ octahedral-tetrahedral framework, whereas the groups of the second type are disordered in the framework cavities (Fig. 33i). **Kainite**, $K_4[Mg_4(H_2O)_{10}(SO_4)_4](H_2O)Cl_4$, is based on kröhnkite-like $[M(TO_4)_2\emptyset_2]$ chains (Fig. 22e) linked into sheets parallel to (100) (Fig. 33j);

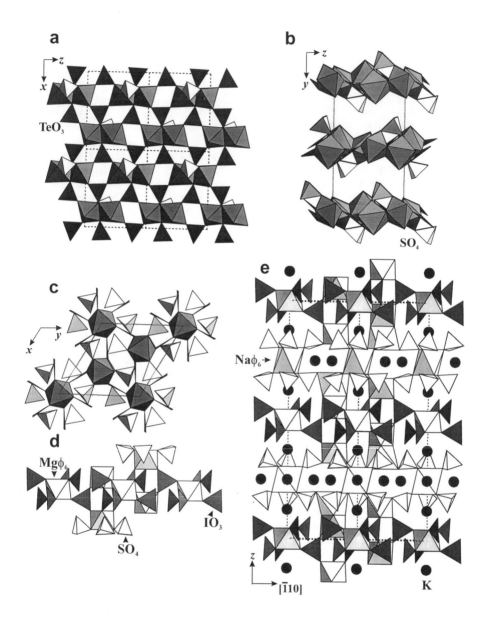

Figure 32. (a) The sheet of (FeØ$_6$) octahedra, (SO$_4$) tetrahedra, and (TeO$_3$) triangles in poughite; (b) the arrangement of sheets in poughite; (c,d) the octahedral–tetrahedral–triangle sheet in fuenzalidaite viewed (c) down [001], and (d) in the plane of the sheet; (e) the structure of fuenzalidaite projected down [110].

Table 12. Sulfate minerals with frameworks of (SO_4) tetrahedra and $(M\phi_6)$ octahedra.

Name	Formula	a (Å)	b (Å)	c (Å)	β (°)	S. G.	Fig.	Ref.
bonattite	$[Cu^{2+}(SO_4)(H_2O)_3]$	5.592	13.029	7.341	97.1	Cc	33a	[1]
kieserite	$[Mg(SO_4)(H_2O)]$	6.891	7.624	7.645	117.7	$C2/c$	33b,c	[2]
dwornikite	$[Ni(SO_4)(H_2O)]$	6.824	7.594	7.457	117.8	$C2/c$	33b,c	[3]
szmikite	$[Mn^{2+}(SO_4)(H_2O)]$	7.116	7.667	7.92	118.1	$C2/c$	33b,c	[3]
gunningite	$[Zn(SO_4)(H_2O)]$	6.925	7.591	7.635	118.2	$C2/c$	33b,c	[3]
szomolnokite	$[Fe^{2+}(SO_4)(H_2O)]$	7.078	7.549	7.773	118.7	$C2/c$	33b,c	[3]
poitevinite[1]	$[Cu^{2+}(SO_4)(H_2O)]$	5.120	5.160	7.535	107.4	$P\bar{1}$	33b,c	[4]
millosevichite	$[Al_2(SO_4)_3]$	8.025	a	21.357	–	$R\bar{3}$	33d,e	[5]
mikasaite	$[Fe^{3+}_2(SO_4)_3]$	8.14	a	21.99	–	$R\bar{3}$	33d,e	(6)
langbeinite	$K_2[Mg_2(SO_4)_3]$	9.919	a	a	–	$P2_13$	33f	[7]
manganolangbeinite	$K_2[Mn^{2+}_2(SO_4)_3]$	10.114	a	a	–	$P2_13$	33f	(6)
efremovite	$(NH_4)_2[Mg_2(SO_4)_3]$	9.99	a	a	–	$P2_13$	33f	(6)
löweite[2]	$Na_{12}[Mg_7(H_2O)_{12}(SO_4)_9](SO_4)_4(H_2O)_3$	11.769	a	–	–	$R\bar{3}$	33i	[8]
chalcocyanite	$[Cu^{2+}(SO_4)]$	8.409	6.709	4.833	–	$Pnma$	33g,h	[9]
zincosite	$[Zn(SO_4)]$	8.604	6.746	4.774	–	$Pnma$	33g,h	[9]
kainite	$K_4[Mg_4(H_2O)_{10}(SO_4)_4](H_2O)Cl_4$	19.720	16.230	9.53	94.9	$C2/m$	33j,k	[10]
voltaite	$K_2[Fe^{2+}_5Fe^{3+}_3(H_2O)_{12}(SO_4)_{12}][Al(H_2O)_6]$	27.254	a	a	–	$Fd\bar{3}c$	33l,m	[11]
zincovoltaite	$K_2[Zn_5Fe^{3+}_3(H_2O)_{12}(SO_4)_{12}][Al(H_2O)_6]$	27.18	a	a	–	$Fd\bar{3}c$	33l,m	(6)
sulfoborite	$[Mg_3(OH)F(SO_4)(B(OH)_4)_2]$	10.132	12.537	7.775	–	$Pnma$	34a,b	[12]
vlodavetsite	$[Ca_2AlF_2Cl(H_2O)_4(SO_4)_2]$	6.870	a	13.342	–	$I4/m$	34c,d	[13]
tychite	$Na_6[Mg_2(CO_3)_4](SO_4)$	13.930	a	a	–	$Fd\bar{3}$	34e,f	[14]
ferrotychite	$Na_6[Fe^{2+}_2(CO_3)_4](SO_4)$	13.962	a	a	–	$Fd\bar{3}$	34e,f	[15]
manganotychite	$Na_6[Mn^{2+}_2(CO_3)_4](SO_4)$	13.995	a	a	–	$Fd\bar{3}$	34e,f	(6)
philolithite	$[Pb^{2+}_2O]_6[Mn^{2+}Mg_2Mn^{2+}_4Cl_4(OH)_{12}(SO_4)(CO_3)_4]$	12.627	a	12.595	–	$P4_2/nnm$	35a,b,c	[16]
brochantite	$[Cu^{2+}_4(OH)_6(SO_4)]$	13.087	9.835	6.015	103.3	$P2_1/a$	36a,b	[17]
antlerite	$[Cu^{2+}_3(OH)_4(SO_4)]$	8.244	6.043	11.987	–	$Pnma$	36c,d	[18]
mammothite	$Pb^{2+}_6[Cu^{2+}_4AlSbO_2(OH)_{16}Cl_2](SO_4)_2Cl_2$	18.390	7.330	11.350	112.4	$C2/m$	36e,f	[19]
caminite	$[Mg_4(SO_4)_3(OH)_2(H_2O)]$	5.242	a	12.995	–	$I4_1/amd$	36g-j	[20]
connellite	$[Cu^{2+}_{19}(OH)_{32}Cl_4](SO_4)(H_2O)_3$	15.780	a	9.100	–	$P\bar{6}2c$	35d	[21]

[1] $\alpha = 107.1°$, $\gamma = 92.7°$; [2] rhombohedral cell: $\alpha = 106.5°$

References: [1] Zahrobsky and Baur (1968), [2] Hawthorne et al. (1987). [3] Wildner and Giester (1991), [4] Giester et al. (1994), [5] Dahmen and Gruehn (1993), (6) Gaines et al. (1997), [7] Mereiter (1979), [8] Fang and Robinson (1970b), [9] Wildner and Giester (1988), [10] Robinson et al. (1972), [11] Mereiter (1972), [12] Giese and Penna (1983), [13] Starova et al. (1995), [14] Shiba and Watanabe (1931), [15] Malinovskii et al. (1979), [16] Moore et al. (2000), [17] Helliwell and Smith (1997), [18] Hawthorne et al. (1989), [19] Effenberger (1985b), [20] Keefer et al. (1981), Hochella et al. (1983), [21] McLean and Anthony (1972)

the sheets are further linked into a framework by corner-sharing through $(MgØ_6)$ octahedra (Fig. 33k). The framework in **voltaite**, $K_2[Fe^{2+}_5Fe^{3+}_3(H_2O)_{12}(SO_4)_{12}]\{Al(H_2O)_6\}$, is a three-dimensional polymerization (Fig. 33l) of complex octahedral–tetrahedral chains (Fig. 33m), with K in the interstices.

Sulfoborite, $[Mg_3(OH)F(SO_4)(B(OH)_4)_2]$, is based on the complex sheets of $(MgØ_6)$ octahedra and (SO_4) tetrahedra; the structure of the sheet is shown in Figure 34b. The amarantite-like octahedral tetramers (Fig. 34a) are polymerized to form octahedral chains parallel to [100] and linked by (SO_4) tetrahedra. The linkage between sheets is provided by $[B(OH)_4]$ tetrahedra (Fig. 34a). **Vlodavetsite**, $[Ca_2AlF_2Cl(H_2O)_4(SO_4)_2]$, consists of a framework of corner-sharing $(CaØ_6)$, $(AlØ_6)$ octahedra and (SO_4) tetrahedra (Fig. 34c). The basis of the framework is the sheet of corner-sharing $(CaØ_6)$ octahedra and (SO_4) tetrahedra shown in Figure 34d. The structure of **ferrotychite**, $Na_6[Fe^{2+}_2(CO_3)_4](SO_4)$, is based on the octahedral–triangular framework formed by corner sharing of (FeO_6) octahedra and (CO_3) triangles (Fig. 34e). The main structural element of

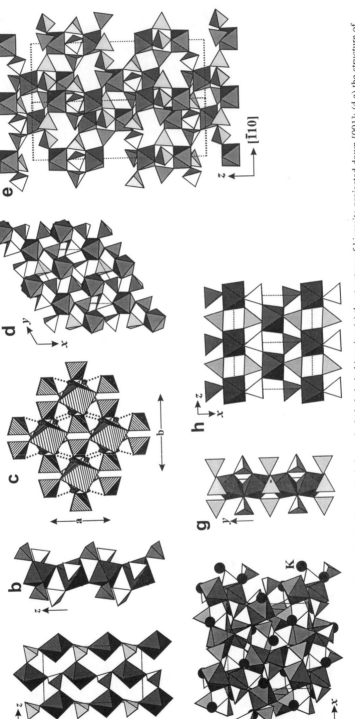

Figure 33. (a) The structure of bonattite; (b) the octahedral–tetrahedral chain in chalcocyanite; (c) the structure of kieserite projected down [001]; (d,e) the structure of millosevichite projected down (d) [001], and (e) [110]; (f) the structure of langbeinite; (g) The octahedral–tetrahedral chain in chalcocyanite; (h) the arrangement of chains in chalcocyanite.

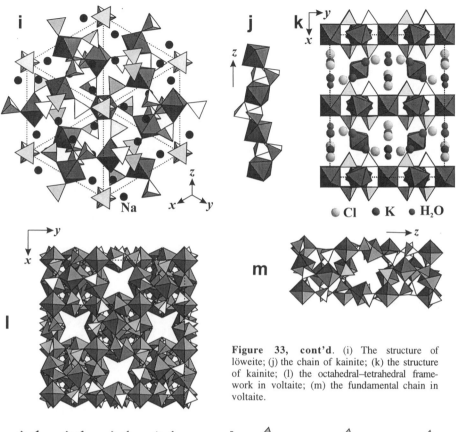

Figure 33, cont'd. (i) The structure of löweite; (j) the chain of kainite; (k) the structure of kainite; (l) the octahedral–tetrahedral framework in voltaite; (m) the fundamental chain in voltaite.

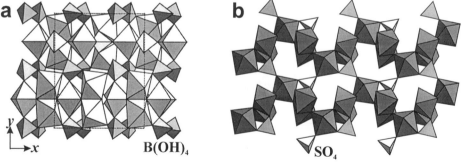

Figure 34. (a) The structure of sulfoborite viewed along [001]; (b) the octahedral–tetrahedral sheet in sulfoborite projected ~5° from [010].

the framework is the octahedral-triangular chain shown in Figure 34f; the (SO_4) groups are in the framework cavities.

The structure of **philolithite**, $[Pb^{2+}_2O]_6[Mn^{2+}(Mg,Mn^{2+})_2(Mn^{2+},Mg)_4Cl_4(OH)_{12}(SO_4)(CO_3)_4]$, consists of a trellis-like open framework based on the complex octahedral-tetrahedral-triangular chain shown in Figure 35a. The chains extend in the [110] and [$\bar{1}$10] directions and cross-link to form a trellis-like framework (Fig. 35b). The large channels of

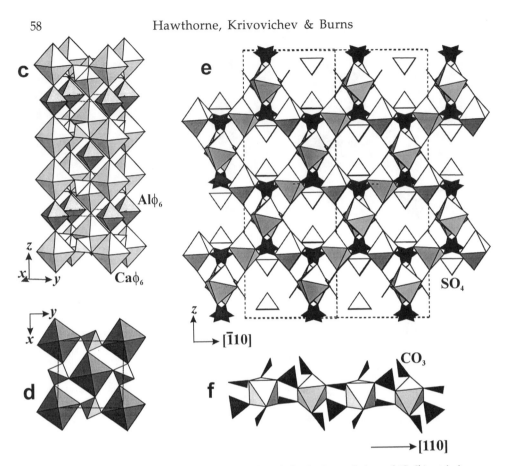

Figure 34, cont'd. (c) the structure of vlodavetsite; (d) the fundamental sheet of (CaØ$_6$) octahedra and (SO$_4$) tetrahedra in vlodavetsite; (e) the structure of ferrotychite projected down [110]; (f) the chain of (MgØ$_6$) octahedra and (CO$_3$) triangles that is an important motif in the framework of ferrotychite.

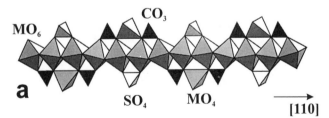

Figure 35. (a) The fundamental [MØ$_4$] chain decorated by (CO$_3$) triangles, (SO$_4$) tetrahedra, and (MO$_4$) tetrahedra in philolithite.

the framework are occupied by [OPb$_2$] chains of edge-sharing (OPb$_4$) tetrahedra, and Cl anions (Fig. 35c). The structure of **connellite**, [Cu$^{2+}_{19}$(OH)$_{32}$Cl$_4$](SO$_4$)(H$_2$O)$_3$, is based on a complex framework of (Cu^{2+}Ø$_6$) octahedra (Ø = OH, Cl) with cavities occupied by disordered (SO$_4$) tetrahedra (Fig. 35d).

Brochantite, [Cu$^{2+}_4$(OH)$_6$(SO$_4$)], and **antlerite**, [Cu$^{2+}_3$(OH)$_4$(SO$_4$)], are based on

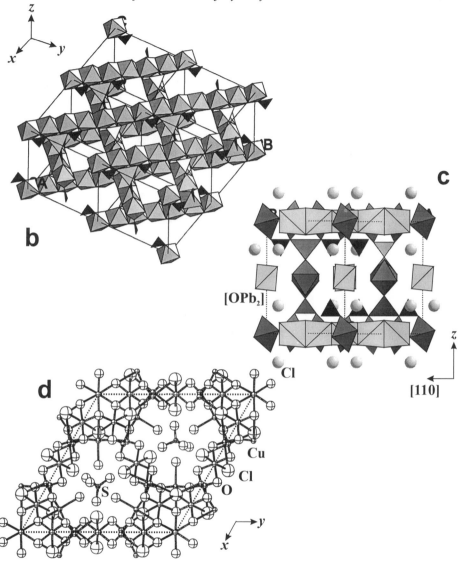

Figure 35, cont'd. (b) The trellis-like framework resulting from the intersection of the chains in (a); (c) the structure of philolithite projected down [$\bar{1}10$]; the [OPb$_2$] chains of edge-sharing (OPb$_4$) tetrahedra are shown as lightly shaded; (d) the structure of connellite projected down [001]; the (SO$_4$) groups are disordered in the channels of the framework of (CuØ$_6$) octahedra.

double and triple chains of (Cu^{2+}Ø$_6$) octahedra, respectively (Fig. 36a,c). The chains are cross-linked by (SO$_4$) tetrahedra to form frameworks (Figs. 36b,d). The structure of **mammothite**, Pb$^{2+}_6$[Cu$^{2+}_4$AlSb^{5+}O$_2$(OH)$_{16}$Cl$_2$](SO$_4$)$_2$Cl$_2$, consists of an octahedral framework based on the open-branched chain of (Cu^{2+}Ø$_6$) octahedra shown in Figure 36e. The chains extend parallel to [010] and are linked by (AlØ$_6$) and (Sb^{5+}Ø$_6$) octahedra in the [100] and [001] directions, respectively (Fig. 36f). The large channels of the framework are occupied by Pb^{2+}, Cl and (SO$_4$) tetrahedra.

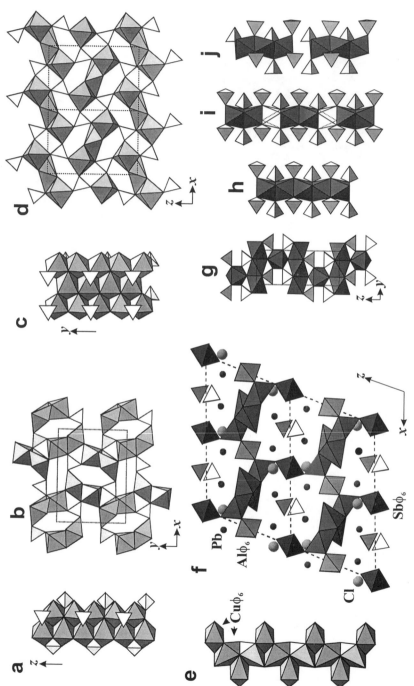

Figure 36. The double (a) and triple (c) chains of (CuØ₆) octahedra and (SO₄) tetrahedra and the frameworks in (b) brochantite and (d) antlerite; (e) the octahedral chain in mammothite; (f) the octahedral framework in mammothite; (g) the octahedral–tetrahedral framework in caminite; (h) the fundamental chain in caminite; (i) the ordered arrangement of occupied and empty octahedra in the caminite chain (as suggested by Keefer et al. 1981); (i) the ordered arrangement of occupied and empty octahedra in synthetic Mg₃(SO₄)₂(OH)₂ (Fleet and Knipe 1997).

Caminite, [$Mg_4(SO_4)_3(OH)_2(H_2O)$], is the only known sulfate mineral with face-sharing divalent-metal octahedra. The ($MgØ_6$) octahedra in caminite share faces to produce chains that extend along [010] (Fig. 36g). The chains are decorated by (SO_4) tetrahedra (Fig. 36h) that provide linkage between the chains. Only two-thirds of the octahedra in the chain shown in Figure 36h are populated by Mg. Keefer et al. (1981) suggested that, in the ordered arrangement, occupied ($MgØ_6$) octahedra are associated in pairs that alternate with unoccupied octahedra (as shown in Fig. 36i). However, Fleet and Knipe (1997) determined the structure of synthetic $Mg_3(SO_4)_2(OH)_2$ with the same stoichiometry as caminite but with a different unit cell and symmetry. In this structure, the face-sharing ($MgØ_6$) octahedra occur as linear ternary groups rather than as octahedral dimers (Fig. 36j).

Table 13. Calcium-sulfate minerals.

Name	Formula	a (Å)	b (Å)	c (Å)	$β$ (°)	S. G.	Fig.	Ref.
anhydrite	[$CaSO_4$]	6.993	6.995	6.245	–	*Amma*	37a-d	[1]
gypsum	[$CaSO_4(H_2O)_2$]	5.679	15.202	6.522	118.4	I2/c	37e,f,h	[2]
ardealite*	[$Ca_2(HPO_4)(SO_4)(H_2O)_4$]	5.721	30.992	6.250	117.3	Cc	37g	[3]
rapidcreekite	[$Ca_2(SO_4)(CO_3)(H_2O)_4$]	15.517	19.226	6.165	–	*Pcnb*	38a,b	[4]
bassanite	[$CaSO_4(H_2O)_{0.5}$]	6.937	a	6.345	–	$P3_12$	38c-f	[5]
syngenite	$K_2[Ca(SO_4)_2(H_2O)]$	6.225	7.127	9.727	104.2	$P2_1/m$	38g,h,i	[6]
görgeyite	$K_2[Ca_5(SO_4)_6(H_2O)]$	17.519	6.840	18.252	113.3	C2/c	39a,b,c	[7]
orschallite	[$Ca_3(SO_3)_2(SO_4)(H_2O)_{12}$]	11.350	a	28.321	–	$R\bar{3}c$	39d-g	[8]
glauberite	[$CaNa_2(SO_4)_2$]	10.129	8.306	8.533	112.2	C2/c	40a-c	[9]
ternesite	[$Ca_5(SiO_4)_2(SO_4)$]	6.863	15.387	10.181	–	*Pnma*	40d-g	[10]

* synthetic material

References: [1] Hawthorne and Ferguson (1975a), [2] Pedersen and Semmingsen (1982), [3] Sakae et al. (1978), [4] Cooper and Hawthorne (1996a), [5] Abriel and Nesper (1993), [6] Bokii et al. (1978), [7] Mukhtarova et al. (1980), Smith and Walls (1980), [8] Weidenthaler et al. (1993), [9] Araki and Zoltai (1967), [10] Irran et al. (1997)

STRUCTURES WITH NON-OCTAHEDRAL CATION-COORDINATION POLYHEDRA

Calcium-sulfate minerals

Divalent Ca has a relatively large ionic radius (1.00 Å and 1.12 Å for [6]Ca and [8]Ca, respectively; Shannon 1976) and is usually coordinated by more than six anions: in sulfate minerals, Ca is most often coordinated by eight or nine anions. Crystallographic parameters for Ca-sulfate minerals are given in Table 13. It is notable that most of the minerals with Ca as the dominant interstitial cation have structures closely related to that of anhydrite, $CaSO_4$.

The structure of **anhydrite**, $CaSO_4$, is based on chains of alternating edge-sharing (SO_4) tetrahedra and (CaO_8) dodecahedra (Fig. 37a,i). These chains extend in the c direction and are linked by edge-sharing between adjacent (CaO_8) dodecahedra and by corner-sharing between the (SO_4) tetrahedra and (CaO_8) dodecahedra (Fig. 37b). The structure contains sheets of edge-sharing chains (Fig. 37c,d) that are parallel to ($\bar{1}$10). Similar sheets occur in the structures of **gypsum**, $CaSO_4(H_2O)_2$, and **ardealite**, $Ca_2(HPO_4)(SO_4)(H_2O)_4$. Figures 37e,f show the structure of gypsum, and Figure 37g shows the structure of ardealite. In gypsum, all sheets have the same orientation, whereas in ardealite, adjacent sheets are rotated by 62.7° relative to each other. Thus, the edge-

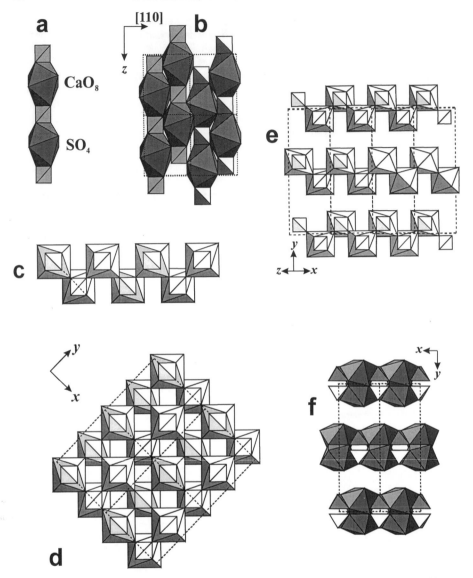

Figure 37. (a) The chain of edge-sharing CaO$_8$ polyhedra and (SO$_4$) tetrahedra in anhydrite; (b) linking of chains into layers parallel to ($1\bar{1}0$) in anhydrite; (c) the layer viewed along [001] in anhydrite; (d) the structure of anhydrite viewed along [001]; (e,f) the structure of gypsum viewed along (e) [101] and (f) [001].

sharing chains of (SO$_4$) and (CaØ$_8$) polyhedra are parallel to [101] in gypsum (Fig. 37h), and to both [001] and [101] in ardealite. The chain repeats are 6.25 Å in anhydrite, 6.27 Å in gypsum, and 6.25 and 6.24 Å in ardealite.

The structure of **rapidcreekite**, Ca$_2$(SO$_4$)(CO$_3$)(H$_2$O)$_4$, can be derived from that of gypsum by twinning along alternate rows of (SO$_4$) groups (Fig. 38a,b). At each alternate row of (SO$_4$) tetrahedra, the structure is twinned by rotation of 180° about [100]. This

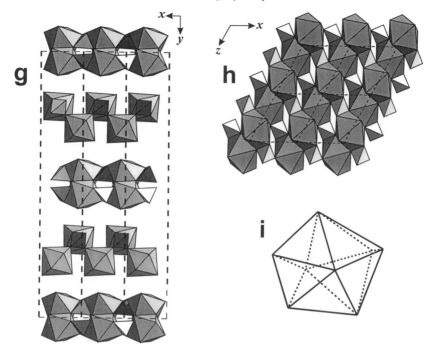

Figure 37, con't. (g) The structure of ardealite viewed along [001]; (h) the layer of chains of edge-sharing polyhedra in gypsum; (i) the (CaØ$_8$) coordination polyhedra in gypsum and ardealite.

process reverses the cant of (CaO$_8$) dodecahedra in adjacent chains, resulting in triangles of O atoms at the twin plane. Occupancy of these triangles by C atoms produces the (CO$_3$) group and the structural unit in rapidcreekite. Detailed work by Hochella et al. (1983) showed that the naturally occurring hydroxy-hydrated Mg sulfate from active hydrothermal vents on the East Pacific Rise has the general chemical composition MgSO$_4$·{Mg(OH)$_2$}$_x$·(H$_2$O)$_{1-2x}$ where $0 \leq x \leq 0.5$. Thus the 'end-member' compositions are MgSO$_4$(H$_2$O) and Mg$_3$(SO$_4$)$_2$(OH)$_2$. Keefer et al. (1981) solved the crystal structure of Mg$_4$(SO$_4$)$_3$(OH)$_2$(H$_2$O) ($x = 0.33$), showing that it is tetragonal, $I4_1/amd$, $a = 5.242$, $c = 12.995$ Å. Haymon and Kastner (1986) described the material as caminite, focusing specifically on the composition Mg$_7$(SO$_4$)$_5$(OH)$_4$(H$_2$O)$_2$ ($x = 0.4$), tetragonal, $I4_1/amd$, $a = 5.239$, $c = 12.988$ Å. Presuming that one end of the 'series', Mg(SO$_4$)(H$_2$O), corresponds to kieserite, the composition with $x = 0.5$ should correspond to the end-member composition for caminite: Mg$_3$(SO$_4$)$_2$(OH)$_2$. Fleet and Knipe (1997) have described a synthetic material similar in stoichiometry to caminite, but with a different structure.

In the structure of **bassanite**, CaSO$_4$(H$_2$O)$_{0.5}$, the (CaØ$_9$) polyhedron has nine vertices and can be described as a triaugmented trigonal prism (Fig. 38f). However, the dominant motif of the structure is a chain of alternating edge-sharing (CaØ$_8$) and (SO$_4$) polyhedra (Fig. 38d). These chains extend along [001] and are connected to form a framework (Fig. 38c,e). The repeat unit of the chain consists of one (CaØ$_9$) and one (SO$_4$) polyhedra; the period of the chain is 6.345 Å, similar to the chain-repeat distances in gypsum, anhydrite, and ardealite.

In the structure of **syngenite**, K$_2$Ca(SO$_4$)$_2$(H$_2$O), bassanite-like chains are linked into sheets parallel to [100] (Fig. 38g). The sheets are linked by K in the interlayer (Fig. 38i,j).

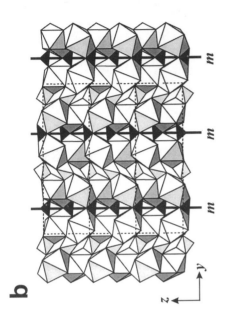

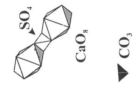

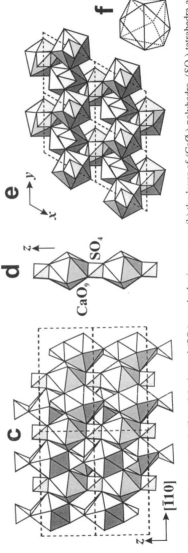

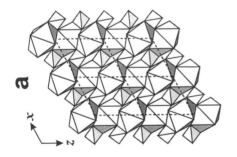

Figure 38. (a) The layer of (CaØ$_8$) polyhedra and (SO$_4$) tetrahedra in gypsum; (b) the layer of (CaØ$_8$) polyhedra, (SO$_4$) tetrahedra and (CO$_3$) triangles in rapidcreekite; the layer in rapidcreekite can be produced from that of gypsum by twinning along the mirror plane; (c) the structure of bassanite viewed along [110]; (d) the chain of edge-sharing (CaØ$_9$) polyhedra and (SO$_4$) tetrahedra in bassanite; (e) the structure of bassanite viewed along [001]; (f) the (CaØ$_9$) coordination polyhedron in bassanite.

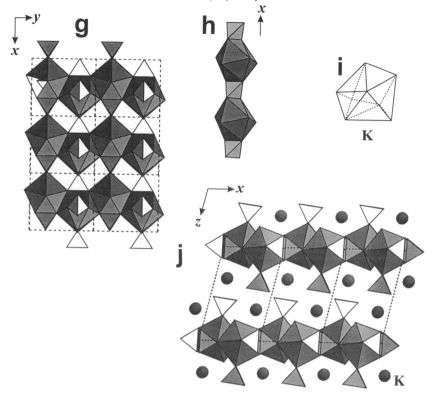

Figure 38, cont'd. (g) the structure of syngenite viewed along [001]; (h) the chain of edge-sharing (CaØ₉) polyhedra and (SO₄) tetrahedra in syngenite; (i) the (KØ₈) coordination polyhedron in syngenite; (j) the arrangement of layers of (CaØ₉) polyhedra and (SO₄) tetrahedra with interstitial K in syngenite.

The repeat distance of the chains of edge-sharing polyhedra (Fig. 38h) is 6.225 Å, very close to the repeat distances of similar chains in other Ca-sulfate structures.

The structure of **görgeyite**, $K_2Ca_5(SO_4)_6(H_2O)$, is an interesting example of combinatorics of structural subunits in this class of sulfate minerals. The structure consists of slabs parallel to (010) and extending along [001] (Fig. 39a). These slabs can be built from finite bassanite-like chains consisting of five (CaO₉) polyhedra and six (SO₄) tetrahedra. These finite chains are linked by edge-sharing of (CaO₉) polyhedra to form slabs. The slabs are linked into a framework (Fig. 39b) by additional edge-sharing of coordination polyhedra (Fig. 39c). Interstitial K cations are located in channels through the framework.

The degree of condensation of edge-sharing chains in anhydrite, gypsum, ardealite, rapidcreekite, bassanite, syngenite, and görgeyite is related to the number of (H₂O) groups per Ca atom, as well as to the presence of additional alkali-metal cations. Where there are zero or 0.5 (H₂O) groups per Ca atom (anhydrite and bassanite), the chains are linked into three-dimensional frameworks. Where there are two (H₂O) groups per Ca atom (gypsum and ardealite), the result is sheet structures. With the addition of K cations, the chains condense into sheets (syngenite) or frameworks (görgeyite).

The crystal structure of **orschallite**, $Ca_3(SO_3)_2(SO_4)(H_2O)_{12}$, a rare Ca-sulfite-

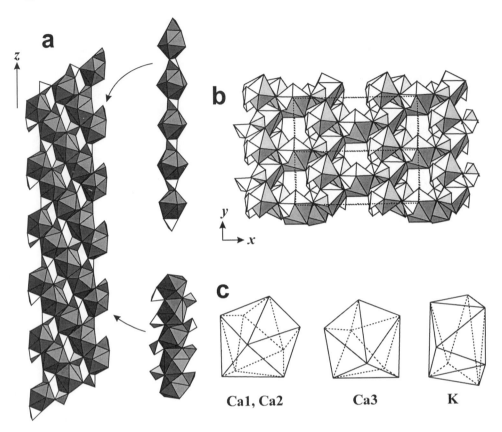

Figure 39. (a) The slab of (CaO$_9$) polyhedra and (SO$_4$) tetrahedra in görgeyite; (b) the arrangement of slabs into a framework in görgeyite; (c) coordination polyhedra of Ca and K in görgeyite.

sulfate-hydrate mineral, is based on finite [Ca$_3$(SO$_3$)$_2$(H$_2$O)$_{12}$]$^{2+}$ clusters consisting of three (CaØ$_8$) polyhedra and two (SO$_3$) groups (Fig. 39d). These clusters are arranged into sheets parallel to (001) (Fig. 39f,g) and are linked *via* hydrogen bonds to (SO$_4$) groups. The (SO$_4$) tetrahedra are held in the structure solely by hydrogen bonds and are disordered, showing two possible orientations (Fig. 39e).

The structure of **glauberite**, CaNa$_2$(SO$_4$)$_2$, is based on chains of edge-sharing (CaO$_8$) polyhedra decorated by edge- and corner-sharing (SO$_4$) tetrahedra (Fig. 40a). The chains are linked into a framework in which the channels are filled by interstitial Na cations (Fig. 40b).

The structure of **ternesite**, Ca$_5$(SiO$_4$)$_2$(SO$_4$), consists of rods of edge-sharing (CaO$_7$) polyhedra augmented trigonal prisms, decorated by (SiO$_4$) and (SO$_4$) tetrahedra and extending along [100] (Fig. 40d). The arrangements of these rods is shown in Figure 40f. The rods are linked by additional (CaØ$_n$) polyhedra into a complex framework (Fig. 40g).

Alkali-metal- and NH$_4$-sulfate minerals

Glaserite-related structures. The glaserite structure-type can be selected as a parent structure-type for many sulfates, chromates, phosphates, and silicates (Table 14)

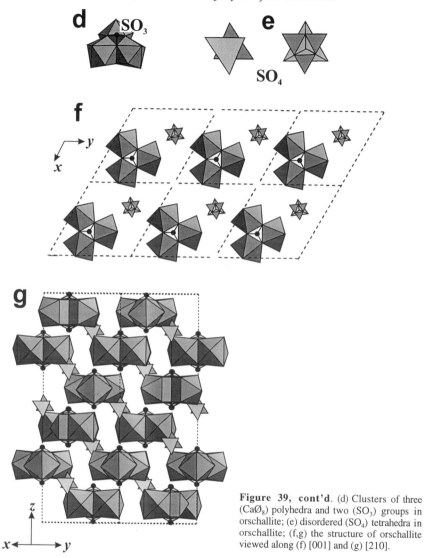

Figure 39, cont'd. (d) Clusters of three (CaØ₈) polyhedra and two (SO₃) groups in orschallite; (e) disordered (SO₄) tetrahedra in orschallite; (f,g) the structure of orschallite viewed along (f) [001] and (g) [210].

(Eysel 1973; Moore 1973, 1976, 1981). According to Moore (1973), the dominant structural subunit (= fundamental building block) of **glaserite**, $K_3Na(SO_4)_2$, is a pinwheel consisting of an (NaO_6) trigonal antiprism (= octahedron) that shares its six corners with (SO_4) tetrahedra (Fig. 41a). The pinwheels are linked through (SO_4) tetrahedra into layers perpendicular to [001] (Fig. 41b). The K atoms are located between the layers, and have two distinct coordinations. Both of these polyhedra are derivatives of the cuboctahedron shown in Figure 41g. The relatively large ionic radius of K allows it to form mixed cation-anion closest (or almost closest) packings with O^{2-} anions. The coordination polyhedron about K(1) in glaserite can be derived from the cuboctahedron by rotation of opposing triangular faces by 30°. Figure 41h shows the resulting polyhedron, a distorted icosahedron with ideal point symmetry $m\overline{3}5$. The $K(1)O_{12}$ icosahedra form sheets

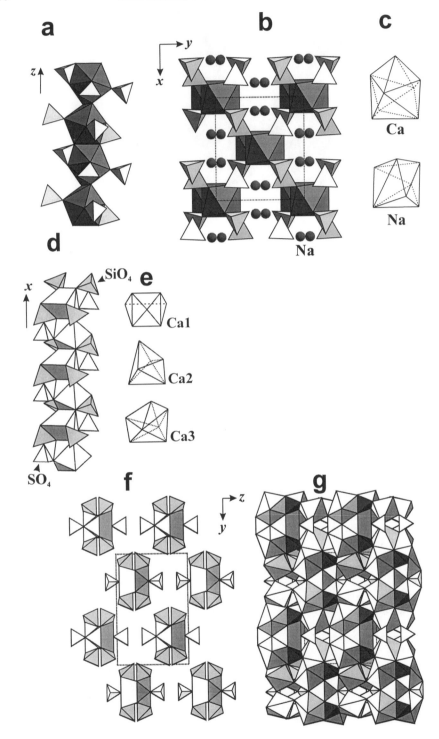

← ← **Figure 40 (opposite page).** (a) The chain of edge-sharing (CaØ$_8$) polyhedra decorated by (SO$_4$) groups in glauberite; (b) linking of chains into a framework in glauberite, viewed along [001]; (c) coordination polyhedra of Ca and Na in glauberite; (d) the chain of edge-sharing {Ca(1)O$_7$} polyhedra decorated by (SO$_4$) and (SiO$_4$) tetrahedra in ternesite; (e) coordination polyhedra of Ca in ternesite; (f) the arrangement of chains of {Ca(1)O$_7$} polyhedra and tetrahedra in ternesite; (g) the structure of ternesite viewed along [100].

Table 14. Alkali-metal- and NH$_4$-sulfate mineral structures.

Name	Formula	a (Å)	b (Å)	c (Å)	β (°)	S. G.	Fig.	Ref.
glaserite	K$_3$Na(SO$_4$)$_2$	5.680	a	7.309	–	$P\bar{3}m$	41	[1]
arcanite	β-K$_2$(SO$_4$)	7.476	10.071	5.763	–	$Pnam$	41	[2]
mascagnite	(NH$_4$)$_2$(SO$_4$)	7.747	10.593	5.977	–	$Pnam$	41j,k	[3]
palmierite[1]	K$_2$Pb(SO$_4$)$_2$	7.560	a	–	–	$R\bar{3}m$	41c,d	[4]
kalistrontite	K$_2$Sr(SO$_4$)$_2$	a	a	–	–	$R\bar{3}m$	–	(5)
lecontite	Na(NH$_4$)(SO$_4$)(H$_2$O)$_2$	8.216	12.854	6.232	–	$P2_12_12_1$	42a	[6]
matteuccite	Na(HSO$_4$)(H$_2$O)	7.811	7.823	8.025	117.5	Cc	42b	[7]
mirabilite	Na$_2$(SO$_4$)(H$_2$O)$_{10}$	11.512	10.370	12.847	107.8	$P2_1/c$	42c,d	[8]
darapskite	Na$_3$(NO$_3$)(SO$_4$)(H$_2$O)	10.564	6.911	5.194	102.8	$P2_1/m$	42e-g	[9]
thenardite	Na$_2$SO$_4$	9.829	12.302	5.868	–	$Fddd$	43a-c	[10]
hanksite	KNa$_{22}$[Cl(CO$_3$)$_2$(SO$_4$)$_9$]	10.490	a	21.240	–	$P6_3/m$	43d-f	[11]
burkeite	Na$_4$(SO$_4$)$_{1.39}$(CO$_3$)$_{0.61}$	5.170	9.217	7.058	–	$Pmnm$	43g,h	[12]
mineevite-(Y)	Na$_{25}$Ba(REE)$_2$(CO$_3$)$_{11}$(HCO$_3$)$_4$ (SO$_4$)$_2$F$_2$Cl	8.811	a	37.030	–	$P6_3/m$	44	[13]
mercallite	K(HSO$_4$)	8.429	9.807	18.976	–	$Pbca$	45a-d	[14]
letovicite	(NH$_4$)$_3$H(SO$_4$)$_2$	15.435	5.865	10.170	101.8	$C2/c$	45e,f	[15]

[1] rhombohedral; $\alpha = 42.5°$

References: [1] Okada and Ossaka (1980), [2] McGinnety (1972), [3] Hasebe (1981), [4] Bachmann (1953), (5) Gaines et al. (1997), [6] Corazza et al. (1967), [7] Catti et al. (1975), [8] Levy and Lisensky (1978), [9] Sabelli (1967), [10] Hawthorne and Ferguson (1975b), [11] Kato and Saalfeld (1972), [12] Giuseppetti et al. (1988), [13] Yamnova et al. (1992), [14] Payan and Haser (1976), [15] Leclaire et al. (1985)

perpendicular to the c direction[1]. The K(2)O$_{10}$ polyhedron (Fig. 41i) can be obtained from the cuboctahedron by collapsing one of the opposing triangular faces into one vertex.

Moore (1973, 1976, 1981), Eysel (1973), and Egorov-Tismenko et al. (1984) reviewed the crystal chemistry of the glaserite-related structures in detail. Moore (1973) noted that the key to understanding the relation of glaserite to other structures is the disposition of the tetrahedra about the octahedron in the pinwheel. He suggested a procedure to derive all possible pinwheels from the thirteen combinatorially distinct *bracelets*, in this case, projections of octahedra with "u" or "d" symbols attached to their

---

[1] Moore (1976, 1981) pointed out that, due to the icosahedral geometry, the glaserite structure-type represents a super-dense-packed oxide. He noted that the glaserite-related minerals and compounds are 5 to 15% more dense that their cubic or hexagonal close-packed counterparts. The only examples he gave are Ca$_2$(SiO$_4$) polymorphs. However, in the structure of 'hexagonal close-packed' (hcp) γ-Ca$_2$(SiO$_4$) ('calcio-olivine'), the Ca atoms are in octahedral coordination (= fill octahedral voids in hcp), which indicates that they expand oxygen hcp packing, and therefore the structure is not ideally close-packed. In contrast, in α- and β-Ca$_2$(SiO$_4$) ('silico-glaserite' and larnite, respectively), Ca participates in mixed cation–anion close packings.

vertices. The **u** and **d** designate tetrahedra pointing either up or down, respectively. Each bracelet can be symbolized as a $(u + d)_r$ combination, where r is a symbol used to identify topologically distinct bracelets. The bracelet for the glaserite structure, shown in Figure 41a, corresponds to the $(3 + 3)c'$ type of Moore (1973). In contrast, the structure of **palmierite**, $K_2Pb(SO_4)_2$, is based on the bracelet $(3 + 3)c$, which is the *complement* of the glaserite bracelet (Fig. 41c) (two bracelets are complementary if the symbols **u** and **d** are interchanged). Whereas in glaserite all pinwheel-layers are located under one another, the pinwheel-layers in palmierite form a more complex three-layer packing (Fig. 41d).

According to Moore (1973), the structure of **arcanite**, β-$K_2(SO_4)$, is based on a $(4 + 2)a$ pinwheel. However, Egorov-Tismenko et al. (1984) proposed a different description. Figure 41j shows the $\{K(2)O_{10}\}$ polyhedron in glaserite, decorated by edge- and corner-sharing (SO_4) tetrahedra to form a $[K(SO_4)_5]$ cluster (Fig. 41l,m). In contrast, Figures 41n,o show a similar cluster formed about the K(1) site (Fig. 41k) in arcanite. The difference between the two configurations is the orientation of one (SO_4) tetrahedron that shares an equatorial edge of the K polyhedron. The $[K(SO_4)_5]$ clusters in glaserite form a continuous layer perpendicular to the c direction, whereas in arcanite, the clusters initially form chains (Fig. 41q), then double chains (Fig. 41r) and, finally, a framework (Figs. 41s,t); the K(2) atoms occur in the interstices of this framework.

The description of the crystal structure of arcanite as an 'anion-stuffed' alloy has been suggested by Smirnova et al. (1967, 1968), O'Keeffe and Hyde (1985), and Hyde and Andersson (1989). They considered several sulfate-mineral structures (e.g. arcanite, barite, langbeinite) as derivatives of simple structure types, such as alloys, simple borides, and simple sulfides. For example, they noted that the arrangement of K and S cations in arcanite is similar to that of the atoms in the alloys Ca_2Sn and Ca_2Si with O atoms in the interstices. It is interesting that the high-pressure modification of K_2S is of the same arrangement, and thus arcanite can be considered as an anion-stuffed high-pressure K_2S polymorph (O'Keeffe and Hyde 1985).

Na- and NH₄-sulfates. Due to its relatively small size, Na can fill octahedral interstices in close-packed oxygen arrays, and the $(Na\emptyset_6)$ octahedron is the most common Na coordination polyhedron in mineral structures. The crystallographic parameters for sulfate minerals in which Na plays a dominant role are given in Table 14. The structures of these minerals can be described as based on structural units consisting of condensed $(Na\emptyset_6)$ octahedra.

The structure of **lecontite**, $Na(NH_4)(SO_4)(H_2O)_2$, is based on infinite chains of face-sharing $(Na\emptyset_6)$ octahedra decorated by (SO_4) tetrahedra that share corners with the $(Na\emptyset_6)$ octahedra (Fig. 42a). The chains are linked *via* NH_4 groups. The structure of **matteuccite**, $Na(HSO_4)(H_2O)$, consists of corner-sharing $(Na\emptyset_6)$ octahedra that are linked into a framework *via* (SO_4) tetrahedra (Fig. 42b). In **mirabilite**, $Na_2(SO_4)(H_2O)_{10}$, edge-sharing of $\{Na(H_2O)_6\}$ octahedra form infinite zigzag chains extending along [001] (Fig. 42c). Together with (SO_4) tetrahedra and additional (H_2O) groups, these chains form layers parallel to (100) (Fig. 42d). The structure of **darapskite**, $Na_3(NO_3)(SO_4)(H_2O)$, is based on chains of octahedra in which face- and edge-sharing alternates (Fig. 42e). These chains are linked *via* (SO_4) tetrahedra and other Na atoms into layers (Fig. 42f) that are linked into a three-dimensional structure by hydrogen bonds through (NH_4) and (H_2O) groups (Fig. 42g).

The structure of **thenardite**, $Na_2(SO_4)$, consists of a framework of (NaO_6) octahedra. Two (NaO_6) octahedra share an edge, forming $[Na_2O_{10}]$ dimers. The dimers are linked *via* edge-sharing with (SO_4) tetrahedra to form chains extended in the $[0\overline{1}2]$ direction. The chains are joined *via* corner-sharing of polyhedra to form sheets parallel to (100)

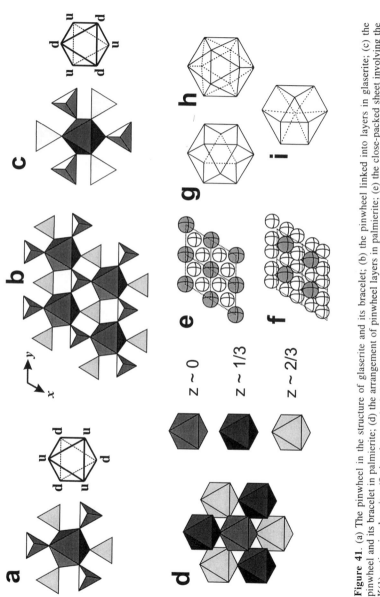

Figure 41. (a) The pinwheel in the structure of glaserite and its bracelet; (b) the pinwheel and its bracelet in palmierite; (d) the arrangement of pinwheel layers in palmierite; (e) the K(1) cation in glaserite; (f) the close-packed sheet involving the K(2) cation in glaserite; (g) ideal coordination polyhedron for a sphere in cubic closest packing; (h,i) coordination polyhedra of (h) K(1) and (i) K(2) in glaserite as derivatives of (g).

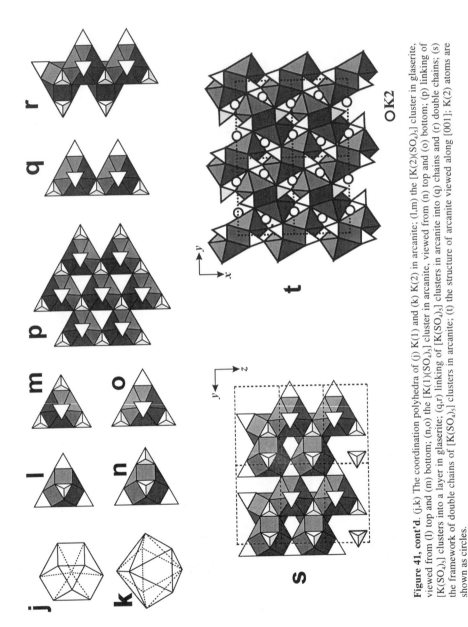

Figure 41, cont'd. (j,k) The coordination polyhedra of (j) K(1) and (k) K(2) in arcanite; (l,m) the [K(2)(SO₄)₃] cluster in glaserite, viewed from (l) top and (m) bottom; (n,o) the [K(1)(SO₄)₅] cluster in arcanite, viewed from (n) top and (o) bottom; (p) linking of [K(SO₄)₅] clusters into a layer in glaserite; (q,r) linking of [K(SO₄)₅] clusters in arcanite into (q) chains and (r) double chains; (s) the framework of double chains of [K(SO₄)₅] clusters in arcanite; (t) the structure of arcanite viewed along [001]; K(2) atoms are shown as circles.

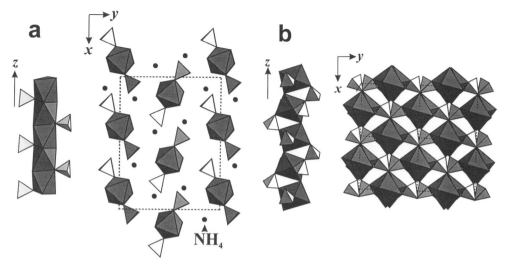

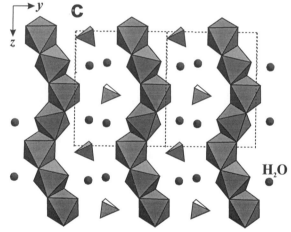

Figure 42. The structures of (a) lecontite and (b) matteuccite based on chains of face-sharing and corner-sharing (NaØ$_6$) octahedra, respectively; (c,d) the structure of mirabilite projected along (c) [100] and (d) [001]; (e) the chain with alternating of face- and edge-sharing (NaØ)$_6$ octahedra in darapskite.

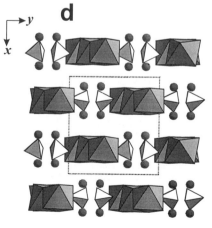

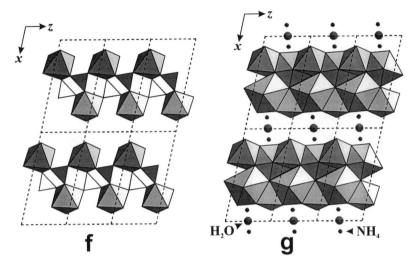

Figure 42, cont'd. (f) The arrangement of chains in darapskite to form a framework;
(g) the structure of darapskite viewed along [010].

(Fig. 43a). The further linkage of the sheets into a framework results in formation of chains of edge-sharing (NaO_6) octahedra (Fig. 41c) along [100] (Fig. 43b).

The crystal structure of **hanksite**, $KNa_{22}(Cl(CO_3)_2(SO_4)_9)$, can be described as rods of face-sharing Na- and K-polyhedra-sulfate pinwheels (Fig. 43d). The orientation of the rods and their stacking are shown in Figure 43e. The remaining cation polyhedra link the rods together (Fig. 43f). The structure of **burkeite**, $Na_4(SO_4)_{1.39}(CO_3)_{0.61}$, is highly disordered, with (SO_4) tetrahedra replaced by (CO_3) triangles and vice-versa. The structure consists of complex chains of (NaO_n) polyhedra (with $n \approx 6$) (Fig. 43g) that are linked into walls parallel to (010) (Fig. 43h). The crystal structure of **mineevite-(Y)**, $Na_{25}Ba(REE)_2$ $(CO_3)_{11}(HCO_3)_4 (SO_4)_2F_2Cl$, is very complex. It consists of rods of face-sharing cation polyhedra [(NaO_6) trigonal prisms, $(REE)O_9$ polyhedra, and (BaO_{12}) icosahedra] girdled by (CO_3) triangles (Fig. 44a). The arrangement of the rods is shown in Figure 44b. Yamnova et al. (1992) described the structure as based on sheets of cation polyhedra. The sheets are of three types (A, B, and C; Figs. 44c,d,e, respectively). The sequence of the sheets is ...*ABBACABBAC*... or (*ABBAC*) (Fig. 44f).

The structure of **mercallite**, $K(HSO_4)$, consists of (SO_4) tetrahedra linked *via* hydrogen bonds into two different substructures. The first substructure is the chain of $(HSO_4)^-$ groups that are linked *via* O⋯H bonds of 1.6–1.8 Å into infinite $[H(HSO_4)]^0$ chains parallel to [100] (Fig. 45a). The second substructure is composed of the finite $[H_2(SO_4)_2]^{2-}$ cluster that consists of two $(HSO_4)^-$ groups linked *via* additional O⋯H bonds (Fig. 45b). The hydrogen-bonded (SO_4) groups form sheets that are parallel to (001) (Fig. 45c), and these sheets are linked by K atoms, the coordinations of which are shown in Figure 45d. The structure of **letovicite**, $(NH_4)_3H(SO_4)_2$, is based on $[H(SO_4)_2]^{3-}$ groups

Figure 43. (a) The layer of edge-sharing $(NaØ_6)$ octahedra and (SO_4) tetrahedra in thenardite; (b) the structure of thenardite viewed along [010]; (c) the coordination polyhedron of Na in thenardite; (d) the rod of face-sharing pinwheels in hanksite; (e) the arrangement of rods in hanksite viewed along [001]; (f) the structure of hanksite viewed along [001]; (g) the complex $[NaØ_n]$ polyhedral chain decorated by (SO_4) tetrahedra and (CO_3) triangles in burkeite; (h) the structure of burkeite viewed along [100].   **Next page → →**

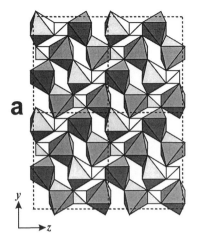

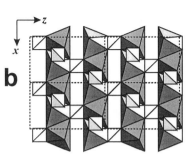

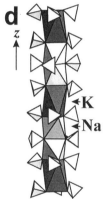

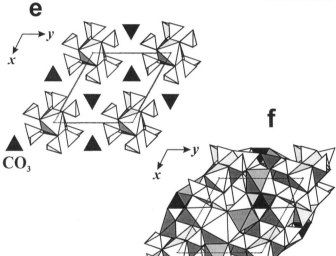

K
Na

CO₃

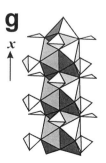

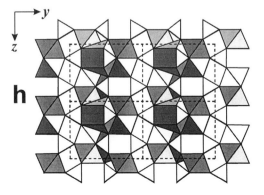

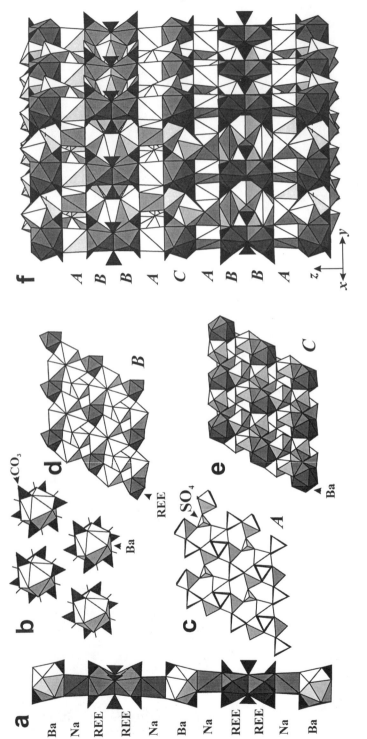

Figure 44. (a) The rod of face-sharing cation polyhedra decorated by (CO₃) triangles in mineevite-(Y); (b) the arrangement of rods viewed down [001] in mineevite-(Y); (c,d,e) *A*, *B*, and *C* sheets of cation-coordination polyhedra in mineevite-(Y), respectively; (f) the structure of mineevite-(Y) viewed along [210] (the sequence of *A*, *B*, and *C* sheets is indicated).

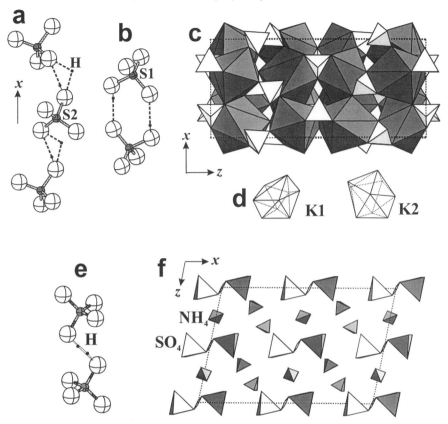

Figure 45. (a) The chain of H-bonded $(HSO_4)^-$ groups in mercallite; (b) $[(HSO_4)_2]^{2-}$ dimers in mercallite; (c) the structure of mercallite viewed along [010]; (d) the coordination polyhedra of K in mercallite; (e) the $[(SO_4)H(SO_4)]^{3-}$ cluster with the disordered H atom in letovicite; (f) the structure of letovicite viewed along [010].

consisting of two (SO_4) tetrahedra linked *via* disordered H atoms (Fig. 45e). As in mercallite, hydrogen-bonded (SO_4) groups form sheets parallel to (001) and are linked by (NH_4) groups (Fig. 45f).

Structures related to sulphohalite. There are four Na-sulfate minerals that occur as multiple salts in the system Na_2SO_4–NaF–$NaCl$: sulphohalite, schairerite, galeite, and kogarkoite (Table 15) (Pabst and Sharp 1973). The structure of **sulphohalite**, $Na_6ClF(SO_4)_2$, consists of a framework of clusters of $(Na\emptyset_6)$ octahedra ($\emptyset$ = O, F, Cl) (Fig. 46a,b). Each octahedron in this cluster shares four of its faces with four adjacent octahedra and one corner with a fifth octahedron (Fig. 46b). By analogy with other star-like clusters (*stella tetrangula* and *stella octangula*, O'Keeffe and Hyde 1996), this cluster should be called *stella hexangula*, a 'star with six angles.' In sulphohalite, the stellae hexangulae are linked by corner-sharing with (SO_4) tetrahedra into a three-dimensional framework (Fig. 46c).

The structures of the other three salts in this system cannot be clearly represented in terms of $(Na\emptyset_6)$ octahedra. A more appropriate description is based on their relation to the halite structure. Both NaF and $NaCl$ crystallize with the halite structure in which each Na

Table 15. Sulfate mineral structures related to sulphohalite, apatite, and barite.

Name	Formula	a (Å)	b (Å)	c (Å)	β(°)	γ(°)	S. G.	Fig.	Ref.
	Structures related to sulphohalite (Na$_2$(SO$_4$) – NaF – NaCl *system*)								
sulphohalite	Na$_6$ClF(SO$_4$)$_2$	10.071	a	a	–	–	$Fm\overline{3}m$	46a-c,e	[1]
kogarkoite	Na$_3$(SO$_4$)F	18.079	6.958	11.443	107.7	–	$P2_1/m$	46d,f	[2]
galeite	Na$_{15}$(SO$_4$)$_5$F$_4$Cl	12.197	a	13.955	–	–	$P31m$	46d,g	[3]
schairerite	Na$_{21}$(SO$_4$)$_7$F$_6$Cl	12.197	a	19.359	–	–	$P31m$	46d,h	[4]
	Structures related to apatite								
caracolite	Na$_3$Pb$^{2+}_2$(SO$_4$)$_3$Cl	9.810	a	7.140	–	–	$P6_3/m$	47a-e	[5]
cesanite	Ca$_{1.31}$Na$_{4.32}$(OH)$_{0.94}$(SO$_4$)$_3$	9.446	a	6.895	–	–	$P6_3/m$	47f	[6]
ellestadite	Na$_3$Ca$_2$(SO$_4$)$_3$(OH)	9.526	9.506	6.922	–	120.0	$P2_1$	–	[7]
chloroellestadite	Ca$_5$(SiO$_4$)$_{1.5}$(SO$_4$)$_{1.5}$Cl	9.530	a	6.914	–	–	$P6_3/m$	–	(8)
fluorellestadite	Ca$_5$(SiO$_4$)$_{1.5}$(SO$_4$)$_{1.5}$F	9.485	a	6.916	–	–	$P6_3/m$	–	(9)
hydroxylellestadite	Ca$_5$(SiO$_4$)$_{1.5}$(SO$_4$)$_{1.5}$(OH)	9.522	9.527	6.909	–	119.9	$P2_1/m$	· –	[10]
mattheddleite	Pb$^{2+}_5$(SiO$_4$)$_{1.5}$(SO$_4$)$_{1.5}$Cl	9.963	a	7.464	–	–	$P6_3/m$	–	(11)
	Structures related to barite								
barite	[Ba(SO$_4$)]	7.154	8.879	5.454	–	–	$Pbnm$	47g-j	[12]
celestine	[Sr(SO$_4$)]	6.867	8.355	5.346	–	–	$Pbnm$	47g-j	[12,13]
anglesite	[Pb^{2+}(SO$_4$)]	6.955	8.472	5.397	–	–	$Pbnm$	47g-j	[12]
hashemite	[Ba(CrO$_4$)]	9.113	5.536	7.340	–	–	$Pnma$	47g-j	[14]
olsacherite	[Pb$^{2+}_2$(SO$_4$)(SeO$_4$)]	8.42	10.96	7.00	–	–	$P22_12$	47g-j	(15)

References: [1] Sakamoto (1968), [2] Fanfani et al. (1980), [3] Fanfani et al. (1975a), [4] Fanfani et al. (1975b), [5] Schneider (1967), [6] Deganello (1983), [7] Organova et al. (1994), [8] Rouse and Dunn (1982), (9) Chesnokov et al. (1987a), [10] Hughes and Drexler (1991), (11) Livingstone et al. (1987), [12] Jacobsen et al. (1998), [13] Hawthorne and Ferguson (1975b), [14] Pasero and Davoli (1987), (15) Hurlbut and Aristarain (1969)

atom is octahedrally coordinated by halogen atoms, X, and vice-versa. Thus, the halite structure can be considered as built from cation-centered (NaX_6) octahedra or anion-centered (XNa$_6$) octahedra. In terms of (XNa$_6$) octahedra, the structure of sulphohalite represents a corner-linked octahedral framework (Fig. 46c). In the structures of **kogarkoite**, Na$_3$(SO$_4$)F, **galeite**, Na$_{15}$(SO$_4$)$_5$F$_4$Cl, and **schairerite**, Na$_{21}$(SO$_4$)$_7$F$_6$Cl, there are clusters of three face-sharing (XNa$_6$) octahedra (Fig. 46d). The structures of these minerals are built from layers of *single* (FNa$_6$) octahedra (S_F layers), layers of *single* (ClNa$_6$) octahedra (S_{Cl} layers), and layers of *triple* octahedra (T layers). The layers are linked into frameworks by corner-sharing of single and triple octahedra. The sequences of layers are shown in Figures 46eBh. Sulphohalite has a sequence ...$S_F S_{Cl} S_F S_{Cl}$... (Fig. 46e), and thus is built from single layers only, whereas the sequence ...TT... (Fig. 46f) in kogarkoite indicates that it consists only of layers of triple octahedra. The structures of galeite (...$S_{Cl} S_F T$...; Fig. 46g) and schairerite (...TTS_{Cl}...; Fig. 46h) are built from both single and triple octahedra. As each Na atom in X-centered octahedra belongs to two such octahedra in all structures, the ratio Na : X is invariably 3:1.

Figure 46. (a) (NaO$_6$) octahedra and (SO$_4$) tetrahedra in the structure of sulphohalite; (b) a *stella hexangula* formed by linkage of six (NaO$_6$) octahedra; (c) sulphohalite viewed as built from (XNa$_6$) octahedra (X = F, Cl); (d) triplet of octahedra from the structures of kogarkoite, galeite, and schairerite; (e,f,g,h) the structures of (e) sulphohalite; (f) kogarkoite; (g) galeite; (h) schairerite, as based on a framework of (XNa$_6$) octahedra (X = F, Cl) (see text for details).   **Next page** $\rightarrow$ $\rightarrow$

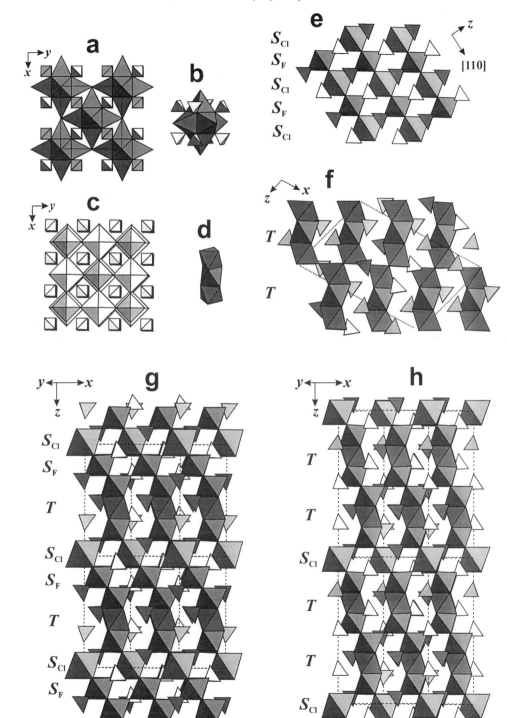

Apatite-like structures

The apatite-related sulfate minerals (Table 15, above) consist of frameworks of infinite rods of face-sharing cation polyhedra (Fig. 47b) that are decorated by (SO_4) tetrahedra (Fig. 45a). These rods are linked into frameworks that contain large channels (Fig. 47c), typically extending along [001]. These channels can be occupied by a variety of cations and anions. In the structure of **caracolite**, $Na_3Pb_2(SO_4)_3Cl$, the channels are occupied (Fig. 47e) by infinite chains of face-sharing (ClA_6) octahedra (A = Na, Pb) (Fig. 47d), whereas in **cesanite**, $Ca_{1.31}Na_{4.32}(OH)_{0.94}(SO_4)_3$, **ellestadite**, $Ca_4(Ca_{5.94}(OH)_{1.2}O_{0.5}Cl_{0.32})(SiO_4)_2$-$(SO_4)_2(Si_{0.5}S_{0.5}O_4)_2)$, and **hydroxylellestadite**, $Ca_{10}(SiO_4)_3(SO_4)_3(F_{0.16}Cl_{0.48}(OH)_{1.36})$, the channels are occupied by (XCa_3) triangles (X = Cl, OH) (Fig. 47f). The channel species are usually disordered and can be exchanged with other cations and/or anions.

Sulfates with the barite structure

Anhydrous sulfate minerals with the **barite**, $BaSO_4$, structure are listed in Table 15. The Ba cations are in [12]-coordination, forming irregular coordination polyhedra that are linked into a framework (Fig. 47g). Considering only the six nearest O neighbors, the Ba coordination can be described as a distorted octahedron and the structure as built from chains of edge-sharing octahedra linked by (SO_4) groups into sheets (Fig. 47h). These sheets are linked into a framework by corner-sharing between (BaO_6) polyhedra and (SO_4) tetrahedra from adjacent sheets (Fig. 47i). The presence of sheets results in the perfect cleavage of barite parallel to (001). An alternative description of the barite structure-type has been given by Smirnova et al. (1967) and O'Keeffe and Hyde (1985). The BaS cation array in barite is the same as that of the alloy FeB, with Ba in place of Fe, and S in place of B; O atoms are 'inserted' into SBa_3 tetrahedra of the BaS array, and barite can be regarded as an O-stuffed BaS array of the Fe–B type.

$Pb_4(SO_4)(CO_3)_2(OH)_2$ polymorphs

The structures of the $Pb^{2+}_4(SO_4)(CO_3)_2(OH)_2$ polymorphs (Table 16: **susannite**, **leadhillite**, and **macphersonite**) are based on the complex $[Pb^{2+}_4(OH)_2(CO_3)_2]^{2+}$ layer. The basis of this layer is the close-packed arrangement of Pb^{2+} atoms (Fig. 48a). Four of these cation sheets form a layer that is filled by (OH) and (CO_3) triangles (Fig. 48b). The resultant $[Pb^{2+}_4(OH)_2(CO_3)_2]^{2+}$ layers are separated in the structures by a single layer of (SO_4) tetrahedra (Figs. 48c,d,e). The difference between polymorphs is primarily in the relative positions of the (SO_4) tetrahedra (Steele et al. 1999).

Uranyl sulfates

There are three uranyl-sulfate minerals (Table 16) in the structural hierarchy for uranyl minerals given by Burns et al. (1996) and Burns (1999). In this scheme, **schröckingerite**, $NaCa_3[(U^{6+}O_2)(CO_3)_3](SO_4)F(H_2O)_{10}$, consists of a complex sheet of $(U^{6+}O_2)(CO_3)_3$ uranyl-carbonate clusters, $(NaØ_6)$ octahedra, and trimers of $(CaØ_8)$ polyhedra capped by (SO_4) groups (Fig. 49a). Note the similarity of the Ca-polyhedral trimer to the clusters in despujolsite (Fig. 16f) and orschallite (Fig. 39f). Linkage between sheets is provided by hydrogen bonding involving interlayer (H_2O) groups (Fig. 49b). The structure of **johannite**, $Cu^{2+}[(U^{6+}O_2)_2(OH)_2(SO_4)_2](H_2O)_8$, is based on an open sheet of $\{(U^{6+}O_2)O_5\}$ dimeric pentagonal bipyramids (Fig. 49c). Cross-linking between the sheets is via Jahn-Teller-distorted $(Cu^{2+}Ø_6)$ octahedra and hydrogen bonds involving interlayer (H_2O) groups (Fig. 49d). **Zippeite**, $K[(U^{6+}O_2)_2(SO_4)(OH)_3](H_2O)$, and its analogs $M^{2+}_2[(UO_2)_6(SO_4)_3(OH)_{10}](H_2O)_{16}$, M^{2+} = Co, Mg, Ni, Zn (Burns 1999), are based on the sheet shown in Figure 49e. The structure consists of chains of edge-sharing $\{(U^{6+}O_2)O_5\}$ pentagonal bipyramids linked by corner sharing with (SO_4) tetrahedra (Fig. 49f). Note that, in johannite and zippeite, (SO_4) tetrahedra and uranyl polyhedra polymerize only by corner-sharing.

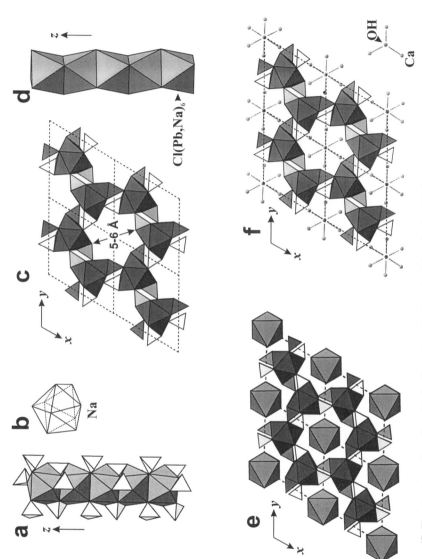

Figure 47. The structures of apatite-related sulfate minerals: (a) rods of face-sharing (NaØ₉) polyhedron in caracolite; (b) the (NaØ₉) polyhedron in caracolite; (c) the arrangement of rods in caracolite; note the large channels parallel to [001]; (d) the chain of face-sharing [Cl(Pb,Na)₆] octahedra in caracolite; (e) the structure of caracolite viewed along [001]; (f) the structure of cesanite viewed along [001]; the structures of ellestadite and hydroxylellestadite closely resemble that of cesanite.

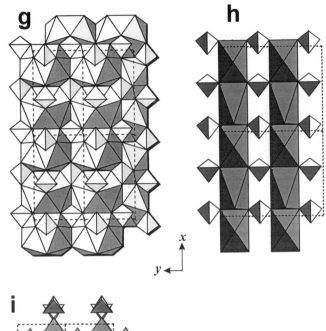

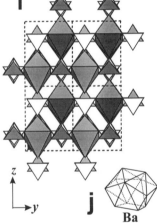

Figure 47, cont'd. (g) the structure of barite viewed along [001]; (h) 'truncated' $(Ba\emptyset_6)$ coordination provides layers of chains of edge-sharing $(Ba\emptyset_6)$ octahedra linked by (SO_4) tetrahedra; (i) the arrangement of layers in barite viewed along [100]; (j) the coordination polyhedron of Ba in barite.

Sulfates with non-sulfate tetrahedral sheets or frameworks

The crystallographic parameters for sulfate minerals of this class are given in Table 16. The structure of **heidornite**, $Na_2Ca_3[B_5O_8(OH)_2](SO_4)_2Cl$, consists of $[B_5O_8(OH)_2]$ sheets of $(B\emptyset_3)$ triangles and $(B\emptyset_4)$ tetrahedra (Fig. 50c). The $(Na\emptyset_6)$ octahedra, $(Ca\emptyset_8)$ polyhedra, and (SO_4) tetrahedra are in the interlayer (Fig. 50a,b). The structures of **tuscanite**, $KCa_{5.8}[(Si,Al)_{10}O_{22}](SO_4)_{1.9}$, **latiumite**, $KCa_3[(Al,Si)_5O_{11}](SO_4)_{0.7}(CO_3)_{0.3}$, and **quietite**, $Pb^{2+}_4[Zn_2(SiO_4)(Si_2O_7)](SO_4)$, are based on the $[T_{10}O_{22}]$ double-tetrahedral sheet shown in Figure 50e. Some large cations (K, Ca, Pb^{2+}) occur within the sheet, whereas others are in the interlayer, together with (SO_4) tetrahedra (Fig. 50d). The dominant structural motif of the structure of **roeblingite**, $Pb^{2+}_2Ca_6(SO_4)_2(OH)_2$-$[Mn^{2+}(Si_3O_9)_2](H_2O)_4$, is the $[Mn^{2+}(Si_3O_9)_2]$ sheet formed by polymerization of $(Mn^{2+}O_6)$ octahedra and (Si_3O_9) silicate rings (Fig. 50g). In heidornite, tuscanite, latiumite, and

Table 16. Sulfate minerals with non-octahedral cation coordination polyhedra.

Name	Formula	a (Å)	b (Å)	c (Å)	β (°)	S. G.	Fig.	Ref.
susannite	$Pb^{2+}_4(SO_4)(CO_3)_2(OH)_2$	9.0718	a	11.570	–	$P3$	48a,b,c	[1]
leadhillite	$Pb^{2+}_4(SO_4)(CO_3)_2(OH)_2$	9.110	20.820	11.590	90.5	$P2_1/a$	48a,b,d	[2]
macphersonite	$Pb^{2+}_4(SO_4)(CO_3)_2(OH)_2$	9.242	23.050	10.383	–	$Pcab$	48a,b,e	[3]
schröckingerite[1]	$NaCa_3[(U^{6+}O_2)(CO_3)_3](SO_4)F(H_2O)_{10}$	9.634	9.635	14.391	92.3	$P\bar{1}$	49a,b	[4]
johannite[2]	$Cu^{2+}[(U^{6+}O_2)_2(OH)_2(SO_4)_2](H_2O)_8$	8.903	9.499	6.812	112.0	$P\bar{1}$	49c,d	[5]
zippeite	$K[(U^{6+}O_2)_2(SO_4)(OH)_3](H_2O)$	8.755	13.987	17.730	104.1	$C2/c$	49e,f	[6]
heidornite	$Na_2Ca_3[B_5O_8(OH)_2](SO_4)_2Cl$	10.210	7.840	18.790	93.5	$C2/c$	50a,b,c	[8]
tuscanite	$KCa_{5.8}[(Si,Al)_{10}O_{22}](SO_4)_{1.9}$	24.030	5.110	10.880	106.9	$P2_1/a$	50d,e	[9]
latiumite	$KCa_3[(Al,Si)_5O_{11}](SO_4)_{0.7}(CO_3)_{0.3}$	12.060	5.080	10.810	106.0	$P2_1$	–	[10]
queitite	$Pb^{2+}_4[Zn_2(SiO_4)(Si_2O_7)](SO_4)$	11.362	5.266	12.655	108.2	$P2_1$	–	[11]
roeblingite	$Pb^{2+}_2Ca_6(SO_4)_2(OH)_2[Mn^{2+}(Si_3O_9)_2]$ $(H_2O)_4$	13.208	8.287	13.089	106.7	$C2/m$	50f,g	[12]
klebelsbergite	$Sb^{3+}_4O_4(OH)_2(SO_4)$	5.766	11.274	14.887	–	$Pca2_1$	51a,b	[16]
peretaite	$CaSb^{3+}_4O_4(OH)_2(SO_4)_2(H_2O)_2$	24.665	5.601	10.185	96.0	$C2/c$	51c,d	[17]
cannonite	$Bi^{3+}_2O(OH)_2(SO_4)$	7.692	13.870	5.688	109.0	$P2_1/c$	51e,f	[18]
d'ansite	$(Na_4Cl)_3[(Na_{0.75}Zn_{0.25})_2(SO_4)_5]_2$	15.913	a	a	–	$I\bar{4}3d$	52a,b,c	[7]
coskrenite-(Ce)[3]	$Ce_2(SO_4)_2(C_2O_4)(H_2O)_8$	6.007	8.368	9.189	105.6	$P\bar{1}$	52d,e	[13]
zircosulfate	$[Zr(SO_4)_2(H_2O)_4]$	25.920	11.620	5.532	–	$Fddd$	52f,g	[14]
vonbezingite	$Ca_6Cu^{2+}_3(SO_4)_3(OH)_{12}(H_2O)_2$	15.122	14.358	22.063	108.7	$P2_1/c$	52h	[15]

[1] $\alpha = 91.4°$, $\gamma = 120.3°$; [2] $\alpha = 109.9°$, $\gamma = 100.4°$; [3] $\alpha = 99.9°$, $\gamma = 107.7°$.

References: [1] Steele et al. (1999), [2] Giuseppetti et al. (1990), [3] Steele et al. (1998), [4] Mereiter (1986), [5] Mereiter (1982), [6] Vochten et al. (1995), [7] Lange and Burzlaff (1995), [8] Burzlaff (1967), [9] Mellini and Merlino (1977), [10] Canillo et al. (1973), [11] Hess and Keller (1980), [12] Moore and Shen (1984), [13] Peacor et al. (1999b), [14] Singer and Cromer (1959), [15] Dai and Harlow (1992), [16] Menchetti and Sabelli (1980a), [17] Menchetti and Sabelli (1980b), [18] Golic et al. (1982)

Table 17. Framework-silicate minerals with (SO_4) tetrahedra in cavities.

Name	Formula	a (Å)	c (Å)	S. G.	Fig.	Ref.
	Cancrinite group					
liottite	$(Ca_{11}Na_9K_4)[(Al_{18}Si_{18})O_{72}](SO_4)_4(CO_3)_2Cl_3(OH)_4(H_2O)_2$	12.870	16.096	$P\bar{6}$	50h	[1]
microsommite	$(Na,Ca,K)_{7-8}[(Al,Si)_{12}O_{24}](Cl,SO_4,CO_3)_{2-3}$	22.138	5.248	$P6_3$	–	[2]
pitiglianoite	$Na_6K_2[Al_6Si_6O_{24}](SO_4)(H_2O)_2$	22.121	5.221	$P6_3$	–	[3]
davyne	$CaNa_{4.26}K_{1.74}[Al_6Si_6O_{24}]Cl_{2.67}(SO_4)_{0.67}$	12.705	5.368	$P6_3/m$	–	[4]
afghanite	$(Na,K,Ca)_8(Si,Al)_{12}O_{24}(SO_4,Cl,CO_3)_3(H_2O)$	12.801	21.412	$P31c$	–	[5]
franzinite	$(Na,Ca)_7[(Si,Al)_{12}O_{24}](SO_4,CO_3,OH,Cl)_3(H_2O)$	12.861	26.45	?	–	(6)
giuseppettite	$(Na,Ca)_{7-8}[(Si,Al)_{12}O_{24}](SO_4,Cl)_{1-2}$	12.850	42.22	?	–	(6)
sacrofanite	$(Na,Ca,K)_9[Si_6Al_6O_{24}](OH,SO_4,CO_3Cl)_4(H_2O)_n$	12.865	74.240	$P6_3/mmc$	–	(6)
tounkite	$(Na,Ca,K)_8[Si_6Al_6O_{24}](SO_4)_2Cl(H_2O)$	12.843	32.239	$P6_222$	–	(6)
vishnevite	$(Na,K)_8[Si_6Al_6O_{24}](SO_4)(H_2O)_2$	12.685	5.179	$P6_3$	–	[7]
	Sodalite group					
haüyne	$(Na,K,Ca)_8[Si_6Al_6O_{24}](SO_4)_2$	9.118	a	$P\bar{4}3n$	50i	[8]
nosean	$Na_8[Si_6Al_6O_{24}](SO_4)(H_2O)$	9.084	a	$P\bar{4}3n$	50i	[9]

References: [1] Ballirano et al. (1996), [2] Klaska and Jarchow (1977), [3] Merlino et al. (1991), [4] Bonaccorsi et al. (1990), [5] Ballirano et al. (1997), (6) Gaines et al. (1997), [7] Hassan and Grundy (1984), [8] Evsyunin et al. (1996), [9] Hassan and Grundy (1989)

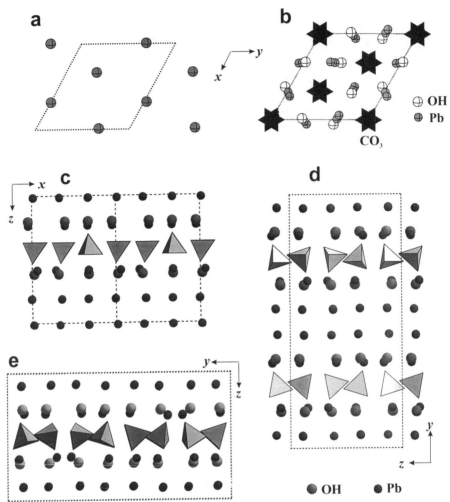

Figure 48. (a) The sheet of eutactically arranged Pb atoms in the $Pb_4(OH)_2(SO_4)(CO_3)_2$ polymorphs; (b) the $[Pb_4(OH)_2(CO_3)_2]$ layer present in all $Pb_4(OH)_2(SO_4)(CO_3)_2$ polymorphs; (c,d,e) the structures of (c) susannite, (d) leadhillite, and (e) macphersonite viewed along the extension of the sheets. The polymorphs differ in the positions of the interlayer (SO_4) tetrahedra.

quietite, the other structural subunits [Ca and Pb^{2+} cations, (SO_4) groups, and (OH) groups] act as connectors of the sheets in three dimensions (Fig. 50f).

Table 17 (above) lists framework silicate minerals that contain (SO_4) tetrahedra encapsulated in cavities. Two examples, **liottite**, $(Ca_{11}Na_9K_4)[(Al_{18}Si_{18})O_{72}](SO_4)_4(CO_3)_2Cl_3$-$(OH)_4(H_2O)_2$, and **haüyne**, $(Na,K,Ca)_8[Si_6Al_6O_{24}](SO_4)_2$, are shown in Figures 50h and 50i, omitting the interframework cations for clarity.

Basic sulfates of Sb^{3+} and Bi^{3+}

Table 16 lists crystallographic parameters for the minerals of this group. The structures of **klebelsbergite**, $Sb^{3+}_4O_4(OH)_2(SO_4)$, **peretaite**, $CaSb^{3+}_4O_4(OH)_2$-$(SO_4)_2(H_2O)_2$, and **cannonite**, $Bi^{3+}_2O(OH)_2(SO_4)$, are strongly influenced by the s^2 lone-

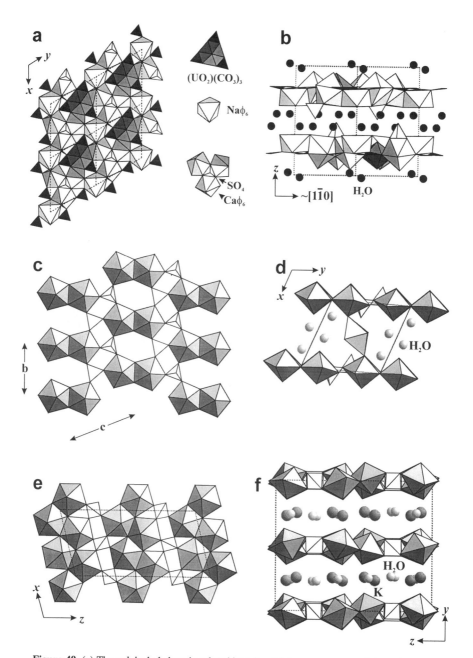

Figure 49. (a) The polyhedral sheet in schröckingerite; (b) the arrangement of sheets in the structure of schröckingerite; (c,d,e,f) the sheets of $\{(UO_2)O_5\}$ pentagonal bipyramids and (SO_4) tetrahedra and the structures of (c,d) johannite and (e,f) zippeite.

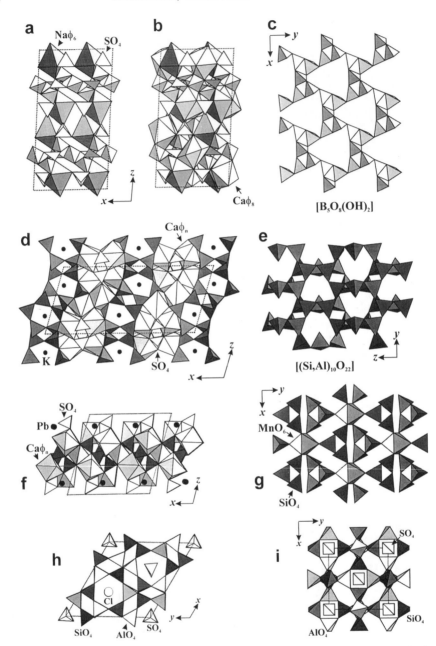

Figure 50. (a) The linkage of [B$_5$O$_8$(OH)$_2$] sheets, (SO$_4$) tetrahedra, and (NaØ$_6$) octahedra in heidornite viewed down [010]; (b) the same view of heidornite with the addition of (CaØ$_8$) polyhedra; (c) the [B$_5$O$_8$(OH)$_2$] sheet in heidornite viewed down [001]; (d) the structure of tuscanite viewed along [010]; (e) the aluminosilicate tetrahedral sheet in tuscanite viewed down [100]; (f) the structure of roeblingite viewed down [010]; (g) the [Mn(Si$_3$O$_9$)$_2$] sheet in roeblingite viewed down [100]; (h,i) the structures of framework silicates with (SO$_4$) tetrahedra in cavities: (h) liottite and (i) haüyne; interframework cations are omitted for clarity.

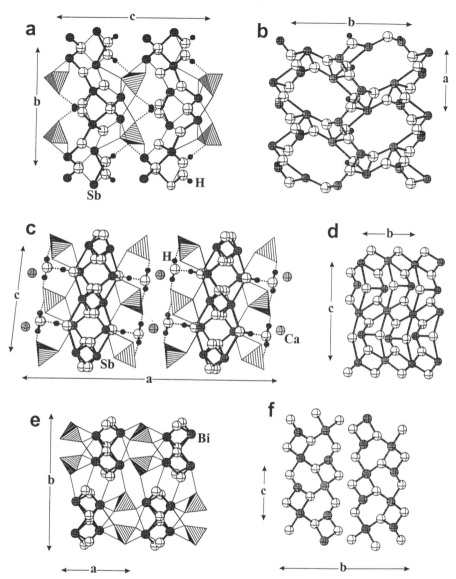

Figure 51. The structures and basic elements of the Sb and Bi sulfates: (a) klebelsbergite viewed along [100]; (b) klebelsbergite viewed down [001]; (c) peretaite viewed along [010]; (d) peretaite viewed down [100]; (e) cannonite viewed down [001]; (f) cannonite viewed down [100].

electron pairs on Sb^{3+} and Bi^{3+}. The coordination polyhedra of these cations are highly irregular and the structures are better understood in the terms of motifs of metal–oxygen bonds rather than in terms of coordination polyhedra. **Klebelsbergite** is based on the $[Sb^{3+}_4O_4(OH)_2]$ sheet parallel to (001); bonding within the sheet is between Sb^{3+} cations and O and (OH) anions not linked to S. The structure of the sheet is shown in Figure 51b. The two Sb^{3+} cations are [3]- and [5]-coordinated, respectively; O anions are [3]-coordinated and (OH) anions are [2]-coordinated (Fig. 51a). The structure of **peretaite** consists of sheets of the same stoichiometry as the sheet in klebelsbergite, but with a rather different structure (Fig. 51d) and linkage (Fig. 51c). In **cannonite** (Fig. 51e,f), Bi^{3+}, O and (OH) form $[Bi^{3+}_2O(OH)_2]$ chains extending along [001] (Fig. 51f). In all three structures, (SO_4) tetrahedra are between the structural units, participating in long cation–oxygen bonds to the trivalent cations (Figs. 51a,c,e). The structures of klebelsbergite, peretaite, and cannonite are examples of the strong tendency of cations with lone-electron pairs to form complexes with O and (OH) anions. These complexes should play an essential role not only in minerals, but also as complexes involved in fluid transport of metals in natural processes.

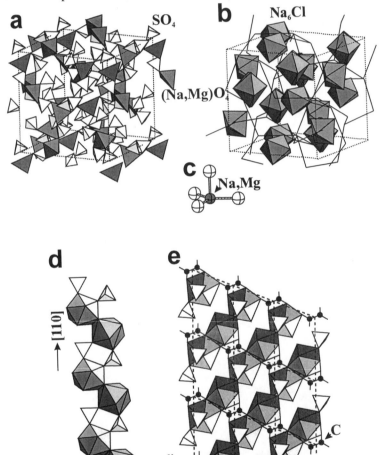

Figure 52. (a) The framework of $[(Na,Mg)O_4]$ trigonal pyramids and (SO_4) tetrahedra in d'ansite; (b) the positions of (Na_6Cl) octahedra in the framework of d'ansite, with the latter shown as a net; (c) coordination of the (Na,Mg) position in d'ansite; (d) the chain of $(REE)O_9$ polyhedra and (SO_4) tetrahedra in coskrenite-(Ce); (e) the chains in coskrenite-(Ce) cross-linked by (C_2O_4) groups to form a sheet.

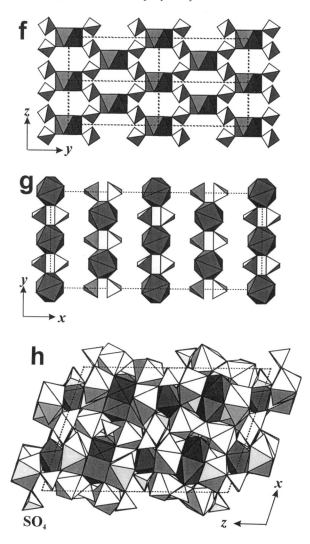

Figure 52, cont'd. (f) The sheet of $(ZrØ_8)$ polyhedra and (SO_4) tetrahedra in zircosulfate; (g) packing of the sheets in the structure of zircosulfate; (h) the structure of vonbezingite projected down [010].

Miscellaneous sulfates

D'ansite, $(Na_6Cl)_3[(Na_{0.75}Zn_{0.25})_2(SO_4)_5]_2$, is based on a framework of corner-sharing $[(Na,Mg)O_4]$ trigonal pyramids and (SO_4) tetrahedra (Figs. 52a,b,c). The cavities in the framework are occupied by $(ClNa_6)$ octahedra.

The structure of **coskrenite-(Ce)**, $(Ce,Nd,La)_2(SO_4)_2(C_2O_4)(H_2O)_8$, the only sulfate-oxalate mineral of known structure, is based on the chain shown in Figure 52d. Edge-sharing dimers of $(REE)Ø_9$ polyhedra are linked by corner-sharing with (SO_4) tetrahedra, forming a chain extending along [110]. These chains are further linked through oxalate anions, $(C_2O_4)^{2-}$, into sheets parallel to (001) (Fig. 52e). Other sulfate–oxalate

Table 18. Sulfate minerals with anion-centered $[XA_4]$ tetrahedra. *

Name	Formula	a (Å)	b (Å)	c (Å)	β (°)	S. G.	Graph	Fig.	Ref.
euchlorine	$NaK[Cu^{2+}{}_3O](SO_4)_3$	18.410	9.430	14.210	113.7	$C2/c$	53a	–	[1]
fedotovite	$K_2[Cu^{2+}{}_3O](SO_4)_3$	19.037	9.479	14.231	111.0	$C2/c$	53a	54a	[2]
kamchatkite	$K[Cu^{2+}{}_3O]Cl(SO_4)_2$	9.741	12.858	7.001	–	$Pna2_1$	53b	54b	[3]
vergasovaite	$[Cu^{2+}{}_3O][(Mo,S)O_4SO_4]$	7.421	6.754	13.624	–	$Pnma$	53b	54c	[4]
piypite	$K_4[Cu^{2+}{}_4O_2](SO_4)_4MCl$	13.600	a	4.980	–	$I4$	53c	54d	[5]
klyuchevskite	$K_3[Cu^{2+}{}_3(Fe,Al)O_2](SO_4)_4$	18.667	4.940	18.405	–	$I2$	53c	–	[6]
alumoklyu-chevskite	$K_3[Cu^{2+}{}_3AlO_2](SO_4)_4$	–	–	–	–	$I2$	53c	–	(7)
lanarkite	$[Pb^{2+}{}_2O](SO_4)$	13.769	5.698	7.079	115.9	$C2/m$	53c	54e	[8]
sidpietersite[1]	$[Pb^{2+}{}_4O_2](OH)_2(SO_3S)$	7.455	6.496	11.207	89.65	$P\overline{1}$	53d	–	[9]
synthetic[2]	$[Pb^{2+}{}_4O_2](OH)_2(SO_4)$	6.378	7.454	10.308	79.37	$P\overline{1}$	53d	–	[10]
nabokoite	$KCu^{2+}[Cu^{2+}{}_6TeO_4](SO_4)_5Cl$	9.833	a	20.591	–	$P4/ncc$	53e	55a	[11]
atlasovite	$KFe^{3+}Cu^{2+}{}_6Bi^{3+}O_4(SO_4)_5Cl$	9.86	a	20.58	–	$P4/ncc$	53e	–	(12)
dolerophanite	$[Cu^{2+}{}_2O](SO_4)$	9.370	6.319	7.639	122.3	$C2/m$	53f	55b	[13]
grandreefite	$[Pb^{2+}{}_2F_2](SO_4)\cdot$	8.667	4.442	14.242	107.4	$A2/a$	53g	55c	[14]
kleinite	$[Hg_2N](SO_4,Cl,H_2O)$	6.762	a	11.068	–	$P6_3/m$	53h	–	[15]
mosesite	$[Hg_4N_2](SO_4)(H_2O)$	9.503	a	a	–	$F\overline{4}3m$	53i	–	[16]
gianellaite	$[Hg_4N_2](SO_4)$	9.521	a	a	–	$F\overline{4}3m$	53i	–	(7)
schuetteite	$[Hg_3O_2](SO_4)$	7.044	a	10.000	–	$P3_121$	53j	–	[17]

* X = O, F, N; A = Cu^{2+}, Pb^{2+}, Hg, Te, Fe^{3+}, Al; M = Na, K; [1] α = 114.33°, γ = 88.69°; [2] α = 75.26°, γ = 88.16°.

References: [1] Scordari and Stasi (1990), [2] Starova et al. (1991), [3] Varaksina et al. (1990), [4] Berlepsch et al. (1999), [5] Effenberger and Zemann (1984), [6] Gorskaya et al. (1992), (7) Gaines et al. (1997), [8] Sahl (1970), [9] Cooper and Hawthorne (1999), [10] Steele et al. (1997), [11] Pertlik and Zemann (1988), (12) Popova et al. (1987), [13] Effenberger (1985a), [14] Kampf (1991), [15] Giester et al. (1996), [16] Airoldi and Magnano (1967), [17] Nagorsen et al. (1962)

minerals, **levinsonite-(Y)**, $(Y,Nd,Ce)Al(SO_4)_2(C_2O_4)\cdot12H_2O$, and **zugshunstite-(Ce)**, $(Ce,Nd,La)Al(SO_4)_2(C_2O_4)\cdot12H_2O$, are known (Peacor et al. 1999b), but complete descriptions and crystal-structure data are not yet published. The crystal structure of **zircosulfate**, $[Zr(SO_4)_2(H_2O)_4]$, is based on sheets of corner-sharing $(ZrØ_8)$ tetragonal antiprisms and (SO_4) tetrahedra (Fig. 52f). The sheets are parallel to (100) and are held together by hydrogen bonds only (Fig. 52g). **Vonbezingite**, $Ca_6Cu^{2+}{}_3(SO_4)_3(OH)_{12}(H_2O)_2$, consists of thick heteropolyhedral slabs parallel to (001) (Fig. 52h).

STRUCTURES WITH ANION-CENTERED TETRAHEDRA

The structures of most of the minerals described in the preceding sections are based on *cation-centered* coordination polyhedra. However, application of the hierarchical principle (Hawthorne 1983a) to some sulfate minerals shows that they should be considered as based on *anion-centered* polyhedra rather than on cation-centered polyhedra (Krivovichev and Filatov 1999a). Recently, Krivovichev et al. (1998a) reported a structural hierarchy of minerals and inorganic compounds based on oxocentered (OM_4) tetrahedra and described their general crystal-chemical features. The importance of anion-centered tetrahedra in minerals has been noted in a number of previous works (e.g. Bergerhoff and Paeslack 1968; O'Keeffe and Bovin 1978; Effenberger 1985a; Hyde and Andersson 1989), but has not been extensively developed until recently.

Table 18 lists sulfate minerals based on anion-centered metal tetrahedra; Figure 53 shows various types of anion-centered tetrahedral units occurring in sulfate minerals. By analogy with other minerals, those in Table 18 can be subdivided into finite-cluster, infinite-chain, infinite-sheet and infinite-framework minerals.

The structures of **euchlorine**, $NaK[Cu^{2+}_3O](SO_4)_3$, and **fedotovite**, $K_2[Cu^{2+}_3O](SO_4)_3$, are based on the $[O_2Cu^{2+}_6]$ edge-sharing tetrahedral dimer linked through (SO_4) groups into infinite sheets parallel to (100) (Fig. 54a). **Kamchatkite**, $K[Cu^{2+}_3O]Cl(SO_4)_2$ (Fig. 54b), and **vergasovaite**, $[Cu^{2+}_3O][(Mo,S)O_4SO_4]$ (Fig. 54c), are based on the $[O_2Cu^{2+}_6]$ infinite chain of corner-sharing (OCu^{2+}_4) tetrahedra of the type shown in Figure 53b; more details about the crystal chemistry of minerals and synthetic compounds based on this chain are given by Krivovichev et al. (1998b). The structures of **piypite**, $K_4[Cu^{2+}_4O_2](SO_4)_4 \cdot MCl$, M = (Na,K), **klyuchevskite**, $K_3[Cu^{2+}_3(Fe^{3+})O_2]$-$(SO_4)_4$, and **lanarkite**, $[Pb^{2+}_2O](SO_4)$, are based on $[O_2M_4]$ chains of *edge*-sharing $[OM_4]$ tetrahedra (Fig. 53c). In piypite (Fig. 54d) and klyuchevskite, these chains are connected through (SO_4) groups into frameworks, whereas in lanarkite (Fig. 54e), they form infinite sheets parallel to ($\bar{1}$01). **Nabokoite**, $KCu^{2+}[Cu^{2+}_6TeO_4](SO_4)_5Cl$, consists of $[O_4Cu^{2+}_6Te]$ sheets based on the tetrahedral tetramers shown in Figure 53e. The sheets are decorated by (SO_4) tetrahedra and are linked together through $(Cu^{2+}O_4Cl)$ tetragonal pyramids and interlayer K atoms (Fig. 55a). **Dolerophanite**, $[Cu^{2+}_2O](SO_4)$, is based on the $[O_2Cu^{2+}_4]$ sheet (Fig. 53f); the sheets are cross-linked by (SO_4) tetrahedra to form a three-dimensional framework (Fig. 55b). The structure of **grandreefite**, $[Pb^{2+}_2F_2](SO_4)$, consists of $[F_2Pb^{2+}_2]$ sheets of the type observed in the structure of tetragonal PbO (Fig. 53g). The (SO_4) tetrahedra are located between the sheets, linking them together (Fig. 55c).

The $[NHg_2]$ frameworks of the nitrogen-centered (NHg_4) tetrahedra in **kleinite**, $[Hg_2N](SO_4,Cl,H_2O)$ (Fig. 53h), and **mosesite**, $[Hg_2N_2](SO_4)(H_2O)$ (Fig. 53i), show striking similarity to the $[SiO_2]$ frameworks in tridymite and cristobalite, respectively. It should be noted that N-centered tetrahedra are well-known in trivalent rare-earth compounds (Schleid 1996). The $[O_2Hg_3]$ framework in **schuetteite**, $[Hg_3O_2](SO_4)$, is built by corner-sharing of the $[O_2Hg_6]$ tetrahedral dimers of the euchlorine-fedotovite type (Fig. 53j).

Inspection of the list of the sulfate minerals with anion-centered tetrahedra (Table 18) shows that their most common metals are Cu, Pb, and Hg. It is of interest that all the Cu minerals possessing (OCu^{2+}_4) tetrahedra are of fumarolic origin, occurring in the exhalation deposits of the Vesuvio (Italy) and Tolbachik (Russia) volcanoes. There are several other examples of fumarolic Cu-chloride, vanadate, arsenate, and selenite minerals with (OCu^{2+}_4) tetrahedra (Krivovichev et al. 1998a; Krivovichev and Filatov 1999a,b). Filatov et al. (1992) suggested that oxocentered tetrahedra play an important role as a form of Cu transport in fumarolic processes. Both Pb^{2+} and Hg^{2+} show a strong tendency to form complex polycations with O anions; these also could act as agents in metal transport in natural and anthropogenic processes.

THIOSULFATE MINERALS

There are a few thiosulfate minerals known (Table 19). Only the structure of sidpietersite is known, although bazhenovite has a synthetic orthorhombic polymorph, the structure of which has been determined (Chesnokov et al. 1987b). Thiosulfates are not highly represented among minerals, but they are of potential environmental importance because they can form as weathering products of anthropogenic materials such as slags (e.g. Braithwaite et al. 1993).

Sidpietersite, $Pb^{2+}_4(S^{6+}O_3S^{2-})O_2(OH)_2$, is a lead hydroxy-thiosulfate, the structure

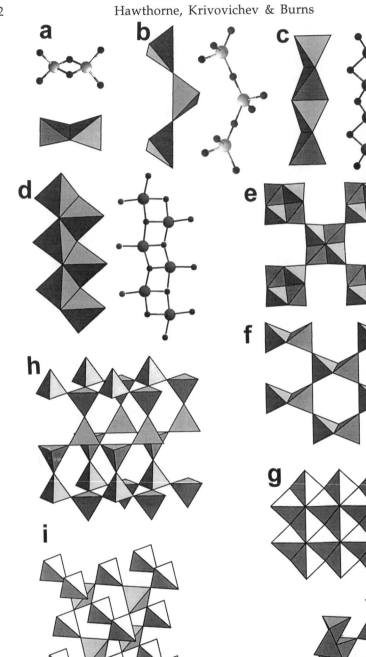

Figure 53. Types of structural units based on anion-centered tetrahedra in structures of the sulfate minerals (see Table 18).

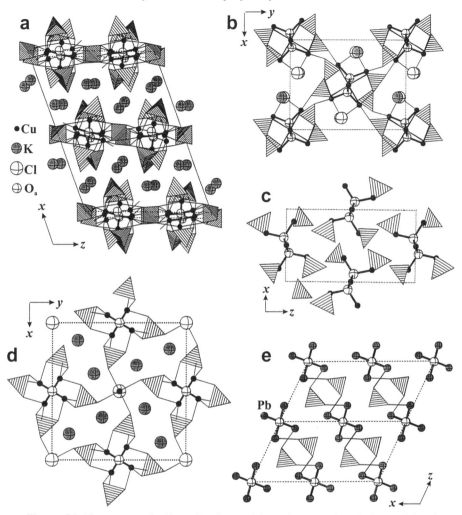

Figure 54. The structure of sulfate minerals containing anion-centered tetrahedra: (a) fedotovite; (b) kamchatkite; (c) vergasovaite; (d) piypite; (e) lanarkite. The anion atoms in the centers of tetrahedra are shown as large unshaded circles.

of which is shown in Figure 56. A ladder of Pb [$Pb(1)$ and $Pb(4)$] and O atoms extends along the *a* axis; note that this is a motif from the PbO structure. This ladder is decorated on either side by a staggered arrangement of Pb^{2+} [$Pb(2)$ and $Pb(3)$] and O atoms to form a ribbon that extends along the *a* axis. These ribbons are linked in the *c* direction by ($S^{6+}O_3S^{2-}$) groups such that the S^{2-} anions [$S(2)$] form an almost linear array in projection (Fig. 56a). This linkage forms thick slabs orthogonal to [001]. In Figure 56b, the '*Pb–O*' ladders are seen 'end-on'; they are linked into slabs orthogonal to [001] by thiosulfate groups and *Pb–O* bonds. These slabs are linked in the *c* direction by long weak *Pb–S(2)* bonds; note that, in this orientation, the $Pb(2)$, $Pb(3)$ and $S(2)$ atoms form a ladder resembling the *Pb–O* ladder in Figure 56a.

Sidpietersite is a unique structure both with regard to minerals and to synthetic

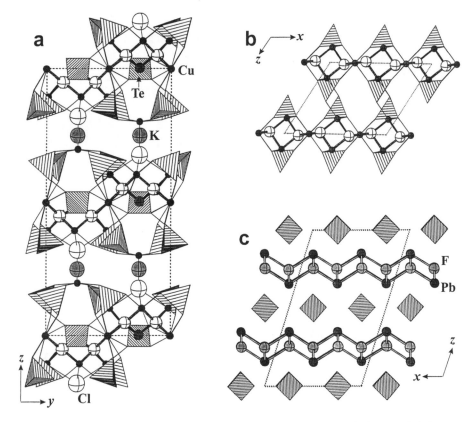

Figure 55. The structure of sulfate minerals containing anion-centered tetrahedra: (a) nabokoite; (b) dolerophanite; (c) grandreefite. The anion atoms in the centers of tetrahedra are shown as large unshaded circles.

compounds. Inspection of Table 2 shows that all the known synthetic thiosulfates have cation:S_2O_3 ratios between 1 and 2, and all of the structural arrangements involve thiosulfate tetrahedra in a network of weak cation–oxygen bonds, usually involving H or alkali cations. Sidpietersite has a cation:S_2O_3 ratio of 4, and contains relatively strongly bonded and polymerized fragments of the PbO structure. However, there does not seem to be anything particularly exotic about the structure of sidpietersite, except for the occurrence of the thiosulfate group in a mineral, and there seems to be no intrinsic reason why structures with cation:S_2O_3 ratios higher than 2 should not be common.

SULFITE MINERALS

The few sulfite minerals known are listed in Table 19 and are illustrated in Figure 57. **Hannebachite**, $Ca(S^{4+}O_3)(H_2O)_{0.5}$, is based on chains of edge-sharing $(Ca\emptyset_8)$ polyhedra, decorated by $(S^{4+}O_3)$ triangles, that extend along the z axis (Fig. 57a). Seen 'end-on' in Figure 57b, these chains are arranged in a checkerboard fashion, linking through shared edges of the $(Ca\emptyset_8)$ polyhedra. **Scotlandite**, $Pb^{2+}(S^{4+}O_3)$, consists of sheets of edge-sharing $(Pb^{2+}\emptyset_9)$ polyhedra decorated by $(S^{4+}O_3)$ triangles and parallel to (001) (Fig. 57c). The $(Pb^{2+}\emptyset_9)$ polyhedra from adjacent sheets share edges to form a fairly dense framework (Fig. 57d). **Gravegliaite**, $Mn^{2+}(S^{4+}O_3)(H_2O)_3$, consists of chains of $(Mn^{2+}\emptyset_6)$ octahedra

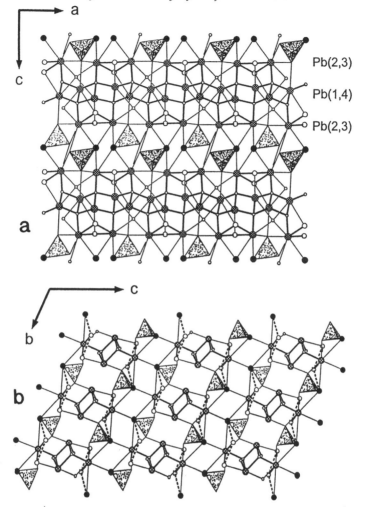

Figure 56. The structure of sidpietersite: (a) viewed down an axis 5° from [010]; (b) viewed down an axis 3° from [100]; the thiosulfate anion is random-dot-shaded, the S^{2-} anion is shown as a large black circle, Pb^{2+} cations are shown as cross-hatched circles, OH anions and O anions are shown as medium-sized and small unshaded circles, respectively, and H bonds are shown as broken lines. After Cooper and Hawthorne (1999).

and $(S^{4+}O_3)$ triangles (Fig. 57e). The chain can be considered as being built from $[M_2(TO_3)\emptyset_6]$ clusters which link by sharing corners between triangles and octahedra; the graph of this chain is shown in Figure 57f. Viewed along [010] (Fig. 57g), these chains are arranged, end-on, at the vertices of a centered plane square lattice, with adjacent chains canted to each other. The chemical formula of the chain is the same as that of the mineral, and the chains are linked only by hydrogen bonds. **Orschallite**, $Ca_3(S^{4+}O_3)_2(S^{6+}O_4)\cdot(H_2O)_{12}$, is a mixed sulfite–sulfate mineral. The structure is discussed in the section on calcium-sulfate minerals. **Abenakiite-(Ce)**, $Na_{26}Ce_6(SiO_3)_6(PO_4)_6(CO_3)_6(S^{4+}O_3)$, has a structure similar to that of mineevite-(Y) (Fig. 44) and hanksite (Figs. 43d,e,f). The structure is based on complex chains and columns that are arranged at and around the

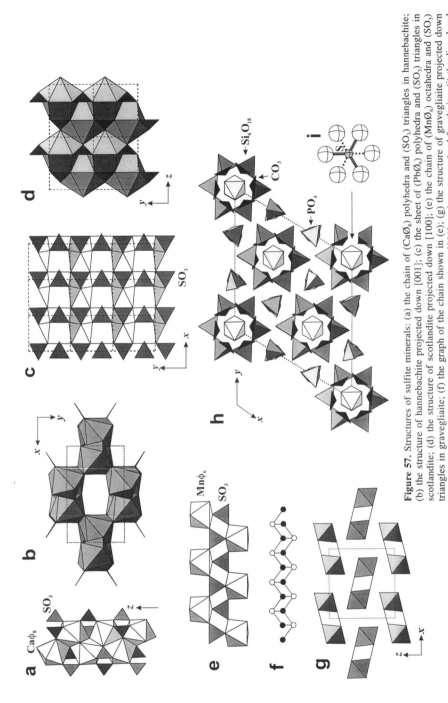

Figure 57. Structures of sulfite minerals: (a) the chain of (CaØ$_8$) polyhedra and (SO$_3$) triangles in hannebachite; (b) the structure of hannebachite projected down [001]; (c) the sheet of (PbØ$_n$) polyhedra and (SO$_3$) triangles in scotlandite; (d) the structure of scotlandite projected down [100]; (e) the chain of (MnØ$_6$) octahedra and (SO$_3$) triangles in gravegliaite; (f) the graph of the chain shown in (e); (g) the structure of gravegliaite projected down [010]; (h) the structure of abenakiite-(Ce) projected down [001]; cations are omitted for clarity; (i) the disordered (S^{4+}O$_3$) group in abenakiite-(Ce); the two preferential positions are shown as solid and dashed, respectively.

Table 19. Thiosulfate, sulfite, and fluorosulfate minerals.

Mineral	Formula	a (Å)	b (Å)	c (Å)	S. G.	Fig.	Ref.
	Thiosulfate minerals						
sidpietersite	$Pb^{2+}_4(S^{6+}O_3S^{2-})O_2(OH)_2$		see Table 18			56a,b	[1]
bazhenovite	$CaS_5Ca(S^{6+}O_3S^{2-})Ca_6$ $(OH)_{12}(H_2O)_{20}$					–	(2)
viaeneite	$(Fe,Pb)(S_2)_{11}(S^6O_3S^{2-})$					–	(3)
	Sulfite minerals						
hannebachite	$Ca(S^{4+}O_3)(H_2O)_{0.5}$	10.664(1)	6.495(1)	9.823(1)	*Pbcn*	57a,b	[4]
scotlandite	$Pb^{2+}(S^{4+}O_3)$	4.505(2)	5.333(2)	6.405(6)	$P2_1/m$	57c,d	[5]
gravegliaite	$Mn^{2+}(S^{4+}O_3)(H_2O)_3$	9.763(1)	5.635(1)	9.558(1)	*Pnma*	57e,f,g	[6]
orschallite	$Ca_3(S^{4+}O_3)_2(SO_4)(H_2O)_{12}$	11.350(1)	a	28.321(2)	$R\bar{3}c$	39d-g	[7]
abenakiite-(Ce)	$Na_{26}Ce_6(SiO_3)_6(PO_4)_6$ $(CO_3)_6(S^{4+}O_3)$	16.018(2)	a	19.761(4)	$R\bar{3}$	57h,i	[8]
	Fluorosulfate minerals						
reederite-(Y)	$Na_{15}Y_2(CO_3)_9(SO_3F)Cl$	8.763(1)	a	10.736(2)	$P\bar{6}$	58a,b,c	[9]

$*\beta = 106.24(3)°$

References: [1] Roberts et al. (1999), Cooper and Hawthorne (1999), (2) Chesnokov et al. (1987b), (3) Kucha et al. (1996), [4] Matsuno et al. (1984), Schröpfer (1973), [5] Pertlik and Zemann (1985), [6] Basso et al. (1991), [7] Weidenthaler et al. (1993), [8] McDonald and Chao (1994), [9] Grice et al. (1995)

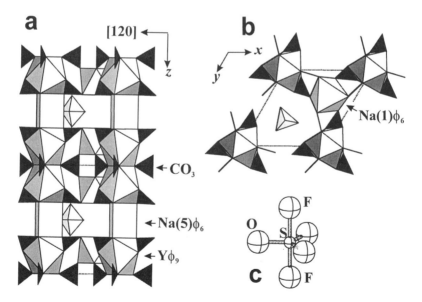

Figure 58. The structure of the fluorosulfate mineral reederite-(Y): (a) projected along [100]; (b) projected along [001]; the Na(2), Na(3) and Na(4) cations are omitted for clarity; (c) the disordered (SO_3F) fluorosulfate anion is shown.

vertices of a plane hexagonal net (Fig. 57h). Prominent six-membered rings of (SiO_4) tetrahedra (Fig. 57h) are linked in the [001] direction by rings of ($NaØ_7$) and ($CeØ_9$) polyhedra to form hollow tubes; within these tubes are (CO_3) groups, ($NaØ_6$) octahedra, and orientationally disordered ($S^{4+}O_3$) groups (Fig. 57i). These filled tubes are linked in the (001) plane by chains of ($NaØ_7$) and (PO_4) polyhedra.

FLUOROSULFATE MINERALS

Only one fluorosulfate mineral is known (Table 19). **Reederite-(Y)**, ideally $Na_{15}Y_2(CO_3)_9(SO_3F)Cl$, is layered on {001}, similar to many other REE carbonates (Grice et al. 1994). The Y occupies a [9]-coordinated site that links to (CO_3) groups arranged both parallel to {001} and at an oblique angle to [001] (Fig. 58a). The ($YØ_9$) and ($NaØ_6$) groups form columns parallel to [001] that occur at the nodes of a plane hexagonal net (Fig. 58b), and are linked by columns of ($NaØ_n$) polyhedra and (SO_3F) groups (Fig. 58c). Note that the ideal formula has one excess positive charge; it is not clear how this is actually compensated in the observed structure.

ACKNOWLEDGMENTS

This work was funded by Natural Sciences and Engineering Research Council of Canada Grants to FCH, by the Environmental Management Sciences Program of the United States Department of Energy grant DE-FG07-97ER14820 to PCB, and by the National Science Foundation NATO Fellowship in Science and Engineering DGE99-03354 to SVK and PCB.

APPENDIX

Index of mineral names and the table numbers in which they appear

Appendix cont'd

lautenthallite (9)	motukoreaite (9)	retgersite (5)	tuscanite (16)
lawsonbauerite (9)	mountkeithite (9)	rhomboclase (11)	tychite (12)
leadhillite (16)	nabokoite (18)	roeblingite (16)	uklonskovite (8)
lecontite (14)	namuwite (9)	roemerite (7)	ungemachite (7)
leightonite (7)	natroalunite (10)	rozenite (7)	vanthoffite (11)
leonite (7)	natrochalcite (11)	sarcofanite (17)	vergasovaite (18)
letovicite (14)	natrojarosite (10)	schairerite (15)	vishnevite (17)
linarite (8)	nickelblödite (7)	schaurteite (6)	viaeneite (19)
liottite (17)	nickelboussingaultite (6)	schlossmacherite (10)	vlodavetsite (12)
lonecreekite (6)	niedermayrite (9)	schröckingerite (16)	voltaite (12)
löweite (12)	nickelhexahydrite (5)	schuetteite (18)	vonbenzingite (16)
macphersonite (16)	nosean (17)	schulenbergite (9)	walthierite (10)
magnesioaubertite (5)	olsacherite (15)	scotlandite (19)	wermlandite (9)
magnesiocopiapite (8)	orschallite (13,19)	serpierite (9)	wherryite (8)
mallardite (5)	orthoserpierite (9)	shigaite (9)	woodhouseite (10)
mammothite (12)	osarizawaite (10)	sideronatrite (8)	woodwardite (9)
manganolangbeinite (12)	palmierite (14)	siderotil (8)	wroewolfeite (9)
manganotychite (12)	parabutlerite (8)	sidpietersite (18,19)	wupatkiite (7)
mascagnite (14)	paracoquimbite (7)	slavikite (10)	xitieshanite (7)
matteuccite (14)	pentahydrite (8)	sodium alum (6)	yavapaite (11)
melanterite (5)	peretaite (16)	spangolite (9)	zincobotryogen (8)
mendozite (6)	philolithite (12)	starkeyite (7)	zincocopiapite (8)
mercallite (14)	pickeringite (7)	sturmanite (6)	zincosite (12)
mereiterite (7)	picromerite (6)	svanbergite (10)	zincovoltaite (12)
metasideronatrite (8)	pitiglianoite (17)	sulphohalite (15)	zippeite (16)
metavoltine (7)	piypite (18)	susannite (16)	zircosulfate (16)
microsommite (17)	plumbojarosite (10)	svyazhinite (5)	zinc-melanterite (5)
mikasaite (12)	poitevinite (12)	syngenite (13)	
millosevichite (12)	polyhalite (7)	szmikite (12)	
minamiite (10)	posnjakite (9)	szomolnokite (12)	
minasragrite (7)	potassium alum (6)	tamarugite (6)	
mineevite-(Y) (14)	poughite (11)	ternesite (13)	
mirabilite (14)	queitite (16)	thaumasite (6)	
mohrite (6)	quenstedtite (7)	thenardite (14)	
mooreite (9)	ramsbeckite (9)	torreyite (9)	
moorhouseite (5)	ransomite (11)	tounkite (17)	
morenosite (5)	rapidscreekite (13)	tschermigite (6)	
mosesite (18)	reederite-(Y) (19)	tsumebite (8)	

REFERENCES

Abdeen AM, Will G, Schaefer W, Kirfel A, Bargouth MO, Recker K, Weiss A (1981) X-ray and neutron diffraction study of alums: III. The crystal structure of ammonium aluminium alum. Z Kristallogr 157:147-166

Abriel W, Nesper R (1993) Bestimmung der Kristallstruktur von $CaSO_4(H_2O)_{0.5}$ mit Röntgenbeugungsmethoden und mit Potentialprofil-Rechnungen. Z Kristallogr 205:99-113

Adams DC, Birchall T, Faggiani R, Gillespie RJ, Vekris JE (1991) The ^{119}Sn Mössbauer and solid-state NMR and the crystal and molecular structure of tin(II) bisfluorosulfate, $Sn(OSO_2F)_2$. Can J Chem 69:2122-2126

Adiwidjaja G, Friese K, Klaska K-H, Schlüter J (1996) The crystal structure of christelite $Zn_3Cu_2(SO_4)_2(OH)_6\cdot4(H_2O)$. Z Kristallogr 211:518-521

Adiwidjaja G, Friese K, Klaska K-H, Schlüter J (1997) The crystal structure of gordaite $NaZn_4SO_4(OH)_6Cl_6(H_2O)$. Z Kristallogr 212:704-707

Airoldi R, Magnano G (1967) Sulla struttura del solfato (di)mercurioammonico. Rassegna Chimica 5: 181-189

Aka Y, Arma_an N, Uraz AA (1980) An X-ray diffraction study of barium thiosulphate monohydrate, $BaS_2O_3\cdot H_2O$. Z Kristallogr 151:61-66

Allmann R (1977) Refinement of the hybrid layer structure $(Ca_2Al(OH)_6)^+(0.5SO_4\cdot3H_2O)^-$. N Jahrb Mineral Monatsh 136-144

Allred AL, Rochow E (1958) A scale of electronegativity based on electrostatic force. J Inorg Nucl Chem 5:264-268

Araki T, Zoltai T (1967) Refinement of the crystal structure of a glauberite. Am Mineral 52:1272-1277

Bachmann HG (1953) Beiträge zur Kristallchemie natürlicher und künstlicher Schwermetallvanadate: III. Das Bleiorthovanadat, seine Darstellung, Kristallstruktur und Isotypiebeziehungen. N Jahrb Mineral Monatsh 209-223

Bacon GE, Titterton DH (1975) Neutron diffraction studies of $CuSO_4(H_2O)_5$ and $CuSO_4(D_2O)_5$. Z Kristallogr 141:330-341

Baggio S, Pardo MI, Baggio R, Gonzalez O (1997) Cadmium thiosulfate dihydrate. Acta Crystallogr C53:1521-1523

Balic Zunic T, Moëlo Y, Loncar Z, Micheelsen H (1994) Dorallcharite, $Tl_{0.8}K_{0.2}Fe_3 (SO_4)_2(OH)_6$, a new member of the jarosite-alunite family. Eur J Mineral 6:255-263

Ballhorn R, Brunner H, Schwab RG (1989) Artificial compounds of the crandallite type: a new material for separation and immobilization of fission products. Sci Basis Nucl Waste Manag 12:249-252

Ballirano P, Merlino S, Bonaccorsi E, Maras A (1996) The crystal structure of liottite, a six-layer member of the cancrinite group. Can Mineral 34:1021-1030

Ballirano P, Bonaccorsi E, Maras A, Merlino S (1997) Crystal structure of afghanite, the eight-layer member of the cancrinite-group: evidence for long-range Si, Al ordering. Eur J Mineral 9:21-30

Bartell LS, Su LS, Yow H (1970) Lengths of phosphorus-oxygen and sulfur-oxygen bonds. An extended Hückel molecular orbital examination of Cruickshank's d–p picture. Inorg Chem 9:1903-1912

Bartlett N, Wechsberg M, Jones GR, Burbank RD (1972) The crystal structure of xenon(II) fluoride fluorosulfate, $FXeOSO_2F$. Inorg Chem 11:1124-1127

Bartmann K, Mootz D (1990) Structures of two strong broensted acids: (I) fluorosulfuric acid and (II) trifluoromethanesulfonic acid. Acta Crystallogr C46:319-320

Basso R, Lucchetti G, Palenzona A (1991) Gravegliaite, $MnSO_3\cdot2H_2O$, a new mineral from Val Graveglia (Northern Apennines, Italy). Z Kristallogr 197:97-106

Baur WH (1960) Die Kristallstruktur von $FeSO_4\cdot4H_2O$. Naturwiss 47:467

Baur WH (1962) Zur Kristallchemie der Salzhydrate. Die Kristallstrukturen von $MgSO_4\cdot4H_2O$ (Leonhardtit) und $FeSO_4\cdot4H_2O$ (Rozenit). Acta Crystallogr 15:815-826

Baur WH (1964) On the crystal chemistry of salt hydrates: III. The determination of the crystal structure of $FeSO_4\cdot7H_2O$ (melanterite). Acta Crystallogr 17:1167-1174

Baur WH, Rolin JL (1972) Salt hydrates. IX. The comparison of the crystal structure of magnesium sulfate pentahydrate with copper sulfate pentahydrate and magnesium chromate pentahydrate. Acta Crystallogr B28:1448-1455

Bayliss P, Atencio D (1985) X-ray powder-diffraction data and cell parameters for copiapite-group minerals. Can Mineral 23:53-56

Bergerhoff G, Paeslack (1968) Sauerstoff als Koordinationszentrum in Kristallstrukturen. Z Kristallogr 126:112-123

Berlepsch P, Armbruster Th, Brugger J, Bykova EY, Kartashov PM (1999) The crystal structure of vergasovaite $Cu_3O[(Mo,S)O_4SO_4]$, and its relation to synthetic $Cu_3O[MoO_4]_2$. Eur J Mineral 11: 101-110

Birchall T, Denes G, Faggiani R, Frampton CS, Gillespie RJ, Kapoor R, Vekris JE (1990) Crystal and molecular structure and Raman and ^{127}I Mössbauer spectra of iodine(III) bis(fluorosulfate) iodide $I(OSO_2F)_2I$. Inorg Chem 29:1527-1530

Blades AT, Kebarle P (1994) Study of the stability and hydration of doubly charged ions in the gas phase: SO_4^{2-}, $S_2O_6^{2-}$, $S_2O_8^{2-}$, and some related species. J Amer Chem Soc 116:10761-10766

Bokii GB, Gorogotskaya LI (1965) The crystal structure of chukhrovite. Dokl Akad Nauk SSSR 163: 183-185 (in Russian)

Bokii GB, Gorogotskaya LI (1969) Crystal chemical classification of sulfates. Zhurnal Strukt Khim 10:624-632 (in Russian)

Bokii GB, Pal'chik NA, Antipin MYu (1978) More precise determination of syngenite crystal structure. Trudy Inst Geol Geofiz Akad Nauk SSSR, Sibir Otd 385:3-7 (in Russian)

Boldyrev A, Simons J (1994) Isolated SO_4^{2-} and PO_4^{3-} do not exist. J Phys Chem 98:2298-2300

Bonaccorsi E, Merlino S, Pasero M (1990) Davyne: Its structural relationships with cancrinite and vishnevite. N Jahrb Mineral Monatsh 97-112

Bookin AS, Drits VA (1993) Polytype diversity of the hydrotalcite-like minerals. I. Possible polytypes and their diffraction features. Clays Clay Minerals 41:551-557

Bookin AS, Cherkashin VI, Drits VA (1993) Polytype diversity of the hydrotalcite-like minerals. II. Determination of the polytypes of experimentally studied varieties. Clays Clay Minerals 41:558-564

Borene J (1970) Structure cristalline de la parabutlerite. Bull Soc franc Minéral Cristallogr 93:185-189

Bowen NL (1928) The evolution of igneous rocks. Princeton University Press, Princeton, New Jersey

Bragg WL (1930) The structure of silicates. Z Kristallogr 74:237-305

Braithwaite RSW, Kampf AR, Pritchard RG, Lamb RPH (1993) The occurrence of thiosulfates and other unstable sulfur species as natural weathering products of old smelting slags. Mineral Petrol 47:255-261

Breidenstein B, Schlüter J, Gebhard G (1992) On beaverite: new occurrence, chemical data and crystal structure. N Jahrb Mineral Monatsh 213-220

Brese NE, O'Keeffe M (1991) Bond-valence parameters for solids. Acta Crystallogr B47:192-197

Brown ID (1981) The bond-valence method: an empirical approach to chemical structure and bonding. *In* O'Keeffe M, Navrotsky A (Eds) Structure and bonding in crystals. Academic Press, New York, 2:1-30

Brown ID, Shannon RD (1973) Empirical bond strength—bond length curves for oxides. Acta Crystallogr A29:266-282

Burdett JK, Hawthorne FC (1993) An orbital approach to the theory of bond valence. Am Mineral 78: 884-892

Burns PC (1999) The crystal chemistry of uranium. Rev Mineral 38:23-90

Burns PC, Hawthorne FC (1994) The crystal structure of humberstonite, a mixed sulfate-nitrate mineral. Can Mineral 32:381-385

Burns PC, Grice JD, Hawthorne FC (1995) Borate minerals: I. Polyhedral clusters and fundamental building blocks. Can Mineral 33:1131-1151

Burns PC, Miller ML, Ewing RC (1996) U^{6+} minerals and inorganic phases: a comparison and hierarchy of structures. Can Mineral 34:845-880

Burns PC, Roberts AC, Nikischer AJ (1998) The crystal structure of $Ca[Zn_8(SO_4)_2(OH)_{12}Cl_2](H_2O)_9$, a new phase from slag dumps at Val Varenna, Italy. Eur J Mineral 10:923-930

Burzlaff H (1967) Die Struktur des Heidornit, $Ca_3Na_2Cl(SO_4)_2B_5O_8(OH)_2$. N Jahrb Mineral Monatsh 157-169

Calleri M, Gavetti A, Ivaldi G, Rubbo M (1984) Synthetic epsomite, $MgSO_4(H_2O)_7$: Absolute configuration and surface features of the complementary (111) forms. Acta Crystallogr B40:218-222

Cannillo E, dal Negro A, Rossi G (1973) The crystal structure of latiumite, a new type of sheet silicate. Am Mineral 58:466-470

Catti M, Ferraris G, Franchini-Angela M (1975) Hydrogen bonding in the crystalline state. $NaHSO_4 H_2O$ (matteuccite), a pseudo-symmetric crystal structure. Atti Accad Sci Torino Fis Matem Natur 109:531-545

Chesnokov BV, Bazhenova LF, Bushmakin AF (1987a) Fluorellestadite $Ca_{10}[(SO_4)_3(SiO_4)_3]_6F_2$—a new mineral. Zapiski Vses Mineral Obshch 116:743-746 (in Russian)

Chesnokov BV, Polyakov VO, Bushmakin AF (1987b) Bazhenovite $CaS_5 \cdot CaS_2O_3 \cdot 6Ca(OH)_2 \cdot 20H_2O$: A new mineral. Zapiski Vses Mineral Obshch 116:737-743 (in Russian)

Christidis PC, Rentzeperis PJ (1985) Experimental charge density distribution in polythionate anions I. X-ray study of electron density distribution in potassium trithionate, $K_2S_3O_6$. Z Kristallogr 173:59-74

Christidis P, Rentzeperis PJ, Kirfel A, Will G (1983) X-ray determination of the electron density distribution in monoclinic ferric sulfate. Z Kristallogr 164:219-236

Collins MJ, Gillespie RJ, Sawyer JF, Schrobilgen GJ (1986) Decaselenium bis(fluorosulfate). Acta Crystallogr C42:13-16

Cooper MA, Hawthorne FC (1994) The crystal structure of wherryite, $Pb_7Cu_2(SO_4)_4(SiO_4)_2(OH)_2$, a mixed sulfate–silicate with $(M(TO_4)Ø_2)$ chains. Can Mineral 32:373-380

Cooper MA, Hawthorne FC (1996a) The crystal structure of rapidcreekite, $Ca_2(SO_4)(CO_3)(H_2O)_4$, and its relation to the structure of gypsum. Can Mineral 34:99-106

Cooper MA, Hawthorne FC (1996b) The crystal structure of shigaite, $(AlMn^{2+}_2(OH)_6)_3(SO_4)_2$ $Na(H_2O)_6(H_2O)_6$, a hydrotalcite-group mineral. Can Mineral 34:91-97

Cooper MA, Hawthorne FC (1999) The structure topology of sidpietersite, $Pb^{2+}_4(S^{6+}O_3S^{2-})O_2(OH)_2$, a novel thiosulfate structure. Can Mineral 37:1275-1282

Corazza E, Sabelli C, Giuseppetti G (1967) The crystal structure of lecontite, $NaNH_4SO_4(H_2O)_2$. Acta Crystallogr 22:683-687

Cromer DT, Kay MI, Larson AC (1967) Refinement of the alum structures. II. X-ray and neutron diffraction in $NaAl(SO_4)_2 \cdot (H_2O)_{12}$, γ-alum. Acta Crystallogr 22:182-187

Cruickshank DWJ (1961) The role of 3d-orbitals in p-bonds between (a) silicon, phosphorus, sulfur, or chlorine to (b) oxygen or nitrogen. J Chem Soc 1961:5486-5504

Cruickshank DWJ (1985) A reassessment of d-p bonding in the tetrahedral oxyanions of second-row atoms. J Mol Struct 130:177-91

Cruickshank DWJ, Eisenstein M (1985) The role of d functions in *ab initio* calculations. Part 1. The deformation densities of sulfamic acid and SO_3^-. J Mol Struct 130:143-56

Cruickshank DWJ, Eisenstein M (1987) The role of d functions in *ab initio* calculations. II. The deformation densities of sulfur dioxide, nitrogen dioxide, and their ions. J Comput Chem 8:6-27

Dahmen T, Gruehn R (1993) Beitrage zum thermischen Verhalten von Sulfaten. IX. Einkristallstrukturverfeinerung der Metall(III)-sulfate $Cr_2(SO_4)_3$ und $Al_2(SO_4)_3$. Z Kristallogr 204:57-65

Dai Y, Harlow GE (1992) Description and crystal structure of vonbezingite, a new Ca-Cu-SO_4-H_2O mineral from the Kalahari manganese field, South Africa. Am Mineral 77:1292-1300

Deganello S (1983) The crystal structure of cesanite at 21 and 263°C. N Jahrb Mineral Monatsh 305-313

Duesler EN, Foord EE (1986) Crystal structure of hashemite, $BaCrO_4$, a barite structure type. Am Mineral 71:1217-1220

Eby RK, Hawthorne FC (1993) Structural relations in copper oxysalt minerals. I. Structural hierarchy. Acta Crystallogr B49:28-56

Effenberger H (1985a) $Cu_2O(SO_4)$, dolerophanite: refinement of the crystal structure with a comparison of $OCu(II)_4$ tetrahedra in inorganic compounds. Monatsh Chem 116:927-931

Effenberger H (1985b) The crystal structure of mammothite, $Pb_6Cu_4AlSbO_2(OH)_{16}Cl_4(SO_4)_2$. Tschermaks mineral petrogr Mitt 34:279-288

Effenberger H (1987) Crystal structure and chemical formula of schmiederite, $Pb_2Cu_2(OH)_4(SeO_3)(SeO_4)$, with a comparison to linarite $PbCu(OH)_2(SO_4)$. Mineral Petrol 36:3-12

Effenberger H (1988) Ramsbeckite, $(Cu,Zn)_{15}(OH)_{22}(SO_4)_4(H_2O)_6$: Revision of the chemical formula based on a structure determination. N Jahrb Mineral Monatsh 38-48

Effenberger H, Zemann J (1984) The crystal structure of caratiite. Mineral Mag 48:541-546

Effenberger H, Pertlik F, Zemann J (1986) Refinement of the crystal structure of krausite: a mineral with an interpolyhedral oxygen-oxygen contact shorter than the hydrogen bond. Am Mineral 71:202-205

Egorov-Tismenko YuK, Sokolova EV, Smirnova NL, Yamnova NA (1984) Crystal chemical features of minerals related to the glaserite structure type. Mineral Zhurnal 6(6):3-9

Elerman Y, Uraz AA, Armagan N (1978) An X-ray diffraction study of nickel thiosulphate hexahydrate. Acta Crystallogr B34:3330-3332

Elerman Y, Bats JW, Fuess H (1983) Deformation density in complex anions: IV. Magnesium thiosulfate hexahydrate, $MgS_2O_3 \cdot 6H_2O$. Acta Crystallogr C39:515-518

Evsyunin VG, Sapozhnikov AN, Rastsvetaeva RK, Kashaev AA (1996) Crystal structure of the potassium-rich hauyne from Arissia (Italy). Kristallografiya 41:659-662 (in Russian)

Eysel W (1973) Crystal chemistry of the system Na_2SO_4-K_2SO_4-K_2CrO_4-Na_2CrO_4 and of the glaserite phase. Am Mineral 58:736-747

Fanfani L, Zanazzi PF (1967) Structural similarities of some secondary lead minerals. Mineral Mag 36:522-529

Fanfani L, Nunzi A, Zanazzi PF (1970) The crystal structure of roemerite. Am Mineral 55:78-89

Fanfani L, Nunzi A, Zanazzi PF (1971) The crystal structure of butlerite. Am Mineral 56:751-757

Fanfani L, Nunzi A, Zanazzi PF, Zanzari AR (1973) The copiapite problem: the crystal structure of a ferrian copiapite. Am Mineral 58:314-322

Fanfani L, Nunzi A, Zanazzi PF, Zanzari AR, Sabelli C (1975a) The crystal structure of schairerite and its relationship to sulphohalite. Mineral Mag 40:131-139

Fanfani L, Nunzi A, Zanazzi PF, Zanzari AR (1975b) The crystal structure of galeite, $Na_{15}(SO_4)_5F_4Cl$. Mineral Mag 40:357-361

Fanfani L, Giuseppetti G, Tadini C, Zanazzi PF (1980) The crystal structure of kogarkoite, Na_3SO_4F. Mineral Mag 43:753-759

Fang JH, Robinson PD (1970a) Crystal structures and mineral chemistry of hydrated ferric sulfates. I. The crystal structure of coquimbite. Am Mineral 55:1534-1540

Fang JH, Robinson PD (1970b) Crystal structures and mineral chemistry of double-salt hydrates. I. The crystal structure of loeweite. Am Mineral 55:378-386

Fang JH, Robinson PD (1972) Crystal structures and mineral chemistry of double-salt hydrates: II. The crystal structure of mendozite, $NaAl(SO_4)_2(H_2O)_{11}$. Am Mineral 57:1081-1088

Farkas L, Pertlik F (1997) Crystal structure determinations of felsöbányaite and basaluminite, $Al_4(SO_4)(OH)_{10} \cdot 4H_2O$. Acta Mineral Petrogr Szeged 38:5-15

Figgis BN, Kepert CJ, Kucharski ES, Reynolds PA (1992) Charge-density study of deuterated ammonium ferrous Tutton salt at 85 K and comparison with Cr(II) and Cu(II) salts. Acta Crystallogr B48:753-761

Filatov SK, Semenova TF, Vergasova LP (1992) Types of polymerization of $[OCu_4]^{6+}$ tetrahedra in compounds with 'additional' oxygen atoms. Proc Acad Sci USSR 322:536-539 (in Russian)

Fischer W, Hellner E (1964) Über die Struktur des Vanthoffits. Acta Crystallogr 17:1613

Fleet ME, Knipe SW (1997) Structure of magnesium hydroxide sulfate $[2MgSO_4 \cdot Mg(OH)_2]$ and solid solution in magnesium hydroxide sulfate hydrate and caminite. Acta Crystallogr B53:358-363

Gaines RV, Skinner HCW, Foord EE, Mason B, Rosenzweig A, King VT (1997) Dana's new mineralogy. Wiley, New York

Galy J, Jaud J, Pulou R, Sempere R (1984) Structure cristalline de la langite, $Cu_4(SO_4(OH)_6H_2O)H_2O$. Bull Minéral 107:641-648

Gaudefroy C, Granger MM, Permingeat F, Protas J (1968) La despujolsite, une nouvelle espece minerale. Bull Soc franc Minéral Cristallogr 91:43-50

Giacovazzo C, Menchetti S (1969) Sulla struttura della amarantite. Rend Soc Ital Mineral Petrol 25:399-406

Giacovazzo C, Menchetti S, Scordari F (1973) The crystal structure of caledonite, $CuPb_5(SO_4)_3CO_3(OH)_6$. Acta Crystallogr B29:1986-1990

Giacovazzo G, Scordari F, Todisco A, Menchetti S (1976a) Crystal structure model for metavoltine from Sierra Gorda. Tschermaks mineral petrogr Mitt 23:155-166

Giacovazzo C, Scandale E, Scordari F (1976b) The crystal structure of chlorothionite $CuK_2Cl_2SO_4$. Z Kristallogr 144:226-237

Gibbs GV (1982) Molecules as models for bonding in solids. Am Mineral 67:421-450

Gibbs GV, Hamil MM, Louisnathan SJ, Bartell LS, Yow H (1972) Correlations between SiBO bond length, SiBOBSi angle and bond overlap populations calculated using extended Hückel molecular orbital theory. Am Mineral 57:1578-1613

Giese RF Jr, Penna G (1983) The crystal structure of sulfoborite, $Mg_3SO_4(B(OH)_4)_2(OH)F$. Am Mineral 68:255-261

Giester G, Rieck B (1995) Mereiterite, $K_2Fe(SO_4)_2 \cdot 4(H_2O)$, a new leonite-type mineral from the Lavrion Mining District, Greece. Eur J Mineral 7:559-566

Giester G, Rieck B (1996) Bechererite, $(Zn,Cu)Zn_2(OH)_{13}((S,Si)(O,OH)_4)_2$, a novel mineral species from the Tonopah-Belmont mine, Arizona. Am Mineral 81:244-248

Giester G, Zemann J (1987) The crystal structure of the natrochalcite-type compounds $(Me^+)Cu_2(OH)(ZO_4)_2 \cdot H_2O$ (Me^+ = Na, K, Rb; Z = S, Se), with special reference to the hydrogen bonds. Z Kristallogr 179:431-442

Giester G, Lengauer CL, Redhammer G (1994) Characterization of the $FeSO_4 \cdot (H_2O)$–$CuSO_4 \cdot (H_2O)$ solid-solution series, and the nature of poitevinite, $(Cu,Fe)SO_4 \cdot (H_2O)$. Can Mineral 32:873-884

Giester G, Mikenda W, Pertlik F (1996) Kleinite from Terlingua, Brewster County, Texas: investigations by single crystal X-ray diffraction, and vibrational spectroscopy. N Jahrb Mineral Monatsh 49-56

Giester G, Rieck B, Brandstatter F (1998) Niedermayrite, $Cu_4Cd(SO_4)_2(OH)_6 \cdot 4H_2O$, a new mineral from the Lavrion Mining District, Greece. Mineral Petrol 63:19-34

Gillespie RJ, Schrobilgen GJ, Slim DR (1977) Crystal structure of mue-fluorosulphato-bis(fluoroxenon(II)-hexafluoroarsenate(V). J Chem Soc Dalton Trans 1977:1003-1006

Gillespie RJ, Kent JP, Sawyer JF, Slim DR, Tyrer JD (1981a) Reactions of S_4N_4 with $SbCl_5$, SbF_5, AsF_5, PF_5 and HSO_3F. Preparation and crystal structures of salts of the $S_4N_4^{2+}$ cation. Inorg Chem 20:3799-3812

Gillespie RJ, Kent JP, Sawyer JF (1981b) Monomeric and dimeric thiodithiazyl cations, $S_3N_2^{2+}$ and $(S_6N_4)^{2+}$: Preparation and crystal structure of $(S_3N_2)(AsF_6)$, $(S_6N_4)(S_2O_6F)_2$ and $(S_6N_4)(SO_3F)_2$. Inorg Chem 20:3784-3799

Ginderow D, Cesbron F (1979) Structure cristalline de l'aubertite, $AlCuCl(SO_4)_2(H_2O)_{14}$. Acta Crystallogr B35:2499-2502

Giuseppetti G, Tadini C (1980) The crystal structure of osarizawaite. N Jahrb Mineral Monatsh 401-407

Giuseppetti G, Tadini C (1983) Structural analysis and refinement of Bolivian creedite, $Ca_3Al_2F_8(OH)_2(SO_4)\cdot(H_2O)_2$. The role of the hydrogen atoms. N Jahrb Mineral Monatsh 69-78

Giuseppetti G, Tadini C (1987) Corkite, $PbFe_3(SO_4)(PO_4)(OH)_6$, its crystal structure and ordered arrangement of the tetrahedral cations. N Jahrb Mineral Monatsh 71-81

Giuseppetti G, Mazzi F, Tadini C (1988) The crystal structure of synthetic burkeite. N Jahrb Mineral Monatsh 203-221

Giuseppetti G, Mazzi F, Tadini C (1990) The crystal structure of leadhillite: $Pb_4(SO_4)(CO_3)_2(OH)_2$. N Jahrb Mineral Monatsh 255-268

Gjerrestad K, Marøy K (1973) The crystal structure of barium telluropentathionate trihydrate. Acta Chem Scand 27:1653-1666

Golic L, Graunar M, Lazarini F (1982) Catena-di-mue-hydroxo-$_3$-oxo-dibismuth(III) sulfate. Acta Crystallogr B38:2881-2883

Gorskaya MG, Filatov SK, Rozhdestvenskaya IV, Vergasova LP (1992) The crystal structure of klyuchevskite, $K_3Cu_3(Fe,Al)O_2(SO_4)_4$, a new mineral from Kamchatka volcanic sublimates. Mineral Mag 56:411-416

Graeber EJ, Rosenzweig A (1971) The crystal structures of yavapaiite, $KFe(SO_4)_2$, and goldichite, $KFe(SO_4)_2(H_2O)_4$. Am Mineral 56:1917-1933

Granger MM (1969) Determination et etude de la structure cristalline de la jouravskite $Ca_3Mn(SO_4)$ $(CO_3)(OH)(H_2O)_{12}$. Acta Crystallogr B25:1943-1951

Grice JD, Van Velthuizen J, Gault RA (1994) Petersenite-(Ce), a new mineral from Mont Saint-Hilaire, and its structural relationship to other REE carbonates. Can Mineral 32:405-414

Grice JD, Gault RA, Chao GY (1995) Reederite-(Y), a new sodium rare-earth carbonate mineral with a unique fluorosulfate anion. Am Mineral 80:1059-1064

Grice JD, Burns PC, Hawthorne FC (1999) Borate minerals: II. A hierarchy of structures based upon the borate fundamental building block. Can Mineral 37:731-762

Griffen DT, Ribbe PH (1979) Distortions in the tetrahedral oxyanions of crystalline substances. Neues Jahrb Mineral Abh 137:54-73

Groat LA (1996) The crystal structure of namuwite, a mineral with Zn in tetrahedral and octahedral coordination, and its relationship to the synthetic basic zinc sulfates. Am Mineral 81:238-243

Groat LA, Hawthorne FC (1986) Structure of ungemachite, $K_3Na_8Fe^{3+}(SO_4)_6(NO_3)_2(H_2O)_6$ a mixed sulfate-nitrate mineral. Am Mineral 71:826-829

Hasebe K (1981) Studies of the crystal structure of ammonium sulfate in connection with its ferroelectric phase transition. J Phys Soc Japan 50:1266-1274

Hassan I, Grundy HD (1984) The character of the cancriniteBvishnevite solid-solution series. Can Mineral 22:333-340

Hassan I, Grundy HD (1989) The structure of nosean, ideally $Na_8(Al_6Si_6O_{24})SO_4H_2O$. Can Mineral 27:165-172

Hawthorne FC (1979) The crystal structure of morinite. Can Mineral 17:93-102

Hawthorne FC (1983a) Graphical enumeration of polyhedral clusters. Acta Crystallogr A39:724-736

Hawthorne FC (1983b) The crystal structure of tancoite. Tschermaks mineral petrogr Mitt 31:121-135.

Hawthorne FC (1984) The crystal structure of stenonite, and the classification of the aluminofluoride minerals. Can Mineral 22:245-251

Hawthorne FC (1985a) Towards a structural classification of minerals: The $^{VI}M^{IV}T_2O_n$ minerals. Am Mineral 70:455-473

Hawthorne FC (1985b) Refinement of the crystal structure of blödite: structural similarities in the $[^{VI}M(^{IV}TØ_4)_2Ø_n]$ finite-cluster minerals. Can Mineral 23:669-674

Hawthorne FC (1986) Structural hierarchy in $^{IV}M_x{}^{III}T_yØ_z$ minerals. Can Mineral 24:625-642

Hawthorne FC (1990) Structural hierarchy in $M^{[6]}T^{[4]}Ø_n$ minerals. Z Kristallogr 192:1-52

Hawthorne FC (1992) The role of OH and H_2O in oxide and oxysalt crystals. Z Kristallogr 201:183-206

Hawthorne FC (1994) Structural aspects of oxide and oxysalt crystals. Acta Crystallogr B50:481-510

Hawthorne FC (1997) Structural aspects of oxide and oxysalt minerals. *In* Merlino (Ed) Modular aspects of minerals. EMU (European Mineralogical Union) Notes in Mineralogy 1:373-429

Hawthorne FC, Ferguson RB (1975a) Anhydrous sulfates. II. Refinement of the crystal structure of anhydrite. Can Mineral 13:289-292

Hawthorne FC, Ferguson RB (1975b) Anhydrous sulphates. I. Refinement of the crystal structure of celestite with an appendix on the structure of thenardite. Can Mineral 13:181-187

Hawthorne FC, Ferguson RB (1975c) Refinement of the crystal structure of kroehnkite. Acta Crystallogr B31:1753-1755

Hawthorne FC, Groat LA (1985) The crystal structure of wroewolfeite, a mineral with $[(Cu_4(OH)_6(SO_4)(H_2O)]$ sheets. Am Mineral 70:1050-1055

Hawthorne FC, Schindler MS (2000) Topological enumeration of decorated $[Cu^{2+}\emptyset_2]_N$ sheets in hydroxy-hydrated copper oxysalt minerals. Can Mineral 38 (in press)

Hawthorne FC, Groat LA, Raudsepp M, Ercit TS (1987) Kieserite, $MgSO_4(H_2O)$, a titanite-group mineral. N Jahrb Mineral Abh 157:121-132

Hawthorne FC, Groat LE, Eby RK (1989) Antlerite, $Cu_3SO_4(OH)_4$, a heteropolyhedral wallpaper structure. Can Mineral 27:205-209

Hawthorne FC, Kimata M, Eby RK (1993) The crystal structure of spangolite, a complex copper sulfate sheet mineral. Am Mineral 78:649-652

Hawthorne FC, Burns PC, Grice JD (1996) The crystal chemistry of boron. In Grew ES, Anobitz LM (Eds) Boron: Mineralogy, petrology and geochemistry. Rev Mineral 33:41-115

Haymon RM, Kastner M (1986) Caminite: A new magnesium-hydroxide-sulfate-hydrate mineral found in a submarine hydrothermal deposit, East Pacific Rise, 21EN. Am Mineral 71:819-825

Helliwell M, Smith JV (1997) Brochantite. Acta Crystallogr C53:1369-1371

Hess H, Keller P (1980) Die Kristallstruktur von Queitit, $Pb_4Zn_2(SO_4)(SiO_4)(Si_2O_7)$. Z Kristallogr 151:287-299

Hess H, Keller P, Riffel H (1988) The crystal structure of chenite, $Pb_4Cu(OH)_6(SO_4)_2$. N Jahrb Mineral Monatsh 259-264

Hesse W, Leutner B, Boehn KH, Walker NPC (1993) Structure of a new sodium thiosulfate hydrate. Acta Crystallogr C49:363-365

Hexiong Y, Pingqiu F (1988) Crystal structure of zincobotryogen. Kuangwu Xuebao 8:1-12

Hill RJ (1980) The structure of mooreite. Acta Crystallogr B36:1304-1311

Hill FC, Gibbs GV, Boisen MBJr (1997) Critical point properties of electron density distributions for oxide molecules containing first and second row cations. Phys Chem Minerals 24:582-596

Hochella MFJr, Keefer KD, de Jong BHWS (1983) The crystal chemistry of a naturally occuring magnesium hydroxide sulfate hydrate, a precipitate of heated seawater. Geochim Cosmochim Acta 47:2053-2058

Hoenle W (1991) Crystal structure of $H_2S_2O_7$ at 298 K. Z Kristallogr 196:279-288

Hoffmann C, Armbruster T, Giester G (1997) Acentric structure ($P3$) of bechererite, $Zn_7Cu(OH)_{13}$ $(SiO(OH)_3SO_4)$. Am Mineral 82:1014-1018

Hughes JM, Drexler JW (1991) Cation substitution in the apatite tetrahedral site: Crystal structures of type hydroxylellestadtite and type fermorite. N Jahrb Mineral Monatsh 327-336

Hyde B, Andersson S (1989) Inorganic crystal structures. Wiley, New York

Hurlbut CS Jr, Aristarain LF (1969) Olascherite, $Pb_2(SO_4)(SeO_4)$, a new mineral from Bolivia. Am Mineral 54:1519-1527

Irran E, Tillmanns E, Hentschel G (1997) Ternesite, $Ca_5(SiO_4)_2SO_4$, a new mineral from the Ettringer Bellerberg/Eifel, Germany. Mineral Petrol 60:121-132

Iskhakova LD, Dubrovinskii LS, Charushnikova IA (1991) Crystal structure, calculation of parameters of atomic interaction potential and thermochemical properties of $NiSO_4 \cdot nH_2O$ ($n = 7,6$). Kristallografiya 36:650-655 (in Russian)

Jacobsen SD, Smyth JR, Swope RJ, Downs RT (1998) Rigid-body character of the SO_4 groups in celestine, anglesite and barite. Can Mineral 36:1053-1060

Jambor JL (1999) Nomenclature of the alunite supergroup. Can Mineral 37:1323-1341

Jambor JL, Owens DR, Grice JD, Feinglos MN (1996) Gallobeudantite, $PbGa_3[(AsO_4),(SO_4)](OH)_6$, a new mineral species from Tsumeb, Namibia, and associated new gallium analogues of the alunite-jarosite family. Can Mineral 34:1305-1315

Jarosch D (1985) Kristallstruktur des Leonits; $K_2Mg(SO_4)_2(H_2O)_4$. Z Kristallogr 173:75-79

Jolly JH, Foster HL (1967) X-ray diffraction data of aluminocopiapite. Am Mineral 52:1220-1223

Kampf AR (1991) Grandreefite, $Pb_2F_2SO_4$: Crystal structure and relationship to the lanthanide oxide sulfates, $Ln_2O_2SO_4$. Am Mineral 76:278-282

Kannan KK, Viswamitra MA (1965) Crystal structure of magnesium potassium sulfate hexahydrate $MgK_2(SO_4)_2(H_2O)_6$. Z Kristallogr 122:161-174

Kato T (1977) Further refinement of the woodhouseite structure. N Jahrb Mineral Monatsh 54-58

Kato T, Miura Y (1977) The crystal structure of jarosite and svanbergite. Mineral J (Japan) 8:419-430

Kato K, Saalfeld H (1972) The crystal structure of hanksite, $KNa_{22}(Cl(CO_3)_2(SO_4)_9)$ and its relation to the K_2SO_4 I structure type. Acta Crystallogr B28:3614-3617

Keefer KD, Hochella MFJr, de Jong BHWS (1981) The structure of magnesium hydroxide sulfate hydrate $MgSO_4 \cdot 1/3Mg(OH)_2 \cdot 1/3H_2O$. Acta Crystallogr B37:1003-1006

Kellersohn T (1992) Structure of cobalt sulfate tetrahydrate. Acta Crystallogr C48:776-779

Kellersohn T, Delaplane RG, Olovsson I (1991) Disorder of a trigonally planar coordinated water molecule in cobalt sulfate heptahydrate, $CoSO_4 \cdot 7D_2O$ (bieberite). Z Naturforsch B46:1635-1640

Kemnitz E, Werner C, Trojanov SI (1996a) Structural chemistry of alkaline metal hydrogen sulfates. A review of new structural data. Part I. Synthesis, metal and sulfur polyhedra. Eur J Solid State Inorg Chem 33:563-580

Kemnitz E, Werner C, Trojanov SI (1996b) Structural chemistry of alkaline metal hydrogen sulfates. A review of new structural data. Part II. Hydrogen bonding systems. Eur J Solid State Inorg Chem 33:581-596

Kemnitz E, Werner C, Trojanov SI (1996c) Reinvestigation of crystalline sulfuric acid and oxonium hydrogensulfate. Acta Crystallogr C52:2665-2668

Kirfel A, Will G (1980) Charge density distribution in anhydrite, $CaSO_4$, from X-ray and neutron diffraction measurements. Acta Crystallogr B36:2881-2890

Klaska R, Jarchow O (1977) Synthetischer Sulfat-hydrocancrinit vom Mikrosommit-Typ. Naturwiss 64:93

Kolitsch U, Tiekink ERT, Slade PG, Taylor MR, Pring A (1999) Hinsdalite and plumbogummite, their atomic arrangements and disordered lead sites. Eur J Mineral 11:513-520

Konnert JA, Evans HTJr, McGee JJ, Ericksen GE (1994) Mineralogical studies of the nitrate deposits of Chile: VII. Two new saline minerals with the composition $K_6(Na,K)_4Na_6Mg_{10}(XO_4)_{12}(IO_3)_{12} \cdot 12H_2O$: fuenzalidaite (X = S) and carlosruizite (X = Se). Am Mineral 79:1003-1008

Krivovichev SV, Filatov SK (1999a) Structural principles for minerals and inorganic compounds containing anion-centered tetrahedra. Am Mineral 84:1099-1106

Krivovichev SV, Filatov SK (1999b) Metal arrays in structural units based on anion-centered metal tetrahedra. Acta Crystallogr B55:664-676

Krivovichev SV, Filatov SK, Semenova TF (1998a) Types of cationic complexes on the base of oxocentered tetrahedra $[OM_4]$ in crystal structures of inorganic compounds. Russ Chem Rev 67: 137-155

Krivovichev SV, Filatov SK, Semenova TF, Rozhdestvenskaya IV (1998b) Crystal chemistry of inorganic compounds based on chains of oxocentered tetrahedra. I. Crystal structure of chloromenite, $Cu_9O_2(SeO_3)_4Cl_6$. Z Kristallogr 213:645-649

Kucha H, Osuch W, Elsen J (1996) Viaeneite, $(Fe,Pb)_4S_8O$, a new mineral with mixed sulphur valences from Engis, Belgium. Eur J Mineral 8:93-102

Ladd MFC, Palmer RA (1994) Structure determination by X-ray crystallography. Plenum Press, New York

Lange J, Burzlaff H (1995) Single-crystal data collection with a Laue diffractometer. Acta Crystallogr A51:931-936

Larson AC, Cromer DT (1967) Refinement of the alum structures: III. X-ray study of the α-alums, K, Rb and $NH_4Al(SO_4)_2(H_2O)_{12}$. Acta Crystallogr 22:793-800

Leclaire A, Ledesert M, Monier JC, Daoud A, Damak M (1985) Structure du disulfate acide de triammonium. Une redetermination. Relations des chaines de liaisons hydrogene avec la morphologie et la conductivite eletrique. Acta Crystallogr B41:209-213

Lengauer CL, Giester G, Irran E (1994) $KCr_3(SO_4)_2(OH)_6$: Synthesis, characterization, powder diffraction data, and structure refinement by the Rietveld technique and a compilation of alunite-type compounds. Powder Diffraction 9:265-271

Levy AH, Lisensky GC (1978) Crystal structures of sodium sulfate decahydrate (Glauber's salt) and sodium tetraborate decahydrate (borax). Redetermination by neutron diffraction. Acta Crystallogr B34:3502-3510

Li G, Peacor DR, Essene EJ, Brosnahan DR, Beane RE (1992) Walthierite, $Ba_{0.5}\square_{0.5}[Al_3(OH)_6(SO_4)_2]$, and huangite, $Ca_{0.5}\square_{0.5}[Al_3(OH)_6(SO_4)_2]$, two new minerals of the alunite group from the Coquimbo region, Chile. Am Mineral 77:1275-1284

Li JJ, Zhou JL, Dong W (1990) The structure of amarillite. Chinese Sci Bull 35:2073-2075

Liebau F (1985) Structural chemistry of silicates. Springer-Verlag, Berlin

Lima-de-Faria J (1994) Structural mineralogy. Kluwer, Dordrecht, the Netherlands

Lima-de-Faria J, Hellner E, Liebau F, Makovicky E, Parthé E (1990) Nomenclature of inorganic structure types. Acta Crystallogr A46:1-11

Lindsay CG, Gibbs GV (1988) A molecular orbital study of bonding in sulfate molecules: implications for sulfate crystal structures. Phys Chem Mineral 15:260-270

Liu Tiegeng, Gong Guohong, Ye Lin (1995) Discovery and investigation of zinc-melanterite in nature. Acta Mineral Sinica 15:286-290 (in Chinese)

Livingstone A, Ryback G, Fejer EE, Stanley CJ (1987) Mattheddleite, a new mineral of the apatite group from Leadhills, Strathclyde region. Scottish J Geol 23:1-8

Louisnathan SJ, Gibbs GV (1972a) The effect of tetrahedral angles on SiBO bond overlap populations for isolated tetrahedra. Am Mineral 57:1614-1642

Louisnathan SJ, Gibbs GV (1972b) Bond length variation in TO_4^{n-} tetrahedral oxyanions of the third row elements: T = Al, Si, P, S and Cl. Mat Res Bull 7:1281-1292

Louisnathan SJ, Hill RJ, Gibbs GV (1977) Tetrahedral bond length variations in sulfates. Phys Chem Minerals 1:53-69

Lynton H, Truter MR (1960) An accurate determination of the crystal structure of potassium pyrosulfate. J Chem Soc 5112-5118

Malinovskii YuA, Baturin SV, Belov NV (1979) The crystal structure of the Fe-tychite. Dokl Akad Nauk SSSR 249:1365-1368 (in Russian)

Marøy K (1971) The crystal structure of potassium pentathionate hemitrihydrate. Acta Chem Scand 25:2580-2590

Maslen EN, Ridout SC, Watson KJ, Moore FH (1988) The structures of Tuttons's salts. I. Diammonium hexaaqua magnesium(II) sulfate. Acta Crystallogr C44:409-412

Mathew M, Takagi S, Waerstad KR, Frazier AW (1981) The crystal structure of synthetic chukhrovite, $Ca_4AlSi(SO_4)F_{13}(H_2O)_{10}$. Am Mineral 66:392-397Matsuno T, Takayanagi H, Furuhata K, Koishi M, Ogura H (1984) The crystal structure of calcium sulfite hemihydrate. Bull Chem Soc Japan 57: 1155-1156

McDonald AM, Chao GY (1994) Abenakiite-(Ce), a new silicophosphate carbonate mineral from Mont Saint-Hilaire, Quebec: description and structure determination. Can Mineral 32:843-854

McGinnety JA (1972) Redetermination of the structures of potassium sulphate and potassium chromate: the effect of electrostatic crystal forces upon observed bond lengths. Acta Crystallogr B28:2845-2852

McKee ML (1996a) Computational study of the mono- and dianions of SO_2, SO_3, SO_4, S_2O_3, S_2O_4, S_2O_6 and S_2O_8. J Phys Chem 100:3473-3481

McKee ML (1996b) Additions and corrections: Computational study of the mono- and dianions of SO_2, SO_3, SO_4, S_2O_3, S_2O_4, S_2O_6 and S_2O_8. J Phys Chem 100:16444

McLean WJ, Anthony JW (1972) The disordered, zeolite-like structure of connellite. Am Mineral 57: 426-438

Medenbach O, Gebert W (1993) Lautenthalite, $PbCu_4[(OH)_6|(SO_4)_2]\cdot 3H_2O$, the Pb analogue of devillite: A new mineral from the Harz mountains, Germany. N Jahrb Mineral Monatsh 401-407

Mellini M, Merlino S (1977) The crystal structure of tuscanite. Am Mineral 62:1114-1120

Mellini M, Merlino S (1978) Ktenasite, another mineral with $((Cu,Zn)_2(OH)_3O)^{1-}$ octahedral sheets. Z Kristallogr 147:129-140

Mellini M, Merlino S (1979) Posnjakite. $(Cu_4(OH)_6(H_2O)O)$ octahedral sheets in its structure. Z Kristallogr 149:249-257

Menchetti S, Sabelli C (1974) Alunogen. Its structure and twinning. Tshchermaks mineral petrogr Mitt 21:164-178

Menchetti S, Sabelli C (1976a) Crystal chemistry of the alunite series: Crystal structure refinement of alunite and synthetic jarosite. N Jahrb Mineral Monatsh 406-417

Menchetti S, Sabelli C (1976b) The halotrichite group: The crystal structure of apjohnite. Mineral Mag 40:599-608

Menchetti S, Sabelli C (1980a) The crystal structure of klebelsbergite $Sb_4O_4(OH)_2SO_4$. Am Mineral 65:931-935

Menchetti S, Sabelli C (1980b) Peretaite, $CaSb_4O_4(OH)_2(SO_4)_2(H_2O)_2$: Its atomic arrangement and twinning. Am Mineral 65:940-946

Menchetti S, Sabelli C (1982) Campigliaite, $Cu_4Mn(SO_4)_2(OH)_6(H_2O)_4$, a new mineral from Campiglia Marittima, Tuscany, Italy. Am Mineral 67:385-393

Mereiter K (1972) Die Kristallstruktur des Voltaits, $K_2Fe^{2+}_5Fe^{3+}_3Al(SO_4)_{12}(H_2O)_{18}$. Tschermaks mineral petrogr Mitt 18:185-202

Mereiter K (1974) Die Kristallstruktur von Rhomboklas $(H_5O_2)^+(Fe(SO_4)_2(H_2O)_2)$. Tschermaks mineral petrogr Mitt 21:216-232

Mereiter K (1976) Die Kristallstruktur des Ferrinatrits, $Na_3Fe(SO_4)_3(H_2O)_3$. Tshchermaks mineral petrogr Mitt 23:317-327

Mereiter K (1979) Refinement of the crystal structure of langbeinite $K_2Mg_2(SO_4)_3$. N Jahrb Mineral Monatsh 182-188

Mereiter K (1982) Die Kristallstruktur des Johannits, $Cu(UO_2)_2(OH)_2(SO_4)_2(H_2O)_8$. Tschermaks mineral petrogr Mitt 30:47-57

Mereiter K (1986) Crystal structure and crystallographic properties of a schroeckingerite from Joachimsthal. Tschermaks mineral petrogr Mitt 35: 1-18

Merlino S, Mellini M, Bonaccorsi E, Pasero M, Leoni L, Orlandi P (1991) Pitiglianoite, a new feldspathoid from southern Tuscany, Italy: chemical composition and crystal structure. Am Mineral 76:2003-2008

Merlino S, Pasero M, Sabelli C, Trosti-Ferroni R (1992) Crystal structure refinements of spangolite, a hydrated basic sulphate of copper and aluminium, from three different occurrences. N Jahrb Mineral Monatsh 349-357

Moore AE, Taylor HFW (1970) Crystal structure of ettringite. Acta Crystallogr B26:386-393

Moore PB (1965) A structural classification of Fe–Mn orthophosphate hydrates. Am Mineral 50:2052-2062

Moore PB (1970) Structural hierarchies among minerals containing octahedrally coordinated oxygen. I. Stereoisomerism among corner-sharing octahedral and tetrahedral chains. N Jahrb Mineral Monatsh 163-173

Moore PB (1973) Bracelets and pinwheels: A topological-geometrical approach to the calcium orthosilicate and alkali sulfate structures. Am Mineral 58:32-42

Moore PB (1976) The glaserite, $K_3Na[SO_4]_2$, structure type as a 'super' dense-packed oxide: Evidence for icosahedral geometry and cation-anion mixed layer packings. N Jahrb Mineral Abh 127:187-196

Moore PB (1981) Complex crystal structures related to glaserite, $K_3Na(SO_4)_2$: Evidence for very dense packings among oxysalts. Bull Minéral 104:536-547

Moore PB, Shen J (1984) Roeblingite, $Pb_2Ca_6(SO_4)_2(OH)_2(H_2O)_2(Mn(Si_3O_9)_2)$: Its crystal structure and comments on the lone pair effect. Am Mineral 69:1173-1179

Moore PB, Kampf AR, Sen Gupta PK (2000) The crystal structure of philolithite, a trellis-like open framework based on cubic closest-packing of anions. Am Mineral 85:810-816

Mootz D, Bartmann K (1991) Hydrate der Fluorsulfonsaeure: Das Schmelzdiagramm des Systems FSO_3H-H_2O und die Kristallstruktur des Monohydrats $(H_3O)FSO_3$. Z Anorg Allg Chem 592:171-178

Mukhtarova NN, Kalinin VR, Rastsvetaeva RK, Ilyukhin VV, Belov NV (1980) The crystal structure of görgeyite $K_2Ca_5(SO_4)_6H_2O$. Dokl Akad Nauk SSSR 252:102-105

Mumme WG, Sarp H, Chiappero PJ (1994) A note on the crystal structure of schulenbergite. Arch Sci Geneve 47:117-124

Nagorsen G, Lyng S, Weiss A, Weiss A (1962) Zur Konstitution von $HgSO_4(HgO)_2$. Angew Chem 74:119

Okada K, Ossaka J (1980) Structures of potassium sodium sulphate and tripotassium sodium disulphate. Acta Crystallogr B36:919-921

Okada K, Hirabayashi J, Ossaka J (1982) Crystal structure of natroalunite and crystal chemistry of the alunite group. N Jahrb Mineral Monatsh 534-540

O'Keeffe M, Bovin JO (1978) The crystal structure of paramelaconite, Cu_4O_3. Am Mineral 63:180-185

O'Keeffe M, Hyde B (1985) An alternative approach to non-molecular crystal structures with emphasis on the arrangements of cations. Struct Bond 61:77-144

O'Keeffe M, Hyde B (1996) Crystal structures: I. Patterns and symmetry. Mineralogical Society of America, Washington, DC

O'Keeffe M, Domenges B, Gibbs GV (1985) Ab initio molecular orbital calculations on phosphates: comparison with silicates. J Phys Chem 89:2304-2309

Organova NI, Rastsvetaeva RK, Kuz'mina OV, Arapova GA, Litsarev MA, Fin'ko VI (1994) Crystal structure of low-symmetry ellestadite in comparison with other apatite-like structures. Kristallografiya 39:278-282 (in Russian)

Ossaka J, Hirabayashi J, Okada K, Kobayashi R, Hayashi T (1982) Crystal structure of minamiite, a new mineral of the alunite group. Am Mineral 67:114-119

O'Sullivan K, Thompson RC, Trotter J (1970) Crystal structure of ammonium fluorosulfate. J Chem Soc A:1814-1817

Otto HH (1975) Die Kristallstruktur von Fleischerite, $Pb_3Ge(OH)_6(SO_4)_2(H_2O)_3$ sowie kristallchemische Untersuchungen an isotypen Verbindungen. N Jahrb Mineral Abh 123:160-190

Pabst A, Sharp WN (1973) Kogarkoite, a new natural phase in the system Na_2SO_4–NaF–NaCl. Am Mineral 58:116-127

Pasero M, Davoli P (1987) Structure of hashemite, $Ba(Cr,S)O_4$. Acta Crystallogr C43:1467-1469

Payan F, Haser R (1976) On the hydrogen bonding in potassium hydrogen sulphate. Comparison with a previous crystal structure determination. Acta Crystallogr B32:1875-1879

Peacor DR, Rouse RC, Coskren TD, Essene EJ (1999a) Destinezite ('diadochite'), $Fe_2(PO_4)(SO_4)(OH)$·$6H_2O$: its crystal structure and role as a soil mineral at Alum Cave Bluff, Tennessee. Clays Clay Minerals 47:1-11

Peacor DR, Rouse RC, Essene EJ, Lauf RJ (1999b) Coskrenite-(Ce), $(Ce,Nd,La)_2(SO_4)_2(C_2O_4)$·$8H_2O$, a new rare-earth oxalate mineral from Alum Cave Bluff, Tennessee: characterization and crystal structure. Can Mineral 37:1453-1462

Pedersen BF, Semmingsen D (1982) Neutron diffraction refinement of the structure of gypsum, $CaSO_4(H_2O)_2$. Acta Crystallogr B38:1074-1077

Pertlik F (1971) Die Kristallstruktur von Poughit, $Fe_2(TeO_3)_2(SO_4)(H_2O)_3$. Tschermaks mineral petrogr Mitt 15:279-290

Pertlik F, Zemann J (1985) The crystal structure of scotlandite, $PbSO_3$. Tschermaks mineral petrogr Mitt 34:289-295

Pertlik F, Zemann J (1988) The crystal structure of nabokoite, $Cu_7TeO_4(SO_4)_5KCl$: The first example of a $Te(IV)O_4$ pyramid with exactly tetragonal symmetry. Mineral Petrol 38:291-298

Pollmann H, Witzke T, Kohler H (1997) Kuzelite $(Ca_4Al_2(OH)_{12})((SO_4)\cdot6H_2O)$, a new mineral from Maroldweisach/Bavaria, Germany. N Jahrb Mineral Monatsh 423-432

Popova VI, Popov VA, Rudashevsky NS, Glavatskikh SF, Polyakov VO, Bushmakin AF (1987) Nabokoite, $Cu_7TeO_4(SO_4)_5\cdot KCl$, and atlasovite, $Cu_6Fe^{3+}Bi^{3+}O_4(SO_4)_5\cdot KCl$: new minerals of volcanic exhalations. Zap Vses Mineral Obshch 116:358-367 (in Russian)

Ptasiewicz-Bak H, McIntyre GJ, Olovsson I (1983) Structure of monoclinic nickel sulphate hexadeuterate, $NiSO_4(D_2O)_6$. Acta Crystallogr C39:966-968

Pushcharovsky DYu, Lima-de-Faria J, Rastsvetaeva RK (1998) Main structural subdivisions and structural formulas of sulphate minerals. Z Kristallogr 213:141-150

Ramondo F, Bencivenni L, Caminiti L, Sadun C (1991) Ab initio SCF study on lithium perchlorate and lithium sulfate molecules: geometries and vibrational frequencies. Chem Phys 151:179-186

Rastsvetaeva RK, Pushcharovsky DYu (1989) Kristallokhimiya sulfatov (= Crystal chemistry of sulfates). Itogi Nauki I Tekhniki, ser Kristallokhimiya Vol 23 (= Advances in Science and Technology, ser Crystal Chemistry) (in Russian)

Rius J, Allmann R (1984) The superstructure of the double layer mineral wermlandite $(Mg_7AlFe(OH)_{18})$ $(Ca(H_2O)_6)(SO_4)_2(H_2O)_6$. Z Kristallogr 168:133-144

Rius J, Plana F (1986) Contribution to the superstructure resolution of the double layer mineral motukoreaite. N Jahrb Mineral Monatsh 263-272

Roberts AC, Cooper MA, Hawthorne FC, Criddle AJ, Stanley CJ, Key CL, Jambor JL (1999) Sidpietersite, $Pb^{2+}_4(S^{6+}O_3S^{2-})O_2(OH)_2$, a new thiosulfate mineral from Tsumeb, Namibia. Can Mineral 37:1269-1273

Robinson DJ, Kennard CHL (1972) Potassium hexa-aquacopper(II) sulfate, $CuH_{12}K_2O_{14}S_2$. Cryst Struct Comm 1:185-188

Robinson PD, Fang JH (1969) Crystal structures and mineral chemistry of double-salt hydrates. I. Direct determination of the crystal structure of tamarugite. Am Mineral 54:19-30

Robinson PD, Fang JH (1971) Crystal structures and mineral chemistry of hydrated ferric sulphates: II. The crystal structure of paracoquimbite. Am Mineral 56:1567-1571

Robinson PD, Fang JH (1973) Crystal structures and mineral chemistry of hydrated ferric sulphates. III. The crystal structure of kornelite. Am Mineral 58:535-539

Robinson PD, Fang JH, Ohya Y (1972) The crystal structure of kainite. Am Mineral 57:1325-1332

Rouse RC, Dunn PJ (1982) A contribution to the crystal chemistry of ellestadite and the silicate sulfate apatites. Am Mineral 67:90B96

Ruben H, Zalkin A, Faltens MO, Templeton DH (1974) Crystal structure of sodium gold(I) thiosulfate dihydrate, $Na_3Au(S_2O_3)_2\cdot2H_2O$. Inorg Chem 13:1836-1839

Rumanova IM, Malitskaya GI (1959) Revision of the structure of astrakhanite by weighted phase projection methods. Kristallografiya 4:510-525 (in Russian)

Sabelli C (1967) La struttura della Darapskite. Atti della Accad Naz Sci Fis Mat Natur Rend Ser 8 42: 874-887

Sabelli C (1980) The crystal structure of chalcophyllite. Z Kristallogr 151:129-140

Sabelli C (1985a) Refinement of the crystal structure of jurbanite, $Al(SO_4)(OH)(H_2O)_5$. Z Kristallogr 173:33-39

Sabelli C (1985b) Uklonskovite, $NaMg(SO_4)F(H_2O)_2$: New mineralogical data and structure refinement. Bull Minéral 108:133-138

Sabelli C, Ferroni T (1978) The crystal structure of aluminite. Acta Crystallogr B34:2407-2412

Sabelli C, Trosti-Ferroni R (1985) A structural classification of sulfate minerals. Per Mineral 54:1-46

Sabelli C, Zanazzi PF (1968) The crystal structure of serpierite. Acta Crystallogr B24:1214-1221

Sabelli C, Zanazzi PF (1972) The crystal structure of devillite. Acta Crystallogr B28:1182-1189

Sahl K (1970) Zur Kristallstruktur von Lanarkit, $Pb_2O(SO_4)$. Z Kristallogr 132:99-117

Sakae T, Nagata H, Sudo T (1978) The crystal structure of synthetic calcium phosphate sulfate hydrate, $Ca_2(HPO_4)(SO_4)(H_2O)_4$, and its relation to brushite and gypsum. Am Mineral 63:520-527

Sakamoto Y (1968) The size, atomic charges, and motion of the sulfate radical of symmetry 4-3m in the crystal of sulfohalite, $Na_6ClF(SO_4)_2$. J Sci Hiroshima Univ A32:101-108

Sandomirsky PA, Belov NV (1984) Crystal chemistry of mixed anionic radicals. Nauka, Moscow (in Russian)

Schindler M, Hawthorne FC, Baur WH (2000a) Crystal chemical aspects of vanadium: Polyhedral geometries, characteristic bond-valences and polymerization of (VO_n) polyhedra. Chem Mater 12:1248-1259

Schindler MC, Hawthorne FC, Baur WH (2000b) A crystal-chemical approach to the composition and occurrence of vanadium minerals. Can Mineral (accepted)

Schlatti M, Sahl K, Zemann A, Zemann J (1970) Die Kristallstruktur des Polyhalits, $K_2Ca_2Mg(SO_4)_4(H_2O)_2$. Tschermaks mineral petrogr Mitt 14:75-86

Schleid Th (1996) [NM$_4$] tetrahedra in nitride sulfides and chlorides of the trivalent lanthanides. Eur J Solid State Inorg Chem 33:227-240

Schlüter J, Klaska K-H, Gebhard G (1999) Changoite, Na$_2$Zn(SO$_4$)$_2$·4H$_2$O, the zinc analogue of blödite, a new mineral from Sierra Gorda, Antofagasta, Chile. N Jahrb Mineral Monatsh 97-103

Schneider W (1967) Caracolit, das Na$_3$Pb$_2$(SO$_4$)$_3$Cl mit Apatitstruktur. N Jahrb Mineral Monatsh 284-289

Schröpfer L (1973) Strukturelle Untersuchungen an Ca(SO$_3$)(H$_2$O)$_{0.5}$. Z Anorg Allg Chem 401:1-14

Scordari F (1977) The crystal structure of ferrinatrite, Na$_3$(H$_2$O)$_3$(Fe(SO$_4$)$_3$) and its relationship to Maus's salt, (H$_3$O)$_2$K$_2$(K$_{0.5}$(H$_2$O)$_{0.5}$)$_6$(Fe$_3$O(H$_2$O)$_3$(SO$_4$)$_6$)(OH)$_2$. Mineral Mag 41:375-383

Scordari F (1978) The crystal structure of hohmannite, Fe$_2$(H$_2$O)$_4$((SO$_4$)$_2$)(H$_2$O)$_4$ and its relationship to amaranite, Fe$_2$(H$_2$O)$_4$((SO$_4$)$_2$O)(H$_2$O)$_3$. Mineral Mag 42:144-146

Scordari F (1981a) Fibroferrite: a mineral with a (Fe(OH)(H$_2$O)$_2$(SO$_4$)) spiral chain and its relationship to Fe(OH)(SO$_4$), butlerite and parabutlerite. Tschermaks mineral petrogr Mitt 28:17-29

Scordari F (1981b) Sideronatrite: a mineral with a (Fe$_2$(SO$_4$)$_4$(OH)$_2$) guildite type chain? Tschermaks mineral petrogr Mitt 28:315-319

Scordari F (1981c) Crystal chemical implications on some alkali hydrated sulphates. Tschermaks mineral petrogr Mitt 28:207-222

Scordari F, Stasi F (1990) The crystal structure of euchlorine, NaKCu$_3$O(SO$_4$)$_3$. N Jahrb Mineral Abh 161:241-253

Scordari F, Stasi F, Milella G (1982) Concerning metasideronatrite. N Jahrb Mineral Monatsh 341-347

Scordari F, Stasi F, Schingaro E, Comunale G (1994) Analysis of the (Na$_{1/3}$,(H$_2$O)$_{2/3}$)$_{12}$-(NaFe^{3+}$_3$O(SO$_4$)$_6$(H$_2$O$_3$)) compound: Crystal structure, solid-state transformation and its relationship to some analogues. Z Kristallogr 209:43-48

Shannon RD (1976) Revised effective ionic radii and systematic studies of interatomic distances in halides and chalcogenides. Acta Crystallogr A32:751-767

Shiba H, Watanabe T (1931) Les structures des cristaux de northupite, de northupite bromee et de tychite. Comptes Rend Hebd Acad Sci 193:1421-1423

Singer J, Cromer DT (1959) The crystal structure analysis of zirconium sulphate tetrahydrate. Acta Crystallogr 12:719-723

Sliznev VV, Solomonik VG (1996) The structure and vibrational spectra of M$_2$SO$_4$ (M = Li, Na, K) molecules: ab initio SCF MO LCAO calculations with effective core potentials and with account taken of all electrons. Russ J Coord Chem 22:655-660

Smirnova NL, Akimova NV, Belov NV (1967) Crystal chemistry of the sulfates. J Struct Chem 8:65-68

Smirnova NL, Akimova NV, Belov NV (1968) Crystal chemistry of the sulfates. II. J Struct Chem 9:724-726

Smith GW, Walls R (1980) The crystal structure of görgeyite K$_2$(SO$_4$)(Ca(SO$_4$))$_5$(H$_2$O). Z Kristallogr 151:49-60

Stadnicka K, Glazer AM, Koralewski M (1987) Structure, absolute configuration and optical activity of α-nickel sulfate hexahydrate. Acta Crystallogr B43:319-325

Starova GL, Filatov SK, Fundamensky VS, Vergasova LP (1991) The crystal structure of fedotovite, K$_2$Cu$_3$O(SO$_4$)$_3$. Mineral Mag 55:613-616

Starova GL, Filatov SK, Matusevich GL, Fundamensky VS (1995) The crystal structure of vlodavetsite, AlCa$_2$(SO$_4$)$_2$F$_2$Cl·4H$_2$O. Mineral Mag 59:159-162

Steele IM, Pluth JJ, Richardson JW Jr (1997) Crystal structure of tribasic lead sulfate (3PbO·PbSO$_4$·H$_2$O) by X-rays and neutrons: An intermediate phase in the production of lead acid batteries. J Solid State Chem 132:173-181

Steele IM, Pluth JJ, Livingstone A (1998) Crystal structure of macphersonite (Pb$_4$SO$_4$(CO$_3$)$_2$(OH)$_2$): Comparison with leadhillite. Mineral Mag 62:451-459

Steele IM, Pluth JJ, Livingstone A (1999) Crystal structure of susannite, Pb$_4$SO$_4$(CO$_3$)$_2$(OH)$_2$): A trimorph with macphersonite and leadhillite. Mineral Mag 11:493-499

Süsse P (1967) Crystal structure of amarantite. Naturwiss 54:642-643

Süsse P (1968) Die Kristallstruktur des Botryogens. Acta Crystallogr B24:760-767

Süsse P (1970) The crystal structure of copiapite. N Jahrb Mineral Monatsh 286-287

Süsse P (1973) Slavikit: Kristallstruktur und chemische Formel. N Jahrb Mineral Monatsh 93-95

Szymanski JT (1985) The crystal structure of plumbojarosite Pb(Fe$_3$(SO$_4$)$_2$(OH)$_6$)$_2$. Can Mineral 23:659-668

Szymanski JT (1988) The crystal structure of beudantite, Pb(Fe,Al)$_3$((As,S)O$_4$)$_2$(OH)$_6$. Can Mineral 26:923-932

Tachez M, Theobald F, Watson KJ, Mercier R (1979) Redetermination de la structure du sulfate de vanadyle pentahydrate VOSO$_4$(H$_2$O)$_5$. Acta Crystallogr B35:1545-1550

Tahirov TH, Lu T-H, Huang C-C, Chung C-S (1994) A precise structure redetermination of nickel ammonium sulfate hexahydrate, Ni(H$_2$O)$_6$·2NH$_4$·2SO$_4$. Acta Crystallogr C50:668-669

Teng ST, Fuess H, Bats JW (1984) Refinement of sodium thiosulfate, $Na_2S_2O_3$ at 120K. Acta Crystallogr C40:1785-1787

Thomas JN, Robinson PD, Fang JH (1974) Crystal structures and mineral chemistry of hydrated ferric sulfates: IV. The crystal structure of quenstedtite. Am Mineral 59:582-586

Tossell JA, Vaughan DJ (1992) Theoretical geochemistry. Oxford University Press, New York

Treiman AH, Peacor DR (1982) The crystal structure of lawsonbauerite, $(Mn,Mg)_9Zn_4(SO_4)_2$ $(OH)_{22} \cdot 8(H_2O)$, and its relation to mooreite. Am Mineral 67:1029-1034

Uraz AA, Arma_an N (1977) An X-ray diffraction study of sodium thiosulfate pentahydrate, $Na_2S_2O_3 \cdot 5H_2O$. Acta Crystallogr B33:1396-1399

Vairavamurthy A, Manowitz B, Luther GW, Jeon Y (1993) Oxidation state of sulfur in thiosulfate and implications for anaerobic energy metabolism. Geochim Cosmochim Acta 57:1619-1623

Varaksina TV, Fundamensky VS, Filatov SK, Vergasova LP (1990) The crystal structure of kamchatkite, a new naturally occuring oxychloride sulphate of potassium and copper. Mineral Mag 54:613-616

Vochten R, Van Haverbeke L, Van Springel K, Blaton N, Peeters OM (1995) The structure and physicochemical characteristics of synthetic zippeite. Can Mineral 33:1091-1101

Wan C, Ghose S, Rossman GR (1978) Guildite, a layer structure with a ferric hydroxy-sulphate chain and its optical absorption spectra. Am Mineral 63:478-483

Wang C-Q, Willner H, Bodenbinder M, Batchelor RJ, Einstein FWB, Aubke F (1994) Formation of cis-bis(carbonyl)palladium(II) fluorosulfate, $cis-Pd(CO)_2(SO_3F)_2$, and its crystal and molecular structure. Inorg Chem 33:3521-3525

Wang C-Q, Lewis AR, Batchelor RJ, Einstein FWB, Willner H, Aubke F (1996) Synthesis, molecular structure, and vibrational spectra of mer-tris(carbonyl)iridium(III) fluorosulfate, $mer-Ir(CO)_3(SO_3F)_3$. Inorg Chem 35:1279-1285

Wang X-B, Ding C-F, Nicholas JB, Dixon DA, Wang L-S (1999) Investigation of free singly and doubly charged alkali metal sulfate ion pairs: $M^+(SO_4^{2-})$ and $[M^+(SO_4^{2-})]_2$ (M = Na, K). J Phys Chem A103:3423-3429

Weidenthaler C, Tillmanns E, Hentschel G (1993) Orschallite, $Ca_3(SO_3)_2(SO_4) \cdot 12(H_2O)$, a new calcium-sulfite-sulfate-hydrate mineral. Mineral Petrol 48:167-177

Wildner M, Giester G (1988) Crystal structure refinements of synthetic chalcocyanite ($CuSO_4$) and zincosite ($ZnSO_4$). Mineral Petrol 39:201-209

Wildner M, Giester G (1991) The crystal structures of kieserite-type compounds: I. Crystal structures of $Me(II)SO_4 \cdot H_2O$ (Me = Mn, Fe, Co, Ni, Zn). N Jahrb Mineral Monatsh 296-306

Willner H, Rettig SJ, Trotter J, Aubke F (1991) The crystal and molecular structure of gold tris(fluorosulfate). Can J Chem 69:391-396

Witzke T (1999) Hydrowoodwardite, a new mineral of the hydrotalcite group from Königswalde near Annaberg, Saxony/Germany and other localities. N Jahrb Mineral Monatsh 75-86

Wood MM (1970) The crystal structure of ransomite. Am Mineral 55:729-734

Yamnova NA, Pushcharovskii DYu, Vyatkin SV, Khomyakov AP (1992) Crystal structure of a new natural sulfatocarbonate $Na_{25}BaTR_2(CO_3)_{11}(HCO_3)_4(SO_4)_2F_2Cl$. Kristallografiya 37:1396-1402 (in Russian)

Zahrobsky RF, Baur WH (1968) On the crystal chemistry of salt hydrates. V. The determination of the crystal structure of $CuSO_4 \cdot 3H_2O$ (bonattite). Acta Crystallogr B24:508-513

Zak Z, Kosicka M (1978) The crystal structure of lithium fluorosulphate $LiSO_3F$. Acta Crystallogr B34: 38-40

Zalkin A, Ruben H, Templeton DH (1964) The crystal structure and hydrogen bonding of magnesium sulfate hexahydrate. Acta Crystallogr 17:235-240

Zemann J, Zobetz E (1981) Do the carbonate groups in thaumasite have anomalously large deviations from coplanarity? Kristallografiya 26:1215-1217 (in Russian)

Zhang D, Rettig SJ, Trotter J, Aubke F (1996) Superacid anions: Crystal and molecular structures of $(H_3O)(Sb_2F_{11})$, $Cs(SO_3F)$, $Cs(H(SO_3F)_2)$, $Cs(Au(SO_3F)_4)$, $Cs_2(Pt(SO_3F)_6)$, and $Cs(Sb(SO_3F)_6)$. Inorg Chem 35:6113-6130

Zhou J, Li J, Dong W (1988) The crystal structure of xitieshanite. Kexue Tongbao 33:502-505

Zoltai T (1960) Classification of silicates and other minerals with tetrahedral structures. Am Mineral 45:960-973

2   X-ray and Vibrational Spectroscopy of Sulfate in Earth Materials

Satish C. B. Myneni

Department of Geosciences
Princeton University
Princeton, New Jersey 08544
and
Earth Sciences Division
Ernest Orlando Lawrence Berkeley National Laboratory
Berkeley, California 94720

INTRODUCTION

Sulfate is one of the most abundant inorganic ligands in the lithosphere and hydrosphere. It plays a major role in mediating mineral dissolution and precipitation, crystal growth, mineral–water and air–water interfacial reactions, aerosol chemistry and global climate, and biogeochemical cycling of several elements including inorganic and organic toxic contaminants. For several years, macroscopic methods and thermodynamic models have been used in understanding and predicting the geochemistry of sulfate in a variety of systems. As shown by several recent studies, molecular chemistry of chemical species cannot be identified by the macroscopic methods alone (Sposito 1990, Brown et al. 1999). Although researchers have been using different spectroscopic methods (especially infrared spectroscopy, IR) for examining sulfate molecular chemistry, sulfate in geologic materials has not been probed as extensively as some of the other oxoanions, such as chromate, selenate and arsenate. Details of different spectroscopic methods and their applications in geochemical studies are discussed by Hawthorne (1988).

The molecular properties of sulfate, such as coordination environment (types of connecting atoms, their number and bond distances), electronic state, and symmetry, dictate how sulfate reacts in a geochemical system. Several electronic states exist in a molecule, with each electronic state containing several vibrational states, and each vibrational state containing several rotational states (Fig. 1). The energies and probabilities of transitions between different electronic, vibrational, or rotational states can be measured and used in the identification of molecules and their chemical states. However, transitions among all of these states in a molecule are not feasible and the molecule symmetry determines whether a particular transition is allowed or forbidden (Cotton 1971). A variety of electron, X-ray, and optical spectroscopic methods can be used in studying the core (innermost) and valence (outermost) electronic transitions, and infrared and Raman spectroscopic methods can be used to obtain information on the vibrational transitions. All of these methods provide complementary information on the structural environment of sulfate. Information on sulfate coordination can also be obtained indirectly using the sulfate-complexed ions or molecules that are sensitive to light in the visible region, such as Fe^{3+} in iron sulfates (Rossman 1975, Huang et al. 2000). Rotational modes and their applications to studying sulfates are not discussed, as they have not been used in probing sulfates in geological materials.

In many cases, samples must be modified or placed in an environment different from their original conditions to enable exploration of sulfate molecular behavior using certain spectroscopic methods. As discussed later, information from such ex-situ methods may not represent the true coordination environment of sulfate. Hence, researchers should be

1529-6466/00/0040-0002$05.00

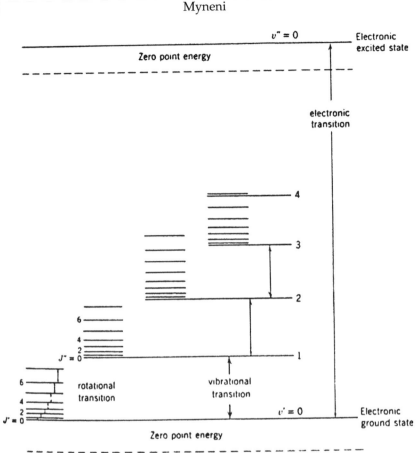

Figure 1. Energy levels of a diatomic molecule (modified from Nakamoto 1984). The letters 'J' and 'v' represent rotational and vibrational quantum numbers, respectively, and the superscripts i, ii,... represent different energy levels. The energy differences between the adjacent rotational intervals increases as J increases, and the energy differences between vibrational intervals decreases as v increases. The vibrational fine structure is also evident in the electronic spectra of several diatomic molecules.

cautious in choosing the appropriate set of spectroscopic methods for their investigation. In addition, several complementary techniques must be used to obtain complete information on the coordination environment of sulfate in heterogeneous matrices.

This chapter is focused primarily on X-ray absorption and vibrational spectroscopic methods, as these methods are widely used in exploring the chemistry of sulfate in geological materials. These methods can also be used to obtain *in situ* information on geological samples in their native state. Structural identification of crystalline sulfate minerals using X-ray diffraction is discussed by Hawthorne et al. (Chapter 1, this volume), and hence this method is not considered here, except for its applications in studying liquids. Also provided here is information on other spectroscopic methods that can be used to probe sulfate minerals directly, yielding information complementary to the above techniques.

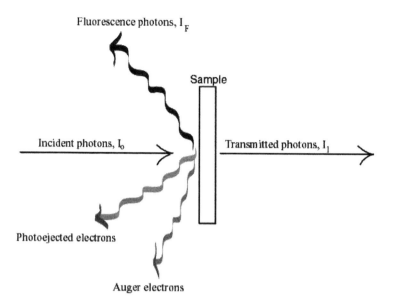

Figure 2. X-ray interactions with sample. X-ray spectra of sample are collected by placing ion chambers or detectors at different locations around the sample to detect X-ray photons or electrons (modified after Brown et al. 1988). For example, the ratio of I_F to I_o gives the X-ray absorption spectrum in fluorescence detection mode. The samples must be placed in vacuum to examine the photo-ejected or Auger electrons.

X-RAY SPECTROSCOPY

The chemical behavior of molecules is determined by their electronic structure. Information on the energy levels and symmetries of the occupied and unoccupied molecular orbitals, where electrons reside in a molecule, is critical for predicting the molecular chemistry. A combination of X-ray spectroscopic techniques can offer this unique information about molecules present in a variety of structural environments. These techniques are based on detecting: (1) absorption of X-rays by samples (X-ray absorption spectroscopy), (2) emitted fluorescence photons from samples (X-ray emission spectroscopy), and (3) ejected photoelectrons from the core or valence levels of X-ray-absorbing atoms in samples (X-ray photoelectron spectroscopy) (Fig. 2). These X-ray methods can be used for studying all electronic states or the electronic transitions of sulfate by selecting suitable energy. All of the above mentioned methods can be used for studying samples under *in situ* conditions, but photoelectron spectroscopy is applicable to samples in high vacuum (recent developments allow measurements under a few torrs of pressure, as is discussed later). Because of different selection rules for different electronic transitions and processes, these methods offer complementary information on the molecular orbital structure of sulfate. In addition, different depth and spatial sensitivities can be achieved in samples by using suitable sample geometry, and by choosing the right photon energy and X-ray spectroscopic method. Detailed discussions of these methods and their applications to geological systems (italicized citations) are available in the following reviews:

X-ray absorption spectroscopy: Stern 1974, Lytle et al. 1982, Teo 1986, Konings-berger and Prins 1988, *Hawthorne 1988,* Stöhr 1992, *Schulze et al. 1999, Brown et al. 1999.*

X-ray photoelectron spectroscopy: Baker and Betteridge 1972, Briggs 1977, Briggs and Seah 1983, *Hawthorne 1988, Hochella 1988, Hochella and White 1990, Perry 1990, Brown et al. 1999.*

A brief introduction to these methods and a review of their application to studies on sulfate geochemistry are presented in this chapter. The primary focus is on X-ray absorption spectroscopy because it has been used extensively for studying minerals. To the author's knowledge, there are no published reports on the molecular chemistry of sulfate in geological samples using X-ray emission or Auger electrons, and hence these are not discussed. However, studies of dilute heterogeneous geological samples in ambient conditions are feasible now using these methods and high-brightness synchrotron X-ray sources.

X-ray absorption spectroscopy (XAS)

The XAS technique has been used widely in the past 15 years to examine the oxidation state and coordination environment of a variety of geochemical species in minerals and at their surfaces, and in soils and aqueous solutions at dilute concentrations. This technique is element-specific and, with some exceptions, has low interference from the sample matrix. Samples can be examined under a wide range of conditions (temperature, pressure, and water content) using this technique. The detection limits are on the order of $<10^{-6}$ μM concentration, especially for heavy elements. Interested readers are referred to Waychunas and Brown (1984), Waychunas (1986), Calas and Manceau (1986), Brown et al. (1988), Brown (1990), Schulze et al. (1999), and Brown et al. (1999) and articles cited therein for details on XAS studies of geological samples.

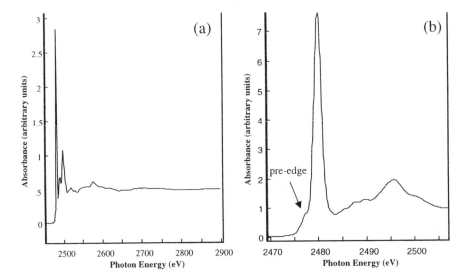

Figure 3. X-ray absorption spectrum of sulfate: (a) EXAFS spectrum of S in alunite; (b) XANES spectrum of S in ferric sulfate (Myneni, unpublished data). Pre-edge region of the XANES spectrum that represents transitions from $1s \rightarrow 3p$ orbitals (hybridized with metal $3d$ orbitals) is shown in (b).

Photon interactions with matter: XANES & EXAFS spectroscopy. When high-energy X-rays are absorbed by a molecule, the innermost electrons from the $1s$, $2s$, $2p$, etc. orbitals of the X-ray-absorbing atom (absorber) are ejected into unoccupied molecular orbitals and continuum state of a molecule (Figs. 3, 4). Such transitions occur

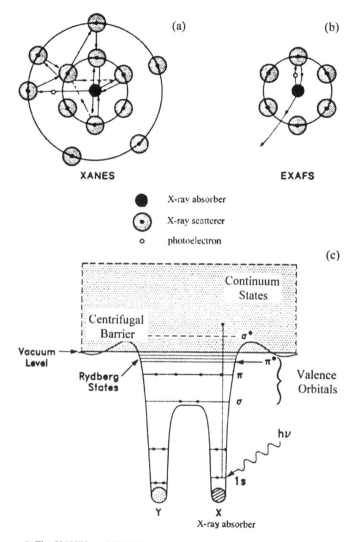

Figure 4. The XANES and EXAFS regions of XAS spectra are explained by multiple and single scattering of photoelectrons by the neighboring atoms of the absorber, respectively (a, b; modified after Brown et al. 1988). The XANES spectrum can also be explained by the electronic transitions from the core hole to the unoccupied or partly filled orbitals. For example the electronic transitions in a diatomic molecule, such as CO is shown in 'c'. The symbols, σ and π, and σ^* and π^*, represent the occupied and unoccupied molecular orbitals, respectively (modified from Stöhr 1992).

with the highest probability when the energy of incident X-ray photons is equal to the binding energy of core electrons of the absorber. In an X-ray absorption spectrum (absorption plotted as a function of incident X-ray energy), such transitions produce a sharp peak, which is commonly referred to as the white line or absorption edge. However, the electronic transitions have low probabilities on either side of the absorption edge. Localized electronic transitions from the core hole to vacant or partly filled atomic

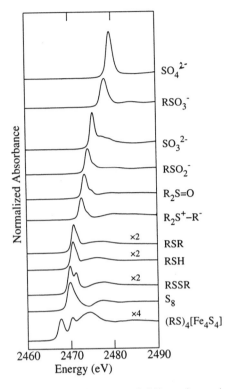

Figure 5. XANES spectra of different inorganic and organic sulfur compounds, and 'R' represents an organic molecule. Spectra of other relevant compounds are also shown in Figure 6. (modified from Pickering et al. 1998).

and molecular orbitals can occur at energies lower than the absorption edge, thereby causing peaks in the absorption spectrum (referred to as the pre-edge; Brown et al. 1988). The spectroscopy that deals with near- and pre-edge part of X-ray absorption spectrum is referred to as X-ray absorption near-edge structure spectroscopy (XANES). Some researchers use 'NEXAFS' (near-edge X-ray absorption fine-structure spectroscopy) to refer the same energy region of the X-ray absorption spectrum. As the incident photon energy is increased well above the absorption edge, the electronic transitions occur with lower probability and the released photoelectrons from the core hole interact with, and are scattered by, the atoms surrounding the absorber. This process produces oscillatory structure in the X-ray absorption spectrum, which is commonly referred to as the extended X-ray absorption fine structure (EXAFS) spectrum (typically includes features ~50 eV above the absorption edge; Fig. 3).

The spectral features in the XANES region can be explained by the localized electronic transitions within the atom, and also by the multiple scattering of released photoelectrons by the atoms surrounding the absorber (Fig. 4). With certain differences, both of these theories can explain the XANES spectra, and are used by researchers to calculate the XANES spectra theoretically from the atomic coordinates. The EXAFS part of the spectrum has been traditionally explained using single and multiple scattering of the released photoelectron by neighboring atoms of the absorber (Sayers et al. 1970, Stern 1974, Sayers 1975, Lee and Pendry 1975, Stöhr 1984, Rehr et al. 1994, Natoli 1995, Rehr and Albers 2000). Currently, the scattering theories are used by researchers to analyze the EXAFS spectra theoretically.

The XANES and EXAFS spectra contain significant chemical information about the X-ray absorber (Teo 1986, Brown et al. 1988, Stöhr 1992). The XANES region can provide information on the oxidation state and coordination environment of the absorber. The energy of the absorption edge is dependent on the oxidation state of the absorber, which is manifested by the type of ligand or metal to which the absorber is bonded, and by the energy levels of the unoccupied molecular orbitals of the absorber. For instance, the absorption edge of $1s$ electrons (also referred to as the K absorption edge, as in K-, L-, and M-electronic states in atoms) shifts to high-energy and its intensity increases with an increase in the oxidation state of the absorber (Figs. 5, 6). In addition, the intensity of the edge is affected by the absorber coordination environment and the extent of mixing of atomic orbitals and their symmetry in a molecule (e.g. s-p mixing, details discussed later;

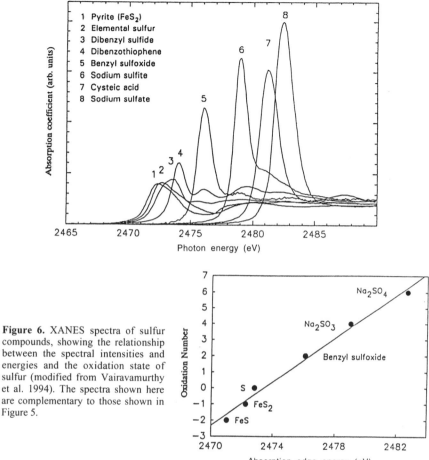

Figure 6. XANES spectra of sulfur compounds, showing the relationship between the spectral intensities and energies and the oxidation state of sulfur (modified from Vairavamurthy et al. 1994). The spectra shown here are complementary to those shown in Figure 5.

Teo 1986, Brown et al. 1988, Stöhr 1992). The EXAFS region is sensitive to absorber-backscatterer distances (frequency of oscillations increases with an increase in bond distance), the number of backscatterers (EXAFS amplitude increases with an increase in the number of backscatterers), and the atomic number of backscatterers (changes both amplitude and frequency). A detailed discussion of the L absorption edges is presented later in this chapter.

The XANES and EXAFS spectra of unknown samples can be analyzed by comparing their spectra with known spectra of structurally well-defined compounds, or by using either the single and multiple scattering theory or density functional theory (for the XANES region) for the hypothesized structures of molecules and clusters. Different computer programs have been developed for performing these theoretical calculations and interested readers are referred to special volumes published on X-ray spectroscopy (*Physica B*, 1995, v 208,209; *Journal de Physique IV*, 1997, v 7; and references therein). Experimental XAS studies can be conducted with laboratory X-ray sources, but it is difficult to obtain high photon flux and tunable monochromatic beams. Synchrotron X-ray sources permit spectral collection in a relatively short time and at high-energy resolution on extremely dilute samples. A list of synchrotron sources and their energy limitations is provided in several review articles on synchrotron research (e.g. Brown et al. 1988). An update on some of these facilities in which X-ray spectroscopic studies of sulfur compounds have been conducted is provided in Table 1.

XAS at the sulfur K absorption edge. The strong absorption of X-ray photons by air in the soft X-ray region (low-energy X-rays) makes XAS studies at the S absorption edge (2472 eV, Vaughan 1986) more complicated than collecting the heavy element absorption edges in the hard X-ray region (high-energy X-rays), which are routinely performed on geological samples. Low fluorescence yields of light elements in the soft X-ray region also limit the detection of S to the order of millimolar concentrations. For these reasons, EXAFS spectroscopic studies were not conducted at the S absorption edge on dilute samples, and most previous XAS studies of S were limited to the edge region and to samples in vacuum conditions. However, with the availability of high-flux and high-brightness beams at synchrotron X-ray sources (e.g. ALS, APS; Table 1) and multi-element solid-state detectors, EXAFS studies on dilute samples under ambient conditions are now possible. As well, several research groups have recently built experimental chambers for studying aqueous, wet, and biological samples by enclosing the X-ray beam path and sample compartment in He gas. Such experimental chambers are common at the SSRL and the ALS for routine XAS studies at the S absorption edge (Table 1, Fig. 7). Several reports are published on experimentation with XAS using hard X-rays. However, a short discussion on experimental facilities and spectral collection is presented here because XAS studies at the S-absorption edge are different from routine hard X-ray methods.

The X-ray absorption spectra are collected directly by monitoring the energy-dependent sample transmission and electron yield, or indirectly by examining changes in sample fluorescence (which is directly proportional to the sample absorbance for dilute samples). Of these, the electron-yield methods are more surface sensitive and are limited to sampling in vacuum conditions (Teo 1986, Brown et al. 1988, Stöhr 1992). Transmission methods are commonly used for concentrated samples, and the X-ray absorption spectra are collected by using ionization chambers behind the sample chamber (Fig. 7). The sample fluorescence can be monitored using ionization chambers, photodiodes, and energy-resolvable solid-state detectors. For examining dilute samples, the solid-state detectors are well-suited because the background sample fluorescence and scattered radiation can be eliminated by collecting the sample fluorescence only (provided that the detector has an energy resolution sufficient to separate fluorescence and scattered photons). The XAS data can also be collected by measuring the photoelectron–induced ionization of gas (e.g. He) present in the sample chamber (similar to electron yield), but this method is not as surface sensitive as are the direct electron yield methods. The surface sensitivity of flat samples can also be enhanced by examining the samples in grazing incidence mode (Brown et al. 1988).

Transmission XAS studies of S–containing samples can only be conducted on thin films because the X-rays can penetrate <10 μm at 2472 eV. Although the fluorescence yield is small at the sulfur K absorption edge (<10%), fluorescence detection is the most commonly used method for studying dilute samples (Vaughan 1986, Brown et al. 1988). With the help of currently available fluorescence detectors, such as the Lytle detector (Lytle et al. 1984) and photodiodes, researchers have successfully examined sulfur at concentrations of a few hundred micromolar at the SSRL, and at millimolar concentrations at the NSLS. Because the fluorescence detectors are placed on the same side of the sample as the incident photon beam (Fig. 7), the sample thickness need not be maintained precisely. However, fluorescence detection schemes pose problems for studying concentrated samples because of self-absorption of fluorescence photons by the

Table 1. Synchrotron X-ray sources optimized for conducting X-ray spectroscopy on S–bearing samples

Country	Synchrotron Light Source	Comments
France	European Synchrotron Radiation Facility (ESRF) & other facilities	X-ray spectromicroscopy beamline ID-21 is optimized for conducting XAS of small samples and for spectromicroscopy studies. LURE also has the capabilities, but the beamlines are not optimized for S-XAS studies. EXAFS SA32 beamline at the super-SCO synchrotron (Orsay) also has the capabilities to conduct S-XAS.
Germany	ANKA	ANKA-XAS is optimized for XANES studies. In addition some of the beamlines at HASY Lab (Hamburg), and SYL1 at the Physics Institute of Bonn University can be used with minor modifications to the existing facilities
	BESSY	KMC-2 is optimized for vacuum studies.
India	INDUSI	Some of the beamlines can be used for conducting XAS of S; however, the beamline information is currently not available.
Japan	Photon Factory, Springate	X-ray spectroscopy studies at the S-absorption edge have been conducted at these facilities; but information on these beamlines is not available.
Sweden	MAX Lab	I-811 is under construction and this facility can be used for conducting S-XAS studies on dilute samples.
Taiwan	Synchrotron Radiation Research Center	BL15B is optimized for S studies, and other beamlines such as SL 3B can also be used for S-XAS studies.
United Kingdom	Daresbury	Stations 3.4 and 6.3 can be used for studies in vacuum. Perhaps some of the other high-flux soft X-ray beamlines can be optimized for S XAS studies.
USA	Advanced Light Source (ALS), Berkeley, CA	Currently BL 9.3.1 and 6.3.1 are optimized for conducting XAS in vacuum. SXEER and Klien's chamber are optimized for studying S in aqueous samples and at the interfaces under ambient conditions. Spectromicroscopy studies with a spatial resolution of 0.8 µm can be performed on wet samples at beamline 10.3.1. With the help of SXEER, L-edge spectroscopy studies can be conducted under ambient conditions on several high-flux beamlines such as 7.0.1 and 8.0.1, and the bend-magnet beamlines 6.3.2 and 9.3.2 (for concentrated samples). STXM endstation at the beamline 7.0.1 can be used for spectromicroscopy studies at the L-edges of S (spatial resolution ~ 0.1 µm).
	Advanced Photon Source (APS), Chicago, IL	S-XAS studies can be conducted at several beamlines, and these are in developmental stages. Spectromicroscopy studies can be performed on samples of size <0.1 µm at some of these beamlines. Currently no beamline is optimized to conduct S L-edge spectroscopy studies.
	National Synchrotron Light Source (NSLS), Brookhaven, NY	X-19A and X-11B are optimized for conducting in-situ studies of S compounds in aqueous solutions or at interfaces. Such studies can also be conducted at other intermediate-energy beamlines after small modifications. S L-edges can be examined at the STXM beamline (X-1A).
	Center for Advanced Microstructures & Devices (CAMD), Baton Rouge, LA	Port 5B is optimized for conducting S-NEXAFS of concentrated samples

Table 1, continued

| Stanford Synchrotron Light Source (SSRL), Stanford, CA | High flux wiggler beamline, 6-2 is optimized for conducting XANES of S at submillimolar concentrations, and EXAFS of concentrated samples. Beamline 2-3, and wiggler beamlines can be used to examine S compounds at millimolar concentrations under ambient conditions, and 3.3 in vacuum |
| Synchrotron Radiation Center (SRC), Madison, WI | Ports 122 and 023 are optimized for vacuum studies |

There are also synchrotron light sources either in planning stages, in construction, or operational in Brazil, China, Korea, Russia, and Thailand, but no information is available. A third-generation source is being built in Canada (Canadian Light Source). Although the storage ring at Aarhus, Denmark (ASTRIO) is optimized for soft X-ray studies, it has no beamlines to go to energy as high as S absorption edges. The sulfur L-edge can be examined at several beamlines at the above facilities, however, only few of them can be used to examine aqueous samples.

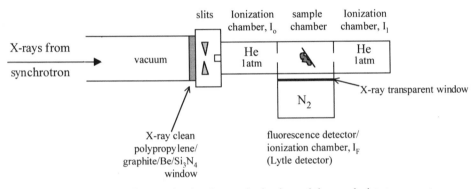

Figure 7. Schematic diagram showing the sample chamber and the sample-detector geometry for collecting the XAS spectra of S under ambient conditions. Helium is used in the beam path and in the sample chamber to prevent photon absorption by air. Nitrogen is used in the ion chambers to capture the fluorescence or transmitted photons.

sample. Hence concentrated samples are diluted with KBr or BN, or thin sample smears are used to minimize these effects. Several research groups have also used theoretical calculations for correcting the experimental spectrum of these self-absorption effects (Waldo et al. 1991, Vairavamurthy et al. 1994). Although electron yield and transmission methods can be used for collecting XAS data of concentrated samples, these methods can be affected by charging, and saturation, respectively (Brown et al. 1988).

Spectral resolution. The intrinsic spectral resolution of the absorption edge of any element is primarily determined by the lifetime of its core hole electrons (width of the core hole), which is approximately 0.4 eV for sulfur (Brown and Doniach 1980). Spectral resolution is also determined by the type of monochromator crystal that is used and by the size of the photon source (e.g. slit openings, or the vertical divergence of the beam). Commonly, a Si(111) double crystal monochromator is used for conducting the XAS of sulfur at several synchrotron facilities. Using Si(111) monochromator crystals and proper source size, the best achieved spectrometer resolution is around 0.5 eV at the sulfur absorption edge. Although other monochromator crystals, such as Ge(111) and InSb(111) can also be used, Si(111) crystals offer better energy resolution. The source size can significantly improve the spectral resolution for collimated beams (Brown et al. 1988). Better spectral resolution can be achieved by reducing the slit openings in the spectrometer, although this severely reduces the photon flux to the sample. At third-generation synchrotron facilities, better spectral resolution is achievable because of the small source size and high photon flux.

Data collection. X-ray absorption spectra can be collected using one of the detection schemes and a set of monochromator crystals as described above. An XANES spectrum typically begun several tens of eV below the absorption edge and extends up to 100-150 eV above the edge (with a step size of a few eV). At the absorption edge, samples are scanned with an edge step of 0.1 eV or less (e.g. Rompel et al. 1998, Frank et al. 1999). Although the best achievable spectral resolution is about 0.5 eV, small changes in the energy of the absorption edge (commonly less than 0.5 eV in chemically similar compounds) and the appearance of shoulders on the absorption edge (e.g. associated with metal complexation) can be detected easily with a smaller step size. For collecting the EXAFS spectra, step sizes of 1-3 eV are sufficient and the spectra must be collected up to several hundred eV above the absorption edge. Samples with a high Cl content can pose problems for long EXAFS scans because of the absorption edge around 2840 eV for Cl. As well, the monochromator has to be detuned over 90% (from maximum incident energy) below the sulfur absorption edge to examine samples with high Fe concentrations. This is done to minimize the third-order component of the incident beam, which has the energies of the Fe absorption edge.

Sulfur XAS data are usually calibrated by comparing with the absorption edge of elemental sulfur (2472 eV). However, as different research groups have used different energies for the absorption edge of elemental sulfur, care must be exercised when spectral comparisons between different reports are made. For example, the absorption edge of sulfur in sodium sulfate is reported to have different energy positions in this chapter, and these are not inherent to samples and data-collection schemes. This difference is present because different calibrations have been used by various research groups, and it is necessary to look at the original articles for the energy calibrations. Researchers often use sulfur compounds that exhibit absorption features sharper than elemental sulfur (e.g. thiosulfate and sulfate) for routine spectrometer calibration during data collection.

XAS of sulfur compounds. The majority of XAS studies conducted on sulfur have focused on a variety of organo-sulfur compounds because of their role in biological systems (proteins, enzymes, cells) and their common occurrence in coals and petroleum hydrocarbons, and as humic substances in soil and sediment organic matter (Spiro et al. 1984, Gorbaty et al. 1990, Waldo et al. 1991, George et al. 1991, Vairavamurthy 1998, Pickering et al. 1998, Sarret et al. 1999). Relatively little work has been done on the XAS of sulfate and sulfide minerals, except for the studies done by Bancroft, Vairavamurthy, Waychunas, Myneni, and their research groups. The difficulties associated with collecting XANES and EXAFS spectra at the sulfur absorption edge, and poor availability of synchrotron facilities optimized for sulfur studies, have also contributed to the small number of sulfur-XAS studies. However, this is expected to change in the near future as several new facilities optimized for sulfur-XAS are under construction.

The energies of maximum absorbance for sulfur in different oxidation states is separated by several eV, and hence the X-ray spectra of sulfates can be distinguished from other sulfur compounds (Spiro et al. 1984, Frank et al. 1987, George and Gorbaty 1989, Waldo et al. 1991, Huffman et al. 1991, 1995; Morra et al. 1997, Vairavamurthy 1998, Xia et al. 1998, Pickering et al. 1998) (Figs. 5, 6). Various studies have indicated that the absorption edge shifts to higher–energy with an increase in the formal oxidation state of sulfur (Figs. 5, 6). As well, the intensity of the absorption edge increases with an increase in the sulfur oxidation state (Fig. 6), because the number of vacancies in the sulfur valence orbitals (primarily sulfur $3p$ character) increases (Waldo et al. 1991,

Vairavamurthy 1998, Pickering et al. 1998). This effect is also commonly found in the X-ray absorption spectra of other elements (Brown et al. 1988). The EXAFS spectroscopic investigation of different forms of sulfate was started only recently, and results have not yet been published. However, some published information is available on the EXAFS of amorphous sulfides (Warburton et al. 1992, Hibble et al. 1999, and references therein).

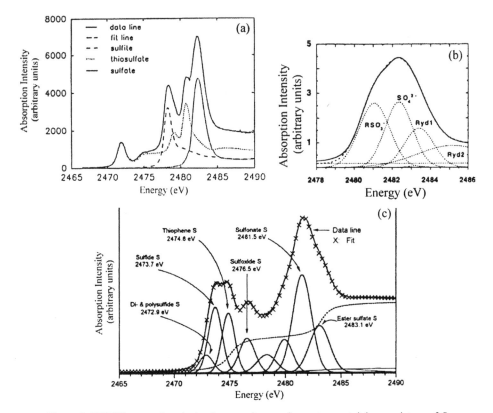

Figure 8. XANES spectral analysis of composite sample spectra containing a mixture of S compounds: (a) composite spectrum fit with a linear combination of sulfate, sulfite and thiosulfate spectra (modified from Vairavamurthy et al. 1994); (b) sulfur in blood cells of tunicate, *Ascidia ceratodes*, fit with a combination of several Gaussian profiles (modified from Frank et al. 1994); and (c) sulfur in humic substances fit with several Gaussian profiles after the spectral component of core hole to continuum transitions is subtracted from the XAS spectrum using an arc tangent function (modified from Vairavamurthy 1998).

Spectral analysis. In samples containing more than one form of sulfur, the XANES spectra can be used to distinguish and to estimate the proportion of each form. Discrimination is possible because sulfur compounds in different oxidation states are separated by several eV, and significant spectral differences exist between compounds of similar form (Figs. 6, 8). Several different techniques are used to fit the XAS data of mixtures of sulfur compounds. Reviews of fitting procedures are provided by Hawthorne and Waychunas (1988), Waldo et al. (1991), George et al. (1991), Vairavamurthy et al. (1994), Frank et al. (1994), and Vairavamurthy (1998). One of the common methods is to fit the compound spectrum with a linear combination of individual component spectra of different models. All of the model and sample spectra are normalized for their spectral intensities several eV above the absorption edge before they are fit with different components (Waldo et al. 1991, George et al. 1991, Vairavamurthy et al. 1994, Frank et al. 1994, Rompel et al. 1998) (Fig. 8). In another approach, the spectral contributions of the core hole excitations into the continuum are subtracted using an arctangent function (Fig. 8), and the remaining peaks ($1s \rightarrow 3p$) of the composite spectrum are fit with a mixture of Gaussian and Lorentzian peaks representing different forms of sulfur (Huffman et al. 1991, Vairavamurthy 1998, Xia et al. 1998). These methods are summarized by Vairavamurthy et al. (1994), and the problems with the fitting routines are discussed in detail by Hawthorne and Waychunas (1988) and by George and Steele (1995). Principal component analyses can be used to obtain information on the number and the probable type of components in the compound spectrum, and details are reported in several recent papers (George and Steele 1995, Wasserman et al. 1999, Ressler et al. 2000). All of these methods are useful for deconvoluting the compound spectrum of an unknown sample, provided that the sample consists of a mechanical mixture of pure compounds. This is a very important caveat in the XANES analysis.

Several researchers have used XAS as a fingerprinting technique to identify the presence of the different sulfur compounds in complex matrices, and other researchers, as mentioned above, have used a series of simple structural models to quantify the fractions of the different forms of sulfur. Although quantification is not absolute, XAS has provided important *in situ* information on the ratio of oxidized and reduced forms of sulfur in coals and humic substances of different origin, which no other non-destructive sampling method has been able to provide. However, quantification of different sulfur forms should be approached with caution because recent studies have shown that changes in protonation and coordination of the sulfur functional groups can cause significant changes in the sulfur spectrum (Frank et al. 1999, Myneni and Martinez, 1999, Myneni et al. 2000b). Hence a small list of structural models may not be sufficient for the accurate determination of a mixture of compounds and their concentrations in complex unknown samples (which may not be simple mechanical mixtures).

Several steps are involved in EXAFS analysis, and these include (in sequence): (1) subtraction of background from the sample X-ray absorption spectrum, (2) fitting a spline function to the X-ray absorption spectrum and extraction of EXAFS, (3) converting the energy of the EXAFS oscillations into the modulus of wave vector, k (Å^{-1}) (4) computing the Fourier transform of EXAFS in distance-space, and fitting the Fourier transform or the EXAFS with the phase- and amplitude-functions derived theoretically or experimentally for different absorber-backscatterer pairs. Details of EXAFS analysis and its limitations are given by Teo (1986), Brown et al. (1988), and Rehr and Albers (2000).

XANES spectra of sulfate and electronic transitions. X-ray absorption spectra of sulfur-bearing salts were collected in the late 1960s and efforts were made to correlate the spectra with the electronic structure of sulfate and its bonding interactions (Nefedov and Formichev 1968). A comprehensive study of this kind was that of Sekiyama et al. (1986). The sulfate ion in tetrahedral (T_d) symmetry has primarily two unoccupied antibonding molecular orbitals: a_1^* of $3s$ character and t_2^* of $3p$ character. Sekiyama et al. (1986) have assigned the intense feature in the XANES spectrum of solid sodium sulfate to the $1s \rightarrow t_2^*$ orbitals (labeled as A in Fig. 9a lower part; spectral assignments in Table 2). The $1s \rightarrow a_1^*$ transitions are dipole-forbidden because of the s-character in a_1^* orbitals, and do not appear as intense features in the X-ray absorption spectrum. The high-energy features above the absorption edge (labeled as B and C in Fig.9a, lower part) are assigned to the $1s \rightarrow$ unoccupied $3d$-like e^* and t_2^* orbitals. Similar assignments were made for sulfate in

Table 2. Assignments of the absorption spectra of S K-edge of Na_2SO_4, Na_2SO_3, $Na_2S_2O_3$, and $Na_2S_2O_5$ (modified from Sekiyama et al. 1986). The alphabetical labels shown in the third column correspond to the spectral features marked in Figure 9.

Molecule	Energy (eV)	Label	Transition
SO_4^{2-}	2479.9	A	S $1s \rightarrow t_2^*$
	2488.7	B	d-type shape resonance
	2495.6	C	d-type shape resonance
SO_3^{2-}	2475.5	A	S $1s \rightarrow e^*$
	2477.5	B	S $1s \rightarrow a_1^*$
	2478.9	C	d-type shape resonance
	2487.6	D	d-type shape resonance
	2494.6	E	d-type shape resonance
$S_2O_3^{2-}$	2469.2	A	terminal S $1s \rightarrow a_1^*$ (terminal, central S $3p\sigma$)
	2476.4	B	central S $1s \rightarrow a_1^*$ (terminal, central S $3p\sigma$)
	2478.0	C	S $1s \rightarrow e^*$ (central S $3p\pi$)
	2479.8	D	S $1s \rightarrow a_1^*$ (central S $3s$ and $3p\pi$)
	2483.2	E	d-type shape resonance
	2493.2	F	d-type shape resonance
$S_2O_5^{2-}$	2475.8	A	S $1s \rightarrow 3p$ (-SO_2)
	2480.2	B	S $1s \rightarrow 3p$ (-SO_3)
	2489.2	C	d-type shape resonance
	2495.9	D	d-type shape resonance

$CuSO_4 \cdot 5H_2O$ (Tyson et al. 1989). The t_2^* orbitals are considered to be the bound-state orbitals because the energy for the $1s \rightarrow t_2^*$ transition is below the ionization potential. The high-energy broad feature is assigned to the $1s \rightarrow$ continuum state transitions, as this feature is above the ionization potential. However, the transitions labeled at B and C in Fig. 9a (lower part) were not observed in the X-ray absorption spectrum of aqueous sulfate, but appear instead as one broad feature (discussed later in this chapter).

Changes in the symmetry of sulfate from T_d to C_{3v}, C_{2v}, or any other low-symmetry group associated with sulfate complexation, causes the splitting of the triply degenerate, unoccupied t_2^* molecular orbitals into either a_1^* and e^*, or its degeneracy may be completely eliminated. The energy difference among these orbitals is modified primarily by the strength of bonding between sulfate and the complexing atoms (or the overlap of sulfate molecular orbitals with those of the complexing atom). Hence, information on the energy levels of molecular orbitals of sulfate and its complexes can be obtained directly from the XANES spectra. The X-ray absorption spectra of other structurally relevant sulfur compounds in different bonding environments are shown in Figure 9 and Table 2. The electronic transitions and spectral features of different oxoanions, such as SiO_4^{4-}, PO_4^{3-}, and ClO_4^-, are expected to exhibit the same trends (Okude et al. 1999).

Theoretical studies to understand the molecular orbital structure of sulfates have not been conducted extensively when compared to other inorganic systems. The majority of the previous theoretical studies on sulfur have focused on organo-S compounds, such as

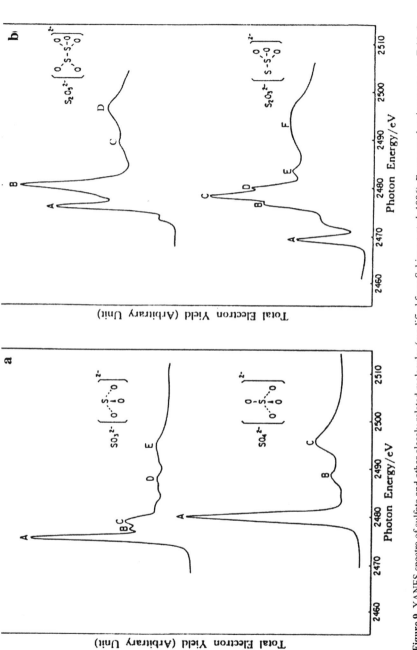

Figure 9. XANES spectra of sulfate and other closely related molecules (modified from Sekiyama et al. 1986). For spectral assignments, see Table 2.

cysteine and thiophene, using multiple scattering X-alpha (MS-Xα), Hatree Fock static-exchange, and density functional theory (Hitchcock et al. 1986, Tyson et al. 1989, Mochizuke et al. 1999, Fernandez-Ramos et al. 2000). The theoretical MS-Xα calculations of the XANES spectra of sulfate in different compounds reproduced the main absorption edge accurately, but not the other spectral features in the vicinity of the absorption edge (Tyson et al. 1989). Lindsay and Gibbs (1988, and references therein) examined the molecular orbital structures of different gas–phase sulfur–oxide molecules, including H_2SO_4, using the 6-31G** bases set. Their vibrational spectral calculations of these molecules were in excellent agreement with the experimental values. However, theoretical studies on solvated species are necessary to understand the influence of solvation on the molecular orbital structure of sulfates and their complexes, which can significantly help in the interpretation of the XANES spectra. Several research groups are currently working on these issues.

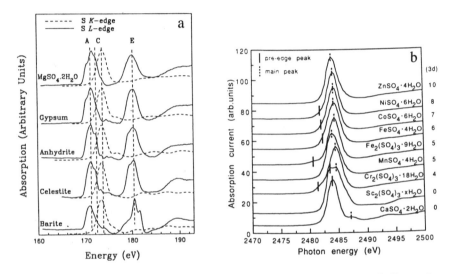

Figure 10. XANES spectra of alkaline earth and transition-metal sulfates: (a) alkaline earth sulfates. Both L-, and K-edge spectra are shown here. The assignments A, C, and E correspond to different L-edge transitions in sulfates (modified from Li et al. 1995); and (b) the pre-edge and absorption edge positions of different transition-metal sulfates (modified from Okude et al. 1999).

XAS of sulfate in solids. XAS studies of sulfate minerals are not as common as those reported for sulfides. Detailed studies on the XANES spectra of alkaline-earth mineral sulfates (Li et al. 1995; Myneni, unpublished data), transition-metal sulfates including several ferric sulfate minerals (Li et al. 1995, Okude et al. 1999, Myneni et al. 2000b, Waychunas, unpublished data), and Al-sulfates (Myneni, unpublished data) were recently conducted using high-resolution spectrometers. Li et al. (1995) compared the K-, and L-edge XANES spectra of the sulfates of Mg (sanderite), Ca (gypsum, anhydrite), Sr (celestine) and Ba (barite), and observed that the K absorption edge is shifted to lower energy and the spectral features above the main absorption edge become more complex as the atomic number of the cation increases. These changes have been attributed to the stronger backscattering of the high-atomic-number elements beyond the first-shell coordination around sulfur (Fig. 10).

The XANES spectra of transition-metal sulfates indicate that the edge position, splitting in the absorption peak, and the energy of deconvoluted peaks are dependent on

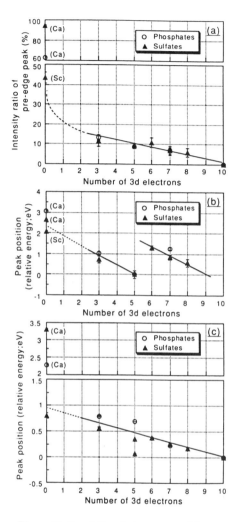

Figure 11. Correlations between spectral features of transition-metal sulfates and phosphates and the number of *d* electrons in the transition metal: (a) intensity ratio of the S pre-edge peak; (b) energy of pre-edge peak; and (c) energy of the absorption edge. (modified from Okude et al. 1999).

the number of *d*-electrons in the transition metal (Figs. 10-12). All of the transition-metal sulfates, with the exception of $ZnSO_4 \cdot 4H_2O$ (boyleite) show a pre-edge feature whose intensity decreases with an increase in the number of *d*-electrons (Fig. 10; Okude et al. 1999). The pre-edge peak overlaps the main absorption edge peak and does not appear as a distinct feature in spectra for some of the transition–metal sulfates (e.g. $FeSO_4 \cdot 7H_2O$, melanterite). An increase in the number of *d* electrons in the transition metal causes their pre-edge to shift to lower energies, with a gap between d^5 and d^6 metal sulfates (Figs. 11, 12). These results indicate that the pre-edge features in these spectra originate from the hybridization of metal 3*d* orbitals with sulfur 3*p* states (Shadle et al. 1995, Rose Williams et al. 1997, Okude et al. 1999). Mineral sulfides and transition-metal chloro salts also exhibit the same behavior (Shadle et al. 1995, Rose Williams et al. 1997, Wu et al. 1997).

Detailed XANES spectroscopic studies are in progress on the naturally occurring Fe-sulfate minerals common to acid mine drainage environments (Myneni and Alpers, unpublished data; Waychunas, unpublished data). These include minerals such as melanterite, jarosite, copiapite, coquimbite, römerite, rhomboclase, voltaite, and halotrichite, some of which are shown in Figure 12. The structure and chemical composition of these minerals are discussed by Hawthorne et al. (Chapter 1, this volume), and Jambor et al. (Chapter 6, this volume). These XANES studies of S in Fe-sulfates indicate that the spectra are sensitive to sulfate coordination to Fe polyhedra and the oxidation state of Fe. There are primarily two effects associated with complexes of Fe: (1) shift in the absorption maximum to high-energy by about 0.5 eV, and (2) appearance of pre-edge features when compared to that of aqueous sulfate. These changes are noticed only where there is a direct linkage of sulfate group to the Fe polyhedra. The pre-edge feature in ferrous salts overlaps the sulfur absorption edge and appears as a low-energy shoulder (Figs. 11, 13). In addition, the pre-edge feature in Fe^{3+} sulfates is present at about 2 eV lower energy than that of the Fe^{2+} salt, halotrichite. Such shifts are also noticed in the K absorption spectra for Cl in Fe^{3+}- and Fe^{2+}-chloride and thiolate salts (Shadle et al. 1995, Rose Williams et al. 1997). Also, the pre-edge

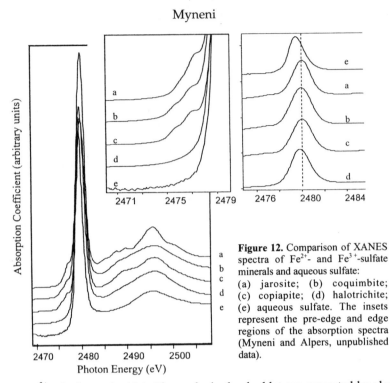

Figure 12. Comparison of XANES spectra of Fe^{2+}- and Fe^{3+}-sulfate minerals and aqueous sulfate: (a) jarosite; (b) coquimbite; (c) copiapite; (d) halotrichite; (e) aqueous sulfate. The insets represent the pre-edge and edge regions of the absorption spectra (Myneni and Alpers, unpublished data).

feature in Fe^{3+} salts has a doublet. The peaks in the doublet are separated by about 2.0 eV, with the highest energy feature 3.0 eV below the absorption maximum (Fig. 12). The energy of the pre-edge feature in Fe^{3+}-sulfate salts is also sensitive to the number of Fe polyhedra connected to each sulfate polyhedron. In coquimbite and copiapite, each sulfate group is connected to two Fe atoms, and in jarosite each sulfate is connected to three Fe atoms (Myneni et al. 2000b). Accordingly, coquimbite and copiapite exhibit pre-edge features at the same energy, approximately 0.5 eV below that of jarosite. However,

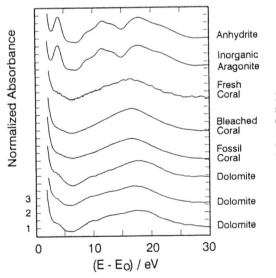

Figure 13. XANES spectra of sulfur in biogenic carbonates (modified from Pingitore et al. 1995). The spectral components just above the S absorption edge are shown.

these pre-edge features disappear when direct interactions between sulfate and the Fe^{3+} polyhedra are absent, or when H-bonds are present between Fe-coordinated water and sulfate, such as in ferric ammonium sulfate (Myneni et al. 2000b). Theoretical studies are necessary to understand the correlation between the changes in the energy levels of molecular orbitals and the change in the linkages in the Fe-sulfate polyhedra. In addition to these pre-edge features, electronic transitions to the continuum state, or the multiple scattering features 10-15 eV above the absorption edge, also vary with the sulfate coordination to cations (Fig. 12). Salts in which sulfate is not bound to any cations have smooth features (e.g. sulfate in aqueous solutions, ettringite, ferric ammonium alum) compared to those of the cation-complexed sulfates. Although Al sulfates, such as alunite, alunogen, basaluminite, and ettringite, follow the trends described above, these minerals do not exhibit any pre-edge features (Myneni unpublished data).

Using the XANES spectroscopy of S in modern and ancient corals, Pingitore et al. (1995) suggested that the form of sulfur in these samples is present almost exclusively as sulfate (Fig. 13). Spectral comparison of coral samples with those of Ca sulfates, such as anhydrite and gypsum, and sulfate in dolomite indicated that sulfate in biogenic carbonate samples is not present in the form of Ca sulfate minerals. Sulfate coordination is similar in all of the aragonite corals Pingitore et al. (1995) examined. Based on this information, they proposed that sulfate substitutes for carbonate in biogenic calcite, aragonite and dolomite and thus influences sulfate chemistry in oceans.

There are no published reports on the EXAFS spectroscopic studies of sulfate salts or sulfate minerals. Recently, EXAFS studies have been conducted to examine sulfate coordination in Fe^{3+} sulfates and Al hydroxy sulfates commonly found in acidic mine drainage environments, CaAl hydroxysulfates in cements, and sulfate complexes in aqueous solutions and at the mineral water interface (Waychunas, Myneni, unpublished data). Because several Al- and Fe hydroxysulfates are poorly crystalline (see Bigham and Nordstrom, this volume), EXAFS studies of sulfur in these materials are useful in understanding their structure and geochemistry.

XANES spectra of aqueous sulfates are reported by Vairavamurthy et al. (1994), Frank et al. (1994), and Myneni et al. (2000a); a detailed analysis of these spectra is presented by the last authors. Sulfate exists in T_d symmetry in aqueous solutions if the hydrated water is excluded from symmetry predictions (see later discussion on vibrational spectra). The XANES spectrum of aqueous SO_4^{2-} has a sharp absorption edge, corresponding to the $1s \rightarrow t_2^*$ transitions, and its energy is the same as that of solid Na- and K-sulfate salts (Figs. 12, 14). A broad feature 10 eV above the absorption edge in the XANES spectrum of SO_4^{2-} corresponds to continuum-state transitions. However, when SO_4^{2-} is complexed with protons or metals, the energy levels of molecular orbitals change and the degeneracy of orbitals is lost. Such changes in SO_4^{2-} coordination may significantly influence its XANES spectral features, and these are shown for sodium salts of SO_4^{2-}, HSO_4^-, $COSO_3$ (organo-sulfates), and $S_2O_8^{2-}$ (persulfate) in Figure 14. With the exception of sodium sulfate, the sulfate group in all other salts is bonded to a proton, carbon, or another oxygen, and accordingly the main absorption edge is split into two peaks. The amount of splitting varies with the type of atom connected to the sulfate group and it is highest in persulfate (Fig. 14).

Protonation of a sulfate polyhedron causes changes in the SO bond length, which increases from 1.473 Å in SO_4^{2-} (uncomplexed S–O) to 1.58 Å in HSO_4^- (SOH). The XANES spectrum of aqueous HSO_4^- ion (e.g. aqueous solutions of $NaHSO_4$) exhibits two overlapping features, which are separated by approximately 1.5 eV, with the peak maximum about 0.5 eV higher than that of aqueous sulfate (Frank et al. 1994, Myneni and Martinez, 1999). The shift to higher energy is caused by the splitting of triply

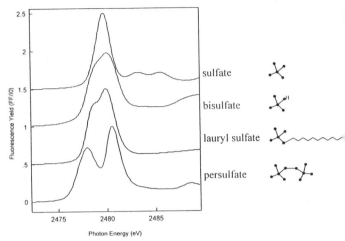

Figure 14. XANES spectra of sulfur in covalently bonded sulfate compounds. The sulfate spectrum shown at the top is that of sulfur in solid potassium sulfate. All other salts are Na salts (from Myneni and Martinez 1999).

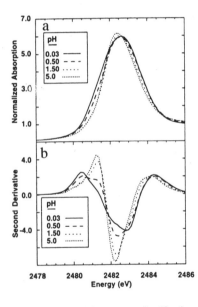

Figure 15. XANES spectra of sulfur in sulfate at different pH values: (a) raw XANES spectra; and (b) second derivative of spectra shown in (a) (modified from Frank et al. 1994).

degenerate t_2^* orbitals into a_2^* and e^* associated with the protonation of SO_4^{2-} (Figs. 14, 15). The two peaks in the XANES spectrum correspond to electronic transitions to these two unoccupied states. It should be noted that these a_2^* orbitals have significant sulfur $3p$ character, and hence the $1s \rightarrow a_2^*$ transitions appear as an intense feature (in contrast to the $1s \rightarrow a_1^*$ transitions discussed above). The XANES spectra of organo-sulfates and metal-complexed sulfates are expected to show the same behavior and are discussed later in this chapter. Frank et al. (1994) titrated a solution of sulfuric acid to its second pK_a (= -$\log K_a$ ~ 2.0, where K_a is the dissociation constant), and their results indicate that the XANES spectra are broader and the peak maximum shifts to higher energy by almost 0.5 eV as the protonation of sulfate increases in aqueous solutions (Fig. 15). The peak broadening is caused by the splitting of the absorption edge and the formation of HSO_4^- close to the second pK_a of sulfuric acid.

The spectroscopic studies of Frank et al. (1994) on vanadium(III)-sulfate complexes in acid solutions showed that the absorption edge of aqueous sulfate becomes broader, with distinct low- and high-energy features (0.7 eV above the peak maximum of aqueous SO_4^{2-}), as the concentration of vanadium(III) increases in aqueous sulfate solutions (1 M sulfate + dilute sulfuric acid, Fig. 16). Frank et al. (1994) assigned these spectral changes to the formation of a VSO_4^+ complex. Additions of Mn^{2+} and NH_4^+ ions to the same solutions (in the absence of vanadium) did not produce any changes in the XANES spectrum of SO_4^{2-}, which indicate that the Mn^{2+} and NH_4^+ ions do not form inner-sphere complexes with SO_4^{2-}.

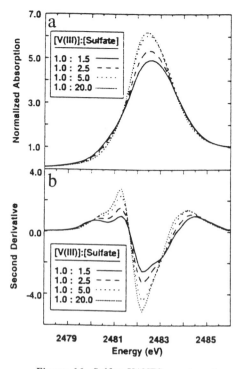

Figure 16. Sulfur XANES spectra of aqueous V(III)-sulfate solutions: (a) raw XANES spectra; (b): second derivative of spectra shown in (a) (modified from Frank et al. 1994).

Myneni et al. (2000a) examined sulfate complexation with Fe^{3+} ions in acid solutions. The X-ray absorption spectra of sulfur for mixtures of Na-sulfate and Fe^{3+}-chloride aqueous solutions (at different ratios and pH values) show a shift in the peak position of SO_4^{2-} to higher energy, and a small pre-edge feature appears when compared to that of aqueous sulfate. These features are similar to those reported earlier for Fe^{3+}-sulfate complexes in ferric sulfate salts discussed above (Myneni et al. 2000b). These studies indicate that Fe^{3+} forms complexes with sulfate in aqueous solutions, but do not distinguish monomeric- and polymeric-Fe^{3+} sulfate complexes (note that Fe^{3+} polymerizes in acidic solutions; Stumm and Morgan 1982, Masion et al. 1997, and references therein). Hence, the percentages of SO_4^{2-} bound to different Fe^{3+} forms and of free SO_4^{2-} in aqueous solutions are difficult to estimate from XAS studies alone. Although vibrational spectroscopic studies support the XAS observations, the question related to the percentage of free and complexed sulfate in aqueous solutions still remains (Hug 1997, Myneni et al. 2000b). Detailed XANES studies together with EXAFS spectroscopy are necessary to evaluate the Fe^{3+} complexes in dilute acidic sulfate solutions. When the pH of Fe^{3+}-containing solutions is increased, Fe oxides and sulfates precipitate from solution, and the nature of aqueous and colloidal Fe^{3+}-sulfate complexes can not be examined unambiguously above pH 4.0.

The above-mentioned studies indicate that the formation of aqueous sulfate complexes and their electronic structure can be studied directly using XAS spectroscopy. Such information on the electronic structure of sulfate, and its modifications associated with protonation and metal complexation, is useful for predicting the geochemistry of sulfate complexes and their redox reactions in aquatic systems.

XAS of organo-sulfates. Sulfate molecules bonded to C atoms in organic molecules, such as organo-sulfates and sulfate-esters, can exist at high concentrations in biological molecules and in the form of humus in soils and aquatic systems (Sposito 1989, Stevenson 1994). Organo-sulfates have been identified as one of the dominant functional groups in humic substances that were collected from different sources around the world (Morra et al. 1997, Vairavamurthy et al. 1997, Xia et al. 1998, Myneni and Martinez 1999). Because of the high aqueous solubility of organo-sulfates and their ability to form micelles, several of these compounds are used as cleaning agents in industrial and household applications. For these reasons, several organo-sulfates are commonly found in wastewater from industrial and residential areas. Organo-sulfur compounds, especially the oxidized forms of sulfur, play a major role in atmospheric chemistry and are also expected to represent a significant organic fraction in aerosol particles (Warneck 1988).

By using spectral deconvolution methods and the spectral features of structurally simple organo-sulfates, several organo-sulfates and sulfonates have been identified as the dominant oxidized forms of sulfur in soils and sediments (Figs. 8, 17). However, several of the previous studies were conducted at low resolution or failed to identify a low-

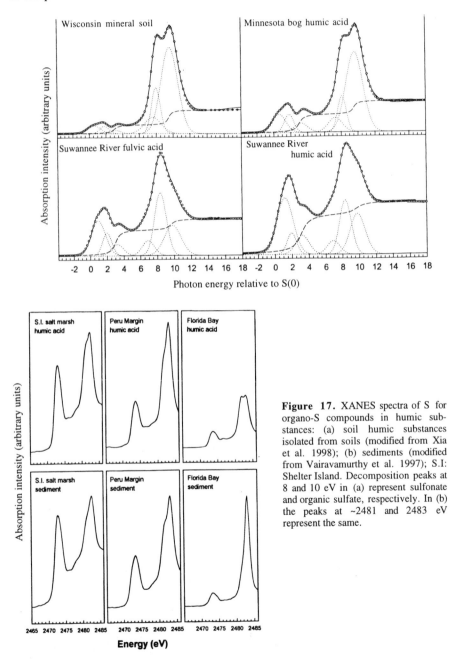

Figure 17. XANES spectra of S for organo-S compounds in humic substances: (a) soil humic substances isolated from soils (modified from Xia et al. 1998); (b) sediments (modified from Vairavamurthy et al. 1997); S.I: Shelter Island. Decomposition peaks at 8 and 10 eV in (a) represent sulfonate and organic sulfate, respectively. In (b) the peaks at ~2481 and 2483 eV represent the same.

energy feature below the absorption edge of sulfate in organo-sulfates (Fig. 14). Myneni and Martinez (1999) reported that this feature arises from the bonding of the SO_4^{2-} group to a carbon atom. The XANES spectral features of organo-sulfates are similar to those of protonated sulfates, and the sulfur absorption edge is split into two distinct features (Fig. 17). Attachment of a carbon atom to one of the oxygen atoms of the sulfate group, as in –C–OSO$_3$ (e.g. lauryl sulfate) changes sulfate symmetry from T_d to C_{3v}, and the $1s \rightarrow t_2^*$ characteristic of uncomplexed SO_4^{2-} splits into $1s \rightarrow a_2^*$ and e^* in –C–OSO$_3$ (Fig. 14). Of particular interest is that the energies of the absorption edge of sulfonate and the low-energy feature in organo-SO_4^{2-} are similar. A detailed examination of several surfactant molecules, such as decyl sulfate, dodecyl sulfate, laureth sulfate, and dodecyl benzene sulfonic acid, also indicated splitting in the absorption edge of sulfur (Myneni, unpublished data). Hence, the low-energy feature (~2478 eV) adjacent to the sulfate peak in humic materials and coals, identified as sulfonate by previous researchers (Morra et al. 1997, Vairavamurthy et al. 1997, Xia et al. 1998), may not necessarily represent sulfonates entirely (Fig. 17). Although researchers used different chemical means (e.g. by treating the samples with Ba-salts, such as barium trifluoroacetate) to separate sulfonate and organo-sulfate, they are not easily distinguishable using the X-ray absorption spectra alone.

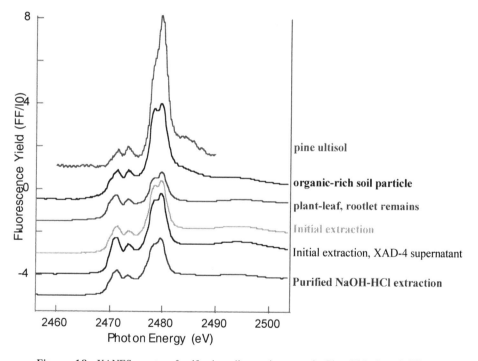

Figure 18. XANES spectra of sulfur in soil organic matter in Pine Ultisol, and different isolated humic substance fractions (from Myneni and Martinez 1999).

In contrast to isolated humic materials, fresh and undisturbed humic substances in soil exhibit different behavior for the ratio of oxidized to reduced sulfur species (Fig. 18). The Pine Ultisol from Puerto Rico (Myneni and Martinez 1999) showed that the percentage of the oxidized forms of sulfur is much higher than that of the reduced forms (Fig. 18). The XANES studies also indicated that the oxidized forms of sulfur (sulfonate, organo-sulfate, inorganic sulfates) are lost during extraction of humic substances from soils and that the isolated fraction preferentially retains the reduced-sulfur forms (Fig. 18; Myneni and Martinez 1999). The reported high concentrations of reduced-sulfur forms relative to sulfate and sulfonate in soils of oxidizing environments may be attributed to this preferential isolation.

XAS of sulfate at interfaces. When sulfate adsorbs on mineral surfaces, it can form (1) outer-sphere complexes, in which the sulfate and mineral surface interact through intermediate water molecules; and (2) inner-sphere complexes, in which sulfate bonds with cation polyhedra directly on mineral surfaces (Sposito 1994, Brown et al. 1999). The inner-sphere sulfate-cation complexes can be monodentate-mononuclear, monodentate-binuclear, bidentate-mononuclear, and bidentate-binuclear complexes. For several years, researchers have been using a variety of molecular methods to identify the types of complexes at the mineral–water interface and in aqueous solutions (Vaughan and Pattrick 1995, Brown et al. 1999). The results indicate that the coordination chemistry of sulfate complexes on surfaces is not understood well, and is widely debated in the case of several mineral surfaces. Although XAS has been used successfully to resolve uncertainties for several complexes (Brown et al. 1999), only recently has this technique been used to study the coordination chemistry of sulfate at the interfaces (Myneni et al. 2000a). The vibrational spectroscopy portion of this chapter contains complementary information on sulfate reactions at interfaces.

Using XAS, Myneni et al. (2000b) have examined sulfate complexes on goethite, ferrihydrite, and hematite at different pH values and sulfate concentrations. These studies indicate that the XANES spectrum of S in sulfate adsorbed on goethite at pH > 3.5 is similar to that of aqueous SO_4^{2-}. When solution pH is <3.5, the absorption maximum shifts to higher energy by about 0.2-0.3 eV, and a pre-edge appears approximately 3 eV below the absorption edge. The pre-edge contains two overlapping features, and their energies are the same as those reported for coquimbite and copiapite (Fig. 19). These results indicated that SO_4^{2-} interacts primarily with goethite surfaces as outer-sphere and

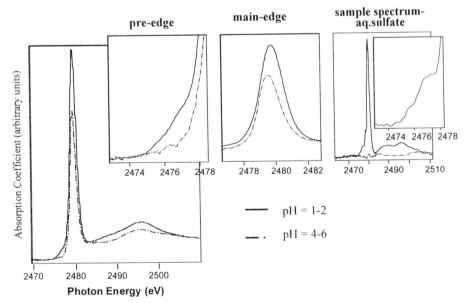

Figure 19. S-XANES spectra of sulfate sorbed on goethite at different pH values (from Myneni et al. 2000b, unpublished data).

H-bonded complexes at pH > 3.5. Below this pH, SO_4^{2-} forms both inner- and outer-sphere complexes on goethite surfaces. From partial least-squares fit of these sample spectra, the inner-sphere complexes were around 35% of total sorbed sulfate at pH ~ 2.0. Although evidence for inner-sphere complexes was not found at higher pH, their presence may not be ruled out completely, as their concentration may be too low to be detected with XAS. In contrast to goethite, sulfate adsorbed on hematite exhibits significant inner-sphere complexation at neutral pH (Myneni et al. 2000b). Although the energies of pre-edges of adsorbed SO_4^{2-} are similar to that reported for copiapite and coquimbite, a detailed investigation is necessary to understand the nature and origin of the pre-edge features and their sensitivity to the sulfate coordination. XANES studies have also been conducted to examine the interactions of sulfur oxide and its oxidation to sulfite and sulfate on several catalyst surfaces (Rodriguez et al. 1999a,b, and references therein).

X-ray absorption spectroscopy at the sulfur L-edge

Electronic transitions from 2s and 2p states ($2p_{1/2}$, $2p_{3/2}$) to unoccupied molecular orbitals, similar to XAS at the K-edge associated with 1s electrons, can also be studied. These transitions occur at 230.9, 163.6, 162.5 eV for elemental sulfur, and these edges are termed L_1, L_2, and L_3, respectively. The L_2 and L_3 edges are commonly studied in addition to the K-edge because of different selection rules for the electronic transitions. The features in the sulfur L_2- and L_3-edge spectra are assigned to the electronic transitions of sulfur from 2p to unoccupied 4s or 3d states. Both of these transitions are dipole allowed and produce significant structure in the L-edge spectra. In addition, the intrinsic core hole lifetime broadening is much smaller for L-edges than for K-edges, and this makes the spectral resolution better than 0.1 eV at the L-absorption edge of sulfur (Brown et al. 1988). Because the X-ray penetration into samples is very small at the L-absorption edge of sulfur (<100 nm in concentrated samples), very high surface sensitivity can be achieved using L-edge spectroscopy. Kasrai et al. (1996) indicated that sampling depth is around 5 nm at the L-edge, compared to 70 nm at the K absorption edge of sulfur in electron yield mode.

The synchrotron instrumentation used to examine L-edges is different from that used for K-edges. Until recently, L-edge spectroscopic studies were limited to vacuum conditions. Typically, diffraction gratings are used instead of monochromator crystals (as at the K-edges) for obtaining monochromatic light from synchrotrons. As the L-edge for S has extremely small fluorescence yield relative to its K-edge, the L-edge XAS studies are done in electron-yield mode. Stöhr (1992) gives more details on synchrotron experimentation at the L-edges (detection schemes, detectors, data collection). Several soft X-ray synchrotron facilities have beamlines that can provide photons at the L-absorption edge of S, but only few beamlines are optimized to do this under ambient conditions (Myneni et al. 2000a). The low fluorescence yield at the L-edges of S, and photon absorption by air in the soft X-ray region restrict the applications of L-edge spectroscopy to aqueous solutions and wet mineral and biological samples. However, recent developments at spectromicroscopy facilities at the ALS and NSLS (Table 1) allow L-edge spectroscopy studies on thin liquid films, mineral pastes, and biological samples under ambient conditions. Specialized beamline end-stations, such as the Soft X-ray Endstation for Environmental Research (SXEER) at the ALS, are optimized to examine dilute samples in electron- and fluorescence-yield modes and in transmission (Myneni et al. 2000a).

Sulfur L-edge spectra of a series of compounds indicate that the L-edges shift to higher energies with increase in the oxidation state (Fig. 20). Significant spectral differences also exist among chemical compounds that are closely related, especially for

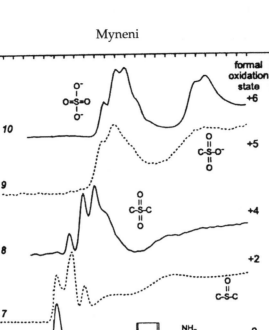
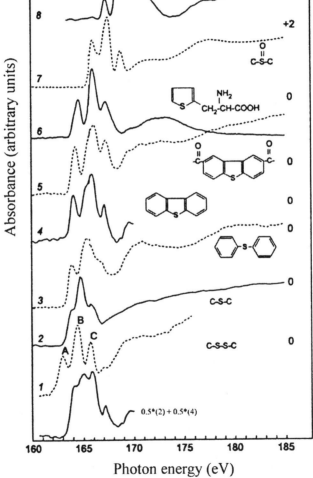

Figure 20. L-edge XANES spectra of S in different redox states and coordination environments (modified from Sarret et al. 1999). The bottom spectrum is a combination of spectra (2) and (4).

reduced-sulfur forms (Kasrai et al. 1994, Sarret et al. 1999). A majority of the previous sulfur L-edge studies were done on transition-metal sulfide minerals and on SO_x reactions on catalyst surfaces (Li et al. 1994, 1995, and references therein). The only study that has reported L-edge spectra of mineral sulfates was conducted by Li et al. (1995), who indicated that the L-edge spectra are highly sensitive to the local coordination in sulfates. In the case of alkaline-earth sulfate minerals, the L-edges shift to lower energies with an increase in the atomic number of the complexing atom (Fig. 20). In transition-metal sulfates, all have L-edges at 171.5 (±0.1) eV, except for $MnSO_4$ (171.8 eV; Li et al. 1995). To the author's knowledge, sulfur L-edge spectroscopic studies in aqueous samples have not been conducted. However, with the availability of the third-generation synchrotron sources, and end-stations optimized to examine samples under ambient conditions, new information is expected in the near future.

X-ray photoelectron spectroscopy

In X-ray photoelectron spectroscopy (XPS), the spectral properties of ejected photoelectrons from a photon-, or electron-absorbing atom are studied (Fig. 2). Although incident X-rays can penetrate much farther than electrons at the same incident energy, released photoelectrons have a much shorter mean-free path and hence XPS can detect only the photoelectrons from surface layers. XPS provides information on the density of occupied electronic states in an atom, and is complementary to the XAS techniques discussed above. When XPS is used as a core hole spectroscopy, the XPS spectra are characteristic of photon-absorbing atoms (as in XAS). XPS spectra also include information on the absorber coordination, as the coordinated atoms can modify the energy of the core hole electrons and the scattering of the released photoelectrons. Because the spectral properties of an electron are investigated in this method, these studies are conducted in high vacuum ($\sim 10^{-6}$ torr or less). However, the availability of high-flux bright X-ray sources at the third-generation synchrotron sources (e.g. the ALS) has led to the development of relatively high-pressure XPS chambers and differentially pumped sample chambers that operate at pressures of 5-10 torr (Ogletree et al. 2000). Some of the preliminary work on XPS of less volatile liquids at moderate pressures ($\sim 10^{-5}$ torr) was conducted by Siegbahn in the late 1970's. The interpretation of the XPS spectra of unknown compounds is typically done by comparing the sample spectra with those of known models. Libraries of XPS spectra of different compounds are available for comparison. However, the interpretation is difficult if the materials have open-shell ions.

Although XPS has not been used in studying mineral sulfates or their interfaces, it has been used for examining the surface chemistry of sulfides. The oxidation of SO_x and sulfate interactions with catalyst surfaces have also been investigated extensively using XPS (e.g. Jirsak et al. 1999, Rodriguez et al. 1999a,b; and references therein). These studies indicate that the $2p$ spectra of S in different oxidation states are separated by several eV; thus, the oxidation state and the bonding environment of different S groups can be identified easily using XPS. Watanabe et al. (1994) used XPS and Fourier transform infrared (FTIR) spectroscopic methods to examine sulfate interactions with surfaces of hematite and maghemite. The $2p$ spectra of S exhibit two overlapping features separated by approximately 2 eV (Fig. 21). The $2p$ spectrum of S in sulfate on maghemite (γ-Fe_2O_3) surfaces are 0.6 eV lower than on hematite (α-Fe_2O_3), which indicates differences in the sorption sites of these mineral surfaces (Fig. 21). Watanabe et al. (1994) attributed the doublet in the XPS spectra of sulfur to sulfate bonded to hematite and maghemite surfaces at two distinctly different sites, and suggested that these sites are less polar in the case of hematite when compared to those of maghemite.

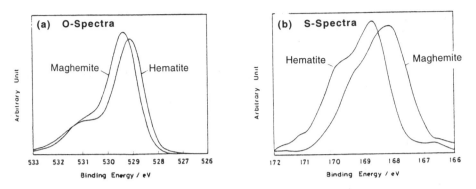

Figure 21. XPS spectra of sulfate sorbed on hematite and maghemite: (a) O $1s$ XPS spectra; (b) $2p$ XPS spectra of sulfur in sulfate (from Watanabe et al. 1994).

X-ray imaging and spectromicroscopy

Obtaining spatially resolved spectroscopic information selectively on different regions of interest in a heterogeneous sample and conducting microscopy based on chemical (or spectral) contrast of different regions of a heterogeneous sample are useful in the examination of spatial heterogeneity in geological samples (Schulze et al. 1999, Myer-Ilse et al. 2000). Such spectromicroscopy studies facilitate the examination of dilute samples and detailed micro-structures, which are otherwise not detectable by conventional molecular probes. Although spectromicroscopy studies of sulfur have not yet been reported for geological samples, the recent availability of these facilities at the third-generation synchrotron sources will help in the examination of a variety of samples of geological and biogeochemical significance. These facilities are operational for the sulfur L-edge spectromicroscopy at the ALS and NSLS, and are under construction for the sulfur K-edge spectroscopy at the ALS and APS (Table 1). The spectromicroscopy facilities are based on focusing the X-ray beam to spot sizes smaller than a micrometer using mirrors, or to better than 100 nm using zone-plate optics (Warwick et al. 1998). For studying the L-edges, the microscopes with zone-plate optics are the only available option. The detectors and other optics are the same as those discussed for XAS above.

Spectromicroscopy studies of heterogeneous samples can also be conducted using XPS, and these are well developed at the synchrotron sources (e.g. ALS). At the third-generation synchrotrons, spatial resolution better than one micrometer can be obtained on mineral surfaces. Researchers have been using the bright beams of the synchrotron sources to develop photoelectron emission microscopy (PEEM), to obtain a spectral resolution better than 50 nm (Anders et al. 1999). This is a photon-in, electron-out technique and is sensitive to surfaces. Research is in progress, and future developments in PEEM may improve the spatial resolution to better than 20 nm. Although these methods are relatively new, they have potential applications in studying the sulfate geochemistry of heterogeneous geological surfaces.

VIBRATIONAL SPECTROSCOPY

Vibrational spectroscopy was developed at the beginning of the 1900s and has been used ever since to study the structural properties of materials. Several books have been written on the theory and interpretation of vibrational spectroscopy, and on the fundamental vibrations of several molecules (e.g. Farmer 1974, Nakamoto 1986, Colthup et al. 1990, Ferraro and Nakamoto 1994, Suetaka 1995). Vibrational transitions in

Table 3. Comparisons of IR and Raman spectroscopy
(modified from Fadini and Schnepel 1991)

Parameter	IR spectroscopy	Raman spectroscopy
Photon interactions	Absorption	Scattering
Vibrational excitations	Polychromatic IR radiation	Monochromatic radiation
Energy measurement	Absolute	Relative to excitation energy
Requirement for the activity of a vibration	Change in dipole moment, $\partial\mu/\partial Q \neq 0$	Change in polarizability, $\partial\alpha/\partial Q \neq 0$
Intensity of the bands	$I \propto (\partial\mu/\partial Q)^2$	$I \propto (\partial\alpha/\partial Q)^2$
Representation of the spectrum	Absorption, logarithmic 'downwards'	Scattering intensity, linear 'upwards'
Preferred technique for	Routine gas analysis	Aqueous solutions, single crystals, polymers

molecules can be studied using infrared (IR) and Raman spectroscopic methods, which are complementary. Together they provide complete information on vibrational transitions. The fundamental differences between these methods are shown in Table 3 and Figure 22.

In IR spectroscopy, molecules absorb photons when the energy of the incident light is equal to the energy difference between the vibrational ground state and the high-energy states. The probability of these transitions is higher when there is a change in the dipole moment of a molecule (Fig. 23). In Raman spectroscopy, vibrational transitions are examined by measuring light that is inelastically scattered by the sample. When high-energy light is incident on a molecule, a majority of it is scattered elastically (Rayleigh scattering) by the molecule. A small fraction (<0.001) of the incident light is scattered inelastically by the sample and causes small features in the energy spectrum of the scattered light. These transitions can represent either Stokes or anti-Stokes conditions (Fig. 22). Such vibrational transitions occur in Raman spectra when there is a change in the polarizability of molecules (Fig. 23).

The total number of vibrations in a molecule is $3n-6$, where n is the total number of atoms in a molecule ($3n-5$ for linear molecules). Although IR and Raman spectroscopic methods can provide information on the majority of vibrational transitions in a molecule, several transitions are not detected by either method. Selection rules are helpful in determining whether a vibrational transition is allowed or forbidden for either of these methods, and a detailed review of these is provided by Cotton (1971), Farmer (1974), Nakamoto (1986), Colthup et al. (1990), and in several books on molecular spectroscopy. Vibrational frequencies of a molecule are related to their bond strength and atomic masses, and the observed shifts in vibrational spectra represent changes in either of the two or in both of these variables. Changes in spectral intensity are related to the concentration of molecules with identical geometry, and to changes in molecule dipole moment (in the case of IR) or polarizability (in Raman spectroscopy). Applications of IR and Raman techniques to geological materials are discussed in detail by McMillan and Hess (1988), McMillan and Hofmeister (1988), and Johnston (1990).

Theoretical vibrational spectra can be calculated using quantum mechanics and semi-empirical methods (Stewart 1989a,b; Seeger et al. 1991, Tossell and Vaughan 1992). The calculated spectra are useful in identifying the experimental spectra of sulfate in different environments.

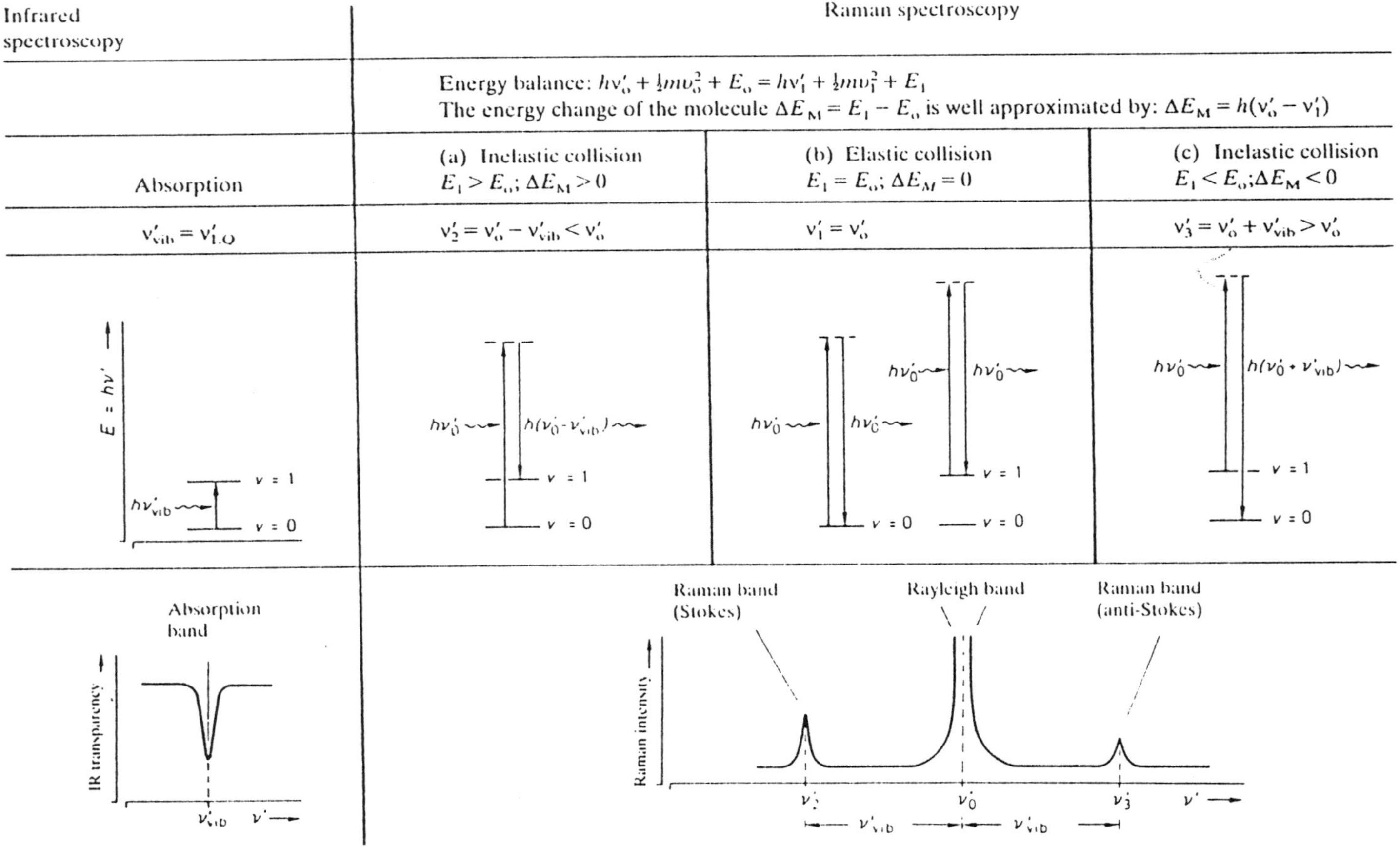

Figure 22. Vibrational transitions and the origin of vibrational spectra in infrared (IR) and Raman spectroscopy. $v = 0$ vibrational ground state, $v = 1$ first excited state (modified from Fadini and Schnepel 1989).

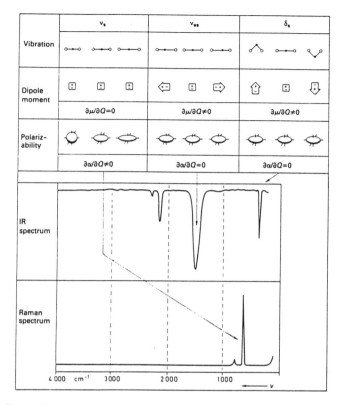

Figure 23. Raman and infrared spectra of CS_2. Changes in molecule dipole moment and polarizability for each of the vibration are also shown with their peak intensities in Raman and IR (from Fadini and Schnepel 1989).

Symmetry and vibrational modes of sulfate and its complexes

The vibrational spectral features, nomenclature and interpretation for arsenate is discussed in detail by Myneni et al. (1998b); parts of that discussion are presented here to clarify the spectral interpretation on sulfates. Sulfate is a tetrahedral (T_d) molecule with an average S–O bond length of 1.473 Å (Hawthorne et al., Chapter 1, this volume). Variation in solvation (coordination of water molecules at different oxygen atoms of sulfate), metal complexation, and protonation of sulfate can modify the S–O bond lengths, and cause changes in symmetry from T_d to either C_{3v}/C_3 (monodentate, corner-sharing), C_{2v}/C_2 (edge-sharing, bidentate binuclear), or C_1/C_s (corner-sharing, edge-sharing, bidentate binuclear, multidentate) (Table 4). Such changes in the bond length and the symmetry of sulfate molecules may shift the vibrational bands to different energies and cause the degenerate vibrations to become nondegenerate. Changes in the vibrational spectrum of sulfate associated with symmetry changes can be interpreted using group theory (Table 4; Cotton 1971, Farmer 1974, Nakamoto 1986).

In the infrared region, SO_4^{2-} exhibits nine normal modes: an A_1 (symmetric stretch, v_1), an E (symmetric bending, v_2), and two F (asymmetric stretching and bending, v_3 and v_4, respectively); the symbols A_1, E, and F correspond to non-, doubly- and triply-degenerate vibrations, respectively. In physical chemistry and spectroscopy literature, both of these notations (A, E, F; v_1, v_2,...) are commonly used. The subscript (e.g. v_1) is

Table 4. Normal modes of sulfate in different symmetries (modified from Farmer 1974)

Site symmetry of sulfate	Vibrational Mode & symmetry class			Activity		No. of IR active bands
	ν_1	ν_2	ν_3, ν_4	R active only	IR and R active	
T_d	A_1	E	F_2	A_1, E	F_2	2
D_{2d}	A_1	A_1+B_1	B_2+E	A_1, B_1	B_2, E	4
S_4	A	A+B	B+E	A	B, E	5
D_2	A	2A	$B_1+B_2+B_3$	A	B_1, B_2, B_3	6
C_{2v}	A_1	A_1+A_2	$A_1+B_1+B_2$	A_2	A_1, B_1, B_2	8
C_2	A	2A	A+2B	-	A, B	9
T	A	E	F	A, E	F	2
C_{3v}	A_1	E	A1+E	-	A_1, E	6
C_3	A	E	A+E	-	A, E	6
C_s	A'	A'+A''	2A'+A''	-	2A', A''	9

used to distinguish different vibrations, and has no relation to the symmetry of a molecule. As the symbol 'ν' has been used by several researchers to represent stretching vibrations only, some of the interpretations in the literature may be confusing unless the notation is clearly specified. The bending vibrations occur at low wavenumbers where several molecules exhibit both vibrational and rotational modes and hence it is difficult to distinguish sulfate-bending vibrations from others in heterogeneous geological materials. Although several researchers have assigned different bands to bending vibrations of sulfate in complex matrices, readers should be cautious about these assignments and interpretations. For these reasons, only the stretching vibrations are focused in this Chapter, and the symbols, ν_s and ν_{as} have been chosen to represent the symmetric and asymmetric stretching vibrations (Myneni et al. 1998b). The vibrating atom-pair has also been specified to avoid misinterpretations (e.g. ν_s of S−OH represents the symmetric stretching vibrations of the S−OH bond).

In association with changes in SO_4^{2-} symmetry and coordination, the A_1 band may shift to different wavenumbers, and the degenerate E and F modes may give rise to several new A_1, B_1, and/or E vibrations (Table 4). In many instances the apparent structural symmetry of SO_4^{2-} species in solutions, crystals, or in the vicinity of interfaces may not follow closely the above-described symmetry predictions. Such discrepancies may be caused by: (1) extensive H-bonding of the non-bonding O atoms of SO_4^{2-} with several H_2O molecules, (2) SO_4^{2-} polydentate complexation with cations, or (3) tilt of the SO_4^{2-}-complexed cation (or proton) away from the S−O−M axis (typically considered to be linear in symmetry analysis). For these reasons, the presence of high-symmetry species in samples may not be excluded if the vibrational spectra exhibit all SO_4^{2-}

fundamental vibrations. Similarly, the appearance of single, unsplit peaks for degenerate SO vibrations do not completely rule out the presence of low-symmetry species in the samples. The symmetry of an SO_4^{2-} group can be very low, in fact as low as C_1, but not result in enough peak splitting to be identified. In summary, disparities between the symmetry predictions and experimental data are caused by the finite spectral resolution and our commonsense perspective on molecular symmetry. However, such deviations in apparent symmetry can be identified with chemical shift information from A_1 vibrational modes and peak splitting of degenerate vibrations.

For ions in crystals, the applications of site-symmetry and factor group analysis, and correlation methods can offer complete information on the vibrational spectra of the ions (Bhagavantam and Venkataraidu 1969, Fateley et al. 1971, Cotton 1971). These methods provide symmetry information for sulfate groups on the basis of their location and the local coordination inside the crystal. Interpretation of molecular spectra of aqueous solutions and interfaces is more complex than that of solids. In aqueous solutions, several sulfato complexes can exist at different concentrations, and the vibrational spectrum of an aqueous sample is a composite spectrum of all of these complexes. The same is true for sulfato complexes on mineral surfaces, where a variety of surface complexes can exist, e.g. the outer-sphere, inner-sphere, and H-bonded complexes discussed above. Inner-sphere complexes can also exist in several forms (discussed above), and identification of these complexes from the vibrational spectra should be approached with caution. The selection rules for the vibrational transitions of surface complexes are also different when compared to those in solutions and solids (Urban 1993, Suetaka 1995). A variety of outer- and inner-sphere complexes can also exist in aqueous solutions, and their identification and relative concentrations are difficult. The challenge in vibrational spectroscopy is to identify the molecular chemistry of these complexes from the energy shifts in v_1 vibrations, and from the amount of splitting in degenerate vibrations (v_2, v_3, v_4), which are directly related to the strength of SO_4^{2-} complexes.

Data collection and analysis

The methods of data collection for vibrational spectra from samples in different environments are well-developed and widely available. A variety of data-collection methods have been developed to obtain IR and Raman spectra of dilute species in aqueous solutions and solids, at mineralwater interfaces, and for spatial resolution of micrometer-size domains. A brief description of these methods is provided here; more detailed discussion of these methods is available in the references cited on IR and Raman spectroscopy.

Infrared spectroscopy. Dry, solid samples can be examined in transmission, diffuse-reflection, and external-reflection modes. For examination in transmission and diffuse-reflection modes, the samples are commonly diluted with IR-transparent KBr or KCl. The diffuse-reflection and external-reflection methods are more surface-sensitive than the transmission methods (assuming that the species of interest is present on surfaces and in bulk). When extreme surface sensitivity is required for solids, a grazing incidence or external reflection method with a high angle of incidence-reflection (80°) is typically used. The surface sensitivity can also be varied by changing the angle of the incident beam (Harrick 1987). Liquid samples are examined in the transmission and attenuated-total-reflection (ATR) modes. For examining liquid samples that have different refractive indices and pH, several different ATR crystals are available (Harrick 1987; also refer to the Spectra-Tech web page, http://www.spectratech.com). The ATR cells are also used for collecting the spectra of wet pastes, suspensions, and interfaces of colloids and fluids. Details of these methods are discussed by Farmer (1974), Nakamoto, (1986), Harrick (1987), Ferraro and Nakamoto (1994), and Suetaka (1995). The optics of the infrared spectrometer are different for different energy regimes (far-, mid-, near-IR), and suitable optics must be used to get the best sensitivity.

Raman spectroscopy. Raman spectra of samples can be collected by exciting them at different wavelengths using different lasers, such as Kr^+ (647.1 nm), He-Ne (632.8 nm), or Ar^+ (488.0, 514.5 nm). Laser wavelength selection depends on the sample type and background fluorescence. Energy-tunable lasers have been developed recently, and these can be used to excite at any wavelength that is best suited for the sample.

Spectral analysis. Spectral identification of unknown samples can be conducted by comparing their spectra with those of several known compounds, or by conducting theoretical calculations. For identifying the fundamental vibrations, researchers typically use spectral libraries that are published by different groups, such as the Sadtler IR Digital Spectra Libraries, the American Society for Testing and Materials (ASTM) Database, and the Canadian Scientific Numerical Database Service (Search Program for IR Spectra, and several others). In practice, the normal modes of vibration of a chemical species cannot be identified unambiguously without examining a series of structural models having similar composition, or by conducting theoretical investigations. As discussed earlier, changes in the bond strength of a molecule can cause significant shifts in its vibrational spectra, and prior knowledge is essential for identifying these changes and for distinguishing the spectral features of a molecule from the spectral features of other molecules.

The presence of molecules in different structural states with closely related spectral features in geological matrices might cause their vibrational spectra to be broad. Overlap of vibrations of molecules of interest with those of other molecules can also contribute to such broadening. For deconvoluting such composite spectra, researchers use spectral second derivatives and principal-component analysis for identifying the number of peaks, and then fit the composite spectrum with a mixture of Laurentzian and Gaussian peaks (Maddoms 1980, Gillette et al. 1982, Hawthorne and Waychunas 1988, Perkins et al. 1991, Hasenoehrl and Griffiths 1993).

Vibrational spectra of sulfate in solids

Reviews of sulfate coordination in a variety of minerals, and their vibrational spectra, were presented by Moenke (1962) and Ross (1974). The collection of vibrational spectra reported in the "Infrared Spectra of Minerals" is extensive and has been widely used by geologists and soil scientists for the identification of a variety of minerals (Farmer 1974). Ross (1974) presented the vibrational spectra of several minerals, with their site symmetries, but did not give a detailed discussion of sulfate bonding in crystals or of the corresponding shifts in sulfate vibrational frequencies. Except for a handful of minerals, a detailed discussion of sulfate minerals is lacking even in the more recent literature. Although there are more recent reviews on the IR and Raman spectra of minerals, they do not include a large a collection of data for sulfate minerals (e.g. Karr 1975, Gadsden 1975, Ferraro 1982, Pechar 1988, Salisbury 1992, Cejka 1999). The vibrational spectra in the mid infrared region of two naturally occurring sulfate minerals are shown in Figure 24. Detailed analysis of energies of sulfate vibrations, together with the complementary information from the vibrational energies of other groups in minerals, is useful to interpret the relationship between the vibrational spectra of sulfate in minerals and its chemical bonding with the neighboring cations (for example, studies of Siedl and Knop (1969) on water in gypsum). Theoretical vibrational spectroscopy studies are also helpful in interpreting the complex vibrational spectra of solid sulfates.

Examination of cation complexes of a series of alkali sulfates, alkaline-earth sulfates, transition-metal sulfates and selected heavy-metal sulfates indicates that the symmetric

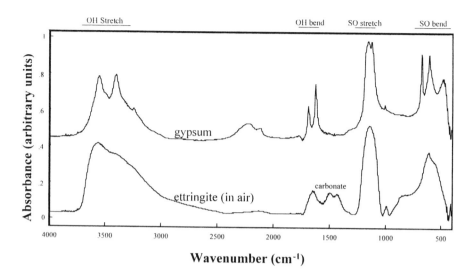

Figure 24. Infrared spectra of gypsum and ettringite (Myneni, unpublished data).

stretching $v_{s(S-OM)}$ (M = cation) vibrations do not exhibit any strong trends with the type of cation that is bonded to the SO_4^{2-} group. The $v_{s(S-OM)}$ vibrations (range: 960-1020 cm^{-1}) do not shift significantly away from those of uncomplexed sulfate (e.g. as in aqueous solutions; ~980 cm^{-1}, discussed later). The largest shift is found in protonated sulfate salts, with the $v_{s(S-OM)}$ vibrations around 890 cm^{-1}. Although definite trends were not observed for the spectral shift of $v_{s(S-OM)}$ vibrations to lower energies when compared to those of uncomplexed sulfate (such as alkali sulfates; Rull and Ohtaki 1997) (Fig. 31c). Changes in cation coordination around sulfate in structurally similar compounds cause shifts in the $v_{s(S-OM)}$ vibrations. If spectral features are resolved and understood, they can provide important information on the environment of sulfate bonding and can offer clues on the structural environment of sulfate in unknown compounds. Using such changes in $v_{s(S-O)}$ vibrations, Myneni et al. (1998a) identified three different types of sulfate groups and their coordination to water in ettringite.

Although information on the vibrational spectroscopy of crystalline and well-characterized minerals is critical for understanding and interpreting the chemical bonding and molecular chemistry of sulfate in complex, heterogeneous geological samples, few recent attempts have been made to correlate the crystal chemistry of sulfates with their vibrational spectral features. Some recent studies have focused on sulfate complexes in aqueous solutions, and at the mineralwater interface (as discussed later), but a majority of these studies have not used structurally similar and well-characterized minerals for interpreting the coordination chemistry of sulfate in complex systems. Detailed vibrational spectroscopic investigations of sulfate minerals can assist in interpreting the sulfate-coordination environment.

Vibrational spectra of sulfate in aqueous solutions

In dilute aqueous solutions of pH >3.0, SO_4^{2-} occurs primarily as a tetrahedral species and exhibits symmetric stretching (v_1) and bending (v_2), and asymmetric stretching (v_3) and bending (v_4) at 983, 450, 1105, and 611 cm^{-1}, respectively (Nakamoto 1986; Table 5). Of these, the asymmetric stretching and bending vibrations are IR active

Table 5. Normal vibrational modes of SO_4^{2-} and its protonated forms in aqueous solutions. Subscripts: IR: Infrared spectra, R: Raman spectra; s: symmetric stretching or bending vibrations, as: asymmetric stretching or bending vibrations.

Chemical Species	v_1	v_2	v_3	v_4
			(cm^{-1})	
SO_4^{2-}				
Nakamoto (1986)	983	450	1105	611
Myneni et al. (1998a)	980_{IR}	--	1098	--
Hug (1997)	--	--	1100_{IR}	--
Persson & Lövgren (1996)	--	--	1100_{IR}	--
Max et al. (2000)	982_{IR}		1099	--
Degenhardt & Mcquillan (1999)	977_{IR}		1099	--
Rull & Ohtaki (1997)				
Rudolph (1998a)				
Faguy et al. (1996)	980.8_R	--	--	--
	981.4_R	452	1110	617
	980_R	--	1100	--
HSO_4^-				
Ataka & Osawa (1998)	$876_{IR\ (S-OH)}$	--	$1050_{(S-O)s}$, $1190_{(S-O)as}$	--
Hug (1997)	$891_{IR\ (S-OH)}$	--	$1051_{(S-O)}$,	--
Max et al. (2000)	$887_{IR\ (S-OH)}$	--	$1194_{(S-O)}$	--
Faguy et al. (1996)	$885_{R(S-OH)s}$	--	$1051_{(S-O)s}$,	--
Rudolph (1996)	$898_{R\ (S-OH)}$	$422_{(S-OH)}$	$1198_{(S-O)as}$ $1050_{(S-O)s}$, $1200_{(S-O)as}$ $1052_{(S-O)s}$, $1202_{(S-O)as}$	--
$H_2SO_4^{\circ}$				
Horn & Sully (1999)	$902_{IR(S-OH)s}$	--	$1158_{(S-O)s}$	--
	$957_{(S-OH)\ as}$	--	$1359_{(S-O)as}$	--

(because of changes in dipole moment), and appear as strong bands in the IR spectrum. However, moderately concentrated (several mM) aqueous solutions of SO_4^{2-} exhibit a weak absorption feature corresponding to the $v_{s(S-O)}$ vibrations at about 980 cm^{-1} (e.g. Myneni et al. 1998a). Appearance of the $v_{s(S-O)}$ band in the vibrational spectrum of SO_4^{2-} suggests that its symmetry may not be perfectly tetrahedral. If the structure of water around SO_4^{2-} is also considered in symmetry analysis, aqueous SO_4^{2-} may not be tetrahedral because of the differences in water solvation and H-bonding at the O atoms of SO_4^{2-}. The sulfate-solvated water is continuously exchanged with the bulk water, and the water-exchange rates should also be considered and compared with the time scale of vibrational transitions ($\sim 10^{-12}$ s). The sample holder used to collect the spectrum of aqueous SO_4^{2-} can also cause symmetry distortions at the interfaces of the IR crystal and water.

In the case of Raman spectra, all of the fundamental modes of SO_4^{2-} are active in tetrahedral geometry and the $v_{s(S-O)}$ vibrations produce an intense band when compared to the other sulfate vibrations (because of changes in polarizability). Hence, several researchers studying aqueous complexes of SO_4^{2-} have focused on the intensity, asymmetry, and energies of this band. The Raman spectra of dilute aqueous solutions of SO_4^{2-} show a strong feature corresponding to the $v_{s(S-O)}$ band of SO_4^{2-}, without changes in other band positions. This suggests T_d symmetry for sulfate (Faguy et al. 1996, Rull and Ohtaki 1997, Rudolph 1998a,b). It should also be noted that small changes in H-bonding of water around solvated sulfate (such as those discussed above) might not cause significant differences in the degenerate (v_2, v_3, v_4) vibrations. Although these studies indicated that the symmetry of sulfate in aqueous solutions is debatable, the symmetry can be considered to be close to T_d geometry.

The vibrational spectra of SO_4^{2-} have been described by several researchers, but only selected results are cited in Table 5 to illustrate the small variation in reported values. It should be noted that these spectra were collected under different experimental conditions, a range of SO_4^{2-} concentrations, and in the presence of different counter-ions (e.g. Na^+, K^+, NH_4^+).

Nature of sulfate-solvated water. When ions are added to aqueous solutions, they disturb the H-bonding environment in liquid water, and form either H-bonds or strong complexes with water molecules (Pauling 1960, Ohtaki and Radnai 1993). However, ionic charge dissipates away from the ions, and theoretical studies indicate that the first two solvation shells of water may neutralize more than 90% of the ionic charge. Water in the solvation shell of sulfate can be probed directly by studying its OH vibrations (Myneni et al. 1998b; Fig. 25). The water molecules exhibit symmetric and asymmetric stretching vibrations above 3000 cm^{-1}, and bending vibrations around 1640 cm^{-1} (Nakamoto 1986). The studies of Myneni et al. (1998b) on a series of tetrahedral oxoanions (SO_4^{2-}, CrO_4^{2-}, MoO_4^{2-}, SeO_4^{2-}, $HAsO_4^{2-}$, HPO_4^{2-}) showed that vibrational spectral features of water (after subtracting the bulk water contributions) shift to different energies in accordance with the type of oxoanion and its protonation state in water. The results indicate that the oxoanions with the strongest H-bonding between the oxoanions and water exhibit weak OH vibrations and thus shift to lower wavenumbers. The strongly hydrating ions also have high pK_a values and narrower metal-O vibrational bands. The sulfate ion has a low pK_a compared to those of the other examined oxoanions (Brookins 1988), and the OH stretching vibrations are at relatively high wavenumbers (Fig. 25). Similar results were also reported for sulfate at different pH values, and for several monovalent anions in HDO (Bergström et al. 1991, Brooker and Tremaine 1992, Ataka and Osawa 1998).

Protonation effects on sulfate vibrations. Protonation and cation complexation of aqueous SO_4^{2-} can modify its vibrational spectral features and these are useful in identifying the SO_4^{2-} coordination and bonding environment. If carefully evaluated, the vibrational spectra can be used to obtain the stability constants of proton-sulfate and cation-sulfate complexes (Rudolph et al. 1998a,b; Max et al. 2000 and references therein).

Addition of a proton to SO_4^{2-} modifies the symmetry of sulfate from T_d to C_{3v} (considering the proton is along the axis of the SO bond). The addition of a proton to SO_4^{2-} also distorts the molecule significantly, and the SO bond length increases from 1.473 Å in uncomplexed SO_4^{2-} to 1.58 Å for SOH in HSO_4^- (Cruickshank and Robinson 1966, Taesler and Olovsson 1969, Hawthorne et al., Chapter 1, this volume). Accordingly, the bond length of uncomplexed SO shortens to 1.46 Å. These changes modify the SO_4^{2-} vibrational spectra completely, and the $v_{s(S-OH)}$ vibrations appear around

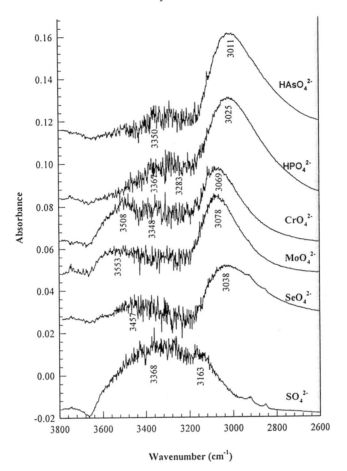

Figure 25. OH stretching vibrations of water coordinated to different aqueous oxoanions (Myneni et al. 1998b). The spectra were collected using ZnSe ATR-FTIR spectroscopy.

890 cm^{-1} (Table 5, Fig. 26). The corresponding $\nu_{s(S-O)}$, $\nu_{as(S-O)}$ vibrations of the SO$_3$ group appear at about 1050 and 1200 cm^{-1}, respectively. These vibrations originate from the splitting of triply-degenerate asymmetric stretching vibration (Tables 4 and 5; Fig. 26). In association with these changes, the degenerate bending vibrations also split accordingly (Table 4; Rudolph 1996, Max et al. 2000). A detailed titration of sulfuric acid by Max et al. (2000) showed these spectral variations clearly (Fig. 26). Max et al. used the spectral energies and intensities to estimate the distribution of HSO$_4^-$ and SO$_4^{2-}$ in aqueous solutions at different pH values and the corresponding pK$_a$ value, which is in close agreement with the experimental values. Proton association with sulfate also changes dramatically with a change in temperature, and the $\nu_{s(S-OH)}$ band shifts to lower energy with increases in temperature (Rudolph 1996; Fig. 27). Based on Raman spectral changes, Rudolph calculated changes in the pK$_a$ of HSO$_4^-$ with temperature, which showed an increase with an increase in temperature (Fig. 27).

Addition of another proton to HSO$_4^-$ to form H$_2$SO$_4^0$ results in the relaxation of the

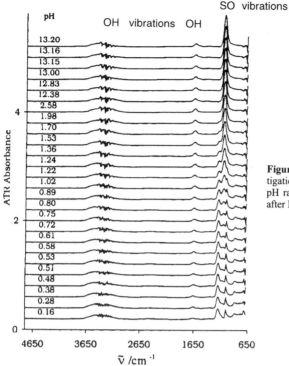

Figure 26. ATR-FTIR spectral investigation of sulfuric acid titration in the pH range of 0.16 to 13.20 (modified after Max et al. 2000).

first SOH bond to 1.54 Å (from 1.58 Å in HSO_4^-), and in further shortening of the uncomplexed SO bond lengths to 1.43 Å (Cruickshank and Robinson 1966, Myneni et al. 1998b). The symmetry of $H_2SO_4^o$ is considered to be C_{2v}, provided that the H atoms are along the SO bond-axis. The degenerate vibrations split accordingly and the IR-inactive bands become active (Table 5). Note that the $v_{s(S-OH)}$ vibrations of $H_2SO_4^o$ shift to higher energies (relative to HSO_4^-) with the addition of protons to HSO_4^- (and with reduction in the SOH bond length). Such bond length and vibrational spectral changes have been observed for other oxoanions such as AsO_4^{3-}, SeO_4^{2-}, and PO_4^{3-} (Cruickshank and Robinson 1966, Myneni et al. 1998a,b; Persson et al. 1996). Theoretical vibrational spectral calculations of H_2SO_4 molecules using the $6\text{-}31G^{**}$ basis set showed an excellent agreement with the experimental results discussed above (Lindsay and Gibbs 1988). Although previous studies focused on gas phase molecules, calculations on solvated species of protonated and metal-complexed sulfates would be useful in understanding the vibrational spectral changes with changes in chemical bonding.

The differences in the vibrational spectra of SO_4^{2-} and HSO_4^- have been used by geochemists in interpreting paleoenvironmental conditions. For example micro-Raman spectroscopic investigation of fluid inclusions in halite were used by Benison et al. (1998) to infer the depositional environment of red beds of Permian age. The spectra of sulfate in the fluid inclusions showed two distinct features, which were assigned to the $v_{s(S-O)}$ vibrations of SO_4^{2-} and HSO_4^-. The chemistry of fluids in inclusions is assumed to represent the chemical conditions of the formation waters, and the study by Benison et al. suggested that the pH of these solutions may have been <1.0 (Fig. 28).

Cation complexation effects on sulfate vibrations. Examination of cation

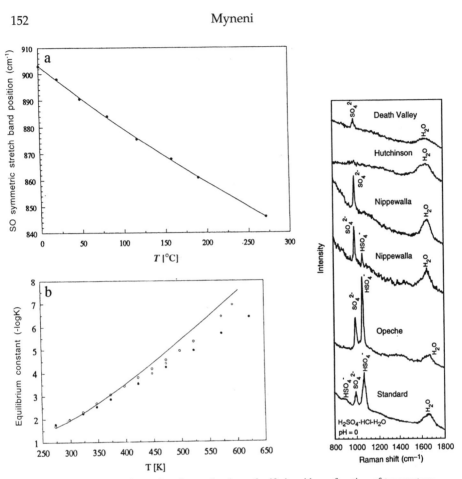

Figure 27. (a) Spectral energies of $\nu_{s(S-O)}$ vibrations of sulfuric acid as a function of temperature; (b) the second pK (K = dissociation constant) of H_2SO_4. (modified from Rudolph 1996).

Figure 28 (right). Micro-Raman spectra of sulfate in fluid inclusions in Permian halite samples (Hutchinson, Nippewalla and Opeche formations) and modern Death Valley halite (modified from Benison et al. 1998).

complexation of sulfate in aqueous solutions, primarily using Raman spectroscopy since the 1960s, has indicated that the $\nu_{s(S-O)}$ vibrations are sensitive to cation complexation. These changes are similar to those reported in the case of protonated sulfates, but the spectral differences between cation-complexed sulfates and aqueous sulfate are much smaller. The cation complexation to sulfate polyhedra does not cause significant changes in the SOM (M = cation) bond distances (relative to the protonated complexes), and this is evident for several of the sulfate salts discussed by Hawthorne et al. (Chapter 1, this volume). Correspondingly, the $\nu_{s(S-OM)}$ vibrations of cation-complexed sulfates are close to those of uncomplexed SO_4^{2-}, and are at a higher wavenumber compared to that of protonated sulfates (Tables 5, 6). Fortunately, this makes it easier to distinguish the vibrational spectral features of protonated- and cation-complexed sulfates. However, the unequivocal identification of uncomplexed and cation-complexed sulfates is difficult.

Another problem with the identification of cationic complexes is related to the presence of several different types of cation-sulfato complexes and uncomplexed SO_4^{2-} in

Table 6. Vibrational modes of different aqueous cation sulfate complexes

Cationic Complexes of SO_4^{2-}	v_1	v_2	v_3	v_4	Source
			(cm^{-1})		
Na^+	980_R	450	1105	625	a
NH_4^+	985_R	460	1120	620	a
Tl^+	985_R	452	1100	625	a
Mg^{2+}	985_R	455	1120	618	a
	982_R				
	$995_R^{(S-O-Mg)}$	450	1115	616	b, c
	985_{ir}				b, c
Ca^{2+}	978_{ir}		1107		h
Fe^{2+}	$981, 989_R$		1104, 1064		d
Cu^{2+}	980_R	450	1100	620	a
Zn^{2+}	985_R	455	1112	618	a
	990.8_R	monoden			e
	1053_R	bident.			a
Cd^{2+}	985_R	451	1105	615	a
	$983, 990_R$				f, g
Fe^{3+}	980_{ir}		1042, 1126		i
Al^{3+}	985_R	455	1100	625	a
Ga^{3+}	985_R	475	1200	600	a
In^{3+}	980_R	450	1105	620	a
	$1000_R^{(S-O-In)}$		1125		

The subscripts IR and R indicate that the data were collected using IR or Raman systems, respectively. Sources of data: a, Hester and Plane (1964); b, Frantz et al. (1994); c, Davis and Oliver (1973); d, Rudolph et al. (1997); e, Rudolph et al. (1999); f, Rudolph et al. (1994); g, Rudolph (1998a); h, Myneni et al. (1998a); i, Hug (1997).

aqueous solutions (Smith and Martell 1976). The nature of sulfate interactions with cations may also change with time, such as the SO_4^{2-} in cationic complexes changing between outer-sphere and inner-sphere complexes (Stumm and Morgan 1982, Stumm 1992). Solutions of sulfate-salts may contain a variety of these species, and the vibrational spectra of each of the component species may overlap each other and produce broad spectral features (if the species are closely related). It is very challenging to distinguish different cationic complexes based on their vibrational spectral features. If the cation-sulfato complexes are understood well, then information can be obtained on the water-exchange rates of cation-SO_4^{2-} complexes and the residence time of SO_4^{2-} in the inner- and outer-sphere complexes.

The spectral contributions of aqueous inner-sphere cation-sulfato complexes in Raman spectra are identified primarily from the asymmetry in the intense peak around 983 cm^{-1}, which corresponds to the $_{s(S-O)}$ vibrations of uncomplexed or outer-spherically bound SO_4^{2-} (Davis and Oliver 1973, Rudolph and Irmer 1994, Rudolph et al. 1997,

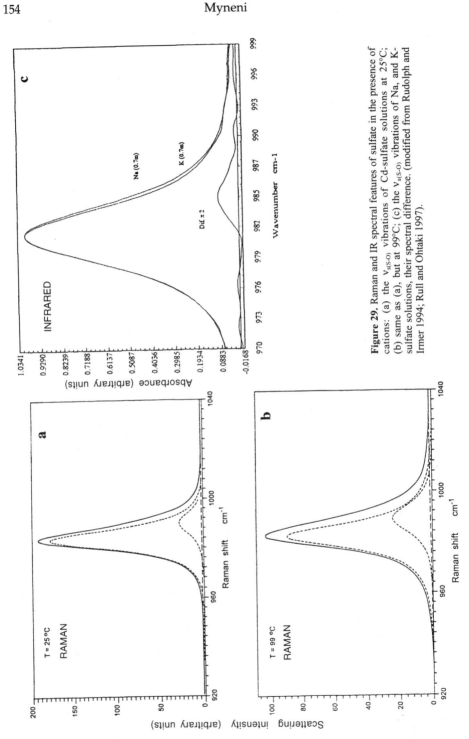

Figure 29. Raman and IR spectral features of sulfate in the presence of cations: (a) the $\nu_{s(S-O)}$ vibrations of Cd-sulfate solutions at 25°C; (b) same as (a), but at 99°C; (c) the $\nu_{s(S-O)}$ vibrations of Na, and K-sulfate solutions, their spectral difference. (modified from Rudolph and Irmer 1994; Rull and Ohtaki 1997).

Rudolph 1998a,b; and references therein) (Fig. 29). Rudolph (1998a,b) observed that this peak asymmetry increases with increases in temperature and cation concentration (Figs. 29, 30). Rudolph and his coworkers used the increases in peak asymmetry to identify the nature of the sulfate-metal aqueous complex and its fraction in aqueous solutions (Fig. 30).

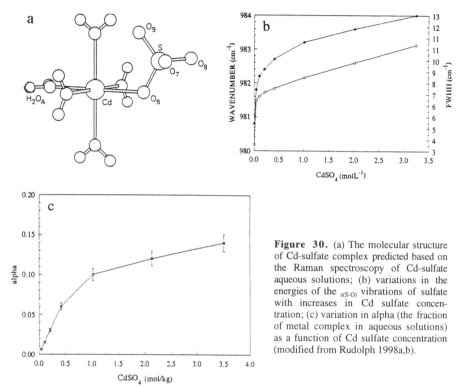

Figure 30. (a) The molecular structure of Cd-sulfate complex predicted based on the Raman spectroscopy of Cd-sulfate aqueous solutions; (b) variations in the energies of the $\nu_{s(S-O)}$ vibrations of sulfate with increases in Cd sulfate concentration; (c) variation in alpha (the fraction of metal complex in aqueous solutions) as a function of Cd sulfate concentration (modified from Rudolph 1998a,b).

Although Rudolph and others have attributed the asymmetry in the Raman peak at 983 cm^{-1} to an inner-sphere cation-sulfato complex, this interpretation has been questioned (Nomura et al. 1980, Rull and Ohtaki 1997). One of the primary reasons is that the solutions of Cu^{2+}- and Al^{3+}-sulfate salts exhibit direct cation-sulfate inner-sphere complexes but have a very symmetric $\nu_{s(S-O)}$ band; moreover, the presence of inner-sphere complexes in these solutions is also supported by neutron and X-ray scattering studies. By contrast inner-sphere cation-sulfato complexes were not detected (by X-ray scattering) in Na$^+$- and K$^+$-salt solutions, but Na-sulfate solutions exhibit distinct asymmetry in the peak at 983 cm^{-1} (Fig. 29). Rull and Ohtaki (1997) suggested that the high asymmetry may be caused by solvation and ionic strength effects rather than by complex formation. Their results support the earlier findings of Nomura et al. (1980), who linked this high-energy asymmetry to the differences between the rotational correlation time of water molecules around cations and sulfate. Complementary spectroscopic information from other techniques is needed to understand the nature of aqueous complexes and the origin of the vibrational spectral features discussed above.

The studies by Rull and Ohtaki (1997) on Li$^+$, Na$^+$, K$^+$, Rb$^+$, and Cs$^+$ sulfate solutions and solids suggest that the $\nu_{s(S-O)}$ vibrations shift linearly with changes in salt concentration (Fig. 31, Table 7). Whereas the slope of the lines plotted for $\nu_{s(S-O)}$ and salt

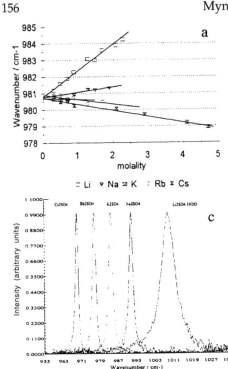

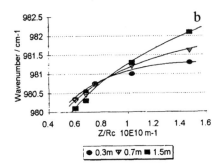

= Li ▽ Na ⊠ K ∷ Rb ⨯ Cs

Figure 31. The $v_{s(S-O)}$ vibrations of sulfate in alkali sulfate salts and their aqueous solutions: (a) variation in the energies of $v_{s(S-O)}$ vibrations of aqueous sulfate as a function of salt concentration; (b) variation in $v_{s(S-O)}$ vibrations with changes in charge to radius ratio of monovalent ion (Z/Rc) (c) Raman spectras of solid salts. Modified from Rull and Ohtaki (1997).

Table 7. Raman spectra of alkali salt solutions at room temperature (modified from Rull and Ohtaki 1997). The energies of symmetric stretching S-O vibrations are shown.

Salt	Concentration (m)	v_{max} (cm^{-1})	v_{av} (cm^{-1})
Li$_2$SO$_4$	0.3	981.2	981.1
	0.7	981.6	981.7
	1.5	982.1	982.4
	2.0	982.6	983.4
Na$_2$SO$_4$	0.3	980.8	980.8
	0.7	980.9	980.5
	1.5	981.3	981.6
	2.0	981.4	981.8
K$_2$SO$_4$	0.3	980.8	980.8
	0.7	981.0	980.7
Rb$_2$SO$_4$	0.3	980.9	980.8
	0.7	980.8	980.7
	1.5	980.6	980.6
Cs$_2$SO$_4$	0.3	980.7	980.4
	0.7	980.6	980.3
	1.5	980.4	980.3
	2.0	980.1	979.7

a 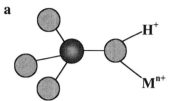 **b**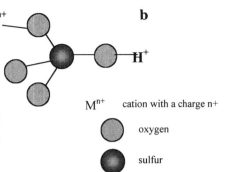

M^{n+} cation with a charge n+

oxygen

sulfur

Figure 32. Metal complexation of HSO_4^-: (a) metal complexation at the protonated oxygen, (b) metal complexation at the terminal oxygen atom.

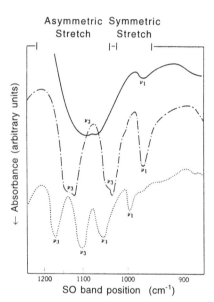

Figure 33. IR spectra of Co-sulfate complexes in different coordination environments (modified from Nakamoto 1986).

concentration showed a positive trend for Li^+ and Na^+ and a negative trend for Rb^+ and Cs^+, the slope was relatively flat for K^+. One of the notable features is that increases in aqueous salt concentration shifted the $\nu_{s(S-O)}$ vibrations of sulfate in the direction as the peaks corresponding to the respective solids (Fig. 31). This suggests that the cation directly influence the sulfate vibrations, either through inner-sphere or outer-sphere complex formation. The significant differences between the energies of $\nu_{s(S-O)}$ vibrations in salts and their aqueous solutions indicate that the cation-anion interactions are relatively weak in aqueous solution. Another notable result from this study is that the $\nu_{s(S-O)}$ band shifts to lower energy with an increase in the cation size in concentrated solutions (Table 7, Fig. 31)

Cation complexation of HSO_4^- ions at low pH has not been investigated thoroughly. Raman studies of $FeSO_4–H_2SO_4$ systems by Rudolph et al. (1997) indicated that Fe^{2+} does not form inner-sphere complexes with the HSO_4^- ion in the temperature range of 25 to 303°C. The vibrational features of the HSO_4^- ion were not perturbed when Fe^{2+} ions were added to the system. Similar results have been reported for arsenate complexes (Myneni et al. 1998b). Based on bond-valence calculations, Myneni et al. (1998a) predicted that the protonated $As–OH/S–OH/P–OH$ groups of $HAsO_4^{2-}/HSO_4^-/HPO_4^{2-}$, respectively, cannot form cation complexes in the form shown in Figure 32. This is because of the excessive charge at the bonding O atoms (>2.0, from bond-valence estimates, Brown and Altermatt 1985, Myneni et al. 1998a). However, cations can bind to free-terminal O atoms of HSO_4^-. As discussed above, protonation-induced perturbations in sulfate polyhedron have more effect than the complexation of sulfate to cations. For this reason, small changes in bond length that are associated with cation complexation of HSO_4^- may not be sufficient to cause changes in the vibrational modes of the HSO_4^- ion. Hence, an absence of changes in the HSO_4^- vibrational spectra does not completely rule out the existence of cation-HSO_4^- complexes. Thus, complexation of cations with HSO_4^- ions in solutions or at mineral–water interfaces should be approached with caution.

Whereas studies of Raman spectra have focused on the $\nu_{s(S-OM)}$ vibrations, infrared studies have focused on spectral splitting in the $\nu_{as(S-O)}$ vibrations. This spectral splitting in IR spectra was demonstrated by Nakamoto (1986) for Co-sulfate complexes, in which sulfate exhibits T_d, C_{3v} and C_{2v} symmetries (also refer Hezel and Ross 1968). Based on the number of bands displayed by the $\nu_{as(S-O)}$ vibrations of these samples, Nakamoto assigned different symmetries for sulfate in these compounds (Fig. 33, Table 4). For example, tetrahedral symmetry was assigned to sulfate in $[Co(NH_3)_6(SO_4)]\cdot5H_2O$ because the bands corresponding to $\nu_{as(S-O)}$ vibrations are not split, and $\nu_{s(S-O)}$ vibrations are weak.

However, the $\nu_{as(S-O)}$ vibrations shift to higher energy by 30-40 cm^{-1} relative to aqueous sulfate. Such shifts may be caused by strong H-bonding of sulfate with the ligands in the coordination shell. Similar results were also reported for sulfate interactions in ettringite, and on goethite surfaces (Persson and Lövgren 1996, Myneni et al. 1998a). For the molecules in C_{2v} symmetry, Nakamoto (1986) suggested that splitting in $\nu_{as(S-O)}$ vibrations is greater in a chelate than in a bridging-bidentate complex. Several researchers have used the number of bands, and their relative energies, to identify the coordination complex of sulfate in unknown samples. Although such assignments are valid, some of the problems associated with such assignments are discussed in the next section.

Vibrational spectra of sulfate at the interfaces

For several years, soil chemists and geochemists have used the results of adsorption measurements to predict the nature of sulfate molecular interactions with different minerals to determine partition coefficients and to develop surface complexation theories (see Nordstrom and Bigham, this volume). As pointed out by Sposito (1990), Brown (1990), and Brown et al. (1999), molecular information on the sorption complex cannot be deduced from macroscopic results alone. For the past two decades, vibrational spectroscopy (primarily IR spectroscopy) has been used extensively to study the nature of sulfate complexes on mineral surfaces, specifically on Fe oxides and oxyhydroxides. The differences in sample preparation, data collection, and interpretation by various authors have led to controversies on the nature of sulfate interactions with mineral surfaces. Although *in situ* studies have been conducted on selected mineral surfaces recently, several critical questions remain unanswered. These are related to: (1) the nature of sulfate coordination on mineral surfaces: i.e. inner- versus outer-sphere complexes, (2) variations in the nature of the surface complex with changes in solution chemistry, and (3) the types of all sorbed complexes and their concentrations. A summary of vibrational spectral information of sulfate interactions at mineral–water, metal–water, and other interfaces is presented below.

Sulfate complexation on Fe-oxide surfaces. Many of the early vibrational

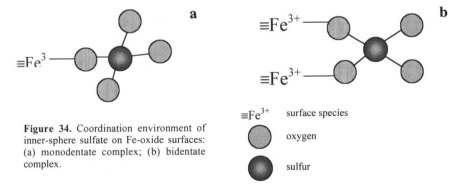

Figure 34. Coordination environment of inner-sphere sulfate on Fe-oxide surfaces: (a) monodentate complex; (b) bidentate complex.

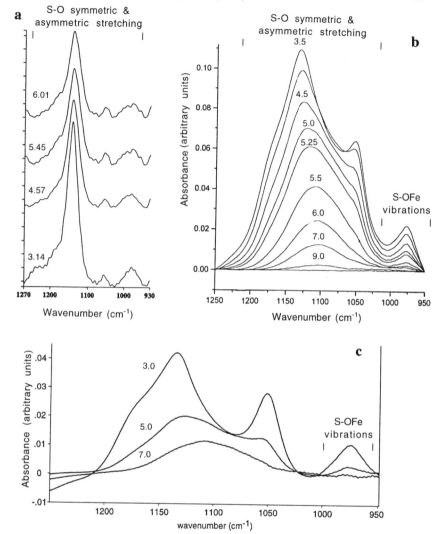

Figure 35. IR spectra of sulfate sorbed on goethite: (a) diffuse-reflectance spectroscopic investigation (modified from Persson and Lövgren 1996); (b) ATR-FTIR spectra (modified from Peak et al. 1999); (c) ATR-FTIR spectra (modified from Ostergren et al. 2000).

spectroscopy studies were done on pellets or powder samples (prepared by mixing with KBr) using transmission, diffuse-reflection and external-reflection sampling geometry (Parfitt and Smart 1978, Harrison and Berkheiser 1982, Turner and Kramer 1991, Watanabe et al. 1994, Persson and Lövgren 1996) (Figs. 35-37). These *ex situ* studies indicated that the asymmetric stretching vibrations of sulfate are split into more than two peaks, and it was therefore suggested that sulfate formed inner-sphere complexes on the surfaces of Fe oxides. All these researchers, with the exception of Persson and Lövgren (1996), suggested that the sorbed sulfate is in C_{2v} symmetry and forms a bidentate bridging complex (Nakamoto 1986), as shown in Figure 34. A summary of these observations are reported in Table 8.

Splitting in the asymmetric stretching frequencies was more pronounced when the samples were kept in vacuum or heated (Fig. 37). The observations of Li et al. (1993) on sulfate reactions with hematite surfaces at high temperatures are in support of these results. The XPS results of sulfate on hematite and maghemite also indicated that sulfate interacts with two chemically distinct sites (Watanabe et al. 1994; Fig. 21).

Table 8. Molecular chemistry of sulfate adsorbed on Fe oxide surfaces.

Authors	*Sample Information*	*Comments on Sulfate Interactions with Fe Oxide Surfaces*
Sulfate on Goethite		
Parfitt & Russell (1977)	Changes in OH/OD vibrations; Sampling on dry systems	H_2SO_4 reactions with goethite surface; Bidentate-bridging complexes
Persson & Lovgren (1996)	Diffuse reflectance; 1145, 1045, 970 cm^{-1}; 1220 for samples with pH<5	Sample pH > 3.0, outer-sphere, HSO_4^- on surfaces above the pKa of $H_2SO_4^\circ$.
Peak et al. (1999)	pH > 6.0: 1104, 975; pH < 6.0: 1170 (wk), 1133, 1055, 976 cm^{-1}; sampling on wet systems	Outer-sphere complex at pH > 6.0. Inner-sphere complex at pH < 6.0, monodentate, some of the sorbed complexes may be present as HSO_4^-
Ostergren et al. (2000)	pH = 3.0; ~1180, ~1140, ~1050 cm^{-1}; pH = 5.0; 1123, 1057 cm^{-1}; pH = 7.0; 1105 cm^{-1}; sampling on wet systems	Outer-sphere complex at pH > 5.0. Inner-sphere complex at pH < 5.0, "combination of complexes with less than C_{3v} symmetry".
Sulfate on Hematite		
Parfitt & Smart (1978)	Dry: 1200, 1128, 1040, 970 cm^{-1} Vacuum: 1245, 1131, 1030, 950 cm^{-1}	Inner-sphere; bidentate bridging complex
Turner & Kramer (1991)	Dry: 1255, 1130, 1030, 950 cm^{-1}; Interpretation is based on IR, and titration of released surface hydroxyls	Inner-sphere; mononuclear & bidentate bridging; mononuclear increases with a decrease in pH
Watanabe et al. (1994)	FTIR and XPS of sulfate on hematite & maghemite; data collection on dry samples	Inner-sphere, bidentate-bridging, sulfate binds at two non-equivalent sites
Hug (1997)	Wet: 1128, 1060, 978 cm^{-1} Dry: 1195, 1135, 1047, 970 cm^{-1}	Sample pH 3-5, monodentate. Bidentate-bridging complex/HSO_4^- in dried samples
Eggleston et al. (1998)	FTIR and STM Observations	FTIR: same as Hug (1997) STM: inner-sphere; spectrum of inner- & outer-sphere; bidentate or bisulfate sorption. Lifetimes are similar to that of aqueous $FeSO_4^+$ complexes.
Sulfate on Other Fe-Oxides		
Parfitt & Smart (1978)	Amorphous Fe-oxides, data collection on dry samples	Bidentate-bridging
Harrison & Berkheiser (1982)	On Freshly prepared Fe-oxides Dry: 1170, 1125, 1050, 970 cm^{-1} Vacuum: 1215, 1125, 1040, 970 cm^{-1}	Bidentate-bridging

Persson and Lövgren (1996) used diffuse-reflectance spectroscopy to examine the nature of sulfate complexes on goethite surfaces. On the basis of the absence of splitting in sulfate $v_{as(S-O)}$ vibrations, they proposed that sulfate formed outer-sphere complexes with goethite surfaces in the pH range of 3-8 (Fig. 35). They also suggested that goethite surfaces stabilize outer-sphere HSO_4^- on solid surfaces in the acidic pH range (for stability of goethite; see Bigham and Nordstrom, Chapter 7, this volume). It should be noted that their spectral features show that the $v_{as(S-O)}$ vibrations of sulfate on goethite are shifted to wavenumbers approximately 40 cm^{-1} above those of aqueous sulfate (Fig. 35). Small peaks are also found at low and high energies of this band. Recent *in situ* ATR-FTIR spectroscopic studies have shown the same spectral features, which suggests that the structure of sorbed sulfate deviates from T_d symmetry even above pH 5.0 (Peak et al. 1999, Ostergren et al. 2000). Although Persson and Lövgren (1996) suggested outer-sphere complexation for sulfate sorbed on goethite, in this author's opinion, sulfate sorbed on goethite is not similar to aqueous sulfate. It is in a different configuration such as a H-bonded complex, as indicated by the shift in the $v_{as(S-O)}$ vibrations (Myneni et al. 2000b).Hug (1997) and Peak et al. (1999) criticized the sample-preparation and data-collection techniques used by previous researchers in studying the nature of sulfate complexes at Fe-oxide–water interfaces. Conversion of wet samples into powders and pellets for IR investigations may alter the sulfate coordination in solids. Drying and diluting the samples with KBr may also cause problems with the water content of the samples. Examination of samples in vacuum condition was also questioned because of the potential loss of structural water and corresponding changes in sulfate coordination (Fig. 37), as has been shown in recent studies (Hug, 1997, Peak et al. 1999).

Hug (1997) conducted a detailed examination of the shifts in vibrational spectra of

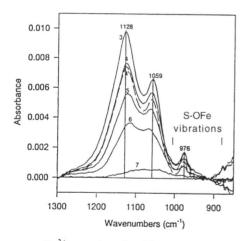

Figure 36. ATR-FTIR spectra of sulfate sorbed on hematite. The pH conditions are shown on each of the spectra (modified from Hug 1997).

aqueous Fe^{3+}-complexed sulfate and HSO_4^-, and suggested that sulfate forms inner-sphere, monodentate complexes on hematite surfaces in the pH range of 3-5. The proposed structure for sulfate on hematite surfaces is shown in Figure 34a. Although the work of Hug (1997) was on hematite, these results are not consistent with the model of tetrahedral outer-sphere sulfate complexes on goethite surfaces as proposed by Persson and Lövgren (1996). The ATR-FTIR spectra of Hug (1997) showed that the $v_{as(S-O)}$ vibrations of sulfate exhibit two bands, which are indicative of C_{3v} symmetry and not C_{2v} as reported in earlier studies (Fig. 36). The spectra of air-dried samples showed a third peak around 1200 cm^{-1}; based on the number of peaks Hug (1997) suggested that the symmetry of sorbed sulfate in these samples may be C_{2v} and that sulfate may be in the form of either bidentate-bridging complex or HSO_4^- (Fig. 36). Recent studies by Eggleston et al. (1998), Peak et al. (1999), and Ostergren et al. (2000) are in support of Hug's interpretations. However, Peak et al. (1999) and Ostergren et al. (2000) reported that sulfate forms outer-sphere complexes on goethite surfaces at relatively high pH and forms an inner-sphere complex at pH values less than 5.0 (Fig. 35). By analyzing the scanning tunneling microscope scan frequency, Eggleston et al. (1998) concluded that the lifetimes of sulfate complexes on hematite surfaces vary from several tens to hundreds of milliseconds. Based on such a large variation in the lifetimes, it was suggested that characterization of adsorbates into inner- and outer-sphere complexes on the basis of macroscopic studies may be an oversimplification.

A summary of these studies indicates that: (1) sulfate coordination on mineral surfaces is affected by sample drying, and studies on dry samples may not represent the coordination chemistry of sulfate in wet conditions, (2) sulfate forms inner-sphere complexes on hematite at pH < 5, and (3) sulfate on goethite forms inner-sphere complexes in the acidic pH range, and outer-sphere complexes at near-neutral pH.

Some critical issues have not been considered by previous researchers in evaluating the nature of sulfate complexes on surfaces. These issues are: (1) all of the previous *in situ* studies had shown similar spectral features with small variations in energies of the $v_{as(S-O)}$ vibrations, and the pH at which splitting in $v_{as(S-O)}$ vibrations begins is different for goethite and hematite (Hug 1997, Peak et al. 1999); (2) the energy of the $v_{s(S-O)}$ mode, which is a strong indicator of the presence of bisulfate complexes was not considered in the previous studies (Myneni et al. 1998b); (3) the simultaneous occurrence of more than one type of surface complex, such as inner- and outer-sphere complexes, was not considered; (3) spectral comparisons of Fe-sulfate salts, in which sulfate forms bidentate-bridging complexes with ferric iron, has not been conducted; and (5) occurrence of Fe sulfate precipitates (such as jarosite) in samples of acidic pH also has not been considered. Such detailed comparisons or theoretical studies will help in identifying the nature of sulfate surface complexes.

Sulfate complexation on Cr-oxide surfaces. The insoluble Cr^{3+} oxides and oxyhydroxides are common in Cr-rich soil profiles, waste environments, and radioactive-waste-disposal sites. Cr-bearing steel is used extensively in industry because of its high resistance to corrosion. Recent IR studies of sulfate interactions with Cr-hydroxide colloids indicate that sulfate geometry changes after its adsorption on Cr-oxide surfaces (Degenhardt and McQuillan 1999). However, the $v_{as(S-O)}$ vibrations at ~1100 cm^{-1} did not exhibit distinct splitting, and the $v_{s(S-O)}$ vibrations appeared at 982 cm^{-1} for sulfate sorbed on Cr-oxide surfaces in the pH range of 3.2 to 8.0 (Fig. 38). Sulfate sorption at pH <4.0 showed a distinct shoulder to the peak corresponding to the $v_{as(S-O)}$ vibrations, and the peak maximum shifted from 1099 to 1116 cm^{-1}. Drying of sulfate-sorbed samples enhanced the peak asymmetry and showed two distinct features with increases in drying time (Fig. 38). Based on these changes Degenhardt and McQuillan (1999) concluded that sulfate forms an outer-sphere complex on Cr-hydroxide surfaces in wet conditions, and that the peak asymmetry is caused by electrostatic interactions between the surface and interfacial sulfate. For dry samples, the formation of inner-sphere monodentate sulfate complexes was proposed. The formation of weak outer-sphere complexes of sulfate on Cr oxides may contribute to pitting on stainless-steel surfaces.

Sulfate complexation on metal surfaces. Sulfate interactions on metal surfaces (e.g. Cu, Ag, Pt, Au) of different orientation have been examined by several researchers because of the importance of these metals in electrochemistry and batteries (for a review, see Bockris and Khan 1993). Several investigations have used Raman and IR spectroscopy for exploring sulfate reactions with metal surfaces at different pH values.

The results indicate that some metal surfaces interact with sulfate through three of its oxygen atoms, and do not interact with bisulfate, e.g. on Au(111) and Cu surfaces (Brown and Hope 1995, Ataka and Osawa 1998). In other cases, e.g. Pt(111) (Faguy et al. 1996), the sulfate-H_3O^+ ion pair interacts with metal surfaces, and the nature of the surface

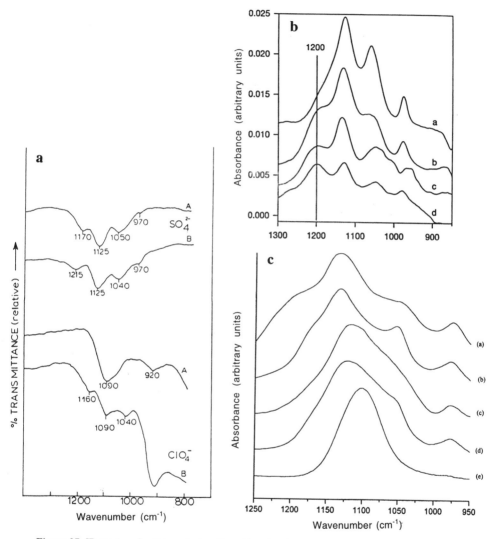

Figure 37. IR spectra of sulfate sorbed on Fe oxides, showing the influence of drying of sample on the vibrational spectra of sulfate: (a) sulfate and perchlorate sorbed on freshly prepared Fe-oxyhydroxides; spectra were collected using the transmission of IR beam through the dried samples; A: air dried samples, and B: evacuated samples (modified from Harrison and Berkheiser 1982), (b) FTIR spectra of sulfate sorbed on hematite. a: sulfate-reacted hematite in wet conditions (ATR-FTIR spectra), b: the same sample without water and after drying in air, c: after drying with N_2, and d: after reaction with water and 0.1 M HCl (modified from Hug 1997); (c) FTIR spectra of sulfate. a: schwertmannite in diffuse reflectance mode (air-dried), b: sulfate sorbed on goethite, pH 3.5, c: schwertmannite in ATR mode (wet), d: sulfate sorbed on goethite, pH 5.0, and e: 100 mM aqueous sulfate (modified from Peak et al. 1999).

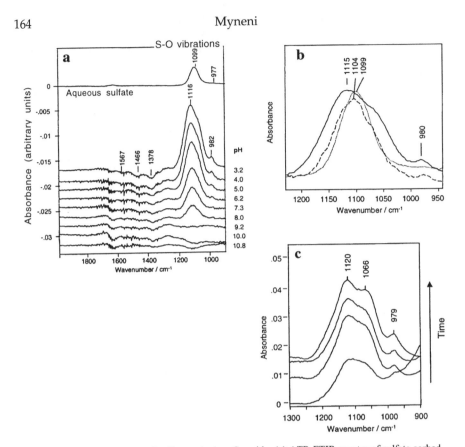

Figure 38. FTIR spectra of sulfate sorbed on Cr oxide: (a) ATR-FTIR spectra of sulfate sorbed on Cr oxide at different pH values; (b) spectral differences between aqueous sulfate (...), and Cr oxide reacted with sulfate at low (----) and high (—) coverages; (c) spectra showing the influence of sample drying on changes in the environment of sulfate coordination. The lowermost spectrum is that of wet sample, and the upper curves are for progressively dried samples. (modified from Degenhardt and McQuillan 1999).

complex is intermediate between sulfate and bisulfate with the H_3O^+ species stabilizing the surface complex. Although these systems are not directly relevant to geological environments, important information on the vibrational modes of sulfate and their changes associated with metal-surface interactions are useful for interpreting samples of geochemical importance.

Spectromicroscopy of sulfates

Although specific spectromicroscopy investigations have not been conducted on sulfate in minerals or other geological materials, such studies are possible. Using laboratory-based micro-IR facilities, a spatial resolution of a few tens of micrometers can be achieved. The synchrotron-based IR spectromicroscopy facilities at the ALS and NSLS can offer a spatial resolution of less than 5 μm because of the higher brightness of the synchrotron beams relative to the beams from the glowbar-source in commercially available IR instruments. In addition, surface-sensitive micro-FTIR studies can be conducted using grazing incidence and ATR objectives. The laboratory-based micro-Raman facilities can offer a resolution of less than one micrometer. Melendres (1998) provides more information on these techniques.

COMPLEMENTARY SPECTROSCOPIC METHODS

Scattering methods

Scattering or diffraction of neutrons, X-rays, and electrons by materials has been widely used for studying the structure of sulfate in various forms. The application of X-ray diffraction in studying crystalline sulfates is discussed by Hawthorne et al. (Chapter 1, this volume). X-ray and neutron scattering methods are also used in studying the coordination chemistry of sulfates in aqueous solutions. These methods are most applicable for concentrated systems, and they have provided important information on the solvation of sulfate in aqueous solutions. Because a neutron beam is scattered by the atomic nuclei, and X-rays are scattered by the electron cloud around nuclei, these methods offer complementary information (Bats 1976, Bacon 1977). In X-ray scattering, the scattering power of atoms is a function of momentum transfer and the atomic number of the elements (electron density), whereas neutron scattering is dependent on the composition of the nucleus of the atom (isotope). The applications of these methods to the study of sulfate solutions are summarized by Magini et al. (1988), Ohtaki and Radnai (1993) and references therein.

The previous studies have indicated that it is difficult to evaluate the solvation of anions because the distance of $SO_4^{2-} \cdot H_2O_w$ overlaps strongly with that of $H_2O_w H_2O_w$ (O_w: oxygen of water molecules; subscript 'w' is used to distinguish the O of water molecules from that of sulfate). For metalsulfate inner-sphere complexes, this problem is even worse because SM (M = cation) distances overlap with those of SO_w. The average SO_w distances in concentrated NH_4^+, Mg^{2+}, Mn^{2+}, Ni^{2+}, Zn^{2+}, Cd^{2+}, Cr^{3+}, and In^{3+} sulfate solutions is approximately 3.80 nm (range: 3.70-3.93 nm; Ohtaki and Radnai 1993), and the hydration number (number of water molecules in the first solvation shell) is about 8.0 (range: 7-12). Correlations among these bond distances and the cation size, charge, and coordination number are not observed (Ohtaki and Radnai 1993).

Infrared emission spectroscopy

Compared to other infrared spectroscopy methods described above, infrared emission spectroscopy has been used infrequently, but it can offer information on *in situ* dehydration and high-temperature phase transitions in minerals. The IR emission spectra are collected by heating the samples to different temperatures, and the IR radiation emitted by the samples is analyzed by IR detectors. Vassallo and Finnie (1992) examined a few sulfate minerals using this technique, and the results indicated that the emission and transmission spectra of these minerals are similar.

Optical spectroscopy

Optical spectroscopy [ultraviolet-visible (UV-VIS) spectroscopy, sum-frequency generation spectroscopy] is not widely used in probing the molecular chemistry of sulfates in geological materials. The studies of Gaft et al. (1985) indicate that anhydrite, barite, celestite, and anglesite are luminescent and exhibit a distinct band around 360 nm. The luminescence was suggested to originate from the crystal structure rather than from physical impurities. However, when luminescent centers (e.g. Mn^{2+}, Gd^{3+}) were added to the crystal in the form of impurities, the luminescence disappeared. Several Fe^{3+} sulfates and other sulfate salts have also been investigated using the optical properties of cations that associate with sulfates (e.g. Rossman 1975, 1976; Wan et al. 1978, and references therein). Similarly, probe molecules have been used to examine the molecular orientation and bonding interactions of organosulfates on clean metal or oxide substrates (Huang et al. 2000, and references therein). Although they give indirect information on sulfate interactions with the optically active centers in crystals, these methods do not provide direct information on sulfate coordination and bonding. Sum-frequency generation spectroscopy has been used recently to investigate the nature of water and organic molecules at the interfaces (Miranda and Shen 1999, and references therein). In the past, the non-availability of lasers at low energies has limited the applications of this technique to studies on sulfates. But the availability of tunable lasers and the broad-band synchrotron beamlines optimized for infrared spectroscopy should assist in these studies in near future.

SUMMARY AND FUTURE DIRECTIONS

(1) Vibrational spectroscopy has been used extensively in studying crystalline sulfates, but very few studies have compared the spectral features of sulfate with its coordination environment. Many previous studies have cited the energies of vibrations without proper band assignments. This has severely limited the identification of sulfate molecular chemistry in unknown crystalline samples and aqueous solutions, and at interfaces. Theoretical studies combined with a detailed investigation of different crystalline sulfates would be useful for researchers in several disciplines.

(2) The number of synchrotron X-ray spectroscopic studies on sulfates and other related S compounds is increasing. The lack of synchrotron beamlines optimized for studying sulfate and other S compounds in geological materials and beamline oversubscription have limited the use of these methods. Availability of suitable multi-element solid-state detectors together with the beamline facilities will make the XANES and EXAFS studies possible on a variety of extremely dilute geological samples. These techniques could provide new and complementary information on the molecular chemistry of sulfates at low concentrations.

(3) Sulfate reactions at the mineralwater interfaces of Fe oxides and certain hydrous Fe oxides in aqueous solutions have been well explored using X-ray and vibrational spectroscopic methods. Such information on other minerals and airwater interfaces is absent. Molecular information on such systems would be useful to predict the nature of redox transformations, trace-element mobility in soil and sediment systems, and gas transfer at the interfaces.

(4) Theoretical studies of sulfate in different coordination environments and calculated electronic and vibrational spectra are useful in interpreting the spectral features of unknown samples. Such studies have been conducted for only a few selected species. Theoretical XANES calculations at the S-absorption edge have not yet been performed.

(5) Because of inconsistencies regarding the nature of aqueous complexation of sulfate in the published literature, detailed studies are necessary to evaluate the nature of complex formation between sulfate and cations and their stability in aqueous solutions. Such information is essential in evaluating the relevant thermodynamic information data that are used extensively in geochemistry.

ACKNOWLEDGMENTS

I thank Dr. C.N. Alpers for putting together this volume and providing valuable comments at every stage in the preparation of this manuscript, and Dr. P.H. Ribbe for helping with the preparation of the final document. Reviews by Drs. G.R. Rossman, P. O'Day and J.L. Jambor helped improve the presentation and technical discussions. I thank Ms. R. Reina and Ms. Jyothi for editorial comments, and Ms. Mayuri for helping with the figures. This work is supported by funding from the Basic Energy Sciences, DOE (Geosciences) and a faculty start-up grant from Princeton University.

REFERENCES

Anders SA, Padmore HA, Duarte RM, Renner T, Stammler T, Scholl A, Scheifein MR, Stöhr J (1999) Photoemission electron microscope for the study of magnetic materials. Rev Sci Instruments 70: 3973-3981

Ataka K, Osawa M (1998) In situ infrared study of water-sulfate coadsorption on gold(111) in sulfuric acid solutions. Langmuir 14:951-959

Bacon GE (1977) Neutron Scattering in Chemistry. Butterworths, Boston

Baker AD, Betteridge D (1972) Photoelectron Spectroscopy: Chemical and Analytical Aspects. Pergamon Press, New York

Bats JW (1976) X-ray Diffraction and Chemical Bonding. Twente University of Technology, Enschede

Benison KC, Goldstein RH, Wopenka B, Burruss RC, Pasteris JD (1998) Extremely acid Permian lakes and ground waters in North America. Nature 392:911-914

Bergström P, Lindgren J, Kristiansson O (1991) An IR study of ClO_4^-, NO_3^-, I^-, Br^-, Cl^-, and SO_4^{2-} anions in aqueous solution. J Phys Chem 95:8575-8580

Bhagavantam S, Venkataraidu T (1969) Theory of Groups and Its Application to Physical Problems. Academic Press, New York, 279 p

Bockris JOM, Khan SUM (1993) Surface Electrochemistry: A Molecular Level Approach. Plenum Press, New York

Briggs D (1977) Handbook of X-ray and Ultraviolet Photoelectron Spectroscopy. Heyden, London

Briggs D, Seah MP (1983) Practical Surface Analysis by Auger and X-ray Photoelectron Spectroscopy. Wiley, New York

Brooker MH, Tremaine PR (1992) Raman studies of hydration of hydroxy complexes and the effect on standard partial molar heat capacities. Geochim Cosmochim Acta 56:2573-2577

Brookins DG (1988) Eh-pH Diagrams for Geochemistry. Springer-Verlag, New York

Brown GE Jr (1990) Spectroscopic studies of chemisorption reaction mechanisms at oxide-water interfaces. *In* Mineral-Water Interface Geochemistry. Hochella MF, White AF (eds) Rev Mineral 23:309-363

Brown GE Jr, Calas G, Waychunas GA, Petiau J (1988) X-ray absorption spectroscopy and its applications in mineralogy and geochemistry. *In* Spectroscopic Methods in Mineralogy and Geology. Hawthorne FC (ed) Rev Mineral 18:431-512

Brown GE Jr, Henrich VE, Casey WH, Clark DL, Eggleston C, Felmy A, Goodman DW, Gratzel M, Maciel H, McCarthy MI, Nealson KH, Sverjensky DA, Toney MF, Zachara JM (1999) Metal oxide surfaces and their interactions with aqueous solutions and microbial organisms. Chem Rev 99:77-174

Brown GM, Hope GA (1995) *In situ* spectroscopic evidence for the adsorption of SO_4^{2-} ions at a copper electrode in sulfuric acid solution. J Electroanal Chem 382:179-182

Brown GS, Doniach S (1980) The principles of X-ray absorption spectroscopy. *In* Synchrotron Radiation Research. Winick H, Doniach S (eds) Plenum Press, New York

Brown ID, Altermatt D (1985) Bond-valence parameters obtained from a systematic analysis of the inorganic crystal structure database. Acta Crystallogr B 41:244-247

Calas G, Manceau A (1986) X-ray absorption spectroscopy of geologic materials. J Physique 47:813-818

Cejka J (1999) Infrared spectroscopy and thermal analysis of uranyl minerals. *In* Uranium: Mineralogy, Geochemistry and the Environment. Burns PC, Finch R (eds) Rev Mineral 38:521-622

Colthup NB, Daly LH, Wiberly SE (1990) Introduction to infrared and Raman spectroscopy. Academic Press, Boston

Cotton FA (1971) Chemical Applications of Group Theory. Wiley Eastern Ltd., New York.

Cruickshank DWJ, Robinson EA (1966) Bonding in orthophosphates and orthosulfates. Spectrochim Acta 22:555-563

Davis AR, Oliver BG. (1973) Raman spectroscopic evidence for contact ion pairing in aqueous magnesium sulfate solutions. J Phys Chem 77:1315-1316

Degenhardt J, McQuillan AJ (1999) *In situ* ATR-FTIR spectroscopic study of adsorption of perchlorate, sulfate, and thiosulfate ions onto chromium(III) oxide hydroxide thin films. Langmuir 15:4595-4602

Eggleston CM, Hug S, Stumm W, Sulzberger B, Afonso MDS (1998) Surface complexation of sulfate by hematite surfaces: FTIR and STM observations. Geochim Cosmochim Acta 62:585-593

Fadini A, Schnepel F-M (1989) Vibrational Spectroscopy. Ellis Horwood Ltd., New York

Faguy PW, Marinkovic NS, Adzic RR (1996) An *in situ* infrared study on the affect of pH on anion adsorption at Pt(111) electrodes from acid sulfate solutions. Langmuir 12:243-247

Farmer VC (1974) The Infrared Spectra of Minerals. Mineralalogical Society, London

Fateley WG, McDevitt NT, Bentley FF (1971) Infrared and Raman selection rules for lattice vibrations: The correlation method. Appl Spectr 25:155-173

Fernandez-Ramos A, Cabaleiro-Lago E, Hermida-Ramon JM, Martinez-Nunez E, Pena-Gallego A (2000) DFT conformational study of cysteine in gas phase and aqueous solution. J Mol Structure (Theochem) 498:191-200

Ferraro JR (1982) The Sadtler Infrared Spectra Handbook of Minerals and Clays. Sadtler, Philadelphia

Ferraro JR, Nakamoto K (1994) Introduction to Raman Spectroscopy. Academic Press, Boston

Frank P, Hedman B, Carlson RMK, Tyson TA, Roe AL, and Hodgson KO (1987) A large reservoir of sulfate and sulfonate resides within plasma sells from *Ascidia ceratodes*, revealed by X-ray absorption near-edge structure spectroscopy. Biochemistry 26:4975-4979

Frank P, Hedman B, Carlson RMK, Hodgson KO (1994) Interaction of vanadium and sulfate in blood cells from the tunicate *Ascidia ceratodes*: Observations using X-ray absorption edge structure and EPR spectroscopies. Inorg Chem 33:3794-3803

Frank P, Hedman B, Hodgson KO (1999) Sulfur allocation and vanadium-sulfate interactions in whole blood cells from the tunicate *Ascidia ceratodes*, investigated using X-ray absorption spectroscopy. Inorg Chem 38:260-270

Franz JD, Dubessy J, Mysen BO (1994) Ion-pairing in aqueous $MgSO_4$ solutions along an isochore to 500°C and 11 kbar using Raman spectroscopy in conjunction with the diamond anvil cell. Chem Geol 116:181-188

Gadsden JA (1975) Infrared spectra of minerals and related inorganic compounds. Butterworths, London

Gaft ML, Bershov LV, Krasnaya AR, Yaskoko VY (1985) Luminescence centers in anhydrite, barite, celestite and their synthetic analogs. Phys Chem Minerals 11:255-260

George GN, Gorbaty ML (1989) Sulfur K-edge X-ray absorption spectroscopy of petroleum asphaltenes and model compounds. J Am Chem Soc 111:3182-3186

George GN, Gorbaty ML, Kelemen SR, Sansone M (1991) Direct determination and wuantification of sulfur forms in coals from the Argonne Premium Sample Program. Energy & Fuels 5:93-97

George WO, Steele D (ed) (1995) Computing Applications in Molecular Spectroscopy. The Royal Society of Chemistry, Cambridge, UK

Gillette PC, Lando JB, Koenig JL (1982) Band shape analysis of Fourier transform infrared spectra. Appl Spectr 36, 401-404

Gorbaty ML, George GN, Kelemen SR (1990) Direct determination and quantification of sulfur forms in heavy petroleum and coals: 2. The sulfur K edge X-ray absorption spectroscopy approach. Fuel 69:945-950

Harrick NJ (1987) Internal Reflection Spectroscopy. Harrick Scientific Corporation, New York

Harrison J.B, Berkheiser V (1982) Anion interactions with freshly prepared hydrous iron oxides. Clays and Clay Minerals 30:97-102

Hasenoehrl EJ, Griffiths PR (1993) Classification of condensed-phase infrared spectra by stuctures using principal component analysis. Appl Spectrosc 47:643-653

Hawthorne FC (ed) (1988) Spectroscopic Methods in Mineralogy and Geology. Reviews in Mineralogy, Vol 18

Hawthorne FC, Waychunas GA (1988) Spectrum fitting methods. *In* Spectroscopic Methods in Mineralogy and Geology. Hawthorne FC (ed) Rev Mineral 18:63-96

Hester RE, Plane RA (1964) A Raman spectrophotometric comparison of interionic association in aqueous solutions of metal nitrates, sulfates and perchlorates. Inorg Chem 3:769-770

Hezel A, Ross SD (1968) Forbidden transitions in the infra-red spectra of tetrahedral anions: IV. The vibrational spectra (4000-400 cm^{-1}) of some cobalt(III) sulphato- and phosphato-complexes. Spectrochim Acta 24A:985-992

Hibble SJ, Walton RI, Feaviour MR, Smith AD (1999) Sulfur-sulfur bonding in the amorphous sulfides WS3, WS5, and Re2S7 from sulfur K-edge EXAFS studies. J Chem Soc, Dalton Trans 2877-2883

Hitchcock AP, Horseley JA, Stohr J (1986) Inner-shell excitation of thiophene and thiolane: Gas, solid, and monolayer states. J Chem Phys 85:4835-4848

Hochella MF Jr (1988) Auger electron and photoelectron spectroscopies. *In* Spectroscopic Methods in Mineralogy and Geology. Hawthorne FC (ed) Rev Mineral 18:573-630

Hochella MF Jr, White AF (eds) (1990) Mineral-Water Interface Geochemistry. Reviews in Mineralogy Vol 23, Mineral Soc Am, Washington, DC

Horn AB and Sully KJ (1999) ATR-FTIR spectroscopic studies of the formation of sulfuric acid and sulfuric acid monohydrate films. Phys Chem Chem Phys 1:3801-3806

Huang MH, Dunn BS, Zink JI (2000) *In situ* luminescence probing of the chemical and structural changes during formation of dip-coated lamellar phase dodecyl sulfate sol-gel thin films. J Am Chem Soc 122:3739-3745

Huffman GP, Mitra S, Huggins FE, Shah NS, Vaidya N, Lu F (1991) Quantitative analysis of all major forms of sulfur in coal by X-ray absorption fine structure spectroscopy. Energy & Fuels 5:574-581

Huffman GP, Shah NS, Huggins FE, Stock LM, Chatterjee K, Kilbane JJ, Chou M, Buchanan DH (1995) Sulfur speciation of desulfurized coals by XANES spectroscopy. Fuel 74:549-555

Hug S (1997) In situ Fourier transform infrared measurements of sulfate adsorption on hematite in aqueous solutions. J Colloid Interface Sci 188:415-422

Jirsak T, Rodriguez JA, Hrbek J (1999) Chemistry of SO_2 on Mo(110), MoO_2/Mo(110) and Cs/Mo(110) surfaces: effects of O and Cs on the formation of SO_3 and SO_4 species. Surface Science 426:319-335

Johnston CL (1990) Fourier transform infrared and Raman spectroscopy. *In* Instrumental surface analysis of geologic materials, Perry DL (ed), VCH Publishers, New York.

Karr C (1975) Infrared and Raman spectroscopy of lunar and terrestrial minerals. Academic Press, New York

Kasrai M, Bancroft GM, Brunner RW, Jonasson RG, Brown JR, Tan KH, Feg X (1994) Sulfur speciation in bitumens and asphaltenes by X-ray absorption fine structure spectroscopy. Geochim Cosmochim Acta 58:2865-2872

Kasrai M, Lennard WN, Brunner RW, Bancroft GM, Bardwell JA, Tan KH (1996) Sampling depth of total electron and fluorescence measurements in Si L-and K-edge absorption spectroscopy. Appl Surf Sci 99:303-312

Koningsberger DC, Prins R (eds) (1988) X-ray Absorption: Principles, Applications, Techniques of EXAFS, SEXAFS, and XANES. John Wiley & Sons, New York

Lee PA, Pendry JB (1975) Theory of extended X-ray absorption fine structure. Phys Rev B 11:2795-2811

Li RS, Chen JF, Yang H, Zhang WY, Wei Q, Jin MZ, Zheng YG (1993) Interaction between ammonium sulfate-iron oxide shown by ESR and Mössbauer spectroscopy. Catal Lett 18:317-322

Li D, Bancroft GM, Kasrai M, Fleet ME, Yang BX, Tan K, Peng M (1994) Sulfur K-, and L-edge X-ray absorption spectroscopy of sphalerite, chalcopyrite and stannite. Phys Chem Minerals 20:489-499

Li D, Bancroft GM, Kasrai M, Fleet ME, Feng XH, Tan K (1995) S K- and L-edge X-ray absorption spectroscopy of metal sulfides and sulfates: Applications in mineralogy and geochemistry. Can Mineral 33:949-960

Lindsay CG, Gibbs GV (1988) A molecular orbital study of bonding in sulfate molecules: Implications for sulfate crystal structures. Phys Chem Minerals 15:260-270

Lytle FW, Sayers DE, Stern EA (1982) The history and modern practice of EXAFS spectroscopy. *In* Advances in X-ray Spectroscopy. Bonnelle C, Mande C (eds) Pergamon, Oxford

Lytle FW, Greegor RB, Sandstrom DR, Marques EC, Wong J., Spiro CL, Huffman GP, and Huggins FE (1984) Measurement of soft X-ray absorption spectra with a fluorescent ion chamber detector. Nucl Instru Meth 226:542-548

Maddoms WF (1980) The scope and limitations of curve fitting. Appl Spectr 34:245-267

Magini M, Licheri G, Paschina G, Piccaluga G, Pinna G (1988) X-ray Diffraction of Ions in Aqueous Solutions: Hydration and Complex Formation. CRC Press, New York

Masion A, Rose J, Bottero JY, Tchoubar D, Elmerich P (1997) Nucleation and growth mechanisms of iron oxyhydroxides in the presence of PO_4 ions: 3. Speciation of Fe by small angle X-ray scattering. Langmuir 13:3882-3885

Max JJ, Menichelli C, Chapados C (2000) Infrared titration of aqueous sulfuric acid. J Phys Chem 104:2845-2858

McMillan PF, Hess AC (1988) Symmetry, group theory and quantum mechanics. *In* Spectroscopic Methods in Mineralogy and Geology. Hawthorne FC (ed) Rev Mineral 18:11-61

McMillan PF, Hofmeister AM (1988) Infrared and Raman spectroscopy. *In* Spectroscopic Methods in Mineralogy and Geology. Hawthorne FC (ed) Rev Mineral 18:99-159

Melendres CA (1998) Synchrotron infrared spectromicroscopy of electrode surfaces and interfaces. Synch Rad News 11:39-46

Miranda PB, Shen Y-R (1999) Liquid interfaces: A study by sum-frequency vibrational spectroscopy. J Phys Chem B103:3292-3307

Mochizuki Y, Agren H, Pettersson LGM, Carravetta V (1999) A theoretical study of sulphur K-shell X-ray absorption of cysteine. Chem Phys Let 309:241-248

Moenke H (1962) Mineralspektren I. Akademie-Verlag, Berlin

Morra MJ, Fendorf SE, Brown PD (1997) Speciation of sulfur in humic and fulvic acids using X-ray absorption near-edge structure (XANES) spectroscopy. Geochim Cosmochim Acta 61:683-688

Myer-Ilse W, Warwick T, Attwood D (2000) X-ray Microscopy. Proc 6th Intern Conf, Berkeley, California (1999), American Institute of Physics, New York

Myneni SCB, Martinez GA (1999) P and S functional group chemistry of humic substances. SSRL Activity Reports–1998, 364-368

Myneni SCB, Traina SJ, Waychunas GA, Logan TJ (1998a) Vibrational spectroscopy of functional group chemistry and arsenate coordination in ettringite. Geochim Cosmochim Acta 62: 3499-3514

Myneni SCB, Traina SJ, Waychunas GA, Logan TJ (1998b) Experimental and theoretical vibrational spectroscopic evaluation of arsenate coordination in aqueous solutions, solids and at mineral-water interfaces. Geochim Cosmochim Acta 62:3285-3300

Myneni SCB, Luo Y, Naslund LA, Ojamae L, Ogasawara H, Pelmenshikov A, Vaterlain P, Heske C, Pettersson LGM, Nilsson A (2000a) Spectroscopic evidence for unique hydrogen bonding structures in water (unpublished data, submitted for publication)

Myneni SCB, Waychunas GA, Traina SJ, Brown GA (2000b) Molecular structure of sulfate on Fe-oxide surfaces (unpublished data, manuscript to be submitted)

Nakamoto K. (1986) Infrared and Raman Spectra of Inorganic and Coordination Compounds. John Wiley & Sons, New York

Natoli CR (1995) XAS, MCD, and PED interpreted in the unifying framework of effective MS theory. Physica B 208 & 209:5-10

Nefedov VI and Formichev VA (1968) J Struct Chem 9:107 (cited by Sekiyama et al. 1986).

Nomura H, Koda S, Miyahara Y (1980) Water and metal cations in biological systems. *In* Proc Symp 1978, Pullman B, Yuji K (eds) Scientific Societies Press, Tokyo.

Ogletree DF (2000) Photoelecron spectroscopy at ten torr. Abstracts of the 8th Intern Conf on Electron Spectroscopy and Structure (conference proceedings will appear in a special issue of J Elctron Spectr Relat Phenom), Berkeley, CA

Ohtaki H, Radnai T (1993) Structure and dynamics of hydrated ions. Chem Rev 93:1157-1204

Okude N, Nagoshi M, Noro H, Baba Y, Yamamoto H, Sasaki TA (1999) P and S K-edge XANES of transition-metal phosphates and sulfates. J Electron Spectrosc Relat Phenom 101-103:607-610.

Ostergren JD, Brown GE, Parks GA, Persson P (2000) Inorganic ligand effects on Pb(II) sorption on goethite (-FeOOH). J Colloid Interface Sci 225:483-493

Parfitt RL, Smart RStC (1978) Mechanism of sulfate adsorption on iron oxides. Soil Soc Am J 42:48-50

Pauling L (1960) The Nature of the Chemical Bond. Cornell Univ Press, Ithaca, NY

Peak D, Ford RG, Sparks DL (1999) An *in situ* ATR-FTIR investigation of sulfate bonding mechanisms on goethite. J Colloid Interface Sci 218:289-299

Pechar F (1988) Infrared Reflection Spectra of Selected Minerals. Academia, Praha

Perkins JH, Hasenoehrl EJ, Griffiths PR (1991) Anal Chem 63:1738

Perry DL (ed) (1990) Instrumental Surface Analysis of Geologic Materials. VCH Publishers, New York

Persson P, Lövgren L (1996) Potential and spectroscopic studies of sulfate complexation at the goethite-water interface. Geochim Cosmochim Acta 60:2789-2799

Persson P, Nilsson N, Sjöberg S (1996) Structure and bonding of orthophosphate ions at the iron oxide-aqueous interface. J Colloid Interface Sci 177:263-275

Pickering IJ, Prince RC, Divers T, George GN (1998) Sulfur K-edge X-ray absorption spectroscopy for determining the chemical speciation of sulfur in biological systems. FEBS Letters 441:11-14

Pingitore NE, Meitzner G, Love KM (1995) Identiifcation of sulfate in natural carbonates by X-ray absorption spectroscopy. Geochim Cosmochim Acta 59:2477-2483

Rehr JJ, Booth CH, Bridges F, Zabinsky SI (1994) X-ray absorption fine structure in embedded atoms. Phys Rev 49:12347-12350

Rehr J and Albers RC (2000) Theoretical approaches to X-ray absorption fine structure. Rev Mod Phys 72:621-654.

Ressler T, Wong J, Roos J, Smith I (2000) Quantitative speciation of Mn-bearing particles emitted from autos burning (methylcyclopentadienyl) manganese tricarbonyl-added gasolines using XANES spectroscopy. Environ Sci Technol 34:950-958

Rodriguez JA, Jirsak T, Chaturvedi S, Kuhn M (1999a) Reaction of SO_2 with ZnO (000-1)-O and ZnO powders: Photoemission and XANES studies on the formation of SO_3 and SO_4. Surf Sci 442:400-412

Rodriguez JA, Chaturvedi S, Hanson J, Brito JL (1999b) Reaction of H_2 and H_2S with $CoMoO_4$ and $NiMoO_4$: TPR, XANES, time-resolved XRD, and molecular orbital studies. J Phys Chem B 103: 770-781

Rompel A, Cinco RM, Latimer MJ, McDermott AE, Guiles RD, Quintanilha A, Krauss RM, Sauer K. Yachandra V, Klein MP (1998) Sulfur K-edge X-ray absorption spectroscopy:A spectroscopic tool to examine the redox state of S-containing metabolites in vivo. Proc Natl Acad Sci 95:6122-6127

Rose Williams K, Hedman B, Hodgson KO, Solomon EI (1997) Ligand K-edge X-ray absorption spectroscopic studies:metal-ligand covalency in transition metal thiolates. Inorg Chim Acta 263:315-321

Ross SD (1974) Phosphates and other oxy-anions of group V. *In* The Infrared Spectra of Minerals. Farmer VC (ed) Mineral Soc London, p 383-422

Rossman GR (1975) Spectroscopic and magnetic studies of ferric iron hydroxy sulfates: Intensification of color in ferric iron clusters bridged by a single hydroxide ion. Am Mineral 60:698-704

Rossman GR (1976) Spectroscopic and magnetic properties of ferric iron hydroxy sulfates: The series $Fe(OH)SO_4 \cdot nH_2O$ and jarosites. Am Mineral 61:398-404

Rudolph WW (1994) Raman and infrared spectroscopic investigation of contact ion pair formation in aqueous cadmium sulfate solutions. J Soln Chem 23:663-683

Rudolph WW (1996) Structure and dissociation of the hydrogen sulfate ion in aqueous solution over a broad temperature range: A Raman study. Z Phys Chemie 194:73-95

Rudolph WW (1998a) Hydration and water-ligand replacement in aqueous cadmium (II) sulfate solution. A Raman and infrared study. J Chem Soc Faraday Trans 94:489-499

Rudolph WW (1998b) A Raman spectroscopic study of hydration and water-ligand replacement reaction in aqueous cadmium(II) sulfate solution: Inner-sphere and outer-sphere complexes. Ber Bunsenges Phys Chem 102:183-196

Rudolph WW, Irmer G (1994) Raman and infrared spectroscopic investigation of contact ion pair formation in aqueous cadmium sulfate solutions. J Soln Chem 23:663-683

Rudolph WW, Brooker MH, Tremaine PR (1997) Raman spectroscopic investigation of aqueous $FeSO_4$ in neutral and acidic solutions from 25°C to 303°C: Inner- and outer-sphere complexes. J Soln Chem 26:757-777

Rudolph WW, Brooker MH, Tremaine PR (1999) Raman spectroscopy of aqueous $ZnSO_4$ solutions under hydrothermal conditions: Solubility, hydrolysis, and sulfate ion pairing. J Soln Chem 28:621-630

Rull F, Ohtaki H (1997) Raman spectral studies on ionic interaction in aqueous alkali sulfate solutions. Spectrochim Acta 53A:643-653

Salisbury JW (1992) Infrared (2.1-25mm) spectra of minerals. Johns Hopkins University Press, Baltimore

Sarret G, Connan J, Kasrai M, Bancroft GM, Charrie-Duhaut A, Lemoine S, Adam P, Albrecht P, Eybert-Berard L (1999) Chemical forms of sulfur in geological and archaeological asphaltenes from the Middle East, France, and Spain determined by sulfur K-, and L-edge X-ray absorption near-edge structure spectroscopy. Geochim Cosmochim Acta 63:3767-3779

Sayers DE (1975) Extended X-ray absorption fine structure technique: III. Determination of physical parameters. Phys Rev B11: 4836-4845

Sayers DE, Lytle FW, Stern EA (1970) Point scattering theory of X-ray K absorption fine structure. Advan X-ray Anal 13:248-271

Schulze DG, Stucki JW, Bertsch PM (eds) (1999) Synchrotron X-ray Methods in Clay Science. The Clay Minerals Society, Boulder, Colorado

Seeger DM, Korzeniewski C, Kowalchyk W (1991) Evaluation of vibrational force fields derived by using semiempirical and *ab initio* methods. J Phys Chem 95:6871-6879

Sekiyama H, Kosugi N, Kuroda H, Ohta T (1986) Sulfur K-edge absorption spectra of Na_2SO_4, Na_2SO_3, $Na_2S_2O_3$, and $Na_2S_2O_x$ (x = 5-8). Bull Chem Soc Jpn 59:575-579

Shadle SE, Hedman B, Hodgson KO, Soloman EI (1995) Ligand K-edge X-ray absorption spectroscopic studies: Metal-ligand covalency in a series of transition metal tetrachlorides. J Am Chem Soc 117: 2259-2272

Siedl V, Knop O (1969) Infrared studies of water in crystalline hydrates: Gypsum, $CaSO_4 \cdot 2H_2O$. Can J Chem 47:1362-1368

Smith R, Martell AE (1976) Critical Stability Constants: Vol 4. Inorganic Complexes. Plenum Press, New York

Spiro CL, Wong J, Lytle FW, Greegor RB, Maylotte DH, Lamson SH (1984) X-ray absorption spectroscopic investigation of sulfur sites in coal: Organic sulfur identification. Science 226:48-50

Sposito G (1989) The Chemistry of Soils. Oxford University Press, New York

Sposito G (1990) Molecular models of ion adsorption on mineral surfaces. *In* Mineral-Water Interface Geochemistry. Hochella MF, White AF (eds) Rev Mineral 23:261-279

Sposito G (1994) Chemical Equilibria and Kinetics in Soils. Oxford University Press, New York

Stern EA (1974) Theory of extended X-ray absorption fine structure. Phys Rev B 10:3027-3037

Stevenson FJ (1994) Humus Chemistry. John Wiley, New York

Stewart JJP (1989a) Optimization of parameters for semiempirical methods: I. Method, J Comp Chem 10:209-220

Stewart JJP (1989b) Optimization of parameters for semiempirical methods: II. Applications, J Comp Chem 10:221-264

Stöhr J (1984) Surface crystallography by means of SEXAFS and NEXAFS. *In* Chemistry and Physics of Solid Surfaces. Vanselow VR, Howe R (eds) Springer-Verlag, New York

Stöhr J (1992) NEXAFS Spectroscopy. Springer-Verlag, Berlin

Stumm W (1992) Chemistry of the Solid-Water Interface. John Wiley, New York

Stumm W, Morgan JJ (1982) Aquatic Chemistry, 2nd edn. Wiley Interscience, New York

Suetaka W (1995) Surface infrared and Raman spectroscopy: Methods and applications. Plenum Press, New York

Taesler I, Olovsson I (1969) Hydrogen bond studies. XXXVII. The crystal structure of sulfuric acid dihydrate $(H_3O^+)_2SO_4^{2-}$. J Chem Phys 51:4213-4219

Teo BK (1986) EXAFS: Basic principles and data analysis. Inorganic chemistry concepts 9. Springer-Verlag, Berlin

Tossell JA, Vaughan DJ (1992) Theoretical Geochemistry: Application of Quantum Mechanics in the Earth and Mineral Sciences. Oxford University Press, New York

Turner LJ, Kramer JR (1991) Sulfate ion binding on goethite and hematite. Soil Sci 152:226-230

Tyson TA, Roe AL, Frank P, Hodgson KO, Hedman B (1989) Polarized experimental and theoretical K-edge X-ray absorption studies of SO_4^{2-}, ClO_3^-, $S_2O_3^{2-}$, and $S_2O_6^{2-}$. Phys Rev B. 39:6305-6315

Urban MW (1993) Vibrational spectroscopy of molecules and macromolecules on surfaces. John Wiley & Sons, New York

Vairavamurthy A (1998) Using X-ray absorption to probe sulfur oxidation states in complex molecules. Spectrochim Acta 54A:2009-2017

Vairavamurthy A, Manowitz B, Zhou W, Jeon Y (1994) Determination of hydrogen sulfide oxidation products by sulfur K-edge X-ray absorption near edge structure spectroscopy. In Environmental Geochemistry of Sulfide Oxidation. Alpers CN, Blowes DW (eds) Am Chem Soc Symp Series 550:412-430, Am Chem Soc, Washington, DC

Vairavamurthy A, Maletic D, Wang S, Manowitz B, Eglinton T, Lyons T (1997) Characterization of sulfur-containing functional groups in sedimentary humic substances by X-ray absorption near-edge structure spectroscopy. Energy Fuels 11:546-553

Vassallo AM, Finnie KS (1992) Infrared emission spectroscopy of some sulfate minerals. Appl Spectrosc 46:1477-1482

Vaughan D (1986) X-ray Data Booklet. Lawrence Berkeley Laboratory Publication, Berkeley, CA

Vaughan DJ, Pattrick RAD (eds) (1995) Mineral Surfaces. Mineralogical Society Series, Chapman & Hall, New York

Waldo GS, Carlson RMK, Moldowan JM, Peters KE, Penner-Hahn JE (1991) Sulfur speciation in heavy petroleums: Information from X-ray absorption near-edge structure. Geochim Cosmochim Acta 55:801-814

Wan C, Ghose S, Rossman GR (1978) Guildite, a layer structure with a ferric hydroxy-sulfate chain and its optical absorption spectra. Am Mineral 63:478-483

Warburton DR, Purdie D, Muryn CA, Prakash NS, Prabhakaran K, Thronton G, Pattrick RAD, Norman D (1992) Transferability of phase shifts for extended X-ray absorption fine structure studies of metal sulfides and sulfur on nickel surfaces. Phys Rev B 45:12043-12049

Warneck P (1988) Chemistry of the Natural Atmosphere. Academic Press, San Diego

Warwick T, Ade H, Cerasari S, Denlinger J, Franck K, Gracia A, Hayakawa A, Hitchcock A, Kikuma J, Kortright J, Meigs G, Moronne M, Myneni SCB, Rightor E, Rotenberg E, Seal S, Shin H-J, Steele R, Tyliszczak T, Tonner B (1998) A scanning transmission X-ray microscope for materials science spectromicroscopy at the Advanced Light Source. Rev Sci Instr 69:2964-2973

Wasserman SR, Allen PG, Shuh DK, Bucher JJ (1999) EXAFS and principal component analysis: A new shell game. J Sync Radiation 6:284-284

Watanabe H, Gutleben CD, Seto J (1994) Sulfate ions on the surface of maghemite and hematite. Solid State Ionics 69:29-35

Waychunas GA, Brown GE (1984) Applications of EXAFS and XANES spectroscopy to problems in mineralogy and geochemistry. In EXAFS and Near-Edge Structure: III. Hodgson KO et al. (eds) Springer-Verlag, New York

Waychunas GA, Brown GE Jr., Apted MJ (1986) X-ray K-edge absorption spectra of Fe minerals and model compounds: II. EXAFS. Phys Chem Minerals 13:31-47

Wu ZY, Ouvrard G, Moreau P, Natoli CR (1997) Interpretation of pre-edge features in the Ti and S K-edge X-ray absorption near edge spectra in the layered disulphides TiS_2, and TaS_2. Phys Rev B55: 9508-9513

Xia K, Weesner F, Bleam W, Bloom PR, Skyllberg UL, Helmke PA (1998) XANES studies of oxidation states of sulfur in aquatic and soil humic substances. Soil Sci Soc Am J 62:1240-1246

3   Sulfate Minerals in Evaporite Deposits

Ronald J. Spencer

Department of Geology and Geophysics
The University of Calgary
2500 University Drive N.W.
Calgary, AB T2N 1N4 Canada

Evaporite deposits preserve a wealth of information on Earth's past surface conditions. The deposits are sensitive indicators of depositional environment and climate, recorded in a wide variety of mechanically and chemically produced sedimentary structures and fabrics. Unique to evaporite deposits is the record they carry of the hydrochemistry of surface waters, including seawater. Sulfate minerals are important components in understanding the hydrochemistry of ancient surface waters.

Sedimentological aspects of evaporite deposits were compiled by Melvin (1991). In the compilation, sedimentology specific to Ca-bearing sulfates was described by Warren (1991); Lowenstein and Hardie (1985), Handford (1991), and Smoot and Lowenstein (1991) discussed the sedimentology of evaporites in general. These general principles can be applied specifically to the sulfate minerals in evaporite deposits. Sedimentological aspects of sulfate minerals in evaporite deposits are not discussed here. For this information, the reader is referred to the above references.

Hardie et al. (1985) and Spencer and Lowenstein (1988) discussed the petrography of evaporites and the interpretation of fabrics of evaporite minerals. Examples of petrographic aspects of sulfate minerals in evaporites are presented here along with references specific to the sulfate minerals. For a more complete understanding of these aspects the reader is referred to the above references.

The major focus of this chapter is on the chemical aspects of the common sulfate minerals that occur in evaporite deposits. The discussion begins with an explanation of the concept of chemical divides and the determination of evaporation paths, and the evolution of brines in accordance with mineral solubility and the composition of dilute inflow waters. Determination of the precipitation pathways of the more soluble salts is more complex. Examples of sulfate-mineral assemblages in evaporation systems, calculated via thermochemical modelling, are given for both marine and non-marine systems.

SOLUBILITY CONTROLS ON MINERAL PRECIPITATION AND PATHS OF EVAPORATION

The mineralogy of evaporite deposits is dependent on the chemical composition of the source waters. Hardie and Eugster (1970) developed the concept of "chemical divides" to explain the evolution of brine compositions in non-marine saline lakes. These "chemical divides" are based on the early precipitation of relatively insoluble minerals such as calcite and gypsum. The concept is used to determine the ultimate chemical signature of the brines and the mineralogy of the evaporites formed from them. This simple concept is used successfully in accounting for the composition of lacustrine brines, where mixing of multiple source waters results in a variety of mineral assemblages. Spencer and Hardie (1990) also used this concept to explain seawater chemistry.

1529-6466/00/0040-0003$05.00

Evaporite deposits are formed from the evaporation of surface waters. The major dissolved cations in surface waters are Na^+, Ca^{2+}, Mg^{2+} and K^+, which are derived through chemical weathering of the crust. The abundance of these cations is dependent on their relative abundance in crustal rocks, and on the solubility of the minerals containing them.

The minerals formed by combining the above cations with sulfate are thus potential candidates for evaporite deposits. Which minerals are formed depends primarily on the ratios of these cations in solution and on the relative solubility of the sulfate minerals that contain these cations.

Solubility of Na-K-Ca-Mg-bearing sulfate minerals

The chemical composition and solubility of several Na-K-Ca-Mg sulfate salts are given in Table 1. There are a limited number of minerals in this chemical system. Minerals range from the simple anhydrous salts anhydrite, arcanite and thenardite, to simple hydrous salts such as gypsum, mirabilite and a series of variably hydrated Mg sulfates from kieserite to epsomite. There are mixed salts for most of the cation pairs with sulfate (all except Ca-Mg), as well as polyhalite, which contains Ca, K, and Mg with sulfate. Relatively few salts have sulfate in combination with either carbonate or chloride.

Table 1. Common Ca-, K-, Mg-, Na-bearing sulfate salts, arranged in
order of solubility for dissolution of mineral in water.

Solubility of minerals in molality and grams per kilogram of water are calculated using the thermochemical model of Harvie et al. (1984). The same abbreviations are used in Figures 1 through 8.

Mineral	Abbrev.	Component formula	Standard formula	Molality	(gm/kg H_2O)
Gypsum	gy	$CaSO_4\ 2H_2O$	$CaSO_4\ 2H_2O$	0.0153	2.08
Anhydrite	an	$CaSO_4$	$CaSO_4$	0.0229	3.11
Syngenite	sy	$K_2SO_4\ CaSO_4\ H_2O$	$K_2Ca(SO_4)_2\ 2H_2O$	0.1528	23.68
Polyhalite	po	$2(CaSO_4)\ K_2SO_4\ MgSO_4\ 2H_2O$	$K_2Ca_2Mg(SO_4)_4\ 2H_2O$	0.3913	55.37
Glauberite	gl	$Na_2SO_4\ CaSO_4$	$Na_2Ca(SO_4)_2$	0.8534	118.20
Arcanite	ar	K_2SO_4	K_2SO_4	0.6927	120.53
Aphthitalite	ap	$.5(Na_2SO_4)\ 1.5\ (K_2SO_4)_2$	$NaK_3(SO_4)_2$	1.118	185.74
Eugsterite	eu	$2(Na_2SO_4)\ CaSO_4\ 2H_2O$	$Na_4Ca(SO_4)_3\ 2H_2O$	1.927	268.50
Mirabilite	mi	$Na_2SO_4\ 10H_2O$	$Na_2SO_4\ 10H_2O$	1.940	273.54
Picromerite	pi	$K_2SO_4\ MgSO_4\ 6H_2O$	$K_2Mg(SO_4)_2\ 6H_2O$	1.959	287.97
Epsomite	ep	$MgSO_4\ 7H_2O$	$MgSO_4\ 7H_2O$	2.986	358.32
Leonite	le	$K_2SO_4\ MgSO_4\ 4H_2O$	$K_2Mg(SO_4)_2\ 4H_2O$	2.514	369.56
Hexahydrite	hx	$MgSO_4\ 6H_2O$	$MgSO_4\ 6H_2O$	3.565	427.80
Pentahydrite	pe	$MgSO_4\ 5H_2O$	$MgSO_4\ 5H_2O$	4.320	518.40
Thenardite	th	Na_2SO_4	Na_2SO_4	3.687	519.87
Blödite	bl	$Na_2SO_4\ MgSO_4\ 4H_2O$	$Na_2Mg(SO_4)_2\ 4H_2O$	4.004	522.52
Starkeyite	st	$MgSO_4\ 4H_2O$	$MgSO_4\ 4H_2O$	5.032	603.84
Kieserite	ks	$MgSO_4\ H_2O$	$MgSO_4\ H_2O$	5.619	674.28
Mixed-anion salts					
Burkeite	bu	$2(Na_2SO_4)\ Na_2CO_3$	$Na_6(SO_4)_2CO_3$		
Kainite	ka	$KCl\ MgSO_4\ 3H_2O$	$KMgClSO_4\ 3H_2O$		

The most common of these are the Na-sulfate-carbonate salt burkeite, and the K-Cl-Mg-sulfate kainite.

The minerals in Table 1 are arranged according to the total dissolved ions at saturation for dissolution of the mineral in water. The common simple Ca sulfate salts, gypsum and anhydrite, are the least soluble of these minerals; only a few grams can be dissolved in a kilogram of water. In contrast, more than 120 g of the K sulfate mineral arcanite, the next least soluble of the simple salts, can be dissolved in the same amount of water. The Na and Mg sulfates mirabilite and epsomite are even more soluble. The solubility of each mixed-cation salt generally falls between those of the simple salts for the cations involved. The least soluble of the mixed salts are the Ca-bearing minerals syngenite, polyhalite, and glauberite; the most soluble is the Na-Mg-sulfate mineral blödite.

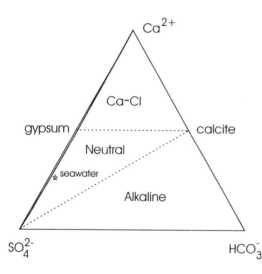

Figure 1. Ternary phase diagram (in equivalents) for Ca–SO$_4$–HCO$_3$ illustrating the concept of chemical divides. The bulk of the diagram is the primary stability field for calcite; the gypsum stability field is along the Ca–SO$_4$ join. Chemical divides from calcite to sulfate and from calcite to gypsum (dashed lines) separate the diagram into three distinct fields. The fields are for brines of the alkaline, neutral, and Ca–chloride type.

Chemical divides in the system Ca^{2+}-SO_4^{2-}-HCO_3^-

The concept of chemical divides is illustrated in Figure 1, which is a ternary phase diagram for the system Ca^{2+}-SO_4^{2-}-HCO_3^-. The units used for this diagram are equivalents, placing the composition of calcite at the midpoint of the Ca^{2+}-HCO_3^- join and gypsum at the midpoint of the Ca^{2+}-SO_4^{2-} join. Primary stability fields for calcite and gypsum were calculated using the thermochemical model of Harvie et al. (1984), with pH controlled using atmospheric values for carbon dioxide. Because calcite is much less soluble than gypsum, the bulk of the body of this diagram is in the primary stability field for calcite. The stability field for gypsum is restricted to a narrow band along the Ca^{2+}-SO_4^{2-} join. The primary stability field determines which mineral is first to precipitate from a given water composition. Therefore, calcite is the first mineral to precipitate from waters whose composition falls within the bulk of this compositional triangle.

Once a mineral begins to precipitate, the composition of the fluid moves directly away from the composition of that mineral on the diagram. This produces two chemical divides within the system Ca^{2+}-SO_4^{2-}-HCO_3^-. One is along the calcite-SO_4^{2-} join, and the other along the calcite-gypsum join. These chemical divides separate the diagram into three distinct fields.

Alkaline waters or brines fall within the calcite-SO_4^{2-}-HCO_3^- sub-triangle; the equivalents of HCO_3^- in solution are greater than the Ca^{2+} equivalents. Alkaline waters precipitate calcite and become depleted in Ca^{2+} as they migrate toward the HCO_3^--SO_4^{2-} join, resulting in Ca free (or very low Ca) sulfate-bearing brines. Average river waters generally lie within the alkaline field. Alkaline brines do not precipitate the Ca-bearing sulfate salts.

Ca-chloride brines fall in the Ca^{2+}-calcite-gypsum sub-triangle; the equivalents of Ca^{2+} in solution are greater than the combined SO_4^{2-} and HCO_3^- equivalents. Precipitation of calcite leads waters to the univariant curve dividing the primary stability fields for calcite and gypsum. After intersecting the univariant curve, waters precipitate both calcite and gypsum as they move toward the Ca^{2+} corner of the diagram. After the precipitation of calcite and gypsum, Ca-chloride brines are depleted in both SO_4^{2-} and HCO_3^-. These brines only yield gypsum and anhydrite as primary sulfate minerals; mixed-cation sulfate salts such as syngenite, polyhalite, and glauberite may form from back reaction of the gypsum and anhydrite with evolved brines.

Neutral brines fall in the third sub-triangle of the diagram; the equivalents of Ca^{2+} in solution are less than the combined SO_4^{2-} and HCO_3^- equivalents, but the equivalents of Ca^{2+} in solution are greater than HCO_3^- equivalents. As in the case of Ca-chloride brines, precipitation of calcite leads waters to the univariant curve dividing the primary stability fields for calcite and gypsum. However, waters in the neutral field move toward the SO_4^{2-} corner of the diagram as both calcite and gypsum precipitate. After the precipitation of calcite and gypsum, neutral brines are depleted in both Ca^{2+} and HCO_3^-. These brines yield not only gypsum and anhydrite as primary sulfate minerals, but also yield more soluble sulfate salts; mixed-cation salts such as syngenite, polyhalite, and glauberite may also form from back reaction of the gypsum and anhydrite with evolved brines. Modern seawater falls within the neutral field.

Precipitation sequences of Na-K-Mg-bearing sulfates

Prediction of the precipitation of sulfate minerals beyond gypsum and anhydrite is more difficult because of the number of possible salts with similar solubility and the complex solution chemistry. For a given water chemistry, sequences of mineral precipitation can be predicted with a thermochemical model suitable for brines. Several ternary diagrams for the system Na^+-Mg^{2+}-K^+-SO_4^{2-}, calculated using the thermochemical model of Harvie et al. (1984), are presented to illustrate some of the variability in this system. Figures 2 through 4 display a sequence from lower to higher Ca^{2+}, and illustrate the influence of higher Ca levels on mineral stability fields. Figures 4 and 5 illustrate the influence of varying chloride levels.

Figure 2 is a ternary Na-K-Mg-sulfate phase diagram constructed for alkaline brines in which the equivalents of HCO_3^- are greater than Ca^{2+}, and SO_4^{2-} and HCO_3^- are equal. This is an example of an alkaline brine system which is very low in Ca^{2+}. Calcite is present in all assemblages, and there is no primary stability field for Ca-bearing sulfate minerals. The primary stability fields for the K-bearing sulfate minerals arcanite and aphthitalite occupy the bulk of the diagram. Primary stability fields for the Na-bearing carbonate minerals gaylussite and pirssonite are present near the Na corner. The primary stability field for mirabilite is along the Na-Mg join, and blödite, picromerite, and epsomite all have primary stability fields near the Mg corner. There is a single eutectic invariant point involving blödite, picromerite, and epsomite. All of the remaining six invariant points are all peritectic reaction points.

The ternary system Na-K-Mg-sulfate in which the equivalents of Ca^{2+} in solution are equal to the HCO_3^- equivalents (along the chemical divide between neutral and alkaline

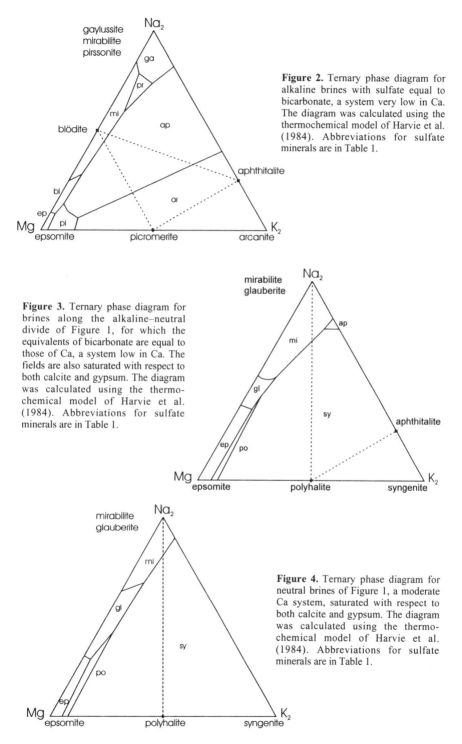

Figure 2. Ternary phase diagram for alkaline brines with sulfate equal to bicarbonate, a system very low in Ca. The diagram was calculated using the thermochemical model of Harvie et al. (1984). Abbreviations for sulfate minerals are in Table 1.

Figure 3. Ternary phase diagram for brines along the alkaline–neutral divide of Figure 1, for which the equivalents of bicarbonate are equal to those of Ca, a system low in Ca. The fields are also saturated with respect to both calcite and gypsum. The diagram was calculated using the thermochemical model of Harvie et al. (1984). Abbreviations for sulfate minerals are in Table 1.

Figure 4. Ternary phase diagram for neutral brines of Figure 1, a moderate Ca system, saturated with respect to both calcite and gypsum. The diagram was calculated using the thermochemical model of Harvie et al. (1984). Abbreviations for sulfate minerals are in Table 1.

brines) is shown in Figure 3. Gypsum and calcite are present in all assemblages. The relative Ca^{2+} concentration in this system is higher than in Figure 2. The primary stability field for the mixed K-Ca-sulfate mineral syngenite occupies the bulk of the diagram, replacing the arcanite and aphthitalite fields from the very low Ca system shown in Figure 2 (a small aphthitalite field remains). The primary stability field for mirabilite occupies the region near the Na corner. Primary stability fields for glauberite and polyhalite are also present. There is a single eutectic involving epsomite, glauberite, and polyhalite. The remaining three invariant points are peritectic reaction points.

The ternary phase diagrams for the chloride-free Na-K-Mg-sulfate system at calcite and gypsum saturation (Fig. 4) is used to depict phase relations for neutral brines near the gypsum-calcite join. This system contains relatively more Ca^{2+} than those shown in Figures 2 and 3. The topology of Figures 3 and 4 is similar. However, the aphthitalite field is not present in Figure 4, and the primary stability fields for the Ca-bearing sulfates glauberite and polyhalite are slightly larger in Figure 4 than in Figure 3. The single eutectic involves epsomite, glauberite, and polyhalite. The remaining two invariant points are peritectic reaction points.

The chloride-saturated Na-K-Mg-sulfate system at calcite and gypsum (or anhydrite) saturation near the gypsum-calcite join is shown in Figure 5. The univariant curves for halite-sylvite, halite-carnallite and sylvite-carnallite are shown for reference. The primary stability field for the K-Ca-bearing sulfate mineral syngenite occupies the bulk of the diagram. The remaining topology of Figure 5 differs significantly from the chloride-free system shown in Figure 4. No primary stability field for mirabilite is present in the chloride-saturated system, and the glauberite and polyhalite fields are much larger than in the chloride-free system. The epsomite field is smaller in the chloride-saturated system, and there are also primary stability fields for blödite, hexahydrite, and kieserite.

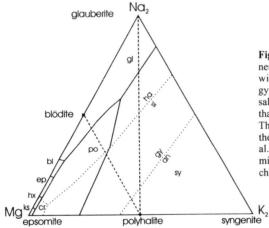

Figure 5. Ternary phase diagram for neutral brines of Figure 1, saturated with respect to both calcite and gypsum (or anhydrite) and a chloride salt. Figure 5 differs from Figure 4 in that it is for a high–salinity system. The diagram was calculated using the thermochemical model of Harvie et al. (1984). Abbreviations for sulfate minerals are listed in Table 1, and for chloride minerals in Table 2.

Although the sulfate mineralogy of evaporites can be fairly complex, certain generalizations can be made. Because the Ca-bearing sulfate minerals are relatively insoluble, the relative concentration of Ca^{2+} in evaporating waters has a strong influence on the sulfate minerals that may form. In high-Ca^{2+} systems with Ca-chloride-type waters, the only sulfate minerals expected are gypsum and anhydrite. In low-Ca^{2+} systems with alkaline waters, Ca-bearing sulfate minerals are not expected. Evaporation of these waters is likely to result in precipitation of simple Na, K, and Mg sulfates with or without mixed K-Mg sulfates. Evaporation of waters with moderate Ca^{2+} contents, neutral brines, is likely to produce Ca-bearing sulfate salts, along with mixed-cation Ca-bearing sulfate salts containing Na, K, and Mg.

Evaporation paths and mineralogy of marine evaporites

Sulfate is the second most abundant anion, after chloride, in modern seawater. Complete evaporation of modern seawater yields a significant quantity of sulfate salts, primarily as Mg-bearing sulfate minerals. Table 2 (Column B) shows the sequence of minerals predicted by the thermochemical model of Harvie et al. (1984) to form from the evaporation of modern seawater. Gypsum is the first sulfate mineral to form; thermodynamically, gypsum may alter to anhydrite just prior to the initiation of halite precipitation. Glauberite is the next sulfate mineral to form, as a reaction product with anhydrite, and is then replaced by anhydrite. The third sulfate mineral to form is polyhalite, again, as an alteration product from anhydrite. Finally, Mg sulfate precipitates, initially as epsomite, followed by hexahydrite and kieserite. The invariant mineral assemblage contains the Ca-sulfate salt anhydrite and the Mg-sulfate salt kieserite.

The mineralogy of evaporite deposits can potentially yield the composition of ancient surface waters. This depends on the extent of evaporation and the minerals precipitated and preserved. For instance, almost any surface water will precipitate calcite (Fig. 1), and both Ca-chloride and neutral waters will precipitate calcite, followed by gypsum. The presence of more soluble salts further constrains the composition of the initial water. As a result, the mineralogy of evaporite deposits that contain the more soluble salts, or bitterns, containing K and/or Mg, is of particular interest. The sulfate-mineral assemblage is an important key in understanding compositions of ancient waters.

The mineralogy of the Permian Stassfurt Series from the German Zechstein II closely follows the predicted modern seawater sequence given above. Harvie et al. (1980) showed that both the sequence and quantity of minerals in these Permian rocks match the prediction. Permian evaporites from the Salado Formation of New Mexico also contain a similar sequence of minerals (Lowenstein 1983). Similar mineral sequences which contain Mg-bearing sulfate salts also occur in Late Precambrian, Late Palaeozoic to Early Mesozoic, and Late Cenozoic evaporites (see compilations by Hardie 1984, 1990). However, most other evaporite deposits formed during the last 600 Ma do not contain Mg-bearing sulfate salts.

Most ancient evaporites that contain bittern salts do not follow the sequence given above for precipitation from modern seawater. It has long been recognized that most ancient evaporites are "deficient" in the Mg-bearing sulfate salts, which are major components of the modern seawater evaporation sequence. This discrepancy, between the observed and the predicted mineralogy, has been explained in numerous ways:

- 1. Metastable equilibrium or non-equilibrium crystallization of minerals from a seawater source during evaporation (Valyashko 1972);

- 2. Syndepositional or post-depositional alteration of the original mineralogy (Stewart 1963, Borchert and Muir 1964, Wardlaw 1968, Evans 1970, Braitsch 1971, Holser 1979);

- 3. Syndepositional modification of seawater chemistry, including biological processes such as sulfate reduction (Borchert and Muir 1964, Braitsch 1971, Wardlaw 1972, Sonnenfeld 1984);

- 4. Syndepositional modification of seawater chemistry by mixing with non-marine inflow, including meteoric (D'Ans 1955, Stewart 1963, Wardlaw 1972, Valyashko 1972), diagenetic, volcanogenic or hydrothermal inflow (Hardie 1984);

Table 2. Calculated precipitation sequences for the evaporation of seawater with different proportions of river water and mid-ocean ridge brines (Spencer and Hardie 1990). A is for mixtures of river water with 96% of the modern mid-ocean ridge input, B is for modern mid-ocean ridge input, C, D, and E are for mixtures of river water with 5, 10, and 25% more mid-ocean ridge brine than in modern seawater.

Abbreviations are as in Table 1, except that abbreviations for chloride minerals are shown in box; mineral names are given in full upon initial appearance, and are abbreviated subsequently.

A. 0.96	B. 1.0	C. 1.05	D. 1.10	E. 1.25
calcite	calcite	calcite	calcite	calcite
cc+gypsum	cc+gypsum	cc+gypsum	cc+gypsum	cc+gypsum
cc+anhydrite	cc+anhydrite	cc+anhydrite	cc+anhydrite	cc+anhydrite
cc+an+glauberite	cc+an+halite	cc+an+halite	cc+an+halite	cc+an+halite
cc+gl	cc+an+ha+glauberite	cc+an+ha+polyhalite	cc+an+ha+sylvite	cc+an+ha+sylvite
cc+gl+halite	cc+an+ha	cc+an+ha+po+sylvite	cc+an+ha+sl+carnallite	cc+an+ha+sl+carnallite
cc+ha+an	cc+an+ha+polyhalite	cc+an+ha+po+sl+carnallite	cc+an+ha+cr	cc+an+ha+cr
cc+gl+ha+an	cc+an+ha+po+epsomite	cc+an+ha+po+cr	cc+an+ha+cr+bischofite	cc+an+ha+cr+antarcticite
cc+ha+an+polyhalite	cc+an+ha+po+hexahydrite	cc+an+ha+po+cr+kieserite	cc+an+ha+cr+bi+tachyhydrite	cc+an+ha+cr+at+tachyhydrite
cc+ha+po	cc+an+ha+po+kieserite	cc+an+ha+cr+ks+bischofite		
cc+ha+po+epsomite	cc+an+ha+po+ks+carnallite			
cc+ha+po+hexahydrite	cc+an+ha+ks+carnallite			
cc+ha+po+kieserite	cc+an+ha+ks+cr+bischofite			
cc+ha+po+ks+kainite				
cc+ha+po+ks+ka+an				
cc+ha+ks+an+carmallite				
cc+ha+ks+an+cr+bischofite				

Mineral formulae and abbreviations not given in Table 1		
antarcticite	$CaCl_2 \cdot 6H_2O$	at
bischofite	$MgCl_2 \cdot 6H_2O$	bi
calcite	$CaCO_3$	cc
carnallite	$KMgCl_3 \cdot 6H_2O$	cr
halite	$NaCl$	ha
sylvite	KCl	sl
tachyhydrite	$CaMg_2Cl_6 \cdot 12H_2O$	tc

- 5. Non-marine parent waters (Hardie 1984);
- 6. Changes in the composition of seawater through time (Spencer and Hardie 1990, Hardie 1996).

The first option above, metastable or non-equilibrium crystallization, is difficult to assess. However, as most of the common evaporite minerals do seem to form readily when the parent solutions are concentrated by evaporation, metastable or non-equilibrium crystallization is an unlikely reason for the absence of Mg-bearing sulfates in most ancient evaporites. The second option, of syndepositional or post-depositional alteration of the original mineralogy, has been interpreted for certain evaporites. On the other hand, Lowenstein and Spencer (1990) used data from fluid inclusions in primary halite to show that the precipitation of potash salts, without Mg-bearing sulfates, in the Oligocene Rhine Graben Formation, Permian Salado Formation, and Devonian Prairie Formation occurred from evaporation of the primary surface brines, and not as a result of alteration of pre-existing minerals.

The remaining options above rely on compositional differences of the evaporating waters to explain the discrepancies between the modern seawater evaporation sequence and the observed mineralogy of ancient evaporites. Systematic changes in evaporites globally through time seem to point toward the final option (6, above) as a likely explanation for the mineralogy of many ancient evaporite deposits. This option is discussed further.

Spencer and Hardie (1990) presented a simple model for the composition of modern seawater as a steady-state mixture of river water with brines from mid-ocean ridges. This mixing model is shown graphically in Figure 6, with the two source waters mixing in a proportion such that, after Ca-carbonate precipitates, modern seawater composition is obtained. If the proportion of the waters in the mixture changes, the composition of seawater changes. A higher mid-ocean ridge flux drives the chemistry of seawater toward the Ca corner of the diagram, whereas a lower flux drives it away from that corner (Fig. 6). The changes in seawater composition result in differences in the predicted sequence of mineral precipitation as shown in Table 2.

Seawater flux through the mid-ocean ridge is a function of heat flow through the ridge. This has almost certainly changed through time as the continents diverged and converged at different rates. Hardie (1996) employed the model of Spencer and Hardie (1990) to track the changes in seawater composition and evaporation sequence through

Figure 6. Ternary phase diagram for seawater constructed as in Figure 1. This diagram shows the steady–state composition of seawater for various mixtures of river water and mid–ocean ridge brines. Values are relative to the modern system of 1; the 0.96 indicates proportionally less mid–ocean ridge brine than the modern system; numbers greater than 1 indicate a relatively larger mid-ocean ridge flux than the modern system (see Spencer and Hardie 1990). Calculated mineral precipitation sequences for these mixtures are given in Table 2.

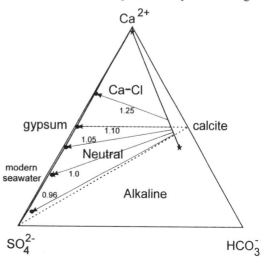

time. Hardie (1996) used secular variations in the production of ocean crust to estimate the changes in mid-ocean ridge flux, and compared the observed and predicted mineralogy of evaporites through geologic time.

The mid-ocean ridge and river-water mixtures for seawater predicted by the Hardie (1996) and Spencer and Hardie (1990) models for the last 600 Ma, along with the type of evaporite minerals expected, are displayed in Figure 7. Evaporite deposits are divided into four types:

- 1. Mg-sulfate evaporites, such as from modern seawater, at low mid-ocean ridge input;
- 2. Mixed Mg-sulfate and K-chloride evaporites at intermediate mid-ocean ridge input;
- 3. K-chloride evaporites, without Mg-bearing sulfates, at high mid-ocean ridge input;
- 4. Ca-chloride evaporites at extreme mid-ocean ridge input.

The type of marine evaporites, along with the periods of dominance of calcite and aragonite in marine carbonates (Stanley and Hardie 1998) are also shown in Figure 7. In general, there is good agreement between the predicted and the observed mineralogy of evaporites.

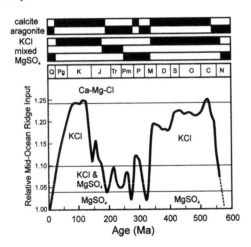

Figure 7. Predicted secular changes in seawater chemistry are shown in terms of the relative mid–ocean ridge input during the last 600 Ma (see Hardie 1996). These are compared with observations of calcite versus aragonite marine intervals (Stanley and Hardie 1998) and mineralogy of marine evaporites (Hardie 1996).

The dominance of the simple Ca-bearing sulfates, gypsum and anhydrite, observed in many ancient evaporites seems to result from changes in seawater composition through time. Only during periods of relatively low mid-ocean ridge input, generally during the existence of super continents, are the Mg-bearing sulfates found as a significant component of marine evaporites. Modern seawater composition does not seem to be a good indicator of past composition.

Evaporation paths and mineralogy of non-marine evaporites

The compositions of non-marine surface brines and the evaporite minerals formed from them are highly variable. Although there is a tendency to think of non-marine brines as simple alkaline waters, which eventually precipitate Na-bearing carbonates, there are numerous modern examples of neutral, and even Ca-chloride-type non-marine brines. Examples of eight modern, non-marine, closed-basin lakes in western North America are displayed in Figure 8. These waters were chosen because they show some of the variability of non-marine systems.

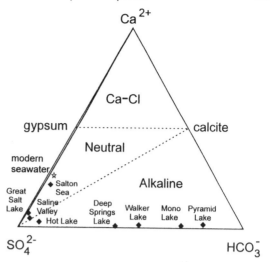

Figure 8. Ternary phase diagram constructed as in Figure 1, showing the composition of the various lakes whose mineral precipitation sequences are given in Table 3.

The waters range in composition from the bicarbonate-carbonate-rich alkaline systems of Pyramid and Mono lakes, to the near-equal sulfate and bicarbonate-carbonate systems of Walker and Deep Springs lakes, to the sulfate-rich, alkaline Hot Lake system. Saline Valley, Great Salt Lake, and the Salton Sea are included as examples of neutral brines. No Ca-chloride-type waters are shown, but such systems as at Bristol Lake, California, do exist. All of these systems receive at least some alkaline-type inflow, as a result of chemical weathering reactions involving meteoric water. The observed compositional differences are largely the result of additional hydrothermal inflow. For instance, Great Salt Lake receives a significant portion of solutes from hydrothermal springs (Spencer et al. 1985), and the Salton Sea is fed in part by a hydrothermal system. Thus, these systems are similar to seawater in that they are mixtures of hydrothermal waters and meteoric waters. The differences in composition result from variation in the proportion of mixing of these source waters.

Evaporation paths were calculated for the waters shown in Figure 8 using the thermochemical model of Harvie et al. (1984) and are presented in Table 3. A variety of sulfate-bearing minerals is predicted to form from these waters, as shown in Table 3. The alkaline waters from Pyramid and Walker lakes (both part of the Pleistocene Lake Lahontan system) are predicted to precipitate the sulfate-bearing salts burkeite, aphthitalite, thenardite, and blödite. Mono Lake, which is also highly alkaline, is predicted to precipitate thenardite, aphthitalite, and blödite, and Deep Springs Lake is predicted to precipitate thenardite, aphthitalite, and burkeite. Jones (1965a) reported the minerals calcite, aragonite, dolomite $[CaMg(CO_3)_2]$, gaylussite, nahcolite $[NaHCO_3]$, thenardite, burkeite, trona and halite from the Deep Springs Playa. This mineralogy compares fairly well with that calculated by the Harvie et al. (1984) model. Hot Lake is predicted to form epsomite, blödite, picromerite, leonite, kainite, and kieserite as sulfate-bearing minerals. None of the alkaline waters described above is predicted to form Ca-bearing sulfate salts.

The neutral brines from Saline Valley are predicted to form the sulfates glauberite, thenardite, aphthitalite, and blödite. Hardie (1968) observed the sequence of Ca-carbonate followed by gypsum, glauberite, halite, and mirabilite or thenardite at Saline Valley. The early sequence of sulfate minerals predicted from Great Salt Lake is similar to that of precipitates from modern seawater, with gypsum, anhydrite, glauberite, and polyhalite forming and back-reacting with the brines. The later predicted sequence

Table 3. Calculated sequences of mineral precipitation for the evaporation of selected lake waters in the alkaline and neutral fields (Fig. 8). The lakes receive different proportions of river water and hydrothermal brines. Mineral abbreviations are as in Tables 1 and 2; except that abbreviations for carbonate minerals are given in full upon initial appearance, and are abbreviated subsequently.

Pyramid Lake Nevada	Mono Lake California	Saline Valley California	Salton Sea California	Walker Lake Nevada	Deep Springs Playa California	Hot Lake Washington	Great Salt Lake Utah
aragonite	aragonite	aragonite	aragonite	aragonite	aragonite	aragonite	aragonite
ag+pirssonite	ag+gaylussite	ag+halite	ag+gypsum	ag+pirssonite	ag+pirssonite	ag+epsomite	ag+gypsum
pr	ga	ag+ha+glauberite	ag+gy+glauberite	pr	pr	ag+ep+blödite	ag+anhydrite
pr+halite	ga+pirssonite	ag+ha+gl+thenardite	ag+gl	pr+trona	pr+trona	ag+ep+bl+picromerite	ag+an+glauberite
pr+ha+trona	pr	ag+ha+gl+th+aphthitalite	ag+gl+halite	pr+tr+thenardite	pr+tr+thenardite	ag+ep+bl+pi+leonite	ag+an+gl+halite
pr+ha+tr+burkeite	pr+trona	ag+ha+gl+th+ap+blödite	ag+gl+ha+blödite	pr+tr+th+halite	pr+tr+th+halite	ag+ep+bl+le	ag+gl+ha
pr+ha+tr+bu+aphthitalite	pr+tr+thenardite		ag+gl+ha+bl+polyhalite	pr+tr+th+ha+burkeite	pr+tr+th+ha+aphthitalite	ag+bl+le+hx+kainite	ag+gl+ha+polyhalite
pr+ha+tr+bu+ap+thenardite	pr+tr+th+halite		ag+gl+ha+bl+po+anhydrite	pr+tr+ha+bu	pr+tr+th+ha+ap+burkeite	ag+bl+hx+ka+kieserite	ag+ha+po
pr+ha+tr+ap+th	pr+tr+th+ha+aphthitalite		ag+ha+bl+po+an	pr+tr+ha+bu+aphthitalite			ag+ha+po+blödite
pr+ha+tr+ap+th+blödite	pr+tr+th+ha+ap+blödite		ag+ha+bl+po+an+epsomite	pr+tr+ha+bu+ap+blödite			ag+ha+po+bl+leonite
							ag+ha+po+le
							ag+ha+po+le+epsomite
							ag+ha+po+ep+kainite
							ag+ha+po+ka+hexahydrite
							ag+ha+po+ka
							ag+ha+po+ka+carnallite
							ag+ha+po+ka+cr+kieserite
							ag+ha+po+cr+ks
							ag+ha+po+cr+ks+anhydrite
							ag+ha+cr+ks+an
							ag+ha+cr+ks+an+bischofite

Mineral formulae and abbreviations not given in Table 1 or Table 2

aragonite	$CaCO_3$	ag
gaylussite	$Na_2Ca(CO_3)_2 \cdot 5H_2O$	ga
pirssonite	$Na_2Ca(CO_3)_2 \cdot 2H_2O$	pr
trona	$Na_3HCO_3CO_3 \cdot 2H_2O$	tr

from Great Salt Lake differs from that of modern seawater and includes blödite, leonite, epsomite, kainite, hexahydrite, and kieserite as sulfate-bearing minerals. The predicted Salton Sea sulfate mineral sequence includes gypsum, glauberite, blödite, polyhalite, anhydrite, and epsomite. All of these neutral brines are predicted to precipitate Ca-bearing sulfate minerals.

MINERAL TEXTURES AND FABRICS

Hardie et al. (1985) and Spencer and Lowenstein (1988) presented more exhaustive discussions of petrographic aspects of evaporites. Only a few examples of textures and fabrics specific to sulfate minerals in evaporites are given here.

Criteria for syndepositional features

Syndepositional features in evaporite deposits include (a) crystalline framework textures and fabrics that are produced as chemically precipitated minerals grow *in situ*; (b) mechanical sedimentary structures with detrital textures and fabrics; and (c) evidence for dissolution, reprecipitation, and cementation in the surface, or near-surface environment. Criteria for identifying these are summarized below. More complete discussions are given by Hardie et al. (1985) and Spencer and Lowenstein (1988).

Crystalline framework fabrics are diagnostic of *in situ*, open-space crystal growth, and form as crystals grow from the substrate on the bottom of brines. The textures are commonly preserved in both modern and ancient deposits of gypsum (Schaller and Henderson 1932, Stewart 1949, 1951; Hardie and Eugster 1971, Schreiber and Kinsman 1975, Caldwell 1976, Schreiber and Schreiber 1977, Arakel 1980, Warren 1982, Lowenstein 1982). In essence, these crystals grow as void-filling cements. The fabrics produced are a result of competitive crystal growth into saturated brine and are illustrated in Figure 9. This growth results in vertically oriented, elongate crystals. The dominant fabric is a layered, syntaxially grown crystalline framework in which crystals tend to coarsen and widen upward.

Because the crystalline framework fabrics are produced by the same mechanism as are cements, i.e. competitive growth of crystals into a void, the same criteria documented by Bathurst (1975) for the recognition of cementation of carbonates can be used to identify these fabrics in evaporites. Although these criteria can be applied to any of the sulfate minerals, the following are most commonly found in gypsum:

- Vertically oriented prisms with euhedral crystal terminations at the top (Fig. 9a). For gypsum, "swallow tail" twins are common;
- Upward coarsening and widening of crystals where the bases of the crystals abut on the substrate, and the sides of the crystals abut on one another along compromise boundaries (Fig. 9b). Therefore, crystal growth is most rapid on upward-directed faces, and crystals oriented in this direction out-compete others and become enlarged upward;
- Crystals that originate from a common, laterally continuous substrate (Fig. 9c). For gypsum, the crystalline framework commonly nucleates on a detrital substrate;
- Sediment overlying the crystalline framework that thickens into the low points, and thins over the crystal high points (Fig. 9d), demonstrating that the crystals grew prior to the deposition of the overlying sediment;
- The presence of dissolution or erosion surfaces that truncate the crystalline framework, and act as a substrate for further crystal growth (Fig. 9e).

Mechanical sedimentary structures, typical of clastic deposition as a result of

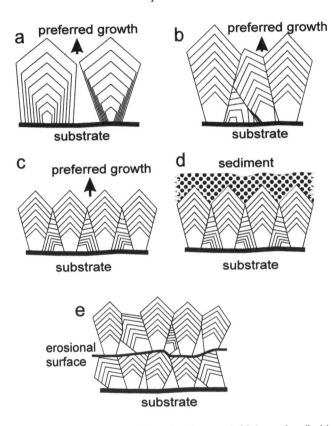

Figure 9. Schematic diagrams of crystalline framework fabrics as described in the text. The diagrams illustrate (a) vertically oriented prisms with euhedral crystal terminations at the top; (b) Upward coarsening and widening of crystals where the base of the crystals abut on the substrate, and the sides of the crystals abut on one another along compromise boundaries; (c) crystals that originate from a common, laterally continuous substrate; (d) sediment overlying the crystalline framework that thickens into the low points, and thins over the crystal high points; and (e) dissolution or erosion surfaces that truncate the crystalline framework.

hydrodynamic processes, are also common in evaporite deposits. This is particularly true for the Ca-sulfate minerals gypsum and anhydrite (Dellwig 1968, Dellwig and Evans 1969, Hardie and Eugster 1971, Schreiber et al. 1976, Schreiber and Hsu 1980, Lowenstein 1988). The most common features are stratification and bed-form structures, but other sedimentary features are also present. The identification of these sedimentary structures is not only important in determination of the primary aspect of the evaporites, but also for information pertaining to the processes involved in deposition and to the interpretation of the environment of deposition.

Detrital framework textures, consisting of a framework of grains with point contacts, such as those typically associated with siliciclastic sediments, can be used to establish the primary detrital nature of evaporite deposits. These textures, which are common in gypsum deposits, result from the erosion, reworking, and hydrodynamic sorting of euhedral to abraded crystals. These fabrics are preserved in many gypsum deposits, and may be seen as remnants in some instances where gypsum has been replaced by anhydrite or mixed Ca-sulfate salts such as glauberite or polyhalite.

Dissolution-reprecipitation features indicate fluctuations in surface-water composition as a mineral grows or layers accumulate. These features are common in gypsum-anhydrite deposits (see Hardie and Eugster 1971, Schreiber and Kinsman 1975, Warren 1982). Lowenstein and Hardie (1985) discussed evidence for syndepositional dissolution and reprecipitation. There are two types of features found. One type consists of sharp, smooth truncations along bedding overlain by either mud, crystal cumulates, or vertically oriented crystals. The other consists of rounding of euhedral crystal terminations, followed by syntaxial overgrowth.

Criteria for burial alteration features

In reconstructing past hydrochemical information from evaporite deposits, it is important to distinguish between primary and secondary features. Many evaporite minerals are susceptible to alteration caused by increases in temperature and pressure during burial. Particularly susceptible are the hydrous minerals that may alter as waters of hydration are released. Given the many hydrous sulfate salts, this problem is of particular interest for sulfate minerals. Metasomatism by invading subsurface waters is also important to consider. Many of the criteria must be inferred rather than obtained directly from comparative studies of active modern systems. Nonetheless, Hardie et al. (1985) put together a tentative working set of criteria, which are given below. In the main, the criteria involve textures and structures that record disruption, deformation, and destruction of primary features.

Disruption and destruction of sedimentary structures is indicated by the absence of primary bedding or the presence of only scattered remnants of primary bedding or other sedimentary structures in massive crystalline mosaic salts. This texture points to destructive secondary crystal growth, early stages of which are seen as irregular, isolated crystalline mosaic patches that crosscut primary sedimentary structures (e.g. Fig. 18 in Schreiber et al. 1976.). This criterion is most persuasive of burial alteration when observed in monomineralic rocks, where the crystalline mosaics must surely have resulted from destructive recrystallization. Such crystalline mosaics have sutured or polygonal grain boundaries. The mosaics are distinct from porphyroblasts and isolated displacive, incorporative, or poikilotopic growth of euhedra and nodules typical of syndepositional intrasediment crystallization.

Sutured mosaic textures in which grain boundaries are sutured in the manner of neomorphic spar of carbonate rocks, as discussed by Bathurst (1975), clearly indicate a secondary grain origin. However, evaporite minerals may undergo "neomorphic" changes more readily than carbonates.

Polygonal mosaic textures have long been documented in experimentally treated metals and metallic ore minerals (Buckley 1951, Burgers 1963, Stanton and Gorman 1968). Similar textures are typical of metamorphic rocks (Voll 1960). The consensus is that these polygonal mosaic textures are a product of annealing recrystallization, whereby grains "optimize" their size, shape, and orientation to minimize the energy in the manner of bubbles in foam (Stanton and Gorman 1968). Evaporites, because of the ease with which they are altered at the temperatures and pressures of even modest burial, are highly susceptible to such annealing recrystallization.

Deformation features include the obvious macroscopic deformation structures such as folds, fractures, veins, flow banding, and pressure solution surfaces (e.g. Balk 1949, Goldman 1952, Muller et al. 1981). There are also more subtle microscopic deformation features visible in thin section or in etched, polished slabs of evaporites (Schlichta 1968).

Stanton and Gorman (1968) discussed the main types of deformation textures recognized in metallurgical studies and applicable to rocks (see also Voll 1960, 1976). These deformation textures are (1) deformation twins, (2) slip lines and slip bands, (3) 3-dimensional distortion of crystal structure, such as undulose extinction, bent cleavages, etc., (4) sub-grain domains, and (5) flattening of grains, thereby leading to the presence of foliation.

Ambiguous features

Numerous features can form either syndepositionally or after burial. In the absence of other criteria to aid interpretation, the time of origin of such features is uncertain. Among the ambiguous features are (1) mineral growth, as euhedra or nodules, within the sediment, (2) pseudomorphous replacements, (3) coarse crystalline fabrics, and (4) certain cavity cements.

Intrasediment growth of euhedra is common in evaporite minerals, including sulfates. Crystals that grow within sediment commonly occur as randomly oriented, isolated or interlocking crystals with euhedral or subhedral habit. Such crystals may incorporate sediment during growth or they may form displacively by pushing aside the host sediment. The Devonian Prairie Formation contains examples of gypsum (or anhydrite) with displacive and incorporative growth habits (Fig. 10 in Spencer and Lowenstein 1988). The Permian Salado evaporite of New Mexico (Lowenstein 1982, 1983, 1988) carries many good examples of this style of intrasediment growth by gypsum, glauberite, and the high-temperature K-Mg sulfate mineral langbeinite (Figs. 11, 12, and 13 in Spencer and Lowenstein 1988). Incorporative growth provides unequivocal evidence of post-depositional crystallization. However, it is dificult to distinguish between early, syndepositional growth and later metamorphic porphyroblastic growth. Intrasediment growth of sulfate minerals such as gypsum, anhydrite, glauberite, and mirabilite as euhedral crystals is a common near-surface feature in many modern playas and sabkhas (Hardie 1968, Butler 1970a, Thompson 1968, Logan et al. 1970). However, porphyroblastic growth might also occur on burial, and could produce similar textures. On the other hand, intrasediment growth of euhedra of minerals such as langbeinite and loewite, which form only at elevated temperature, probably indicates burial crystallization.

Intrasediment growth of nodules is also common in modern sediments. The discovery of nodules of anhydrite in Holocene sediments of the Persian Gulf (Curtis et al. 1963, Shearman 1966) rekindled interest in evaporites among sedimentologists. An astonishing number of published papers have since called on nodular anhydrite, or its presumed pseudomorphs, as a criterion for syndepositional growth of evaporites in a sabhka environment. However, nodules of anhydrite can form under a variety of conditions, such as from the Red Sea hot brines in the Atlantis II Deep under 2000 m of water (see Degens and Ross 1969, color plate between p. 366 and p. 367). Dehydration of gypsum crystals to make masses of anhydrite laths is a major mechanism for the generation of anhydrite nodules. This conversion may just as easily occur on deep burial as during primary surface deposition. The same reasoning holds for the dehydration of all other hydrous evaporite minerals that have anhydrous, or less hydrous, conterparts. Without independent evidence, nodules may not give unambiguous information about their time, mode, or environment of formation.

Pseudomorphs are unequivocal evidence of replacement. However, pseudomorphous replacement may be of syndepositional or later burial origin (Stewart 1963). Recognition of pseudomorphous replacement rests on the identification of the crystal morphology of the parent mineral. Such replacements are widely observed for gypsum in evaporites (Schaller and Henderson 1932, Stewart 1949, 1951, 1963; Borchert and Baier 1953, Kerr and Thompson 1963, Jones 1965b, Nurmi and Friedman 1977, Handford 1981, Presley and McGillis 1982, Crawford and Dunham 1982, Lowenstein 1982, 1988).

Several different types of pseudomorphous replacements in evaporite deposits have been recognized. These are (1) "dehydration pair" replacement such as gypsum-anhydrite, (2) "reaction pair" replacement involving a common component such as gypsum-glauberite, or anhydrite-polyhalite, and (3) "non-reaction pair" replacement wherein the original and replacment minerals have no common component such as gypsum-halite. Although Borchert and Muir (1964) considered all types of pseudo-morphous replacement to be metamorphic, both dehydration and replacements of the reaction-pair type are observed and are predicted thermodynamically as a result of syn-depositional interaction between early-formed minerals and evolving brines. This type of replacement is especially common for Ca-bearing sulfate minerals.

Temperature-dependent salts that have restricted temperature stability ranges are an attractive option in deciding a syndepositional or burial origin. There is a large overlap in temperatures experienced by primary depositional environments and secondary burial environments. Many brines in modern evaporite settings remain at temperatures below 40°C the year around (such as the Persian Gulf sabkha; Butler 1970a). However, temperatures near 50°C due to solar heating have been measured in shallow stratified brine ponds (Hudec and Sonnenfeld 1974). Stewart (1963) reported temperatures in the 60-70°C range for some shallow brine bodies, and values of >100°C are reached in hot-spring brine pools (Holwerda and Hutchinson 1968). Temperatures of surface sediments on salt flats commonly exceed 50°C in the summer (Butler 1970b). Under a normal geothermal gradient, temperatures of 50°C are reached at about 900-m depth, and 100°C at about 2400-m depth. This overlap in environment temperatures means that "high temperature" salts such as langbeinite [$K_2Mg_2(SO_4)_3$], löweite [$Na_{12}Mg_7(SO_4)_{13} \cdot 15H_2O$], and vanthoffite [$Na_6Mg(SO_4)_4$], stable only at temperatures greater than 40 to 50°C (Stewart 1963) could be of either syndepositional or burial origin.

SUMMARY

The sulfate mineralogy of evaporite deposits is of key importance in understanding the hydrochemistry of ancient surface waters. The sulfate mineral assemblage may be complex, and a variety of dehydration or replacement reactions may occur. However, careful examination of mineral textures and fabrics allows determination of the crystallization sequence during evaporation. The complex nature of sulfate mineral assemblages in evaporite deposits places constraints on the chemical composition of the parent waters that evaporated. Modern thermochemical models can be used to formulate and test hypotheses regarding the composition of ancient waters responsible for the deposits.

REFERENCES

Arakel AV (1980) Genesis and diagenesis of Holocene evaporitic sediments in Hutt and Leeman lagoons, Western Australia. J Sed Pet 50:1305-1326
Balk R (1949) Structure of Grand Saline salt dome, Van Zandt County, Texas. Am Assoc Petrol Geol Bull 33:1791-1829
Bathurst RGC (1975) Carbonate Sediments and Their Diagenesis. 2nd Edition. Elsevier, Amsterdam, 658 p
Borchert H, Baier E (1953) Zur metamorphose ozeaner Gipsablagerungen. N Jahrb Mineral Abh 86:103-154
Borchert H, Muir RO (1964) Salt Deposits: The Origin, Metamorphism and Deformation of Evaporites. D Van Nostrand, Princeton, New Jersey, 300 p
Braitsch O (1971) Salt Deposits, Their Origin and Composition. Springer-Verlag, New York, 297 p

Buckley HE (1951) Crystal Growth. John Wiley and Sons, New York, 571 p

Burgers WG (1963) Principles of recrystallization. In JJ Gilman (Ed) The Art and Science of Growing Crystals. John Wiley and Sons, New York, p 416-450

Butler GP (1970a) Holocene gypsum and anhydrite of the Abu Dhabi Sabkha, Trucial Coast: An alternative explanation of origin. In Jl Rau, Lf Dellwig (Eds) 3rd Symposium on Salt. Northern Ohio Geol Soc, Cleveland, Ohio, p 120-152

Butler GP (1970b) Secondary anhydrite from a sabkha, northwest Gulf of California, Mexico. In JL Rau, LF Dellwig (Eds) 3rd Symposium on Salt. Northern Ohio Geol Soc, Cleveland, Ohio, p 153-155

Caldwell RH (1976) Holocene Gypsum Deposits of the Bullara Sunkland, Carnarvon Basin, Western Australia. PhD Dissertation, Univ Western Australia, Nedlands, W Aust, 123 p

Crawford GA, Dunham JB (1982) Evaporite sedimentation in the Permian Yates Formation, Central Basin Platform, Andrews County, West Texas. In CA Handford, RG Loucks, GR Davies (Eds) Depositional and Diagenetic Spectra of Evaporites—A Core Workshop. Soc Econ Paleo Mineral, Core Workshop 3, Calgary, Canada, p 238-275

Curtis R, Evans G, Kinsman DJJ, Shearman DJ (1963) Association of dolomite and anhydrite in the Recent sediments of the Persian Gulf. Nature 143:679-680

D'Ans J (1955) Die Losungsgleichgewichte der Systeme der Salze ozeanischer Salzablagerungen. Kali-Forschungsanstalt, Verlagsgesellschaft Ackerbau, 254 p

Degens ET, Ross DA (1969) Hot Brines and Recent Heavy Metal Deposits in the Red Sea. Springer Verlag, New York, 600 p

Dellwig LF (1968) Significant features of deposition in the Hutchinson Salt, Kansas, and their interpretation. In RB Mattox (Ed) Saline Deposits. Geol Soc Am Spec Paper 88:421-426

Dellwig LF, Evans R (1969) Depositional processes in Salina salt of Michigan, Ohio and New York. Am Assoc Petrol Geol Bull 53: 949-956

Evans R (1970) Genesis of sylvite and carnallite-bearing rocks from Wallace, Nova Scotia. In JL Rau, LF Dellwig (Eds) 3rd Symposium on Salt. Northern Ohio Geol Soc, Cleveland, Ohio, p 239-245

Goldman MI (1952) Deformation Metamorphism, and Mineralization in Gypsum-Anhydrite Cap Rock, Sulfur Salt Dome, Louisiana. Geol Soc Am Mem 50, 163 p

Handford CR (1981) Coastal sabkha and salt pan deposition of the Lower Clear Fork Formation (Permian) Texas. J Sed Pet 51:761-778

Handford CR (1991) Marginal marine halite: sabkhas and salinas. In JL Melvin (Ed) Evaporites, Petroleum and Mineral Resources. Elsevier, Amsterdam, p 1-68

Hardie LA (1968) The origin of the Recent non-marine evaporite deposit of Saline Valley, Inyo County, California. Geochim Cosmochim Acta 32:1279-1301

Hardie LA (1984) Evaporites: marine or non-marine? Am J Sci 284:195-240

Hardie LA (1990) Potash evaporites, rifting and the role of hydrothermal brines. Am J Sci 290:43-106

Hardie LA (1996) Secular variations in seawater chemistry: An explanation for the coupled secular variation in the mineralogy of marine limestones and potash evaporites over the past 600 m.y. Geol 24:279-283

Hardie LA, Eugster HP (1970) The evolution of closed-basin brines. Mineral Soc Am Spec Paper 3: 273-290

Hardie LA and Eugster HP (1971) The depositional environment of marine evaporites: A case for shallow, clastic accumulation. Sedimentol 16:187-220

Hardie LA, Lowenstein TK, Spencer RJ (1985) The problem of distinguishing between primary and secondary features in evaporites. In BC Schreiber (Ed) 6th Symposium on Salt. Salt Institute, Alexandria, Virginia, p 1-59

Harvie CF, Moller N, Weare JH (1984) The prediction of mineral solubilities in natural waters: The Na-K-Mg-Ca-H-Cl-SO_4-OH-HCO_3-CO_3-H_2O system to high ionic strengths at 25°C. Geochem Cosmochim Acta 48:723-751

Harvie CF, Weare JH, Eugster HP, Hardie LA (1980) Evaporation of sea water: calculated mineral sequence. Science 208:498-500

Holser WT (1979) Trace elements and isotopes in evaporites. In RG Burns (Ed) Marine Minerals. Rev Mineral 6:295-546

Holwerda JG, Hutchinson RW (1968) Potash-bearing evaporites in the Danakil area, Ethiopia. Econ Geol 63:124-150

Hudec PP, Sonnenfeld P (1974) Hot brines on Los Roques, Venezuela. Science 185:440-442

Jones BF (1965a) The hydrology and mineralogy of Deep Springs Lake, Inyo County, California. U S Geol Surv Prof Paper 502:1-56

Jones CL (1965b) Petrography of evaporites from the Wellington Formation near Hutchinson, Kansas. U S Geol Surv Bull 1201A:70

Kerr SD, Thompson A (1963) Origin of nodular and bedded anhydrite in Permian shelf sediments, Texas and New Mexico. Am Assoc Petrol Geol Bull 47:1726-1732

Logan BW, Davies GR, Read JR, Cebulski DF (1970) Carbonate sedimentation and environments, Shark Bay, Western Australia. Am Assoc Petrol Geol Mem 13:223

Lowenstein TK (1982) Primary features in a potash evaporite deposit, the Permian Salado Formation of west Texas and New Mexico. *In* CA Handford, RG Loucks, GR Davies (Eds) Depositional and Diagenetic Spectra of Evaporites—A Core Workshop. Soc Econ Paleo Mineral, Core Workshop 3, Calgary, Canada, p 276-304

Lowenstein TK (1983) Deposition and Alteration of an Ancient Potash Evaporite: The Permian Salado Formation of New Mexico and West Texas. PhD Dissertation, Johns Hopkins Univ, Baltimore, Maryland, 411 p

Lowenstein TK (1988) Origin of depositional cycles in a Permian "saline giant": The Salado (McNutt Zone) evaporites of New Mexico and Texas. Geol Soc Am Bull 100:592-608

Lowenstein TK, Hardie LA (1985) Criteria for the recognition of salt-pan evaporites. Sedimentol 92: 627-644

Lowenstein TK, Spencer RJ (1990) Syndepositional origin of potash evaporites: Petrographic and fluid inclusion evidence. Am J Sci 290:1-42

Melvin JL (Ed) (1991) Evaporites, Petroleum and Mineral Resources. Elsevier, Amsterdam, 556 p

Muller WH, Schmid SM, Briegel U (1981) Deformation experiments on anhydrite rocks of different grain sizes: Rheology and microfabrics. Tectonophysics 78:527-544

Nurmi R0, Friedman GM (1977) Sedimentology and depositional environments of basin center evaporites, Lower Salina Group (Upper Silurian) Michigan Basin. *In* JH Fisher (Ed) Reefs and Evaporites: Concepts and Depositional Models. Am Assoc Petrol Geol Stud Geol 5:23-52

Presley MW, McGillis KA (1982) Coastal evaporite and tidal flat sediments of the Upper Clear Fork and Glorieta formations, Texas Panhandle. Univ Texas, Austin, Bur Econ Geol Rep Invest 115:50

Schaller WT, Henderson EP (1932) Mineralogy of drill cores from the potash field of New Mexico and Texas. U S Geol Surv Bull 833:124

Schlichta PJ (1968) Growth, deformation and defect-structure of salt crystals. *In* RB Mattox (Ed) Saline Deposits. Geol Soc Am Spec Paper 88:597-617

Schrieiber BC, Friedman GM, Decima A, Schreiber R (1976) Depositional environments of Upper Miocene (Messinian) evaporite deposits of the Sicilian Basin. Sedimentol 23:729-760

Schreiber BC, Hsu KJ (1980) Evaporites. *In* GO Hobson (Ed) Developments in Petroleum Geology 2. Appl Sci Pubs, Barking, Essex, UK, p 87-138

Schreiber BC, Kinsman KJJ (1975) New observations on the Pleistocene evaporites of Montellegro, Sicily, and a modern analogue. J Sed Pet 45:469-479

Schreiber BC, Schreiber E (1977) The salt that was. Geol 5:527-528

Shearman DJ (1966) Origin of marine evaporites by diagenesis. Inst Mining Metal Trans 75:208-215

Smoot JP, Lowenstein TK (1991) Depositional environments of non-marine evaporites. *In* JL Melvin (Ed) Evaporites, Petroleum and Mineral Resources. Elsevier, Amsterdam, p 189-348

Sonnenfeld P (1984) Brines and Evaporites. Academic Press, Orlando, Florida, 613 p

Spencer RJ, Hardie LA (1990) Control of seawater composition by mixing of river waters and mid-ocean ridge hydrothermal brines. *In* RJ Spencer, I-Ming Chou (Eds) Fluid-Mineral Interactions: A Tribute to H.P. Eugster. Geochem Soc Spec Pub 2:409-419

Spencer RJ, Eugster HP, Jones BF (1985) Geochemistry of Great Salt Lake, Utah II: Pleistocene-Holocene evolution. Geochem Cosmochim Acta 49:739-747

Spencer RJ, Lowenstein TK (1988) Evaporites. *In* IA McIlreath, DW Morrow (Eds) Diagenesis. Geosci Canada Rep Ser 4:141-163

Stanley SM, Hardie LA (1998) Secular oscillations in the carbonate mineralogy of reef-building and sediment-producing organisms driven by tectonically forced shifts in seawater chemistry. Palaeogeog Palaeoclimat Palaeoecol 144:3-19

Stanton RL, Gorman H (1968) A phenomenological study of grain boundary migration in some common sulfides. Econ Geol 63:907-923

Stewart FH (1949) The petrology of the evaporites of the Eskdale No. 2 boring, east Yorkshire: Part I. The middle evaporite bed. Mineral Mag 29:445-475

Stewart FH (1951) The petrology of the Eskdale No. 2 boring, east Yorkshire: Part II. The middle evaporite bed. Mineral Mag 29:445-475

Stewart FH (1953) Early gypsum in the Permian evaporites of north-eastern England. Proc Geol Assoc London 64:33-39

Stewart FH (1963) Marine Evaporites. U S Geol Surv Prof Paper 440-Y, 53 p

Thompson RW (1968) Tidal Flat Sedimentation on the Colorado River Delta, Northwestern Gulf of California. Geol Soc Am Mem 107, 133 p

Valyashko MG (1972) Playa lakes—a necessary stage in the development of a salt-bearing basin. *In* G Richter-Bernburg (Geology of Saline Deposits. UNESCO, Earth Sci Series 7:41-51

Voll G (1960) New work on petrofabrics. Liverpool Manchester Geol J 2:503-567

Voll G (1976) Recrystallization of quartz-biotite and feldspars from Eastfeld to the Leventina Nappe, Swiss Alps and its geological significance. Schweiz mineral petrogr Mitt 56:641-647

Wardlaw NC (1968) Carnallite-sylvite relationships in the Middle Devonian Prairie Evaporite Formation, Saskatchewan. Geol Soc Am Bull 79:1273-1294

Wardlaw NC (1972) Unusual marine evaporites with salts of calcium and magnesium chloride in cretaceous basins of Sergipe, Brazil. Econ Geol 67:156-168

Warren JK (1982) The hydrological setting, occurrence and significance of gypsum in late Quaternary salt lakes in South Australia. Sedimentol 29:609-639

Warren, J.K. (1991) Sulfate dominated sea-marginal and platform evaporative settings: Sabkhas and salinas, mudflats and salterns. *In* JL Melvin (Ed) Evaporites, Petroleum and Mineral Resources. Elsevier, Amsterdam, p 69-188

4

Barite–Celestine Geochemistry and Environments of Formation

Jeffrey S. Hanor

Department of Geology and Geophysics
Louisiana State University
Baton Rouge, Louisiana 70803

INTRODUCTION

Minerals in the barite ($BaSO_4$)–celestine ($SrSO_4$) solid solution series, $(Ba,Sr)SO_4$, occur in a remarkably diverse range of sedimentary, metamorphic, and igneous geological environments which span geological time from the Early Archean (~3.5 Ga) to the present. The purpose of this chapter is to review: (1) the controls on the chemical and isotopic composition of barite and celestine and (2) the geological environments in which these minerals form. Some health risks are associated with barite, and these are discussed near the end of this chapter.

Although complete solid solution exists between $BaSO_4$ and $SrSO_4$ most representatives of the series are either distinctly Ba-rich or Sr-rich. Hence, it is convenient to use the term *barite* to refer to not only the stochiometric $BaSO_4$ endmember but also to those $(Ba,Sr)SO_4$ solid solutions dominated by Ba. Similarly, the term *celestine* will refer here not only to the stoichiometric $SrSO_4$ endmember but to solid solutions dominated by Sr. Such usage is in accord with standard mineral nomenclature. The Committee on Mineral Names and Nomenclature of the International Mineralogical Association recognizes "celestine" as the official name for $SrSO_4$. However, the name "celestite" is still commonly used in the literature.

Geological significance of barite and celestine

Most of the barite which exists in the Earth's crust has formed through the mixing of fluids, one containing Ba leached from silicate minerals, and the other an oxidized shallow fluid, such as seawater, which contains sulfate. Large deposits of barite represent areas of focused fluid flow and mineral precipitation and thus aid in the reconstruction of the hydrogeologic history of the Earth's crust. The stability of barite is redox sensitive, and the presence or absence of this mineral helps to constrain interpretation of paleoredox conditions. Pelagic barite, the precipitation of which is biologically mediated, may serve as a paleoproductivity indicator. In addition, the $^{87}Sr/^{86}Sr$, $\delta^{18}O$, and $\delta^{34}S$ isotopic composition and trace-element composition of barite help to identify sources of mineral-forming components, environments of precipitation, and secular variations in global seawater composition.

Celestine has a much more restricted geological distribution than barite, and most major occurrences appear to be the product of the reaction of hypersaline Sr-bearing fluids with gypsum and anhydrite. The presence of celestine reflects special geological environments where there has been preferential concentration of Sr over Ba.

Economic importance

Barite, because of its high density (4.48 g/cm^3), relative abundance, and ease of grinding into powder, has long been used for fillers, extenders, and weighting agents (Brobst 1994). Prior to the passage of the United States Pure Food and Drug Act of 1906, the principal use of barite in the U.S. was as an adulterant in flour and sugar. Today, barite

1529-6466/00/0040-0004$10.00

and to a lesser degree witherite ($BaCO_3$) are the sources of barium chemicals used in glass, ceramics, ferrites and titanites, TV tubes, paint, plastics, green pyrotechnics, and photographic print paper. Kyle (1994) provides a detailed review of manufacturing processes involving barite. United States Pharmacopoeia barite is used in gastrointestinal X-ray examinations. By far the greatest use of barite, however, is as a weighting agent in drilling fluids used in the oil and gas industry. As such, annual production figures for barite closely track oil and gas exploration and development. Of the 2.18×10^6 metric tons (or 2.18 megatons, Mt) of barite used in the United States in 1997, the most recent year for which detailed figures were available at the time of this writing (1999), 2.08 Mt were used in drilling fluids (Searls 1997). Much of this barite was imported from China, the world's largest 1997 producer at 3.5 Mt, and was processed in Louisiana for use in petroleum exploration and development in the Gulf of Mexico sedimentary basin. Total world production was 6.93 Mt at an estimated total value in United States currency of $173,000,000 or $25/t. Nearly 50 countries currently produce barite. The top producers in 1997 were China, United States, Morocco, Kazakhstan, and India.

Celestine is the principal source of strontium chemicals used in ceramics, glass, red pyrotechnics, and metallurgy (Ober 1997). In 1997, more than 75% of the strontium consumed in the United States was used in color-television face plates, which are required by law to contain strontium to block X-ray emissions. Faceplate glass contains approximately 8 wt % SrO. World production of celestine in 1997 was estimated at 0.306 Mt, with a value of $72/t. Seven countries currently mine celestine: the principal producers are Mexico, Spain, Turkey, and Iran. There have been no active celestine mines in the United States since 1959.

Some conventions and terms used in this chapter

Concentration units. A variety of units are used to describe the concentrations of dissolved species in aqueous solutions. Marine chemists and geochemists routinely use mass of solute per kg of seawater. Units of mass per L of solution are typically used in studies of groundwater and basinal waters. Molality, m, moles of solute per kg H_2O, is the concentration unit used in thermodynamic work. Precise conversion between these three types of units requires knowledge of fluid density and total dissolved solutes for each analysis. This information is not always readily available, and no attempt has been made to convert units here. Concentrations are reported in the units given in the original paper cited. It should be noted, however, that units such as mg/L and mg/kg of solution or ppm are generally numerically comparable to two significant figures, i.e. 21 mg/L $\approx$ 21 mg/kg solution.

Bedded barite. Barite disseminated as individual grains and nodules in a layered sediment matrix conformable to bedding. Can include detrital barite, syngenetic barite, and diagenetic nodules. Equivalent in meaning to stratiform barite.

Deep-sea barite. Barite found in deep-sea sediments located away from continental margins.

Diagenetic. Precipitated within the host sediment from autochthonous sources of Ba and sulfate. Examples are barite nodules formed by the local dissolution, mobilization, and reprecipitation of syngenetic barite.

Epigenetic. Precipitated within a previously deposited host sediment from allochthonous sources of Ba and/or sulfate. An example is carbonate-hosted Mississippi Valley type barite. There is considerable overlap in usage between the terms syngenetic and diagenetic and between diagenetic and epigenetic among various workers.

Hydrothermal. Deposited from allochthonous waters of sedimentary, igneous, or metamorphic origin, warmer than the initial temperature of the host rock or sediment.

MOR. Mid-ocean rise or ridge.

Pelagic barite and celestine. Barite and celestine formed in the upper part of the open-ocean water column by any process.

Sedex deposits. Sedimentary-exhalative mineral deposits. Mineral deposits formed by fluids venting into a marine environment where there is no obvious magmatic source of heat.

Syngenetic or synsedimentary. Formed penecontemporaneously with the host sediment, usually by mixing of Ba-rich fluids with marine waters, or by the incorporation of pelagic barite.

VHM deposits. Volcanic-hosted mineral deposits in which magmatic activity drives convective circulation of fluids. Includes volcanic-hosted massive sulfide (VMS or VHMS) deposits.

PHYSICAL CHEMISTRY

Crystal chemistry and solid-phase relations

Crystallography. The barite ($BaSO_4$) structure is orthorhombic, dipyramidal, and has the space group *Pnma*. Celestine ($SrSO_4$) is isostructural with barite. The sulfur and two oxygen atoms of each SO_4 tetrahedron in the $BaSO_4$ structure lie on a mirror plane (Gaines et al. 1997). The other two oxygens are equidistant above and below the plane. The Ba ions lie on the same mirror plane and are in 12-fold coordination with oxygens belonging to seven different sulfate groups.

Barite and celestine sometimes occur as a replacement of the mineral anhydrite ($CaSO_4$). Solid solution between $BaSO_4$ and $SrSO_4$ is complete, but that with $CaSO_4$ is incomplete. The anhydrite ($CaSO_4$) structure is also orthorhombic, but of the space group *Cmcm*. Unlike the barite structure, S and Ca atoms lie on lines of intersection of two sets of mirror planes, and planes containing evenly spaced Ca and SO_4 ions lie parallel to (100) and (010). In contrast to the larger barium ion, which is in 12-fold coordination in the barite structure, Ca is in 8-fold coordination with oxygens belonging to six different sulfate groups. According to Gaines et al. (1997), the anhydrite structure is similar to that of zircon.

A more detailed description of the crystal chemistry of these sulfate minerals is given by Hawthorne et al. (this volume).

Table 1. Ionic radii (6-fold coordination) of selected cations.

Cation	Ionic radius, Ångstroms	Cation	Ionic radius, Ångstroms
K^+	1.33	Pb^{2+}	1.18
Mg^{2+}	0.72	La^{3+}	1.05
Ca^{2+}	1.00	Ce^{3+}	1.01
Sr^{2+}	1.16	Lu^{3+}	0.85
Ba^{2+}	1.36	Eu^{2+}	1.17
Ra^{2+}	1.43		

Source: Huheey (1972)

Solid solutions. The substitution of cations other than Ba^{2+} and Sr^{2+} into barite and celestine is controlled by their degree of similarity in charge, ionic radius, and electronegativity to Ba^{2+} and Sr^{2+}. Table 1 shows the ionic radii (Huheey 1972) for key cations which will be discussed further in this chapter. Values are given for 6-fold coordination. Ionic radii generally increase with increasing coordination number, but the proportional differences in ionic radii between cations are similar with increasing coordination number.

Despite the fairly significant difference in ionic radii between Ba^{2+} and Sr^{2+}, 1.34 Å and 1.18 Å respectively, there is complete solid solution between $BaSO_4$ and $SrSO_4$ at room temperature, at least metastably, as first established by Grahmann (1920). Additional evidence for complete solid solution between barite and celestine has been obtained through X-ray diffraction by Sabine and Young (1954), Boström et al. (1967), and Burkhard (1973, 1978). Boström et al. (1967) further established the existence of complete solid solution between $BaSO_4$, $SrSO_4$, and $PbSO_4$ at 375°C. Takano et al. (1970), however, document a structure gap in the $BaSO_4$-$PbSO_4$ series.

There is more restricted solid solution between $CaSO_4$ and barite and celestine, reflecting the greater differences in ionic radii between Ca^{2+} and Ba^{2+} and the significant differences in crystal structure between barite and anhydrite noted above. Grahmann (1920) showed that at room temperature only approximately 6 mol % $CaSO_4$ can enter the $BaSO_4$ structure, slightly more $CaSO_4$ in strontian barite, and 8 mol % $BaSO_4$ in anhydrite. Grahmann also established that 12 mol % $CaSO_4$ can substitute in celestine and much as 42 mol % $SrSO_4$ in anhydrite. Substitution of K^+, Ra^{2+}, and rare-earth elements (REE) for Ba^{2+} and Sr^{2+} also occurs (e.g. Church 1979, Guichard et al. 1979, Morgan and Wandles 1980, Jebrak et al. 1985). The temperature dependence of the degree of substitution of other cations for Ba^{2+} and Sr^{2+} has not been established.

Chang et al. (1996) have summarized previous analytical work reporting minor substitution of Ba in barite by Fe, Cu, Zn, Ag, Ni, Ra, Hg and V. The minerals and compounds $BaSeO_4$, $PbSeO_4$, $SrSeO_4$, $BaCrO_4$, $KMnO_4$, $KClO_4$, and $(K,Cs)BF_4$ are isostructural with barite, and the anionic groups $[MnO_4]$, $[SeO_4]$, and $[CrO_4]$ have been substituted for sulfate in synthetic material (Chang et al. 1996). Some Archean barites contain Xe produced by cosmic ray bombardment of Ba (Srinivasin 1976).

Thermal stability. Barite undergoes a transition at 1148°C to a monoclinic structure that melts at 1580°C (Weast 1974). $SrSO_4$ melts at 1605°C (Weast 1974). An increase in pressure probably increases these melting temperatures. Hence, it is safe to assume that barite and celestine are intrinsically stable over the entire P-T range of the Earth's crust in the absence of other, reactive components.

Solubility of barite and celestine in aqueous solutions

Both Ba^{2+} and Sr^{2+} have high ionic potentials (the ratio of charge to ionic radius) and can be readily accommodated in aqueous solution as hydrated divalent cations, without the need for exotic aqueous complexing. The solubilities of barium chloride and strontium chloride, for example, are 375 g/L at 26°C and 538 g/L at 20°C, respectively (Weast 1974). The sulfate anionic complex is also readily accommodated in aqueous solution as a hydrated anionic species. For example, the solubility of $Na_2SO_4 \cdot 10H_2O$ is 927 g/L at 30°C (Weast 1974). The reaction between dissolved Ba, Sr, and sulfate, however, produces solids of such low solubility that natural waters can exist which have significant concentrations of either dissolved barium and strontium or of sulfate, but not both.

P-T conditions of natural waters that precipitate barite and celestine. Barite and, to a much lesser extent, celestine have formed over a wide range of tempera-tures and pressures in many different geological settings. As we shall see, most barite has formed by precipitation from aqueous solution. It is thus of interest to examine the aqueous solubility of barite and celestine over a wide range of pressure and temperature conditions. Figure 1 shows typical temperature and pressure ranges for fluids in seawater, sedimentary basins, seafloor vents, and volcanic areas, all environments in which barite has been found. Much of the *P-T* space between 0 and 400°C and 1 to 2000 bars represents potential conditions under which barite can precipitate. Barite has also been found in metamorphic and igneous rocks, although it is not common, and the *P-T* conditions could be extended

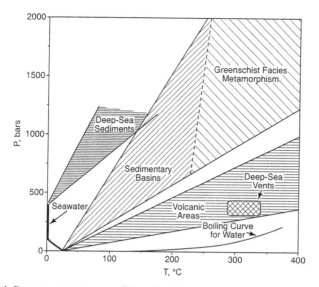

Figure 1. Pressure-temperature conditions for shallow crustal and marine environments.

beyond those shown in Figure 1.

Hydrolysis reactions involving pure endmembers. The mass-action relations for pure barite and celestine and an aqueous phase can be written as

$$BaSO_{4(s)} = Ba^{2+}_{(aq)} + SO_4^{2-}_{(aq)} \tag{1}$$

$$SrSO_{4(s)} + Sr^{2+}_{(aq)} + SO_4^{2-}_{(aq)} \tag{2}$$

For each of these reactions we can define an equilibrium constant, K

$$K_{BaSO_4} = (a_{Ba^{2+}})(a_{SO_4^{2-}})/(a_{BaSO_4}) \tag{3}$$

$$K_{SrSO_4} = (a_{Sr^{2+}})(a_{SO_4^{2-}})/(a_{SrSO_4}) \tag{4}$$

where a_i is the activity of component or species i. If we define standard state conditions such that the activities of pure $BaSO_4$ and pure $SrSO_4$ are unity at any given P and T, then

$$K_{BaSO_4} = (a_{Ba^{2+}})(a_{SO_4^{2-}}) \tag{5}$$

$$K_{SrSO_4} = (a_{Sr^{2+}})(a_{SO4^{2-}}) \tag{6}$$

Based on the calculated values of these equilibrium constants by Bowers et al. (1984), there is a several order-of-magnitude range in K_{BaSO_4} and K_{SrSO_4} over the P-T range of Table 2. At any fixed temperature, there is an increase in K_{BaSO_4}, and thus aqueous solubility, with increasing pressure, reflecting the reduction in volume accompanying the release of ions into solution. At a fixed pressure, there is an increase in K_{BaSO_4} of nearly an order of magnitude from 0 to 100°C, then a progressive decrease with increasing temperature. Celestine, in contrast, has retrograde solubility with increasing temperature over the entire temperature range. The lowest values of K for both phases are at high-T, low-P conditions. The saturation state for pure barite or celestine can be defined by Q/K or log(Q/K), where Q is the measured value of the activity product in Equation (5) or (6). At equilibrium, Q = K and thus Q/K = 1 and log(Q/K) = 0.

Table 2. Log hydrolysis constants for barite and celestine and barite-celestine equilibria at selected temperatures and pressures. Data from Bowers et al. (1984).

log K(barite)   T °C

P, kbar / barite	0	25	50	100	150	200	250	300	350	400
2.0	-9.12	-8.72	-8.50	-8.36	-8.47	-8.74	-9.11	-9.57	-10.11	-10.79
1.5	-9.30	-8.94	-8.72	-8.58	-8.70	-9.00	-9.14	-9.92	-10.57	-11.41
1.0	-9.56	-9.21	-8.98	-8.84	-8.97	-9.29	-9.76	-10.39	-11.25	-12.48
0.5									-15.73	-15.43
sat.*	-10.49	-9.96	-9.68	-9.51	-9.69	-10.16	-10.96	-12.35		

log K (celestite)   T °C

P, kbar	0	25	50	100	150	200	250	300	350	400
2.0	-5.09	-5.13	-5.28	-5.72	-6.26	-6.85	-7.47	-8.12	-8.82	-9.63
1.5	-5.29	-5.38	-5.53	-5.97	-6.51	-7.13	-7.79	-8.49	-9.30	-10.28
1.0	-5.61	-5.67	-5.81	-6.24	-6.80	-7.44	-8.16	-8.99	-10.01	-11.39
0.5									-14.65	-14.46
sat.*	-6.51	-6.43	-6.53	-6.95	-7.56	-8.35	-9.41	-11.03		

log [K(barite)/K(celestite)]   T °C

P, kbar	0	25	50	100	150	200	250	300	350	400
2.0	-4.03	-3.59	-3.22	-2.64	-2.21	-1.89	-1.64	-1.45	-1.29	-1.16
1.5	-4.01	-3.56	-3.19	-2.61	-2.19	-1.87	-1.35	-1.43	-1.27	-1.13
1.0	-3.95	-3.54	-3.17	-2.60	-2.17	-1.85	-1.60	-1.40	-1.24	-1.09
0.5									-1.08	-0.97
sat.*	-3.98	-3.53	-3.15	-2.56	-2.13	-1.81	-1.55	-1.32		

*sat. = saturation vapor pressure

Solubility in non-ideal aqueous solutions. The complex electrostatic interactions that occur in aqueous solutions, particularly in saline waters of high ionic strength, give rise to considerable departures between analytically-determined molal concentrations, m_i, of solute species and their thermodynamic activity, a_i (Nordstrom and Munoz 1994). The non-ideal behavior of sulfate, in particular, is pronounced even at moderate ionic strengths. These deviations can be accounted for by use of conventional free-ion activity coefficients, γ_i

$$a_{Ba^{2+}} = (\gamma_{Ba^{2+}})(m_{Ba^{2+}}) \qquad (7)$$

or stoichiometric activity coefficients, γ^*_i

$$a_{Ba^{2+}} = (\gamma^*_{Ba^{2+}})(m_{Ba\ tot}) \qquad (8)$$

where $m_{Ba\ tot}$ is the molality of total dissolved Ba.

There is an extensive body of experimental work on the solubility of end-member barite (e.g. Strübel 1967, Gundlach et al. 1972, Blount 1977) and celestine (e.g. Strübel 1966, Reardon and Armstrong 1987) in a variety of electrolyte solutions at different temperatures. The information derived both from theoretical considerations and experimental studies has been useful in deriving more generalized equations of state, such as those of Pitzer (1987) which can be used to establish activity–concentration relations for dissolved Ba, Sr, and sulfate over a wide range of fluid compositions (e.g. Monnin and Galinier 1988, Monnin 1999, Ptacek and Blowes, this volume).

We examine briefly here some of the experimental work by Blount (1977) to assess the magnitude of the effects of increasing salinity on barite solubility. Many subsurface saline waters are dominated by Na and Cl as the principal solutes (Hanor 1994). Figure 2 shows the solubility of barite in NaCl solutions of varying NaCl concentration, as plotted from Blount's (1977) data. There is a significant increase in the solubility of barite by over an order of magnitude with increasing concentration of NaCl from 0 to 6 m, reflecting in large part the strongly non-ideal behavior of the sulfate ion in concentrated electrolyte solutions. The solubility curves are convex up relative to the NaCl axis, and mixing of two barite-saturated solutions of differing NaCl concentration would produce an intermediate water which is undersaturated with respect to barite.

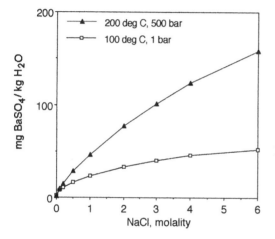

Figure 2. Variation in the solubility of barite in NaCl solutions as a function of the molality of NaCl at two selected *P-T* conditions. Plotted from data of Blount (1977). Mixing of two barite-saturated solutions of differing salinity at a fixed temperature and pressure produces an intermediate solution which is undersaturated.

Aqueous solution–solid-solution equilibria. A detailed account of solubility and thermodynamic properties of sulfate solid solutions is given by Glynn (this volume), and only a very brief overview is presented here. The problem of determining the stability of barite and celestine in natural waters is made complex by the fact that they rarely exist in nature as pure end-member phases, unlike the solids used in much experimental work, but rather as solid solutions. The activities of the components in the solid solution are given by

$$a_{BaSO_4} = (\lambda_{BaSO_4})(X_{BaSO_4}) \tag{9}$$

$$a_{SrSO_4} = (\lambda_{SrSO_4})(X_{SrSO_4}) \tag{10}$$

where λ_i is the activity coefficient and X_i is the mole fraction of the end-member component. In an ideal solid solution, λ_i is unity and $a_i = X_i$ for all components.

In theory, in a aqueous system in equilibrium with respect to both components, $BaSO_4$ and $SrSO_4$, must meet both conditions (9) and (10). In a strictly binary solid solution, these two equations are coupled through the relation.

$$X_{BaSO_4} + X_{SrSO_4} = 1 \tag{11}$$

A variety of graphical and computational techniques have been developed to portray or evaluate equilibrium involving aqueous solutions and binary solid solutions (see Glynn and Reardon 1990, Glynn et al. 1990, Felmy et al. 1993, Glynn this volume). One simple method for predicting the composition of solid solutions that should be formed during precipitation is through the use of partition coefficients in which the activity of sulfate is eliminated as a variable. Subtracting mass-action Equation (2) from Equation (1) we have:

$$BaSO_{4(ss)} + Sr^{2+}_{(aq)} = SrSO_{4(ss)} + Ba^{2+}_{(aq)} \tag{12}$$

The equilibrium constant for this new mass-action relation is

$$K_{Sr-Ba} = (a_{SrSO_4})(a_{Ba^{2+}})/(a_{BaSO_4})(a_{Sr^{2+}}) \tag{13}$$

where

$$K_{Sr-Ba} = K_{BaSO_4}/K_{SrSO_4} \tag{14}$$

At equilibrium we can rewrite Equation (13) as

$$(a_{SrSO_4})/(a_{BaSO_4}) = K_{Sr-Ba} (a_{Sr^{2+}})/(a_{Ba^{2+}}) \tag{15}$$

Thermodynamic partition coefficients for ideal solid solutions. Assuming initially that the barite-celestine solid-solution series behaves ideally and that deviations from ideality of Sr^{2+} and Ba^{2+} are similar, Equation (15) can be used to estimate the $(SrSO_4/BaSO_4)$ activity ratio of a solid that should precipitate out of a solution of a given (Sr/Ba) ratio under equilibrium conditions. Table 2 shows the variation in $K_{Sr-Ba} = (K_{BaSO_4}/K_{SrSO_4})$ over the P-T range of Figure 1. There is a very slight decrease in K_{Sr-Ba} with increasing pressure at a fixed temperature, but there is a nearly three order-of-magnitude increase in K_{Sr-Ba} with increasing temperature, from values of approximately 10^{-4} at 0°C to values approaching 10^{-1} at temperatures of 400°C. These extremely small values of K_{Sr-Ba}, particularly at low temperatures, result in the strong preferential partitioning of Ba into the solid phase and Sr into the aqueous phase.

Partition coefficients from precipitation experiments. A number of laboratory studies have been made of the partitioning of Ba and Sr between coexisting aqueous solutions and crystalline sulfate phases. Here, partitioning is expressed not as a thermodynamic constant, but rather a proportionality term, D_{Sr-Ba}, whose magnitude may reflect non-ideal behavior of the aqueous and/or solid solution, kinetics of precipitation, back reaction or re-equilibration after precipitation, and compositional zoning of the precipitate:

$$(Sr)/(Ba)_{solid} = D_{Sr-Ba} [(Sr^{2+})/(Ba^{2+})]_{aqueous} \tag{16}$$

Experimental studies of Ba-Sr partitioning during the precipitation of $(Ba,Sr)SO_4$ from aqueous solution include those of Goldschmidt (1938) at 18°C, Gordon et al. (1954) at 83°C, Starke (1964) at various temperatures between 20°C and 250°C, Church (1970) at temperatures of between 3 and 55°C, and Blount (1977) at 95°C. Experimental procedures and aqueous media varied widely among these workers. In general, experimentally derived partition coefficients, D_{Sr-Ba}, from most of the above precipitation experiments are one or more orders of magnitude *greater* than thermodynamic exchange constants at the same temperature. For reference, the thermodynamic K_{Sr-Ba} at 25°C and 1 bar is $10^{-3.53}$. Starke (1964) reported a D_{Sr-Ba} of approximately $10^{-1.5}$ at 20°C, Goldschmidt a value of approximately $10^{-2.5}$ at 18°C, and Blount, D_{Sr-Ba} values of between 0.015 and 0.03 at 95°C. Starke's values for D_{Sr-Ba} progressively increase with increasing temperature to values of 0.6 at 250°C.

In his experiments to determine partition coefficients for barite in seawater, Church (1970) mixed dilute sulfate-free seawater containing low concentrations of Ba with large volumes of seawater and observed that slower reaction rates produced lower concentrations of Sr, Ca, and K in the solids precipitated. Church's values of approximately $10^{-4.0}$ for D_{Sr-Ba} and $10^{-6.0}$ for D_{Ca-Ba} at 25°C are closer to thermodynamically predicted values that those of other workers.

Precipitation experiments involving $(Ba,Sr)SO_4$ are complicated because the preferential partitioning of Ba into the precipitate results in a continuing increase in the Sr/Ba ratio of the aqueous solution and hence in the Sr/Ba ratios of solids sequentially precipitated from solution. If earlier-formed solids do not back-react with the fluid, then the progressive change in fluid composition as a result of partitioning can be described by the Doerner-Hoskins relation, originally developed to describe the coprecipitation of Ra with Ba (Gordon and Rowley 1957):

$$\ln (Sr_i/Sr_f) = \lambda_{Sr-Ba} \ln (Ba_i/Ba_f) \qquad (17)$$

where λ_{Sr-Ba} is the logarithmic distribution coefficient, which is usually of the same magnitude as D_{Sr-Ba}, and the subscripts i and f refer to initial and final aqueous solution concentrations. Prieto et al. (1993) provide an in-depth review and discussion.

Partition coefficients from re-equilibration experiments. A second technique for determining equilibrium relations between aqueous and solid solution is to attempt to re-equilibrate a previously formed solid precipitate with an aqueous fluid. This is one of the approaches followed by Church (1970), who attempted to re-equilibrate pure barite and an artificial barite enriched in Sr, Ca, and K with surface seawater at 3°C for 17 months. Church observed a progressive loss with time of Sr, Ca, and K from the artificial barite.

In their re-equilibration experiments involving a wide range of solid compositions over a time span of nearly three years, Felmy et al. (1993) found evidence that small quantities of a phase close in composition to pure barite controlled aqueous Ba concentrations, and that substitution of Sr occurred in a second solid-solution phase in which the Ba component did not attain equilibrium with the aqueous phase.

Evidence for non-ideal behavior of the $BaSO_4$-$SrSO_4$ solid solution. The thermodynamic properties of the $BaSO_4$-$SrSO_4$ solid solution have not yet been rigorously established, but there is evidence for non-ideal behavior. This is not surprising in view of the large difference in ionic radii between Ba^{2+} and Sr^{2+} (Malinin and Urusov 1983). Goldish (1989) found that cell dimensions deviate from strictly linear relations as a function of composition. Burkhard (1973) demonstrated nonlinear variations in barite-celestine refractive indices. Brower and Renault (1971), however, concluded that the solid solution was ideal on the basis of their experimentally determined enthalpy values. See Hanor (1973) for a critical review of their experimental techniques.

Hanor (1966) noted that D_{Sr-Ba} in Starke's (1964) precipitation experiments varied as a function of solid composition at fixed temperature, and attempted from Starke's precipitation data to derive values for B in the following equation relating activity coefficients and solid-solution compositions

$$RT \ln \lambda_1 = B (X_2)^2 \qquad (18)$$

where B is comparable to the term W_G in a symmetrical Margules model solid solution (cf., Nordstrom and Munoz 1994). The results, not surprisingly, show increasing departure from ideal behavior with decreasing temperature. Church (1970) calculated the free energy of non-ideal mixing for K^+, Sr^{2+}, and Ca^{2+} in synthetic marine barite from long-term precipitation-equilibration experiments. He found that K behaved nearly ideally, but that Sr is non-ideal and Ca more strongly non-ideal, consistent with the progressive differences in ionic radii with Ba^{2+} (Table 1). Not discussed by Church (1970) is the fact that the K^+ substitution must in some way be coupled to preserve charge balance. Felmy et al. (1993) reviewed more recent work and some of the experimental problems that make determination of the thermodynamic properties of this solid solution difficult in re-equilibration experiments, including the apparent formation of two solid phases, one nearly pure barite, the other a $(Ba,Sr)SO_4$ solid solution, as noted above.

Kinetics of precipitation and dissolution. The kinetics of precipitation and dissolution of barite and celestine are dealt with in detail by Hina and Nancollas (this volume). Some recent studies include those of Bosbach et al. (1998), Christy and Putnis (1993), Denis and Michard (1983), Fernandez-Diaz et al. (1990), Kornicker et al. (1991), Mohazzabi and Searcy (1976), Prieto et al. (1990), and Putnis et al. (1992, 1995).

As discussed in this chapter, the precipitation of barite occurs in geologic settings which range from the rapid mixing of 350°C Ba-bearing hydrothermal fluids with 2°C deep seawater to the replacement of solid gypsum and anhydrite in probably nearly isothermal settings over extended periods of time. Barite thus exists in a wide range of crystal sizes, morphologies, and textures which in part reflect the kinetics and conditions of its crystallization. Barite, because of its low solubility, resists wholesale aqueous dissolution. Barite can be effectively dissolved, however, in highly reducing systems. Here the rates of dissolution may ultimately be dependent on the kinetics of biologically mediated redox reactions.

Practical problems involving the kinetics of barite and celestine precipitation and dissolution include the formation of barite scale in hydrocarbon wells (Weinritt and Cowan 1967) and the destruction of particles of pelagic barite and celestine that are produced in near-surface seawater and subsequently settle down through undersaturated marine waters (Dehairs et al. 1980).

Stability ranges in multicomponent systems

The thermodynamic stability of barite and celestine depends not only on their solubility in aqueous solution at various P-T and salinity conditions, but also on the presence or absence of other reactive components that can combine with Ba, Sr, and sulfate. Some of these other stability constraints are briefly reviewed.

Redox stability. Both barite and celestine can be destroyed through the inorganic and the biologically mediated reduction of sulfate, as in the following reactions at intermediate pHs, where $SO_4^{2-}{}_{(aq)}$, $HS^-{}_{(aq)}$, and $HCO_3^-{}_{(aq)}$ are predominant species:

$$BaSO_{4(s)} \rightarrow Ba^{2+}{}_{(aq)} + SO_4^{2-}{}_{(aq)} \tag{19}$$

$$SO_4^{2-}{}_{(aq)} + H^+{}_{(aq)} \rightarrow HS^-{}_{(aq)} + 2O_{2(aq\ or\ g)} \tag{20}$$

In low temperature marine environments, organic carbon, here represented by CH_2O, is a common reductant of sulfate:

$$SO_4^{2-}{}_{(aq)} + CH_2O_{(s)} \rightarrow HCO_3^-{}_{(aq)} + HS^-{}_{(aq)} + O_{2(aq\ or\ g)} \tag{21}$$

The reduction of barite and celestine is thus favored by low f_{O2} and high H^+ (low pH) conditions. The low solubility of barite extends the stability field of $BaSO_4$ well into conditions where reduced sulfur, not sulfate, is the predominant aqueous sulfur species, and it is common to find stable coexisting mineral assemblages consisting of barite with pyrite, galena, and sphalerite (Fig. 3). BaS has been found in some anthropogenic reducing environments (Baldi et al. 1996) and is often used in constructing phase diagrams to denote the lower practical limit of barite stability. With progressively decreasing f_{O2}, $CaSO_4$ will be reduced first because of it is more soluble, followed by $SrSO_4$, and then $BaSO_4$.

Potential reducing agents for barite at elevated *P-T* conditions include CH_4, H_2, CO, NH_3, and solid carbon. Theoretical calculations by Kritsotakis and von Platen (1980) show that reduction is favored by temperatures in excess of 200°C, low pH, and high gas fugacities. Despite the results of these theoretical calculations, there is abundant field

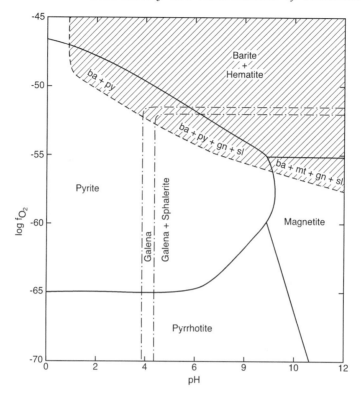

Figure 3. Log f_{O_2}–pH diagram showing the stability fields for barite (shaded), Fe oxides and sulfides, galena, and sphalerite at 100°C and a salinity of 20 wt % in seawater proportions. Total sulfur = 5.5×10^{-4} m. Galena and sphalerite fields drawn for 1 ppm each of Pb and Zn. Barite field drawn for 100 ppm Ba. Modified from Cooke et al. (1998). Note overlap of stability fields for barite and pyrite, galena, and sphalerite.

evidence of the remarkable stability of barite during metamorphism in the presence of reducing agents, as discussed later in this chapter.

Stability in presence of calcium sulfates. Both barite and celestine occur as replacements of gypsum or anhydrite as the result of reaction of Ba- and Sr-bearing fluids with calcium sulfate. Neglecting solid solution, these stability relations can be described by the following reactions:

$$BaSO_{4(s)} + Ca^{2+}_{(aq)} = CaSO_{4(s)} + Ba^{2+}_{(aq)} \qquad (22)$$

$$SrSO_{4(s)} + Ca^{2+}_{(aq)} = CaSO_{4(s)} + Sr^{2+}_{(aq)} \qquad (23)$$

The corresponding ion activity ratios at equilibrium are given by:

$$a_{Ba^{2+}}/a_{Ca^{2+}} = \log (K_{BaSO_4}/K_{CaSO_4}) \qquad (24)$$

$$a_{Sr^{2+}}/a_{Ca^{2+}} = \log (K_{SrSO_4}/K_{CaSO_4}) \qquad (25)$$

For example, based on data of Bowers et al. (1984) at 25°C and 1 bar (Table 3), anhydrite is unstable relative to barite in waters having log activity ratios of Ba^{2+}/Ca^{2+} greater than -5.69 and could be replaced by celestine where waters have Sr^{2+}/Ca^{2+} log activity ratios greater than -2.16. We will pursue this theme further when we discuss replacement of calcium sulfate minerals by barite and celestine.

Table 3. Log hydrolysis constants for barite-anhydrite and celestine-anhydrite equilibria at selected temperatures and pressures. Calculated from data in Bowers et al. (1984).

log [K(barite)/K(anhydrite)]   T deg C

P, kbar	0	25	50	100	150	200	250	300	350	400
2.0	-6.43	-5.78	-5.19	-4.27	-3.59	-3.07	-2.66	-2.34	-2.06	-1.83
1.5	-6.41	-5.74	-5.16	-4.24	-3.55	-3.04	-2.36	-2.30	-2.03	-1.78
1.0	-6.36	-5.71	-5.13	-4.21	-3.53	-3.01	-2.60	-2.26	-1.98	-1.71
0.5									-1.71	-1.52
sat.*	-6.42	-5.69	-5.10	-4.17	-3.47	-2.95	-2.52	-2.13		

log [K(celestite)/K(anhydrite)]   T deg C

P, kbar	0	25	50	100	150	200	250	300	350	400
2.0	-2.40	-2.19	-1.97	-1.63	-1.38	-1.18	-1.02	-0.89	-0.77	-0.67
1.5	-2.40	-2.18	-1.97	-1.63	-1.36	-1.17	-1.01	-0.87	-0.76	-0.65
1.0	-2.41	-2.17	-1.96	-1.61	-1.36	-1.16	-1.00	-0.86	-0.74	-0.62
0.5									-0.63	-0.55
sat.*	-2.44	-2.16	-1.95	-1.61	-1.34	-1.14	-0.97	-0.81		

*sat. = saturation vapor pressure

Carbonate–sulfate equilibria. Witherite, $BaCO_3$, occurs in low-temperature deposits, and its formation is generally described as the replacement of barite (Chang et al. 1996). Witherite may react with sulfuric acid produced by the oxidation of sulfide ores to form barite (Fitch 1931). Strontianite, $SrCO_3$, also is generally formed from alteration of celestine and/or in association with celestine. In the presence of Ca, witherite requires high activities of barium and low activities of sulfate (Fig. 4) to be stable. Such conditions can exist where barite is being dissolved through sulfate reduction.

There are, in addition to witherite, other barium carbonates whose origin may include replacement of barite. These include alstonite, paralstonite, and barytocalcite, all

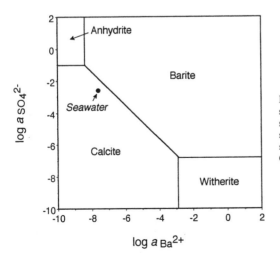

Figure 4. Stability fields for Ba-Ca sulfates and carbonates in aqueous solution at 25°C and 1 atm in aqueous solutions of modern seawater composition. Modified from Maynard and Okita (1991).

polymorphs having the formula $BaCa(CO_3)_2$, and norsethite, $MgBa(CO_3)_2$ (Chang et al. 1996). Benstonite, $(Ba,Sr)_6(Ca,Mn)_6Mg(CO_3)_{13}$, occurs in the Chamberlain Creek sedimentary-exhalative (sedex) bedded barite deposit in Arkansas (Lippman 1962).

Sulfate–silicate equilibria. Several Ba-bearing silicates coexist with barite in some metamorphosed barite deposits and are potential sinks for Ba during metamorphism. These minerals include the zeolite harmotome, $(Ba_{0.5}Ca_{0.5},K,Na)_5Al_5Si_{11}O_{32}\cdot12H_2O$, and the feldspars celsian, $BaAl_2Si_2O_8$, and hyalophane, $(K_{1-x}Ba_x)Al(Si_{3-x}Al_x)O_8$, where x = 0.25 to 0.40, and cymrite, $BaAl_2Si_2(O,OH)_8\cdot H_2O$. The thermodynamic properties of the Ba aluminosilicates have not been well established, so their stability ranges are not well known. However, barite is remarkably refractory during metamorphism in silicate-bearing rocks in the absence of a sink for sulfate, as is described in the section in this chapter on behavior of barite during metamorphism.

Naerh and Bohrmann (1999) described Ba-rich authigenic clinoptilolite in sediments from the Japan sea, which they propose as a sink for dissolved Ba. Removal of Ba in silicate phases may be occuring inhydrothermal systems on the flank of the Juan de Fuca ridge (Bautier et al. in press).

Ba, Sr, AND S IN CRUSTAL ROCKS AND NATURAL WATERS

Most barite and celestine has been precipitated from aqueous fluids which contain Ba and Sr derived from the alteration of silicate, carbonate, and sulfate minerals in sedimentary, metamorphic, and igneous rocks. It is thus useful in understanding the controls on the distribution and composition of barite and celestine to examine briefly the crustal distribution of Ba, Sr, and sulfate both in terms of bulk-rock and bulk-fluid compositions. More extensive reviews on the general geochemistry of Ba and Sr are given by Puchelt (1967, 1972) and Faure (1972).

Crustal abundance and controls on the distribution of Ba and Sr

Barium, Sr, and S are relatively abundant and widely distributed elements in the rocks of the Earth's crust. Estimates of average crustal abundance vary, but Faure (1998) gives the following subequal values: Ba, 250 ppm; Sr, 260 ppm; and S, 310 ppm. The crustal geochemistry of Ba^{2+} and Sr^{2+} is dominated by their similarity in ionic radius and electronegativity to the major rock-forming cations, K^+ and Ca^{2+}, respectively. Most Ba in the Earth's crust exists in solid solution for K in K-bearing minerals, such as K-feldspars and K-micas, including biotite. A lesser amount substitutes for Ca in Ca-silicates. In marked contrast, most strontium exists as a proxy for Ca in Ca minerals such as anorthite, calcite, gypsum, and anhydrite.

Barium and Sr are present in most major igneous and siliciclastic rock types in subequal amounts at concentrations of between 200 and 900 ppm (Table 4, in the section entitled *Barite of Continental-Igneous and Igneous-Hydrothermal Origin,* below). Exceptions are tholeiite mid-ocean-ridge basalts, which have very low concentrations of both Ba and Sr, and carbonatites, which have exceptionally high concentrations. The abundance of Ba and Sr in igneous rocks increases with increasing Si, and hence, K and Ca, concentrations. The Sr/Ba weight ratio of most major rock types varies within the fairly narrow range of between 5 and 0.5, with the important exception of evaporites and carbonates, which can have Sr/Ba ratios in excess of 1000. This significant fractionation of Sr over Ba into Ca-carbonate and Ca-sulfate sediments plays a key role in the formation of celestine deposits, as is discussed later in this chapter. The general lack of celestine deposits of Precambrian age may well reflect the paucity of carbonate and Ca-sulfate sediments of this age and hence the absence of a mechanism for concentrating Sr.

Sulfate geochemistry

Sulfur exists primarily as sulfide, S^{2-}, sulfate, SO_4^{2-}, and as native sulfur, S^o, although other valence states are known. The geochemistry of sulfur in the Earth's crust is controlled largely by redox state. Under reducing conditions sulfur may exist primarily as H_2S and metal sulfides, and in highly oxidizing conditions it exists mainly as dissolved sulfate and the alkaline earth sulfates gypsum, anhydrite, celestine, and barite.

The geochemical cycling of sulfur has evolved significantly over time in Earth history. The Earth's earliest atmosphere was most likely reducing and dominated by CO_2, N_2, and perhaps CH_4, with small amounts of reduced gases such as H_2, CO, and reduced S species (Kasting 1993, Des Marais 1994). The appearance of small amounts of free oxygen in the atmosphere and in surface ocean waters probably did not occur until the Proterozoic during a transition period marked by the appearance of redbeds at 2.3 Ga (Grotzinger and Kasting 1993). By 2.0 Ga, the Earth entered an aerobic stage in which free oxygen pervaded the entire ocean-atmosphere system (Eriksson et al. 1998).

Although minor occurrences of evaporitic gypsum are reported in the Archean (Buick and Dunlop 1990) they seem to reflect local sources of sulfate, perhaps ultimately magmatic in origin. The first extensive sedimentary sulfates known were deposited in the MacArthur basin, Australia from 1.7 to 1.6 Ga (e.g. Warren 1997, Cooke et al. 1998). Halite was commonly deposited immediately above carbonates in Early Proterozoic sediments, without intervening Ca-sulfates. According to Eriksson et al. (1998) the scarcity of Ca-sulfate evaporites may reflect either low concentrations of sulfate in seawater or a bicarbonate/Ca ratio that was sufficiently high to remove Ca by carbonate precipitation before the gypsum stability field was reached (Grotzinger and Kasting 1993). Most gypsum and anhydrite deposits are of Phanerozoic age.

Waters in sedimentary basins

As a result of an extensive body of analyses of waters coproduced with crude oil and natural gas, much is known about the composition of aqueous fluids in deep sedimentary basins. The behavior of Ba, Sr, and S in these waters provides clues as to the probable behavior in aqueous fluids in other shallow crustal rock types as well.

As discussed by Hanor (1979, 1994), the salinity of pore fluids in sedimentary basins varies by nearly five orders of magnitude, from <100 mg/L in shallow meteoric groundwater systems to hypersaline brines containing >400,000 mg/L. Barium and strontium show a similarly wide range in concentrations, from 0.1 to >1000 mg/L each.

In contrast to the anionic composition of saline waters, which is dominated almost entirely by chloride, there is a progressive, systematic change in the cationic makeup of formation waters with increasing salinity (Fig. 5), reflecting rock-buffering of fluid compositions (Hanor 1988). Where the composition of a fluid is rock-buffered by a specific silicate-carbonate mineral assemblage, the activity ratios of individual major cations to hydrogen ion, e.g. Na^+/H^+, $Ca^{2+}/(H^+)^2$, are nearly constant at a fixed P and T and vary only slightly as a result of variations in the activity of H_2O induced by changes in salinity (Hanor 1994). Hanor (1996a) showed that fixed cation-activity ratios and charge-balance constraints in such systems require that the activities of monovalent cations such as H^+, Na^+ and K^+ increase in 1:1 proportion to increases in total anionic charge or salinity, but that activities of Mg^{2+}, Ca^{2+}, and Sr^{2+} increase by a factor of 10:1. These relations for Ca^{2+} and Sr^{2+} versus TDS are reflected in 2:1 slopes on log-log scatter plots for basinal fluids world-wide (Fig. 5). Rather than a single unique phase assemblage responsible for controlling fluid compositions, various assemblages made up of common sedimentary silicate and carbonate minerals impose broadly similar constraints on fluid composition.

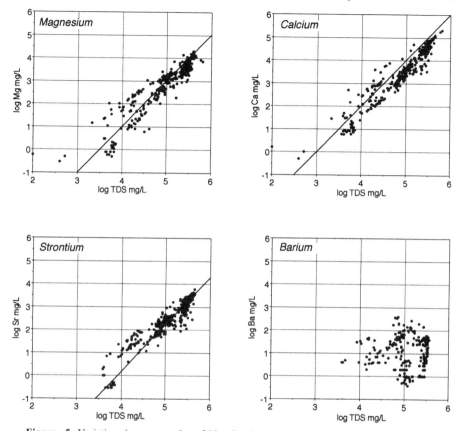

Figure 5. Variations in concentration of Mg, Ca, Sr, and Ba as a function of salinity in basinal waters worldwide. Solid lines for Mg, Ca, and Sr represent a 2:1 log-log slope. Data: Hanor (1994).

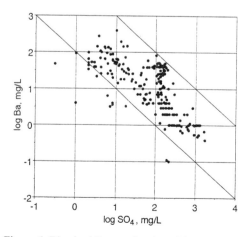

Figure 6. Dissolved Ba as a function of dissolved sulfate in basinal waters world-wide. Solid lines represent a -1:1 log-log slope. Data from Hanor (1994).

Of significance in the discussion of barite and celestine deposits is the dissimilarity of the behavior of Sr and Ba in basinal waters. In contrast to Ca^{2+} and Sr^{2+}, which each show a 10:1 increase in concentration with increasing salinity (Fig. 5), Ba shows a high degree of scatter. There is, however, a rough inverse correlation with a slope of minus 1 between log Ba and log sulfate (Fig. 6), which is consistent with the hypothesis that equilibrium with respect to barite ($BaSO_4$) may be one factor controlling Ba concentrations (see also Macpherson 1989). Calculations by Fisher (1995, 1998) for similar subsurface brines in Texas show the fluids to be close to saturation with respect to barite. The master variable controlling Ba concentrations in many basinal waters thus seems to be sulfate, whose own concentration is controlled by rate-dependent processes of dissolution of gypsum and anhydrite, dispersive fluid mixing, and reduction to sulfide. Not surprisingly, the waters highest in Ba are low in sulfate. Based on presently available data, fluids that would have the potential for transporting >10 mg/L Ba to a site of ore-mineral precipitation would typically have sulfate concentrations of <200 mg/L but could have salinities ranging from 10,000 to 300,000 mg/L.

Meteoric groundwaters

Buffering by barite is an important control on the concentration of barium in some shallow meteoric groundwater systems (Fig. 7). Where there has been bacterial reduction of sulfate, dissolved Ba can exceed the U.S. Environmental Protection Agency drinking water standard of a maximum of 2 mg/L (Gilkeson et al. 1981), even in waters of very low salinity.

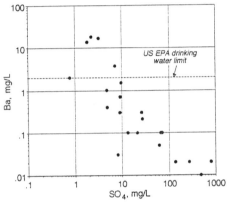

Figure 7. Dissolved Ba as a function of dissolved sulfate in portions of fresh and brackish water aquifers in northeastern Illinois. The waters are close to equilibrium with respect to barite. Plotted from data of Gilkeson et al. (1981).

Seawater

The approximate residence time of Ba in open seawater is a relatively short 5×10^4 y (Chester 1990), which, in the presence of sources and sinks for Ba within the water column, gives rise to significant spatial variations in dissolved Ba. The concentration of Ba in seawater typically varies from approximately 5 to 20 µg/kg of seawater (~5 to 20 µg/L) in open ocean water. The lowest values occur in the surface North Atlantic, the highest are in deep North Pacific waters (Chan et al. 1977, Ostlund et al. 1987, Rhein et al. 1987, Rhein and Schlitzer 1988, Jeandel et al. 1996). Much higher Ba concentrations, up to 60 µg/kg, are found in marine anoxic basins, such as the deep Black Sea (Falkner et al. 1993) and the anoxic hypersaline brines of the Tyro and Bannock basins, eastern Mediterranean (de Lange et al. 1990a,b).

Interest in the use of dissolved Ba as a potential stable proxy for radium led Chow and Goldberg (1960) to make the first studies of the concentration of dissolved Ba in seawater. These authors established that the depth distribution of dissolved Ba is similar to that of Ra (Koczy 1958), namely low in surface waters and higher at depth (Fig. 8). They concluded that Ba is extracted from surface waters by some process and is then released into deeper water. Chow and Goldberg (1960) also made the first calculations of saturation state of barite in seawater, taking into account pressure and temperature, and concluded that surface seawater is undersaturated with respect to barite but that saturation may be reached at greater depths.

The establishment of the GEOSECS program in the early 1970s brought with it large

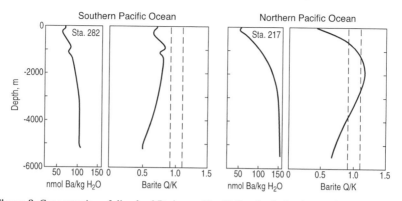

Figure 8. Concentration of dissolved Ba in nmol/kg H$_2$O and calculated saturation state, Q/K, for pure barite for seawater samples from two GEOSECS stations in the Pacific Ocean. A value of Q/K = 1 represents saturation. Modified from Monnin et al. (1999).

numbers of high-precision analyses of dissolved Ba (Church and Wolgemuth 1972; Chan et al. 1977; Chan and Edmund 1987). Church and Wolgemuth (1972) calculated the concentration of Ba in seawater saturated with respect to pure barite as a function of depth, taking into account the concentration of sulfate in seawater and the decreases in temperature and increases in pressure with depth. They compared their calculated values with the then recently measured Ba concentrations from the GEOSECS program and from Ba concentrations in pore waters in deep sea sediments and demonstrated that waters of the eastern Pacific were undersaturated from the surface to the seafloor, but that pore waters in deep-sea sediments were approximately saturated. The high-precision barium analyses of seawater of Chan et al. (1977) and Chan and Edmund (1987) have been utilized in a more recent study by Monnin et al. (1999) to calculate the saturation state of the oceans with respect to pure barite (Fig. 8). The results of these calculations show that while the oceans are generally undersaturated with respect to barite, equilibrium is reached in cold surface waters of the Southern Ocean, intermediate waters of the Pacific and deep waters of the Bay of Bengal. Slight supersaturation occurs in surface waters of the Waddell Gyre. Waters of the Atlantic are undersaturated. The calculated saturation state (Q/K) for pure barite ocean-wide ranges from approximately 0.15 to nearly 1.5.

Strontium is much more abundant in seawater than Ba. Strontium concentrations range from approximately 7.7 to 8.1 mg/kg of seawater, and its residence time is approximately 4×10^6 y (Chester 1990). The residence time of Sr is several orders of magnitude larger than oceanic mixing times of 1,500 years, and hence the isotopic composition of marine Sr is uniform at any given time, although it shows long-term secular variations. There are only minor variations in Sr/salinity ratios in the open ocean. Bernstein et al. (1992) estimated the saturation index of celestine in surface ocean water to be approximately 0.16.

Sulfate is the second most abundant anion in modern seawater at an average concentration of 2,712 mg/kg of seawater, and is surpassed in abundance only by chloride (Chester 1990). The long residence time of sulfate, 8.7×10^6 y, results in uniform sulfate/salinity ratios in the open ocean. In anoxic basins, however, sulfate can be reduced to sulfide. Like Sr, the isotopic composition of marine S is uniform at any given time in open ocean conditions, although it shows long-term secular variations.

Seawater today is sulfate-rich and Ba-poor. Although the oceans are generally slightly undersaturated with respect to pure BaSO$_4$ and biological activity is effective in maintaining this undersaturation, the principal reason the oceans are Ba-poor is the presence of high levels of dissolved sulfate. Is it possible that during the Archean, when there was little free oxygen and sulfate, that the oceans were barium-enriched.

River and estuarine waters

The concentration of Ba in surface ocean water is significantly less than that of average river water. Hanor and Chan (1977) for example, found that dissolved Ba in the lower Mississippi River at the time of their study of the Mississippi River Estuary was 60 µg/L, whereas the Gulf of Mexico had Ba values of only 11 µg/L (~11 µg/kg seawater). Hanor and Chan (1977) also found that the behavior of dissolved Ba in the Mississippi River Estuary is strongly non-conservative, with excess Ba introduced into solution during mixing as a result of Na displacing adsorbed Ba on suspended clays. Waters of intermediate salinity are actually approximately two times supersaturated with respect to barite. Approximately half of the flux of dissolved Ba into the Gulf of Mexico from the Mississippi River was derived by desorption occurring near the mouth of the river. Shaw et al. (1998) have recently documented a diffuse flux of dissolved Ba from groundwaters discharging into the ocean along the U.S. Atlantic coast. Additional information on Ba in river and estuarine systems is given by Edmond et al. (1978), Carroll et al. (1993), Coffey et al. (1997), and Stecher and Kogut (1999).

Waters in crystalline rocks

Data on Ba in pore waters in crystalline rocks of the upper crust are limited. Most of the major cations, however, show trends with salinity and chlorinity similar to those documented for sedimentary basins, and it is probable that minor dissolved species such as Ba and Sr may show trends similar to those observed in sedimentary basins.

Continental rifts

Analyses of waters from the Salton Sea continental rift, California (McKibben and Hardie 1997) provide clues as to the behavior of Ba and Sr in areas of high heat flow and continental hydrothermal activity. Temperatures of waters which have been analyzed range from 200 to >300°C. Most dissolved sulfate values are <100 mg/kg fluid. Under these conditions of high temperature and low sulfate, concentrations of Ba are not buffered by barite. Barium instead shows a roughly 2:1 increase with increasing salinity on a log-log plot, similar to the behavior of Ca and Sr for these same waters (Fig. 9).

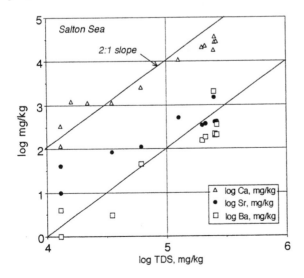

Figure 9. Variation in dissolved Ca, Sr, and Ba in waters from the Salton Sea continental rift. Solid lines show 2:1 slopes on the log-log plot. Compare Ba to Figure 5. Plotted from data of McKibben and Hardie (1997).

Seafloor hydrothermal vents

Hydrothermal vents are a significant source of Ba to the ocean. Edmond et al. (1979) and Von Damm (1995) estimated that the hydrothermal input of Ba is about 1/3 its total input into the ocean. Scott (1997) summarized Ba, Sr, and chlorinity values for modern seafloor hydrothermal vents and noted that Ba values represent minima because of the rapid removal of Ba by precipitation as barite. Barium values range from 0.01 to 0.06 mmol/L, or roughly two orders of magnitude higher than seawater levels of Ba. Strontium ranges from from 0.04 to 0.27 mmol/L, or from approximately half to three times seawater levels. Chlorinities range from slightly less to nearly twice that of seawater.

Review of controls on Sr/Ba ratios in natural waters

The Sr concentration of many crustal waters appears to be controlled largely by buffering by multi-phase silicate-carbonate mineral assemblages, whereas the concentration of Ba in many crustal waters is controlled instead by buffering with respect to barite. This decoupling of the geochemical behavior of Sr and Ba has given rise to a wide range of Sr/Ba ratios in crustal fluids. It is this diverse range of fluid Sr/Ba ratios that is partly responsible for the observed range of Sr/Ba ratios in naturally occurring barite and celestine. Because of the strong preferential partitioning of Ba into the solid sulfate phase during precipitation, only waters of extremely high Sr/Ba ratios are capable of precipitating celestine. The highest Sr/Ba waters today are those of high salinity (Fig. 10) and low Ba/SO$_4$ ratio (Fig. 11), i.e. highly saline waters with significant concentrations of dissolved sulfate.

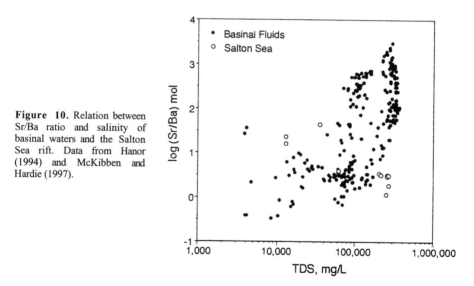

Figure 10. Relation between Sr/Ba ratio and salinity of basinal waters and the Salton Sea rift. Data from Hanor (1994) and McKibben and Hardie (1997).

ENVIRONMENTS OF FORMATION OF BARITE AND CELESTINE: AN OVERVIEW

This section reviews some of the recurring themes discussed in the remainder of this chapter. These include general mechanisms for forming barite and celestine and the global relations between the occurrence of barite and celestine and regional tectonics, hydrogeology, and the secular evolution of the composition of seawater and sedimentary rocks.

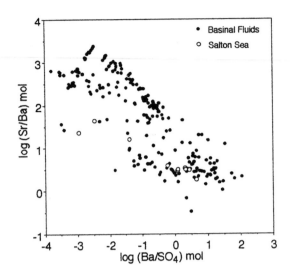

Figure 11. The relation between the Sr/Ba and Ba/SO4 ratios of basinal waters and Salton Sea rift. Data from Hanor (1994) and McKibben and Hardie (1997).

Mechanisms for the precipitation of barite and celestine

Most barite and celestine has formed in one or more of the following ways:

Barium-rich fluids ascending from depth and mixing with seawater. Such deposits are known as *submarine exhalative* deposits. There are two principal end-member types, *submarine volcanic hydrothermal* deposits, whose formation is directly related to fluid convection driven by magmatic heat, and *sedimentary exhalative (sedex)* deposits, for which no obvious magmatic activity is involved. The Ba-rich fluids are derived from the leaching of oceanic or continental rocks or both. The sulfate in the overlying marine fluids may have undergone partial reduction, thus altering the isotopic composition of residual dissolved sulfate to heavier values, which are reflected in the precipitated barite. Sedex deposits include economically important occurrences of Pb and Zn as well as barite, and the submarine volcanic hydrothermal deposits include a wide range of *volcanic-hosted massive sulfide (VMS)* deposits, many of which contain barite, as well.

Biologically mediated precipitation of barite and celestine. A number of marine and freshwater organisms are known to induce the precipitation of barite or celestine from the Ba, Sr, and sulfate present in ambient water (e.g. Dehairs et al. 1980, Bernstein et al. 1992). There is also evidence that release of Ba and/or sulfate through degradation of biologically precipitated materials can create microenvironments favorable for the precipitation of barite.

Reaction of barium- or strontium-rich fluids with non-seawater sulfate. These include reactions involving Ca-sulfate evaporite minerals, sulfate-bearing waters derived from the subaerial evaporation of seawater and/or the dissolution of evaporites, sulfate derived from the oxidation of sulfides, and magmatic sulfate. Examples of barite formed this way include the carbonate-hosted *Mississippi Valley Type* and *Irish Type* ore deposits, vein deposits occurring along continental rifts, and innumerable examples of dispersed barite in the form of concretions and cements. Most of the largest deposits of celestine in the world have formed in specialized environments in which Sr-rich fluids have reacted with sulfate in an evaporate-carbonate environment. Other examples include replacement of gypsum and anhydrite in salt dome caprock by barite and celestine and the

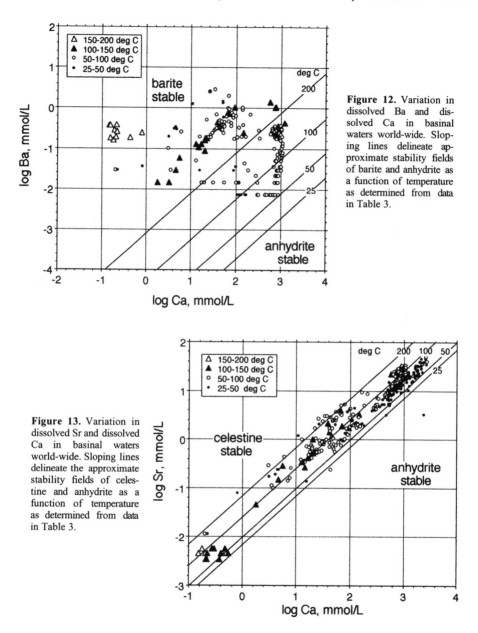

Figure 12. Variation in dissolved Ba and dissolved Ca in basinal waters world-wide. Sloping lines delineate approximate stability fields of barite and anhydrite as a function of temperature as determined from data in Table 3.

Figure 13. Variation in dissolved Sr and dissolved Ca in basinal waters world-wide. Sloping lines delineate the approximate stability fields of celestine and anhydrite as a function of temperature as determined from data in Table 3.

apparent discharge of barium-rich ground waters into sites of active subaerial evaporation of seawater. Figures 12 and 13 show the variations in dissolved Ba and Sr with dissolved Ca in sedimentary basins and the phase relations between barite and celestine with anhydrite calculated from Table 3 for typical basinal P–T conditions (Fig. 1). Composition-temperature-pressure relations for many of these waters favor replacement of anhydrite by barite or celestine. Decreasing temperature and pressure also favors replacement of anhydrite.

Decreasing solubility resulting from changes in fluid pressure and temperature. Precipitation of barite and celestine can result from changes in pressure and temperature. Barite shows a reduction in solubility through lowering of pressure at any temperature, and below 100°C a lowering of temperature. The hydrolysis constant for barite decreases by about an order of magnitude for an ascending fluid following a typical basinal *P-T* trajectory (Fig. 1). Celestine, because of its strongly retrograde solubility, is less likely to be precipitated by ascending fluids undergoing cooling.

Release of Sr during diagenesis of marine carbonates. In the presence of interstitial marine sulfate, release of Sr during diagenetic alteration of a Sr-rich carbonate, such as aragonite, to a carbonate-poor carbonate such as calcite or dolomite, can produce celestine (Baker and Bloomer 1988).

Relation of barite and celestine occurrences to global tectonics and regional hydrogeology

Most major occurrences of barite and celestine have formed in response to the development of tectonic regimes which have driven waters carrying Ba and Sr derived from depth into shallow, oxidizing conditions where sulfate is stable. As reviewed by Garven and Raffensperger (1997), fluids within sedimentary basins and the continental crust are subjected to a variety of forces which can cause large-scale fluid migration. The types of flow associated with these forces are sometimes grouped into topography-driven flow, free convection (e.g. thermohaline overturn), and forced convection (Cathles 1997). Fluid flow in the oceanic crust, in contrast, is dominated by thermal convection induced by high heat flow associated with magmatic activity.

Extensional faulting has played a key role in the localization of many barite deposits. Such faults occur in a wide variety of tectonic settings, including even compressional continental margins. Many barite deposits occur along areas of extension in compressional, passive, and strike-slip continental margins, continental rift basins, and oceanic crust.

Relation of barite and celestine occurrences to the secular evolution of sedimentary rock types

During the early Archean, most marine sulfate was probably derived from local magmatic activity, and barite precipitated in sedimentary-volcanic environments now preserved in greenstone belts.The progressive oxidation of the Earth's atmosphere and oceans during the Precambrian was important in creating large quantities of oxidized sulfur which in turn led to the formation of two reservoirs of sulfate which subsequently have been involved in barite precipitation: sulfate in marine waters and Ca-sulfates in evaporite sequences (e.g. Warren 1997). The development of organisms that could precipitate massive quantities of platform and epicontinental carbonates in the Paleozoic resulted in the formation of thick and regionally-extensive carbonate-evaporite sequences. Some of these carbonates served as regional aquifers during crustal deformation, permitting fluids from depth to react with sedimentary sulfate. The appearance of Ca-rich sediments also provided a mechanism for concentrating Sr and creating high Sr/Ba environments necessary for precipitating celestine. Since the Jurassic, however, the principal locus of carbonate deposition has shifted from the shallow marine coastal environment into the pelagic realm.

Marine versus continental barite

Goldberg et al. (1969) introduced the concept, which continues to be invoked in the literature, that there are demonstrable differences between what they termed marine and continental barite, i.e. barite presently found on or near the seafloor and barite found in rocks that now make up part of the continental crust, respectively. Unfortunately, this classification obscures the sources of Ba and sulfate and the mechanisms by which barite precipitates.

Barite precipitated with seawater sulfate as the principal source of sulfur has several distinct modes of formation in modern marine environments, including: (1) biologically-mediated pelagic precipitation in near-surface waters; (2) precipitation related to oceanic submarine hydrothermal activity; (3) precipitation related to submarine discharge of fluids from continental margins and in epicontinental marine basins. Each type of barite has been referred to in the literature by different authors as "marine barite."

The term "continental barite" has no genetic connotation other than barite which is presently part of a continental terrane, even if it originally precipitated in a marine environment on oceanic crust. Of the four examples of continental barite discussed by Goldberg et al. (1969), one is from a Precambrian metamorphosed sedex deposit that formed in a marine setting, two are from carbonate-hosted MVT deposits, and one is a magmatic-hydrothermal vein deposit. Of the four examples of marine barite discussed by these authors, two are from the California borderland in a tectonic setting which may involve continental or transitional crust. The other two are dispersed barites from Pacific deep-sea sediments.

CHEMICAL AND ISOTOPIC COMPOSITION OF BARITE AND CELESTINE: AN OVERVIEW

Natural barite and celestine rarely exist as end-member phases. There is a wide variation in Sr/Ba and the incorporation of other cations, such as Ca^{2+}, Ra^{2+}, Pb^{2+}, and K^+ in the barite-celestine series. These compositional variations have the potential for helping to constrain interpretation of environments of formation. In addition, there is measurable variation in the Sr, S, O, Ra, Pb, and Nd isotopic compositions of barite and celestine. The use of Sr- and S-isotope geochemistry, in particular, has added greatly to our understanding of sources of components and environments of formation of barite and celestine. Finally, both minerals can contain fluid inclusions, which provide information on the salinity, chemical composition, and temperature of the host fluids. Systematics on the controls on the chemical and isotopic composition of barite and celestine are briefly reviewed to provide a background for the later sections of this chapter.

Frequency distribution of compositions in the barite-celestine series

Although a complete solid solution exists between barite and celestine, apparently even at temperatures approaching 25°C, there is a strongly bimodal distribution to the frequency distribution of compositions in this solid-solution series (Hanor 1968, Prieto et al. 1993, 1997). Most barite contains less than 7 mol % $SrSO_4$ and most celestine contains less than 4 mol % $BaSO_4$ (Fig. 14) (Hanor 1968). Intermediate compositions are known in nature (e.g. Burkhard 1978), but they are uncommon.

Hanor (1968) explained this strongly bimodal distribution through a series of model calculations predicting the compositions in the barite-celestine series that would be expected during the progressive titration of Ba and Sr out of an aqueous solution (Fig. 15). The calculations assumed a closed system with respect to Ba and Sr and precipitation following the Doerner-Hoskins fractionation relation (Eqn. 17), i.e. early-formed precipitates do not re-equilibrate with the aqueous solution. In the model results reproduced here, it was assumed that initial aqueous solution compositions between $Ba_{90}Sr_{10}$ and $Ba_{10}Sr_{90}$ occur with equal frequency and that all of the Ba and Sr is eventually precipitated out of solution. A value for the partition coefficient λ_{Sr-Ba} of 0.03 (Eqn. 17) was used which represents an average value of D_{Sr-Ba} for temperatures between 20 and 80°C and the complete range of solid solution compositions studied by Starke (1964).

During titration of Ba and Sr from a fluid of given initial Sr/Ba content, the first solid to form is highly enriched in Ba, as expected from the small value of the partition

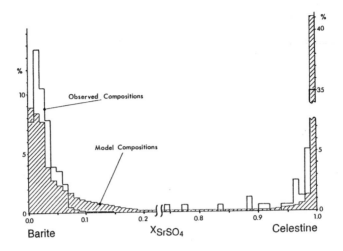

Figure 14. Frequency distribution of observed compositions in the barite-celestine solid-solution series normalized so that the areas under the Ba-rich and the Sr-rich ends of the histogram are the same. Shaded area shows the calculated frequency distribution of compositions formed during closed-system precipitation of Ba and Sr from solution. See text for details. Modified from Hanor (1968).

coefficient. As Ba is removed preferentially over Sr from aqueous solution, the Sr/Ba concentration of the fluid increases, and the Sr/Ba ratio of the next increment of solid to be precipitated has a higher Sr/Ba ratio than the preceding. There is, however a strongly non-linear variation in solid composition with the degree of progress of removal of Ba plus Sr from aqueous solution. There are *three* distinct stages in the evolution of the compositions of the solids as precipitation progresses. In the first stage, a series of Ba-rich, Sr-poor solids is formed (Fig. 15). The Sr content of these solids increases only slightly as precipitation progresses. In the second stage, compositions of the solids swing abruptly from Ba-rich to Sr-rich over a very narrow range of precipitation. In the third stage, the solids precipitated are very nearly pure $SrSO_4$. As a consequence of the strong fractionation of Ba and Sr which occurs during precipitation, most of the solids produced are thus either Ba-rich or Sr-rich. Only a small mass has intermediate compositions. On this basis, Hanor (1968) concluded that the paucity of intermediate compositions in nature does not require the existence of a miscibility gap, but can be simply explained by the strong partitioning of Ba into the solid during precipitation and the eventual formation of a residual fluid having a high Sr/Ba ratio. Subsequent authors have come to the same conclusion (Prieto et al. 1993, 1997).

Nature's modes for titrating Ba and Sr out of solution are far more varied than the above and probably only rarely involve sequential removal of most of the Ba plus Sr from aqueous solution. Submarine exhalative deposits, for example, are systems open with respect to Ba and Sr, where hydrothermal fluids carrying Ba and Sr are injected into an essentially infinite volume of well-mixed sulfate-rich seawater.

It is thus not surprising from the small partition coefficient for Sr incorporation into barite that most barite has only modest concentrations of Sr. More problematic are the venues which can produce celestine in the presence of barium. One mechanism is to start with a fluid whose initial Sr/Ba ratio is high, then precipitate the Ba end-member early in the precipitation history, and then precipitate nearly pure $SrSO_4$, much like the titration curves on the far right of Figure 15.

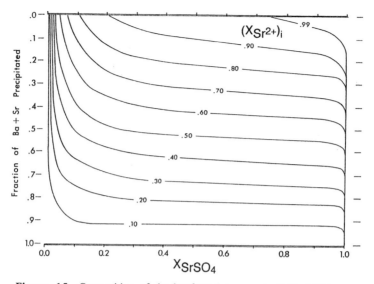

Figure 15. Composition of the last-formed increment of $(Ba,Sr)SO_4$ as a function of the mole fraction of Ba plus Sr precipitated from aqueous solution for a series of initial aqueous solution compositions of varying initial Sr/Ba content. The calculations assume Doerner-Hoskins behavior and $\lambda_{Sr-Ba} = 0.03$. Solid and liquid compositions are expressed as mole fractions, where $X_{Sr}^{2+} + X_{Ba}^{2+} = 1$ and $X_{SrSO_4} + X_{BaSO_4} = 1$. Modified from Hanor (1968).

Ba-Sr zoning in barite-celestine

The partitioning of Ba and Sr described above has the potential for giving rise to compositional zoning that reflects changes in fluid compositions resulting from fractionation. In addition, because the Ba-Sr partition coefficient is temperature dependent, changes in temperature in time and in space may also give rise to compositional variations. Hanor (1966) and many others (e.g. Starke 1964, Burkhard 1978, Breit et al. 1990) have described compositional zoning in barite which exist on scales from the several micron to the 1,000 km.

For example, the Sr content increases from the center outward in some barite concretions, suggesting precipitation from a closed system (Hanor 1966). In contrast, concretions associated with shale-hosted stratiform deposits commonly show reverse zoning, in which Sr decreases from the center outward. This may reflect diffusion-controlled rates of transport of Sr and Ba to sites of precipitation (Okita 1983).

Zoning of Ba and Sr has been described within individual ore districts. At the Julcani Ag-Bi-Pb-S district, central Peru, ore-forming solutions rose vertically and spread laterally from a central zone (Hanor 1966). There is a strong increase in the Sr-content of barite out from this central zone, consistent with partitioning of Ba into the solid phase and an increase in fluid Sr/Ba ratio during fluid flow and precipitation (Hanor 1966). In contrast, Tischendorf (1963) established that the Sr content of barite decreases upward over a vertical distance of 500 m in veins of the Schneckenstein district, Germany. Although Tischendorf does not give an explicit interpretation of this zoning, it could have been produced by a pronounced decrease in temperature of ascending fluids.

Hanor (1966) established that there is a regional zoning in Sr content of barite in the eastern half of North America. The westernmost zone is characterized by barites having values of from 1 to 2 mol % $SrSO_4$. The central zone, which spatially encompasses the Cincinnati and Findlay arches, is characterized by high Sr values in the range of 6-26 mol % $SrSO_4$. The eastern zone is characterized by intermediate Sr values, ranging from 1.5 to 9.0 mol %. The high-Sr central zone encompasses most of the major occurrences of celestine in the eastern half of North America. The boundary between the central and western zones is also a regional boundary for the occurrence of fluorite mineralization. Barite districts west of this boundary contain no fluorite. Nearly every barite district east of this line contains appreciable fluorite. Hanor (1966) proposed that this regional zoning reflects compositional differences in potential Precambrian source rocks below Paleozoic cover. As an alternative hypothesis, this regional Sr zoning may reflect instead a Paleozoic carbonate, high Sr/Ba source to the east and a Paleozoic siliciclastic, low Sr/Ba source to the south and west.

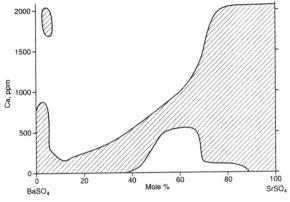

Figure 16. Variation in the Ca content of barite and celestine from the Swiss Alps and Jura Mountains. Data from Burkhard (1978).

Other cations

The elements Ca and K are generally present in quantities of a few hundred to a few thousand ppm in barite (e.g. Church 1970), but other than Burkhard's (1978) extensive documentation of Ca in Alpine barite (Fig. 16), there has been little attempt to date to develop the systematics of their occurrence. Burkhard found higher Ca values to be associated with carbonate host rocks. Lead in the hundreds of ppm range has been reported from some barite, particularly in volcanic-hosted massive sulfide deposits of Japan, and Kajiwara and Honma (1976) have used the Pb content of barite coexisting with galena (PbS) to estimate f_{O_2} conditions prevailing during ore deposition from the equilibrium reaction $PbS_{(s)} + 2O_{2(g)} = PbSO_{4(ss)}$. Takano and Watanuki (1974) have reviewed other aspects of the Pb content of barite.

Radium is close in ionic radius to Ba (Table 1), and has been detected in recent seafloor occurrences in barite (e.g. Moore and Stakes 1990, Paytan et al. 1996), in barite sinter (Cecile et al. 1984), and barite well scale (Fisher 1998). Guichard et al. (1979) described the distribution of rare earth elements of barite, although more recent work suggests some of the REE may be present in a separate phase (Martin et al. 1995).

Strontium isotopic composition

Strontium-isotope analyses have been widely used to constrain the origin of both barite and celestine (e.g. Reesman 1968, Whitford et al. 1992). These minerals are ideal for Sr-isotope work because of their high Sr content and the fact they are depleted in Rb, which eliminates any need for correcting for the in-situ production of radiogenic [87]Sr since the time the barite or celestine precipitated. In addition, the very high Sr content makes these minerals insensitive to contamination during their post-depositional history (Whitford et al. 1992).

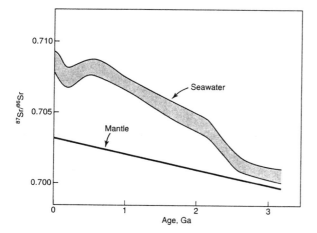

Figure 17. Secular variation in the $^{87}Sr/^{86}Sr$ ratio of the mantle and seawater. Modified from Franklin et al. (1981).

The Sr-isotopic composition of geological materials is expressed as the ratio of $^{87}Sr/^{86}Sr$, which can be measured with great analytical precision. Radiogenic ^{87}Sr is produced by the decay of ^{87}Rb (half-life = 4.88×10^{10} y). Rocks having high initial concentrations of Rb , such as granites, are characterized by high $^{87}Sr/^{86}Sr$ ratios. Rocks derived from materials having low Rb concentrations, such as the mantle (Fig. 17), have low $^{87}Sr/^{86}Sr$ ratios.

Since the Precambrian, the $^{87}Sr/^{86}Sr$ of seawater has fluctuated between approximately 0.7070 and 0.7092 as the result of variations in the rates of input of enriched Sr from continental weathering and depleted Sr from mantle sources (Fig. 17). Sulfate and carbonate minerals ultimately derived from seawater, including pelagic and submarine exhalative barite and celestine, have Sr isotopic compositions which reflect the Sr composition of seawater at the time the minerals or their precursors were formed, unless they have undergone major diagenetic alteration. Silicate minerals, however, have a much wider range of isotopic compositions. For example, some detrital K-feldspar in the North Sea has $^{87}Sr/^{86}Sr$ values in excess of 0.7300, and high Rb/Sr micas can have values in excess of 0.8000 (Whitford et al. 1992). Volcanoclastic sediments and barite derived from volcanic sources, on the other hand, may have $^{87}Sr/^{86}Sr$ ratios lower than contemporaneous seawater if they were sourced from low-Rb rocks having a low initial $^{87}Sr/^{86}Sr$ ratio.

Potential barite-forming fluids in sedimentary basins containing Paleozoic strata commonly have $^{87}Sr/^{86}Sr$ ratios in excess of seawater values contemporaneous or coeval with the depositional age of the current host sediment. This is well illustrated by the data of Connolly et al. (1990) for the Alberta basin, Canada (Fig. 18). The enrichment represents Sr derived from alteration of silicates. Because of the significant increase in $^{87}Sr/^{86}Sr$ with time since the Jurassic, however, some fluids in Cenozoic sedimentary basins actually have $^{87}Sr/^{86}Sr$ ratios lower than contemporaneous seawater values, representing introduction of Sr from older and deeper sedimentary sources (e.g. McManus and Hanor 1988, 1993). Barite derived from such sources, such as barite in recent chimneys in the Gulf of Mexico (Fu 1998), have $^{87}Sr/^{86}Sr$ ratios less than that of coeval seawater.

There is no isotopic fractionation of Sr during sulfate precipitation, and barite-celestine Sr derived from seawater or from the dissolution of a marine evaporite or carbonate phase will thus have an isotopic composition with a marine signature. Many synsedimentary barite and celestine occurrences have Sr isotopic ratios heavier than that of contemporaneous marine water (Fig. 18), representing a partial continental silicate input for Sr and, by extension, Ba.

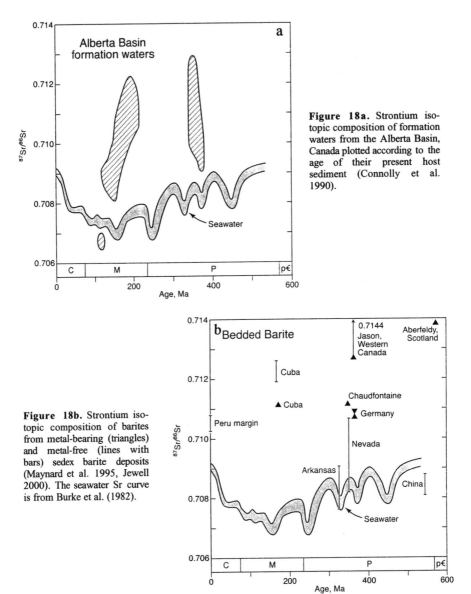

Figure 18a. Strontium isotopic composition of formation waters from the Alberta Basin, Canada plotted according to the age of their present host sediment (Connolly et al. 1990).

Figure 18b. Strontium isotopic composition of barites from metal-bearing (triangles) and metal-free (lines with bars) sedex barite deposits (Maynard et al. 1995, Jewell 2000). The seawater Sr curve is from Burke et al. (1982).

Sulfur isotopic composition

Sources of sulfate in barite and celestine include seawater sulfate; seawater sulfate modified by microbial reduction; calcium sulfate minerals; magmatic sulfate; and sulfate produced by oxidation of reduced sulfur, including sulfide, elemental sulfur; and organically bound sulfur. The $\delta^{34}S$ isotopic composition of barite and celestine has proved invaluable in the identification of sources of sulfate and environments of deposition (e.g. von Gehlen et al. 1962, 1983; Kusakabe and Robinson 1977, Strizhov et al. 1988).

The isotopic systematics of sulfur are complex because sulfur exists in several redox states and a variety of fractionation mechanisms exist (Seal et al. this volume). Below approximately 200°C, the rate of isotopic exchange between dissolved sulfate and sulfide is slow, and isotopic equilibrium between the species is rare (Ohmoto and Lasaga 1982). Kinetic effects therefore dominate the isotopic systematics of sulfur in sedimentary basinal conditions. The principal kinetic effect is associated with the reduction of sulfate to sulfide, which can be achieved both microbially and thermochemically. In general, substantial isotopic fractionations occur when sulfate is reduced to sulfide or sulfur, whereas negligible fractionation occurs during oxidation of sulfide (Seal et al., this volume).

The lighter isotopes of sulfur are preferentially partitioned into sulfide during microbial reduction, thereby producing hydrogen sulfide which is much lighter than the precursor sulfate and a residual sulfate which is heavier. Thus, sulfate reduction drives residual seawater sulfate toward higher values. The major factors controlling the sulfur isotopic composition of seawater are the relative proportion of sulfide and sulfate removed from the oceans, and the $\delta^{34}S$ value of average river sulfate, derived from the continental weathering of sulfate and sulfide bearing rocks. Holland (1984) argued that fluctuations in river sulfate is less of a controlling factor than fluctuations in the removal rates of sulfide and sulfate.

The value of $\delta^{34}S$ in gypsum is only approximately 1.65 permil heavier than the aqueous sulfate in solutions from which it precipitates, and the isotopic composition of gypsum deposits precipitated from seawater during evaporation thus track secular changes in seawater isotopic composition. The isotopic composition of seawater sulfur has fluctuated throughout the Phanerozoic from approximately +10 to +30 permil VCDT, based on analyses of gypsum and anhydrite in the geological record (Fig. 19). Late Proterozoic values of $\delta^{34}S$ range from +15 to +25 (Claypool et al. 1980). Because sulfate has a long oceanic residence time, the isotopic composition of sulfur is assumed to have been uniform throughout the oceans at any given time. Waters in anoxic basins or sulfate-bearing brines in evaporite sedimentary basins may have more positive $\delta^{34}S$ values than contemporaneous seawater as a result of bacterial reduction. Thermochemical reduction of sulfate at higher basinal temperatures, on the other hand, typically produces sulfides similar in isotopic composition to the parent sulfate.

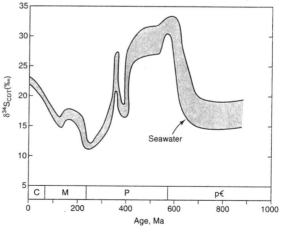

Figure 19. Variation in the sulfur isotopic composition of seawater with time (after Claypool et al. 1980).

Early Archean sulfates in submarine exhalative deposits typically have $\delta^{34}S$ compositions slightly heavier than zero, while coexisting sulfides group around zero. With the later appearance of sulfate-reducing bacteria and the preferential removal of light sulfur, marine sulfate became heavier, and barite compositions reflect this shift. Barite in many sedex deposits has sulfur isotopic compositions similar to that of coeval seawater sulfate (Fig. 19), indicating that much of the barite was formed by mixing of Ba-bearing hydrothermal fluids with seawater. Sulfide minerals, if present, are generally lighter, reflecting that much or most of the sulfide has been derived by bacterial reduction of marine sulfate (Lydon 1995). Sedex barite formed in slightly anoxic basins has an isotopic composition heavier than coeval seawater (Goodfellow 1999). Barite in carbonate-hosted MVT deposits typically has a sulfur isotopic composition inherited from marine evaporites (Kesler 1996).

Oxygen isotopic composition

The $\delta^{18}O$ of present-day dissolved marine sulfate is +8.6 permil VSMOW, far out of equilibrium with the $\delta^{18}O$ value of ocean water (Seal et al. this volume). Holser et al. (1979) have shown that this difference can be explained by the persistence of non-equilibrium between oxygen in seawater and oxygen in the river input of sulfate, which owes its $\delta^{18}O$ value to the combined effect of the solution of evaporite minerals and the oxidation of sulfides during weathering. It is estimated that the isotopic equilibration time between seawater and dissolved sulfate is greater than 10^9 years, far larger than the oceanic mixing time of approximately 2,000 years. An alternative explanation involves equilibrating marine sulfate with seawater oxygen at temperatures of 200°C in mid-ocean ridge hydrothermal systems. The $\delta^{18}O$ of seawater sulfate has varied throughout geological time, with $\delta^{18}O$ values of late Proterozoic and Phanerozoic anhydrites varying generally between 13 and 17 permil, with exception of the Permian when a minimum value of 10 was reached (Holser 1979).

The oxygen isotopic composition of barite is not reported as often as that of $\delta^{34}S$. At most diagenetic conditions of temperature and pH, oxygen in dissolved sulfate and in water are not in isotopic equilibrium because of the sluggish kinetics of isotopic exchange. At temperatures above about 150°C and near-neutral pH conditions, the oxygen isotopic composition of sulfate can be used as a geothermometer and to infer conditions under which the sulfate formed (Ohmoto and Lasaga 1982). Under highly acidic conditions, oxygen exchange between sulfate and H_2O is more rapid (Seal et al. this volume). However, this is probably not a factor in most barite deposits.

Radium

The several naturally-occurring isotopes of radium, ^{224}Ra, ^{226}Ra, and ^{228}Ra, are produced by the radioactive decay of parents in the U and Th decay series (Fig. 20). All of these isotopes are radioactive and have half-lives ranging from 1622 y to 3.64 d. Radium and Ba are of similar ionic radius, and Ra often occurs along with its various daughters in measurable quantities in recently precipitated barite. Radiometric dating schemes based Th-Ra systematics have been used to date recent hydrothermal deposits of bariite and to calculate sedimentation rates (Kadko and Moore 1988, Paytan et al. 1996). Techniques of Th-Ra dating include the assumption that the $^{228}Th/^{228}Ra$ activity ratio is of the same magnitude as $^{232}Th/^{226}Ra$, and that ^{228}Th produced by decay of ^{228}Ra does not migrate.

Fluid inclusions

Coarse-grained barite typically contains fluid inclusions (Roedder and Bodnar 1997). The study of heating and freezing behavior of primary fluid inclusions in barite can provide valuable information on the temperature of fluid entrapment and on fluid salinity, and chemical analysis of fluid inclusions provides direct information on the details of fluid chemistry in minerals as old as the Early Archean (e.g. Rankin and Shepard 1978). It has been found, however, that many fluid inclusions in barite stretch upon heating, leading to homogenization temperatures that are far higher than true entrapment temperatures (Ulrich and Bodnar 1988). These factors may explain why analyses of fluid inclusions in barite by

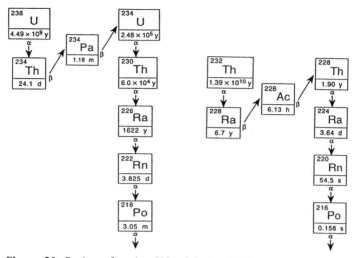

Figure 20. Portions of uranium-238 and thorium-232 decay schemes, showing the Th and Ra isotopes refered to in text. Modified from Fisher (1998).

Ramboz and Charef (1988) for the Les Malines deposit in France yield anomalously high fluid pressures and temperatures.

BARITE IN SUBMARINE VOLCANIC HYDROTHERMAL SYSTEMS

The basic concept of the syngenetic accumulation of massive sulfides and sulfate minerals on the seafloor in direct association with magmatic activity was well established by the late 1950s by Oftedahl (Oftedahl 1958, Franklin et al. 1981, Franklin 1995). The subsequent discovery in the 1970s of presently active seafloor systems has established the general applicability of the model.

Heat from magmatic activity drives deep convective circulation of fluids in the oceanic crust resulting in leaching of Ba from a variety of source rocks. Extensional faults and fractures help to focus upward discharge of hydrothermal fluids onto the seafloor, where they mix with seawater. Principal regions of thermal convection today occur along mid-ocean rises and rifts, seamounts, and back-arc basins (Scott 1997). It is reasonable to include Early Archean barite deposits in this section of the chapter because of their occurrence in volcanic greenstone belts and probable association with volcanic activity, even though their original tectonic setting has been obscured by later geological events.

The basic processes by which volcanic-hosted mineral (VHM) deposits form have been derived both from detailed field examination of ancient deposits exposed through outcrop, the drill bit and mine workings, and from study of modern seafloor examples. The underlying driving force for creating a VHM deposit is the high heat associated with emplacement of magma which generates convective fluid flow in the oceanic crust. Seawater is recharged into the system over a broad area, but the upward discharge is controlled by faults and fractures and is thus much more focused. Seawater is modified in composition as it descends through the crust and is progressively heated. Notable changes in modern seawater systems include the loss of Mg, Sr, Ca, and sulfate as the result of formation of anhydrite, zeolites, and clay minerals (Franklin 1995, Scott 1997). Reaction between modified seawater and basalt at about 385°C causes a lowering of pH and the leaching of metals and sulfide from host rocks (Franklin 1995, Scott 1997). Presumably Ba is leached from source rocks after the loss of sulfate. During the Archean, surface marine waters were low in sulfate, and presumably there were significantly different chemical pathways taken during seawater-volcanic rock interaction.

Sulfides in modern systems are precipitated on or near the seafloor by rapid cooling of the hydrothermal fluid. Some sulfide minerals are dispersed into the overlying water column in distinctive black plumes or smokers. Isotopic studies have established that the metals and sulfide are transported simultaneously to the seafloor and that, unlike sedex Pb-Zn deposits, reduction of seawater sulfate is unimportant as a source of reduced sulfur (Janecky and Shanks 1988). Precipitation of sulfide and sulfate minerals occurs both as chimneys around vent orifices and in plumes which rise from the vents. In modern deposits dominated by metal sulfides, initial venting produces anhydrite chimneys rapidly precipitated from seawater-sourced Ca and sulfate as a result of the retrograde solubility of anhydrite. The chimneys act as a substrate for subsequent deposition of sulfides. With cooling, the anhydrite chimneys eventually dissolve and collapse. Minerals also precipitate around the exterior of sessile worm tubes in the complex community of vent fauna which populate areas of hydrothermal discharge.

According to Franklin (1995) the presence or absence of barite in vent systems is largely dependent on temperature of discharge. In the high-temperature phase of discharge, most barite is dispersed into the overlying water column in smoke plumes. Lower temperature fluids, T <250°C, which are depleted in base metals and sulfur, form barite-rich chimneys over the vents. Barite is precipitated with seawater as the source of sulfate. In ancient deposits, barite probably formed where hydrothermal systems continued to vent after the peak of high-temperature discharge and where seawater contained abundant sulfate. Modern seafloor hydrothermal fluids generally contain abundant silica which is precipitated upon cooling and/or boiling forming white plumes or smokers rising above discharge vents. Ferruginous chert overlies many deposits and is the product of low temperature discharge of silica- and iron-rich waters. Thus, barite in VHM deposits is often associated with silica-rich sediments and ferruginous cherts.

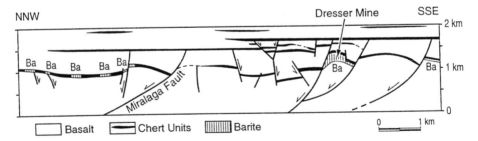

Figure 21. Restored cross-section through the North Pole area, Western Australia, showing the Early Archean North Pole cherts interbedded with basalts, and the relation of the occurrence of barite to synsedimentary extensional faults. Modified from Nijman et al. (1998).

Sulfide-poor barite deposits of Archean age

Barite makes its first documented appearance in the geologic record in a spectacular manner: as large synsedimentary mounds more than 20 m high and 50 m across that were precipitated on the seafloor of the East Pilbara block of Western Australia approximately 3.5 Ga ago (Nijman et al. 1998). The mounds occur as part of a complex of baritic veins and sediments in the North Pole Chert of Early Archean age which crops out over an area of approximately 10 by 15 km. The North Pole Chert itself consists of a series of thin chert units interbedded with thicker basalts in the basal portion of the Warrawoona Group, which in turn overlies older basalts (Fig. 21). It is the lowermost of these cherts which contains extensive barite. This chert also contains some of the oldest documented stromatolite-like structures and contains possible remains of living organisms. Galena from vein barite in the North Pole Chert gives a model lead age of 3,490 Ma (Nijman et al. 1998).

Nijman et al. (1998) documented the existence of synsedimentary extensional growth faults active during the deposition of the North Pole Chert and immediately overlying units (Fig. 21) and coeval with felsic volcanism. Swarms of chert veins fan upward from growth-fault planes in the underlying basalt. The large barite mounds and apparent underlying feeder veins of barite and chert are spatially associated with some of these faults.

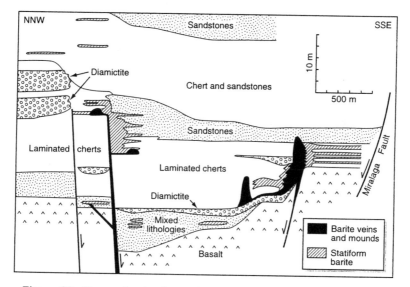

Figure 22. Diagram showing the stratigraphic and structural relations of barite in the northwesternmost part of the cross section in Figure 21. Modified from Nijman et al. (1998).

One very large mound described by Nijman et al. (1998) stands on a base of diamictite, which may have formed by collapse of a submarine fault scarp (Fig. 22). The barite structure is made up of smaller hemispherical mounds of barite deposited in subparallel layers. The layers now consist of bladed crystals of up to 10 cm long. Judging from the diagrams in the paper by Nijman et al. the blades are perpendicular to bedding and may be similar to secondary blades of barite noted in detrital barite deposits of Early Archean age in South Africa (Reimer 1980). Thin chert and Fe-oxide layers separate some successive barite layers. Beds of stratiform barite that are described as sinter occur proximal to some of the mounds (Fig. 22).

There is consensus that the North Pole Chert was deposited within a shallow-water evaporite lagoon behind a barrier of volcanoclastic sand (Buick 1990, Buick and Dunlop 1990). Fluid inclusions in the barite contain CO_2, H_2S, calcite and sphalerite, and there is a large variation in both CO_2 content and gas-bubble size. Nijman et al. (1998) invoke submarine discharge of fluids containing Ba, Si, and sulfide into a 50-m-deep stratified water body. Boiling or degassing and mixing with the overlying water caused rapid oxidation of sulfide into sulfate. Some of the barite occurs as replacement of bladed gypsum thought to be evaporitic in origin (Dunlop and Buick 1990), and some occurs as detrital beds derived from erosion of the mounds.

Barite in similarly massive quantities also occurs in sediment-hosted deposits in Early Archean portions of the Precambrian shield areas of India, South Africa, Canada, and Russia. Much of this barite was apparently deposited in shallow, high-energy aqueous environments, which resulted in its erosion and reworking into sand-size clasts. Deb et al. (1991), for example, described a detrital stratiform barite occurrence in the Indian Shield which is only 1 m in thickness, but extends 10 km along strike. The barite is interlayered with Cr-rich muscovite-bearing quartzite in the upper amphibolite facies of the Sargur Group (>3.3 Ga). The barite sediments have retained their primary sedimentary stratification, with alternating quartzose layers interbedded with micaceous and pyritic layers. The authors interpreted the barite to have been originally derived from a submarine hydrothermal deposit.

Stratiform barite in Early Archean (>3.2 Ga) metasediments of the Barberton Mountain Land, Transvaal, South Africa, may have had a similar origin. Reimer (1980) described several different major sedimentary barite occurrences which include: 12-m-thick zones of barite with finely disseminated pyrite and other sulfides alternating with chert; lens-shaped deposits of barite within chert; and detrital barite sands containing detrital pyrite, chromite, zircon, magnetite, ilmenite, molybdenite, and gold as accessory minerals. The barite sand was presumably eroded from older deposits and mechanically mixed with the other heavy minerals. Transport distances were probably short. Diagenetic recrystallization has led to the formation of radiating, cauliflower-like large crystals of secondary barite that cut across original bedding. Reimer (1980) envisioned the barite as being formed by hydrothermal solutions containing Ba that reacted with seawater sulfate, with little obvious association with evaporites. DeWit et al. (1982) have identified potential submarine vent features in the vicinity of the barite and at approximately the same stratigraphic horizon.

Isotopic composition and sources of sulfate. Archean barite has low $^{87}Sr/^{86}Sr$ ratios, reflecting primitive mantle compositions (Fig. 17), and $\delta^{34}S$ values just slightly greater than magmatic values (Perry et al. 1971, Lambert et al. 1978). For example, the Archean barite of the Sargur Group in India has a $^{87}Sr/^{86}Sr$ ratio of 0.7018 and $\delta^{34}S$ values of 4.02 to 7.45 permil VCDT (Deb et al. 1991).

One of the more problematic aspects of the Early Archean barite deposits is the source of sulfate in an ocean presumably nearly devoid of free oxygen. One possibility is that the sulfate is magmatic in origin. As discussed by Hattori and Cameron (1986), sulfate in igneous rocks can be formed by the oxidation of primary sulfide in silicate magma and by the hydrolysis of SO_2 gas in igneous-hydrothermal fluids. The amount of Fe is much larger than S in typical melts, and f_{O2} is determined by the ambient Fe^{III}/Fe^{II} ratio. In the first process sulfide is oxidized by a reaction such as

$$S_{2(melt)} + 4H_2O_{(melt)} \rightarrow SO_{4(melt)} + 4H_{2(g)} \tag{26}$$

Degassing of H_2 moves reaction to right. In an aqueous phase, sulfate may be produced by reactions such as

$$SO_{2(aq)} + 2H_2O_{(aq)} \rightarrow H_2SO_{4(aq)} + H_{2(aq)} \tag{27}$$

or

$$4SO_{2(aq)} + 4H_2O_{(aq)} \rightarrow H_2S_{(aq)} + 3H_2SO_{4(aq)} \tag{28}$$

Reaction of sulfate with Ca^{2+} or Ba^{2+} to form anhydrite or barite helps to drive these reactions to the right.

Volcanic-hosted massive sulfide (VHMS) deposits

Barite occurs not only in submarine deposits having low concentrations of associated metals, but is also a common mineral in sulfide-bearing VHMS deposits in submarine volcanic rocks of all ages, from pre-3.4 Ga volcanic strata of the Pilbara Block in Australia (Barley 1992) and Archean gold deposits in Canada (Cameron and Hattori 1985) to modern active spreading centers (Franklin et al. 1981, Franklin 1995). The deposits consist of bulbous to tabular stratiform accumulations of massive pyrite and lesser pyrrhotite with Zn, Pb, and Cu sulfides. The deposits are spatially associated with submarine volcanic rocks and associated sedimentary sequences and are underlain by discordant alteration zones which may occupy synvolcanic faults. VHMS deposits occur in present or former areas of high heat flow that result from magmatic activity. Most deposits are mineralogically simple and contain from 50 to more than 80 vol % sulfides. Pyrite is the dominant sulfide, followed by sphalerite, chalcopyrite, and galena.

One of the earliest known VHMS-barite deposit is the Early Archean Hemlo occurrence in Ontario (Cameron and Hattori 1985). The Cenozoic Kuroko deposits in Japan also contain barite and have been studied in detail (Lambert and Sato 1974, Farrell and Holland 1983). Barite is also commonly associated with gold-rich massive sulfide deposits (Poulsen and Hannington 1995). Hannington and Scott (1989) have shown that gold is effectively transported in hydrothermal fluids by bisulfide complexing and quantitatively precipitated by oxidation of the reduced sulfur. At subcritical seafloor depths of approximately 1900 m or less, boiling of hydrothermal fluids can occur, which causes precipitation of Cu, Fe and Zn as sulfides and produces a fluid enriched in Au, Ba, and Pb. Gold is precipitated during boiling where the ambient seawater is oxidizing. For reasons which are not well understood, barite is less common in Late Archean and Proterozoic deposits (Franklin 1995, Lydon 1995). Huston (1999) proposed this may be the result of low total concentrations of sulfur in Precambrian seawater.

Modern submarine hydrothermal barite

Prior to the recognition in the late 1960s of the relation between mid-ocean rises and ridges, seafloor spreading, and plate tectonics, Arrhenius and Bonatti (1965) and Boström and Peterson (1966) had suggested the possible existence of magmatic-hydrothermal activity along the East Pacific Rise. They noted that the areas of maximum barite concentrations in Pacific pelagic sediments do not coincide with areas of biologically high productivity under the Peru current but rather with the East Pacific Rise, and they suggested the possibility of the existence of active magma chambers discharging Ba-rich fluids on the seafloor. Direct observation of seafloor venting of hydrothermal fluids was first made in the mid-1960s in the Atlantis II Deep of the Red Sea (Degens and Ross 1969) where dense warm brines overlie metalliferous muds. With the development of submersible research vessels, hydrothermal venting has since been found at more than 100 oceanic sites (Scott 1997).

Tectonic setting. Barite is a major component of submarine hydrothermal deposits in modern back-arc basins, such as the Lau Basin, the Okinawa trough, and the Mariana trough (Bertine and Keene 1975, Herzig et al. 1993, Scott 1997). Barite is also abundant in the Guaymas Basin in the Gulf of California, where it makes up approximately 15% of the precipitated minerals (Peter and Scott 1988), and on the Galapagos ridge (Varnavas 1987). In contrast, barite is usually only a minor component or is absent from many sediment-starved mid-ocean ridges and from seamounts. Back-arc rocks are commonly more enriched in Ba than are mid-ocean rise tholeiites, which may account for the more common occurrence of barite with the former (Scott 1997).

All of the sites, except for the seamounts, are located in extensional or rifted terranes, and all of the sites, including the seamounts, are characterized by high heat flow. Modern hydrothermal deposits are mineralogically complex and diverse, but basically consist of two general types: sulfide and Fe-Si-Mn oxide deposits. The deposits can consist of a single vent chimney a few centimeters high to coalesced mounds hundreds of meters wide and topped by chimneys several tens of meters high. Most important to the size and composition of deposits and to the geochemistry of the vent fluid is the type and amount of host volcanic rocks and the sediments through which the fluids have passed. These, in turn, are dependent on the tectonic setting of hydrothermal activity (Scott 1997).

Morphology of submarine hydrothermal barite deposits. An example of modern submarine hydrothermal barite deposits is provided by the chimneys of the Mariana trough, which are composed of barite, sphalerite, galena, chalcopyrite, and silica (Scott 1997). Well-formed polyhedra of barite appear to have formed from recrystallization of early fine-grained barite on the outer side of chimneys. The basic framework of the chimneys consists predominately of euhedral lathlike crystals of barite ranging from than 1 mm across to large crystals up to 5 mm across. Crystal size and morphology change rapidly across centimeter-dimension regions of the chimney. Barite crystals commonly form radiating bundles or sector-zoned euhedral minerals that project into open vugs and contain vermicular inclusions of the sulfides sphalerite, chalcopyrite, and galena.There is no petrographic evidence that the barite crystals have been dissolved or leached.

Urabe and Kusakabe (1990) described barite-bearing silica chimneys and crusts in the back-arc basin in the Izu-Bonin arc, northwestern Pacific. The deposits occur on the flank of a rhyolite lava dome along the rift axis. The chimneys are composed of filamentous amorphous silica and minor amounts of barite and amorphous Fe oxide. Barite $\delta^{34}S$ values range from 21.7 to 22.3 permil, and are only slightly higher than seawater sulfate of 20.2 permil. The range in isotopic composition is similar to that barite from the Mariana backarc basin (Shikazono 1994) and the Kuroko deposits (Farrell and Holland 1983). Seawater sulfate is thus the dominant source of sulfate in all three barite deposits.

Thorium-radium systematics. The young age of many of the barite chimneys has lent them to Th-Ra dating (Moore and Stakes 1990, Kadko and Moore 1988). Kadko and Moore calculated the ages of barite chimneys on the Endeavor Ridge to be 2.0 to 3.9 years. Low $^{226}Ra/Ba$ ratios in marine bottom waters and high ratios in the chimneys preclude bottom-water as a major source of Ba.

SEDIMENTARY EXHALATIVE (SEDEX) DEPOSITS OF BARITE

A different type of submarine barite deposit consists of barite formed by ascending barium-rich fluids venting into a marine environment where there is no obvious magmatic source of heat to drive fluid flow. The terms *sedimentary-exhalative* and *sedex*, used here to describe these deposits, are employed in their broadest context and do not have as a requirement the presence of base metals such as Pb, Zn, and Cu. Some of the largest sedex barite deposits in the world, such as the Arkansas deposits, contain only trace amounts of base metals (Howard and Hanor 1987).

Many sedex deposits of barite, particularly those associated with base metals, occur in sedimentary basins controlled by tectonic subsidence associated with intracratonic or epicontinental marine rift systems (Goodfellow 1999). The deposits occur in shallow-water sediments deposited during the stage of thermal subsidence that follows active extension and fill by coarse clastics, turbidites, and volcanics. Where the structural setting can be reconstructed, these deposits are generally spatially associated with synsedimentary faults in small basins and depressions situated within larger rift basins. Although the deposits may have been located in areas of high heat flow, there is little evidence to support direct crustal heating by intrusion of magma.

Other sedex deposits of barite occur within a wide variety of continental margin settings, including collisional margins, passive margins, and strike-slip margins. Many of these barite deposits are devoid of associated base metals. Faulting, however, appears to be a universal requirement for the transport and focusing of subsurface fluid flow. Many of the largest and economically most important deposits of barite in the world occur along what were or are today Phanerozoic continental margins (Murchey et al. 1987, Maynard and Okita 1991) in districts that extend hundreds and even thousands of kilometers along strike. Continental margins, of course, are often the locus of tectonic activity, and Ba-bearing fluids released during tectonic activity can ascend vertically into overlying sulfate-bearing marine waters. Phanerozoic continental margins have also been sites of oceanic upwelling and high biological productivity. Because there is a documented association of pelagic barite with biological productivity during the Cenozoic, some have proposed that barite deposits along older continental margins have had a similar origin (Jewell 2000). We will examine here examples of sedex barite deposits occurring within continental margins in convergent, extensional, and strike-slip tectonic settings.

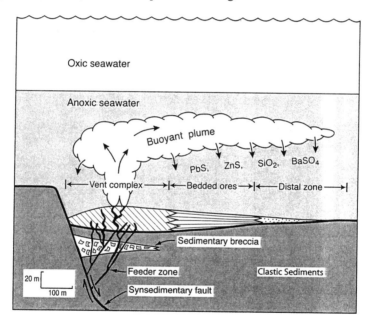

Figure 23. Schematic diagram showing principal features of a sedex sulfide-barite deposit at the time of ore deposition. Based on reconstructions of Lydon (1995) and Goodfellow (1999).

Proterozoic barite

Zn-Pb sedex deposits. The Middle Proterozoic is marked by the first appearance of sedimentary-exhalative sulfide deposits dominated by sphalerite and galena, with barite as a common, though not universally present, accessory. As shown in the schematic illustration modified from Lydon (1995) and Goodfellow (1999) (Fig. 23), who present excellent general reviews of sedex deposits, the Zn-Pb sedex deposits of Proterozoic and younger age typically consist of: (1) a vent complex, which lies above and within a feeder zone of ascending hydrothermal fluids, (2) a zone of bedded ores, which consists of a

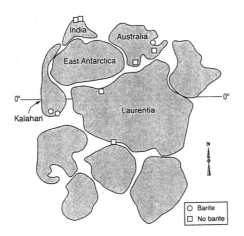

Figure 24. Reconstruction of a mid-Proterozoic supercontintent showing the location of major sedex Pb-Zn and barite deposits. Base from Goodfellow (1999).

compositionally layered apron of hydrothermal precipitates, and (3) a distal facies which lies outside the limits of ore-grade sulfide mineralization. The ratio of barite to heavy metals generally increases laterally away from the vent complex. Barite also occurs at the top of some vent complexes. Sedimentary breccias formed as a result of the collapse of fault scarps. Ore-forming fluids either formed buoyant plumes, as in the example shown, or bottom-hugging brines. Examples of Proterozoic sedex deposits containing barite include the well-documented Lady Loretta deposit, Australia, and several major deposits in the central Bushmanland Group, western Namaqua Province, southern Africa (Lydon 1995). Most Proterozoic Pb-Zn sedex deposits formed during the time period 1650-1700 Ma and seem to be associated spatially (Fig. 24) with possible intracratonic suture systems, which later became continental margins, within a Proterozoic supercontinent (Goodfellow 1999). Many Proterozoic sedex Pb-Zn deposits of northern Australia and elsewhere lack barite, which may reflect transport of metals by brines too high in sulfate to transport significant Ba (Cooke et al. 2000).

Metal-poor sedex barite deposits. There are also metal-poor barite sedex deposits of probable Proterozoic age, including barite and ironstones in Australia (Lottermoser and Ashley 1996). Horton (1989) described massive barite which occurs in layers and pods that range in thickness from a few centimeters to up to 3.7 m in thickness which occur in a 40-km-long zone in quartz-sericite schist and schistose pyroclastic rocks of inferred Late Proterozoic age of the Battleground Formation, Kings Mountain Belt, western South Carolina. The high quartz content and lack of volcanic textures suggest that the schist originated from epiclastic or sedimentary materials and possibly in part from hydrothermally altered volcanic materials. Horton (1989) suggested that the barite may have originated from seafloor hot springs, but was redistributed and locally concentrated in discordant veins during regional metamorphism from greenschist to amphibolite facies.

Phanerozoic convergent continental margins

Some of the largest accumulations of barite in the world, including the extensively studied deposits of Arkansas and Nevada, occur within former Phanerozoic convergent continental margins (Fig. 25). Here, microcrystalline barite occurs as beds and lenses in fine-grained siliceous sedimentary sequences containing shale and some chert (Orris 1986). The more massive, higher-grade beds of barite are commonly laminated. Barite nodules of early diagenetic origin are common in beds containing lower concentrations of barite. Intraformational conglomerates, breccias, and clasts of barite-cemented, fine-grained sediment are common in synsedimentary debris flows within the deposits.

Mississippian Ouachita orogenic belt. The stratiform, siliciclastic-hosted barite deposits of the Ouachita region, Arkansas, USA, were long an economically important source of barite. The Chamberlain Creek deposit alone yielded approximately 25% of the world's production for a total of nearly 8 Mt during the period from 1939 to 1972 (Shelton

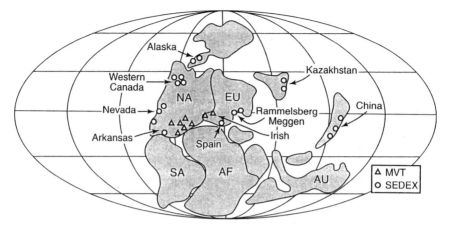

Figure 25. Devonian continental reconstruction showing the location of major Paleozoic sedex (circles) and MVT and Irish-type (triangles) barite deposits. Base from Goodfellow (1999) with additional deposits from the literature.

1989, Brobst 1994). The Fancy Hill deposit in western Arkansas (Okita 1983, Howard and Hanor 1987) still contains large reserves of barite.

Barite in the Ouachitas occurs as conformable, stratiform lenses and disseminations in shales and siltstones of the lowermost Stanley Group of Middle Mississippian age. Individual deposits are lenticular and range from <1 m to 30 m in maximum stratigraphic thickness and from 40 to over 7,500 m in outcrop length. The barite content exceeded 85 wt % in portions of the Chamberlain Creek deposit, but the maximum concentrations of barite are less in most other deposits in the region.

The barite occurs primarily as disseminated single crystals <150 μm in a shale or carbonate matrix; as dense, finely crystalline laminae; and as nodules of spherically radiating crystals. On the basis of the spatial proximity of the Chamberlain Creek deposit to the Magnet Cove alkali intrusive complex, Park and Branner (1932) originally postulated that the deposit was a hydrothermal replacement related to the nearby igneous intrusive complex. Scull (1958) expanded the concept to include all of the Ouachita deposits. In 1964, however, Zimmerman and Amstutz (1964) presented a series of arguments based on pioneering textural and sedimentological studies that convincingly demonstrated that the barite was deposited penecontemporaneously with the host silts and clays. Locally, there are textures reflecting primary precipitation from bottom waters, early diagenetic replacement of other minerals by barite, and penecontemporaneous mechanical transport and redeposition of clasts of barite. Many of these features have been described in further detail by Zimmerman (1970) and Okita (1983).

Hanor and Baria (1977) proposed the following model for the genesis of the Ouachita barite deposits. During Middle Mississippian time, compressive regional deformation and a change in tectonic regime along the Ouachita continental margin (Thomas 1976) resulted in an increased influx of clastic materials into the Ouachita basin and the development of submarine depressions on the seafloor near the base of the continental slope (Fig. 26). Hanor and Baria (1977) in considering potential sources for the Ba, rejected a pelagic source on the basis that the rate of supply would have been too slow. Calculations by Boström et al. (1973) showed that present maximum documented fluxes of Ba to the seafloor are only on the order of 10 mg Ba cm^{-2}/1000 yr. At these rates, it would have required approximately 500×10^6 years to have produced a Chamberlain Creek deposit.

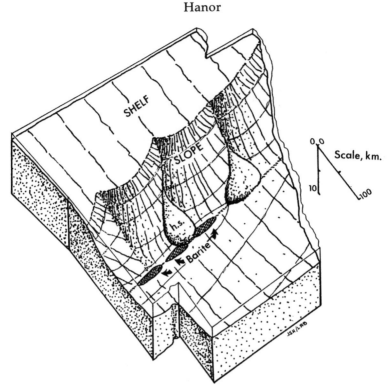

Figure 26. Conceptual model for the origin of the stratiform barite deposits of Arkansas. Ba-rich brines pooled in local depressions near the base of the continental slope between lobes of the Hot Springs sandstone (h.s.). Modified from Hanor and Baria (1977).

Hanor and Baria (1977) thus concluded that it was necessary to invoke a submarine source.

Hanor and Baria then considered mechanisms for the concentration of barite in stratiform lenses and proposed that the tectonically active nature of the Ouachita trough favored the formation of local submarine depressions (Fig. 26). Faults provided conduits for formation water in underlying sediments to migrate upward and mix with overlying marine waters, precipitating barite. Fluids more dense than seawater would pond in these depressions, and much of the precipitation of barite would thus be confined to localized areas. It was thought that injection of a buoyant fluid directly into seawater as a plume would have produced a highly dispersed and economically unimportant precipitate. Removal of barite from a brine pool could be accomplished by precipitation at the brine-seawater interface (Fig. 27). These authors estimated that approximately 1.6×10^{14} L of a normal formation water containing 200 mg/L Ba would have been necessary to form a deposit the size of Chamerlain Creek. The isotopic composition of a barite sample from Chamberlain Creek is $\delta^{18}O = 13.2$ permil (SMOW) and $\delta^{34}S = 18.3$ permil (VCDT) (Hanor and Baria 1977), values consistent with seawater sulfate of Middle Mississippian age, the estimated age of the host sediment. Sr isotopic values range from 0.7077 to 0.7092 (Maynard et al. 1995) or from approximately coeval seawater values to moderately radiogenic levels (Fig. 18).

Ordovician-Devonian Roberts Mountain allochthon, Nevada

The Nevada barite belt has been called the areally largest barite province in the world (Papke 1984, Dube 1986, 1988), but it is really a subpart of a much larger Paleozoic barite

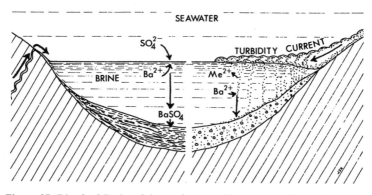

Figure 27. Dissolved Ba in a brine pool reacts with seawater sulfate to form barite. Ba-rich clay beds may have formed as clays settling through the brine. Modified from Hanor and Baria (1977).

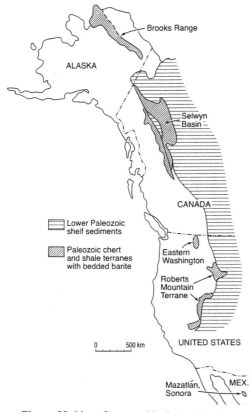

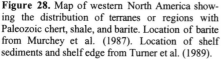

Figure 28. Map of western North America showing the distribution of terranes or regions with Paleozoic chert, shale, and barite. Location of barite from Murchey et al. (1987). Location of shelf sediments and shelf edge from Turner et al. (1989).

province (Fig. 28) which extends from northwestern Mexico through southern Nevada, and northward into Alaska (Murchey et al. 1987). In addition to the deposits discussed by Murchey et al., the bedded barite deposits of eastern Washington State (Mills et al. 1971) should be placed in this group.

As summarized by Dube (1986, 1988), bedded barite in Nevada is hosted by sediments that range in age from Cambrian to Early Mississippian. Most deposits, however, occur in Late Ordovician and Late Devonian host sediments. The barite bodies are lensoid, have thicknesses of up to 75 m, and have high thickness to width (length) ratios (Shawe et al. 1969). Single large deposits contain up to nearly 15 Mt of barite (Dube 1986). Horton (1963) noted that most of the barite deposits of Nevada lie in a belt that spatially coincides with the Antler orogenic belt, which is now known as the Roberts Mountain allochthon.

The $\delta^{34}S$ compositions of massive and laminated barite in East Northumberland Canyon, Nevada range from 20.9 to 28.6 permil, with most values around 25.0 permil (Rye et al. 1978). These values are typical of Upper Devonian marine sulfate, and because of this, seawater is thought to be the source of the sulfate (Rye et al. 1978). Although there is general agreement that the bedded deposits are syngenetic and that the sulfate is marine, there has been disagreement as to the source of Ba.

Some of the criteria used by Dube (1986, 1988) to support a sedex origin for the Nevada deposits are the stratiform, lens-like form of the orebodies: massive to laminated textures; coarse or angular sedimentary units near ore horizons, suggesting faulting; and presence of vent fauna in the form of fossil tube worms and brachiopods. Barite conglomerates are present in some of the deposits, suggesting slumping at proximal fault scarps near feeder zones. The high thickness to length ratio of many of the barite lenses suggests that the fluid from which the barite precipitated may have been ponded against fault scarps, perhaps within minibasins.

Jewell and Stallard (1991) and Jewell (1992) proposed instead that the Nevada bedded barite is ultimately the result of the concentration of Ba by pelagic organisms thriving in a coastal upwelling system in the Late Devonian ocean. These authors have cited the high levels of P (see Graber and Chafetz 1990) and organic carbon in barite-bearing sections relative to non-barite sections, and low Al/(Al+Fe+Mn) and low Fe/Ti bulk-sediment values indicative to them of a terrigenous rather than a hydrothermal source for the Ba. In their model, barite produced by Paleozoic pelagic organisms sank into underlying anoxic waters and then dissolved. Barite was then reprecipitated on the continental slope at an interface between sulfate-rich oxic and Ba-rich anoxic waters. Although this is a hypothesis worth testing further, there are difficulties in reconciling it with the strong field evidence for a sedex origin discussed by Dube (1986, 1988) and others. It is also unclear whether pelagic processes have ever been dynamic enough in the geological past to generate Ba fluxes large enough to produce localized, million metric ton orebodies of barite. The association of barite with phosphorites and cherts in some deposits may simply reflect deposition of barite in elevated concentrations during times of low background clastic sedimentation.

Further evidence cited by Jewell (2000) to support a biological origin for Pb- and Zn-free bedded barite in general, including deposits in Nevada, Arkansas, and China, consists of "...$^{87}Sr/^{86}Sr$ analyses that are comparable to contemporaneous seawater...". However, while some analyses do approach those of coeval seawater, most of the Sr-isotopic signatures (Maynard et al 1995, Jewell 2000) of the Nevada bedded barites and other bedded barite deposits are not comparable. Many of the Arkansas and Nevada analyses show enrichment in radiogenic Sr, and the Chinese barite is less radiogenic than coeval seawater (Fig. 18). The Pb- and Zn-free bedded barite from Cuba is highly radiogenic (Maynard et al. 1995).

Murchey et al. (1987) proposed a hybrid origin for the barite deposits along the western North America Paleozoic continental margin. They have suggested that the barite deposits are sedex, but that the source sediments from which Ba was derived by hydrothermal leaching were enriched in Ba derived from pelagic processes at the time of deposition.

Cenozoic convergent margin: Peru. The present convergent margin of Peru is characterized by an extensional tectonic regime and the lack of a well-developed accretionary prism. What have been described as "massive" barite deposits actually consist of crusts of barite a few millimeters thick, concretions, and chimneys up to 15 cm in height at two active vent sites (Dia et al. 1993, Torres et al. 1996a, Aquilina et al. 1997). Venting fluids sampled during deep submersible dives show an enrichment in Na, Cl, Ba, and radiogenic Sr ($^{87}Sr/^{86}Sr = 0.710151$ to 0.711186) relative to modern seawater (Fig. 18) (Aquilina et al. 1997). Values for $\delta^{34}S$ range from 10.7 to 29.7 permil. One site is close to sea water in sulfur isotopic composition and the otheris enriched in heavy sulfur. The venting fluids and barite have similar isotopic ratios of S, O, and Sr. Torres et al. (1996a) proposed that the Ba was derived by remobilization of non-detrital barite deposited in area of high biological productivity along the continental margin. Aquilina et al. (1997), in contrast, proposed that the Ba was derived from mainly continental sources.

Phanerozoic passive continental margins and epicontinental rifts

Selwyn basin, Canada: Cambrian-Mississippian. The Selwyn basin is an elongated epicontinental marine rift basin that extends from northeastern British Columbia, Canada, to Alaska. It is one of the best known sedex mineral districts because of excellent outcrop and the extensive and detailed work done there by the Geological Survey of Canada and others. Much of the development of general conceptual models for sedex deposits by Lydon (1995), Goodfellow (1999), and others stems from the study of these deposits.

The Selwyn basin was initiated in the Cambrian as result of extension along the northwest American passive margin and was terminated during the Mississippian by infilling with clastics. Stratiform Pb-Zn deposits were formed within the basin during three major periods, early(?) Cambrian, early Silurian, and middle to late Devonian. Sulfide deposits in early Cambrian and Devonian sediments are typically interbedded with barite and are often commonly capped by barren barite. Early Silurian deposits, in contrast, contain no barite. Stratiform barite deposits without significant metals commonly occur in rocks of Devonian and Mississippian age. Goodfellow (1999) noted that an important characteristic of sedex-bearing rifts in general is that they were long-lived and have a sedimentary record that spans hundreds of millions of years. Rift-bounding faults were periodically reactivated (Fig. 29), resulting in several hydrothermal events within the same sedimentary basin.

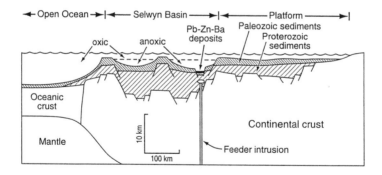

Figure 29. Schematic cross section of the Selwyn basin showing location of anoxic bottom waters and sedex sulfide-barite deposits. Intrusive activity may have been involved in the generation in ore-forming fluids but not in driving fluid expulsion. After reconstructions by Goodfellow et al. (1993) and Goodfellow (1999).

All major sedex deposits in the Selwyn basin formed during periods of anoxic bottom-water conditions (Figs. 23, 29). The $\delta^{34}S$ values for barite are consistently 10 to 15 permil more positive than the global values for marine evaporites indicating partly restricted circulation between the Selwyn basin and open seawater and/or vertical stratification within the water column (Cecile et al. 1983). Isotopic values of sulfide minerals are generally lighter, reflecting that much or most of the sulfide has been derived by bacterial reduction of marine sulfate (Goodfellow et al. 1993, Lydon 1995). The Sr isotopic composition of the barite is highly radiogenic, reflecting a sialic crustal source for the ore-forming fluids.

On the basis of fluid-inclusion and field data Goodfellow (1999) concluded that hydrothermal fluids in the Selwyn basin vented into sea water both as buoyant plumes (Fig. 23) and as dense bottom-hugging brines (cf. Fig. 27). Deposits formed from plumes are mound-shaped and are characterized by rapid changes in thickness and mineralogy away from the discharge vents. In contrast, deposits formed from dense brines tend to be more uniform in thickness, are laterally widespread, and are delicately bedded. Heating by magma injection into the lithosphere (Fig. 29) may have played an important role in generating ore-forming fluids, but was not responsible for the driving forces which caused fluid expulsion.

Sedex deposits of late Paleozoic age occur in the Alaska basin (Murchey et al. 1987). Most completely studied have been the Red Dog sedex Pb-Zn-Ba deposits of Carboniferous age that occur in northwesternmost Alaska. The Upper Carboniferous to Middle Jurassic Etivuk Group includes baritic shales which record a change to low sedimentation rates and oxygenated conditions and the continuation of low-temperature hydrothermal activity long after the formation of the Red Dog deposits (Edgerton 1997).

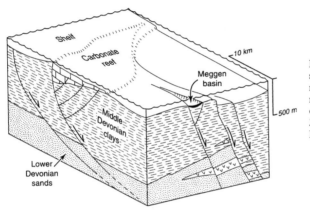

Figure 30. Block diagram showing a paleo-structural reconstruction of the Meggen region, Germany, at the time of submarine-hydrothermal Zn-Pb-Ba mineralization. Modified from Krebs (1981).

Meggen and Rammelsberg, Devonian, Germany. Meggen and Rammelsberg, Germany are two important sedex sulfide-barite deposits in Devonian host sediments and have been studied since the mid-1850s. Meggen is noteworthy for having been the first barite deposit in which the use of sulfur isotopes established Devonian seawater as the source of sulfur in the barite and hence a synsedimentary origin for the deposit (Buschendorf et al. 1963). According to Krebs (1981) the deposit formed when warm metalliferous basinal brines were driven toward syndepositional growth faults and ascended to the seafloor. Metal and barite precipitation occurred in restricted, fault-bounded sub-basins (Fig. 30). The Rammelsberg ore bodies are similarly associated with a major growth fault (Hannak 1981). Sawkins and Burke (1980) noted the relation of the occurrence of mid-Paleozoic massive sulfide deposits in Europe with extensional tectonic features.

China and Kazakhstan, Cambrian; Spain, Silurian. Large bedded barite deposits of apparent sedex origin exist over an area of approximately a million square kilometers in Lower Cambrian strata in China (Wang and Li 1991) and in Cambrian sediments in central Kazakhstan (Vinogradov et al. 1978). The Kazakhstan barite has fluid inclusions which yield homogenization temperatures of 150 to 250°C, which are interpreted by Vinogradov et al. to preclude a synsedimentary origin. It cannot be determined from the published description of these deposits, however, whether the inclusions exist in coarse, secondary barite, and whether stretching and/or the presence of dissolved methane has contributed to high homogenization temperatures. In other respects the Kazakhstan deposits closely resemble sedex deposits described above. The Chinese and Kazakhstan deposits constitute some of the largest reserves of barite in the world (Clark et al. 1990). The literature is sparse on these occurrences, but one hopes that additional studies will be performed and published.

Moro and Arribas (1981), in a review of stratiform barite deposits worldwide, included a brief description of several bedded barite districts in Spain. Host rocks are siliciclastic and volcanic sequences of Silurian age. Their field outcrop photos and photomicrographs document sedimentary features very similar to those observed in the Arkansas and Nevada deposits (Zimmerman 1970, Graber and Chafetz 1990).

Northwest Africa, Cretaceous. Dean and Schreiber (1977) described the occurrence of rosettes, lenses and laminae of barite in Lower Cretaceous sediments at two DSDP sites along the continental margin of northwest Africa. Salinities of interstitial porewaters increase downward at both sites in response to suspected underlying evaporites of Jurassic age. Dean and Schreiber suggested three possible sources of sulfate: seawater sulfate, sulfate formed by oxidation of pyrite, and sulfate derived from evaporites. Sulfur isotopic work would help resolve probable sources. They concluded that the main source of Ba was from oxidation of organic matter.

Northern Gulf of Mexico, Recent. Although the northern Gulf of Mexico sedimentary basin is technically a passive continental margin, it is active both tectonically and hydrologically (Hanor and Sassen 1990). Sediments both onshore and offshore are undergoing active deformation as a result of salt tectonics, subsidence, and extensional growth-faulting. The combination of topographically-driven fluid flow, thermohaline instabilities resulting from the dissolution of salt and perturbation of thermal gradients, and overpressuring have given rise to dynamic and complex regional subsurface fluid flow (Hanor and Sassen 1990, Hanor 1999a). It is not surprising, therefore, that brine seeps exist on the seafloor in the Gulf of Mexico and that some of these have barite associated with them.

Fu (1998) described several minor occurrences of barite chimneys and crusts associated with brine seeps in the offshore Louisiana Gulf of Mexico. On the basis of [228]Th/[228]Ra dating the chimneys studied by Fu are recent, from 0.5 to 6.5 years old, and the crusts from 9 to 23 years old at the time of sampling. Chimney $\delta^{34}S$ and $\delta^{18}O$ values are similar to that modern seawater, but the crusts are enriched in heavy S and O, reflecting sulfate that has been derived from seawater that has undergone bacterial reduction (Fu 1998). The [87]Sr/[86]Sr ratio of the barite is lower than that of modern seawater, consistent with strontium derived from anhydrite in Jurassic salt (McManus and Hanor 1993), which forms domes and allochthonous sheets in the offshore Louisiana Gulf of Mexico.

Cenozoic strike-slip margins

Southern California borderland. Occurrences of barite are common in proximity to fault scarps in the California borderland (Lonsdale 1979). The barite occurs as cones, columns and irregular piles of mineral debris on the seafloor. At one location, barite occurs as spherical and cylindrical concretions, the latter appearing to be baritized worm burrows (Hanor, unpublished). Some of the deposits studied by Lonsdale (1979) stand 2 to 3 m above the seafloor, but one mass has an estimated height of 10 m. Dense colonies of tube worms occur around some of these deposits. The barite is thought by Lonsdale and others who have studied similar occurrences (see references in Lonsdale 1979) to be the result of submarine discharge along faults. An active hydrothermal circulation system has been proposed for the San Clemente fault zone on the basis of the pattern of heat flow. Lonsdale proposed convective circulation of seawater within the San Clement fault zone, with sulfate removed by reduction and Ba added by leaching. As cited by Lonsdale, basaltic rocks of the region are notable for their high concentrations of Ba. Widespread submarine fluid expulsion occurs further north along this translational continental margin within Monterey Bay, although the presence of barite has not yet been identified (Orange et al. 1999).

Cortecci and Longinelli (1972) measured oxygen-isotope variations in one of several large barite slabs from this offshore region. These values, and isotopic analyses reported by Goldberg et al. (1969) and Church (1970), indicate the barite is enriched in ^{18}O and ^{34}S relative to seawater sulfate. The relative fractionation of sulfur to oxygen follows the 4 to 1 ratio observed experimentally in the bacterial reduction of dissolved sulfate (Mitzutian and Rafter 1969).

Sedex barite deposits in active marine evaporite settings

Precipitation of barite can apparently occur contemporaneously with seawater evaporation. Such may be the case for the Chaudfontaine barite deposit in Belgium (Dejonghe 1990, Dejonghe and Boulvain 1993), where barite, pyrite, sphalerite, and galena formed in small shallow pools that evolved under evaporitic conditions in a Devonian carbonate bioherm. According to Dejonghe and Boulvain, much of the barite grew at the sediment–water interface from marine S and non-marine Ba and Sr. Similar deposits may occur in Argentina (Brodkorb et al. 1982). Given the low concentration of Ba in normal seawater, however, it is unlikely that significant amounts of barite could ever be precipitated by direct evaporation only.

The question of metal-bearing and metal-free bedded barite deposits

There has been considerable interest in determining why some bedded barite deposits, such as Meggen, Rammelsberg, and many of the Selwyn basin deposits, are spatially associated with significant amounts of metal sulfides, whereas other barite deposits, such as the Arkansas, Nevada, and many Chinese deposits are not. Maynard and Okita (1991) attempted to explain this difference on the basis of differences in tectonic setting. They proposed a dual classification of "continental margin type" and "cratonic rift type" barite deposits. In the continental margin type deposit barite was deposited without associated metal sulfides in small ocean basins lying between subduction zones and passive continental margins. Sulfide-bearing deposits, in contrast, formed in cratonic rifts and have a continental geochemical signature. Turner (1992) noted, however, that metal-free barite deposits occur in the Selwyn basin along with Pb-Zn-Ba deposits. Turner further noted that most Cambrian to Devonian age stratiform Pb-Zn-Ba deposits world-wide formed in a continental margin setting rather within cratonic rifts.

Maynard et al. (1995) presented an extensive compendium of Sr-isotopic analyses of bedded barite and concluded that barite associated with Pb and Zn is much more radiogenic than coeval seawater than Pb-Zn-free barites (Fig. 18). They suggested that these differences in Sr-isotopic composition of barite could be used to screen exploration targets for the possible occurrence of heavy metals. They further explained the occurrence of barite-only deposits in intracratonic rift settings by the exhalation of ore-forming fluids into sulfide-free oxic seawater.

Other variables to consider in the relation between barite and metal sulfides include temperature and salinity. Elevated temperature and salinity facilitate the mobilization and transport of lead and zinc through the formation of chloro-complexes (Hanor 1996b, 1999b), whereas the mobilization of Ba under basinal conditions is more critically dependent on the presence of low concentrations of sulfate (Fig. 6). Terranes or sedimentary sequences with halite may be more likely to generate metal-bearing sedex fluids than those without. As noted above, an alternative explanation for the origin of Pb- and Zn-free bedded barite deposits has been proposed by Jewell (2000), who has advocated a biological origin for the Ba.

CENOZOIC PELAGIC BARITE AND DISPERSED BARITE IN DEEP SEA SEDIMENTS

The final types of barite to be discussed which have seawater as the principal source of sulfate are: pelagic barite, a term restricted in this chapter to barite produced in the shallow part of the water column of the open ocean, and fine-grained barite regionally dispersed in deep-sea sediments in environments spatially removed from localized submarine hydrothermal vents. The term *pelagic*, derived from the Greek word *pelagos*, referring to the level surface of the sea, is used here in its strict sense. Barite dispersed in deep-sea sediments may include pelagic barite, barite precipitated at the sediment–water interface, and particulate barite derived from distant hydrothermal sources. As discussed earlier, the term *marine barite* by itself is ambiguous and includes a wide variety of barite deposits. Some have coined the term *biobarite* to refer to pelagic barite thought to originate from biologically mediated processes, but this is not appropriate because some benthic organisms can also precipitate barite.

Probably no types of barite have received more attention in the recent literature than pelagic barite and barite in deep-sea sediments, because of their potential as paleo-productivity indicators and because of their potential use in tracking secular changes in the isotopic composition of seawater. The literature on these types of barite and on Ba in seawater is large, and some highlights are reviewed here.

Barite in seawater

Particulate barite is nearly ubiquitous in seawater. Dehairs et al. (1980) found barite particles in all water samples they investigated in an areally extensive sampling of the Atlantic, Antarctic, and Pacific oceans. Barite concentrations ranged from 10 to 100 ng/kg of seawater, with most water samples containing several tens of ng/kg. Barite is particularly abundant in waters underlying areas of high surface productivity.

The barite studied by Dehairs et al. (1980) consists of aggregates of submicrometer grains with or without a crystalline habit, sub-spherical particles, and particles with a distinct crystalline habit. Dehairs et al. reported a wide range of Sr concentrations, with 67% of the barite samples containing <10 mol % $SrSO_4$, 22% containing from 10 to 50% $SrSO_4$, and 11% containing over 50% $SrSO_4$. Barium-free, Sr- and S-rich particles are also present as biogenic Acantharian celestine debris or as smaller particles with no obvious biogenic morphology. The suspended barite particles contain minor amounts of K.

Stroobants et al. (1991) found that barite particles in the upper 10-20 m of the water column occurred as bioaggregates without distinct crystalline habit. Below these surface waters, however, barite existed in bioaggregrates containing microparticles with a crystalline habit. At water depths below a few hundred meters barite crystals, were present as free discrete particles, possibly as a result of breakup of aggregates by bacterial activity. The free microcrystals settle much more slowly than the carrier-aggregates, and local concentration of suspended barite within the water column at mid-depth can result (Dehairs et al. 1992). Bertram and Cowen (1997) observed three distinct crystalline forms of barite in plankton tows and net tows and on artificial seafloor substrates in the central Pacific. Microcrystalline aggregates of barite contained no detectable Sr, whereas ovoid and hexagonal-type crystals contained variable, and some high, Sr concentrations. The microcrystalline aggregates may be formed passively within abiotic microenvironments, and certain hexagonal crystals are precipitated by benthic forams.

Mode(s) of precipitation of pelagic barite

The exact mechanisms of formation of pelagic barite in the upper water column are still unknown and a subject of debate. The following mechanisms have been proposed.

Inorganic precipitation. Most authors argue that highly variable Sr/Ba ratios of pelagic barite are inconsistent with inorganic precipitation from surface seawater, which has a restricted range of Sr/Ba ratios (e.g. Dehairs et al. 1980). Inorganic precipitation has been deemed unlikely because most near-surface seawater is also undersaturated with respect to pure $BaSO_4$ (Fig. 8) (Monnin et al. 1999). Bernstein et al. (1992) estimated saturation indices for celestine and barite in ocean surface waters to be 0.16 and 0.14 respectively. Calculations by Hanor (1969), based on the Sr-Ba partition coefficients determined by Starke (1964), however, showed that Sr-enriched barite is significantly less soluble than pure $BaSO_4$, and that seawater could be nearly saturated with respect to a strontian-barite containing 33 mol % $SrSO_4$.

Precipitation in sulfate-enriched microenvironments. Bishop (1988) proposed that barite forms in microenvironments enriched in sulfate released from decaying diatom organic matter. According to Bishop, the C:Si ratio of a diatom is 2:1, and the C:S ratio is approximately 100:1. Therefore the Si:S ratio of a diatom is approximately 50:1. Dissolved Si and Ba covary in seawater in a ratio of 1300:1. According to Bishop, the reaction of organically bound S with dissolved Ba in seawater need only be 4% efficient if diatoms are the only source of S for barite. In addition, other plankton groups can contribute to the pool of S. Fresh diatom silica surfaces may adsorb Ba better than other biogenic organic phases (Bishop 1988) thus enhancing the formation of microenvironments conducive for barite precipitation.

Incorporation of Ba into skeletons of siliceous plankton. Because of the excellent correlation between dissolved Ba and Si in the marine water column, it has been suggested that the distribution of both elements is governed by the dissolution of Ba-enriched siliceous frustrules. However, plankton samples from the Pacific described by Martin and Knauer (1973) had low Ba levels.

Biogenic precipitation of barite by other planktonic organisms. A number of freshwater and marine organisms are known to precipitate barite. These include the benthic marine Rhizopoda of the class Xenophyophorida (Arrhenius 1963, Tendal 1972, Gooday and Nott 1981, Gooday et al. 1995, Hopwood 1997), benthic foraminifera (Dugolinsky et al. 1977), marine algae in the Mediterranean (Fresnel et al. 1979, Gayral and Fresnel 1979), freshwater protozoa (Hubert et al. 1975, Finlay et al. 1983, Fenchel and Finlay 1984), and freshwater desmid algae (Brook et al. 1980, 1988; Wilcock et al. 1989). Finlay et al. (1983) speculated that the intracellular precipitation of ~4 μm size spherical barite granules by the freshwater protozoa of the genus *Loxodes* in the English Lake District of England, may be instrumental in gravitropic responses. A related marine form, *Remanella*, precipitates celestine, however, not barite (Bishop 1988). Production of pelagic barite by direct intracellular biological precipitation has not yet been demonstrated.

Dissolution of acantharian debris. Acantharians are an abundant planktonic species of radiolaria which secrete shells or tests of celestine (see section in this chapter on celestine). Bernstein et al. (1992, 1998) argued that the uptake of Ba in the celestine acantharian shells and its subsequent release during ingestion of acantharians by zooplankton may result in the precipitation of barite. Despite the abundance of acantharians in the euphotic zone, celestine remains have apparently not been reported in either the gut contents or fecal material of zooplankton. However, siliceous remains are common. Bernstein et al. speculated that the absence of acantharian debris may be the result of rapid dissolution of celestine skeletons and cysts. The authors calculated that the dissolution of a single acantharian or acantharian cyst of 1 μg mass containing 5.4×10^{-9} mol $SrSO_4$ would saturate a 10-μL volume of water with respect to $SrSO_4$, while supersaturating the water with respect to $BaSO_4$ by seven times.

Fate of barite in the water column

Undersaturation of seawater with respect to $BaSO_4$ almost everywhere in the water column (Fig. 8) (Monnin et al. 1999) induces the dissolution of suspended barite. Nearly all suspended barite particles are affected by dissolution to varying degrees (Dehairs et al. 1980). The edges of euhedral particles are rounded in the process, and particles become progressively ellipsoidal and spherical. There is also etching of crystal surfaces. Some pelagic barite dissolves completely as it sinks through the water column, and there is a decrease in the number of suspended barite particles with water depth. Most of the barite, however, is transported rapidly within fecal pellets to the seafloor where it accumulates in bottom sediments. The interstitial pore water of these sediments rapidly becomes saturated with respect to barite, inhibiting further dissolution. Bertram and Cowen (1997) observed that both size and Sr content of suspended barite particles decrease from surface water to the seafloor in the central Pacific, suggesting that the particles undergo dissolution and preferential removal of Sr-rich barite.

High concentrations of dissolved Ba possibly derived from the dissolution of pelagic barite are found in marine anoxic basins, including the Cariaco trench, Framvaen Fjord, Black Sea (Falkner et al. 1993), and the Mediterranean (de Lange et al. 1990a,b). Barium distribution is not strongly affected by adsorption or uptake on Mn or Fe oxyhydroxides formed at the redox interfaces of these basins. However, microbial activity just above the O_2/H_2S interface in the Black Sea promotes breakdown of settling particulate matter and the release of barite. Dissolution of barite in the marginal sediments of these basins also probably contributes to the small Ba maxima observed at the redox interface.

Barite in deep-sea sediments

Occurrence. Barite is a common constituent of those deep-sea sediments whose mineralogy is dominated by non-terrigenous sources. The high content of Ba in east equatorial Pacific sediments was first described by Revelle (1944), who suggested Ba may have been deposited in biogenic carbonate on the basis of the correlation between the abundance of Ca carbonate and Ba. Arrhenius et al. (1957) identified the existence of small (1 μm) crystals of high refractive index and high alpha activity as possibly barite in north equatorial Pacific sediments. Since that study, barite has been identified in deep-sea sediments ranging in age from Recent to Cretaceous (Dean and Schreiber 1977).

Arrhenius and Bonatti in their 1965 paper on neptunism and volcanism in the oceans showed that maximum concentrations of barite, 7 to 9 wt % on a carbonate-free basis, coincide spatially with the East Pacific Rise and not with the area of high productivity along the Peru current to the east (Fig. 31). According to Church (1970, 1979) barite occurs in three different types of deep-sea sediments, those characterized by: (1) abundant calcareous and siliceous biological debris; (2) abundant Mn and Fe phases in nodular and dispersed forms, and (3) altered volcanic debris with montmorillonite, palygorskite, and clinoptilolite. The order of frequency of occurrence of these three types of barite is about 10:2:1. Euhedral crystals typically 1 to 2 μm in size can comprise up to 2% of the sediments underlying productive areas of the eastern equatorial Pacific. Barite crystals ranging in size from 25 to 100 μm are occasionally found in deep-sea sediments containing altered volcanic debris or manganiferous phases. In their survey of barite reported in sediment samples collected during the Deep Sea Drilling Project, Dean and Schreiber (1977) noted the common occurrence of barite both in the 2 to 20 μm and <2 μm grain size fractions.

Chemical composition. The first detailed and extensive study of the chemical composition of barite in deep-sea sediments was performed by Church (1970). In contrast to pelagic barite, which has a wide range of Sr concentrations, Sr in deep-sea barite from

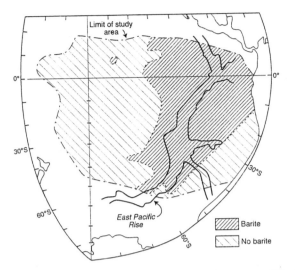

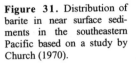

Figure 31. Distribution of barite in near surface sediments in the southeastern Pacific based on a study by Church (1970).

the Pacific ranges from 0.2 to 3.4 wt % and averages 1%. Calcium and K are present in levels of 0.01 to 0.1 wt %, with no correlation to each other, Sr content, or crystal size. Church found no evidence for high levels (100-1000 ppm) of Cr, Zn, Pb, and Zr earlier reported by Arrhenius (1963).

Sr-isotopic composition. The relatively high concentrations of Sr in barite has permitted use of very small samples of deep-sea barite for Sr-isotope studies. Goldberg et al. (1969) determined that barite crystals from Pacific surface sediments had Sr isotopic compositions similar to that of present day seawater. With the advent of more precise Sr-isotopic techniques there has been considerable refinement in the understanding of the Sr-isotopic compositions of pelagic barite (Paytan et al. 1996). Martin et al. (1995) found that all samples from near the core tops in eastern equatorial Pacific sediments have values close to modern-day seawater (0.709175). The $^{87}Sr/^{86}Sr$ ratios for barite from Late Miocene to Pliocene sediments deeper in the cores agree very well with published Sr-isotope data for carbonates over the past 36 Ma. The crystal morphology of barite over these depth intervals shows no noticeable change, and the barite has not re-equilibrated with pore fluids in terms of Sr isotopic composition. Martin et al. (1995) thus concluded that marine barite may be useful for Sr-isotopic stratigraphy and correlations in epochs for which the seawater curve for Sr is known and in times or environments where Ca carbonate has either undergone diagenetic alteration or is absent. However, barite separates in two sites north of the productivity maximum had consistently slightly lower $^{87}Sr/^{86}Sr$ ratios than coexisting foraminifera and fish teeth, thus suggesting the presence of small amounts of oceanic hydrothermal barite in these sediments (Martin et al. 1995).

Neodymium isotopic compositions and REE. Guichard et al. (1979) reported relatively high Nd concentrations (5-100 ppm) for barite in deep-sea sediments, which would make it a useful mineral for Nd-isotope studies of seawater. However, Martin et al. (1995) observed Nd isotopic values of barite in pelagic sediments to be significantly different than values for contemporary seawater. The Nd values appear instead to be dominated by an unidentified aeolian contaminant having a high Nd concentration and low $^{143}Nd/^{144}Nd$ ratios. Martin et al. (1995) concluded that while deep-sea barite is potentially useful as a recorder of Nd paleo-seawater compositions, its utility for both REE and Nd studies may be hampered by the presence of various refractory contaminants, such as rutile, anatase, zircon, and titanite, which are not easily removed using standard chemical separation techniques.

Radium and thorium. Some of the results of early studies of Ra, Pb, Th, and U in barite samples from pelagic sediments suggested rapid exchange and remobilization of these elements after burial (Church and Bernat 1972, Borole and Somayajulu 1977). Church and Bernat (1972) concluded from the high $^{228}Th/^{232}Th$ they measured in barite near core tops that ^{228}Ra in barite was in rapid exchange equilibrium with pore waters. Borole and Somayajulu (1977) measured excess ^{226}Ra in barite samples too old to contain excess ^{226}Ra, and concluded that barite must be continually growing in the sediment column. However, there was no increase in size in the barite crystals with depth in the sediment.

More recent work by Paytan et al. (1996) sheds new light on Th and Ra systematics in barite in pelagic sediments. Concentrations of U in the barite are less than 0.5 ppm, negligible in comparison to that of Th, indicating essentially all the ^{230}Th in barite is unsupported. $^{230}Th/^{232}Th$ activity ratios range between 20 and 190, with a mean of ~100, which is consistent with the bulk sediment ratio. The exponential decrease in ^{226}Ra of barite in the upper 25 cm of the sediment suggests that barite behaves as a closed system and is not affected by exchange or recrystallization. The lack of detectable ^{228}Th, ^{228}Ra, and ^{224}Ra activities in any of the barite samples indicates that no significant barite growth is occurring below the bioturbated zone.

Paytan et al. (1996) ascribed problems in the earlier results for Ra and Th in marine barite to the use of milder sequential leaching procedures that did not remove all detrital material and oxyhydroxide coatings from the barite. An extreme HF leach step is necessary to remove detrital components, and special precautions are needed to remove surface coatings of Fe-Mn oxyhydroxides from barite surfaces.

Origin of barite in deep-sea sediments. There are still uncertainties regarding the origin(s) of this type of barite, despite the large amount of work which has been done on it. A spatial association between the occurrence of Ba in deep-sea sediments and areas of high biological productivity was made early on. The high content of Ba in east equatorial Pacific sediments was first described by Revelle (1944), who suggested that Ba may have been deposited in biogenic carbonate, because of the correlation between the abundance of Ca carbonate and Ba. Arrhenius (1952) tested the hypothesis that rate of accumulation of Ba is a function of the rate of organic production in the euphotic zone by measuring variations in BaO/TiO_2 ratios with depth in sediment cores, and assuming that the rate of TiO_2 accumulation is uniform in space and time. Arrhenius concluded that the rate of Ba accumulation is approximately 20 times higher under the equatorial divergence high productivity zone between 10°N and 10°S than outside this zone. Arrhenius (1952) also established that there has been a marked variation in the rates of Ba accumulation with time.

Church (1979) proposed that the flux of particulate barite down through the water column could account for approximately half the sedimentary barite he observed. According to Church, the remainder appears to grow in surface sediments as a consequence of further organic degradation. Church (1970) suggested that barite deposition on the East Pacific Rise is correlated with high concentrations of $CaCO_3$ and organic matter. However, Boström et al. (1973) indicated that these $CaCO_3$ data do not agree with earlier studies on the areal distribution of carbonates.

Although it is sometimes stated in the literature that barite is particularly abundant both in the water column and underlying sediments in areas of high biological productivity, Arrhenius and Bonatti (1965) showed that maximum concentrations of barite, 7 to 9% on a carbonate-free basis, coincide spatially with the East Pacific Rise and not with the area of high productivity along the Peru current. Boström et al. (1973) determined the concentration of Ba in more than one thousand sediment samples from the Pacific, Indian, South Atlantic and Arctic oceans and determined Ba accumulation rates in the Pacific (Fig. 32). Some barite may form by release of Ba during alteration of volcanics. There is some correlation of Ba accumulation rates with accumulation rates of opaline silica (Fig. 32), particularly in the equatorial Pacific, but there is remarkably little Ba in sediments underlying high productivity regions off the coast of Japan and South America. All high Ba values are located on the active ridge system in the Indian and Pacific oceans. Regions of high productivity in the north Pacific, eastern Pacific, and east Indian oceans lack Ba-rich sediments.

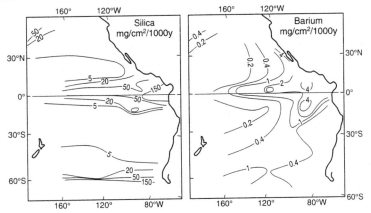

Figure 32. Accumulation rates for (a) opaline silica and (b) Ba in sediments of the Pacific. Both diagrams modified from Boström et al. (1973).

An unknown, but possibly significant amount of barite is generated by xenophyophores, a group of bottom-dwelling protozoans closely related to foraminifera (Tendal 1972, Tendal and Gooday 1981). Pictures of the seafloor taken from submersibles in the Atlantic ocean suggest that these protozoans are quite common. Gingele and Dahmke (1994) concluded that the barite secreted by the xenophyophores cannot be distinguished from barite dispersed in sediments in terms of size and morphology. The quantitative contribution of the xenophyophores to the barite budget remains unclear. Perhaps these organisms ingest rather than secrete barite.

There would thus appear to be multiple origins for the barite dispersed in deep-sea sediments. Among these origins are the more Ba-rich of the pelagic barites which settle intact through the water column; barite precipitated at the sediment-water interface, perhaps by dissolution of other forms of biogenic minerals or recrystallization of pelagic barite; particulate barite derived from distant hydrothermal sources; and barite secreted by benthic organisms.

Paleoenvironmetal significance. A number of authors have concluded that barite in deep-sea sediments has utility as a paleoenvironmental indicator (Schmitz 1987, Andrew et al. 1990, Dymond and Collier 1996). Dymond et al. (1992) developed an algorithm for computing absolute paleoproductivity rates from rates of barite accumulation. Barite accumulation rates show cyclical variations in which maxima correspond to glacial and minima correspond to interglacial stages (Gingele and Dahmke 1994).

The association between barite accumulation rates and primary productivity seems well established for Pacific equatorial waters, but as noted above, there are areas where there is no correlation. The accuracy of the calculations is further limited by the precision in distinguishing between biogenic barite and the background barium which increases with increasing concentration of terrigenous matter. In addition, secular variations in the concentration of Ba in the water column which could effect barite production and dissolution rates. That absolute concentrations of Ba in seawater have fluctuated in the past is evidenced by variations in Ba/Ca ratios of benthic foraminifera. Schroeder et al. (1997) concluded that total Ba in equatorial Pacific carbonate sediment is not a good proxy for barite because of the association of Ba with terrigenous input. It is possible that the non-barite Ba is adsorbed on grain surfaces. McManus et al. (1998) noted that barite can be reduced under suboxic conditions. They proposed that the accumulation of authigenic U may serve as an indicator of when the Ba record may be unreliable.

EPIGENETIC BARITE DEPOSITS AND EVAPORITES

This section explores a diverse group of barite deposits in which sulfate is sourced not from seawater, as in the previous examples, but from marine evaporites and/or buried evaporative marine waters.

Carbonate-hosted barite deposits

The evolution in sedimentary rock types over geological time manifests itself in the Late Proterozoic and Paleozoic with the appearance of laterally extensive biochemical continental margin and epicontinental platform carbonate sediments, many of which have served as hosts for ore deposits. The Paleozoic thus marks the development of new types of carbonate-hosted ore deposits, the Mississippi Valley type (MVT) and Irish type deposits, which have no known Archean or Lower Proterozoic counterparts. During low stands of sea level, the platform carbonates formed karsts and transformed into aquifers of regional extent which then served to focus fluid flow during large-scale deformation of the platform margins. Marine evaporites that are commonly associated with these carbonates provided potential sources of S for the formation of both barite and associated metal sulfides.

Mississippi Valley type deposits occur as epigenetic cavity fill and replacements by galena, sphalerite, fluorite, and barite in lithified and diagenetically cemented limestones and dolomites deposited in platform environments. Most MVT deposits were formed from 10 to 100 Ma after deposition of the host sediment (Hitzman 1999). Most deposits are related spatially to regional unconformities, and the timing of mineralization is related to large-scale orogenic events. Barite commonly occurs late, after the main stage of base metal mineralization.

Irish type Pb-Zn-Ba deposits occur in non-argillaceous, shallow marine carbonates and include syndiagenetic to epigenetic mineralization (Hitzman and Beaty 1996). A distinguishing feature of the Irish type deposits is their occurrence in carbonate environments that have undergone extension. The deposits are located adjacent to normal faults that served as fluid conduits, and mineralization occurred within 1 to 10 Ma after deposition of the host rocks. Barite is typically early, in contrast to the late deposition in many MVT deposits. Ore fluids were typically hot, 120-280°C, and had salinities of 10-25 wt % NaCl equivalent. Deposits apparently formed from the mixing of two fluids, one metal rich and hot with moderate salinities, and the other of lower temperature and salinity and which may have contained the sulfur.

Since the Jurassic, the principal locus of carbonate deposition has shifted from the shallow marine environment into the pelagic realm. There are no sediments being deposited today that are strictly comparable to those that host the MVT and Irish type deposits.

An extensive review of carbonate hosted lead-zinc-barite deposits world-wide is contained in the volume edited by Sangster (1996). Representative examples of MVT and

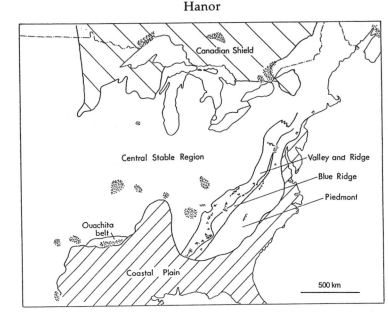

Figure 33. Map showing the distribution of barite deposits (stippled areas) in the central and eastern United States and Canada . From Hanor (1966).

Irish type deposits of eastern North America are described below.

Carbonate-hosted MVT deposits of the Appalachian orogen. The Appalachian orogen of eastern North America has had a complex tectonic history with three extended periods of orogeny in the Cambrian-Ordovician (Taconic, Penobscottian), Devonian (Acadian) and Mississippian through Permian (Alleghanian). There followed a period of rifting during the Triassic and Jurassic. Kesler (1996) provided a comprehensive review of the continental-scale MVT mineralization which extends throughout the Appalachian orogen as exposed in the Valley and Ridge province (Fig. 33) from Alabama through Quebec (Williams-Jones et al. 1992, Paradis and Lavoie 1996) to Newfoundland. The mineral deposits contain various combinations of sphalerite, galena, barite, fluorite, dolomite, calcite, and quartz and are most abundant in thicker stratigraphic sections along the strike of the orogen. Each of the four units which host significant MVT mineralization (Fig. 34) were important paleoaquifers that focussed fluid flow during proximal tectonic events.

Barite, largely without significant amounts of sulfides, occurs in the Upper-Precambrian, Lower Cambrian Clastic Sequence with major deposits in the Cartersville District of Georgia. The Late Cambrian-Early Ordovician Knox-Beekmantown-St. George groups constitute the most important MVT paleoaquifer in the Appalachians and host large deposits of sphalerite. Barite occurs with or without fluorite in districts, such as the Sweetwater district, Tennessee, peripheral to zinc deposits. Farther to the west, MVT mineralization, including barite, is present in the Knox Group along the crest of the Cincinnati arch in central Tennessee and central Kentucky.

Possible sedimentary exhalative barite occurs in the Athens Shale in the Tennessee embayment of the Upper Ordovician Clastic Sequence and in foreland basin shales in the Pennsylvania embayment (Nuelle and Shelton 1986, Clark and Mosier 1989). MVT mineralization in Pennsylvanian and Permian sediments is confined to nodules and veinlets of sulfide and local barite, commonly associated spatially with coal layers. MVT

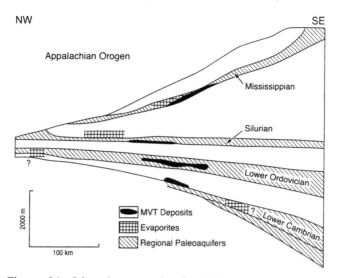

Figure 34. Schematic cross-section through the southern and central Appalachians showing the location of MVT deposits and evaporites in relation to principal paleoaquifers. Based on Kesler (1996).

mineralization is zoned on a regional scale throughout the Appalachians, with barite much more common in the southern part of the orogen. Most mineralization in the southern Appalachian orogen seems to have formed during Devonian time, presumably in response to the Acadian orogeny.

Information on the nature of the mineralizing fluids has been obtained from fluid-inclusion studies. Freezing temperatures yield fluid salinities in sphalerite of 20 to 26 equivalent wt % NaCl, and homogenization temperatures from <100 to >200°C. Some homogenization temperatures for Appalachian barite range from 200 to 300°C, unusually high relative to the other minerals. The temperatures may be artifacts produced by either stretching of the inclusions (Ulrich and Bodnar 1988) or the presence of dissolved methane (Hanor 1980).

Kesler (1996) presented chemical evidence for the presence of several distinctly different mineralizing brines in the orogen. Fluid-inclusion leachates from east Tennessee have very low Na/Cl and Cl/Br ratios, typical of brines that have undergone extreme evaporation to the point of precipitation of sylvite and other non-Na salts, but not typical of brines that have formed as the result of dissolution of halite. Kesler et al. (1995) thus concluded that the high salinities observed in the inclusions are the result of seawater evaporation rather than the subsurface dissolution of halite. An alternative explanation to that of Kesler et al. (1995) would be the dissolution of bittern salts high in Br.

Sulfur-isotope studies indicate that the most likely sources of S in both sulfide and sulfate minerals are buried seawater sulfate and evaporite sulfate. Sulfate S in barite was derived directly from these marine sources with a fractionation of only a few permil. Sulfide sulfur was derived almost entirely from H_2S produced by bacterial and thermochemical reduction of marine sulfate. Studies of most MVT deposits show that $^{87}Sr/^{86}Sr$ increases with decreasing age in the paragenetic sequence in a given deposit. Some ratios exceed that of enclosing carbonate host rocks and probably reflecting the introduction of Sr from a siliciclastic source. In the Sweetwater district of Tennessee, however, $^{87}Sr/^{86}Sr$ values for early fluorite and barite are actually less than those for enclosing rocks and coeval seawater (Kesler 1996). Schrijver et al. (1994) have presented Pb-isotopic evidence supporting a siliciclastic source for both Pb and Ba in Taconic Orogen of Quebec.

Intracratonic MVT deposits. More problematic in origin than the MVT deposits of the Paleozoic Appalachian orogen are the carbonate-hosted MVT deposits of the intracratonic area of North America (Fig. 33), including the eponymous MVT district of the Upper Mississippi Valley region of southeastern Wisconsin and adjacent areas of Illinois and Iowa, the MVT districts of central and southeast Missouri, central Kentucky, and central Tennessee, and the Pine Point deposit of northwest Canada. These deposits are very similar in mineralogy and host rock setting to MVT deposits of the Appalachian orogen, but problematic in origin because of their greater distance from orogenic belts whose deformation might have driven ore-fluid transport (Garven and Raffensperger 1997). Some districts, such as the Cave-in-Rock fluorite district, which also contains barite, may be related to intracratonic igneous activity (Ruiz et al. 1988). Barite typically occurs as a late-stage mineral whose presence may not be related to the deposition of earlier base metals (Misra et al. 1996).

One of the more completely documented barite districts in the continental interior of North America is that of southeast Missouri (Kaiser et al. 1987). Lead-Zn-Ba-quartz mineralization is localized in flat-lying, elongate structures called runs, which represent high permeability zones formed at the intersection of near-vertical faults and areas of intense jointing within the subhorizontal Upper Cambrian Potosi and Eminence cherty dolomites. Fluids moved through these areas of high permeability depositing sulfides and barite in the central core areas of the runs and additional barite in sediments peripheral to the runs. Barite becomes progressively finer-grained and plumose laterally away from these core areas.

Fluid-inclusion studies indicate that the main stage of sulfide mineralization was deposited from warm saline Na-Ca-Cl fluids having temperatures from 70 to 105°C, salinities of 20 to 26 equivalent wt % NaCl, and carrying isotopically light S. Galena and sphalerite have $\delta^{34}S$ values which range from 3 to 16 permil.

In contrast, main-stage barite was deposited following sulfide deposition, from fluids far less saline (0 to 14 equivalent wt % NaCl). Temperatures of barite deposition are difficult to determine because most of the inclusions are single phase and the few two-phase inclusions analyzed apparently stretched upon heating. Barite $\delta^{34}S$ values range from 21 permil near the faults to approximately 32 permil laterally away from the faults. Fluid-inclusion and O-isotope evidence suggests that the barite was precipitated when a warm saline fluid flowing through the runs mixed with a cooler, less saline ambient fluid. Sulfate was derived from two distinctly different sources, oxidized H_2S from near the runs, and a heavier sulfate derived from the host Cambro-Ordovician carbonates. Uplift of the Ouachitas in late Paleozoic time apparently initiated MVT mineralization in the Ozark region through development of a continental-scale fluid-flow regime in which topographically-driven flow was initiated (Garven and Raffensperger 1997).

Leach (1980) determined that barite in the Central Missouri district was deposited as open-space filling in carbonates from fluids having salinities ranging from 4 to 10 wt % NaCl at temperatures of less than 40 to 50°C. Both salinities and Sr-content fluctuated during the growth of many of the barite crystals examined by Leach, who proposed that the barite was precipitated by mixing of a barium-rich and sulfate-rich fluid in near-surface environments affected by the seasonal introduction of sulfate-bearing water.

Hanor (1966) noted that major occurrences of barite in the mid-continent region of North America are spatially related to domal uplifts (Fig. 35). There is no exposed dome in

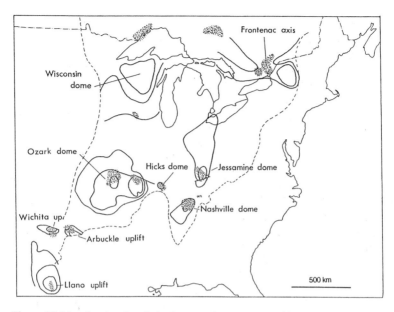

Figure 35. Map showing the relation between the occurrence of barite deposits (stippled areas) and domal uplifts in the central interior of the United States (Hanor, 1966).

this region which does not have some barite mineralization associated with it. Nearly all of the host rocks are Upper Cambrian and Ordovician carbonates. Carbonate solution features associated with the uplifts provided channelways for transport and sites of deposition of Ba. Barite may have been formed in the latter stages of hydrothermal activity when there was greater influx of ambient oxygenated shallow groundwaters circulating through the domal uplifts.

Irish Type deposits. Hitzman (1999) has provided an overview of Irish type base-metal and barite deposits world-wide. In the Walton and Gays River districts of Nova Scotia, Canada, Irish-type sulfide and barite deposits are restricted to areas of synsedimentary faults. Base metals and barite replace host rocks and occlude porosity. Sulfides and barite form stratiform bodies at Walton and discordant bodies at Gays River. Mineralization apparently occurred just prior to the Appalachian orogeny ~320 to 300 Ma when high-temperature high-salinity ore fluids discharged from depth and mixed with cooler, lower-salinity waters. Local evaporites provided the source of sulfur (Mossman and Brown 1986, Sangster et al. 1998, Chagon et al. 1998, Hitzman 1999) (Fig. 36).

Figure 36. Schematic cross-section through the Gays River Zn-Pb-barite deposit, Nova Scotia, showing the relation of ore to the presence of evaporites. Based on a reconstruction of Chagon et al. (1998).

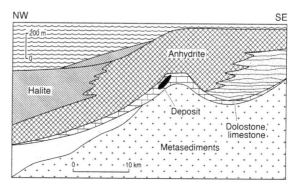

Continental rifts

Not all barite deposits associated with continental rifts occur in the immediate proximity to continental margins or active marine basins. The continental rifts discussed here are distinguished from the epicontinental marine rift basins discussed earlier in that the barite was deposited in epigenetic veins rather than in a syndepositional marine setting.

As noted by Robinson and Woodruff (1988), the margins of continental rift basins commonly host base-metal and barite vein deposits. Examples cited by Robinson and Woodruff include: the Late Proterozoic rift basins in the Lake Superior region of Ontario, Canada; the Cretaceous Benue Trough, Nigeria; the Paleozoic Midland Valley, Scotland; rift basins associated with the Rhine Graben, Germany; the Grenville Front, Canada; and rifts proximal to the Rhodesian craton, South Africa. Two of the more completely described settings are the Mesozoic rift basins of the eastern United States (Robinson and Woodruff 1988) and the Rio Grande rift (McLemore and Lueth 1996).

Mesozoic rift basins of the eastern United States. The Mesozoic rift basins of the eastern United States include 92 ore deposits in 14 districts extending from New Hampshire to northern Virginia (Fig. 37). This is a tectonic environment characterized by basinal subsidence and crustal extension in a continental setting. The vein deposits are spatially associated with regional faults.

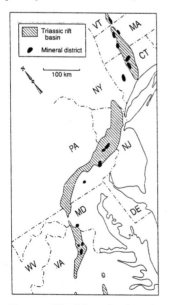

Mineralization is characterized by quartz, galena, sphalerite, chalcopyrite, minor pyrite and tetrahedrite, and local barite, fluorite, and carbonate minerals. Barite occurs as open-space fill in high-angle faults and fractures. Host rocks for the barite range from Precambrian through Jurassic and represent a wide range of lithologies, including gneiss, schist, diabase, sandstone, shale, granite, and basalt. Fluid-inclusion evidence establishes that vein mineralization was deposited from moderately saline brines having from 10 to 16 equivalent wt % NaCl with a temperature range of 100 to 250°C.

The early Mesozoic basins of the eastern United States containing red-bed sediments may be important as sources of both saline brines and leachable metals (Zielinski et al. 1983). The deposits appear to be the result of transport of brines from within the basins and adjacent basement rocks to shallow sites of mineral precipitation. Fluid migration may have been be the consequence of dynamic and changing stresses that developed during the Middle Jurassic separation of North America from Africa. Seismic pumping may have been important as a drive for fluid flow. Principal mechanisms of deposition (Robinson and Woodruff 1988) appear to have been cooling, sulfate reduction, wallrock reaction, and possibly fluid mixing. Pre-existing fractures and faults were apparently important in localizing vein development.

Figure 37. Map showing the location of sulfide-barite deposits in relation to Triassic rifts of the eastern United States. Modified from Robinson and Woodruff (1988).

Rio Grande rift deposits. According to McLemore and Lueth (1996) barite deposits of the Rio Grande rift in New Mexico (Fig. 38) formed from basinal brines having slightly higher temperatures but lower salinities than those responsible for typical MVT deposits. Paleozoic carbonate rocks host veins, breccia cements, and cavity fillings adjacent

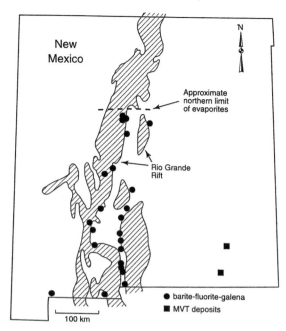

Figure 38. Map showing the distribution of Rio Grande Rift barite-fluorite-galena deposits. After McLemore and Lueth (1996).

to or within faults related to the Basin and Range. The barite deposits occur in the southern two-thirds of the rift, where there are evaporites in the section. Lead-isotope data indicate that the metals were derived from upper crustal sediments. Brines were ejected along fractures, faults, and unconformities. Fluid-inclusion homogenization temperatures and salinities vry widely. For the region as a whole, temperatures range from 95 to 341°C and salinities from 0 to 18 equivalent wt % NaCl; however, the ranges are typically more constrained within a given deposit. At the Hansonburg barite deposit, for example, homogenization temperatures range from 125-210°C and salinities from 10-18 wt % NaCl. McLemore and Lueth (1996) stated that there is no obvious association of the barite deposits with magmatic or volcanic activity, but some of the high homogenization temperatures must be related to high regional heat flow associated with extension of the rift.

Barite in late-stage thrust belts

Numerous studies have been done on barite in European orogenic belts, where the barite commonly exists in vein deposits (e.g. Burkhard 1973, Frimmel 1990, Frimmel and Papesch 1990). For example, Barbieri et al. (1982) documented the Sr geochemistry of late-orogenic epithermal barite deposits of late Triassic age in northern Tuscany, Italy. The $^{87}Sr/^{86}Sr$ values for barite range from 0.7101 to 0.7102, reflecting a large contribution of radiogenic ^{87}Sr derived from a continental source, identified by Barbieri et al. as the Verrucano Formation in the Paleozoic Basement. Phyllites of the Verrucano Formation of Paleozoic age have an initial isotopic composition of 0.7100. These metasediments also have Ba contents of up to 7000 ppm. Formation of barite is attributed to the interaction of ascending Ba-bearing solutions with sulfate-bearing evaporitic rocks. According to Barbieri et al. (1982) mineralization is genetically related to a granitic pluton that intruded sedimentary terranes during the late stage of the Apenninian orogeny. The high $^{87}Sr/^{86}Sr$ values rule out precipitation of barite either from coeval seawater Sr in which $^{87}Sr/^{86}Sr$ ranged from 0.7073 to 0.7090, or from waters that leached marine limestones of similar Sr isotopic composition. Additional Sr isotopic information on barite deposits in Sardinia and

Sicily is given by Barbieri et al. (1984, 1987).

Burkhard (1973, 1978), Frimmel (1990), and Frimmel and Papesch (1990) described the occurrence and composition of barites and celestines from the Alpine fold belt and Jura Mountains.

Dispersed cements and nodules of barite

Barite is common in minor quantities in proximity to marine evaporites, either as a direct replacement of gypsum or anhydrite or as a reaction product formed as a result of mixing of Ba-rich waters with waters derived from the dissolution of Ca sulfate. Variants on this theme are examined briefly in this section.

Dispersed barite cements, Colorado Plateau. Barite is dispersed over a 250 by 250 km area in Late Jurassic sandstones of the Morrison Formation, Colorado Plateau (Breit et al. 1990). Barite abundance ranges from <0.2 g/kg of rock to >4 g/kg. The region of highest concentration of barite coincides approximately with the areal extent of bedded evaporites in underlying Hermosa Formation of Pennsylvanian age, 2 km below. The $\delta^{34}S$ values of the barite range from 8 to 14 permil, similar to the isotopic composition of gypsum and anhydrite in the Hermosa. Breit et al. postulated that sulfate-bearing fluids moved up through faults, altered the overlying arkosic sandstone which contained Ba adsorbed on clays and in carbonate cements, and precipitated barite. Breit et al. (1990) found high Sr contents (2 to 6 wt %) in barite near faults, and low Sr (~0.4 wt %) in the western area of the Colorado Plateau where barite is sparse. In the area of the Dolores fault zone, in western Colorado, a principal feeder for ascending fluids, the combination of Sr content and Sr isotopic composition of the barite supports the hypothesis that sulfate-rich, high-Sr fluids having $^{87}Sr/^{86}Sr$ values close to 0.7100 mixed with a low-Sr source having an $^{87}Sr/^{86}Sr$ value of 0.7080 over a distance of nearly 12 km laterally from the fault zone.

Gluyas et al. (1997) documented, on the basis of the S and O isotopic composition of sulfates, the precipitation of barite and anhydrite cements in the southern North Sea, where inversion of normal faults promoted mixing of sulfate-rich waters derived from the Zechstein evaporites of the Upper Permian with Ba-rich waters derived from underlying Carboniferous coal measures. Precipitation of these cements has resulted in pervasive cementation of the Rotliegend Sandstone, causing degradation of reservoir quality and poor production of natural gas. Considerable volumes of fluid may have been involved. Savage has calculated that as much as from 2.6 km^3 to 26 km^3 of fluid containing from 5000 to 500 mg/L Ba is required to account for the barite cements found in the Amethyst Field, North Sea (David Savage, pers. comm. 2000).

Other examples of barite dispersed in sediments in the form of nodules and cements are provided by Bogoch et al. (1987), Khalaf and El Sayed (1989), Hill (1992), and Stamatakis and Hein (1993).

Salt domes. Barite is a minor, but common mineral in salt dome caprock. The concentration of Ba in most primary marine evaporites is low, and the Ba near salt has presumably been transported in from other sources. For example, Esch (1995) documented pseudomorphs of barite-celestine after anhydrite in the Eugene Island 128 salt dome, offshore Louisiana. Replacement by Sr appears to occur first, followed by conversion to nearly pure $BaSO_4$. Bedded sands surrounding the dome contain barite, among other diagenetic phases. Samuelson (1992) described pervasive barite in the caprock over the top of the Bolling salt dome in Texas. There is an abrupt boundary between the spatial distribution of barite and occurrence of hauerite (MnS_2), sphalerite, and galena, which probably marks a paleoredox front.

Table 4. Abundance of Ba and Sr in common igneous rocks and sediments.

Material	Ba, ppm	Sr, ppm	Ref.*	Material	Ba, ppm	Sr, ppm	Ref.*
Igneous Rocks				*Sediments*			
anorthosite	171	639	*1*	chert	450	250	*1*
kimberlite	847	906	*1*	phosphorite		600	*1*
gabbro	246	293	*1*	gypsum		850	*1*
diabase		175	*1*	anhydrite		2,000	*1*
diorite	714	472	*1*	halite		100	*1*
granodiorite	811	457	*1*	Holocene carbonate		6,000	*1*
granite	732	147	*1*	deep sea sed		1,000	*1*
carbonatite	3,800	2,350	*1*	Paleozoic carbonate		400	*1*
tholeiite, ridge	15	124	*1*	carbonate	90		*1*
tholeiite, island	246	329	*1*	shale	546		*1*
alkali basalt	613		*1*	sandstone	316		*1*
basalt		393	*1*	pelagic clay, Atlantic	1,260		*1*
andesite	703	442	*1*	pelagic clay, Pacific	4,160		*1*
olivine basalt	613	626	*1*				
ultramafic rocks	1	6	2	shale	580	300	2
basalt	315	452	2	sandstone	10	20	2
high-Ca granite	420	440	2	carbonate	10	610	2
low-Ca granite	840	100	2	deep sea clay	2,300	180	2

*References: *1* = Puchelt (1972), Faure (1972); 2 = Faure (1998)

BARITE OF CONTINENTAL IGNEOUS AND IGNEOUS-HYDROTHERMAL ORIGIN

Barite has been described as an accessory mineral in carbonatites, dacites, and rhyolites. Barite is also common in a wide variety of metal deposits related to continental igneous activity. Some examples are reviewed here.

Carbonatites

Both barite and celestine are common in carbonatites, which have the highest average bulk Ba and Sr content of common igneous rocks (Table 4). Carbonatitic barite is late magmatic to more commonly hydrothermal (Heinrich and Vian 1967). Barite is most common in late-generation carbonatites that are mineralogically and texturally complex, particularly those containing REE carbonates and phosphate minerals, fluorite, sulfides, and quartz. Heinrich and Vian (1967) list 23 occurrences of carbonatitic barite, including those at Magnet Cove, Arkansas, the Arkansas River area of Fremont County, Colorado, and Mountain Pass, California.

The Sr/Ba ratio of barite in carbonatites varies widely, from 0.01 to 0.43. The barite is deposited relatively late, largely after magmatic calcite and dolomite, both of which are typically both rich in Ba and Sr. The deposition of barite was followed in a few deposits by deposition of celestine and strontianite. Carbonatitic celestine occurs, for example, at Mountain Pass, California; Nano-Vara, former USSR; and Tamazeret, Morocco. Heinrich and Vian (1967) proposed the following geochemical history of Ba and Sr during formation of a carbonatite: some Ba is removed early in the crystallization history by early K-silicates, such as feldspar and micas; there is then subsequent removal of large amounts of both Ba and Sr by carbonates with preferential removal of Sr. This increases the Ba/Sr ratio of the fluid. As barite with trace amounts of Sr is deposited, the Ba/Sr ratio is reversed, leading to late-stage precipitation of celestine.

Other magmatic examples

Fine grains of barite-celestine occur in feldspar porphyries in the Late Archean Kirkland Lake alkaline igneous province (Hattori 1989). Most grains contain from 15 to 20 mol % $BaSO_4$. The $\delta^{34}S$ values are low (6.0 to7.0 permil), reflecting a magmatic source of S, and coexisting sulfides are lower. Values for $^{87}Sr/^{86}Sr$ range from 0.70271 to 0.70403, consistent with the Late Archean age of the rocks. Estimated temperatures of formation exceed 400°C. Hattori concluded that barite-celestine formed instead of anhydrite because of low bulk Ca/Sr and Ca/Ba ratios of the host intrusive and the high P_{CO_2} of the fluid. Other examples include magmatic barite in vesicles in a rhyolite porphyry in Precambrian rocks at Pilot Knob, Missouri (Frank and Moynihan 1947), in a high-K dacite from Bulgaria (Marchev 1991), and in a diorite at Ratagain, Kintail district, Scotland (May et al. 1963).

Hydrothermal barite in the Cordilleran of the western United States

Small deposits of vein barite occur widely through much of the western United States, some in association with porphyry Cu and Cu-Mo deposits, with polymetallic vein deposits (Tooker 1991), and with W vein deposits (Worl 1991). Barite deposits in California lie in a belt that is spatially coincident with the Nevadan batholithic and metamorphic belt. The time of deposition of most of the barite occurrences has not yet been established. Weber (1963), however, has suggested that the deposits may be temporally and genetically related to batholithic intrusions.

In the Rocky Mountain region, barite is largely confined to Laramide orogenic belts. Although barite shows a preference for Tertiary uplifts in the easternmost belt, not all uplifts are mineralized. The tectonic setting of the western Laramide belt has been largely obscured by complex Cenozoic subsidence and faulting. Dunham and Hanor (1967) noted that barite deposits in Laramide Tertiary uplifts are preferentially spatially associated with areas characterized by diorite-granodiorite and latite-monzonite associations, and are generally absent from basaltic provinces, the extensive trachyte-phonolite and alkaline provinces, and the spilite-keratophyre province of the western United States. Barite deposits are not spatially associated with the most Ba-rich igneous rocks. Dunham and Hanor (1967) proposed that these relations may reflect the general geochemical behavior of Ba during the crystallization of melts of various bulk compositions.

If plagioclase is the first feldspar to crystallize from a melt, then Ba is enriched in the residual melt. As long as the melt becomes saturated with water before K-feldspar or biotite begin to crystallize, much of the Ba present in the initial melt should be available for incorporation into a fluid phase. On the other hand, if K-feldspar crystallizes early and before the melt becomes saturated with water then a large portion of Ba present in the initial melt will be captured in the earliest fraction of K-feldspar to crystallize. The behavior of barium during crystallization is thus dependent on the following characteristics of the initial bulk composition of the magma: (1) Ca-content of magma. High Ca is generally associated with high Na/K ratios. Hence plagioclase crystallizes before K-feldspar. (2) Na/K ratio. If the Na/K ratio is low and Ca is low, K-feldspar crystallizes early, capturing Ba. (3) Water content. Water content and confining pressure will determine the stage at which magma becomes saturated with water and a vapor phase is evolved.

In general, most deposits of barite associated with igneous activity in the western United States are small. Some of the deposits included by Dunham and Hanor (1967) in their analysis are probably not directly related to igneous activity. For example McLemore and Lueth (1996) stated that there is no obvious association of major barite deposits in New Mexico with magmatic or volcanic activity, although some of the high homogenization temperatures in fluid inclusions may be related to high regional heat flow associated with the extension of the Rio Grande rift (Fig. 38) or with deep circulation of the ore-forming fluids. The metaliferous deposits in New Mexico that are associated with volcanic rocks include typical Pb-Zn-Cu-Ag Laramide skarn and replacement deposits formed by mixing of meteoric waters with local magmatic fluids derived from Late Cretaceous to Miocene calc-alkaline or Oligocene to Miocene alkaline igneous rocks. Barite in these deposits is only minor at best.

Barite from outer space

As cited by Chang et al. (1996), barite has been identified in chondritic interplanetary dust particles by Rietmejer (1990).

BEHAVIOR OF BARITE DURING WEATHERING, DIAGENESIS, AND METAMORPHISM

Barite is stable over a wide range of *P-T-X* and redox conditions. In the absence of a sink for sulfate, say by reduction to sulfide, barite is refractory during weathering, diagenesis, and metamorphism, although it can undergo recrystallization and remobilization. Barite is readily destroyed, however, in reducing environments. Some examples are presented below.

Formation of barite in soil environments

Barite has been identified as an authigenic mineral in some soils (Stoops and Zavaleta 1978, Darmody et al. 1989, Davis et al. 1990). Hanor et al. (1995) described the occurrence of authigenic barite in contaminated soils at a hazardous waste disposal site in southwestern Louisiana.

Subaerial weathering of barite

Not surprisingly, barite is highly resistant to subaerial weathering in oxidizing environments because of its low solubility. During weathering of carbonate-hosted MVT deposits, it is not uncommon to have near-surface host limestone and dolomite in temperate climates completely removed by dissolution, leaving behind a residuum of red clay, chert, and barite. In some instances, the concentration of residual barite is sufficient to create an economically-viable ore deposit. Well-known examples are found in southeastern Missouri, in the Cartersville District, Georgia, and in the Sweetwater District, Tennessee (Brobst 1994). Oxidation of coexisting sulfide minerals, such as pyrite, galena, and sphalerite, produces sulfuric acid, which not only enhances destruction of carbonate minerals but helps to stabilize barite through an increase in the activity of sulfate ion.

Some of the Ba dissolved during weathering may be preferentially adsorbed by clays making the Ba available for reprecipitation as secondary barite if a source of sulfate, for example by the oxidation of pyrite, is available. Strontium can be preferentially lost in the process (Hanor 1966).

Solution and reprecipitation of barite in diagenetic redox fronts

Barite is unstable in strongly reducing environments and can dissolve during diagenesis. For example, Bjorlykke and Griffin (1973) reported pseudomorphs of hyalophane after barite in carbonaceous Lower Ordovician shales of the Oslo region, Norway. Sedex deposits of massively bedded barite commonly contain beds of nodular or concretionary barite of early diagenetic origin. Examples include the Arkansas and Nevada bedded sedex deposits (Okita 1983, Graber and Chavetz 1990). The nodules typically contain pyrite as an accessory mineral, and it is likely that these nodules represent the reductive dissolution of barite by bacterial respiration and subsequent reprecipitation under more oxidizing conditions.

Clarke and Mosier (1989) described similar nodules in Devonian shale and mudstone in western Virginia. Instead of a submarine source for the Ba, however, they invoked a pelagic source, where settling particles of barite(?) were dissolved in anoxic bottom waters. Barite nodules then precipitated at the interface between the anoxic waters and overlying dysaerobic waters. The high S and O isotopic compositions of these nodules relative to Devonian seawater (Nuelle and Shelton 1986) reflect a reduced marine sulfate source.

Redox fronts have been reported in rapidly accumulating hemipelagic sediments containing high concentrations of organic matter. Torres et al. (1996b) determined that microcrystalline barite recovered by deep-sea drilling on the Peru Margin and in the Japan Sea is enriched in heavy sulfate up to a $\delta^{34}S$ value of +84 permil. Concentrations of barite of between 10 and 50% are locally present as layers within the cores. On the basis of depth profiles of dissolved barium and sulfate in interstitial pore waters, it was proposed by Torres et al. that barite remobilization is operating at several sites along the continental margins of the circum-Pacific. In the absence of advective flow, upward diffusion of Ba into the overlying sulfate-reducing zone leads to in-situ formation of barite where there is sufficient downward diffusion of sulfate to cause supersaturation. Torres et al. considered this barite to be an analog of barite in ancient rocks of similar lithology, including some sedex deposits. A diagenetic layer of barite was found by Gingele and Dahmke (1994) in sediments of the Angola slope, Africa, southwestern Atlantic ocean. McManus et al. (1998) further discuss the remobilization of Ba in continental margin sediments.

Unusually high concentrations of Ba have been reported from streams draining barite and witherite mining areas (Klein 1959). Bolze et al. (1974) showed through experiment that microbial reduction of barite can occur in systems where barite is the only source of sulfate available, and it was proposed that sulfate reduction may account for the increase in dissolved Ba content of streams that drain barite mining areas.

Behavior of barite during metamorphism

Recrystallization. During metamorphism barite commonly recrystallizes to equigranular textures resembling marble. There has been little study of chemical changes barite might undergo during diagenetic or metamorphic alteration. One exception is barite in the deposits at the Cartersville district, Georgia (Hanor 1966). Carbonate-hosted barite and surrounding sediments have been metamorphosed to amphibolite facies (Kesler and Kesler 1971) and inclusion-free grains of barite have been broken, rotated, and partly replaced by fine-grained, inclusion-rich, equigranular barite. The older barite contains from 7 to 9 mol % $SrSO_4$ (Hanor 1966). The younger barite, in contrast, contains 1.6 to 2.0 mol % $SrSO_4$. It is thus possible that Sr was preferentially removed during recrystallization.

Remobilization. Massive barite occurs in layers and pods ranging in thickness from a few centimeters to up to 4 m thick in 40-km-long zone in quartz-sericite schist and schistose pyroclastic rocks of greenschist to amphibolite facies in the Battleground Formation, Kings Mountain belt, western South Carolina. Horton (1989) suggested the deposits may have originated from seafloor hot springs, but that barite was redistributed and locally concentrated in discordant veins during regional metamorphism.

Stability in silicate systems. Potential reducing agents for barite at elevated *P-T* conditions include CH_4, H_2, CO, NH_3, and solid carbon. Calculations by Kritsotakis and von Platen (1980) showed that reduction of barite is favored by temperatures in excess of 200°C, low pH, and high gas fugacities. Despite the results of these theoretical calculations, there is abundant field evidence of the remarkable stability of barite during metamorphism.

As an example, metal-rich metasediments which include layers and lenses of massive barite, occur in graphitic quartz-muscovite schist of the Ben Eagach Schist of Lower

Cambrian age in Aberfeldy, Scotland (Coats et al. 1980, Fortey and Beddoe-Stephens 1982). The sequence was subjected to amphibolite grade metamorphism during the Caledonian orogeny. The area has been of interest because associated with the barite are barium silicates, among which are the Ba-K feldspars celsian (>80% cn) and hyalophane (25 to 40% cn), cymrite, $BaAl_2Si_2(O,OH)_8 \cdot H_2O$, and a barian muscovite.

The barite typically occurs as anhedral or polygonal to irregular crystals 1-2 mm across with textures reminiscent of marble. Coats et al. (1980) found no evidence for net input of Ba into the mineralized zone during metamorphism and no evidence for reaction between barite and other Ba-bearing rocks. The high thermal stability of barite and favorable $f_{O2}-f_{S2}$ conditions could explain the survival of barite during amphibolite grade metamorphism

Coats et al. (1990) considered the deposit to be synsedimentary exhalative and to have formed by the introduction of Ba, Fe, Zn, Pb, Mn, and Si from a brine into a black-shale environment. Reaction produced an area of sulfide-rich carbonate deposition, precipitation of barite by mixing with sulfate-bearing seawater, and either the formation of a Ba-Al-Si rich gel under reducing conditions or the precipitation of silica with a Ba zeolite. Additional reaction of Ba-rich pore fluids with clay minerals on the periphery of the deposit formed barian mica. In contrast to many sedex deposits in which barite is the dominant Ba mineral, the barite at Aberfeldy represents only a minor amount of the total Ba present in the deposit. Barium feldspars and muscovite are the product of metamorphic crystallization of Ba-rich clays and or a Ba-bearing chert formed from a Ba-Al-Si gel.

Kribek et al. (1996) described barite-hyalophane sulfide ores in the Bohemian Massif, Czech Republic, which are metamorphosed equivalents of black-shale-hosted submarine exhalative deposits. Despite high temperatures of metamorphism, 600-700°C, it is possible to reconstruct environments of both formation and metamorphism. Kribek et al. (1996) compared these deposits to submarine hydrothermal deposits of the Guaymas basin, Gulf of California, where barite, carbonates, and silica have been deposited in organic-rich sediments (Scott 1997).

CELESTINE IN SEDIMENTARY ENVIRONMENTS

The number of occurrences and modes of occurrence of celestine is far more limited than those of barite. Because of the strong preferential partitioning of Ba into the barite-celestine solid solution, celestine can form only where some process discriminates against the co-precipitation of barium or from sources that have extremely high Sr/Ba ratios. We examine here examples of both: the biological precipitation of celestine in which the organisms involved discriminate against Ba and sedimentary environments that produce fluids with high Sr/Ba ratios.

Pelagic celestine

Bernstein et al. (1992) presented an excellent discussion of the marine production of celestine by Acantharians, from which much of the following description is based. Acantharians are marine planktonic protozoans and are the only marine organism that uses $SrSO_4$ as a major structural component. Other marine organisms apparently can contain celestine, because Bernstein et al. (1992) observed numerous $SrSO_4$ crystals 10 µm long inside a colonial radiolarian from the family *Collosphaeridae*. Hollande and Martoja (1974) and Hughes et al. (1989) also described celestine occurring in radiolaria.

Acantharian life cycle. Both the skeletons (Wilcock et al. 1988) and cysts of acantharians consist of celestine. Prior to reproduction certain acantharians resorb their spines and undergo active $SrSO_4$ metabolism to produce a $SrSO_4$ cyst that surrounds the acantharian cytoplasm. The cytoplasm within the cyst eventually differentiates into numerous flagellated spores or gametes approximately 5 μm long. These subsequently exit the cyst through a pore. Bernstein et al. (1992) proposed that each spore/gamete contains one or more $SrSO_4$ granules, which possibly serve as nuclei for subsequent skeletal growth or buoyancy adjustment.

Acantharians are low on the food chain, and one mode of their demise is ingestion by other zooplankton. Celestine remains have apparently not been reported in either zooplankton gut contents or fecal material. However, siliceous remains are common. Bernstein et al. have speculated that acantharian skeletons and cysts dissolve rapidly within the predator after ingestion. Fecal material would then contain fluid having elevated concentrations of Sr, Ba, and SO_4, which might serve as a microenvironment for precipitating barite.

Abundance and fate in water column. Acantharians and their cysts are present in most of the world's oceans, and the acantharians are frequently more abundant in the euphotic zone than either radiolaria or pelagic foraminifera. Bernstein et al. (1992) found that the greater than 63 μm size fraction of particulate matter in the Sargasso Sea was composed almost exclusively of acantharian cysts. The <63 μm fraction contained numerous discrete particles of $SrSO_4$ approximately 1 to 3 μm in length which may be fragments of spores or gametes. Beers et al. (1975) reported 7200-16,000 celestine fragments/m³. Sea water is undersaturated with respect to celestine, and rapid dissolution of acantharian celestine with depth limits the export of Sr beyond intermediate depths. The Sr released is thus rapidly recycled.

Composition. The reported molar Ba/Sr ratio in acantharians ranges from 3×10^{-3} to 4×10^{-3} (Bernstein et al. 1992). These ratios are approximately an order of magnitude higher than the surface seawater molar ratio of 3.3×10^{-4}. Because Sr/Cl depletions in upper ocean are as large as 5%, there is thus the possibility of significant acantharia-mediated Ba depletion in surface waters as well (Bernstein et al. 1992).

Carbonate sediments

Carbonate sediments of pelagic origin consist primarily of calcitic forams and coccoliths and aragonitic pteropods. These sediments thus contain several times less Sr than typical shallow water marine carbonates, where Sr-enriched aragonite is commonly the predominant carbonate. Nonetheless, the submarine dissolution of pelagic carbonates and precipitation of inorganic calcite do release significant Sr into pore fluids (Sayles and Manheim 1975). Rates of release are often slow because of low rates of reaction, low rates of accumulation, or because of Sr uptake by underlying basement rocks. Where carbonate accumulation rates are high, a higher fraction of reactive carbonate reaches the seafloor. Under some conditions, dissolved interstitial Sr/Ca reaches a plateau value that is approximately 25 times the Sr/Ca ratio of the biogenic calcite (Baker and Bloomer 1988)

Baker and Bloomer (1988) have described celestine nodules of unspecified size and morphology recovered at several sites on DSDP Leg 90 from the Lord Howe Rise in the southwest Pacific. The host sediments are predominately very pure calcareous oozes and chalks of Middle Miocene to Early Pliocene age which have undergone relatively rapid burial diagenesis. This has resulted in the expulsion of Sr from biogenic calcite into the ambient interstitial waters. As a result of microbial sulfate reduction, dissolved sulfate decreases down-core. The down-core increase in Sr, however, is proportionally higher than the sulfate decrease, and celestine precipitates at depths of approximately 100 m below the seafloor at each of four sites. The nodular celestine contains 4 to 8 mol % $BaSO_4$, 2-6 mol % $CaSO_4$, 1100-2,600 ppm Al and 400-750 ppm K. The isotopic composition of this celestine has not yet been determined. Calculations of saturation state showed that the porewaters are saturated with respect to celestine and are undersaturated with respect to barite.

According to Kinsman (1969), celestine is a common early diagenetic mineral of coastal sabkas and is abundant in areas of intense dolomitization. The celestine forms during replacement of aragonite (7000-8000 ppm Sr) by dolomite (600-700 ppm Sr). Examples include the Persian Gulf, Baja California, and shallow lagoons in southwestern Australia.

Celestine and salt domes. Celestine is a common accessory mineral in the caprocks of salt domes (Saunders et al. 1988), where it can exist as a replacement of anhydrite or gypsum. Esch (1995), in a study of diagenesis at the salt-sediment interface in an offshore Louisiana salt dome, observed celestine replacing anhydrite within a matrix of halite. Some of the celestine pseudomorphs are partly replaced by barite.

Celestine in coastal carbonate-evaporite sequences

General features. The largest and most massive occurrences of celestine are invariably in coastal carbonates and evaporites, usually with up-dip or overlying clastics (Scholle et al. 1990). Host rocks range in age from Silurian to Miocene. There is variable evidence of associated hydrothermal activity. Lower Permian deposits in the former USSR and Cretaceous deposits in Mexico have associated barite, but most of others do not. Six of the largest known celestine deposits have proven reserves of 0.5 to 8 Mt. These include (Scholle et al. 1990): Montevive, Spain; Molkabade, Iran; San Augustin, Mexico; Cifteagil, Turkey; Yate, England (Nickless et al. 1975, Wood and Shaw 1976); and Lake Enon, Nova Scotia, Canada (Andrews and Collins 1991). Current principal producers are Spain, Turkey, Mexico, Iran, and the United Kingdom.

Example: East Greenland. Scholle et al. (1990) provided an instructive description of a major celestine deposit in central eastern Greenland. An estimated 25-50 Mt of celestine-bearing sediment having an average grade of 50-60% SrSO$_4$ occur in Upper Permian algally laminated limestones, limestone breccias, and karstified units that surround an isolated gypsum deposit of probable salina origin. This deposit has not yet been mined because of its remote location.

Deposition of the host sediments took place in shallow subtidal to supratidal environments including evaporitic hypersaline lagoons and algal shoals. The main occurrences of celestine are confined to limestone in the proximity of the evaporite lagoon. Here, celestine replaces calcite, dolomite, and gypsum in algally laminated sediment and as replacement of algally laminated limestone clasts as cements, veins, and cave fillings in overlying conglomerate and karst sediments. Celestine almost always appears to be the last stage of replacement of rocks that are dominated by early diagenetic features.

Strontium isotopic ratios of the celestine range from 0.7130 to 0.7137, and all celestine formed roughly at same time and from same pore fluids. Seawater coeval with the host sediments had an [87]Sr/[86]Sr ratio of 0.7068 to 0.7070, and the highly radiogenic Sr cannot be the result of diagenetic alteration of associated aragonitic limestone. Gypsum also shows slight enrichment in radiogenic Sr (0.7074 to 0.7082) relative to contemporaneous seawater values, possibly the result of diagenetic alteration.

The δ^{34}S isotopic composition of celestine is 15.0 to 19.6 permil and is considerably heavier than that of the spatially associated gypsum deposits, which range from 9.1 to 11.4 permil. These values seem to be beyond the range of the effects of simple isotopic fractionation expected during dissolution of gypsum and precipitation of celestine. Isotopically heavier S has also been noted in Mexican celestine deposits by Kesler and Jones (1981), who invoked possible inorganic fractionation or limited enrichment of gypsum sulfate by bacterial reduction. Halas and Mioduchowski (1978) have invoked bacterial reduction of sulfate in similar celestine deposits.

Scholle et al. (1990) call on precipitation of celestine by mixing of near-surface meteoric waters which had been involved in calcitization or dissolution of evaporites, and thus were rich in sulfate, with rising basinal waters rich in Sr. Such a mechanism was also invoked by Kesler and Jones to explain the more radiogenic Mexican deposits. Other workers have invoked the continental groundwater alteration of limestones to supply Sr.

A radically different mechanism for forming celestine deposits was proposed by Müller (1962), who called on direct evaporation of seawater for the origin of the Himmelte West celestine deposit of Germany, which contains several Mt of celestine. According to Muller, saturation with respect to celestine is reached during progressive evaporation of sea water near the boundary between precipitation of carbonates and gypsum. Brodtkorb et al. (1982) discussed other examples of the possible synsedimentary precipitation of celestine by evaporation of seawater. The difficulty here would seem to be accounting for rock celestine/gypsum ratios significantly higher than seawater Sr/Ca ratios. Additional information on celestine deposits is provided by Carlson (1987), Kushnir (1986). Mitchell and Pharr (1961), Brodtkorb (1989), and Ramos and Brodtkorb (1989, 1990).

Sedimentary venues for forming waters of high Sr/Ba ratios and celestine deposits. The association of large celestine deposits with coastal carbonate-evaporite sequences most likely reflects the difference in geochemical behavior of Sr and Ba in subsurface fluids. As noted earlier in this chapter, the concentration of Sr in subsurface sedimentary fluids is silicate/carbonate-buffered such that the concentration of Sr increases expotentially with increases in salinity. The concentration of Ba, in contrast, is controlled by the concentration of sulfate. The fluids of highest Sr/Ba ratio, and thus those most likely to form celestine rather than barite, are those of high salinity and high sulfate relative to Ba (Figs. 9, 10). Such waters could be produced in coastal evaporitic settings where residual evaporated seawater is saline with appreciable levels of dissolved sulfate. Such waters, if refluxed into underlying sediments, would leach appreciable Sr as the fluids attempted to achieve equilibrium with the host sediments. If such fluids were then discharged back up into shallow gypsum or anhydrite beds, the precipitation of celestine could result. As noted earlier, decreasing temperature and pressure favors celestine stability over anhydrite (Fig. 13).

ENVIRONMENTAL ASPECTS

Barium in potable water supplies

Barium toxicity. High levels of Ba in the human body are potentially toxic. According to Gregus and Klaassen (1996), Ba^{2+} interferes with signaling mechanisms for neurotransmitters by inhibiting Ca^{2+}-activated K^+ channels in cell membranes. This is accompanied by potentially lethal neuroexcitatory and spasmodic effects. Poisoning associated with ingestion of soluble barium salts has resulted in gastroenteritis, muscular paralysis, decreased pulse rate, and ventricular fibrillation and extra-systoles (Goyer 1974). Stokinger and Woodward (1958) calculated a safe threshold of 2.0 mg/L of dissolved Ba in drinking water on the basis of threshold limiting values in industrial atmospheres, an estimate of the amount absorbed into the blood stream, and daily consumption of two liters of water. However, to provide a safety factor and to allow for possible accumulation in the body, a limit of 1.0 mg/L was intially recommended. The limit has recently been reset to 2.0 mg/L (Fig. 7).

Barium sulfate is sufficiently insoluble that it is routinely used as a radioplaque aid to X-ray diagnosis. Occupational poisoning by Ba is uncommon, but benign pneumoconiosis (baritosis) may result from inhalation of Ba sulfate (Goyer 1974). Baritosis is not incapacitating and is usually reversible with cessation of exposure.

Strontium apparently does not have the same toxic effects as Ba, and in fact, strontium chloride is routinely used to desensitize teeth in humans. More serious is the potential uptake of radioactive ^{90}Sr by bone. This isotope has a half life of 21.8 years, however, and has not been reported from naturally occurring celestine deposits, although it could coprecipitate with recent barite.

Barium in drinking water and in municipal wastes. Buffering by barite is an important control on the concentration of Ba in drinking-water supplies (Fig. 7). Dissolved Ba in concentrations of up to 18 mg/L occur in portions of freshwater aquifers of northeastern Illinois (Gilkeson et al. 1981). The area is anomalously depleted in dissolved sulfate as a result of reduction under anaerobic conditions. The reduction of sulfate is accompanied by the presence of dissolved H_2S, increased alkalinity and enrichment of residual sulfate in ^{34}S.

Sewage sludge in Florence, Italy, contains high levels of total Ba (0.4 to 1.6 mg/g) as a result of disposal of barite in hospital wastes (Baldi et al. 1996). Under the reducing conditions, the barite is solubilized and the Ba is partly reprecipitated, probably as witherite ($BaCO_3$) and as BaS.

Barite and the uranium industry

Waters from U mine wastes typically contain an array of radioactive nuclides, including ^{230}Th, ^{226}Ra, ^{210}Pb, and ^{210}Po. According to Paige et al. (1998) ^{226}Ra is the radionuclide responsible for more than 75% of the radioactivity originally present in the ore. The precipitation and the dissolution of Ra-bearing barite are controlling factors in the behavior and mobility of Ra in uraniferous mine wastes. Precipitation of Ba-Pb sulfates on mica and quartz grains is a process which influences the distribution and mobility of ^{210}Pb (Paige et al. 1998).

Beard et al. (1980) described the use of $BaCl_2$ in some U milling operations in Colorado, New Mexico, and Utah to reduce the Ra content of mine and processing wastewaters by coprecipitation with barite. Removal efficiencies ranged from 25 to 99.98% of the dissolved Ra originally present.

Margaritz et al. (1990) used the distribution of dissolved Ba and Sr as proxies of radioactive isotopes to study the potential behavior of nuclear wastes disposed of in the unsaturated zone and/or near the water table. They found a complex distribution of Ba and Sr and a transition from waters saturated to undersaturated with respect to barite over a vertical interval of only 50 cm.

Problems related to oil and gas production

Problems with scaling by barite. Clogging and plugging of production casing by barite scale is a common phenomenon in the oil and gas industry (Weintritt and Cowan 1967). Barite precipitation can be induced during water-flooding operations, particularly if seawater, which contains high concentrations of sulfate, mixes with native formation waters containing high concentrations of Ba. Other mechanisms for precipitating barite include mixing of native formation waters of varying Ba and sulfate concentrations and decreases in temperature and pressure that accompany transport of fluid to the surface. There has been considerable interest in the development of techniques for inhibiting the formation of barite scale, and much of the work done on nucleation and the kinetics of crystal growth of barite (Ali and Nancollas, this volume) stems from these problems.

Radioactive barite scale. Water produced from oil, gas, and geothermal wells can have radioactivity levels approaching those in U mill tailings (Fisher 1995, 1998). The chief contributors to this naturally occurring radioactive material (NORM) are ^{226}Ra and ^{228}Ra, which are ultimately derived from the decay of ^{238}U and ^{232}Th, respectively (Fig. 20) and which occur as Ra^{2+} in solution. Because of the similarity in ionic radius and electronegativity to Ba^{2+}, some Ra^{2+} is removed from aqueous solution during the precipitation of barite scale in oil field production and processing equipment. In an extensive study of NORM wastes in the hydrocarbon producing fields in Texas, Fisher (1995, 1998) found that ^{226}Ra levels typically ranged from approximately one to several thousand picocuries per liter. The levels of ^{228}Ra are comparable. Highest dissolved Ra values were found in waters whose salinity exceeded 35,000 mg/L or were derived from reservoirs in which volcanic rock fragments are abundant. Well scale contained from 600 to 2700 pCi/g ^{226}Ra and 400 to 4000 pCi/g ^{232}Th. Radium activity correlates with the abundance of Ba in the well scale. Pardue and Guo (1998) reported on the biochemistry of ^{226}Ra in contaminated bottom sediments in oil-field waste pits.

Produced water. Produced water released into the environment from oil and gas operations is likely to have substantial concentrations of dissolved Ba and Ra. Mixing of produced waters with ambient groundwaters containing sulfate has the potential for coprecipitating Ba and Ra as radioactive barite and thus creating local areas of elevated radioactivity and radon emission (Fig. 20).

Disposal of barite drilling muds. Dial and Huff (1989) reported levels of dissolved Ba above background in water wells spatially associated with oil-field activities in Ascension Parish, Louisiana. Maximum levels of dissolved Ba, however, were 800 µg/L. These authors cite reports of contamination within two years of when drilling operations began. It is not known whether the increased Ba was due to disposal of produced waters high in Ba or from dissolution of barite drilling muds or cuttings.

Because of the current low cost of barite drilling muds and the cost of separating barite from rock fragments in the cuttings produced by drilling, much barite used in drilling is not recycled. Most wastes in offshore environments are discharged directly into seawater.

In an extensive U.S. Environmental Protection Agency study of barite drilling mud disposal pits at a site in south Louisiana the sole potential risk identified during the site assessment was to future residential children who might ingest the barite wastes on a regular basis. Because potential risks associated with agricultural use of the property were not quantified, it was determined that the property can not be used for farming or grazing purposes. This means that future use of the property is restricted to industrial purposes.

An example of a "bad rap" for barite

Many scientists and some of the general public are aware of the potential toxicity of Ba. Many are also aware that barite is used in oil and gas operations. Fewer are aware, however, that Ba is between the 12th or 15th most abundant element in the Earth's crust, and that the presence of Ba in rocks and sediments at levels of several hundred ppm is not surprising.

Isphording (1982) provides an instructive example of a case where the public, the oil industry, and state and Federal governments became involved in protracted controversy resulting from the discovery of Ba at levels of several hundred ppm in offshore sediments near the mouth of Mobile Bay, Alabama. This Ba was attributed by some to spillage of barite drilling mud used in offshore oil and gas operations, and concern was raised regarding pollution, wetlands damage, and the destruction of marine life. Subsequent determination (Isphording 1982) of the partitioning of Ba between various minerals of the sediment, however, showed the Ba occurred associated with clay minerals and as a trace element in biogenic carbonate shells. No barite was detected by X-ray diffraction.

CONCLUDING REMARKS

The author's Ph.D. thesis was entitled "The Origin of Barite" (Hanor 1966). The conclusions of that work were based on thermodynamic information available at the time, a new field-sampling and geochemical study of barite deposits in the midwestern and eastern United States, and a synthesis and re-interpretation of the available literature on barite deposits in North America. Because of the major advances in geology which have taken place since 1966, it is now possible in the year 2000 to make much more definitive interpretations regarding the diverse origins of barite and celestine.

These advances include the development of conceptual models and quantitative techniques in plate tectonics, deep-basin hydrogeology, Sr-isotope systematics, compositional systematics and thermodynamics of subsurface brines, ore-mineral age dating, and marine geochemistry, and the direct observation and sampling of deep-sea hydrothermal sites. Critically important also has been the general acceptance among geologists since the mid-1960s of the basic concepts of synsedimentary ore deposition. It is hoped that this chapter has provided an overview of current thought on the properties, occurrence, and origin of barite and celestine as of the year 2000.

Some key areas for future work on the occurrence and origin of barite and celestine include: the conditions under which Ba and Sr are leached from shallow crustal rocks and sediments, the generation and expulsion of barite- and celestine-forming fluids in extensional terranes, the potential significance of the occurrence of major concentrations of barite and celestine as paleo-focal points for discharge of deep fluids, mass-balance constraints to determine much fluid was involved, the factors that result in metal-rich and or metal-poor barite deposits, the mechanisms and rates of precipitation of pelagic barite today and in the geological past, importance of redox fronts in concentrating massive quantities of barite, what Precambrian barite and celestine occurrences can tell us about the early history of the Earth, the role of barite equilibria in determining the concentrations of Ba and Ra in oil and gas exploration and production sites, and the development of new exploration strategies for the location of additional deposits of these economically important minerals.

ACKNOWLEDGMENTS

I thank editors Charlie Alpers and John Jambor and external reviewers George Breit, Christophe Monnin, and Fred DeHairs for their helpful reviews of this chapter. My professional interest in barite goes back to my graduate student days at Harvard, where Ray Siever, Ulrich Peterson, Cliff Frondel, and Bob Garrels helped to guide me along. Further insight was provided by Kurt Boström, Gustav Arrhenius, and Tom Church during a subsequent NSF postdoctoral fellowship at Scripps. My LSU faculty colleague, Lui Chan, and my former graduate students Larry Baria, Larry Bunting, Pat Okita, Kevin Howard, Karen Graber, and Lee Esch have helped me sustain an active interest in heavy divalent cations during my tenure at Louisiana State University. My more recent undergraduate and graduate student research assistants, Kathleen (KT) Moran, Katie Cooper, Romie Coronado, and Kimberly Noble, helped me track down references and put things in order during the preparation of this paper. The recent work on fluids in sedimentary basins reported on here was supported in part by NSF grants EAR 98-05446 and EAR 98-05459.

It is a pleasure to be able to dedicate this chapter to my wife, Leslie, who has been with me since the beginning of all of this.

REFERENCES

Andrew JE, Funnel BM, Jickells TD, Shackleton NJ, Swallow JE, Williams AC, Young KA (1990) Preliminary assessment of cyclic variations in foraminifers, barite, and cadmium/calcium ratios in early

Pleistocene sediments from Hole 709C (equatorial Indian ocean). Proc ocean Drilling Program, Scientific Results 115:611-619

Andrews PRA, Collins RK (1991) Celestine in Canada. Can Inst Mining Metall Bull 84:36-39

Aquilina L, Dia AN, Boulegue J, Bourgois J, Fouillac AM (1997) Massive barite deposits in the convergent margin off Peru: Implications for fluid circulation. Geochim Cosmochim Acta 61:1233-1245

Arrhenius G (1952) Sediment cores from the East Pacific. Rept Swedish Deep Sea Exp, 1947-48, 5:1-120

Arrhenius G (1963) Pelagic sediments. In The Sea, vol 3. MN Hill (Ed) Interscience, New York, p 655-727

Arrhenius G, Bonatti E (1965) Neptunism and vulcanism in the ocean. In Progress in Oceanography. M Sears (Ed) Pergamon Press, New York, p 7-22

Arrhenius G, Bramlette M, Picciotto E (1957) Localization of radioactive and stable heavy nuclides in ocean sediments. Nature 180:5-86

Baker PA, Bloomer SH (1988) The origin of celestite in deep-sea carbonate sediments. Geochim Cosmochim Acta 52:335-339

Baldi F, Pepi M, Burrini D, Kniewald G, Scali D, Lanciotti E (1996) Dissolution of barium from barite in sewage sludges and cultures of Desulfovibrio desulficans. Appl Environ Microbiol 62:2398-2404

Barbieri M, Masi U, Tolomeo L (1982) Strontium geochemistry in the epithermal barite deposits from the Apuan Alps (Northern Tuscany, Italy). Chem Geol 35:351-356

Barbieri M, Masi U, Tolomeo L (1984) Strontium geochemical evidence for the origin of the barite deposits from Sardinia, Italy. Econ Geol 79:1360-1365

Barbieri M, Ballanca A, Neri R, Tolomeo L (1987) Use of strontium isotopes to determine the sources of hydrothermal fluorite and barite from northwestern Sicily (Italy). Chem Geol 66:273-278

Barley ME (1992) A review of Archean volcanic-hosted massive sulfide and sulfate mineralization in Western Australia. Econ Geol 87:855-872

Bautier M, Monnin C, Fruh-Green G, Karpoff AM (in press) Fluid-sediment interactions related to hydrothermal circulation in the Eastern flank of the Juan de Fuca ridge. Chem Geol

Beard, HR Salisbury, HB and Shirts, MB (1980) Absorption of radium and thorium from New Mexico uranium mill tailing solutions. US Bureau of Mines Rept Inv 8463, 14 p

Beers JR, Reid FMH, Stewart J (1975) Microplankton of the north Pacific central gyre. Population structure and abundance, June 1973. Int'l Rev Hydrobiology 60:607-638

Bernstein RE, Byrne RH, Betzer PR, Greco AM (1992) Morphologies and transformations of celestine in seawater: The role of acantharians in strontium and barium geochemistry. Geochim Cosmochim Acta 56:3273-3279

Bernstein RE, Byrne RH, Schijf J (1998) Acantharians: A missing link in the oceanic biogeochemistry of barium. Deep-Sea Res 45:491-505

Bertine KK, Keene JB (1975) Submarine barite-opal rocks of hydrothermal origin. Nature 188:150-152

Bertram MA, Cowen JP (1997) Morphological and compositional evidence for biotic precipitation of marine barite. J Mar Res 55:577-593

Bishop JKB (1988) The barite-opal-organic carbon association in oceanic particulate matter. Nature 332:341-343.

Bjorlykke KO, Griffin WL (1973) Barium feldspars in Ordovician sediments, Oslo region, Norway. J Sed Pet 43:461-465

Blount CW (1977) Barite solubilities and the thermodynamic quantities up to 300 degrees Celsius and 1400 bars. Am Mineral 62:942-957

Bogoch R, Buchbinder B, Nielsen H (1987) Petrography, geochemistry, and evolution of barite concretions in Eocene pelagic chalks from Israel. J Sed Pet 57:522-529

Bolze CE, Malone PG, Smith MJ (1974) Microbial mobilization of barite. Chem Geol 13:141-143

Borole DV, Somayajulu (1977) Radium and lead-210 in marine barite. Marine Chem 5:291-296

Bosbach D, Haal C, Putnis A (1998) Mineral Precipitation and dissolution in aqueous solution: in-situ microscopic observation on barite (001) with atomic force microscopy. Chem Geol 151:143-160

Boström K, Peterson MNA (1966) Precipitates from hydrothermal exhalations on the East Pacific Rise. Econ Geol 61:1258-1265

Boström K, Frazer J, Blankenburg J (1967) Subsolidus phase relations and lattice constants in the system $BaSO_4$-$SrSO_4$-$PbSO_4$. Arkiv Mineral Geol 4:477-485

Boström K, Joensuu O, Moore C, Bostrom B, Dalziel M, Horowits A (1973) Geochemistry of barium in pelagic sediments. Lithos 6:159-174

Bowers TS, Jackson KJ, Helgeson HC (1984) Equilibrium Activity Diagrams. Springer-Verlag, Berlin, 397 p

Breit G, Goldhaber MB, Martin B, Shawe D, Simmons EC (1990) Authigenic barite as an indicator of fluid movement through sandstones within the Colorado Plateau. J Sed Pet 60:884-896

Brobst DA (1994) Barium minerals. *In* Industrial Minerals and Rocks (6th ed.). DD Carr (Ed) Soc Mining Metall Explor, Littleton, Colorado, p 125-134

Brodtkorb de MK (1989) Celestite: worldwide classical ore fields. *In* Nonmetalliferous Stratabound Ore Fields. MK de Brodtkorb (Ed) Van Nostrand, New York, p 17-39

Brodtkorb de MK, Ramos V, Barbieri M, Ametrano S (1982) The evaporitic celestite-barite deposits of Neuquen, Argentina. Mineral Dep 17:423-426

Brook AJ, Fotheringham A, Bradly J, Jenkins A (1980) Barium accumulation by desmids of the genus *Closterium* (Zygnemaphyceae). British Phycology J 15:261-264

Brook AJ, Grime GW, Watt F (1988) A study of barium accumulation in desmids using the Oxford scanning proton microprobe (SPM). Nuclear Instr Meth Physics Res B30:372-377

Brower E, Renault J (1971) Solubility and enthalpy of barium-strontium sulfate solid solution series. New Mexico Bureau of Mines and Mineral Resources Circular 116, 21 p

Buick R (1990) Microfossil recognition in Archean rocks: an appraisal of spheroids and filaments from a 3500 m.y. old chert barite unit at North Pole, Western Australia. Palaios 5:411-459

Buick R, Dunlop JSR (1990) Evaporitic Sediments of Early Archean age from the Warrawoona Group, Western Australia. Sedimentology 37:247-278

Burke WH, Denison RE, Hetherington EA, Koepnik RB, Nelson HF, Otto JB (1982) Variation of seawater $^{87}Sr/^{86}Sr$ throughout Phanerozoic time. Geology 10:516-519

Burkhard A (1973) Optische und röntgenographische Untersuchungen am System $BaSO_4$-$SrSO_4$ (Baryt-Celestin). Schweiz mineral petrogr Mitt 53:185-197

Burkhard A (1978) Baryt-Celestin und ihre Mischkristalle aus Schweizer Alpen und Jura. Schweiz mineral petrogr Mitt 58:1-95

Buschendorf F, Nielsen H, Puchelt H, Ricke W (1963) Schwefel-isotopen-Untersuchen am Pyrit-Sphalerit-Baryt-Lager, Meggen/Lenne (Deutschland) und an verschiedenen Devon-Evaporiten. Geochim Cosmochim Acta 27:501-523

Cameron EM, Hattori K (1985) The Hemlo gold deposit: A geochemical and isotopic study. Geochim Cosmochim Acta 49:2041-2050

Carlson EH (1987) Celestine replacements of evaporites. Sed Geol 54:93-112

Carroll JL, Falkner KK, Brown ET, Moore WS (1993) The role of the Ganges-Brahmaputra mixing zone in supplying Ba and ^{226}Ra to the Bay of Bengal. Geochim Cosmochim Acta 57:2981-2990

Cathles LM (1997) Thermal aspects of ore formation. *In* Geochemistry of Hydrothermal Ore Deposits (3rd Edn) HL Barnes (Ed) Wiley, New York, p 191-228

Cecile MP, Goodfellow WD, Jones LD, Krouse HR, Shakur MA (1984) Origin of radioactive sinter, Flybye springs, Northwest Territories, Canada. Can J Earth Sci 21:383-395

Cecile MP, Shakur MA, Krouse HR (1983) The isotopic composition of western Canadian barites and the possible derivation of oceanic sulphate $\delta^{34}S$ and $\delta^{18}O$ age curves. Can J Earth Sci 20:1528-35

Chagon A, St.-Antoine P, Savard MM, Heroux Y (1998) Impact of Pb-Zn sulfide precipitation on the clay mineral assemblage in the Gays River deposit, Nova Scotia, Canada Econ Geol 93:779-792

Chan LH, Edmond JH (1987) Barium. *In* GEOSECS Atlantic, Pacific and Indian Ocean Expeditions, Vol 7, Shore-based Data and Graphics. H Ostlund, H Craig, W Broecker, D Spencer D (Eds) Nat'l Sci Fdn, Washington, DC, p 13

Chan LH, Drummond D, Edmond JH, Grant B (1977) On the barium data from the Atlantic GEOSECS expedition. Deep Sea Res 24:613-649

Chang LLY, Howie RA, Zussman J (1996) Rock-forming Minerals, Second Edition, Vol 5B, Non-silicates: Sulfate, Carbonates, Phosphates, Halides. Longman Group, Harlow, UK, 383 p

Chester R (1990) Marine Geochemistry. Chapman & Hall, London, 698 p

Chow TJ, Goldberg ED (1960) On the marine geochemistry of barium. Geochim Cosmochim Acta 20:192-198

Christy AG, Putnis A (1993) The kinetics of barite dissolution and the precipitation in water and sodium chloride brines at 44-85 degrees Celsius. Geochim Cosmochim Acta 57:2161-2168

Church TM (1970) Marine Barite. PhD dissertation, University of California San Diego, 100 p

Church TM (1979) Marine barite. Rev Mineral 6:175-209

Church TM, Bernat M (1972) Thorium and uranium in marine barite. Earth Planet Sci Lett 14:139-144

Church TM, Wolgemuth K (1972) Marine barite saturation. Earth Planet Sci Lett 15:35-44

Clark SHB, Mosier EL (1989) Barite nodules in Devonian shale and mudstone of western Virginia. US Geol Survey Bull 1880, 30 p

Clark SHB, Gallagher MJ, Poole FG (1990) World barite resources: A review of recent production patterns and genetic classification. Trans Inst Mining Metall B-Appl 99:B125-B132

Claypool GE, Holser, WT, Kaplan, IR, Sakai, H, Zak, I (1980) The age curves of sulfur and oxygen in marine sulfate and their mutual interpretations. Chem Geol 28:199-260

Coats JS, Smith CG, Fortey NJ, Gallegher MJ, May F, McCourt WJ (1980) Strata-bound barium-zinc mineralization in Dalradian schist near Aberfeldy, Scotland. Trans Inst Mining Metall 89:B110-B122

Coffey M, Dehairs F, Collette O, Luther G, Church T, Jickells T (1997) The behaviour of dissolved barium in estuaries. Estuarine Coastal Shelf Sci 45:113-121

Connolly CA, Walter LM, Baadsgaard H, and Longstaffe FJ (1990) Origin and evolution of formation waters, Alberta Basin, Western Canada sedimentary basin. Appl Geochem 5:397-413

Cooke DR, Bull SW, Donovan S, Rogers JR (1998) K-metasomatism and the base metal depletion in volcanic rocks from the McArthur basin, Northern Territory—implications for base metal mineralization. Econ Geol 93:1237-1265

Cooke DR, Bull SW, Large RR, McGoldrick PJ (2000) The importance of oxidized brines for the formation of Australia Proterozoic stratiform sediment-hosted Pb-Zn (sedex) deposits. Econ Geol 95:1-18

Cortecci G, Longinelli A (1972) Oxygen isotope variations in a barite slab from the sea bottom off southern California. Chem Geol 9:113-117

Darmody RG, Hardin, SD, Hassett JJ (1989) Barite authigenesis in surficial soils of mid-continental United States. In Water-Rock Interaction (WRI-6). DL Miles (Ed) Balkema, Rotterdam, 183-186

Davis GR, Hale M, Dixon CJ, Bush PR, Wheatley MR (1990) Barite dispersion in drainage sediments in arid climatic regimes. Trans Inst Mining Metall 99:15-20

Dean WE, Schreiber BC (1977) Authigenic barite, Leg 41 Deep Sea Drilling Project. Initial Rept Deep Sea Drilling Proj 41:915-931

de Lange GJ, Boelrijk NAIM, Catalano G, Corselli, C, Klinkhammer G (1990a), Sulphate-related equilibria in the hypersaline brines of the Tyro and Bannock Basins, eastern Mediterranean. Mar Chem 31:89-112

de Lange GJ, Middleburg JJ, van der Weijden CH, Catalano G (1990b) Composition of anoxic hypersaline brines in the Tyro and Bannock Basins, eastern Mediterranean. Mar Chem 31:63-80

Deb M, Hoefs J, Baumann A (1991) Isotopic composition of two Precambrian stratiform barite deposits from the Indian Shield. Geochim Cosmochim Acta 55:303-308

Degens ET and Ross DA (1969) Hot brines and recent heavy metal deposits in the Red Sea. Springer-Verlag, New York, 100 p

Dehairs F, Baeyens W, Goeyens L (1992) Accumulation of the suspended barite at mesopelagic depths and export production in the southern-ocean. Science 258:1332-1335

Dehairs F, Chesselet R, Jedwab J (1980) Discrete suspended particles of barite and the barium cycle in the open ocean. Earth Planet Sci Lett 49:528-550

Dejonghe L (1990) The sedimentary structures of barite: examples from the Chaudfantaine ore deposit, Belgium. Sedimentology 37:303-323

Dejonghe L, Boulvain F (1993) Paleogeographic and diagenetic context of a baritic mineralization enclosed within Frasnian peri-reefal formations—case-history of the Chaudfontaine mineralization (Belgium). Ore Geol Rev 7:413-431

Denis J, Michard G (1983) Dissolution d'une solution solides etude theoretique et experimentale. Bull Minéral 106:309-319

Des Marais DJ (1994) The Archean atmosphere: its composition and fate. In Archean Crustal Evolution. KC Condie (Ed) Elsevier, Amsterdam, p 505-523

DeWit MJ, Hart R, Martin A, Abbott P (1982) Archean abiotic and probable biogenic structures associated with mineralized hydrothermal vent systems and regional metasomatism, with implications for greenstone belt studies. Econ Geol 77:1783-1802

Dia A, Aquilina L, Boulegue J, Bourgois J, Suess E, Torres M (1993) Origin of fluids and related barite deposits at vent sites along the Peru convergent margin. Geology 21:1099-1102

Dial DC, Huff GF (1989) Occurrence of minor elements in ground water in Louisiana including a discussion of three selected sites having elevated concentrations of barium. Louisiana Dept Trans Water Res Tech Rept 47, 88 p

Dube TE (1986) Depositional setting of exhalative bedded barite and associated submarine fan deposits of the Roberts Mountains. Geol Soc Am Bull 18:259-267

Dube TE (1988) Tectonic significance of Upper Devonian igneous rocks and bedded barite, Roberts Mountains Allochton, Nevada, USA. Mem Can Soc Pet Geol 14:235-249

Dugolinsky BK, Marhoils SV, Dudley WC (1977) Biogenic influence on the growth of manganese nodules. J Sed Pet 47:428-445

Dunham AC, Hanor JS (1967) Controls on barite mineralization in the western United States. Econ Geol 62:82-94

Dymond J, Collier R (1996) Particulate barium fluxes and their relationships to biological productivity. Deep-Sea Res Part II 43:1283-1308

Dymond J, Suess E, Lyle M (1992) Barium in deepsea sediments: a geochemical proxy for paleo-productivity. Paleooceanography 7:163-181

Edgerton D (1997) Reconstruction of the Red Dog Zn-Pb-Ba ore body, Alaska, implications for the vent environment during the mineralization event. Can J Earth Sci 34:1581-1602

Edmond JM, Boyle ED, Drummond D, Grant B, Mislick T (1978) Desorption of barium in the plume of the Zaire (Congo) River. Netherlands J Sea Res 12:324-328

Edmond JM, Measures C, McDuff RE, Chan LH, Collier R, Grant B, Gordon LI, Corliss, JB (1979) Ridge crest hydrothermal activity and the balance of major and minor elements in the ocean: The Galapagos data. Earth Planet Sci Lett 46:1-18

Eriksson PG, Condie KC Tirsgaard H, Mueller WU, Altermann W, Miall AD, Aspler LB, Catuneanu O, Chiarenzelli C (1998) Precambrian clastic sedimentation systems. Sed Geol 120:5-53

Esch WL (1995) Effects of salt dome dissolution on sediment diagenesis: an experimental and field study. PhD Dissertation, Louisiana State University, Baton Rouge, 219 p

Falkner KK, Klinkhammer GP, Bower TS, Todd JF, Leweis BL, Landing WM, Edmond JM (1993) The behavior of barium in anoxic marine waters. Geochim Cosmochim Acta 57:537-554

Farrell CW, Holland HD (1983) Strontium isotope geochemistry of the Kuroko deposits. Econ Geol Monogr 5:302-319

Faure G (1972) Strontium. *In* Handbook of Geochemistry. KH Wedepohl (Ed) Springer-Verlag, New York, p 38B1-38B35

Faure G (1998) Principles and Applications of Geochemistry. Prentice Hall, Upper Saddle River, New Jersey, 600 p

Felmey AR, Rai D, Moore DA (1993) The solubility of $(Ba,Sr)SO_4$ precipitates: thermodynamic equilibrium and reaction path analysis. Geochim Cosmochim Acta 57:4345-4363

Fenchel T, Finlay BJ (1984) Geotaxis in the ciliated protozoon *Loxodes*. J Exper Biol 110:17-33

Fernandez-Diaz L, Putnis A, Cumberbatch J (1990) Barite nucleation kinetics and the effect of additives. Eur J Mineral 2:495-501

Finlay BJ, Hetherington NB, Davison W (1983) Active biological participation in lacustrine barium chemistry. Geochim Cosmochim Acta 47:1325-1329

Fisher RS (1995) Naturally occurring radioactive materials (NORM) in produced water and scale from Texas oil, gas, and geothermal wells. Texas Bur Econ Geol 95-3, 43 p

Fisher RS (1998) Geologic and geochemical controls on naturally occuring radioactive materials (NORM) in produced water from oil, gas, and geothermal operations. Environ Geosci 5:139-150

Fitch AA (1931) Barite and witherite from near El Portal, Mariposa Co., California. Am Mineral 16:461-468

Fortey NJ, Beddoe-Stevens B (1982) Barium silicates in stratabound Ba-Zn mineralization in the Scottish Dalradian. Mineral Mag 46:63-72

Frank AJ, Moynihan CS (1947) occurrence of the barite at Pilot Knob, Missouri. Am Mineral 32:681-683

Franklin JM (1995) Volcanic-associated massive sulfide base metals. *In* Geology of Canadian Mineral Deposit Types. OR Eckstrand, WD Sinclair, RI Thorpe (Eds) Geol Soc Am, The Geology of North America P-1:158-183

Franklin JM, Lydon JW, Sangster DF (1981) Volcanic-hosted massive sulfide deposits. *In* Economic Geology 75th Anniversary Volume. BF Skinner (Ed) Economic Geology Publishing Co, El Paso, Texas, p 485-627

Fresnel J, Galle P, Gayral P (1979) Résultats de la microanalyse des cristaux vacuolaires ches deux Chromophytes unicellulaires marines: *Exanthemachrysis gayraliae, Pavlova sp.* (Prymnesiophycées, Pavlovacées). Compt Rendus Acad Sciences Paris D 112:823-825

Frimmel HE (1990) Isotopic constraints on fluid rock ratios in carbonate rocks – barite sulfide mineralization in the Schwarz dolomite, Tyrol (Eastern Alps, Austria). Chem Geol 90:195-209

Frimmel HE, Papesch W (1990) Sr, O and C isotope study of the Brixlegg barite deposit, Tyrol (Austria). Econ Geol 85:1162-1171

Fu B (1998) A study of pore fluids and barite deposits from hydrocarbon seeps: deepwater Gulf of Mexico. PhD Dissertation, Louisiana State University, 243 p

Gaines RV, Skinner HCW, Foord EE, Mason B, Rosenzweig A (1997) Dana's New Mineralogy. Wiley, New York, 1819 p

Garven G, Raffensperger JP (1997) Hydrogeology and geochemistry of ore genesis in sedimentary basins. *In* Geochemistry of Hydrothermal Ore Deposits, 3rd edn. HL Barnes (Ed) Wiley, New York, p 797-875

Gayral P, Fresnel J (1979) *Exanthemachrysis gayraliae*, Lepailleur (Prymnesiophyceae, Pavlovales): Ultrastructure et discussion taxinomique. Protistologica 15:217-282

Gilkeson RH, Perry EC, Cartwright K (1981) Isotopic and geologic studies to identify the sources of sulfate in groundwater containing high barium concentration. Illinois State Geol Surv Rept 1981-4, 39 p

Gingele F, Dahmke A (1994) Discrete barite particles and barium as tracers of paleoproductivity in south-atlantic sediments. Paleooceanography 9:151-168

Gluyas J, Jolley L, Primmer T (1997) Element mobility during diagenesis: sulphate cementation of Rotliegeng sandstones, southern North Sea. Marine Petrol Geol 14:1001-1011

Glynn PD, Reardon EJ (1990) Solid–solution aqueous–solution equilibria: thermodynamic theory and representation. Am J Sci 290:164-201

Glynn PD, Reardon EJ, Plummer LN, Busenberg E (1990) reaction paths and equilibrium endpoints in solid–solution aqueous–solution systems. Geochim Cosmochim Acta 54:267-282

Goldberg ED, Somayajulu BLK, Galloway J, Kaplan IR, Faure G (1969) Differences between barites of marine and continental origins. Geochim Cosmochim Acta 33:287-289

Goldish E (1989) X-ray diffraction analysis of barium-strontium sulphate (barite-celestite) solid solutions. Powder Diff 4:214-216

Goldschmidt B (1938) Sur la precipitation mixte des sulfates de baryum et de strontium. Compt Rendu Paris Acad des Sci 206:1110

Gooday AJ, Nott JA (1982) Intracellular barite crystals in two Xenophyophores, *Aschemonella ramuliformis and Galatheammina sp*. (Protozoa: Rhizopoda) with comments on the taxonomy of *A. ramuliformis*. J Mar Biol Assoc UK 62: 595-605

Gooday AJ, Nott JA, Davis D, Mann S (1995) Apatite particles in the test wall of the large agglutinated forminifer *Bathysiphon major* (Protista). J Mar Biol Assoc UK 75:469-481

Goodfellow WD (1999) Sediment-hosted Zn-Pb-Ag deposits of North America. *In* Basins, Fluids and Zn-Pb Ores. O Holm, J Pongratz, P McGoldrick (Eds) Univ Tasmania Centre for Ore Deposits Research (CODES) Spec Pub 2: 59-91

Goodfellow WD, Lydon JW, Turner RW (1993) Geology and genesis of stratiform sediment-hosted (SEDEX) Zn-Pb-Ag sulphide deposits. *In* Mineral Deposit Modeling. RV Kirkham, WD Sinclair, RI Thorpe, JM Duke (Eds) Geol Assoc Canada Spec Paper 40:201-251

Gordon L, Rowley K (1957) Coprecipitation of radium with barium sulfate. Analyt Chem 29: 34-37

Gordon L, Reimer CC, Burit PB (1954) Distribution of strontium within barium sulfate precipitated from homogeneous solution. Analyt Chem 26, 842-846

Goyer RA (1974) Toxic effects of metals. *In* Casarett and Doull's Toxicology: the Basic Science of Poisons. CD Klassen (Ed) McGraw-Hill, New York, p 691-736

Graber KK, Chafetz HS (1990) Petrography and origin of bedded barite and phosphate in the Devonian Slaven Chert of Central Nevada. J Sed Pet 60:897-911

Grahmann W (1920) Uber barytocolestin und das verhaltnis von anhydrit zu colestin und baryt. Neues Jahrb Mineral 1:1-23

Gregus Z, Klaasen CD (1996) Mechanisms of toxicity. *In* Casarett and Doull's Toxicology: The Basic Science of Poisons. CD Klassen (Ed) McGraw-Hill, New York, p 35-74

Grotzinger JP, Kasting JF (1993) New constraints on Precambrian ocean composition. J Geology 101:235-243

Guichard F, Church TM, Treuil M, Jaffrezic H (1979) Rare earths in barites: distribution and effects in aqueous partitioning. Geochim Cosmochim Acta 43:983-987

Gundlach H, Stoppel D, Strubel G (1972) Zur hydrothermalen Löslichkeit von Baryt. Neues Jahrb Mineral 116:321-328

Halas S, Mioduchowski L (1978) Isotopic composition of oxygen in sulfate minerals of calcium and strontium and in water sulfates from various regions of Poland. Ann Univ Mariae Curie-Sklodowska 33:115-130

Hall AJ, Boyce AJ, Fallick AE, Hamilton PJ (1991) Isotopic evidence of the depositional environment of the Proterozoic stratiform barite. Chem Geol 87:99-114

Hannak WW (1981) Genesis of the Rammelsberg ore deposit near Goslar/Upper Harz, Federal Republic of Germany. *In* Handbook of Strata-Bound and Stratiform Ore Deposits, vol 9. K Wolf (Ed) Elsevier, Amsterdam, p 551-642

Hannington, MD and Scott SD (1989) Gold mineralization in volcanogenic massive sulfides: implications of data from active hydrothermal vents on the modern sea floor. Econ Geol Monogr 6:491-506

Hanor JS (1966) The origin of barite. PhD Thesis, Harvard University, 257 p

Hanor JS (1968) Frequency distribution of compositions in the barite-celestite series. Am Mineral 53:1215-1222

Hanor JS (1969) Barite saturation in sea water. Geochim Cosmochim Acta 33:894-898

Hanor JS (1973) Critical comment: The synthesis of barite, celestite, and barium-strontium sulfate solid solution crystal. Geochim Cosmochim Acta 37:2685-2687

Hanor JS (1979) Sedimentary genesis of hydrothermal fluids. *In* Geochemistry of Hydrothermal Ore Deposits, 2nd ed. HL Barnes (Ed) John Wiley and Sons, New York, p 137-168

Hanor JS (1980) Dissolved methane in sedimentary brines: Potential effect on the PVT properties of fluid inclusions. Econ Geol 75:603-609

Hanor JS (1988) Origin and Migration of Subsurface Sedimentary Brines. Society of Sedimentary Geology Short Course No. 21, Tulsa, Oklahoma, 247 p.

Hanor JS (1994) Origin of saline fluids in sedimentary basins. *In* Geofluids: Origin and Migration of Fluids in Sedimentary Basins. J Parnell (Ed) Geol Soc London Spec Pub 78:151-174

Hanor JS (1996a) Variations in chloride as a driving force in siliciclastic diagenesis. *In* Siliciclastic Diagenesis and Fluid Flow: Concepts and Applications. LJ Crossey, R Loucks, MW Totten (Eds) Soc Econ Paleo Mineral Spec Pub 55:3-12

Hanor JS (1996b) Controls on the solubilization of dissolved lead and zinc in basinal brines. *In* Carbonate-hosted Lead-Zinc Deposits. DF Sangster (Ed) Soc Econ Geol Spec Pub 4:483-500

Hanor JS (1999a) Thermohaline porewater trends of southeastern Louisiana revisited. Gulf Coast Assoc Geol Soc Trans 49:273-280

Hanor JS (1999b) Geochemistry and origin of metal-rich brines in sedimentary basins. *In* Basins, fluids, and Zn-Pb ores, 2nd edn. *In* H Holm, J Pongratz, P McGoldrick (Eds) Univ Tasmania, Centre for Ore Deposits Research (CODES) Spec Pub 2:129-146

Hanor JS, Baria LR (1977) Controls on the distribution of barite deposits in Arkansas. *In* Geology of the Quachitas, vol. 2. CG Stone (Ed) Arkansas Geol Commission, Little Rock, p 48-55

Hanor JS, Chan LH (1977) Non-conservative behavior of barium during mixing of Mississippi River and Gulf of Mexico waters. Earth Planet Sci Lett 37:242-250

Hanor JS, Sassen R (1990) Large-scale vertical migration of formation waters, dissolved salt, and hydrocarbons in the Louisiana Gulf Coast. *In* Gulf Coast Oils and gases, Their Characteristics, Origin, Distribution, and Exploration and Production Significance. D Schumacher, BF Perkins (Eds) Proc 9th Ann Res Conf, Gulf Coast Section, Soc Econ Paleo Mineral p 283-296.

Hanor JS, McManus KM, Ranganathan V, and Su S (1995) Mineral buffering of contaminated ground water compositions at a hazardous waste site in southwestern Louisiana: Trans Gulf Coast Assoc of Geol Soc 45:237-243

Hattori K (1989) Barite-celestine intergrowths in Archean plutons: the product of oxidizing hydrothermal activity related alkaline intrusions. Am Mineral 74:1270-1277

Hattori K, Cameron EM (1986) Archaean magmatic sulphate. Nature 319:45-47

Heinrich EW, Vian RW (1967) Carbonatitic barites. Am Mineral 52:1179-1189

Herzig PM, Hannigton MD, Fouquet Y, Vonstackelberg U, Petersen S (1993) Gold-rich polymetallic sulfides from the Lau back-arc and implications for the geochemistry of gold in sea-floor hydrothermal systems of the southwest Pacific. Econ Geol 88:2182-2209

Hill CA (1992) Isotopic values of native sulfur, barite, celestine and calcite: their relationship to sulphur deposits and the evolution of the Delaware Basin (preliminary results). Soc Metall Expl 147-157

Hitzman MW (1999) Characteristics and worldwide occurrence of Irish-type Zn-Pb-(Ag) deposits. *In* Basins, Fluids and Zn-Pb ores, 2nd edn. O Holm, J Pongratz, P McGoldrick (Eds) Univ Tasmania Centre for Ore Deposits Research (CODES) Spec Pub 2: 93-116

Hitzman MW, Beaty DD (1996) The Irish Zn-Pb-(Ba) ore field. *In* Carbonate-Hosted Lead-Zinc Deposits. DF Sangster (Ed) Soc Econ Geol Spec Pub 4:29-57

Holland HD (1984) The Chemical Evolution of the Atmosphere and Oceans. Princeton University Press, Princeton, New Jersey, 582 p

Hollande A, Martoja R (1974) Identification du cristalloide des isospores de radiolaires à un cristal de célestite ($SrSO_4$). Détermination de la constitution du cristalloide par voie cytochimique et à l'aide de le microsonde électronique et du micoanalyseur par émission ionique secondaire. Protistologica 10:603-609

Holser WT (1979) Mineralogy of evaporites. Rev Mineral 6:211-294

Hopwood JD, Mann S, Gooday AJ (1997) The crystallography and possible origin of barium sulphate in deep-sea rhizopod protists (Xenophyophorea). J Mar Biol Assoc UK 77:969-987.

Horton JWJ (1989) Barite in quartz-sericite schist and schistose pyroclastic rock of the Battleground Formation, Kings Mountain Belt. US Geol Survey Prof Paper 1112:115-118

Horton RC (1963) An inventory of barite occurrences in Nevada. Nev Bur Mines Rept 4, 18 p

Howard KW, Hanor JS (1987) Compositional zoning in the Fancy Hill stratiform barite deposit, Quachita Mountains, Arkansas and evidence for lack of associated massive sulfides. Econ Geol 82:1377-1385

Hubert G, Rieder N, Schmitt G, Send W (1975) Bariumanreicherung in den Müllerschen Körperchen der *Loxodidae* (Ciliata, Holotriche). Z Natur 30c:422.

Hughes, NP, Perrry CC, Anderson OR, Williams RJP (1989) Biological minerals formed from strontium and barium sulphates. III. The morphology and crystallography of strontium sulphate crystals from the radiolarian *Sphaerozoum punctatum*. Proc Royal Soc London B238:223-233

Huheey JE (1972) Inorganic Chemistry. Harper & Row, New York, 737 p

Huston DL (1999) Stable isotopes and their signficance for understanding the genesis of volcanic-hosted massive sulfide deposits: A review. Rev Econ Geol 8:157-180

Isphording WC (1982) Misinterpretation of environmental monitoring data—a plague on mankind! Gulf Coast Assoc Geol Soc Trans 32:399-411

Janecky DR, Shanks, WC III (1988) Computational modelling of chemical and sulphur isotopic reaction processes in seafloor hydrothermal systems: chimneys, massive sulfides, and subjacent alteration zones. Can Mineral 26:805-826

Jeandel C, Dupré B, Lebaron G, Monnin C, Minster JF (1996) Longitudinal distributions of dissolved barium, silica and alkalinity in the western and southern Indian ocean. Deep-Sea Res 43:1-31

Jebrak M, Smejkal V, Albert D (1985) Rare earth and isotopic geochemistry of the fluorite-barite vein deposits from the western Rouergue district (France). Econ Geol 80:2030-2034

Jewell PW (1992) Hydrodynamic and chemical controls of Paleozoic phophorite and barite in the western Cordillera of North America. *In* Water-Rock Interaction (WRI-7). Y Kharaka, A Maest (Eds) Balkema, Rotterdam, 1235-1238

Jewell PW (2000) Bedded barite in the geologic record. Soc Econ Paleo Mineral Spec Pub 66:147-161

Jewell PW, Stallard RF (1991) Geochemistry and paleooceanographic setting of central Nevada bedded barites. J Geol 99:151-170

Kadko D, Moore WS (1988) Radiochemical constraints on the crustal residence time of submarine hydrothermal fluids: Endeavor ridge. Geochim Cosmochim Acta 52: 659-668

Kaiser CJ, Kelly WC, Wagner RJ, Shanks WCI (1987) Geologic and geochemical controls of mineralization in the Southeast Missouri barite district. Econ Geol 82:719-734

Kajiwara Y, Honma H (1976) Lead content of barite coexisting with galena—an indicator of oxygen fugacity. Mining Geol (Japan) 25:397-400

Kasting JF (1993) Earth's early atmosphere. Science 259:920-926

Kesler SE (1996) Appalachian Mississippi Valley Type deposits: Paleoaquifers and brine provinces. *In* Carbonate-Hosted Lead-Zinc Deposits. DF Sangster (Ed) Soc Econ Geol Spec Pub 4:29-57

Kesler SE, Jones LM (1981) Sulphur- and strontium-isotopic geochemistry of celestite, barite and gypsum from the Mesozoic Basins of north eastern Mexico. Chem Geol 31:211-24

Kesler, SE, Appold MS, Walter LM, Martini A, Huston TS (1995) Na-Cl-Br systematics of fluid inclusions from MVT deposits. Geology 23:642-644

Kesler TL, Kesler SE (1971) Amphibolites of the Cartersville district, Georgia. Geol Soc Am Bull 82:3163-3168

Khalaf FI, El-Sayed MI (1989) Occurrence of barite-bearing mud balls within the Mio-Pleistocene terrestrial sand (Kuwait Group) in southern Kuwait, Arabian Gulf. Sed Geol 64:197-202

Kinsman D (1969) Reinterpretation of Sr^{2+} concentrations in carbonate minerals and rocks. J Sed Pet 39:512-517

Klein L (1959) River Pollution: 1. Chemical Analyses. Butterworths, London, 206 p

Kornicker WA, Presta PA, Paige CR, Johnson DM, Hileman OE, Snodgrass WJ (1991) The aqueous dissolution kinetics of the barium/lead sulfate solid solution series at 25 degrees Celsius and 60 degrees Celsius. Geochim Cosmochim Acta 55:3531-3541

Koczy FF (1958) Natural radium as a tracer in the ocean. Proc 2nd Int'l Conf on the Peaceful Uses of Atomic Energy 18:351

Krebs W (1981) Geology of the Meggen ore deposit. *In* Handbook of Strata-Bound and Stratiform ore Deposits. K Wolf (Ed) Elsevier, Amsterdam 9:509- 549

Kribek B, Hladikova J, Zak K, Bendl J, Pudilova M, Uhlik Z (1996) Barite-hylophane sulfide ores at Rozna, Bohemian Massif, Czech Republic: Metamorphosed black shale-hosted submarine exhaltive mineralization. Econ Geol 91:14-35

Kritsotakis K, von Platen H (1980) Reduktive Barytmobilisation. Neues Jahrb Mineral Abh 137:282-306

Kusakabe M, Robinson BW (1977) Oxygen and sulfur isotope equilibria in the $BaSO_4$-H_2SO_4-H_2O system from 110 to 350 degrees C and applications. Geochim Cosmochim Acta 41:1033-1040

Kushnir SV (1986) The epigenetic celestine formation mechanism for rocks containing $CaSO_4$. Geochem Int'l 23:1-9

Kyle JR (1994) The barite industry and resources of Texas. Texas Bur Econ Geol, Mineral Res Circ 85, 86 p

Lambert IB, Donelly TH, Dunlop JSR, Groves DI (1978) Stable isotopic compositions of early Archean sulphate deposits of probable evaporitic and volcanogenic origins. Nature 276:808-811

Lambert IB, Sato T (1974) The Kuroko and associated ore deposits of Japan: A review of their features and metallogenesis. Econ Geol 69:1215-1236

Leach DL (1980) Nature of mineralizing fluids in the barite deposits of central and southeast Missouri. Econ Geol 75:1168-1180

Lippmann F (1962) Benstonite, $Ca_7Ba_6(CO_3)_{13}$, and new mineral from the barite deposit in Hot Springs County, Arkansas. Am Mineral 47:585-598

Lonsdale P (1979) A deep-sea hydrothermal site on a strike-slip fault. Nature 282:531-534.

Lottermoser BG, Ashley PM (1996) Geochemistry and explanation of the significance of ironstones and barite-rich rocks in the Proterozoic Willyama Supergroup, Olary Block, South Australia. J Geochem Expl 57:57-73

Lydon JW (1995) Sedimentary exhalative sulfides. *In* Geology of Canadian Mineral Deposit Types. OR Eckstrand, WD Sinclair, RI Thrope (Eds) Geol Soc Am, The Geology of North America P-1:130-152

Macpherson GL (1989) Sources of lithium and barium in Gulf of Mexico formation waters, USA. *In* Water-Rock Interaction (WRI-6). DL Miles (Ed) Balkema, Rotterdam, p 453-456

Malinin SD, Urusov VS (1983) The experimental and theoretical data on isomorphism in the (Ba,Sr)SO$_4$ system in relation to barite formation. Geochem Int'l 20:70-80

Marchev P (1991) Primary barite in high-K dacite from the eastern Rhodope, Bulgaria. Eur J Mineral 3:1005-1008

Margaritz M, Brenner IB, Ronen D (1990) Ba and Sr distribution at the water table: Implications for monitoring ground-water at nuclear waste repository sites. Appl Geochem 5:555-562

Martin EE, McDougall JD, Herbert TD, Paytan A, Kastner M (1995) Strontium and neodymium isotopic analyses of marine barite separates. Geochim Cosmochim Acta 59:1353-1361

Martin JH, Knauer GA (1973) The elemental composition of plankton. Geochim Cosmochim Acta 37:1639-1653

May F, Peacock JD, Smith DL, Barber AJ (1963) Geology of the Kintail district. British Geol Surv Memoir for sheet 72W and part of 71E (Scotland), 77 p

Maynard JB, Okita PM (1991) Bedded barite deposits in the United States. Econ Geol 86:364-376

Maynard JB, Morton J, Valdes-Nodarse EL, Diaz-Carmona, A (1995) Sr isotopes of bedded barites: guide to distinguishing basins with Pb-Zn mineralization. Econ Geol 90:2058-2064

McKibben MA, Hardie, LA (1997) Ore-forming brines in active continental rifts. *In* Geochemistry of Hydrothermal Ore Deposits, 3rd edn. HL Barnes (Ed) Wiley, New York, p 877-936

McLemore VT, Leuth VW (1996) Lead-zinc deposits in carbonate rocks in New Mexico. *In* Carbonate-Hosted Lead-Zinc Deposits. DF Sangster (Ed) Soc Econ Geol Spec Pub 4:264-276

McManus J, Berelson WM, Klinkhammer GP, Kilgore TE, Hammond DE (1998) Geochemistry of barium in marine sediments: Implications for its use as a paleoproxy. Geochim Cosmochim Acta 62:3453-3473

McManus KM, Hanor JS (1988) Calcite and iron sulfide cementation of Miocene sediments flanking the West Hackberry salt dome, southwest Louisiana, U.S.A. Chem Geol 74, 99-112

McManus KM, Hanor JS (1993) Diagenetic evidence for massive evaporite dissolution, fluid flow, and mass transfer in the Louisiana Gulf Coast. Geology 21:727-730.

Mills JW, Carlson CL, Fewkes RH, Handlen LW, Jaypakash GP, Johns MA, Margati JM, Neitzel TW, Ream LR, Sanford SS, Todd SG (1971) Bedded barite deposits of Stevens County, Washington. Econ Geol 66:1157-1163

Misra KC, Gratz JF, Lu C (1996) Carbonate-hosted Mississippi Valley-type mineralization in the Elmwood-Gordonsville deposits, Central Tennessee zinc district: A synthesis. Econ Geol Spec Pub 4:58-73

Mitchell RS, Pharr RF (1961) Celestite and calciostrontianite from Wise County, Virginia. Am Mineral 46:189-185

Mohazzabi P, Searcy AW (1976) Kinetics and the thermodynamics of decomposition of barium sulphate. J Chem Soc Farad Trans 72:290-295

Monnin C (1999) A thermodynamic model for the solubility of barite and celestite in electrolyte solutions and seawater to 200°C and to 1 kbar. Chem Geol 153, 187-209

Monnin C, Galinier C (1988) The solubility of celestite and the barite in electrolyte solutions and natural waters at 25 degrees C: A thermodynamic study. Chem Geol 71:283-296

Monnin C, Jeandel C, Cattaldo T, Dehairs F (1999) The marine barite saturation state of the world ocean. Marine Chem 65:253-261

Moore W, Stakes D (1990) Ages of barite-sulfide chimneys from the Mariana Trough. Earth Planet Sci Lett 100:265-274

Morgan JW, Wandles GA (1980) Rare earth element distribution in some hydrothermal minerals: Evidence for crystallographic control. Geochim Cosmochim Acta 44:973-980

Moro MC, Arribas A (1981) Los yacimientos espanoles de barita estratiforme y su significado metalogenico en el contexto mundial. Tecniterrae 42:18-45

Mossman DJ, Brown MJ (1986) Stratiform barite in sabkha sediments, Walton-Cheverie, Nova Scotia. Econ Geol 81:2016-2021

Müller G (1962) Zur Geochemie des Strontiums in ozeanen Evaporiten unter besonderer Berucksichtigumg der sedimentaren Celestin-Lagerstatten von Hemmelte West. Geologie, Beiheft 35, 90 p

Murchey BL, Madrid RJ, Poole FG (1987) Paleozoic bedded barite associated with chert in western North America. *In* Siliceous Sedimentary Rock-Hosted Ores and Petroleum. JR Hein (Ed) Van Nostrand, p 269-283

Naehr TH, Bohrmann G (1999), Barium-rich authigenic clinoptilolite in sediments from the Japan Sea: a sink for dissolved barium? Chem Geol 158:227-244

Nickless EFP, Booth SJ, Mosley PN (1975) Celestite deposits of the Bristol area. Trans Inst Mining Metall B84:62-64

Nijman W, de Bruijne KH, Valkering ME (1998) Growth fault control of Early Archaean cherts, barite mounds and chert-barite veins, North Pole Dome, Eastern Pilbara, Western Australia. Precamb Res 88:25-52

Nordstrom DK, Munoz JL (1994) Geochemical Thermodynamics. Blackwell, Boston, 493 p

Nuelle LM, Shelton KL (1986) Geologic and geochemical evidence of possible bedded barite deposits in Devonian rocks of the Valley and Ridge Province, Appalachian Mountains. Econ Geol 81:1408-1430

Ober JA (1997) Strontium. US Geol Survey Minerals Inf Circ, 6 p

Oftedahl C (1958) On exhalative-sedimentary ores. Geol Foren Stockholm Forh 80:1-19

Ohmoto H, Lasaga AC (1982) Kinetics of reactions between aqueous sulfates and sulfides in hydrothermal systems. Geochim Cosmochim Acta 46:1727-1745

Okita PM (1983) Petrography of the Fancy Hill (Henderson) stratiform barite deposit, Montgomery County, Arkansas. MS Thesis, Louisiana State University, Baton Rouge, 208 p

Orange DL, Greene HG, Reed D, Martin JB, McHugh CM, Ryan WBF, Maher N, Stakes D, Barry J (1999) Widespread fluid expulsion on a translational continental margin: mud volcanoes, fault zones, headless canyons, and organic-rich substrate in Monterey Bay, California. Geol Soc Am Bull 111:992-1009

Orris GJ (1986) Descriptive model of bedded barite. US Geol Survey Bull 216:112-113

Ostlund H, Craig H, Broecker W, Spencer D (1987) GEOSECS Atlantic, Pacific and Indian ocean expeditions, vol. 7, Shore-based data and graphics, Nat'l Sci Fdn, Washington DC, 200 p

Paige CR, Kornicker WA, Hileman OE, Snodgrass WJ (1998) Solution equilibria for unranium ore processing: the $BaSO_4$-H_2SO_4-H_2O system and the $RaSO_4$-H_2SO_4-H_2O system. Geochim Cosmochim Acta 62:15-23

Papke KG (1984) Barite in Nevada. Nevada Bur Mines Geol Bull 98, 125 p

Paradis S, Lavoie D (1996) Multiple-stage diagenetic alteration and fluid history of Ordovician carbonate-hosted barite mineralization, southern Quebec Appalachians. Sed Geol 107:121-139

Pardue JH, Guo TZ (1998) Biochemistry of Ra-226 in contaminated bottom sediments and oilfield waste pits. J Environ Radioactivity 39:239-253

Park B, Branner JA (1932) A barite deposit in Hot Spring County, Arkansas. Arkansas Geol Surv Inf Circ 1, 52 p

Paytan A, Kastner M, Martin EE, McDougald JD, Herbert T (1993) Marine barite as a monitor of seawater strontium isotope composition. Nature 366:445-449

Paytan A, Moore WS, Kastner M (1996) Sedimentation rate as determined by Ra-226 activity in marine barite. Geochim Cosmochim Acta 60:4313-4319

Perry EC, Monster J, Reimer T (1971) Sulfur isotopes in Swaziland System barites and the evolution of the Earth's atmosphere. Science 171:1015-1016

Peter JM, Scott SD (1988) Mineralogy, composition and fluid-inclusion microthermometry of seafloor hydrothermal deposits in the southern trough of Guymas Basin, Gulf of California. Can Mineral 26:567-587

Pitzer KS (1987) A thermodynamic model for aqueous solutions of liquid-like density. Rev Mineral 17:97-142

Poulsen KH, Hannington MD (1995) Volcanic-associated massive sulfide gold. *In* Geology of Canadian Mineral Deposit Types. OR Eckstrand, W D Sinclair, RI Thorpe (Eds) Geol Soc Am, The Geology of North America P-1:183-196

Prieto M, Putnis A, Fernandez-Diaz L (1990) Factors controlling the kinetics of crystallization; supersaturation evolution in a porous medium: Application to barite crystallization. Geol Mag 127:485-495

Prieto M, Putnis A, Fernandez-Diaz L (1993) Crystallization of solid solutions from aqueous solutions in a porous medium: Zoning in (Ba,Sr)SO_4. Geol Mag 130:289-299

Prieto, M, Fernandez Gonzalez A, Putnis A, Fernandez Diaz L (1997) Nucleation, growth, and zoning phenomena in crystallizing (Ba,Sr)CO_3, Ba(SO_4,CrO_4), (Ba,Sr)SO_4, and (Cd,Ca)CO_3 solid solutions from aqueous solutions. Geochim Cosmochim Acta 61: 3383-3397

Puchelt H (1967) Zur Geochemie des Bariums im exogenen Zyklus. Springer-Verlag, Heidelberg, 287 p

Puchelt H (1972) Barium. *In* Handbook of Geochemistry. KH Wedepohl (Ed) Springer-Velag, New York, p 56B1-56O22

Putnis A, Fernandez-Diaz L, Prieto M (1992) Experimentally produced oscillatory zoning in the (Ba,Sr)SO$_4$ solid solution. Nature 358:743-745

Putnis A, Juntarosso JL, Hochella MF (1995) Dissolution of barite by a chelating ligand - an atomic-force microscopy study. Geochim Cosmochim Acta 59:4623-4632

Ramboz C, Charef A (1988) Temperature, pressure, burial history, and paleohydrogeology of the Les Malines Pb-Zn deposit: Reconstruction from aqueous inclusions in barite. Econ Geol 83:784-800

Ramos VA, Brodtkorb de MK (1989) Celestite, barite, magnesite and fluorspar: Stratabound settings through time and space. *In* Nonmetalliferous Stratabound Ore Fields. MK de Brodtkorb (Ed) Van Nostrand Reinhold, New York, p 297-321

Ramos VA, Brodtkorb de MK (1990) The barite and celestine metalotects of the Neuquen retroarc basin, central Argentina. *In* Stratabound Ore Deposits in the Andes. L Fonbote, GC Amstutz, M Cardozo, E Cedillo, J Frutos (Eds) Springer-Verlag, Berlin, p 599-613

Rankin AH, Sheppard TJ (1978) H$_2$O-bearing fluid inclusions in baryte from the North Pole deposit, Western Australia. Mineral Mag 42:408-410

Reardon EJ, Armstrong DK (1987) Celestine (SrSO$_4$) solubilty in water, seawater and NaCl solutions. Geochim Cosmochim Acta 51:63-72

Reesman RH (1968) Strontium isotopic composition of gangue minerals from hydrothermal veins. Econ Geol 63:731-736

Reimer TO (1980) Archean sedimentary baryte deposits of the Swaziland Supergroup (Barberton Mountain Land, South Africa). Precamb Res 12:393-410

Revelle RR (1944) Marine bottom samples collected in the Pacific ocean by the Carnegie on its seventh cruise. Publ Carneige Inst Washington 556, 2(1), 133 p

Rhein M, Chan LH, Roether W, Schlosser P (1987) [226]Ra and Ba in northeast Atlantic Deep Water. Deep-Sea Res 34:1541-1564

Rhein M, Schlitzer R (1988) Radium-226 and barium sources in the deep East Atlantic. Deep-Sea Res 35:1499-1510

Rietmejer FJM (1990) Salts in two chondritic porous interplanetary dust particles. Meteoritics 25:209-213

Robinson GRJ, Woodruff LG (1988) Characteristics of base-metal and barite vein deposits associated with rift basins, with examples from some early Mesozoic basins of eastern North America. US Geol Survey Bull 1776:377-390

Roedder E, Bodnar RJ (1997) Fluid inclusion studies of hydrothermal ore deposits. *In* Geochemistry of Hydrothermal Ore Deposits, 3rd edn. HL Barnes (Ed) Wiley, New York, Wiley, p 657-698

Ruiz J, Richardson CK, Patchett PJ (1988) Strontium isotope geochemistry of fluorite, calcite, and barite of the Cave-in-Rock fluorite district, Illionois. Econ Geol 83:203-210

Rye RO, Shawe DR, Poole FG (1978) Stable isotope studies of bedded barite at East Northumberland Canyon in Toquima Range, central Nevada. Res US Geol Survey 6:221-229

Sabine PA, Young BR (1954) Cell size and composition of the baryte-celestine isomorphous series. Acta Crystallogr 7:630 (abstract)

Samuelson SF (1992) Anatomy of an elephant: Boling dome. *In* Native Sulfur: Developments in Geology and Exploration. GR Wessel BH Wimberly (Eds) Soc Mining Metall Explor, Littleton, Colorado, p 59-71

Sangster DF (Ed) (1996) Carbonate-hosted lead-zinc deposits. Soc Econ Geol Spec Pub 4, 664 p

Sangster DF, Savard MM, Kontak DJ (1998) A genetic model for mineralization of Lower Windsor (Visean) carbonate rocks of Nova Scotia, Canada: Econ Geol 93:932-952

Saunders JA, Prikryl JD, Posey HH (1988) Mineralogic and isotopic constraints on the origin of the strontium-rich cap rock, Tatum Dome, Mississippi, U.S.A. Mineral Mag 32:63-86

Sawkins FJ and Burke K (1980) Extensional tectonics and mid-Paleozoic massive sulfide occurrences in Europe. Geol Rundschau 69:349-360

Sayles FL, Manheim F (1975) Interstitial solutions and diagenesis in deeply buried marine sediments, results from the deep sea drilling project. Geochim Cosmochim Acta 39:103-127

Schmitz B (1987) Barium, equatorial high productivity, and the northward wandering of the Indian continent. Paleooceanography 2:63-67

Scholle PA, Stemmerik L, Harpoth O (1990) Origin of major karst-associated celestine mineralization in Karstrynggen, Central East Greenland. J Sed Pet 60:397-410.

Schrijver K, Zartman RE, Williams-Jones AE (1994) Lead and barium sources in Cambrian siliciclastics and sediment provenance of a sector of the Taconic orogen: A mixing scenario based on Pb-isotopic evidence. Appl Geochem 9:455-476

Schroeder JO, Murray RW, Leinen M, Pflum RC, Janecek TR (1997) Barium and equatorial Pacific carbonate sediment: Terrigenous, oxide and biogenic associations. Paleooceanography 12:125-146

Scott SD (1997) Submarine hydrothermal systems and deposits. *In* Geochemistry of Hydrothermal Ore Deposits, 3rd edn. HL Barnes (Ed) Wiley, New York, p 797-875

Scull BJ (1958) Origin and occurrence of barite in Arkansas. Ark Geol Conserv Comm Info Circ 18, 101 p

Searls JP (1997) Barite. US Geol Survey Minerals Inf Circ, 7 p

Shaw TJ, Moore WS, Kloepper J, Sochaski K (1998) The flux of barium to the coastal waters of the southeastern USA: the importance of submarine groundwater discharge. Geochim Cosmochim Acta 62:3047-3054

Shawe DR, Poole FG, Brobst DA (1969) Newly discovered barite deposits in East Northumberland Canyon, Nye County, Nevada. Econ Geol 64:245-254

Shelton KL (1989) Mineral deposits and resources of the Ouachita Mountains. In The Appalachian-Ouachita Orogen in the United States. RD Hatcher, WA Thomas, GW Viele (Eds) Geol Soc Am, The Geology of North America F-2:729-737

Shikazono N (1994) Precipitation mechanisms of barite in sulfate-sulfide deposits in back-arc basins. Geochim Cosmochim Acta 58:2203-2213

Srinivasin B (1976) Barites: Anomalous xenon from spallation and neutron-induced reactions. Earth Planet Sci Let 31:129-141

Stamatakis MG, Hein JR (1993) Sedimentary rocks from Lefkas island, Greece. Econ Geol 88:91-103

Starke R (1964) Die strontiumgehälte der baryte. Freiberger Forschungsh C150, 123 p

Stecher HA, Kogut MB (1999) Rapid barium removal in the Delaware estuary. Geochim Cosmochim Acta 63:1003-1012

Stokinger HE, Woodward RL (1958) Toxicological methods for establishing drinking water standards. J Am Water Works Assoc 50:515-530

Stoops GJ, Zavaleta A (1978) Micromorphological evidence of barite neoformation in soils. Geoderma 20:63-70

Strizhov VP, Sval VN, Nokolayev SD, Zhabina NV (1988) Isotope composition of sulfur in barite crusts of the Red Sea. Oceanology 28:83-85

Stroobants N, Dehairs F, Goeyens L, Van Der Heijden N, Van Grieken R (1991) Barite formation in the southern-ocean water column. Mar Chem 35:411-421

Strübel G (1966) Die hydrothermale löslichkeit von colestin im system $SrSO_4$-$NaCl$-H_2O. Neues Jahrb Mineral 99-108

Strübel G (1967) Zur kentniss und genetischen bedetung des systems $BaSO_4$-$NaCl$-H_2O. Neues Jahrb Mineral 233-338

Takano B, Watanuki K (1974) Geochemical implications of the lead content of barite from different origins. Geochem J 8:87-95

Takano B, Yanagisawa M, Watanuki K (1970) Structure gap in $BaSO_4$-$PbSO_4$ solid solution series. Mineral J (Japan) 6:159-171

Tendal OS (1972) A monograph on the Xenophyophoria. In Galathea Report, Scientific Results of the Danish Deep-Sea Expedition Round the World 1950-52. W Torben (Ed) Danish Science Press, Copenhagen, p 6- 90

Tendal OS, Gooday AJ (1981) Xenophoria (Rhizopoda, Protozoa) in the bathyal and abyssal NE Atlantic. Oceanol Acta 4:415-422

Thomas WA (1976) Evolution of the Ouachita–Appalachian continental margin. J Geol 84:323-342

Tischendorf G (1963) Über $SrSO_4$-gehälte im Baryt als ein Kriteium für dessen Bildungs bedingungen. Problems of postmagmatic ore deposition, vol 1, Prague

Tooker EW (1991) Copper and molybdenum deposits in the United States. In Economic Geology, U.S. HJ Gluskoter, DD Rice, RB Taylor (Eds) Geol Soc Am, The Geology of North America P-2:23-42

Torres ME, Bohrmann G, Suess E (1996a) Authigenic barites and fluxes of barium associated with fluid seeps in the Peru subduction zone. Earth Planet Sci Lett 144:469-481

Torres ME, Brumsack HJ, Bohrmann G, Emeis KC (1996b) Barite fronts in continental margin sediments: A new look at barium remobilization in the zone of sulfate reduction and formation of heavy barites in diagenetic fronts. Chem Geol 127:125-139

Turner RJW (1992) Bedded barite deposits in the United States, Canada, Germany, and China: two major types based on tectonic setting: A discussion. Econ Geol 87:198-201

Turner RJW, Madrid RJ, Miller EL (1989) Roberts Mountains allochthon: Stratigraphic comparison with lower Paleozoic outer continental margin strata of the northern Canadian Cordillera. Geology 17:341-344

Ulrich MR, Bodnar RJ (1988) Systematics in stretching of fluids inclusions: II. Barite at 1 atm confining pressure. Econ Geol 83:1037-1046

Urabe T, Kusakabe M (1990) Barite silica chimneys from the Simisu Rift, Izu-Bonin Arc: possible analog to hematitic chert associated with kuruko deposits. Earth Planet Sci Lett 100:283-290

Varnavas SP (1987) Marine barite in sediments from deep sea drilling projects sites 424 and 242A (Galapagos hyrothermal mounds field). Mar Chem 20:245-253

Vinogradov VI, Kheraskova TN, Petrova SN (1978) Formation of stratiform barite deposits in the siliceous units of Kazakhstan. Lith Mineral Res 13:238-247

Von Damm KL (1995) Controls on the chemistry and temporal variability of seafloor hydrothermal fluids. Am Geophys Union Geophys Monogr 91:222-247

von Gehlen K, Nielsen H, Chunnett I, Rosendaal A (1983) Sulphur isotopes in metamorphosed Precambrian Fe-Pb-Zn-Cu sulphides and baryte at Aggeneys and Gamsberg, South Africa. Mineral Mag 47:481-486

von Gehlen K, Nielsen H, Ricke W (1962) S-isotopen-verhaltnisse in baryt und sulfiden aus hydrothermalen gangen im Schwarzwald und jungeren barytgangen in Suddeutschland und ihre genetische bedeutung. Geochim Cosmochim Acta 26:1189-1207

Wang ZC, Li GZ (1991) Barite and witherite deposits in the lower cambrian shales of south China: Stratigraphic distribution and geochemical characterization. Econ Geol 86:354-363

Warren JK (1997) Evaporites, brines and base metals: Fluids, flow and 'the evaporite that was.' Australian J Earth Sci 44:149-183

Weast RC (Ed) (1974) Handbook of Chemistry and Physics, 55th edn. CRC Press, Cleveland, Ohio

Weber, FH (1963) Barite in California. Calif Div Mines Geol Mineral Inf Service 16:1-10

Weinritt DJ, Cowan JC (1967) Unique characteristics of barium sulfate scale deposition. J Petroleum Tech 30:1381-1394

Whitford DJ, Kursch MJ, Solomon M (1992) Strontium isotope studies of barites, implications for the origin of base metal mineralization in Tasmania. Econ Geol 87:953-959

Wilcock JR, Perry CC, Williams RPJ, Brook AJ (1989) Biological minerals formed from strontium and barium sulphates. II. Crystallography and control of mineral morphology in desmids. Proc Roy Soc London B238:203-221

Wilcock JR, Perry CC, Williams RPJ, Mantoura RFC (1988) Crystallographic and morphological studies of the celestite skeleton of the acantharian species *Phyllostaurus siculus*. Proc Roy Soc London B233:393-405

Williams-Jones AE, Schrijver K, Doig R, Sangster DF (1992) A model for epigenetic Ba-Pb-Zn mineralization in the Appalachian thrust belt, Quebec: Evidence from fluid inclusions and isotopes. Econ Geol 87:154-174

Wood MW, Shaw HF (1976) The geochemistry of celestites from the Yate area near Bristol (U.K.). Chem Geol 17:179-193

Worl W (1991) The other metals. *In* Economic Geology, U.S. HJ Gluskoter, DD Rice, RB Taylor (Eds) Geol Soc Am, The Geology of North America P-2:125-151.

Zielinski RA, Bloch S, Walker TR (1983) The mobility and distribution of heavy metals during the formation of first cycle red beds. Econ Geol 78:1574-1589

Zimmerman RA (1970) Sedimentary features in the Meggen barite-pyrite-sphelerite deposit and a comparison with Arkansas barite deposits. Neues Jharb Mineral Abh 113:179-214

Zimmerman RA, Amstutz GC (1964) Die Arkansas-schwerspätzone, neue sedimentpetrographisch Beobachtung und genetische Umdeutung. Z Erz u Metall 7:365-371

5 Precipitation and Dissolution of Alkaline Earth Sulfates: Kinetics and Surface Energy

A. Hina and G. H. Nancollas

Chemistry Department
State University of New York at Buffalo
Buffalo, New York 14260

INTRODUCTION

The precipitation and dissolution of sparingly soluble alkaline earth sulfates play important roles in the establishment of geological equilibria (Davis and Hayes 1986). There is little doubt that there are close relationships among the interfacial tensions between the solid phases and their solutions, the observed solubilities, and the kinetics of crystallization and dissolution. The experimental methods used for investigating the rates of these reactions are reviewed with particular reference to the Constant Composition (CC) technique (Tomson and Nancollas 1978) which enables both crystallization and dissolution reactions to be studied at constant thermodynamic driving forces. Interfacial tensions of the alkaline earth sulfates may be measured using thin-layer wicking methods (Van Oss et al. 1992), and surface-tension components theory yields detailed information concerning the Lifshitz–van der Waals and Lewis acid–base parameters of the surface tension. The mechanisms of the growth and dissolution reactions are discussed from the point of view of kinetics and interfacial energies.

DRIVING FORCES FOR GROWTH AND DISSOLUTION

Supersaturation

Studies of crystallization from solution first require the preparation of metastable, supersaturated solutions in which the concentration of solute exceeds the saturation value at a given temperature and pressure. Some common expressions describing super-saturation are defined by Equations (1)–(3):

$$\Delta C = C - C_s \tag{1}$$

$$S = \frac{C}{C_s} \tag{2}$$

$$\sigma = \frac{C - C_s}{C_s} = S - 1 \tag{3}$$

where C is the solute concentration (moles of solute per liter of solvent, for example), C_s is the solubility value, ΔC is the concentration driving force, S is the supersaturation ratio, and σ is the relative supersaturation.

It is important to note that these definitions of supersaturation assume that the solutions are ideal despite the fact that they consist of mixtures of electrolytes and, in many cases, additional nonelectrolytes. The definitions are correct, therefore, only for very dilute solutions for which it is possible to assume that the activity coefficients are unity. In crystallization studies, one inevitably deals with solutions with high concentrations even for sparingly soluble electrolytes such as those of interest in this chapter. Concentrations must therefore be replaced by activities, and the supersaturations must be replaced as in Equation (4). Another complication arises because the

1529-6466/00/0040-0005$05.00

concentration ratios of the constituent ions in the solution commonly deviate from the stoichiometric values. This is commonly the case for calcium sulfate ($CaSO_4$) systems (Zhang and Nancollas 1992a). For an electrolyte, the relative supersaturation is defined as

$$\sigma = \left(\frac{IP}{K_{SP}}\right)^{1/\nu} - 1 = \left(\frac{\{Ca^{2+}\}\{SO_4^{2-}\}}{K_{SP}}\right)^{1/\nu} - 1 \tag{4}$$

where K_{sp} is the activity solubility product, IP is the activity ion product, $\{\}$ braces enclose aqueous ionic activities and the stoichiometric coefficient ν is the total number of ions per formula unit of the electrolyte (e.g. $\nu = 2$ for $CaSO_4$). It is assumed in Equation (4) that the solid phase activity (a component of the activity ion product, IP) is unity.

When gypsum crystals are brought into contact with an aqueous solution, the tendency to grow or dissolve depends upon the value of the Gibbs free energy of the resulting reaction. For the formation of one mole of gypsum ($CaSO_4 \cdot 2H_2O$), or crystalline calcium sulfate dihydrate (CSD), the change in the Gibbs free energy, ΔG, is given by

$$\Delta G = -RT\ln S = -\frac{RT}{\nu}\ln\left(\frac{IP}{K_{SP}}\right) \tag{5}$$

where R is the gas constant and T is temperature in Kelvin.

DEFINITION AND DETERMINATION OF GROWTH RATE

The linear growth rate, r ($m \, s^{-1}$), of a crystal face is defined as the velocity of advancement of that face in a direction normal to the surface. Growth rates may also be expressed as the mass deposited per unit surface area per unit time, R ($mol \, m^{-2} \, s^{-1}$). Generally, the relationship between r and R is complex, as is discussed in another section.

To elucidate the mechanisms of crystal growth, it is important to study the growth rates at different driving forces. For sparingly soluble salts growing in suspension, the rates can be determined directly by measuring the changes of particle size as a function of time, or indirectly from the measured changes in concentration of constituent ions in the solution. Titration methods may also be used to record the amount of constituent ions that must be added to maintain constant concentrations during the reactions. These methods can be divided into three main categories.

One of the earliest experimental approaches, still in use by many research groups, is that of "free drift" (Davies and Jones 1955, Nancollas 1973) in which the rate of a crystallization reaction is obtained by following concentration changes of the constituent ions as a function of time. With the development of rapid sensors, the concentrations may be recorded continuously by means of potentiometric, conductometric, or photometric methods (Ebrahimpour 1990). If such continuous monitoring methods are not available, the reaction can be followed by sampling the solution suspension, filtering rapidly, and analyzing for constituent ions using conventional analytical techniques.

The next developmental stage in measurement the of seeded growth was the potentiostatic method. For some systems, it is necessary and convenient to investigate the rate of reaction while the activity of one of the species is held constant. In the conventional potentiostatic method, the activity of a single ionic species is controlled by titrant while the concentrations of other constituents are allowed to vary with time. This results in variations in the thermodynamic driving forces during the reactions. Although it would seem that rate data for crystal growth as a function of supersaturation could be obtained from a single experiment, large errors are commonly obtained in the determination of the rates, especially at low supersaturation. It is therefore desirable to use a constant composition (CC) method in which the activities of all species in the solution can be maintained constant during the reactions (Tomson and Nancollas 1978).

The CC method has several advantages as compared with the free drift and conventional potentiostatic techniques: (1) the growth and dissolution rates can be determined more precisely, especially at very low driving forces; (2) experiments can be carried out at precisely known points on a phase diagram in order to avoid side reactions such as undesirable phase transformation (Nancollas et al. 1989); (3) relatively large amounts of material can be grown at a specific driving force for characterization of the solid phases (Tomson and Nancollas 1978); (4) the influence of additives on the reaction rate can be determined over extended periods of time while maintaining constant concentrations of both additives and lattice ions (Budz et al. 1988); (5) changes in reactivity of the crystal surfaces can be investigated as a function of time at a given thermodynamic driving force (Zhang and Nancollas 1990a); (6) growth rates can be determined at different ratios of constituent ion in solution while keeping the supersaturation constant (Zhang and Nancollas 1990b); and (7) crystallization experiments can be carried out in ranges of supersaturation at which no homogeneous nucleation takes place.

Rate determination

One of the advantages of the CC method is that a large extent of reaction can be achieved even at very low driving forces. The rate of crystal growth, R, is expressed as functions of surface characteristics $[f(m)]$ and supersaturation $[f'(\sigma)]$ in Equation (6)

$$R = k_1 f(m) f'(\sigma) \tag{6}$$

where k_1 is an effective rate constant, $f'(\sigma)$ is constant and the rate reflects changes that take place at the crystal surfaces during the reaction.

The effective surface area, s_e, of a crystal suspension at time t, is related to the initial value, s_o, by the empirical Equation (7) (Christoffersen and Christoffersen 1976),

$$s_e = s_o \left(\frac{m}{m_o} \right)^p \tag{7}$$

where m_o and m are the masses of crystals present initially and at time t, respectively. For isotropic three-dimensional growth, $p = 2/3$ (Van Oosterhout and Van Rosmalen 1980), whereas p values of 0 and 1/2 indicate that growth occurs exclusively in one and two directions, respectively (Barone et al. 1983). In general, the rates of reaction may be calculated from the volume of added titrant, V, which can be converted to moles of materials deposited or dissolved as a function of time. Thus, the overall reaction rate is given by Equation (8),

$$R = \frac{C_{eff}}{s_e} \frac{dV}{dt} \tag{8}$$

where dV/dt is the slope of CC titration curve at time t, and the effective concentration, C_{eff}, is the equivalent number of moles precipitated per liter of added titrant. However, as the crystals grow, the effective surface area available for deposition increases, leading to more rapid titrant addition. Thus, the CC titration curves may reflect changes in the crystal habit based on the relation between the effective surface area and the extent of growth. The growth rate, in terms of flux density (mole $min^{-1}m^{-2}$), is given by Equation (9),

$$R = \frac{C_{eff}}{s_o} \frac{dV/dt}{(m/m_o)^p} \qquad (9)$$

Normally, p values may be calculated from Equation (9) by rewriting it as Equation (10),

$$\log\left(\frac{dV}{dt}\right) = \log\left(\frac{R\,s_o}{C_{eff}}\right) + p\,\log\left(\frac{m}{m_o}\right) \qquad (10)$$

and plotting $\log (dV/dt)$ against $\log (m/m_o)$. However, this procedure requires the calculation of a number of dV/dt values. As the titration curve is not always perfectly smooth, especially when a divalent-ion selective electrode is used, significant errors may be involved in the estimation of dV/dt. The following approach avoids the use of derivatives, thereby yielding more accurate values for p and the rate of nucleation, J, which is discussed in a later section.

The total mass of crystallites can be calculated from the volume of titrant addition, V,

$$m = m_o + M\,C_{eff}\,V \qquad (11)$$

where M is the molecular mass of the growing phase. Substitution of Equation (11) into Equation (9) yields

$$R = \frac{C_{eff}}{s_o} \frac{dV/dt}{(1 + MC_{eff}V/m_o)^p} \qquad (12)$$

The integration of Equation (12) with the initial condition $V = 0$ at $t = 0$, gives Equations (13) and (14),

$$R = (C_{eff}/s_o)(V_c/t) \qquad (13)$$

$$V_c = \frac{(1 + MC_{eff}V/m_o)^{(1-p)} - 1}{(1-p)MC_{eff}/m_o} \qquad (p \neq 1) \qquad (14)$$

where V_c is the corrected titrant volume for constant effective surface area. From Equations (13) and (14), both R and p can be evaluated simultaneously using nonlinear least squares. Moreover, Equation (13) suggests that if the correct p value is used, plots of V_c against time t should be linear. It was found that at $p = 0.4 \pm 0.1$, a linear correlation was obtained between the corrected volume from Equation (14) and time t. A typical corrected titrant curve is shown in Figure 1.

The growth kinetics of gypsum (calcium sulfate dihydrate, CSD) has been investigated at equivalent calcium sulfate concentration by a number of workers (Liu and Nancollas 1970, Smith and Sweett 1971, Nancollas 1973, Liu and Nancollas 1973, Van Rosmalen et al. 1981, Christoffersen and Christoffersen 1982, Amathieu and Boistelle 1988, Witkamp et al. 1990). However, in most practical applications, such as scale formation (van Rosmalen et al. 1981), gypsum wallboard manufacture (Ridge 1960), and phosphoric acid production (Stevens 1961), gypsum crystallization proceeds in a non-stoichiometric medium. The nucleation rate of CSD, as demonstrated by Keller et al. (1980), markedly depends on the calcium/sulfate molar ratio at a given supersaturation. The growth kinetics of CSD have been also investigated (Zhang and Nancollas 1992a) using the constant-composition method over a range of calcium/sulfate molar ratios in supersaturated solutions ($0.17 < Ca/SO_4 < 5.0$, at ionic strength, $I = 0.500$ mol L^{-1} with KCl as the supporting electrolyte and $0.08 < Ca/SO_4 < 10.0$ at $I = 1.00$ mol L^{-1} in NaCl solutions). To assess the influence of molar ratio on growth rates, the latter were corrected to a relative supersaturation of 0.3 using Equation (15),

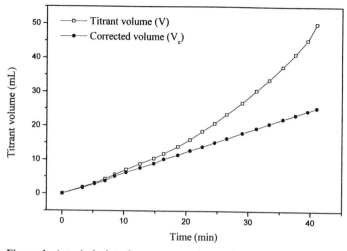

Figure 1. A typical plot of gypsum crystal growth titrant addition and the corrected volume against time at T = 25°C in 0.5 mole L^{-1} NaCl aqueous solution (Zhang and Nancollas 1992a).

$$R_c = R \ (0.3 \ / \ \sigma)^2 \tag{15}$$

which is based on the previously confirmed parabolic rate law for CSD growth (Christoffersen et al. 1982, Nielsen 1984). Plots of the corrected growth rate, R_c, against Ca/SO_4 (Fig. 2) show marked decreases in R_c with increasing Ca/SO_4. Moreover, the rate is considerably smaller in NaCl solutions even though the higher ionic strength may have yielded a higher rate (Witkamp et al. 1990). The inhibitory effect of Na^+ ions, in comparison with K^+, is probably due to their relatively stronger adsorption at the gypsum surfaces. The difference in rates in the two media becomes smaller at higher Ca/SO_4 ratios as the Na concentration decreases. In spite of the constancy of ion-activity product,

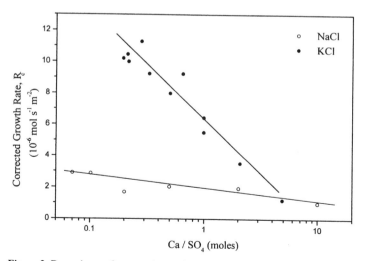

Figure 2. Dependence of corrected growth on the calcium/sulfate molar ratio at $\sigma = 0.3$ (Zhang and Nancollas 1992a).

the rate was found to increase with decreasing Ca/SO_4 molar ratio, indicating that the rate of crystal growth is not merely a function of the thermodynamic driving forces but also depends upon the relative concentrations and characteristics of the individual constituent ions.

CRYSTALLIZATION AND DISSOLUTION KINETICS

Crystallization processes can be divided conveniently into three distinct stages: (1) the achievement of the supersaturated state as a result of chemical reaction or changes in temperature or pressure, or by mixing solutions containing the constituent ions of interest; (2) nucleation or birth of the crystalline phase through the generation of the first minute clusters or nuclei of the crystal; (3) subsequent growth of nuclei to larger crystallites.

Nucleation is traditionally divided into primary homogeneous nucleation, primary heterogeneous nucleation, and secondary nucleation. In primary homogeneous nucleation, clusters (nuclei) are produced in gas, solid, melt, or solution phases in the absence of foreign bodies. Primary heterogeneous nucleation is induced by surfaces, other than those of the crystallizing substance, which reduce, to some extent, the energy barrier for homogeneous nucleation. Secondary nucleation may occur in the presence of crystals of the material being crystallized, and involves the dislodgment of nuclei from the parent crystal at supersaturations at which primary (heterogeneous) nucleation would not occur.

Theories of nucleation have developed along several different lines, among which are the kinetic models of Volmer and Weber (1926), Becker and Döring (1935), Zeldovich (1942), and Frenkel (1946). Other treatments have adopted statistical-mechanical approaches that focus attention on the partition functions of nuclei, how the concentrations of nuclei are distributed with size (Band 1939 a,b; Frenkel 1939 a,b; McDonald 1963), and computer simulations that include two techniques of major importance, namely, the Monte Carlo method (MC), and molecular dynamics (MD) calculations.

Homogeneous nucleation

Homogeneous, or spontaneous, nucleation occurs when nuclei of the solute are formed in perfectly clean solutions in the absence of foreign surfaces. Homogeneous nucleation describes the spontaneous formation of critical nuclei within the supersaturated solution. Statistical fluctuations in the interactions between solute ions or molecules result in the formation of ordered aggregates or embryos (Gibbs 1928) of critical size, after which crystal growth will occur.

An important application of nucleation theory has been for the estimation of solid–liquid interfacial free energies. Briefly, the free-energy change in forming a crystal in a homogeneous solution is usually expressed as Equation (16) (Nielsen 1964):

$$\Delta G = -\alpha \frac{L^3 \Delta\mu}{v_m} + \beta L^2 \gamma_{s\ell} \tag{16}$$

where α and β are the volume and surface shape factors, respectively, $\Delta\mu$ is the chemical potential difference between the two phases, L is the size of the crystal (in nm), v_m the molecular volume, and $\gamma_{s\ell}$ is the interfacial tension between solid and liquid phases. The dependence of the Gibbs free energy on L passes through a maximum corresponding to the critical size L^*. The condition, $(\partial\Delta G/\partial L) = 0$, yields Equation (17)

$$L^* = \frac{2}{3} \frac{\beta \gamma_{s\ell} v_m}{\alpha \Delta G} \tag{17}$$

Substitution of L^* into Equation (16) yields the change in the Gibbs free energy during critical nucleus formation:

$$\Delta G^* = \frac{4}{27} \frac{\beta^3 v_m^2 v_{s\ell}^3}{\alpha^2 k^2 T^2 \ln S} \tag{18}$$

where k is the Boltzmann constant, T is temperature in Kelvins, and S is the degree of supersaturation. According to classical nucleation theory (Nielsen 1964, Walton 1967), the rate of nucleation, J (cm^{-3} s^{-1}), or the number of nuclei formed per unit time and volume, depends primarily on the degree of supersaturation and temperature. It can be expressed in the form of the Arrhenius reaction velocity Equation (19) commonly used for the rate of a thermally activated process:

$$J = A \exp(-\frac{\Delta G^*}{kT}) \tag{19}$$

where A is the pre-exponential factor and k is the Boltzmann constant.

From Equations (18) and (19)

$$J = A \exp\left(-\frac{4}{27} \frac{\beta^3 v_m^2 \gamma_{s\ell}^3}{\alpha^2 k^3 T^3 (\ln S)^2} \right) \tag{20}$$

Equation (20) indicates that three main variables govern the rate of nucleation: temperature, T, degree of supersaturation, S, and interfacial tension, $\gamma_{s\ell}$. Equation (20) can be rewritten in the form

$$\ln J = \ln A - \frac{4}{27} \frac{\beta^3 v_m^2 \gamma_{s\ell}^3}{\alpha^2 k^3 T^3 (\ln S)^2} \tag{21}$$

Therefore, at constant temperature, the interfacial tension, $\gamma_{s\ell}$ can be estimated from linear plots of $\ln J$ against $(\ln S)^{-2}$. Many attempts have been made to obtain the ionic crystal/solution interfacial tension using this approach (Nielsen 1961, 1964; Garten and Head 1973, Möller and Rajagopalan 1975, 1976; Söhnel et al. 1977, Söhnel and Mullin 1982).

In practice, precipitation does not occur instantaneously upon the creation of supersaturation in a solution. Rather, a period of time, the "induction period," τ, elapses before newly created crystals are detected. The induction period, τ, is a complex quantity involving, amongst other factors, both nucleation and growth components. However, if the simplifying assumption is made that τ is mainly concerned with the nucleation process, i.e. that

$$\tau \propto J^{-1} \tag{22}$$

then from the classical nucleation described by Equation (20), we obtain Equation (23)

$$\ln \tau \propto \left[-\ln A + \frac{4}{27} \frac{\beta^3 v_m^2 \gamma_{s\ell}^3}{\alpha^2 k^3 T^3 (\ln S)^2} \right] \tag{23}$$

According to Equation (23), plots of the logarithm of the induction time, $\ln \tau$, as a function of $(\ln S)^{-2}$ are expected to give a straight line if the data represent homogeneous nucleation, the interfacial tension may then be calculated from the slope (Mullin and Osman 1973, Mullin and Ang 1976, Söhnel and Mullin 1978, Joshi and Antony 1979,

Mullin and Zacek 1981).

For liquid systems, the nucleation theory, in general, gives satisfactory agreement with experiments (La Mer and Pound 1949, Higuchi and O'Konski 1960). However, for solid-solution systems, the data are much less reliable (Table 1) and experimental testing is difficult to achieve for the following reasons:

(1) It is difficult to prepare systems free from impurities, which might act as nucleation catalysts, and to eliminate the influence of the walls of the retaining vessel, which commonly catalyze nucleation. An experimentally determined linear relationship between $\ln\tau$ or $\ln J$ and $(\ln S)^{-2}$ is no guarantee that homogeneous nucleation has occurred. From precipitation data for celestine ($SrSO_4$), Malollari et al. (1995) showed that, for a number of solutions maintained at constant supersaturation, kinetic data from log–log plots of the measured initial rates as a function of relative supersaturation defined two linear parts corresponding to low and high supersaturations.

Malollari et al. (1995) suggested that the apparent change in effective order, from $n = 1$ to $n = 4$, reflected a transition from homogeneous to heterogeneous nucleation. From the slope of the plot of $\ln\tau$ against $(\ln S)^{-2}$ corresponding to the homogeneous process, a value of $\gamma_{s\ell} = 88$ mJ m^{-2} was calculated for the interfacial tension of the precipitating $SrSO_4$. This value is within the range of values (82–92 mJ m^{-2}) reported by other authors (Nielsen 1969, Garten and Head 1973, Füredi-Milhofer et al. 1977). A similar change in the growth order has been reported for the spontaneous precipitation of $SrSO_4$ (Doremus 1970). Moreover, data for the seeded growth of $SrSO_4$ (Campbell and Nancollas 1969) showed similar features. In this case, slopes of $n = 2$ and 5 were found for low and high supersaturations, respectively.

(2) In principle, experimental methods used for studying nucleation are very simple and much qualitative work has been done by preparing supersaturated solutions and observing the visible onset of nucleation. However, such experiments cannot be expected to yield reproducible results, and more advanced techniques such as interferometry (Humphreys-Owen 1949), turbidimetry and dilatometry (Allen 1953), and light scattering (Kenyon and La Mer 1949, Garten and Head 1973) were developed. Other advanced techniques involve monitoring the composition of the nutrient medium by methods such as conductometry (Davies and Jones 1949), potentiometry (Van Hook 1940), and the use of radioactive tracers (Enüstün and Turkevich 1960). However, all of these experimental approaches fail to address the real requirements of the nucleation theory. Thus, according to Nielsen (1964), the size of the critical nucleus, n^*

$$n^* = \frac{d\ln J}{d\ln S} = \frac{8\beta^3\, v_m^2 \gamma_{s\ell}^3}{27\alpha^2 (kT\ln S)^3} \qquad (24)$$

which is estimated from Equation (24) by plotting experimental values of $\ln J$ against $\ln S$, may be as small as about 10 molecules, or less than 10 nm. However, the critical sizes obtained by substituting data on the volume of crystallizing molecules into the classical nucleation equations are in most cases more than 100 nm, and some are as large as 50 μm. In Equation (20), J describes the number of nuclei that reach critical size per unit volume per second. However, most experimental data refer to nucleated particles that have already grown appreciably.

(3) The theoretical value of the pre-exponential factor A in Equation (20) is about

10^{33} m^{-3} sec^{-1} (Nielsen 1964, Walton 1969), but according to He et al. (1995) this value could be on the order of 10^{29} to 10^{39} m^{-3} sec^{-1}. Lothe and Pound (1962) reconsidered classical nucleation theory in the light of the quantum-mechanical contributions to the free energy of formation of nuclei, arising from their absolute entropy, and were able to explain the anomalous pre-exponential factors for homogeneous nucleation. Assumptions involved in their derivation are still questionable (Dunning 1969). Nevertheless, as the theoretical values of A do not agree with those obtained experimentally, either the classical nucleation theory is incomplete or the experimental methods are inadequate.

(4) The classical theory for homogeneous nucleation cannot be applied to seeded growth systems (Koutsopoulos et al. 1994).

(5) The geometrical shape factor, β (= $4s^3/27v^2$), where s and v are the surface area and the volume of the nucleus), could be equal to $16\pi/3$, 50, or 4 assuming spherical, rectangular parallelepiped, or cubic-shaped nuclei, respectively). It should be noted that the geometrical factor was assumed to be a constant with or without the presence of inhibitors in the calculation. However, it is well known that additives, such as scale inhibitors, commonly change the crystal morphology (Cowan and Weintritt 1976), therefore, the geometrical factor may be a variable. Thus, uncertainties in the geometrical shape can also introduce errors in the calculated interfacial tension.

Table 1. Estimated results of the interfacial tension, obtained from rate data for crystal growth kinetics.

Mineral	Reaction order	Interfacial tension (mJ m^{-2})	Growth technique	Reference
Gypsum	1.5 – 2.1	100	CC	Christoffersen et al. (1982)
Gypsum	4.6	4	CC	Zhang and Nancollas (1992a)
Gypsum	2.7	7	CC	unpublished
Gypsum	–	45	Turbidity	He et al. (1994a)
Barite	3.2	40	Conductivity	Van der Leeden et al. (1992)
Barite	2.0	25	Temperature	Gardner and Nancollas (1983)
Barite	–	140	Turbidity	He et al. (1994a)
Celestine	–	76	Turbidity	He et al. (1995)
Celestine	–	84	SP[a]	Enüstün and Turkevich (1960)
Celestine	–	103	TN[b]	Mealor and Townshend (1966)
Celestine	–	72	TN[b]	Garten and Head (1973)
Celestine	–	76	Turbidity	He et al. (1995)
Celestine	–	43	Turbidity	He et al. (1995)

SP[a], the interface tension was calculated from experimental data of celestine solubility as a function of particle size.
TN[b], interfacial tensions were calculated using the critical supersaturation for homogeneous nucleation obtained from the total number of crystals and their size.

Heterogeneous nucleation

It has been known for many years that certain solid bodies (containing hetero-geneities, motes, inclusions, etc.), extraneous to the system, promote phase trans-formations, particularly condensation and crystallization. As the presence of a suitable foreign body or surface can induce nucleation at degrees of supersaturation lower than those required for spontaneous homogeneous nucleation, the overall free energy change associated with the formation of a critical nucleus under heterogeneous conditions, ΔG^{het}, must be less than that, ΔG^{hom}, associated with homogeneous nucleation, or

$$\Delta G^{het} = \varphi \Delta G^{hom} \tag{25}$$

where the factor φ, less than unity, can be expressed as (Turnbull 1952)

$$\varphi = \frac{(2+\cos\theta)(1-\cos\theta)^2}{4} \tag{26}$$

where θ is the contact angle formed between the crystalline deposit and the foreign solid surface, corresponding to the angle of wetting in a liquid–solid system (Young's equation, Young 1805). At the limit, $\theta = 180°$, there is a complete mismatch between the crystalline lattices of the substrate and that of the overgrowth, whereas at $\theta = 0°$ a perfect match is implied. However, these two limiting cases do not account for the influence of substrata surface properties on nucleation. Thus, it is impossible to determine interfacial tensions in the case of heterogeneous nucleation. The erroneous application of Young's equation to solid-solution systems has also been criticized from a thermodynamic point of view (Wu and Nancollas 1996).

Determination of interfacial free energy

The concept of surface free energy plays an important role in colloid and surface science, especially in interpreting such factors as its relationship to the rate of crystal nucleation and growth in solutions. The surface tension of liquids (at the interface with air, or vacuum, expressed as a force per unit length, mN m^{-1}, or free energy per unit area, mJ m^{-2}), like that of interfacial tension between two liquids, or between a liquid and a solid (expressed in the same units), is at first sight easy to grasp, and as a rule raises few intuitive problems. The concept of surface tension or energy of a solid, however, is more difficult to visualize, but can be defined by Equation (27) (Adamson and Gast 1997, van Oss 1994):

$$\gamma_i \equiv \left(\frac{\partial F}{\partial s}\right)_{T,V,\mu_i} \tag{27}$$

where F is the Helmholtz free energy and s is the surface area. Equation (27) expresses the reversible work required to create a unit area of the surface at constant temperature, volume, and chemical potential (Johnson 1959, Mullins 1963, Good 1967).

Although equilibrium interfaces may be treated using either mechanical or the mathematically equivalent concepts of surface free energy, the latter holding for all capillary phenomena, it is first necessary to distinguish between surface tension, γ, and specific surface free energy, f. The most convenient thermodynamic definitions make the surface tension strictly equal to the specific surface free energy only in the case of one-component, as distinct from multicomponent, systems. For the latter, the surface free energy is given by Equation (28):

$$f = \left(\frac{\partial F}{\partial s}\right)_{T,V} = \gamma + \sum_{i=1}^{n} \mu_i \Gamma_i \tag{28}$$

where μ_i and Γ_i are the chemical potential and surface excess free energy, respectively, for the ith component. If $n = 1$, then by the Gibbs '$\Gamma = 0$' convention defining the surface, Equation (28) reduces to Equation (27). In a two-component system composed of two condensed phases, there must be some interfacial excess free energy of one or both components, especially where adsorption phenomena are important. For strongly interacting systems, such as a polar liquid on an ionic solid, this interfacial excess free energy must be quite appreciable (Herring 1952, Zisman 1964, Defay and Prigogine 1966, Good 1967).

Although the surface-tension terms, surface free energy and interfacial tension, are used interchangeably in the literature, surface tension and surface free energy commonly refer to a condensed-phase–vapor system, whereas interfacial tension is always used to describe two condensed phases, such as liquid–liquid and solid–liquid interfaces. It is important to keep this in mind when comparing surface-tension or interfacial-tension data obtained using different approaches. Thus, recalculation of the data of Hulett (Hulett 1901, Freundlich 1923), gave a γ_{sl} value for gypsum ($CaSO_4 \cdot 2H_2O$) of 1050 dynes/cm, and for barite ($BaSO_4$), 1300 dynes/cm. These values are greater than those found by Berggren (1914) for the surface tension of amorphous solids. The latter data refer to solid–vapor (air) interfaces. For most solid–liquid systems, the magnitude of the surface free energy is difficult to predict accurately because all models of crystal growth involve multiple parameters, many of which cannot be determined. The majority of experimental studies have been based on measurement of homogeneous nucleation, for which the results may differ appreciably.

Experimental. In potentiostatic seeded-growth and free-drift techniques, the solution supersaturation changes during the reaction, due to the reduction in the concentrations of constituent ions. At each stage, therefore, the supersaturated solutions may be metastable with respect to the different phases that may form and subsequently redissolve as the concentrations decrease (Zhang and Nancollas 1990b). It is impossible to obtain reliable values of the solid/solution interfacial tension because both the rate of crystal growth and the supersaturation change with time. These problems were avoided with the development of the Constant Composition (CC) method (Tomson and Nancollas 1978), in which titrant solutions containing the precipitating constituent ions were simultaneously added to the reaction solution to compensate for their removal by crystal growth. The stoichiometries of the titrants were matched to those of the growing phases while taking into account the concomitant dilution of the reaction solutions that results from the addition of multiple titrants.

Instrumentation for implementing the CC method consisted of thermostatted cell, a potentiometer incorporating ion-selective and reference electrodes, and a potentiostat containing an appropriate computer-interfaced switching device. During the reaction, the output of the potentiometer was constantly compared with a preset value, and the difference, or error signal, was amplified and relayed to an electronic switch. When the error signal exceeded a predetermined threshold value, the switch activated motor-driven titrant burets. The effective titrant concentrations and the number of burets could also be changed to ensure optimal rates of titrant addition. Thus a constant thermodynamic driving force for crystal growth was maintained during an experiment, facilitating the determination of accurate kinetic data with which to calculate interfacial tension.

Possible mechanisms in crystal growth. Once stable nuclei are formed in a supersaturated solution, crystal growth of these nuclei occurs, decreasing the total free energy of the system. The subsequent growth of crystals is governed by at least two processes: diffusion-controlled transport of ions to the growing crystal surface, and surface-mediated processes in which ions are incorporated into the crystal structure. The latter can be divided into the following possible rate-determining steps which take place successively or simultaneously (Burton et al. 1951, Nielsen 1987): (1) transport of a growth unit to the surface by convection; (2) diffusion through the solution to the crystal surface; (3) adsorption at the hydrated adsorption layer; (4) two-dimensional diffusion within the hydrated adsorption layer; (5) adsorption at a step; (6) one-dimensional diffusion along a step; (7) integration at a kink site on a step; and (8) partial or total dehydration of the lattice ion. The first four processes represent transport processes, whereas the last four steps are considered to be surface reactions (Nancollas et al. 1989). The rate equations for dissolution processes may thus be identical to those for corresponding growth processes but with negative values of σ. The mechanisms of crystal growth and dissolution are usually interpreted from measured reaction rates as a function of thermodynamic driving forces, and the data are usually fitted to an empirical rate law such as Equation (29):

$$R = k_1 \sigma^n \tag{29}$$

where k_1 is the effective rate constant, σ is the relative supersaturation with respect to the growing phase, and n is the effective reaction order. The probable mechanism(s), summarized in Table 2, can be deduced recognizing that one or more elementary processes may operate simultaneously (Zhang and Nancollas 1992b). By fitting the kinetic rate data to the appropriate model(s), the interfacial tensions can be calculated.

The effective reaction order (n) can be derived from logarithmic plots of rate, R, against relative supersaturation, σ, using Equation (30):

$$n = \frac{\partial \ln R}{\partial \ln \sigma} \tag{30}$$

From precipitation data for gypsum, linear logarithmic plots of the growth rates of gypsum, $\ln R$ as a function of $\ln \sigma$, shown in Figure 3, gave slope of 3.4 suggesting a polynucleation crystal growth mechanism.

Table 2. Effective reaction order, n, for different mechanisms of crystal growth (Zhang and Nancollas 1990b).

n	Probable mechanisms
1	Volume diffusion; Adsorption; Volume diffusion + Adsorption
1–2	Combined mechanisms such as: Adsorption + Surface diffusion; Adsorption; Integration; Volume diffusion + Surface diffusion; Volume diffusion + Integration; Volume diffusion + Polynucleation
2	Surface diffusion; Integration; Surface diffusion + Integration
>2	Polynucleation growth; Polynucleation + Spiral growth; Polynucleation dissolution; Spiral dissolution controlled by surface diffusion and/or detachment; Polynucleation + Spiral dissolution

Model parameters for crystal growth. Problems arise in defining parameters such as the molar volume of the growth phase (V_m), the size of a growth unit (a), and the geometrical shape factor (β) involving the ratio of the perimeter to surface dimensions. Because the interfacial tension may be proportional to the second power of the size, small errors in size result in very different calculated values of the interfacial tension. It is, therefore, difficult to obtain reliable surface free energy from kinetic data. To be able to compare interfacial values from different sources, the parameters should be equivalent. In practice, the mean diameter and edge length of a growth unit, a, is approximated by Equation (31) (Nielsen 1964, Ohara and Reid 1973, Nyvlt et al. 1985):

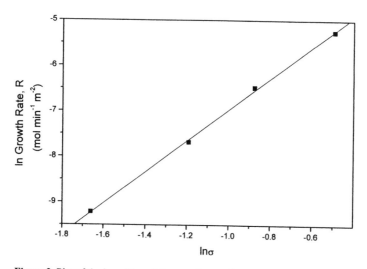

Figure 3. Plot of the logarithm of the growth rate of gypsum as a function of the logarithm of the relative supersaturation (unpublished).

$$a = v_m^{1/3} \tag{31}$$

The volume of a single ion has often been used to represent v_m for calculating the interfacial tension; i.e. $v\, v_m$ = molar volume/Avogadro's number, where v is the number of ions in the chemical formula of the growing crystal (Young 1805, Nielsen 1964, Koutsoukos and Nancollas 1987, Christoffersen et al. 1988, Koutsopoulos et al. 1994). For example, the surface tension of barium sulfate (Nielsen 1964) was found to be 126 mJ m^{-2} by using $2\, v_m$ = the formula weight (crystal density × Avogadro's number)$^{-1}$ = 8.60×10^{-23} or $v_m = 4.30 \times 10^{-23}$ cm^3. The surface tension between strontium sulfate and aqueous NaCl solutions was found to be 75.6 mJ m^{-2} by using the same formula ($v_m = 3.84 \times 10^{-23}$ cm^3) (He et al. 1995). Although Nielsen (1964) gave no explanation for the use of a factor of 2 in calculating v_m for determining the interfacial tension between barium sulfate surface and solution, other researchers have followed his lead.

To calculate the interfacial tension from kinetic data for crystal growth and dissolution, γ_{sl} should represent the interfacial tensions at the crystal/solution interface. The crystallization driving force (or relative supersaturation) depends only on the solution composition and concentration, and not on the values of v_m. However, γ_{sl} is not independent of v_m because, if the ionic volume is used in the calculation, the resulting γ_{sl} represents the interfacial tension between the ion and the solution interface. Equivalent chemical formulae volumes also may not be used to represent v_m in the calculation of interfacial tensions at the crystal/solution interfaces. A single $CaSO_4 \cdot 2H_2O$ molecule does not describe a gypsum crystal, four such molecules repeated in three dimensions make up the gypsum unit cell, which is the smallest unit of the crystal structure. Therefore, in calculating the interfacial tension at a crystal/solution interface, v_m should be taken as the effective volume of the crystallizing solute molecule, i.e. the volume of the unit cell of the crystal (Liu and Nancollas 1996). The selected parameters for several crystal phases (Lide 1995) used in this chapter are listed in Table 3.

Table 3. Cell parameters for gypsum, barite, and celestine.

Crystal	a_o $(10^{-10} m)$	b_o $(10^{-10} m)$	c_o $(10^{-10} m)$	Z	v_m $(10^{-30} m^3)$	v_i $(10^{-30} m^3)$
Gypsum	5.68	15.18	6.29	4	496.1	124.0
Barite	8.88	5.45	7.15	4	346.0	86.5
Celestine	8.36	5.35	6.87	4	307.2	76.8

Note: Z is the number of molecules in the unit cell, v_m the effective volume of the crystallizing solute molecule, and v_i the volume of one molecule (Lide 1995).

Interfacial energies between minerals and aqueous solutions

Calcium sulfate, barium sulfate, and strontium sulfate. Calcium sulfate (gypsum, hemihydrate, and anhydrite), barium sulfate (barite) and strontium sulfate (celestine) are the most common sulfates encountered by the oil and gas industry (Collins 1975, Cowan and Weintritt 1976, Vetter 1976), evaporative desalination plants (Stumm and Morgan 1970), cooling water technology (Cowan and Weintritt 1976), and by other industries involving subsurface waters or brines, such as geothermal energy and brine disposal (Rosenbauer et al. 1992). Sulfate scales deposit and accumulate in formations, in production tubing, and in surface equipment mainly because of changes in temperature, pressure, or both. Once formed, sulfates are very difficult to remove chemically. In addition, scales of barium sulfate have the potential to contain coprecipitated radium, which results in the occurrence of naturally occurring radioactive materials (NORM) (Smith 1987, Gray 1993). Handling and disposal of scales containing NORM are closely regulated by governmental agencies and associated costs can be significant. Therefore, prevention of the deposition of sulfate scales deposition is of practical significance.

The kinetics of crystallization of these minerals have been well studied using different methods (Campbell and Nancollas 1969, Doremus 1970, Smith and Sweett 1971, Nancollas and Reddy 1974, Liu and Nancollas 1975, Gill and Nancollas 1980, Nancollas and Sawada 1982). The most economical method to control scale is to prevent its deposition by inhibiting nucleation and crystal growth using chemical inhibitors (Van Rosmalen 1981, He et al. 1994a,b). These inhibitors are reported to be effective (Rizkalla 1983, Van der Leeden et al. 1992, Amjad 1988, 1994). The performance of scales inhibitors is significantly influenced by many variables, such as the degree of supersaturation, temperature, pH, and the constituent cation/anion molar ratio. Although these parameters have been recognized, the most important factor involved in scale formation is the adherence of precipitates to the scaling surfaces. A knowledge of the precipitation kinetics is needed to predict scale formation and guide the choice of inhibitors. Adding inhibitors, such as hexamethylenediaminetetra (methylene phosphonic) acid (HDTMP) and hydroxyethylene-1,1-diphosphonic acid (HEDP), to supersaturated solutions of calcium sulfate, barium sulfate, or strontium sulfate will increase the interfacial tension between the crystal and aqueous solution, thereby prolonging the nucleation induction period. The increase in the interfacial tension depends upon the nature and concentration of inhibitor. Generally, the interfacial tension increases with increasing concentration of inhibitors. In the case of gypsum inhibition, the presence of 10 mg L^{-1} HDTMP elevated the interfacial tension from 46 to 70 mJ m^{-2} at 25°C, the presence of 10 mg L^{-1} HEDP elevated the interfacial tension of barite from 141 to 248 mJ m^{-2}, and of celestine from 53 to 90 mJ m^{-2} at 25°C, respectively (Table 4). Therefore, the effect of inhibitors on the nucleation rate can be interpreted in terms of the elevated interfacial tension between the crystal and aqueous solution in the presence of inhibitors. The same conclusions were reached for the inhibition of nucleation of other sparingly soluble sulfates of the alkaline-earth metals. Table 4 summarizes the values of the interfacial tension between the sulfate surfaces and the aqueous phases, in the absence and presence of scale inhibitors at 25°C (He et al. 1994a,b; 1995).

The differences between various additives for nucleation inhibition may be caused by changes in the interfacial tension for a given crystal. These changes may depend on the different molecular structures of these inhibitors, such as their molecular weight and size, number of functional groups, and degree of dissociation (Amjad 1994). The affinity between the inhibitor molecule and the crystal surface also plays an important role in determining the efficiency of the inhibition. Although gypsum and barite both consist of alkaline earth metals and sulfate, gypsum is more selective to inhibitors probably because of the occurrence of a water sheet around calcium ions at the nucleus surface, and because of gypsum's much higher aqueous solubility.

The performance of nucleation inhibitors can be affected by many factors other than the degree of supersaturation and temperature. The most important parameters include ionic strength, pH, and the constituent cation/anion molar ratio of the solution. Several studies have discussed the influence of solution pH on the performance of polymeric and non-polymeric inhibitors on the crystal growth of sparingly soluble salts. Griffiths et al. (1979) have shown that the performance of phosphonic acids as inhibitors of $CaSO_4 \cdot 2H_2O$ crystal growth improves with increasing pH, this improvement was attributed to an increase in the degree of ionization of the acids. Leung and Nancollas (1978), using a seeded growth technique, arrived at similar conclusions after studying the effect of benzene polycarboxylic acids on the crystal growth of $BaSO_4$.

Table 4. Estimated values of the interfacial tension between gypsum, barite, or celestine and aqueous solutions, obtained from kinetics rate data in the presence of scale inhibitors at 25°C (He et al. 1994a,b; 1995).

Sample	*NaCl* (mol L^{-1})	Inhibitor	Concentration (mg L^{-1})	Interfacial tension (mJ m^{-2})
Gypsum	3.0	HDTMP	0	45.5
Gypsum	3.0	HDTMP	1.0	49.4
Gypsum	3.0	HDTMP	5.0	55.7
Gypsum	3.0	HDTMP	10.0	69.9
Barite	1.0	HEDP	0	140.5
Barite	1.0	HEDP	0.5	171.2
Barite	1.0	HEDP	1.0	181.0
Barite	1.0	HEDP	5.0	223.6
Barite	1.0	HEDP	10.0	248.2
Celestine	1.0	HEDP	0	53.3
Celestine	1.0	HEDP	0.5	61.9
Celestine	1.0	HEDP	1.0	67.5
Celestine	1.0	HEDP	5.0	80.3
Celestine	1.0	HEDP	10.0	90.2

The cation/anion molar ratio of constituent ions in aqueous solutions (e.g. $[Ca^{2+}]/[SO_4^{2-}]$ or $[Ba^{2+}]/[SO_4^{2-}]$) also influences the inhibition of nucleation of gypsum and barite. He et al. (1994a) observed that the cation/anion molar ratio of the solution had a strong impact on the effectiveness of anionic inhibitors, such as HDTMP and HEDP. The degree of the inhibition of nucleation generally increased with increasing cation/anion ratios. However, when no inhibitors were present, the nucleation induction period appeared to be independent of this ratio in the case of gypsum, or slightly dependent for celestine. The effect of the cation/anion molar ratio on the inhibition of nucleation can be explained in terms of nucleation theory (Gunn and Murthy 1972).

For strontium sulfate, the variation of induction-period with ionic strength can be explained by the variation of $SrSO_4$ solubility in strong electrolytes. The increased solubility of $SrSO_4$ in NaCl and KCl solutions results in a decrease of the interfacial tension between $SrSO_4$ and electrolyte solutions. The increased nucleation rate in the presence of electrolytes is thus a result of the decreased interfacial tension. The same phenomenon was also observed for the nucleation of gypsum and barite (He et al. 1994a,b) in electrolyte solutions. Thus, in the crystallization reactions, the much smaller values of the interfacial tension between aqueous media and the solid phases suggest that these phases would have a greater tendency to be nucleated either on their own or on foreign surfaces. Larger interfacial tensions would result in prolonged induction times, which are exponentially related to $\gamma_{s\ell}$. This effect is illustrated by the results of studies of the nucleation and growth of sparingly soluble minerals (Perez and Nancollas 1988).

CONTACT-ANGLE METHOD

Direct observations and measurements of contact angle have become standard experimental methods to obtain surface energetics (solid–vapor and solid–liquid surface tensions). Thomas Young, in 1805, proposed the following relationship between the contact angle θ and the interfacial tensions γ:

$$\gamma_{sv} = \gamma_{s\ell} + \gamma_{\ell v}\cos\theta \tag{32}$$

where the subscripts sv, sℓ, and ℓv refer to the solid–vapor, solid–liquid, and liquid–vapor interfaces, respectively, and θ is the contact angle formed between a pure liquid and the surface of the solid.

Surface-tension component theory

The surface free energy of a given substance can be represented by the sum of contributions of different types of interactions (Fowkes 1964):

$$\gamma_i = \gamma_i^d + \gamma_i^i + \gamma_i^p + \gamma_i^H \tag{33}$$

In Equation (33), the superscript d refers to dispersion forces, i to induction forces, p to the Keesom force, and H to hydrogen bonding. Generally, γ may be expressed as the sum of only two terms: a dispersive component (superscript d) and a "polar" component, which includes all of the nondispersive forces (superscript AB), such as Debye (induction), Keesom, and hydrogen-bond interaction (Fowkes and Mostafa 1978, Fowkes et al. 1980, Fowkes 1987):

$$\gamma_i = \gamma_i^d + \gamma_i^{AB} \tag{34}$$

In the same year that Fowkes (1987) defined the Lewis acid–base interactions (AB), Van Oss et al. (1987a,b; 1988) showed that the use of a single polar term was empirically inadequate. Two terms are needed: γ^+ and γ^-, which are, respectively, the acidic (or electron acceptor) and basic (or electron donor) parameters of the surface-tension component, γ^{AB}. The component can be used only in conjunction with these two terms:

$$\gamma_i^{AB} = 2\sqrt{\gamma_i^+\gamma_i^-} \tag{35}$$

Using the Lifshitz (1955) approach for van der Waals interactions in condensed media, Chaudhury (1984) showed that the dispersion (d, London), induction (i, Debye) and dipole (p, Keesom) contributions to the Lifshitz–van der Waals (or apolar) surface-tension component γ^{LW}, are simply additive:

$$\gamma^{LW} = \gamma_i^d + \gamma_i^i + \gamma_i^p \tag{36}$$

This can be proved experimentally in a rather simple manner (Good 1992, Van Oss 1994). These studies considerably clarified the treatment of "polar" interactions of hydrogen-bonding origin, without unnecessary concern for putative encroachments of, or overlaps between, the electrodynamic LW forces. The AB component of interfacial (or surface) tension must be added to the LW component,

$$\gamma_i = \gamma_i^{LW} + \gamma_i^{AB} \tag{37}$$

The postulated combining rule for acid–base interactions across an interface is, for the free energy of adhesion,

$$\Delta G_{ij}^{AB} = -2\left[\sqrt{\gamma_i^+\gamma_j^-} + \sqrt{\gamma_i^-\gamma_j^+}\right] \tag{38}$$

For γ_{ij}^{AB}, the combining rule is

$$\gamma_{ij}^{AB} = 2\left[\sqrt{\gamma_i^+\gamma_i^-} + \sqrt{\gamma_j^+\gamma_j^-} - \sqrt{\gamma_i^+\gamma_j^-} - \sqrt{\gamma_i^-\gamma_j^+}\right] \tag{39}$$

or

$$\gamma_{ij}^{AB} = 2\left[\sqrt{\gamma_i^+} - \sqrt{\gamma_j^+}\right]\left[\sqrt{\gamma_i^-} - \sqrt{\gamma_j^-}\right] \tag{40}$$

This form of the combining rule (which is not a geometric mean), first suggested by Small (1953), can be derived from molecular orbital theory (Kollman et al. 1975, Kollman 1977).

The Dupré equation is also applicable to adhesive AB interactions, *in vacuo*. Combination of $\gamma_{ij}^{\cdots}$ and $\gamma_{ij}^{\cdots}$ yields the total interfacial tension between two different condensed phases:

$$\gamma_{ij} = \left[\sqrt{\gamma_i^{LW}} - \sqrt{\gamma_j^{LW}}\right]^2 + 2\left[\sqrt{\gamma_i^+\gamma_i^-} + \sqrt{\gamma_j^+\gamma_j^-} - \sqrt{\gamma_i^+\gamma_j^-} - \sqrt{\gamma_i^-\gamma_j^+}\right] \tag{41}$$

The AB contribution to the surface tension or the interfacial tension between a solid surface and a polar liquid, such as water, is of paramount importance. The contribution of acid–base terms may be negative, but the LW component is always positive, the total interfacial tension can be either negative or positive, depending upon the values of the polar interaction parameters γ^+ and γ.

The adhesive free-energy change for apolar and acid–base components is expressed by

$$\Delta G_{ij} = -2\left[\sqrt{\gamma_i^{LW}\gamma_j^{LW}} + \sqrt{\gamma_i^+\gamma_j^-} + \sqrt{\gamma_i^-\gamma_j^+}\right] \tag{42}$$

The complete contact-angle equation, for a liquid on a solid, is

$$\Delta G_{s\ell} = -\gamma_\ell(1 + \cos\theta) \tag{43}$$

or

$$\gamma_\ell(1 + \cos\theta) = 2(\sqrt{\gamma_s^{LW}\gamma_\ell^{LW}} + \sqrt{\gamma_s^+\gamma_\ell^-} + \sqrt{\gamma_s^-\gamma_\ell^+}) \tag{44}$$

To determine the surface tension component ($\gamma_s^{L^w}$) and parameters (γ_s^+ and γ_s^-) of a solid, the contact angle must be determined with at least three liquids, two of which must be polar (Van Oss et al. 1988, Van Oss 1994, Wu et al. 1995).

Contact-angle measurement and thin-layer wicking

For the determination of the contact angle formed between a liquid (or air, or vapor) and finely divided solid particles, a thin-layer wicking, capillary-rise method may be used (Van Oss et al. 1992, Wu 1994, Liu and Nancollas 1996, Wu and Nancollas 1997). In this method, the measured rate of capillary rise of a liquid through a packed column of powder supported on a glass slide for microscopy was substituted into the Washburn equation (Washburn 1921)

$$h^2 = \frac{t r_{eff} \gamma_L \cos\theta}{2\eta} \tag{45}$$

where r_{eff} is the effective radius of the interstitial pore, γ_L is the surface tension of the liquid, θ is the contact angle of the liquid on the solid, η is the viscosity of the liquid, and h is the rate of capillary rise during given time t. To obtain $\cos\theta$, the value of r_{eff} must be determined independently. This can be done by using low-energy nonpolar liquids, in this case octane, decane, and dodecane (Table 5), which spread over the solid surface without forming a finite contact angle. It was shown that, with such spreading liquids, θ remains exactly equal to zero, so that $\cos\theta = 1$, as the result of the formation of a precursor film. Once r_{eff} is known, the contact angle between the solid and the wicking liquids α-bromo-naphtalene (α-Br), diiodomethane (DIM), water (W), ethylene glycol (EG), and formamide (FO) can be determined (Table 6), permitting Equation (44) to be solved for γ_s^{LW}, γ_s^+, and γ_s^- of the crystal–particle surface.

To prepare thin CSD crystallite layers for measurements of contact angle, 1.5% suspensions in triple-distilled water (TDW) were maintained with continuous magnetic stirring while 5 ml aliquots withdrawn by a pipette were uniformly distributed on horizontal, glass slides (7.5 cm _ 2.5 cm). The water was allowed to evaporate, leaving a uniform, thin deposit of the powdered mineral firmly adhering to the glass.

Table 5. Surface-tension components γ, of liquids used in direct contact-angle determination, or in wicking, and viscosities, η, at 200°C.

Liquid	γ_L (mJ m^{-2})	γ^{LW} (mJ m^{-2})	γ^{AB} (mJ m^{-2})	γ^+ (mJ m^{-2})	γ^- (mJ m^{-2})	η (poise)
Heptane	20.14	20.14	0	0	0	0.00409
Octane	21.62	21.62	0	0	0	0.00542
Decane	23.83	23.83	0	0	0	0.00907
Dodecane	25.35	25.35	0	0	0	0.01493
Tetradecane	26.56	26.56	0	0	0	0.02322
Hexadecane	27.47	27.47	0	0	0	0.03451
α-Bromonaphthalene	44.4	44.4	0	0	0	0.0489
Diiodomethane	50.8	50.8	0	0	0	0.028
Ethylene glycol	48.0	29.0	19.0	1.92	47.0	0.199
Formamide	58.0	39.0	19.0	2.28	39.6	0.0455
Glycerol	64.0	34.0	30.0	3.92	57.4	14.9
Water	72.8	21.8	51.0	25.5	25.5	0.01

Note. Total surface $\gamma_L = \gamma^{LW} + \gamma^{AB}$, and $\gamma^{AB} = 2(\gamma^+ \gamma^-)^{1/2}$. The values of γ_L, γ^{LW}, γ^{AB}, γ^+, γ^-, and η are from van Oss et al. (1992b) and Van Oss (1994).

Thin-layer wicking was performed by immersing the coated glass slides in the vertical position to a depth of about 5 mm in one of the test liquids listed in Table 5, in cylindrical glass containers fitted with gas-tight stoppers of ground glass. Before immersion, the coated slides were placed inside the closed containers, for about 1 hour, to allow the powders to become equilibrated with the vapor of the low-energy spreading liquids. No equilibration was necessary when the test liquid was α-bromonaphthalene, diiodomethane, water, ethylene glycol, or formamide. Following these procedures, the slides were brought into contact with the liquid surfaces, and the vertical movement of the liquid fronts through the layers of powder was observed.

In Table 6, the surface-tension components and parameters of CSD crystallites were calculated by means of Young's Equation (44). The surface-tension component (γ_L^{LW}) and parameters (γ_L^+ and γ_L^-) of test liquids and values of γ_S^{LW} were determined by wicking with 1-bromonaphthalene and diiodomethane, nonpolar liquids that form finite contact angles. Finally, the polar parameters, γ_S^+ and γ_S^-, were obtained by wicking with polar liquids (formamide, ethylene glycol, and water), using the measured γ_S^{LW} values. The interfacial tension between the crystals and aqueous solution was calculated from Equation (41).

Thin-layer wicking is a novel method for the determination of contact angles (θ) on powdered materials. To employ the wicking approach for the determination of cos θ, it is essential to use exceedingly well-packed columns of rather monodispersed particles (Ku et al. 1985). This packing can be achieved by coating the particles of the solid onto glass slides, followed by measurement of the capillary rise of various test liquids in a manner similar to thin-layer-wicking chromatography (Van Oss et al. 1992, Liu and Nancollas 1996, Wu and Nancollas 1997). In general, the surface-tension components and parameters are obtained by first measuring cos θ, and then solving the complete Young's Equation (44).

Table 6. Interfacial tensions, γ, between water and gypsum and surface-energy component, obtained from crystal growth, solubility, and contact-angle methods.

Material	γ^{LW} (mJ m^{-2})	γ^+ (mJ m^{-2})	γ (mJ m^{-2})	Y_θ, γ_{sl} (mJ m^{-2})	GK, γ_{sl} (mJ m^{-2})	S, γ_{sl} (mJ m^{-2})
Gypsum [a]	25.7	2.3	23.9	1.31	7	
Gypsum [a]	36.5	3.0	8.6	+15.9	4	12.2–36.7
Gypsum [b]	43.1	1.1	52.2	-13.8		

Abbreviations: Y_θ, interfacial tension from measurements of contact angle, GK, interfacial tension from growth kinetics, S, interfacial tension from solubility.
[a] Unpublished.　　[b] Data from Wu (1994).

CONCLUSIONS

The precipitation of calcium sulfate, barium sulfate, and strontium sulfate, all surface-controlled, can be described by a (heterogeneous) nucleation and growth model in which the growth rate is determined by two-dimensional surface nucleation. A number of methods, such as turbidity, conductivity, and constant composition, have been used to study the growth rates. For strontium sulfate, kinetic analysis of the pre-nucleation induction times suggested the existence of two regions: in the first, a mainly homogeneous process, and in the second, a predominantly heterogeneous process. The dependence of the measured rates on the relative supersaturation showed that, at low supersaturation, the apparent order corresponded to a surface-control process. At higher extremely difficult, and commonly the calculated free-energy data are seriously in error. Thin-layer wicking seems to offer a promising method for the determination of contact angles and the mean pore size of powdered materials. Hitherto, such studies have been made only for the characterization of calcium sulfate dihydrate (gypsum). It will be very interesting to extend use of the wicking method to the other sparingly soluble sulfates.

Although approximate values of interfacial tension values have been estimated from experimental data on homogeneous and heterogeneous nucleation, including nucleation rates and induction periods as a function of supersaturation, the calculated interfacial energies may be subject to appreciable errors.

NOMENCLATURE

Symbols

A	Constant
a	Area of growth unit
a_o	Cell parameter
b_o	Cell parameter
c_o	Cell parameter
c	Constant
C	Solute concentration
CC	Constant composition
C_s	Solubility concentration
C_{eff}	Equivalent number of moles precipitated per liter of added titrant
F	Helmholtz free energy
f	Specific surface free energy
G	Gibbs free energy
ΔG^*	Change in Gibbs free energy during critical nucleus formation
ΔG_{ii}	Gibbs free energy change of cohesion
ΔG_{ij}	Gibbs free energy change of adhesion
h	Capillary rise
IP	Activity ionic product
J	Rate of nucleation
K_{sp}	Solubility product
k	Boltzmann constant
k_1	Effective rate constant
L	Size of a crystal
L^*	Critical size
m	Mass
M	Molecular weight
n	Order of reaction
n^*	Critical nucleus
n_i	Number of mole of component, i
p	Constant
r	Linear growth rate
R	Growth rate
R_c	Corrected growth rate
R	Gas constant
r_{eff}	Effective interstitial pore radius
S	Supersaturation ratio
s_e	Effective surface area of a crystal suspension at time t
s	Surface area

T	temperature
t	Time
V	Volume
v	Volume of the nucleus
v_i	Volume of one molecule
V_m	Molar volume
v_m	Molecular volume of unit cell
Z	Number of molecules in the unit cell
{ }	Aqueous ionic activity
[ ]	Molar concentration
α	Volume shape factor
β	Surface shape factor
γ_i	Surface tension for component i
γ_{ij}	Interfacial tension between components i and j
θ	Contact angle
μ_i	Chemical potential
η	Viscosity of the liquid
ν	Number of ions per formula unit
σ	Relative supersaturation
τ	Induction period
ϕ	Heterogeneous nucleation factor
Γ_i	Surface excess free energy

Subscripts and superscripts

AB	Lewis acid–base component (polar)
d	Dispersion forces
H	Hydrogen bonding
het	Heterogeneous nucleation
hom	Homogeneous nucleation
i	Induction forces
l, L	Liquid or water
ℓv	Liquid–vapor interface
LW	Lifshitz–van der Waals component (nonpolar)
p	Keesom forces
s	Solid
sℓ	Solid–liquid interface
sv	Solid–vapor interface

REFERENCES

Adamson AW, Gast AP (1997) Physical Chemistry of Surfaces. John Wiley and Sons, New York

Allen JA (1953) The precipitation of nickel oxalate. J Phys Chem 57:715-716

Amathieu L and Boistelle R (1988) Crystallization kinetics of gypsum from dense suspension of hemihydrate in water. J Crystal Growth 88:183-192

Amjad Z (1988) Calcium sulfate dihydrate (gypsum) scale formation on heat exchanger surfaces: The influence of scale inhibitors. J Colloid Interface Sci 123:523-536

Amjad Z (1994) Inhibition of barium sulfate precipitation: Effects of additives, solution pH, and supersaturation. Water Treatment 9:47-56

Band W (1939a) Dissolution treatment of condensing systems. J Chem Phys 7:324-326

Band W (1939b) Dissolution treatment of condensing systems II. J Chem Phys 7:927-931

Barone JP, Nancollas GH, Yoshikawa Y (1983) Crystal growth as a function of seed surface area. J Crystal Growth 63:91-96

Becker VB, Döring W (1935) Kinetische Behandlung der Keimbildung in übersättigten Dämpfen. Ann Physik 24 (5):719-542

Berggren B (1914) A method for determination of the surface tension of amorphous substances. Ann Physik 4: 61-80

Budz JA, LoRe M, Nancollas GH (1988) The influence of high- and low-molecular-weight inhibitors on dissolution kinetics of hydroxyapatite and human enamel in lactate buffers: A constant composition study. J Dental Res 67:1493-1498

Burton WK Cabrera N and Frank FC (1951) The growth of crystals and the equilibrium structure of their surfaces. Phil Trans Roy Soc Lond 242:299-358

Campbell JR, Nancollas GH (1969) The crystallization and dissolution of strontium sulfate in aqueous solution. J Phys Chem. 73:1735-1740

Chaudhury MK (1984) Short-range and long-range forces in colloidal and macroscopic systems. PhD Dissertation, State University of New York at Buffalo, New York, 189 p

Christoffersen J, Christoffersen MR (1976) The kinetiks of dissolution of calcium sulfate dihydrate in water. J Crystal Growth 35:79-88

Christoffersen J, Christoffersen MR, Kibalczyc W, Perdok WG (1988) Kinetics of dissolution and growth of calcium fluoride and effects of phosphate. Acta Odontol Scand 46:325-336

Christoffersen MR, Christoffersen J, Weijnen WGJ, van Rosmalen G (1982) Crystal growth of calcium sulfate dihydrate at low supersaturation. J Crystal Growth 58:585-595

Collins AJ (1975) Geochemistry of Oilfield Waters. Elsevier Scientific, New York, p 367-386

Cowan JC, Weintritt DJ (1976) Water-Formed Scale Deposits. Gulf Publishing Co, Houston, Texas

Davies CW, Jones AC (1949) The precipitation of silver chloride from aqueous solutions. Discuss Faraday Soc 5:103-111

Davies CW, Jones AL (1955) The precipitation of silver chloride from aqueous solutions. Faraday Soc Trans 51:812-829

Davis JA, Hayes KF (Eds) (1986) Geochemical Processes at Mineral Surfaces. American Chemical Society, Chicago, p 2-18

Defay R, Prigogine I (1966) Surface Tension and Adsorption. John Wiley and Sons, New York

Doremus RH (1970) Crystallization of slightly soluble salts from solution. J Phys Chem 74:1405-1408

Dunning WJ (1969) General and Theoretical Introduction. In Nucleation. Zettlemoyer AC (Ed) Marcel Dekker, New York, p 1-61

Ebrahimpour A (1990) Kinetics of dissolution, growth and transformation of calcium salts and the development of spectroscopic and dual constant composition. Ch 7 in PhD thesis, State University of New York at Buffalo

Enüstün BV, Turkevich J (1960) Solubility of fine particle of strontium sulfate. J Am Chem Soc 82:4502-4509

Fowkes FW (1964) Attractive forces at interfaces. Ind Eng Chem 56:40-52

Fowkes FW (1987) Role of acid-base interfacial bonding in adhesion. J Adhesion Sci spreading pressures of liquids on hydrophobic solids. J Colloid Interface Sci 78:Tech 1:7-27

Fowkes FW, MacCarth DC, Mostafa MA (1980) Contact angles and the equilibrium spreading pressures of liquids on hydrophobic solids. J Colloid Interface Sci 78:200-206

Fowkes FW, Mostafa MA (1978) Acid-base interactions in polymer adsorption. Int Eng Chem Prod Res Dev 17:3-7

Frenkel J (1939a) Statistical theory of condensation phenomena. J Chem Phys 7:200-201

Frenkel J (1939b) A general theory of heterophase fluctuations and pretransition phenomena. J Chem Phys 7:538-547

Frenkel J (1946) Kinetic Theory of Liquids. Oxford University Press, London

Freundlich H (1923) Colloid and Capillary Chemistry. EP Dutton and Company, New York

Füredi-Milhofer H, Brecevic L, Komunjer M, Purgaric B, Babic-Ivancic V (1977) The use of precipitation diagrams in the determination of critical supersaturation for homogeneous nucleation. Croat Chem Acta 50:139-154

Gardner GL and Nancollas GH (1983) Crystal growth in aqueous solution at elevated temperatures: Barium sulfate growth kinetics. J Phys Chem 87:4699-4703

Garten VA and Head RB (1973) Nucleation in salts solutions. Faraday Trans 1, Chem Soc London 69:514-520

Gibbs JW (1928) Collected Works. Longmans Green, London

Gill JS and Nancollas GH (1980) Kinetics of growth of calcium sulfate crystals at the heated metal surfaces. J Crystal Growth 48:34-40

Good RJ (1967) Physical significance of parameters γ_c, γ_s and ϕ, that govern spreading on adsorbed films. In Wetting. Soc Chem Ind Monogr 25:328-344

Good RJ (1992) Contact angle, wetting and adhesion: A critical review. J Adhesion Sci Technol 6:1269-1302

Gray PR (1993) NORM contamination in the petrolium industry. J Petrol Technol 45:12-16

Griffiths DW, Roberts SD, Liu ST (1979) Inhibition of calcium sulfate dihydrate crystal growth by phosphoric acids—influence of inhibitor structure and solution pH. Paper No. 7862, Intern'l Symp on Oilfield and Geothermal Chem, Soc Petrol Engineers, Houston

Gunn DJ, Murthy MS (1972) Kinetics and mechanisms of precipitation. Chem Eng Sci 27:1293-1313

He S, Oddo JE, Tomson MB (1994a) The inhibition of gypsum and barite nucleation in NaCl brines at temperature from 25 to 90°C. Appl Geochem 9:561-567

He S, Oddo JE, Tomson MB (1994b) The nucleation kinetics of calcium sulfate dihydrate in NaCl solutions up to 6 molar and 90°C. J Colloid Interface Sci 162:297-303

He S, Oddo JE, Tomson MB (1995) The nucleation kinetics of strontium sulfate in NaCl solutions up to 6 molar amd 90°C with or without inhibitors. J Colloid Interface Sci 174:327-335

Herring C (1952) The use of classical macroscopic concepts in surface energy problems. *In* Structure and Properties of Solid Surfaces. Gomer R, Smith CS (Eds) Univ Chicago Press, Chicago, p 5-72

Higuchi WI, O'Konski CT (1960) A test of the Becker-Doering theory of nucleation kinetics. J Colloid Sci 15:14-49

Hulett GA (1901) Surface tension and solubility. Z Phys Chem 37:385-406

Humphreys-Owen SPF (1949) Crystal growth from solution. Proc Roy Soc A 197:218-237

Johnson RE (1959) Conflicts between gibbsian thermodynamic and recent treatments of interfacial enegies in solid–liquid-vapor systems. J Phys Chem 63:1655-1658

Joshi MS, Antony AV (1979) Nucleation in supersaturated potassium dihydrogen orthophosphate solutions. J Crystal Growth 46:7-9

Keller DM, Massey RE, Hileman OE (1980) Studies on nucleation phenomena occurring in aqueous solutions supersaturated with calcium sulfate. III. The cation:anion ratio. Can J Chem 58:2127-2131

Kenyon AS, La Mer VK (1949) Light scattering properties of monodispersed sulfur sols. I. Monochromatic ultraviolet angular scattering. II. Effect of the complex index of refraction upon transmittance. J Colloid Sci 4:163-184

Kollman P (1977) A general analysis of noncovalent intermolecular interactions. J Am Chem Soc 99: 4875-4894

Kollman P, McKelvey J, Johannson A and Rothenberg S (1975) Theoretical studies of hydrogen-bond dimers. Complexes involving HF, H_2O, NH_3, HCl, H_2S, PH_3, HCN, HNC, HCP, CH_2NH, H_2CS, H_2CO, CH_4, CF_2H, C_2H_2, C_2H_4, C_6H_6, F⁻, and H_3O^+. J Am Chem Soc 97:955-965

Koutsopoulos S, Paschalakis PC, Dalas E (1994) The calcification of elastin in vivo. Langmuir 10: 2423-2428

Koutsoukos PG, Nancollas GH (1987) The mineralization of collagen in vivo. Colloids Surfaces 28: 95-108

Ku CA, Henry JD, Siriwardane R, Roberts L (1985) Particle transfer from a continuous oil to a dispersed water phase: Model particle study. J Colloid Interface Sci 106:377-387

La Mer VK, Pound GM (1949) Surface tension of small droplets from Volmer and Flood's nucleation data. J Chem Phys 17:1337-1338

Leung WH, Nancollas GH (1978) A kinetics study of the seeded growth of barium sulfate in the presence of additives. J Inorg Nucl Chem 40:1871-1875

Lifshitz EM (1955) The theory of molecular attraction forces between solid bodies. Zh Eksperimentalnoi Teareticheskoi Fiziki 29:94-110

Lide DR (1995) CRC Handbook of Chemistry and Physics. CRC Press, Boca Raton, Florida

Liu ST, Nancollas GH (1970) The kinetics of crystal growth of calcium sulfate dihydrate. J Crystal Growth 6:281-289

Liu ST, Nancollas GH (1973) Linear crystallization and induction period studies of the growth of calcium sulfate dihydrate crystals. Talanta 20:211

Liu ST, Nancollas GH (1975) A kinetics and morphological study of the seeded growth of calcium sulfate dihydrate in the presence of additives. J Coll Interface Sci 52: 593-601

Liu Y, Nancollas GH (1996) Fluorapatite growth kinetics and influence of solution composition. J Crystal Growth 165:116-123

Lothe J, Pound GM (1962) Reconsiderations of nucleation theory. J Chem Phys 36:2080-2085

Malollari I Xh, Klepetsanis P, Koutsoukos PG (1995) Precipitation of strontium sulfate in aqueous solutions at 25°C. J Crystal Growth 155:240-246

McDonald JE (1963) Homogeneous nucleation of vapor condensation: II. Kinetics aspects. Am J Phys 31:31-41

Mealor D, Townshend A (1966) Homogeneous nucleation of some sparingly soluble salts. Talanta 13:1069-1074

Möller P, Rajagopalan G (1975) Precipitation kinetics of $CaCO_3$ in presence of Mg^{2+} ions. Z Phys Chem NF 94:297-314

Möller P, Rajagopalan G (1976) Charges of excess free energies in the crystal growth process of calcite and aragonite due to the presence of Mg^{2+} ions in solution. Z Phys Chem NF 99:187-198

Mullin JW, Osman MM (1973) The nucleation and precipitation of nickel ammonium sulphate crystals from aqueous solution. Kristall Technik 8:471-481

Mullin JW, Ang H-M (1976) Nucleation characteristics of aqueous nickel ammonium sulfate solutions. Faraday Soc Disc 61:141-148

Mullin JW, Zacek S (1981) The precipitation of potassium aluminum sulfate from aqueous solution. J Crystal Growth 53:515-518

Mullins WW (1963) Metal Surfaces: Structure, Energetics and Kinetics. American Society for Metals, Metals Park, Ohio, p 17-66

Nancollas GH (1973) The crystal growth of sparingly soluble salts. Croat Chem Acta 45:225

Nancollas GH, Reddy MM (1974) The kinetics of crystallization of scale-forming minerals. J Soc PetrolEng 117-126

Nancollas GH, Sawada K (1982) The formation of scales of calcium carbonate polymorphs: The influence of magnesium ion and inhibitors. J Soc PetrolEng 645-652

Nancollas GH, Perez L, Richardson C, Zawacki SJ (1989) Mineral phases of calcium phosphate. Anatomical Record 224:234-241

Nielsen AE (1961) Homogeneous nucleation in barium sulfate. Acta Chem Scand 15:441-442

Nielsen AE (1964) Kinetics of Precipitation. Pergamon Press, Elmsford, New York

Nielsen AE (1969) Nucleation and growth of crystals at high supersaturation. Krist Technik 4:17-38

Nielsen AE (1984) Electrolyte crystal growth mechanisms. J Crystal Growth 67:289-310

Nielsen AE (1987) Rate laws and rate constants in crystal growth. Croat Chem Acta 60:531-539

Nyvlt J, Söhnel O, Matuchova M, Broul M (1985) The Kinetics of Industrial Crystallization. **XXXXX ??** Amsterdam

Ohara M, Reid RC (1973) Modeling Crystal Growth Rates from Solution. Prentice-Hall, Englewood Cliffs, New Jersey

Perez L, Nancollas GH (1988) The growth of calcium and strontium sulfates on barium s 'fate surfaces. Scanning Electron Micros 2:1437-1443

Ridge MJ (1960) Mechanism of setting of gypsum plaster. Rev Pure Appl Chem 10:243-276

Rizkalla EN (1983) Kinetics of the crystallization of barium sulfate. J Chem Soc Farady Trans 79: 1857-1867

Rosenbauer RJ, Bischoff JL, Kharaka YK (1992) Geochemical effects of deep-well injection of the Paradox Valley brine into paleozoic carbonate rocks. Appl Geochem 7:273-286

Small PA (1953) Some factors affecting the solubility of polymers. J Appl Chem 3:71-96

Smith AL (1987) Radioactive scale formation. J Petrol Technol 39:697-706

Smith BR, Sweett F (1971) The crystallization of calcium sulfate dihydrate. J Colloid Interface Sci 37:612-618

Söhnel O, Garside J, Jancic SJ (1977) Crystallization from solution and the thermodynamics of electrolytes. J Crystal Growth 39:307-314

Söhnel O, Mullin JW (1978) A method for the determination of precipitation induction periods. J Crystal Growth 44:377-382

Söhnel O, Mullin JW (1982) Precipitation of calcium carbonate. J Crystal Growth 60:239-250

Stevens HM (1961) Wet process phosphoric acid. In Phosphorus and Its Compouds. van Wazer JR (Ed) Vol 2, Wiley Interscience, New York, p 1025-1066

Stumm S, Morgan JJ (1970) The regulation of the chemical composition of natural waters. In Aquatic Chemistry: An introduction emphasizing chemical equilibria in natural waters. Stumm S, Morgan JJ (Eds) Wiley-Interscience, New York, p 383-444

Tomson MB, Nancollas GH (1978) Mineralization kinetics: a constant composition approach. Science 200:1059-1060

Turnbull D (1952) Kinetics of solidification of supercooled liquid mercury droplets. J Chem Phys 20: 411-424

Van der Leeden MC, Kashchiew DK, van Rosmalen GM (1992) Precipitation of barium sulfate: Induction time and the effect of an additive on nucleation and growth. J Colloid Interface Sci 152:338-350

Van Hook A (1940) The precipitation of silver chromate. J Phys Chem 44:751-764

Van Oosterhout GW, van Rosmalen GM (1980) Analysis of kinetics experiments on growth and dissolution of crystals in suspension. J Crystal Growth 48:464-471

Van Oss CJ, Chaudury MK, Good RJ (1987) Monopolar surfaces. Adv Colloid Interface Sci 28:35-64

Van Oss CJ, Chaudury MK, Good RJ (1987) Mechanism of partition in aqueous media Separation Sci Tech 22:1515-1526

Van Oss CJ, Chaudury MK, Good RJ (1988) Interfacial Lifshitz–van der Waals and polar interactions in macroscopic systems. Chem Rev 88:927-941

Van Oss CJ, Giese RF, Li Z, Murphy K, Norris J, Chaudury MK, Good RJ (1992) Determination of contact angles and pore sizes of porous media by column and thin layer wicking. J Adhes Sci Technol 6: 413-428

Van Oss CJ (1994) Interfacial Forces in Aqueous Media. Marcel Dekker, New York, 401 p

Van Rosmalen GM, Daudey PJ and Marchée WGJ (1981) An analysis of growth experiments of gypsum-crystals in suspension. J Crystal Growth 52:801-811

Van Rosmalen GM (1981) Scale prevention: Study on the crystallization of calcium sulfate and barium sulfate with and without inhibitors. PhD Dissertation, Delft University Press, The Netherlands

Vetter OJ (1976) Oilfield scale: Can we handle it? J PetrolTech 28:1402-1408

Volmer M, Webber A (1926) Nucleus formation in supersaturated systems. Z Physik Chem (Leipzig) 119:277-301

Walton AG (1967) The formation and properties of precipitates. Interscience, New York

Walton AG (1969) Nucleation in liquids and solutions. *In* Nucleation. Zettlemoyer AC (Ed) Marcel Dekker Inc, New York, p 225-305

Washburn EW (1921) The dynamics of capillary flow. Phys Rev 17:273-283

Witkamp GJ, van der Eerden JP, van Rosmalen GM (1990) Growth of gypsum: I. Kinetics. J Crystal Growth 102:281-289

Wu W (1994) Linkage between ζ-potential and electron donicity of charged polar surfaces: The mechanisms of flocculation and repeptization of particle suspensions. PhD Dissertation, State Univ New York at Buffalo

Wu W, Giese RF, van Oss CJ (1995) Evaluation of the Lifshitz–van der Waals/acid–base approach to determine surface tension components. Langmuir 11:379-382

Wu W, Nancollas GH (1996) Interfacial free energies and crystallization in aqueous media. J Colloid Interface Sci 182:365-373

Wu W, Nancollas GH (1997) Nucleation and crystal growth of octacalcium phosphate on titanium oxide surfaces. Langmuir 13:861-865

Young T (1805) An essay on the cohesion of fluids. Trans R Soc London 95:65-87

Zeldovich J (1942) Zur Theorie der Buildung einer neuen Phase: Die Kavitation. J Exp Theor Phys (URSS) 12:525-538

Zhang J, Nancollas GH (1990a) Kink densities along a crystal surface step at low temperature and under non equilibrium conditions. J Crystal Growth 106:181-190

Zhang J, Nancollas GH (1990b) Mechanisms of growth and dissolution of sparingly soluble salts. Rev Mineral 23:365-396

Zhang J, Nancollas GH (1992a) Influence of calcium/sulfate molar ratio on the growth rate of calcium sulfate dihydrate at constant supersaturation. J Crystal Growth 118:287-294

Zhang J, Nancollas GH (1992b) Interpretation of dissolution kinetics of dicalcium phosphate dihydrate. J Crystal Growth 125:251-269

Zisman WA (1964) Relation of equilibrium contact angle to liquid and solid constitution. *In* Contact Angle, Wettability and Adhesion. Fowkes FM (Ed) Adv Chem 43:1-51

6   Metal-sulfate Salts from Sulfide Mineral Oxidation

John L. Jambor

Department of Earth and Ocean Sciences
University of British Columbia,
Vancouver, British Columbia V6T 1Z4 Canada

D. Kirk Nordstrom

U.S. Geological Survey
Water Resources Division, 3215 Marine Street
Boulder, Colorado 80303

Charles N. Alpers

U.S. Geological Survey
Water Resources Division, 6000 J Street
Sacramento, California 95819

> *"It is plain that almost all kinds of* atramentum *are made of Earth and Water.*
> *At first they were liquid and afterwards solid, and still they can be redissolved,*
> *by heat and moisture."*
>
> Albertus Magnus (1205-1280) *Book of Minerals* (transl. D. Wyckoff, 1967)

The observation of "efflorescences," or the flowering of salts, associated with periods of dryness in soils, in closed-basin lakes, in rock outcrops, and in mines and mine wastes has been noted since early antiquity. The formation of metal-sulfate salts, in connection with the mining of metals, was a phenomenon well known to the early Greek and Roman civilizations. Alum, most commonly potash alum $KAl(SO_4)_2 \cdot 12H_2O$, which is from the Latin *alumen*, was extensively mined and used by goldsmiths, dyers, paper manufacturers, and physicians in ancient civilizations. It forms from the oxidation of pyrite in shales and slates and from oxidation of sulfurous gases in geothermal areas. The Greeks and the Romans described stalactites of *atramentum* (soluble metal-sulfate salts) that formed within mines and along rock faces (Agricola 1546, 1556). Furthermore, the toxic effects of these salts on animals were also noted. For example, in *De Natura Fossilium*, Agricola (1546) stated "....I mention the congealed acid juice which usually produces *cadmia*. It is white, hard, and so acrid that it can eat away walls, grills and even destroy all living matter." *Cadmia* is thought to be derived from the oxidation of zinc, cobalt, and arsenic sulfides, such as cobaltite. He goes on to say that "Pyrite, unless it contains sulphates, is either a golden or silver color, rarely any other, while *cadmia* is black, yellow brown, or gray. The former will cure gatherings while the latter is a deadly poison and will destroy any living substance. It is used to kill grasshoppers, mice and flies." These descriptions suggest the presence of arsenic compounds. The range of colors from white to black commonly is caused by different amounts of admixed pyrite with sulfate minerals. From the days of the Greek philosopher Theophrastus (*ca* 325 BCE) and the Greek physician Dioscorides (first century CE), the efflorescent salts *atramentum sutorium virida* or melanterite (also called *melanteria*) and *atramentum sutorium caeruleum* or chalcanthite were well known to form from the corrosion of pyrite and chalcopyrite by moisture (Agricola 1546, footnotes on p. 47-51). By the time of Pliny the Second (Caius Plinius Secundus, 23-79 CE), the names "green vitriol" for melanterite and "blue vitriol" for chalcanthite were in common use and continued to be used from the Middle Ages to the 20th century.

1529-6466/00/0040-0006$05.00

Today, we know that the formation and dissolution of metal-sulfate salts play an important role in the storage and transport of acids and metals released upon weathering of mineralized rocks, coal deposits, metallic ore deposits, and mine wastes. The composition of the salts reflects the composition of the evaporated waters from which the salts form, including information on whether the waters are acidic, basic, or near neutral. The original composition of the waters reflects water–mineral interactions involving sulfide-mineral oxidation and reactions with non-sulfide gangue minerals, especially carbonates and silicates. Hence, identification of efflorescent salts can provide information about water quality and water–rock interactions.

Oxidation of sulfide minerals in coal and metallic ore deposits typically leads to the formation of both insoluble and water-soluble, metal-bearing sulfates, hydroxysulfates, and hydrous oxides. The reactions generally lead to the generation of acidic solutions. As pH decreases, sulfide oxidation is accelerated because mineral solubilities and metal concentrations increase.

On a global scale, the largest tonnages of metal production are derived from remarkably few primary sulfide minerals. For example, the world's principal sources of Pb, Zn, and Cu are galena PbS, sphalerite $(Zn,Fe)S$, and chalcopyrite $CuFeS_2$, respectively. If bornite Cu_5FeS_4, chalcocite Cu_2S, and covellite CuS are added as important, albeit minor relative to chalcopyrite, sources of Cu, probably >90% of the world's annual production of Cu + Pb + Zn from mineral deposits would be accounted for. These ore minerals, however, are accompanied by gangue minerals. Although most silicate gangue minerals are environmentally benign, they are commonly accompanied, or exceeded in the case of massive sulfide deposits, by iron sulfides such as pyrite FeS_2, pyrrhotite $Fe_{1-x}S$, and marcasite FeS_2, which produce sulfuric acid upon oxidation and hydrolysis. The last two are locally important, but on a global scale nearly all acid rock drainage, whether natural (as in the development of massive or disseminated gossan, the oxidized equivalent of massive or disseminated sulfide, respectively) or related to anthropogenic activities (as in the generation of wastes from mining and mineral processing) can be traced to the oxidation of pyrite. The metal:sulfur ratio in sulfide minerals plays an important role in determining the amount of sulfuric acid that is liberated by oxidation. Pyrite and marcasite are more S-rich than other sulfides, and consequently produce more acid per mole (Blanchard 1967, Blain and Andrew 1977, Alpers and Brimhall 1989, Plumlee 1999). Two important results from the oxidation of pyrite are (a) the generation of low-pH conditions, and (b) the consequent release of heavy metals into surface and ground waters.

Metals liberated by sulfide oxidation may precipitate locally as soluble or relatively insoluble sulfate minerals, with the latter acting as solid-phase controls on dissolved metal concentrations. Soluble sulfates generally act only as temporary 'sinks' for the heavy metals, but deposits of chalcanthite, $CuSO_4 \cdot 5H_2O$, of commercial size have been found in arid regions, as at Copaquire, Chile; chalcanthite was also a minor ore in the oxidized zone at Butte, Montana, and was shipped in large quantities from the Bluestone mine, Nevada (Palache et al. 1951). Many of the soluble minerals have the simplified formula $MSO_4 \cdot nH_2O$, wherein M represents divalent Fe, Mn, Co, Ni, Mg, Cu, or Zn, and n ranges from 1 to 7. Mutual substitution among the divalent cations is common, but complete solid solutions have been documented for only a few of the binary systems. Although the soluble sulfate minerals are generally ephemeral, they provide important clues or direct evidence of the pathways of sulfide oxidation and the alteration of the associated mineral assemblages. Where soluble metals salts are present, they provide information about the reactions that have occurred, the composition of the solutions from which they formed, and the types of primary minerals likely to have weathered.

The purpose of this paper is to summarize the occurrence, geochemical properties, and environmental behavior of metal-sulfate salts, in particular those that are readily water-soluble, because these minerals are important in trace-metal cycling and have often been overlooked or ignored. The relatively insoluble sulfate minerals that result from sulfide oxidation are described in three other chapters of this volume. The poorly crystalline, relatively insoluble, hydroxysulfates of iron and aluminum are described by Bigham and Nordstrom (this volume). The well-crystalline minerals of the alunite supergroup, including alunite, jarosite, and related phases, are described by Dutrizac and Jambor (this volume) and by Stoffregen et al. (this volume).

In the following section, variations in the compositions of the more common soluble metal-sulfate salts are reviewed. The sequence begins with the simple hydrated salts of the divalent cations, progresses to salts of the trivalent cations, and thence to those salts that contain both divalent and trivalent cations and (or) monovalent cations. Additional sections in this chapter describe processes of formation, transformation, and dissolution of the soluble salts, including a summary and synthesis of available data on solubility and stability relationships. The chapter concludes with a discussion of the paragenesis of metal-sulfate salts.

COMPOSITIONS AND CRYSTAL CHEMISTRY OF HYDRATED METAL SALTS

Divalent cations

The simple hydrated salts with divalent cations are of the type $M^{2+}SO_4 \cdot nH_2O$ (Table 1). Numerous experimental studies have shown that n decreases as relative humidity or water activity is decreased, and as temperature or acidity of the mother liquor is increased. The melanterite and epsomite groups are both heptahydrates that are distinguished on the basis of crystal structure. In most compositionally simple solid solutions, the substituting ion is accommodated by various degrees of distortion of the crystal structure of the host mineral. When distortion is too severe, a new structure is formed. For the $M^{2+}SO_4 \cdot nH_2O$ salts, increasing structure distortion follows the sequence $Ni^{2+} < Zn^{2+} < Mg^{2+} < Co^{2+} < Fe^{2+} < Cu^{2+}$ (Aslanian and Balarew 1977).

Melanterite group. Melanterite, $FeSO_4 \cdot 7H_2O$, is one of the most common soluble sulfate minerals formed in nature, whereas the other four minerals in the group (Table 1) are relatively rare. Boothite, $CuSO_4 \cdot 7H_2O$, for example, has been reported from only a few localities (Palache et al. 1951, Skounakis and Economou 1983), and although the supporting morphological and chemical data are good, no modern description has been made. No numerical X-ray diffraction (XRD) data are available for either natural or synthetic material, although Jambor and Traill (1963) reported that a pentahydrate with Cu:Fe:Zn = 70:24:6, from Alameda County, California, gave a melanterite-type X-ray pattern upon artificial hydration. Similarly, the mineral zinc-melanterite, $(Zn,Cu,Fe)SO_4 \cdot 7H_2O$, is known from fewer than five localities, and no XRD data were available prior to the recent description by Tiegeng Liu et al. (1995). Bieberite, $CoSO_4 \cdot 7H_2O$, and mallardite, $MnSO_4 \cdot 7H_2O$, are not as rare, and both are also well known as synthetic compounds.

Solid solution in binary subsystems of the heptahydrate metal salts has been examined by Aslanian et al. (1972), Balarew et al. (1973), and Siebke et al. (1983). In those studies, supersaturated solutions were prepared at 65-70°C, and crystalline precipitates were obtained by cooling the solutions to room temperature and slowly evaporating the solvent. For the melanterite-group minerals, solid solution between melanterite and bieberite was found to be complete. In the Ni-Fe series, the maximum Ni uptake in solid solution is $(Fe_{0.54}Ni_{0.46})SO_4 \cdot 7H_2O$ (Fig. 1a); analyses of natural materials are within the lower part of the allowable range of Ni substitution (Palache et al. 1951, Rutstein 1980).

Table 1. Simple hydrated sulfate salts of the divalent metal cations.

Melanterite Group (monoclinic, $P2_1/c$)[1]		Rozenite Group (monoclinic, $P2_1/n$)[6]	
melanterite	$FeSO_4 \cdot 7H_2O$	rozenite	$FeSO_4 \cdot 4H_2O$
boothite	$CuSO_4 \cdot 7H_2O$	starkeyite	$MgSO_4 \cdot 4H_2O$
bieberite	$CoSO_4 \cdot 7H_2O$	ilesite	$MnSO_4 \cdot 4H_2O$
mallardite	$MnSO_4 \cdot 7H_2O$	aplowite	$CoSO_4 \cdot 4H_2O$
zinc-melanterite	$(Zn,Cu)SO_4 \cdot 7H_2O$	boyleite	$ZnSO_4 \cdot 4H_2O$

Epsomite Group (orthorhombic, $P2_12_12_1$)[2]			
epsomite	$MgSO_4 \cdot 7H_2O$	Bonattite (monoclinic, Cc)[7]	$CuSO_4 \cdot 3H_2O$
morenosite	$NiSO_4 \cdot 7H_2O$		
goslarite	$ZnSO_4 \cdot 7H_2O$	Sanderite	$MgSO_4 \cdot 2H_2O$

Hexahydrite Group (monoclinic, $C2/c$)[3]		Kieserite Group (monoclinic, $C2/c$)[8]	
hexahydrite	$MgSO_4 \cdot 6H_2O$	kieserite	$MgSO_4 \cdot H_2O$
chvaleticeite	$MnSO_4 \cdot 6H_2O$	szmikite	$MnSO_4 \cdot H_2O$
ferrohexahydrite	$FeSO_4 \cdot 6H_2O$	szomolnokite	$FeSO_4 \cdot H_2O$
nickelhexahydrite	$NiSO_4 \cdot 6H_2O$	dwornikite	$NiSO_4 \cdot H_2O$
moorhouseite	$CoSO_4 \cdot 6H_2O$	gunningite	$ZnSO_4 \cdot H_2O$
bianchite	$ZnSO_4 \cdot 6H_2O$		
		Poitevinite (triclinic)[9]	$(Cu,Fe)SO_4 \cdot H_2O$

Retgersite (tetragonal, $P4_12_12$)[4]	
	$NiSO_4 \cdot 6H_2O$

Chalcanthite Group (triclinic, $P\overline{1}$)[5]	
chalcanthite	$CuSO_4 \cdot 5H_2O$
pentahydrite	$MgSO_4 \cdot 5H_2O$
jôkokuite	$MnSO_4 \cdot 5H_2O$
siderotil	$FeSO_4 \cdot 5H_2O$

[1]Baur (1964a); Kellersohn et al. (1991); [2]Baur (1964b); Ferraris et al. (1973); Beevers and Schwartz (1935); [3]Zalkin et al. (1962, 1964); Elerman (1988); Gerkin and Reppart (1988); [4]Beevers and Lipson (1932); [5]Bacon and Titterton (1975); Baur and Rolin (1972); [6]Baur (1962, 1964c); Kellersohn (1992); [7]Zahrobsky and Baur (1968); [8]Le Fur et al. (1966); Hawthorne et al. (1987); Wildner and Giester (1991, 1988); Giester et al. (1994).

In the Zn-Fe series, the maximum Zn uptake is $(Fe_{0.45}Zn_{0.55})SO_4 \cdot 7H_2O$. The consequence is that two heptahydrate minerals of monoclinic structure are possible, one with formula Fe > Zn, which is melanterite, and the other with Zn > Fe, which is zinc-melanterite (Fig. 1b). Analyses of natural melanterite have shown up to 8.92 wt % ZnO (Palache et al. 1951), which corresponds to $(Fe_{0.69}Zn_{0.30})_{\Sigma0.99}SO_4 \cdot 6.8H_2O$, and Alpers et al. (1994) reported an analysis with Fe:Zn:Cu:Mg = 53:29:14:4. Only two complete analyses are available for zinc-melanterite; one (Palache et al. 1951) corresponds to $(Zn_{0.44}Cu_{0.43}Fe_{0.08})_{\Sigma0.95}SO_4 \cdot 6.6H_2O$, and the other (Tiegeng Liu et al. 1995) to $(Zn_{0.57}Fe_{0.35}Mg_{0.10}Ca_{0.01})_{\Sigma1.03}SO_4 \cdot 6.96H_2O$. The latter slightly exceeds the maximum Zn uptake of 55 mol % in the Zn-Fe binary system; however, the Mg content of the mineral is 10 mol %, and Mg slightly extends the range of substitution within which decreases in Fe are tolerated (Fig. 1c).

Compositions of zinc-melanterite are shown along with other available data for natural metal-sulfate heptahydrates in two ternary diagrams, Fe–Cu–(Zn+Mg) (Fig. 2a) and Fe–(Cu+Zn)–Mg (Fig. 2b). The data points in Figure 2 from Jamieson et al. (1999) for heptahydrate salts from Iron Mountain represent Fe:Cu:Zn ratios from electron microprobe analyses, in which Mg was not analyzed. The Mg contents of other melanterite samples from Iron Mountain are about 2 mol % (Alpers et al. 1994). The Richmond deposit at Iron

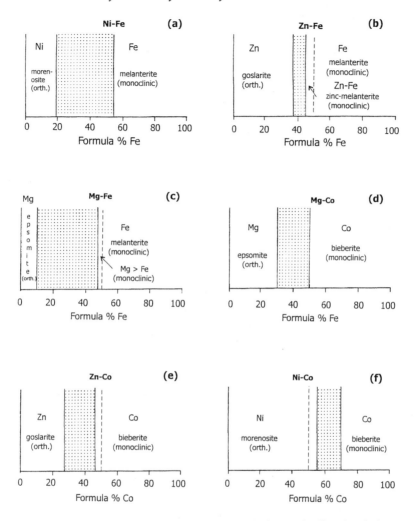

Figure 1. Binary solid solutions among the synthetic metal-sulfate heptahydrates precipitated at room temperature (data from Aslanian et al. 1972, Balarew et al. 1973, and Siebke et al. 1983). Dotted pattern shows the miscibility gaps in the solid-solution series: (a) Ni-Fe; (b) Zn-Fe, vertical dashed line at 45 mol % Fe (55 mol % Zn) represents the maximum solid solution of Zn in monoclinic $M^{2+}SO_4 \cdot 7H_2O$, where M = Fe,Zn. At 0-50 mol % Fe in $(Fe,Zn)SO_4 \cdot 7H_2O$ the equivalent mineral is melanterite, and at >50 mol % Zn in $(Zn,Fe)SO_4 \cdot 7H_2O$ the equivalent mineral is zinc-melanterite; (c) Mg-Fe; (d) Mg-Co; (e) Zn-Co; (f) Ni-Co. Note that, as for zinc-melanterite, narrow fields are available for the Mg-dominant analog of melanterite, the Zn analog of bieberite, and the orthorhombic Co analog of morenosite.

Mountain is a massive sulfide that hosts extremely acidic mine waters, some with negative pH values (Nordstrom and Alpers 1999a,b; Nordstrom et al. 2000). Some of the melanterite from the Richmond mine precipated from mine water with a pH of -0.7 (Alpers et al. 1994, Nordstrom et al. 2000). The Mattie deposit at Iron Mountain is located adjacent to the Richmond deposit, and is accessed by a common tunnel. Mine waters associated with melanterite and zinc-melanterite formation in the Mattie deposit have pH values in the range of 3-4 (Alpers and Nordstrom, unpublished data).

The diagram for Fe-Mg solid solution (Fig. 1c) shows that Mg can be accommodated up to $(Fe_{0.47}Mg_{0.53})SO_4 \cdot 7H_2O$. Theoretically, therefore, the Mg analog of zinc-melanterite could exist as a mineral. Magnesium-rich melanterite was previously called 'kirovite,' a name no longer in good standing. However, an analysis reported by Palache et al. (1951) contains 7.45 wt % MgO and the corresponding formula is $(Mg_{0.48}Fe_{0.47}Al_{0.07}Zn_{0.02}Cu_{0.01}Mn_{0.01})_{\Sigma1.06}SO_4 \cdot 6.80H_2O$, which is (barely) the Mg analog of zinc-melanterite. A Mg-dominant mineral was also described by Pasava et al. (1986b), who obtained the formula $(Mg_{0.48}Fe^{2+}_{0.32}Fe^{3+}_{0.07}Mn_{0.21}Al_{0.02})_{\Sigma1.00}SO_4 \cdot 6.66H_2O$.

Synthetic bieberite (Rohmer 1939, Aslanian et al. 1972) incorporates up to 50 mol % Mg (Fig. 1d), up to 54 mol % Zn (Fig. 1e), and up to 30 mol % Ni (Fig. 1f); solid solution between Co and Fe is complete. Substitution of Co by Mn seems to be limited to about 18 mol % at 20°C (Balarew et al. 1984), and incorporation of up to 32 mol % Cu at 25°C was reported by Crockford and Brawley (1932). Few analyses of natural bieberite are available, and these show little range in solid solution. Some of the few available analyses of mallardite, $MnSO_4 \cdot 7H_2O$, are near the end-member (Palache et al. 1951, Nambu et al. 1979), but Pasava et al. (1986a) obtained an analysis corresponding to $(Mn_{0.48}Fe^{2+}_{0.26}Mg_{0.24})_{\Sigma0.98}SO_4 \cdot 6.52H_2O$, and another analysis has $Mn:Mg:Fe^{3+} = 50:47:2$ (Pasava et al. 1986b).

Most analyses of boothite (Palache et al. 1951) are near that of the end-member. Skounakis and Economou (1983) described an occurrence of boothite with compositions ranging to cuprian melanterite, but analytical results were not reported. Although morphological data seem to indicate that boothite and cuprian melanterite are isostructural, the two minerals are reportedly separated by a miscibility gap (Palache et al. 1951). Cuprian melanterite was previously known as 'pisanite,' a name no longer in good standing. Keating and Berry (1953) described melanterite with $Fe:Cu:Zn = 100:80:3$, thus corresponding to $(Fe_{0.53}Cu_{0.44}Zn_{0.02})SO_4 \cdot 7H_2O$, and a slightly more Cu-rich analysis was given by Palache et al. (1951). A Cu-Zn-rich melanterite with cation ratios $Fe_{0.40}Cu_{0.34}Zn_{0.26}$ was reported by Dristas (1979). The most Cu-rich analysis (18.81 wt % CuO) given by Palache et al. (1951) for 'melanterite' corresponds to $(Cu_{0.68}Fe_{0.34})_{\Sigma1.02}SO_4 \cdot 7.12H_2O$, which is ferroan boothite rather than melanterite. Collins (1923) synthesized the Fe-Cu series at room temperature and concluded that the maximum Cu uptake is to $(Cu_{0.66}Fe_{0.34})SO_4 \cdot 7H_2O$, which is almost identical to the composition of the above-mentioned ferroan boothite. The range in compositions suggests that, if there is a solid-solution gap in the melanterite-boothite series, it must be rather narrow and it may be in the Cu-dominant part of the series. It has yet to be demonstrated unequivocally that boothite is isostructural with the other members of the melanterite group. However, comparison of the cell dimensions of the various members shows that the length of the b axis is almost constant; thus, cell parameters for boothite on the basis of its morphological axial ratios can be calculated to be $a = 13.89$, $b = 6.50$, $c = 10.64$ Å, $\beta = 105.60°$ assuming isomorphism with melanterite. Cell parameters for other members of the melanterite group, and most other sulfate minerals mentioned in this chapter, are given by Hawthorne et al. (this volume).

Epsomite group. The heptahydrates of the epsomite group are orthorhombic, in contrast to those of the melanterite group, which are monoclinic (Table 1). As in the monoclinic melanterite group (Baur 1964a, Kellersohn et al. 1991), one of the water molecules in the orthorhombic epsomite-group structures is not bonded to a metal ion and is readily lost (Beevers and Schwartz 1935, Baur 1964b, Ferraris et al. 1973).

The three minerals in the epsomite group (Table 1) are defined by their predominance of Mg (epsomite), Zn (goslarite), and Ni (morenosite). In the synthetic system, near room temperature, solid solutions of Zn-Ni (Aslanian et al. 1972), Zn-Mg (Balarew et al. 1973), and Ni-Mg (Soboleva 1958, 1960) are apparently complete. Presumably, therefore, the Zn-Ni-Mg ternary system does not contain a miscibility gap.

Substitution of Fe for Mg in epsomite in the synthetic system (Balarew et al. 1973, Siebke et al. 1983) is to a maximum of only 10 mol % (Fig. 1c), but older results (Palache et al. 1951) indicated that Mg:Fe ~5:1 was achievable (Fig. 2). In natural material, the most Fe-rich analysis listed by Palache et al. (7.77 wt % FeO) corresponds to $(Mg_{0.72}Fe_{0.28})_{\Sigma1.00}SO_4 \cdot 6.99H_2O$, which is considerably beyond the limit indicated by syntheses.

Substitution of Mg in epsomite by elements such as Co, Cu, and Mn is possible. For the Mg-Co binary, the limit is 30 mol % Co (Fig. 1d). Solid-solution of Cu in the Mg-Cu binary is apparently limited to a few mol %, beyond which a monoclinic heptahydrate is formed (Balarew and Karaivanova 1975). Neither Co nor Cu seems to have been detected in appreciable quantities in natural epsomite. The extent of substitution of Mg by Mn in natural epsomite is probably large, but supporting data are poor; in synthetic material the reported limit is Mg:Mn ~5:2 (Palache et al. 1951).

For the Zn member, goslarite, solid solutions with Mg and with Ni are complete. Substitution of Zn by Fe can extend to 37 mol % Fe (Fig. 1b), and of Zn by Co can extend to 27 mol % Co (Fig. 1e). Analyses of natural goslarite also indicate that substantial substitution by Cu (up to 15 mol %; Milton and Johnston 1938) and Mn may occur, but the limits are not well-defined.

Few analyses of the Ni member, morenosite, are available because of uncertainties about the identification and homogeneity of the older samples. More recent analyses (e.g. King and Evans 1964, Otto and Schuerenberg 1974, Boscardin and Colmelet 1977) show that substitution of Ni is mainly by Mg, or (Mg+Fe), in agreement with the older analyses. In the synthetic system, substitution of Ni by Fe in the binary system is limited to 19 mol % Fe (Fig. 1a), but Co in the Ni-Co series can exceed Ni and still maintain the orthorhombic structure (Fig. 1f). At 25 °C, substitution of Ni by Cu is limited to about 18 mol % Cu (Jangg and Gregori 1967).

Hexahydrite group. The hexahydrite group consists of monoclinic sulfates of the type $M^{2+}SO_4 \cdot 6H_2O$. The 'hexa' in the name alludes to the water content, and coincidentally, there are six minerals in the group (Table 1). Except for hexahydrite, $MgSO_4 \cdot 6H_2O$, which is known to precipitate in diverse settings such as saline lakes, soils, and weathered mine wastes, the minerals of the group occur sparingly and are mainly found as the oxidation products of sulfide deposits. An indication of the relative sparseness of the group is that only two of the minerals (hexahydrite and bianchite) were discovered prior to the 1960s.

Although little is known about the limits of solid solution in the hexahydrates, the similarity in crystal structure with that of the heptahydrates suggests that comparable levels of substitution can be accommodated. Synthetic bianchite, $ZnSO_4 \cdot 6H_2O$ apparently can contain Fe in solid solution up to Zn:Fe ~2:1, and an analysis of natural material (in Palache et al. 1951) corresponds to $(Zn_{0.65}Fe_{0.33})_{\Sigma0.98}SO_4 \cdot 5.89H_2O$. Semi-quantitative analysis of bianchite from the Sterling mine, New Jersey, gave Zn:Fe:Mn = 55:45:5, with no Mg or Cu detected (Jenkins and Misiur 1994). A cupriferous variety with Zn:Fe:Cu = 64:21:15 has been reported from Bulgaria (Zidarov 1970). Compositions of natural metal-sulfate hexahydrates in the Fe–Cu–(Zn+Mg) system are shown in Figure 3.

Analysis of chvaleticeite, the Mn member so far known only from the type locality

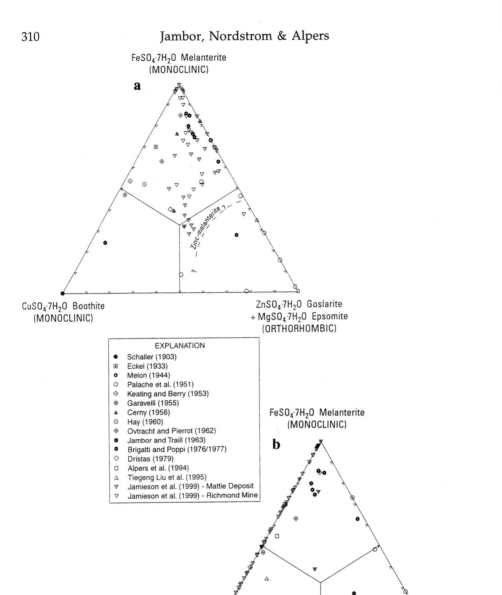

Figure 2. Ternary diagrams showing compositions of natural metal-sulfate heptahydrates: (a) $FeSO_4 \cdot 7H_2O$–$CuSO_4 \cdot 7H_2O$–$(Zn,Mg)SO_4 \cdot 7H_2O$; dashed line indicates possible compositional limit of zinc-melanterite (monoclinic); (b) $FeSO_4 \cdot 7H_2O$–$(Cu,Zn)SO_4 \cdot 7H_2O$–$MgSO_4 \cdot 7H_2O$, symbols are for both (a) and (b).

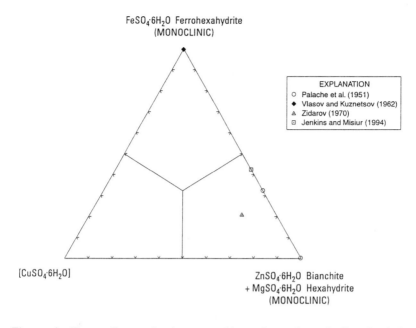

Figure 3. Ternary diagram showing compositions of natural metal-sulfate hexahydrate minerals: $FeSO_4 \cdot 6H_2O$–$CuSO_4 \cdot 6H_2O$–$(Zn,Mg)SO_4 \cdot 6H_2O$; note that $CuSO_4 \cdot 6H_2O$ is not a known mineral species.

(Pasava et al. 1986a), gave $(Mn_{0.57}Mg_{0.40})_{\Sigma 0.97}SO_4 \cdot 6.4H_2O$. For ferrohexahydrite, the original analysis showed Fe^{2+} but no Mg or Zn (Vlasov and Kuznetsov 1962). Although several occurrences of the mineral have since been reported, quantitative compositional data are sparse. The original analyses of nickelhexahydrite (Oleinikov et al. 1965) correspond to $(Ni_{0.78}Mg_{0.16}Fe_{0.10})_{\Sigma 1.04}SO_4 \cdot 5.90H_2O$ and $(Ni_{0.49}Mg_{0.28}Fe_{0.23}Cu_{0.01})_{\Sigma 1.01}SO_4 \cdot 6.04H_2O$. Analyses of the mineral from other occurrences (Karup-Møller 1973, Nawaz 1973, Otto and Scheurenberg 1974) show additional small amounts of Mn, Zn, and Co, but do not extend the range for Ni-Mg-Fe solid solution. For hexahydrite, which is the Mg-dominant member, Ni substitution to Mg:Ni = 73:27 has been found (Janjic et al. 1980). The Co-dominant member of the group, moorhouseite, is known only from a single locality. Analysis of the type material (Jambor and Boyle 1965) gave cation ratios of Co:Ni:Mn:Cu:Fe:Zn = 55:25:12:5:3:1. In the synthetic system, Co-Ni solid solution is complete at 61 °C, but mixed phases appear at lower temperatures (Rohmer 1939). At 50°C, the limit of Fe substitution is about 27 mol % (Balarew and Karaivanova 1976b).

Retgersite. Retgersite, $NiSO_4 \cdot 6H_2O$ is tetragonal, dimorphous with nickel-hexahydrite (Angel and Finger 1988; Table 1). The limits of solid solution are not known, and in natural occurrences (e.g. Frondel and Palache 1949, Fedotova 1967, Sejkora and Rídkosil 1993) only small amounts of substituting elements, especially Fe and Mg, have been detected. A mineral with Ni:Mg:Fe:Zn:Co = 65:19:13:2:1 described by Eliseev and Smirnova (1958) has since been determined to be nickelhexahydrite (Oleinikov et al. 1965), possibly with admixed morenosite (Sejkora and Rídkosil 1993).

Chalcanthite group. The chalcanthite group consists of triclinic pentahydrates of Cu, Mg, Mn, and Fe (Table 1, Fig. 4). The Cu member, chalcanthite, is of common occurrence whereas the Mn member, jôkokuite, is known only from two localities.

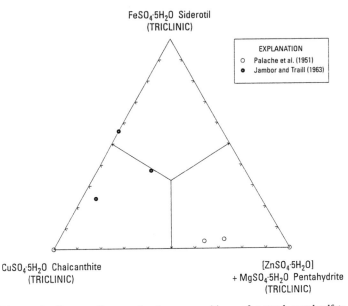

Figure 4. Ternary diagram showing compositions of natural metal-sulfate pentahydrate minerals: $FeSO_4 \cdot 5H_2O–CuSO_4 \cdot 5H_2O–(Zn,Mg)SO_4 \cdot 5H_2O$; note that $ZnSO_4 \cdot 5H_2O$ is not a known mineral species.

The composition of type jôkokuite (Nambu et al. 1978) corresponds to $(Mn_{0.94}Fe_{0.04}Zn_{0.03})_{\Sigma 1.01}SO_4 \cdot 5.07H_2O$, and material from Chvaletice, Czech Republic (Pasava et al. 1986b) has the composition $(Mn_{0.63}Mg_{0.31}Fe^{3+}_{0.02})_{\Sigma 0.96}SO_4 \cdot 5.33H_2O$. Although only small amounts of Mn have been reported as substituting in natural occurrences of the other members of the group, in the synthetic system at 21°C the Mn-Cu solid solution is evidently complete (Mellor 1932).

In synthetic chalcanthite at 25°C, up to 20 mol % Ni substitution for Cu was obtained by Jangg and Gregori (1967). The isomorphous pentahydrates of Ni and Co are known as synthetic products (Hammel 1939, Rohmer 1939) but have not yet been described as minerals. Substitution of 5 mol % Co for Cu in chancanthite was obtained by Crockford and Brawley (1932), and this increased to only 16 mol % Co at 60°C (Balarew and Karaivanova 1976a,c). Solid-solution incorporation of Mg or Zn in chalcanthite at 25°C was ≤6 mol % in the experiments by Balarew and Karaivanova (1975), but up to 19 mol % Zn was incorporated at 40°C (Balarew and Karaivanova 1976a). Mutual Cu-Fe substitution (chalcanthite-siderotil) is extensive in natural and synthetic material, and compositions extending to Cu:Fe near 1:1 have been reported for both minerals (Jambor and Traill 1963).

Relatively few analyses of the Mg member, pentahydrite, are available, and their low Fe and Zn contents of a percent or two are unlikely to reflect the solid-solution limits in natural material. The highest amounts of substitution for Mg are by Cu and Zn (Palache et al. 1951), with one analysis corresponding to $(Mg_{0.63}Cu_{0.35}Fe_{0.03}Mn_{0.01})_{\Sigma 1.02}SO_4 \cdot 4.78H_2O$, and another to $(Mg_{0.53}Cu_{0.26}Zn_{0.16}Fe_{0.04}Mn_{0.01})_{\Sigma 1.00}SO_4 \cdot 4.95H_2O$ (Fig. 4). These results suggest that the extent of mutual substitutions among Cu-Mg-Mn-Fe in the natural pentahydrates may be much larger than has yet been observed.

Rozenite group. The rozenite group consists of five minerals (Table 1), with the Mn member, ilesite, the only species known prior to 1960. Nevertheless, the two minerals that occur the most abundantly and which have the widest distribution are rozenite, $FeSO_4 \cdot 4H_2O$, and starkeyite, $MgSO_4 \cdot 4H_2O$. Ilesite (Mn), aplowite (Co), and boyleite (Zn) are known from only a few localities.

All of the members of the group are readily synthesized, but solid-solution limits, if present, have not been established. In natural occurrences, some analyses of the Mg and Fe members show small amounts of solid solution (e.g. Kubisz 1960a, Brousse et al. 1966, Snetsinger 1973, Baltatzis et al. 1986). Pasava et al. (1986b) reported a manganiferous variety with the composition $(Fe_{0.78}Mn_{0.11}Mg_{0.09})_{\Sigma 0.98}SO_4 \cdot 3.85H_2O$. An analysis of ilesite (Palache et al. 1951) corresponds to $(Mn_{0.70}Zn_{0.16}Fe_{0.13})_{\Sigma 0.99}SO_4 \cdot 3.89H_2O$, and a magnesian variety reported by Pasava et al. (1986b) has the composition $(Mn_{0.62}Mg_{0.40})_{\Sigma 1.02}SO_4 \cdot 4.2H_2O$. A Zn-rich variety devoid of Fe and coexisting with the monohydrate was mentioned by Jambor and Boyle (1962), but no analyses were given. For the Co member, aplowite, cation ratios are Co:Mn:Ni:Cu:Fe:Zn = 50:25:22:1:1:1 (Jambor and Boyle 1965), thus indicating that extensive substitution by Mn and Ni is possible. The amount of Cu in the analysis is small, and the corresponding Cu tetrahydrate is not known in either natural or synthetic material; dehydration of $CuSO_4 \cdot 5H_2O$ leads directly to the trihydrate or, depending on conditions, to the monohydrate. For the Zn member of the group, boyleite, analysis of the type material gave Zn:Mg = 84:16 (Walenta 1978).

Bonattite. Bonattite, $CuSO_4 \cdot 3H_2O$, is the only trihydrate among the $MSO_4 \cdot nH_2O$ minerals (Table 1). The compound crystallizes from solution (Posnjak and Tunell 1929), and forms by dehydration of the pentahydrate, chalcanthite, at elevated temperatures (Hammel 1939, Guenot and Manoli 1969). The mineral is extremely rare in ore deposits, probably in part because it may hydrate to chalcanthite at normal atmospheric humidity. Compositions of the mineral are close to that of the end-member (Garavelli 1957, Jambor 1962). Bonattite has also been observed as an atmospheric weathering product of sculptures (Zachmann 1999) and as a corrosion product of buried bronze artifacts (Nord et al. 1998). The synthetic analog was among several compounds that formed on artificially patinated layers on artistic bronze exposed to laboratory SO_2 contamination (Bastidas et al. 1997).

Sanderite. Sanderite, $MgSO_4 \cdot 2H_2O$, is poorly described but is nonetheless a mineral in good standing. It occurs with hexahydrite, pentahydrite, and starkeyite in marine salt deposits (Berdesinki 1952), and as an efflorescence on Neogene rocks in Greece (Schnitzer 1977), and is a well-defined synthetic phase.

Kieserite group. The kieserite group of monohydrate sulfates is made up of five members (Table 1). The cell volumes of these sulfates, and those of the analogous selenates, have been shown by Le Fur et al. (1966), Giester (1988), Wildner and Giester (1991), and Giester and Wildner (1992) to increase in accordance with the size of the predominant M^{2+} cation (Fig. 5). The most abundant and widely occurring members of the group are kieserite (Mg) and szomolnokite (Fe), with szmikite (Mn) much less common. Gunningite (Zn) is thought to be relatively rare, but reports of new occurrences seem to be increasing rapidly (e.g. Grybeck 1976, Sabina 1977, 1978; Jambor 1981, Avdonin 1984, Perroud et al. 1987, Avdonin et al. 1988, Yakhontova et al. 1988). Dwornkite, the Ni member, is known only from its type locality, Minasragra, Peru (Milton et al. 1982).

Analyses of kieserite, the Mg member, generally give compositions near that of the end-member, to the extent that kieserite is one of the few minerals for which analytical results are not listed in Palache et al. (1951). The stoichiometric compositions probably reflect occurrences in marine salt deposits, wherein high-purity, abundant material is available. Compositions of szomolnokite, the Fe member, are also commonly near that of

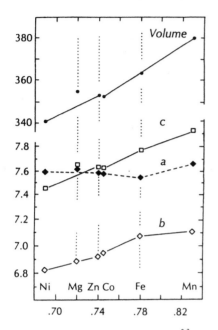

Figure 5. Dependency of cell volume (Å³) and cell-dimensions a, b, c (Å) on ionic radii of the M^{2+} cations in the minerals of the kieserite group. Redrawn from Wildner and Giester (1991).

the end-member, but Kubisz (1960b) analyzed a magnesian variety corresponding to $(Fe_{0.59}Mg_{0.41})SO_4 \cdot H_2O$, and Jamieson et al. (1999) reported szomolnokite compositions with up to 16 wt % Zn (corresponding to 39 mol % Zn) from Iron Mountain, California (Fig. 6). The few quantitative analyses for the Mn member, szmikite, show little substitution other than 1-2 wt % FeO (Palache et al. 1951, Matsubara et al. 1973). Although Jambor and Boyle (1962) reported the occurrence of a Zn-rich variety of szmikite, no analyses were given. In dwornikite, the Ni member, about 10 mol % of the Ni is substituted by Fe (Milton et al. 1982). The compositions of gunningite were found to range from the Zn end-member to Zn:Mn:Cd:Fe = 89:8:1:1 (Jambor and Boyle 1962).

In synthetic products (Jambor and Boyle 1962), up to 39 mol % Mn was substituted in gunningite, and indications were that the Zn-Fe series may be complete. These results must be treated with caution, however, because it is difficult to distinguish by X-ray powder-diffraction patterns the monoclinic $MSO_4 \cdot H_2O$ members from those

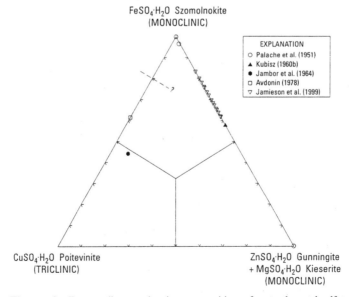

Figure 6. Ternary diagram showing composition of natural metal-sulfate monohydrates: $FeSO_4 \cdot H_2O$–$CuSO_4 \cdot H_2O$–$(Zn,Mg)SO_4 \cdot H_2O$. The dashed line represents possible compositional limit of triclinic and monoclinic phases.

that are triclinic. For example, solid solution between Cu and Fe is limited to 20 mol % Cu, whereupon a triclinic phase (poitevinite) is formed (Giester et al. 1994). Similarly in the Mg-Cu series, a triclinic phase analogous structurally to poitevinite is formed when Cu equals or exceeds 20 mol % (Lengauer and Giester 1995). The situation arises because the M^{2+} cations are distributed into two sites, one of which is more distorted than the other, and Cu is preferentially accommodated within the distorted site; above 20 mol % Cu, this ordering becomes increasingly evident, and at >60 mol % Cu the more distorted site is occupied entirely by Cu (Lengauer and Giester 1995). It is likely that the relatively restricted range of solid solution for most ions in natural occurrences of the kieserite-group minerals (Fig. 6) reflects the small number of quantitative analyses rather than limits imposed by crystal-structure considerations.

Poitevinite. Poitevinite, $(Cu,Fe)SO_4 \cdot H_2O$, has triclinic symmetry, as noted above, thereby distinguishing it from the kieserite group, which is monoclinic. Analysis of the type material corresponds to $(Cu_{0.50}Fe_{0.46}Zn_{0.08})_{\Sigma1.04}SO_4 \cdot 1.2H_2O$ (Jambor et al. 1964). Avdonin (1978) obtained $(Fe_{0.61}Cu_{0.38})SO_4 \cdot 1.2H_2O$ for a second occurrence, and the mineral has since been reported in association with siderotil, bianchite, apjohnite, and other supergene minerals in the Deputatsk tin deposit, Yakutia, Russia (Zhdanov and Solov'ev 1998). The analysis of poitevinite reported by Avdonin (1978), however, has Fe > Cu and is therefore neither poitevinite nor cuprian szomolnokite, but is instead apparently the unnamed Fe-dominant analog of poitevinite (Fig. 6). The situation arises because of Cu-Fe immiscibility in the monohydrate series, with Cu substitution in szomolnokite limited to 20 mol %. Thus, the remainder is a triclinic series extending from $(Fe_{0.80}Cu_{0.20})SO_4 \cdot H_2O$ to $CuSO_4 \cdot H_2O$, wherein the mineral with Fe > Cu is unnamed, and that with Cu > Fe is poitevinite.

Table 2. Simple hydrated sulfate salts of the trivalent ions.

Mineral	Formula	Symmetry	Reference
lausenite	$Fe_2(SO_4)_3 \cdot 6H_2O$	monoclinic	Posnjak and Merwin (1922)
kornelite	$Fe_2(SO_4)_3 \cdot 7H_2O$	monoclinic, $P2_1/n$	Robinson and Fang (1973)
coquimbite	$Fe_2(SO_4)_3 \cdot 9H_2O$	hexagonal, $P\bar{3}c$	Fang et al. (1970)
"paracoquimbite"	$Fe_2(SO_4)_3 \cdot 9H_2O$	rhombohedral, $R\bar{3}$	Robinson and Fang (1971)
quenstedtite	$Fe_2(SO_4)_3 \cdot 11H_2O$	triclinic, $P\bar{1}$	Thomas et al. (1974)
alunogen	$Al_2(SO_4)_3 \cdot 17H_2O$	triclinic, $P\bar{1}$	Fang and Robinson (1976)
meta-alunogen	$Al_4(SO_4)_6 \cdot 27H_2O$	orthorhombic	Naray-Szabo (1969)

"Paracoquimbite", although still a mineral in good standing, has been shown by Fang and Robinson (1974) to be a polytype of coquimbite.

Trivalent cations

The minerals of this type are characterized by the general formula $A_2(SO_4)_3 \cdot nH_2O$, where A is Fe^{3+} or Al, and n ranges from 6 to 17 (Table 2). The anhydrous compounds occur naturally as mikasaite, $(Fe,Al)_2(SO_4)_3$, and millosevichite $(Al,Fe)_2(SO_4)_3$. Both minerals form in fumarolic conditions and are hygroscopic. In mikasaite, Fe:Al is 1.56:0.44 (Miura et al. 1994), and in millosevichite, Al:Fe extends to 1.58:0.42 (Srebrodol'skiy 1974c). The monohydrate $Fe_2(SO_4)_3 \cdot H_2O$ is known as a synthetic product, and an apparently natural occurrence in association with rozenite was noted by Omori and Kerr (1963). Coquimbite and alunogen are by far the most commonly occurring, whereas lausenite is known from only the type locality (Palache et al. 1951) and from a burning coal dump in the Ukraine (Srebrodol'skiy 1974a). The latter has Fe:Al almost 1:1, but total cations are in considerable excess of the formula requirements. Kornelite, quenstedtite, and meta-alunogen, although not abundant, have been reported from several localities (e.g. Palache et al. 1951, Pemberton 1983, Gaines et al. 1997, Nordstrom and Alpers 1999a). In natural settings, the Fe^{3+} minerals represent a more advanced stage in the oxidation sequence insofar as the Fe^{2+} sulfates are the ones that are generally proximal to Fe sulfides.

Table 2 shows that the hydrated Fe^{3+} sulfates span a range of water contents, reaching a maximum of $11H_2O$ in quenstedtite. A higher hydrate, corresponding to $Fe_{3-x}(SO_4)_2 \cdot 14H_2O$ was described by Wang and Lee (1988), but the cell dimensions and matching triclinic symmetry and space group indicate that the mineral is likely lishizhenite, $ZnFe^{3+}_2(SO_4)_4 \cdot 14H_2O$ (Table 3).

Table 3. Mixed divalent – trivalent hydrated sulfate minerals

Mineral	Formula	Symmetry	Reference
ransomite	$Cu(Fe,Al)_2(SO_4)_4 \cdot 6H_2O$	monoclinic, $P2_1/c$	Wood (1970)
römerite	$Fe^{2+}Fe^{3+}_2(SO_4)_4 \cdot 14H_2O$	triclinic, $P\bar{1}$	Fanfani et al. (1970)
lishizhenite	$Zn\,Fe^{3+}_2(SO_4)_4 \cdot 14H_2O$	triclinic, $P\bar{1}$	Li and Chen (1990)
Halotrichite Group		monoclinic, $P2_1/c$	Menchetti and
pickeringite	$MgAl_2(SO_4)_4 \cdot 22H_2O$		Sabelli (1976);
halotrichite	$Fe^{2+}Al_2(SO_4)_4 \cdot 22H_2O$		Lovas (1986)
apjohnite	$Mn^{2+}Al_2(SO_4)_4 \cdot 22H_2O$		
wupatkiite	$Co^{2+}Al_2(SO_4)_4 \cdot 22H_2O$		
dietrichite	$ZnAl_2(SO_4)_4 \cdot 22H_2O$		
bilinite	$Fe^{2+}Fe^{3+}_2(SO_4)_4 \cdot 22H_2O$		
redingtonite	$Fe^{2+}(Cr,Al)_2(SO_4)_4 \cdot 22H_2O$		
Copiapite Group (OH-bearing)		triclinic, $P\bar{1}$	Süsse (1972);
aluminocopiapite	$Al_{2/3}\,Fe^{3+}_4(SO_4)_6(OH)_2 \cdot 20H_2O$		Fanfani et al. (1973);
magnesiocopiapite	$Mg\,Fe^{3+}_4(SO_4)_6(OH)_2 \cdot 20H_2O$		Bayliss and Atencio
calciocopiapite	$Ca\,Fe^{3+}_4(SO_4)_6(OH)_2 \cdot 20H_2O$		(1985)
copiapite	$Fe^{2+}\,Fe^{3+}_4(SO_4)_6(OH)_2 \cdot 20H_2O$		
ferricopiapite	$Fe^{3+}_{2/3}Fe^{3+}_4(SO_4)_6(OH)_2 \cdot 20H_2O$		
cuprocopiapite	$Cu\,Fe^{3+}_4(SO_4)_6(OH)_2 \cdot 20H_2O$		
zincocopiapite	$Zn\,Fe^{3+}_4(SO_4)_6(OH)_2 \cdot 20H_2O$		

Substitution of Fe^{3+} by Al is possible in this group of minerals, but analyses generally show no more than 1-3 wt % Al_2O_3 and lower levels of other metals. However, in coquimbite, $Fe_2(SO_4)_3 \cdot 9H_2O$, higher Al contents have been reported, with formula Fe:Al extending to 58:42 mol % (Palache et al. 1951, Fang and Robinson 1970).

The Al-sulfate salts are represented by alunogen and meta-alunogen; the latter is a lower hydrate per mole of aluminum (Table 2) and is inadequately described, although occurrences have been reported from several localities. The maximum H_2O content of alunogen was determined by Fang and Robinson (1976) to be 17 molecules; however, loss of water can occur without structure breakdown, and the variation is from 17 to 16 H_2O (Taylor and Bassett 1952). For meta-alunogen, the formula in Table 2 is from Mandarino (1999), reflecting the original analysis in which the water content was determined to be 13.5 H_2O. On the basis of the crystal structure of alunogen, Fang and Robinson (1976) concluded that the minimum water content to sustain the structure of alunogen would be 13.5 H_2O per formula unit. In the Powder Diffraction File, however, synthetic meta-alunogen is assigned the formulas $Al_2(SO_4)_3 \cdot 12H_2O$ and $Al_2(SO_4)_3 \cdot 14H_2O$, and the two XRD patterns have distinct differences from one another, and with that of alunogen. It is evident that meta-alunogen, if it is to be retained as a mineral name, is in need of a formal

Mixed divalent–trivalent salts

Many of the mixed divalent–trivalent sulfate salts have the general formula $AR_2(SO_4)_2 \cdot nH_2O$, where A is Mg, Fe^{2+}, Mn^{2+}, Co^{2+}, or Zn, and R is Al, Fe^{3+}, or Cr^{3+} (Table 3). The minerals described in the previous sections can be considered as "simple" salts, on the basis that sulfate is the only anion and that formula OH is absent. In the discussion of the divalent–trivalent category, however, the OH-bearing copiapite group, generally $AR_4(SO_4)_6(OH)_2 \cdot 20H_2O$, is included because its minerals meet the divalent-trivalent criterion and they are among the most commonly observed metal-bearing soluble salts derived from the oxidation of mineral deposits rich in Fe sulfides.

Ransomite, $Cu(Fe,Al)_2(SO_4)_4 \cdot 6H_2O$, is known from only one locality, at which it formed as a result of a fire in a mine. Compounds that form under such conditions are no longer accepted as new minerals (Nickel 1995). In lishizhenite, which is also known only from a single locality, the A position has Zn:Mn:Fe:Mg = 86:9:1:1, and R has Fe^{3+}:Al = 1.96:0.06 (Li and Chen 1990). Compositions of römerite (Palache et al. 1951; also summarized by Van Loan and Nuffield 1959) show formula Fe:Zn:Mg up to 65:30:5, suggesting that a series may extend to lishizhenite. Appreciable substitution of Al for Fe^{3+} in römerite is generally not present, and the maximum that has been observed is Fe:Al = 84:16.

Halotrichite group. In contrast to the preceding minerals, extensive solid solution occurs in the minerals of the halotrichite group. Intermediate compositions in the Mg-Fe^{2+} solid-solution series on the A site (pickeringite-halotrichite) are known (Fe:Mg = 51:49, Martin et al. 1999; Fe:Mg = 63:37, Cody and Grammer 1979), and solid solution seems likely to be complete (Bandy 1938, Palache et al. 1951). For apjohnite (A = Mn^{2+}), compositions extend from near the Mn end-member, to Mn^{2+}:Mg:Zn:Fe^{2+} = 64:28:6:2 (Menchetti and Sabelli 1976) and to Mn:Mg:Fe = 58:31:11 (Paulis 1991); Mn-rich halotrichite with Fe^{2+}:Mn^{2+} = 62:38 is also known, and it is likely that ternary Fe^{2+}-Mg-Mn^{2+} solid solution on the A site is complete (Palache et al. 1951).

Each of dietrichite (A = Zn, R = Al), redingtonite (A = Fe^{2+}, R = Cr,Al), and wupatkiite (A = Co, R = Al) is known from single or no more than two or three localities, and compositional data are accordingly sparse. For dietrichite, reported Zn:Fe^{2+}:Mn^{2+} = 42:33:25, and the Zn end-member apparently has been synthesized (Palache et al. 1951). Wupatkiite, known only from a single locality, has Co:Mg:Ni:Mn:Fe:Ca:Cu = 40:36:6:2:2:2:1 (Williams and Cesbron 1995). Bilinite (A = Fe^{2+}, R = Fe^{3+}) is known from several localities (Palache et al. 1951, Bolshakov and Ptushko 1967, Srebrodol'skiy 1977, Keith and Runnells 1998). Data for redingtonite are incomplete and the mineral has not been described adequately. The extent of substitution in R^{3+} within the mineral group is not well known, but an analysis of halotrichite with Al:Fe^{3+} = 62:38 is recorded in Palache et al. (1951), and Srebrodol'skiy (1974b) gave an analysis with Mg:Fe^{2+} = 85:17 and Al:Fe^{3+} = 120:76.

Thus, some of the sulfate minerals in the divalent–trivalent category are rather rare with regard to data and number of reported occurrences. Halotrichite is the most abundant and widely distributed, followed by pickeringite and römerite. In some cases the distinction between halotrichite and pickeringite is not made, but both are appreciably more abundant than römerite. Römerite has been observed in sulfate deposits and as an oxidation product of coals and massive sulfides. Römerite occurred at the Alcaparrosa sulfate deposit in Chile in sufficient abundance to be mined and processed as an acid source (Bandy 1938) and is locally abundant in the Richmond mine at Iron Mountain (Nordstrom and Alpers 1999a), but most occurrences are in oxidized sulfide deposits in which only minute quantities of the mineral are present.

Copiapite group. The general formula of the copiapite group is $A^{2+}R^{3+}_4(SO_4)_6(OH)_2 \cdot 20H_2O$, wherein R is dominated by Fe^{3+} in all members. The seven minerals of the copiapite group are listed in Table 3, wherein all of the members are assigned 20 formula H_2O in accordance with the conclusions of Bayliss and Atencio (1985). Substitution of Al and Fe^{3+} in the A^{2+} position leads to the peculiarity that the aluminocopiapite and ferricopiapite end-members are trivalent rather than mixed divalent-trivalent salts. Substitution of trivalent ions in A leads to excess positive charge that is accommodated by vacancies, hence the formula is written with $Al_{2/3}$ for aluminocopiapite, and $Fe^{3+}_{2/3}$ for ferricopiapite (see also the discussion by Hawthorne et al., this volume). Because the A-site cations form such a small proportion of the total formula mass, small changes in analytical wt % have a pronounced effect on the mol % of the A-site cations. For example, end-member magnesiocopiapite contains 3.31 wt % MgO, and end-member copiapite contains 5.75 wt % FeO. Small amounts of contamination, or analytical errors, may therefore have a significant effect on the apparent range of solid solution.

With regard to A-site solid solution in the copiapite group, the 42 analyses listed by Berry (1947), and other analyses reported or discussed by Palache et al. (1951), Jolly and Foster (1967), Fanfani et al. (1973), Zodrow (1980), Bayliss and Atencio (1985), and Robinson (1999) strongly suggest that mutual substitution among Mg-Fe^{2+}-Fe^{3+}-Al is complete. Significant amounts of Na and Ca may be present in A, and only 4.54 wt % CaO is required for end-member calciocopiapite. For calciocopiapite, however, occurrences other than that of the type locality (Fleischer 1962) have not been documented.

The range in Cu contents found within the copiapite group (Palache et al. 1951, Escobar and Gifford 1961) is large, and includes compositions near that of cuprocopiapite, the Cu end-member. Type zincocopiapite (in Fleischer 1964) has $(Zn_{0.75}Fe^{2+}_{0.07}Mn^{2+}_{0.06}Ca_{0.04}K_{0.04}Na_{0.01})_{\Sigma 0.97}$ for the A site, and an analysis of the naturally occurring pure end-member is given by Perroud et al. (1987). Manganese contents within the copiapite group are generally <1 wt % MnO, but Pasava et al. (1986b) obtained an unusual composition for magnesiocopiapite that corresponds to $(Mg_{0.93}Mn_{0.07})_{\Sigma 1.00}(Fe^{3+}_{3.50}Fe^{2+}_{0.43}Mn_{0.17})_{\Sigma 4.10}(SO_4)_6(OH)_2 \cdot 19.9H_2O$. Copiapite-group minerals from the Richmond mine at Iron Mountain have two distinct textures and compositions (Fig. 7). The larger platy minerals, 10-50 μm in diameter, are magnesio-

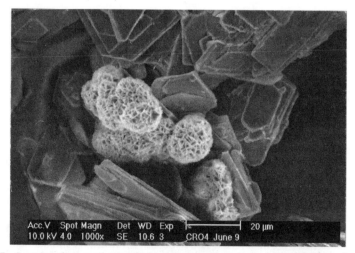

Figure 7. Scanning electron micrograph showing magnesiocopiapite (plates 10-50 μm in diameter) and Al-bearing ferricopiapite (spheroidal rosettes). Reproduced with permission from Robinson (1999).

copiapite, whereas the spheroidal aggregates are Al-bearing ferricopiapite, which may have formed by evaporation of pore waters after sample collection (Robinson 1999, Robinson et al. 2000a). Partial Al-for-Fe^{3+} substitution in R^{3+} has been demonstrated by Berry (1947), but the extent of possible accommodation of divalent metals in the R^{3+} position is not known. The synthesis of all of the minerals in the copiapite group, including calciocopiapite, was reported by Atencio et al. (1996), who also obtained the Mn, Ni, and Co analogs not yet known as minerals.

Other minerals

Two complex, hydrated sulfate salts, fibroferrite, $Fe^{3+}(SO_4)(OH)\cdot5H_2O$, and botryogen $MgFe^{3+}(SO_4)_2(OH)\cdot7H_2O$, particularly the former, are commonly associated with $M^{2+}SO_4\cdot nH_2O$ assemblages. The dimorphs butlerite and parabutlerite have the composition $Fe^{3+}(SO_4)(OH)\cdot2H_2O$; one would expect them to form as dehydration products of fibroferrite, but the authors are not aware of this having been documented.

Among the numerous other sulfates commonly associated with the preceding soluble sulfates in acidic environments, gypsum, $CaSO_4\cdot2H_2O$, occurs almost universally. With the availability of alkalis, as is typical in saline soils, the variety of salts can increase considerably. Alkali sulfates are not discussed in detail in this chapter (see Spencer, this volume, for end-member compositions and paragenetic relations among evaporite minerals).

It is appropriate to mention two K-bearing metal sulfates that form in association with the metal salts described previously: rhomboclase and voltaite. The formula of rhomboclase is variously written as $HFe^{3+}(SO_4)_2\cdot4H_2O$, $(H_3O)Fe^{3+}(SO_4)_2\cdot3H_2O$, or $(H_5O_2)Fe^{3+}(SO_4)\cdot2H_2O$, with the last formula from the crystal-structure determination by Mereiter (1974). A phase related in composition to rhomboclase is goldichite, $KFe^{3+}(SO_4)_2\cdot4H_2O$, which has K instead of H, and which is not isostructural with rhomboclase (Hawthorne et al., this volume). Unlike hydronium jarosite, which has a comparatively low solubility, dissolution of rhomboclase is fairly rapid, and the mineral can be an efficient, albeit temporary, storage place for sulfuric acid because of the extra H^+ in the formula. Several localities for rhomboclase are known, typically for occurrences in oxidized sulfide deposits. Nordstrom and Alpers (1999a) described the formation of rhomboclase stalagmites at Iron Mountain in association with water of pH = -3.6, the lowest value recorded in a field setting (Nordstrom et al. 2000).

Voltaite, $K_2Fe^{2+}_5Fe^{3+}_4(SO_4)_{12}\cdot18H_2O$, occurs in oxidized sulfide deposits and in fumarolic deposits. One of the analyses included for voltaite by Palache et al. (1951) corresponds to that of an unnamed Mg analog (Mg > Fe^{2+}). In the Zn analog, zincovoltaite, which is known only from an oxidized sphalerite-galena-pyrite deposit in China, the divalent Fe site is occupied by $(Zn_{3.69}Fe^{2+}_{0.91}Mn_{0.35})_{\Sigma4.95}$ (Li et al. 1987). Zoned crystals of voltaite, intergrown with szomolnokite, were determined to vary systematically in Zn-for-Fe substitution for samples collected from Iron Mountain, California (Jamieson and Przybylowicz 1997, Jamieson et al. 1999).

PROCESSES OF FORMATION, TRANSFORMATION, AND DISSOLUTION

Pyrite oxidation

The oxidation of pyrite to form sulfate involves a series of reactions, reviewed by Nordstrom (1982a), Lowson (1982), Nordstrom and Southam (1997), and Nordstrom and Alpers (1999b). The overall process is most commonly reported by the following reaction:

$$FeS_{2(s)} + 15/4\ O_{2(aq)} + 7/2\ H_2O_{(l)} \rightarrow Fe(OH)_{3(s)} + 2\ H_2SO_{4(aq)} \tag{1}$$

where $Fe(OH)_3$ is generally regarded to be a surrogate for ferrihydrite, a Fe^{3+} oxyhydroxide whose formula is contentious (see Bigham and Nordstrom, this volume). Other oxyhydroxides, such as goethite, α-FeOOH, and (rarely) lepidocrocite, γ-FeOOH, also form as insoluble precipitates. Reaction (1) forms sulfuric acid, and in acidic conditions ferrous iron can be generated

$$FeS_{2(s)} + 14\ Fe^{3+}_{(aq)} + 8\ H_2O_{(l)} \rightarrow 15\ Fe^{2+}_{(aq)} + 2\ SO_4^{2-}_{(aq)} + 16\ H^+_{(aq)} \qquad (2)$$

Indeed, soluble ferric iron and pyrite cannot coexist for any significant length of time because pyrite rapidly reduces $Fe^{3+}_{(aq)}$. This observation begs the question: how do goethite pseudomorphs after pyrite form? Such pseudomorphs are common for single pyrite crystals of generally large crystal size (≥ 1 mm), relatively discrete occurrence of pyrite, and within rock that contains sufficient buffering capacity to keep the ground water near neutral (or even alkaline) in pH. Hence, the low surface area slows the reaction rate considerably, and the buffered water maintains low concentrations of $Fe^{3+}_{(aq)}$ in association with the formation of insoluble Fe^{3+}-bearing minerals. In this manner, conditions are optimized for direct transformation from pyrite to goethite without the formation of sulfate minerals. However, during formation of gossans, which commonly are the oxidized equivalent of massive sulfide deposits, there are large quantities of pyrite in contact with ground water in systems that do not have the buffering capacity to override the acid production from pyrite oxidation.

In the initial stage of gossan development, oxidation of sulfides leads to the formation of Fe oxyhydroxides and soluble Fe sulfates. The soluble sulfates form in the upper parts of the vadose zone as the acid waters evaporate and dry out. These waters may have extremely low pH values (e.g. Nordstrom and Alpers 1999a,b; Nordstrom et al. 2000). The soluble sulfates, if detectable, initially form within the oxidized rims of the sulfide minerals, with melanterite usually the first to appear (Nordstrom 1982a, Nordstrom and Alpers 1999b). In the pH range of 2.5 to 8 typical of the saturated zone at and below the water table, Fe oxyhydroxides form. By far the most abundant of the oxyhydroxide products is goethite, with smaller amounts of ferrihydrite and possibly the sulfate-bearing oxyhydroxide schwertmannite (see Bigham and Nordstrom, this volume). The Fe oxyhydroxides are fine-grained, and in mineral deposits they may contain large amounts (commonly low percentages) of sorbed elements, including those of the heavy metals, as well as sulfate. Ferrihydrite and schwertmannite are metastable with respect to goethite, and thus goethite attains overwhelming predominance both by direct precipitation and by conversion of metastable Fe phases. As oxidation progresses and an overall acidic domain is established, the initial oxidation scenario moves downward, in concert with the evolving low-pH front. In the zone nearer surface, however, the Fe oxyhydroxide assemblage seems to undergo a recycling, which is probably recrystallization and the concomitant development of a coarser grain size; the attendant effect is that sorption capacity decreases, and thus the bulk of the sorbed elements is lost and moves downward. The near-surface product is a purer oxyhydroxide, most commonly goethite. Over time, the goethite may convert to hematite; there may also be pathways directly from ferrihydrite to hematite, depending on pH and the presence of trace elements (e.g. Alpers and Brimhall 1989; and references therein). The soluble sulfates are dissolved and the low-pH, sulfate-rich pore waters may precipitate jarosite, generally of composition $(K,Na,H_3O)Fe_3(SO_4)_2(OH)_6$ (see Dutrizac and Jambor, this volume; Stoffregen et al., this volume). In sialic rocks lacking carbonates, even minor dissolution of aluminosilicates typically occurs only after acidic conditions have already been established. Among the common rock-forming minerals that are potential sources of the K that is incorporated in jarosite, the trioctahedral micas (e.g. biotite) have been observed to be the most susceptible to alteration (Jambor and Blowes 1998, Malström and Banwart 1997).

With the maturation of a gossan, the near-surface sulfides are depleted and the aluminosilicate minerals are leached, leaving a goethite-rich, siliceous residue (Blanchard 1967, Blain and Andrew 1977). Relatively insoluble sulfate minerals may accumulate at depth, and soluble sulfates and other metal oxides may precipitate at lateral seeps if climatic conditions are appropriate.

In mine settings, contemporary precipitates of soluble sulfates are generally most noticeable on the older walls of open pits and underground mine workings, and at the exposed surfaces of accumulated mine wastes, such as undisturbed tailings impoundments. The walls of open-pit mines that exploit porphyry copper deposits commonly display blue to pale green stains that consist predominantly of chalcanthite and melanterite. Such efflorescent blooms are generally temporary because of the high solubility of the salts and their susceptibility to dissolution by rain or snowmelt. Underground mines not only provide a more sheltered environment for the preservation of water-soluble minerals, but such mines also have yielded a wider spectrum of oxidation minerals because the ore types are commonly more diverse than those extracted by open-pit methods.

Field studies

Metal-sulfide deposits. Metal-sulfate salts occur most commonly in association with the oxidation of metal-sulfide mineral deposits. Many of the most famous deposits of the world (Rio Tinto, Spain; Rammelsberg, Germany; Chuquicamata, Chile; Bingham, USA; Butte, USA; Cornwall, UK; Sain Bel, France; Falun, Sweden) have efflorescent metal-sulfate salts of various types in mine workings, in open pits, and on waste rock and tailings piles. The classic paper on this subject is the work of Bandy (1938), who described in considerable detail the mineralogy of salts at Chuquicamata, Quetena, and Alcaparrosa, Chile.

Numerous efflorescent sulfate minerals have been reported from the Rio Tinto mines in Spain. Among those listed by García García (1996) are chalcanthite, copiapite, coquimbite, botroygen, epsomite, fibroferrite, gypsum, goslarite, halotrichite, mallardite, melanterite, and römerite. At the Nikitov mercury deposits, Donet'sk, Ukraine, oxidation of melanterite was concluded by Bolshakov and Ptushko (1971) to have produced rhomboclase, römerite, bilinite, copiapite, and voltaite. Numerous reports of similar assemblages are available in the Russian literature. Among the many examples, Velizade et al. (1976) identified chalcanthite, melanterite, siderotil, halotrichite, jarosite, alunogen, and slavikite $NaMg_2Fe^{3+}{}_5(SO_4)_7(OH)_6 \cdot 33H_2O$ in the oxidation zone of the Datsdag deposit, and Kravtsov (1971, 1974) reported jarosite, copiapite, römerite, and quenstedtite from cassiterite–sulfide deposits at Sakha, Azerbaijan. The processes of sulfide oxidation and sulfate formation in cryogenic zones have been discussed by Kravtsov (1974), Chernikov et al. (1994), Yurgenson (1997), and others. Permafrost conditions generally limit the availability of aqueous solutions, thereby simulating, in some respects, arid conditions.

Many examples of occurrences of soluble metal salts in the United States could be cited. Milton and Johnston (1938) photographically documented extensive blooms of epsomite and pickeringite associated with gypsum, melanterite, goslarite, pentahydrite, copiapite, voltaite, and rhomboclase at the Comstock Lode, Nevada. At the Dexter Number 7 mine, Utah, chalcanthite, melanterite, copiapite, coquimbite, römerite, fibroferrite, halotrichite, alunogen, voltaite, butlerite, parabutlerite, goldichite, and other minerals cement a talus breccia (Rozenzweig and Gross 1955). Stalactitic material at the San Manuel mine, Pinal County, Arizona, contained the first occurrence of jurbanite, $AlSO_4OH \cdot 5H_2O$, which was associated with epsomite, hexahydrite, starkeyite, pickeringite, and lone-creekite, $(NH_4)_4Fe(SO_4)_2 \cdot 12H_2O$. At the Leviathan mine, California, chalcopyrite and cryptocrystalline pyrite (Pabst 1940) oxidized and produced chalcanthite, halotrichite, melanterite, and römerite (Gary 1939). Nordstrom (1982a) demonstrated, with a color photograph, the conversion of melanterite to rozenite and to copiapite from an efflorescence taken from the face of the Brick Flat massive sulfide at Iron Mountain, California. Nordstrom and Alpers (1999a) observed massive quantities of melanterite, rhomboclase, voltaite, coquimbite, copiapite, halotrichite, and römerite in underground workings of the Richmond Cu-Zn-pyrite mine at Iron Mountain. Acid waters that seemed to be in equilibrium with many of these minerals had, in several instances, negative pH values (Nordstrom et al. 2000). Solubility studies that examined Cu-Zn partitioning in melanterite from Iron Mountain showed a greater incorporation of Cu over Zn during precipitation relative to the starting solutions. These results help to explain the variations in Zn/Cu ratios of the effluent water during the annual wet-dry cycle (Alpers et al. 1994).

Tailings impoundments. The surfaces, channels, and vadose zones in tailings impoundments in arid climates have yielded a rich array of soluble salts. Agnew (1998), for example, reported the presence of several Fe and Mg salts of the type $M^{2+}SO_4 \cdot nH_2O$ at impoundments in Australia. At some sites these were accompanied by rare or unusual minerals such as: blödite $Na_2Mg(SO_4)_2 \cdot 4H_2O$, hydrobasaluminite $Al_4(SO_4)(OH)_{10} \cdot 12-36H_2O$, hemimorphite $Zn_4Si_2O_7(OH)_2 \cdot H_2O$, a mineral in the picromerite group $A_2Mg(SO_4)_2 \cdot 6H_2O$, and possibly wattevillite $Na_2Ca(SO_4)_2 \cdot 4H_2O(?)$. The latter two are from the Ranger uranium mine, at which sulfuric acid used in mineral processing apparently accounts for the acid source. The association between arid climate and elevated temperatures is common but, as has been noted, the metal salts also form in cryogenic (permafrost) zones, as in the Antarctic (Keys and Williams 1981).

Mineralogical studies of tailings impoundments have revealed the presence of many of the sulfate minerals of the types listed in Tables 1, 2, and 3 (Jambor 1994, Jambor and Blowes 1998). Most weathered sulfide-rich impoundment surfaces contain blooms of the $FeSO_4 \cdot nH_2O$ minerals. Blowes et al. (1991) observed that near-surface melanterite was sufficiently abundant to form a hardpan layer at the Heath Steele tailings impoundment in New Brunswick, Canada. In various impoundments, the simple hydrated sulfates (e.g. Tables 1 and 2) are commonly accompanied by those of more complex formulation, such as rhomboclase, copiapite, fibroferrite, and halotrichite (Shcherbakova and Korablev 1998, Jambor et al. 2000). Dagenhart (1980) noted that, during dry-weather conditions, melanterite, rozenite, magnesiocopiapite, aluminocopiapite, halotrichite, pickeringite, and gypsum were abundant on tailings and related wastes at the Sulfur, Boyd Smith, and Arminius Cu-Zn mines along Contrary Creek, Virginia. He also identified lesser amounts of alunogen, ferricopiapite, chalcanthite, ferrohexahydrite, siderotil, szomolnokite, gunningite, bianchite, epsomite, hexahydrite, pentahydrite, rhomboclase, fibroferrite, coquimbite, paracoquimbite, jarosite, the Cu minerals antlerite $Cu_3SO_4(OH)_4$, brochantite $Cu_4(SO_4)(OH)_6$, and serpierite $Ca(Cu,Zn)_4(SO_4)_2(OH)_6 \cdot 3H_2O$, and the Pb minerals anglesite $PbSO_4$ and linarite $PbCu(SO_4)(OH)_2$.

Although the sulfate minerals in tailings impoundments are typically Fe-dominant, the Mg-dominant sulfates have been noted to form abundantly where the gangue minerals contain dolomite. As well, the identification of Zn-dominant minerals such as gunningite, boyleite, and goslarite in mine wastes (Avdonin et al. 1988) serves to emphasize that, although the sulfates are overwhelmingly those of Fe, the compositions of the sulfate salts will reflect those of the oxidizing source materials.

Coal deposits. Soluble metal-sulfate salts commonly occur in coal or form by the oxidation of pyrite and marcasite in coal wastes (Gruner and Hood 1971, Taylor and Hardy 1974, Wagner et al. 1982, Baltatzis et al. 1986, Foscolos et al. 1989, Ward 1991). Sulfate efflorescences in coal deposits commonly contain melanterite (± rozenite and szomolnokite), copiapite, and the pickeringite-halotrichite series (McCaughey 1918,

Nuhfer 1967, Young and Nancarrow 1988, Cravotta 1994, Rose and Cravotta 1998, Querol et al. 1999). Cravotta (1994) observed römerite, copiapite, and coquimbite on mine spoils, coal outcrops, and overburden at the bituminous coalfields of western Pennsylvania. Wiese et al. (1987) showed that melanterite and rozenite were the earliest Fe-sulfate hydrates to form from the oxidation of pyrite and marcasite in Utah and Ohio coalfields; also noted was that szomolnokite and halotrichite appeared to be the most stable phases.

McCaughey (1918) observed the common occurrence of melanterite in bituminous coal mines and described one specimen that formed a 14 kg (30 lb) mass. After storage for a year, the mineral was noted to be transforming to copiapite (and possibly some rozenite-szomolnokite). Similar transformations have been noted for other melanterite specimens stored in museums (Nordstrom, unpublished data). The dehydration of melanterite to rozenite has been observed frequently, beginning with the reports of Kubisz (1960a) and Kossenberg and Cook (1961).

Nuhfer (1967, 1972) studied the occurrence of efflorescent sulfates in the bituminous coal deposits near Morgantown, West Virginia. Melanterite, szomolnokite, pickeringite, copiapite, gypsum, and hexahydrite were observed to be common; thenardite, Na_2SO_4, and possible boothite were uncommon, and the thenardite may have had an anthropogenic source. The following is a summary of Nuhfer's observations and inferences:

1. Melanterite and szomolnokite were the first to form as a result of pyrite oxidation.
2. Copiapite was abundant and likely included significant quantities of ferricopiapite and minor magnesiocopiapite.
3. Epsomite was known to occur in the area, but he detected only hexahydrite.
4. Halotrichite-pickeringite and copiapite occur in sheltered areas in close proximity to pyrite.
5. Sulfate efflorescences reflect the composition of the waters from which they form.
6. Sulfate efflorescences are easily dissolved by rainwater, and the sulfate-rich solutions move into adjacent streams.
7. Formation of hydrated sulfate minerals contributed to fracturing and spallation of rock faces.

The presence of hydrated iron sulfates in the Sydney coalfield of Cape Breton, Nova Scotia, was observed by Zodrow and colleagues (Zodrow and McCandlish 1978, Zodrow et al. 1979, Zodrow 1980), who identified melanterite (as the first mineral formed), rozenite, epsomite, pickeringite, halotrichite, aluminocopiapite, fibroferrite, rhombo-clase, thenardite, sideronatrite $Na_2Fe^{3+}(SO_4)_2(OH)_3 \cdot 3H_2O$, and metasideronatrite $Na_2Fe^{3+}(SO_4)_2(OH) \cdot H_2O$. Zodrow et al. (1979) noted that the sulfate efflorescences induced mechanical stress and rock erosion; thus, the buildup of salts in fractures, with consequent heaving and slippage, was at least partly responsible for some of the pillar collapses in the coal mines. Oxidation of pyrite-bearing shales and mudrocks is a common geotechnical problem because of the ensuing heaving and cracking of many types of structures (Hawkins and Pinches 1997, Cripps and Edwards 1997, Hawkins and Higgins 1997), and because of the rapid erosion of highway material (Byerly 1996).

Zodrow (1980) demonstrated a wide range of substitutions by Al, Mg, Na, Cu, Ni, and Zn in copiapite, and complex transformations of aluminocopiapite on storage:

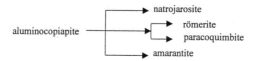

(The formula of amarantite is $Fe^{3+}_2(SO_4)_2O\cdot7H_2O$; Mandarino 1999).

Other occurrences. Metal-sulfate salts can form in diverse environments other than weathering zones of coal and metallic deposits. Hexahydrite has been identified in human concretions (Gibson 1974), and the Mg salts are common in evaporites. For example, kieserite occurs widely in marine evaporite deposits in Germany and several other countries, among which in the United States are the Permian salt deposits of Texas and New Mexico (Palache et al. 1951). Epsomite occurs similarly, and both it and hexahydrite form as precipitates in saline lakes. Both minerals are also present in some soils and their efflorescences (Doner and Lynn 1977, Skarie et al. 1986, Timpson et al. 1986, Ducloux et al. 1994), and these minerals may have formed evaporite deposits on the surface of Europa, one of the satellites of Jupiter (McCord et al. 1998). Melanterite, rozenite, and copiapite have been reported commonly to form at or near the surfaces of active acid sulfate soils (Van Breemen 1982, Wagner et al. 1982, Fanning et al. 1993, Montoroi 1995).

Metal-sulfate salts are common around active crater lakes, fumaroles, and acid hot springs. In these environments, H_2S oxidizes to elemental sulfur, which accumulates and is further oxidized by microbial activity to form sulfuric acid. The acid reacts with the surrounding silicate bedrock and, upon evaporation, can form a variety of metal-sulfate salts. In deposits of native sulphur in the Ukraine, Srebrodol'skiy (1977) determined that alteration of melanterite resulted in two different assemblages. One of the paragenetic associations consisted of ferrohexahydrite, siderotil, rozenite, and szomolnokite, i.e. the 7-, 6-, 5-, 4-, and 1-hydrates of $FeSO_4$. The other association consisted of melanterite, römerite, bilinite, copiapite, and Fe oxyhydroxides. The formation of soluble sulfates in other sulfur deposits and in aluminite deposits has been reported in numerous papers (e.g. Srebrodol'sky 1971, Smirnova 1971, Vdovichenko et al. 1974, Sokolov et al. 1985, Lizalek et al. 1989, Zavalía and Galliski 1995). Among the minerals most commonly observed are melanterite, siderotil, rozenite, szomolnokite, epsomite, basaluminite (felsöbányaite), halotrichite, pickeringite, copiapite, fibroferrite, and voltaite. Charles et al. (1986) included szomolnokite, rozenite, halotrichite, and alunogen in the vapor-produced alteration assemblage in the Sulphur Springs hydrothermal system at Valles Caldera, New Mexico. Delines (1975) observed chalcanthite among the sublimates at Nyamuragira Volcano, Zaire, and Africano and Bernard (2000) reported alunite, schlossmacherite (hydronium-dominant alunite), anhydrite/gypsum, and an Fe-Mg sulfate close to magnesiocopiapite in composition in the fumarolic environment of Usu Volcano, Japan.

There are a number of localities that use the name of Alum Creek or Alum Rock. One striking example is the deposit of alum rock next to the Gila River in Grant County, New Mexico, where an andesitic breccia is highly altered and contains aluminum sulfate salts in addition to large quantities of kaolinite (Hayes 1906). Alunogen and halotrichite are present, and one incrustation of alunogen was described as being more than a meter thick. The weathering of fine-grained pyrite and acid hydrothermal alteration seem to have been responsible for the occurrence of metal-sulfate salts rich in aluminum.

Stoiber and Rose (1974) identified 29 metal-sulfate salts at 14 Central American volcanoes; most of the salts were soluble hydrated minerals and are the same as those at metal-sulfide mines and mineral deposits. In fumarolic incrustations at Mount St. Helens, Washington, Keith et al. (1981) observed that halotrichite and gypsum are common, and also present are anhydrite, melanterite, alunite, thenardite, sal ammoniac NH_4Cl, and glauberite $Na_2Ca(SO_4)_2$. Melanterite and other salts occur as volcanic sublimates in Japan (Ossaka 1965) and Kamchatka, Russia (Vergasova 1983), as authigenic minerals from volcanic emanations in Greece (Stamatakis et al. 1987, Kyriakopoulos et al. 1990), and as pseudofumarolic deposits in burning coal dumps (Lazarenko et al. 1973, Novikov and Suprychev 1986). Martin et al. (1999) found alunogen, meta-alunogen, halotrichite, melanterite, potash alum, mirabilite $Na_2SO_4 \cdot 10H_2O$, and tschermigite $(NH_4)Al(SO_4)_2 \cdot 12H_2O$ as products of the sulfuric-acid alteration of ignimbrite at Te Kopia geothermal field in New Zealand. Gypsum, alunite, jarosite, halite $NaCl$, kalinite $KAl(SO_4)_2 \cdot 11H_2O$, and tamarugite $NaAl(SO_4)_2 \cdot 6H_2O$ also occur in the field (Rodgers et al. 2000). Diverse assemblages that include water-soluble simple sulfates such as rozenite, coquimbite, halotrichite, and alunogen have been observed in other geothermal fields (Zhu and Tong 1987). Among the minerals identified by Minakawa and Noto (1994) as occurring in a tufa deposit at the Myoban hot spring, Oita Prefecture, Japan, are melanterite, halotrichite, copiapite, alunogen, tamarugite, voltaite, and metavoltine $K_2Na_6Fe^{2+}Fe^{3+}_6(SO_4)_{12}O_2 \cdot 18H_2O$.

Shales commonly contain fine-grained pyrite that forms salt efflorescences upon weathering. Badak (1959) and Kubisz and Michalek (1959) detected melanterite and epsomite in weathered oil shales, and Ievlev (1988) reported the occurrence of melanterite, rozenite, halotrichite, copiapite, and coquimbite that formed from the weathering of shale in permafrost conditions. Development of efflorescences is typical during dry periods when evaporation takes place, which may occur in a variety of climates. However, the salts tend to persist for longer periods of time in regions with prolonged dry seasons, i.e. hyper-arid, arid, and semi-arid climates. In the Upper Colorado River Basin, for example, Whittig et al. (1982) observed gypsum, epsomite, hexahydrite, pentahydrite, starkeyite, kieserite, thenardite, mirabilite $Na_2SO_4 \cdot 10H_2O$, löweite $Na_{12}Mg_7(SO_4)_{13} \cdot 15H_2O$, and blödite $Na_2Mg(SO_4)_2 \cdot 4H_2O$ as efflorescences on the Mancos Shale. Their dissolution was determined to be responsible for the high contents of dissolved salts in some of the streams draining into the Upper Colorado River.

As has been emphasized, the compositions of soluble metal-sulfate minerals reflect the compositions of the solutions from which the minerals precipitate. In supergene zones, therefore, these sulfates are closely linked to the compositions of the sulfide minerals that have been oxidized, as well as to the composition of the surrounding minerals that have been susceptible to dissolution by the ensuing low-pH solutions.

Dissolution during rainfall events

Dissolution of Fe sulfates during storm runoff (Dagenhart 1980, Olyphant et al. 1991, Bayless and Olyphant 1993) or after reclamation efforts (Cravotta 1994) can acidify streams, can rapidly increase metal loading to surface waters, and can lead to the development of acidic groundwaters. Such dissolution can also lead to deleterious consequences for remediation efforts that involve the flooding of mine voids by plugging of mine openings (e.g. Cravotta 1994, Nordstrom and Alpers 1999a).

Dissolution of metal-sulfate salts during storm runoff events has been found to cause a spiked increase in the concentrations of dissolved metals and sulfate even as the stream discharge increased (Dagenhart 1980). Figure 8 shows spiked increases in Fe, Cu, and Zn during the rising limb of the discharge in Contrary Creek, Virginia, for a rainstorm event of June 19, 1978. Careful mineralogical examination showed that the spiked increases in solute concentrations were related to the dissolution of efflorescent salts on mine tailings.

Keith and Runnells (1998) and Keith et al. (1999) noted that the dissolution of sulfates, accumulated at base-metal waste sites during the dry season, gave a pronounced hydrogeochemical response following the first storm of the wet season. Rapid response to local climatic conditions has also been noted by Alpers et al. (1994), who observed that dissolution of melanterite, in which solid solutions of other metals were present, led to seasonal variations of Zn/Cu in acidic effluents at Iron Mountain, California. Cyclical changes in metal and sulfate concentrations through storm events (Dagenhart 1980) and cyclical changes in Zn/Cu ratios through seasonal wet-dry cycles (Alpers et al. 1992, 1994)

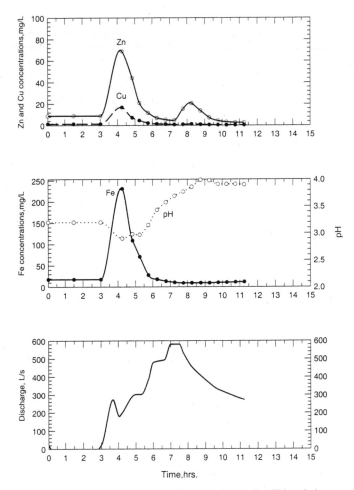

Figure 8. Variations in Zn, Cu, and Fe, and decrease in pH in relation to stream discharge after a rainstorm at Contrary Creek, Virginia (after Dagenhart 1980).

can be attributed to dissolution and reprecipitation of soluble metal-sulfate salts.

One of the environmental consequences of metal concentrations increasing during the rising limb of a stream discharge during storm events is that downstream aquatic life receives a much greater loading of metals than if the rainstorm simply diluted the existing water quality. Indeed, fish kills are commonly associated with some of the early rainstorm events of the wet season in areas with a periodically dry climate.

Plugging of mine adits and the consequent rise in underground water levels can cause the dissolution of soluble salts that have accumulated in the mine workings. For the Richmond mine at Iron Mountain, California, it was estimated by Nordstrom and Alpers (1999a) that about 600,000 m^3 of highly acidic water (pH < 1) containing many grams per liter of dissolved heavy metals could accumulate underground as a result of proposed (and later rejected) plans for remediation by mine plugging. Figure 9 shows the anticipated drop in pH with increasing dissolution of soluble salts that was computed on the basis of the

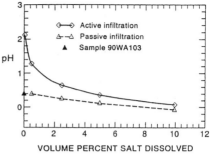

Figure 9. Calculated decrease in pH related to the potential volume of metal-sulfate salts dissolved at Iron Mountain, California (from Nordstrom and Alpers 1999a).

observed mineralogy and composition of salts found underground at Iron Mountain. A plot similar to Figure 9 was produced by Keith et al. (1999) to show the effect on predicted pH by the dissolution of different mass amounts of specific salts.

Laboratory studies

One of the more accessible examples of the development of soluble sulfate minerals is their presence, as a result of spontaneous growth, on laboratory and museum specimens (Workman and Rader 1961, Buurman 1975, Wiese et al. 1987). Such sulfates formed so readily on samples of cored pyritiferous mine tailings that the minerals were included in Jambor's (1994) classification as 'quaternary' minerals to distinguish them from 'tertiary' sulfates that had crystallized from pore waters during drying of the cores. In a study of museum and collectors' specimens of pyrite and marcasite, Blount (1993) identified ten Fe-sulfate minerals, and two other sulfates that were concluded to have formed as alteration products as a result of humidity variations during storage of pyrite and marcasite specimens. Luzgin (1990) noted that, in addition to szomolnokite and chalcanthite, the Co arsenate mineral erythrite had formed during long-term storage of collected specimens. Grybeck (1976) observed that specimens of sphalerite continued to form coatings of gunningite even after repeated washings had cleansed the sphalerite surfaces.

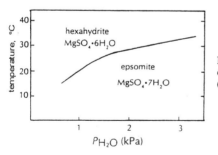

Figure 10. Stability of the simple sulfates of Mg as a function of temperature and water (modified from Keller et al. 1986a).

Many of the simple salts are extremely sensitive to atmospheric conditions and will change their hydration state, in some cases reversibly, in response to the local temperature and humidity conditions (Fig. 10; Keller et al. 1986a,b; Waller 1992, Chou et al. 2000). Preservation of samples in their original, as-collected state is thus a common problem for mineralogical studies. Waller (1992) compiled data on humidity-controlled reactions among hydrous minerals (Table 4) and discussed various strategies for sample conservation.

Solubilities and stability relationships

Ferrous sulfate hydrates. Pure, synthetic forms of three of the naturally occurring ferrous sulfate minerals, melanterite, rozenite, and szomolnokite, have been the subject of

Table 4. Sulfate minerals subject to humidity-related phase transitions. Modified from Waller (1992).

Species	Formula[a]	Reaction[b]	%RH[c]	T[d]	Reference[e]
aluminite	$Al_2(SO_4)(OH)_4 \cdot 7H_2O$	$-nH_2O$			Waller 1992
alunogen	$Al_2(SO_4)_3 \cdot 17H_2O$	deliquesce	88	20	Waller 1992
		$-nH_2O$			
bianchite	$(Zn,Fe^{2+})SO_4 \cdot 6H_2O$	$+1H_2O \to 7$	59	20	Bonnell and Burridge 1935
		$-1H_2O \to 6$	56	20	
bieberite	$CoSO_4 \cdot 7H_2O$	deliquesce	94	20	Waller 1992
		$-1H_2O \to 6$	70	20	Waller 1992
bonattite	$CuSO_4 \cdot 3H_2O$	$+2H_2O \to 5$	33	20	Collins and Menzies 1936
		$-2H_2O \to 1$	22	25	Collins and Menzies 1936
boothite	$CuSO_4 \cdot 7H_2O$	$-nH_2O$			Parsons 1922
boussingaultite	$(NH_4)_2Mg(SO_4)_2 \cdot 6H_2O$	deliquesce	96	25	Waller 1992
		$-2H_2O \to 4$	20	25	Caven and Ferguson 1924
boyleite	$(Zn,Mg)SO_4 \cdot 4H_2O$	$-nH_2O$			Bari et al. 1982
chalcanthite	$CuSO_4 \cdot 5H_2O$	deliquesce	97	25	Collins and Menzies 1936
		$-2H_2O \to 3$	33	25	Collins and Menzies 1936
chalcocyanite	$CuSO_4$	$+1H_2O \to 1$	3	25	Foote and Scholes 1911
coquimbite	$Fe^{3+}_2(SO_4)_3 \cdot 9H_2O$	deliquesce	~73	20	Waller 1992
		$-nH_2O$			Bannister 1937
cyanochroite	$K_2Cu(SO_4)_2 \cdot 6H_2O$	$-2H_2O \to 4$	36	25	Caven and Ferguson 1922
darapskite	$Na_3(SO_4)(NO_3) \cdot H_2O$	deliquesce			Parsons 1922
epsomite	$MgSO_4 \cdot 7H_2O$	deliquesce	91	20	Waller 1992
		$-H_2O \to 6$	83	20	Randall 1930
ettringite	$Ca_6Al_2(SO_4)_3(OH)_{12} \cdot 26H_2O$	$-nH_2O$			Palache et al. 1951
ferrinatrite	$Na_3Fe^{3+}(SO_4)_3 \cdot 3H_2O$	deliquesce			Palache et al. 1951
ferrohexahydrite	$Fe^{2+}SO_4 \cdot 6H_2O$	$+1H_2O \to 7$	61	25	Bonnell and Burridge 1935
fibroferrite	$Fe^{3+}(SO_4)(OH) \cdot 5H_2O$				Palache et al. 1951
goslarite	$ZnSO_4 \cdot 7H_2O$	deliquesce	89	25	O'Brien 1948
		$-nH_2O$			Palache et al. 1951
gunningite	$(Zn,Mn^{2+})SO_4 \cdot H_2O$	$+5H_2O \to 6$	54	25	Foote and Scholes 1911
		$-1H_2O \to 0$	4	25	Foote and Scholes 1911
halotrichite	$Fe^{2+}Al_2(SO_4)_4 \cdot 22H_2O$	$-nH_2O$			Bannister 1937
hanksite	$KNa_{22}(SO_4)_9(CO_3)_2Cl$	deliquesce	~75	RT	Waller 1992
hexahydrite	$MgSO_4 \cdot 6H_2O$	$+1H_2O \to 7$	51	25	Bonnell and Burridge 1935
		$-1H_2O \to 5$	41	25	Foote and Scholes 1911
hydrobasaluminite	$Al_4(SO_4)(OH)_{10} \cdot 15H_2O$ [a]	$-nH_2O \to 5$			Sunderman and Beck 1969
jôkokuite	$Mn^{2+}SO_4 \cdot 5H_2O$	$+2H_2O \to 7$	84	15	Randall 1930
		$-4H_2O \to 1$	74	20	Randall 1930
kainite	$MgSO_4 \cdot KCl \cdot 3H_2O$	deliquesce			
		$-3H_2O \to 0$	3	32	Randall 1930
kieserite	$MgSO_4 \cdot H_2O$	$+3H_2O \to 4$	21	25	Foote and Scholes 1911
		$-1H_2O \to 0$	4	25	Foote and Scholes 1911

Table 4, continued

konyaite	$Na_2Mg(SO_4)_2 \cdot 5H_2O$	$-1H_2O \rightarrow 4$			van Doesburg et al. 1982
langbeinite	$K_2Mg_2(SO_4)_3$	deliquesce			Bannister 1937
leonite	$K_2Mg(SO_4)_2 \cdot 4H_2O$	$+2H_2O \rightarrow 6$			Palache et al. 1951
löweite	$Na_{12}Mg_7(SO_4)_{13} \cdot 15H_2O$	deliquesce			Palache et al. 1951
mallardite	$Mn^{2+}SO_4 \cdot 7H_2O$	$-2H_2O \rightarrow 5$		<9	Linke 1965
mascagnite	$(NH_4)_2SO_4$	deliquesce	81	20	Wexler and Hasegawa 1954
matteuccite	$NaHSO_4 \cdot H_2O$	deliquesce	52	20	O'Brien 1948
		$-1H_2O \rightarrow 0$	16	20	Heitmankova and Cerny 1974
melanterite	$Fe^{2+}SO_4 \cdot 7H_2O$	deliquesce	95	20	Waller 1992
		$-6H_2O \rightarrow 1$	57	20	Waller 1992
mendozite	$NaAl(SO_4)_2 \cdot 11H_2O$	$-5H_2O \rightarrow 6$			Palache et al. 1951
mercallite	$KHSO_4$	deliquesce	86	20	O'Brien 1948
mirabilite	$Na_2SO_4 \cdot 10H_2O$	deliquesce	93	RT	O'Brien 1948
moorhouseite	$(Co,Ni,Mn^{2+})SO_4 \cdot 6H_2O$	$+1H_2O \rightarrow 7$	70	20	Schumb 1923
morenosite	$NiSO_4 \cdot 7H_2O$	deliquesce	93	25	Waller 1992
pentahydrite	$MgSO_4 \cdot 5H_2O$	$+1H_2O \rightarrow 6$	41	25	Randall 1930
		$-1H_2O \rightarrow 4$	37	25	Randall 1930
poitevinite	$(Cu,Fe^{2+},Zn)SO_4 \cdot H_2O$	$+2H_2O \rightarrow 3$	22	25	Collins and Menzies 1936
		$-1H_2O \rightarrow 0$	3	25	Foote and Scholes 1911
quenstedtite	$Fe^{3+}_2(SO_4)_3 \cdot 10H_2O$	$-1H_2O \rightarrow 9$			Bandy 1938
retgersite	$NiSO_4 \cdot 6H_2O$	$+1H_2O \rightarrow 7$	84	20	Bonnell and Burridge 1935
rhomboclase	$(H_5O_2)^+Fe^{3+}(SO_4)_2 \cdot 2H_2O$	$-nH_2O$			Bannister 1937
sanderite	$MgSO_4 \cdot 2H_2O$	$+2H_2O \rightarrow 4$	22	31	Randall 1930
sideronatrite	$Na_2Fe^{3+}(SO_4)_2(OH) \cdot 3H_2O$	$-nH_2O$			Zodrow et al. 1979
sodium alum	$NaAl(SO_4)_2 \cdot 12H_2O$	$-6H_2O \rightarrow 6$	86	20	Hepburn and Phillips 1952
starkeyite	$MgSO_4 \cdot 4H_2O$	$+1H_2O \rightarrow 5$	37	25	Foote and Scholes 1911
szmikite	$Mn^{2+}(SO_4) \cdot H_2O$	$+4H_2O \rightarrow 5$	83	25	Carpenter and Jette 1923
		$-1H_2O \rightarrow 0$	17	20	Randall 1930
szomolnokite	$Fe^{2+}SO_4 \cdot H_2O$	$-1H_2O \rightarrow 0$	11	20	Waller 1992
		$+6H_2O \rightarrow 7$	57		
tamarugite	$NaAl(SO_4)_2 \cdot 6H_2O$	$+6H_2O \rightarrow 12$	86	20	Hepburn and Phillips 1952
thenardite	Na_2SO_4	$+10H_2O \rightarrow 10$	81	25	Wilson 1921
tschermigite	$(NH_4)Al(SO_4)_2 \cdot 12H_2O$	$-6H_2O \rightarrow 6$	7	25	Hepburn and Phillips 1952
uranopilite	$(UO_2)_6(SO_4)(OH)_{10} \cdot 12H_2O$	$-nH_2O$			Palache et al. 1951
voltaite	$K_2Fe^{2+}_5Fe^{3+}_4(SO_4)_{12} \cdot 18H_2O$	$-nH_2O$			Bannister 1937
zaherite	$Al_{12}(SO_4)_5(OH)_{26} \cdot 20H_2O$	$-nH_2O$			Ruotsala and Babcock 1977
zinc-melanterite	$(Zn,Cu,Fe^{2+})SO_4 \cdot 7H_2O$	deliquesce	89	25	O'Brien 1948
		$-1H_2O \rightarrow 6$	65	25	Bonnell and Burridge 1935

[a] Formulas from Waller (1992) or Mandarino (1999) except for hydrobasaluminite (Bigham and Nordstrom, this vol.)
[b] The number of water molecules specified as being gained or lost in a reaction is for the reaction as written with the formula given. If the reaction is deliquescence, then decomposition, oxidation, or hydrolysis may also occur.
[c] The relative humidity at which a reaction occurs is dependent on temperature. The relative humidity given in column 4 applies at the temperature given in column 5. See Waller (1992) for methods of calculation.
[d] Temperatures are given in degrees Celsius. RT = room temperature (normally ~ 22°C).
[e] References as cited by Waller (1992).

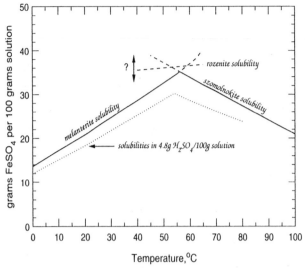

Figure 11. Solubility data for melanterite ($FeSO_4 \cdot 7H_2O$), rozenite ($FeSO_4 \cdot 4H_2O$), and szomolnokite ($FeSO_4 \cdot H_2O$); data of Bullough et al. (1952) and Reardon and Beckie (1987).

numerous laboratory studies of solubility. These studies help to define the range of environmental stability for each mineral. Early data have been compiled by Mellor (1923), International Critical Tables (1927), Gmelins Handbuch (1932), and Linke and Seidell (1958). Phase equilibria and electrolyte properties in the $FeSO_4$–H_2SO_4–H_2O system have been studied more recently by Bullough et al. (1952) and Reardon and Beckie (1987). In anoxic systems the solubility equilibria for the three hydrate minerals are shown in Figure 11. Rozenite appears to be a metastable phase in aqueous solutions according to the carefully reversed solubility measurements of Bullough et al. (1952) and as discussed by Reardon and Beckie (1987). Bursa and Stanisz-Lewicka (1982) found rozenite to be a metastable intermediate during the recrystallization of melanterite with increasing sulfuric acid concentrations at temperatures of 45, 55, and 60°C. Rozenite formed first and then slowly recrystallized to szomolnokite. Hence, the solubility curve of rozenite is estimated to be slightly above the point of intersection of the solubilities for melanterite and szomolnokite. A conclusion to be drawn from these data is that the occurrence of rozenite would require dehydration of melanterite (or rehydration of szomolnokite) in the absence of a solution phase. Rozenite must form by dehydration at <100% relative humidity. For example, Mitchell (1984) determined that melanterite dehydrated to rozenite when the relative humidity is less than 65%. Parkinson and Day (1981) measured an equilibrium relative humidity of 55.8% at 20°C.

The dotted line in Figure 11 shows the effective lowering of the solubilities by the additional of 4.8 wt % sulfuric acid. This concentration of acid is close to the maximum concentration of sulfuric acid commonly observed for acid mine waters, and it would be equivalent to a pH range of 0 to 1. The presence of sulfuric acid produced by pyrite oxidation can be seen to enhance the formation of melanterite and szomolnokite by decreasing their solubility through the common-ion effect. Melanterite commonly contains some divalent metals in solid solution (Fig. 2), which further decreases its solubility (Mellor 1923) and increases its stability (see Glynn, this volume).

The question mark in Figure 11 suggests that there are several inadequacies in the solubility data. Solubility data for rozenite are not well-established, and no data seem to exist for ferrohexahydrite or siderotil to confirm their metastability. The solubility data between 50 and 65°C are in need of detailed study because that is the temperature range in which ferrohexahydrie and siderotil would form, and the existing measurements contain inconsistencies. The data of Reardon and Beckie (1987) produce a melanterite-szomolnokite transition temperature of about 65°C instead of the 56.7°C measured by Bullough et al. (1952), but it is likely that the data cannot be fit much better at this time. Mitchell (1984) was able to prepare rozenite both by recrystallization from solution at 60°C and by hydration of an X-ray amorphous monohydrate in the relative humidity range of >0 to 65%. Above 65% the heptahydrate formed. A dihydrate salt reported rarely in lab studies requires verification and has not been reported to occur in nature.

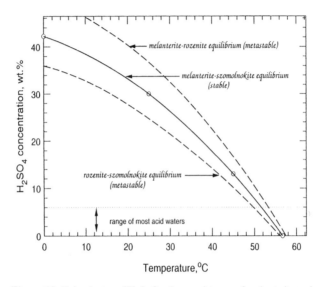

Figure 12. Univariant equilibria for the coexistence of melanterite and szomolnokite as a function of temperature and concentration of H_2SO_4; data sources as in Figure 11.

A number of solubility studies over a range of sulfuric acid concentrations help to define the stability relationships and provide convincing evidence that only melanterite and szomolnokite are stable in contact with saturated solutions. Figure 12 shows the variation in the stable univariant equilibrium for the coexistence of melanterite and szomolnokite as a function of temperature and sulfuric acid concentration. Estimated curves for the metastable extensions of the melanterite-rozenite and rozenite-szomolnokite equilibria in the presence of water (100% relative humidity) are shown as dashed lines in Figure 12. The metastable equilibria intersect the stable equilibrium line at <100% relative humidity. Both the concentrations of sulfuric acid and the relative humidity are independent measurements of the activity of water and can be used to confirm the consistency and accuracy of the stability relations and to recover thermodynamic properties.

Zinc sulfate hydrates. Solubility data for the zinc sulfate hydrates, goslarite ($7H_2O$), bianchite ($6H_2O$), and gunningite ($1H_2O$) are based on the best fits by Linke and Seidell (1958) and the data shown in Figure 166 of Gmelins Handbuch (1956). The solubility curves (Fig. 13) are similar to the ferrous sulfate data except that bianchite, the hexahydrate, is the stable intermediate hydrate between 37.9 and 48.8°C. The enthalpy of

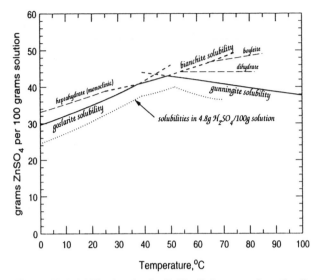

Figure 13. Solubility data for the $ZnSO_4$–H_2O system, from Gmelins Handbuch (1956) and Linke and Seidell (1958).

transition for the reaction goslarite to bianchite plus water in aqueous solution was determined by Grønvold and Meisingset (1982) to be 16.5 kJ mol^{-1} at 38°C. The tetrahydrate, boyleite, and the dihydrate are metastable in aqueous solutions. The hemiheptahydrate, the trihydrate, and the pentahydrate have been reported from lab studies but these are not confirmed as metastable phases from the solubility studies. The dashed lines in Figure 13 show the metastable extensions of each solubility, and the dotted line shows the lowering of the solubility of the stable phases with the addition of 4.8 wt % sulfuric acid. Impurities of divalent metal also lower the solubility of the zinc sulfate hydrates (Mellor 1923, Petlicka 1971).

Solubility data for the zinc sulfate system in sulfuric acid are sufficient to plot the univariant equilibria as a function of temperature (Fig. 14). The quadruple point for the coexistence of goslarite, bianchite, gunningite, and solution does occur for saturated solution conditions (at 21.5°C) although it is outside the acid range of most acid mine waters. These equilibria demonstrate that all three zinc hydrates can form directly from solution and that the temperature and sulfuric acid concentrations for their stability range fall within well-defined limits. Furthermore, boyleite (like rozenite) is metastable and must form by dehydration of bianchite or goslarite (or hydration of gunningite). The solubility studies also predict the metastability of a zinc sulfate dihydrate phase that has not yet been identified as a mineral. There are several examples of synthetic compounds that were later identified as minerals. For example, the solubility of boyleite was known from the 1930s, but the compound was not identified as a mineral until 1962 (Jambor and Boyle 1962, Walenta 1978).

Inconsistencies occur in the phase-equilibria data for the aqueous zinc sulfate system. For example, the equilibrium for the coexistence of bianchite and gunningite has been determined by more than one investigator and, in the absence of sulfuric acid, it occurs at 48.8°C. However, carefully measured solubilities over a range of sulfuric acid concentrations extrapolate to about 56°C at zero sulfuric acid concentration (Fig. 14). This difference is greater than the error of the measurements.

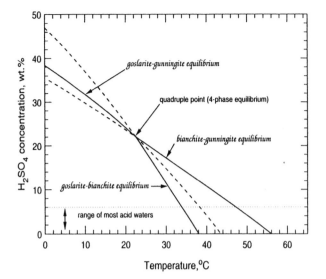

Figure 14. Univariant equilibria for the $ZnSO_4$ system; sources of data as in Figure 13.

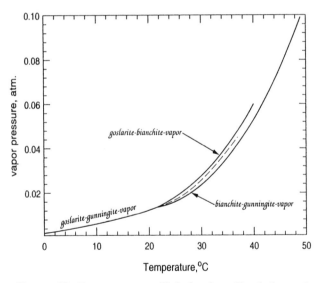

Figure 15. Vapor-pressure equilibria for zinc sulfate hydrates; data from Gmelins Handbuch (1956).

Considerable data on equilibrium vapor-pressure studies exist for the zinc sulfate hydrate thermal transitions. These data are summarized in Gmelins Handbuch (1956) and are plotted here in Figure 15. As in Figure 14, the stable three-phase univariant equilibria are shown by solid lines and the metastable extension for the goslarite-gunningite-vapor line is dashed. The invariant quadruple point (coexisting goslarite, bianchite, gunningite, and water vapor) is located in Figure 15 at 21.5°C, consistent with the solubility diagram (Fig. 14). The vapor-pressure data demonstrate the range of stability for goslarite, bianchite, and gunningite, with boyleite an unmeasured metastable phase.

Copper sulfate hydrates. Solubility and other phase-equilibria studies summarized by Mellor (1923) and in the Gmelins Handbuch (1958) indicate that only the penta-, the tri-, and the monohydrate (chalcanthite, bonattite, and poitevinite, respectively) are stable phases in the $CuSO_4$–H_2O system. Experimental work has demonstrated that pure boothite is very unstable but there are indications that it can be made more stable with the substitution of divalent cations. For example, de Boisbadran (cited in Mellor 1923) was able to synthesize boothite by adding a seed crystal of melanterite to a supersaturated solution of copper sulfate. This observation, combined with the difficulty in sythesizing pure boothite in solubility studies and its rare occurrence in nature, indicates that boothite is unlikely to be found as end-member, monoclinic $CuSO_4 \cdot 7H_2O$.

The transition from chalcanthite to bonattite occurs at 95.9°C (Collins and Menzies 1936) which helps to explain the rare occurrence of bonattite and its tendency to hydrate to chalcanthite under most environmental conditions. The pure copper sulfate monohydrate has an even higher transition temperature of 116.6°C (Collins and Menzies 1936).

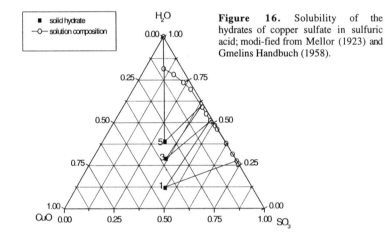

Figure 16. Solubility of the hydrates of copper sulfate in sulfuric acid; modi-fied from Mellor (1923) and Gmelins Handbuch (1958).

The ternary phase diagram for the solubility of copper sulfate hydrates in sulfuric acid solutions is shown in Figure 16. Note that chalcanthite covers a large field of stability which also accounts for the rare occurrence of bonattite and boothite. Recent evaluations of aqueous copper-sulfate electrolyte systems using the Pitzer approach have made refinements on the solubility of chalcanthite for systems involving sulfuric acid (Baes et al. 1993) and copper chloride and sodium sulfate (Christov 2000).

Nickel sulfate hydrates. The relatively evenly spaced solubilities of morenosite ($7H_2O$, orthorhombic), retgersite ($6H_2O$, tetragonal), and nickelhexahydrite ($6H_2O$, monoclinic) as a function of temperature are shown in Figure 17. Dwornikite, the monohydrate, is not stable until the temperature is above 84°C. The solubilities for the first three minerals also have nearly congruent slopes because morenosite and retgersite differ by only one water molecule of hydration, and retgersite and nickelhexahydrite are dimorphs. Four metastable hydrates have been measured at temperatures above 90°C: the di-, tri-, tetra-, and pentahydrate. Reardon (1989) has applied the Pitzer model to the solubilities of morenosite, retgersite, and nickelhexahydrite in water and in sulfuric acid for temperatures of 0 to 70°C.

Cobalt sulfate hydrates. The solubilities of the cobalt sulfate hydrates (Fig. 18)

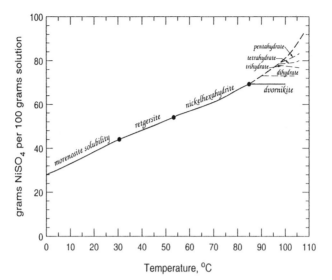

Figure 17. Solubility of the hydrates of nickel sulfate; data from Gmelins Handbuch (1966) and Reardon (1989).

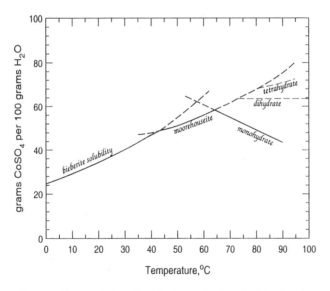

Figure 18. Solubility of the hydrates of cobalt sulfate; data from Gmelins Handbuch (1961).

are similar to, but have a larger range than, those of the zinc sulfate hydrates. The monohydrate is stable at temperatures above 62°C, and the dihydrate and tetrahydrate form two metastable phases.

Magnesium sulfate hydrates. The pattern of solubility for the magnesium sulfate hydrates (Fig. 19) is similar to that for iron, zinc, nickel, and cobalt in that the heptahydrate (epsomite) covers the largest temperature range of stability. With increasing temperature,

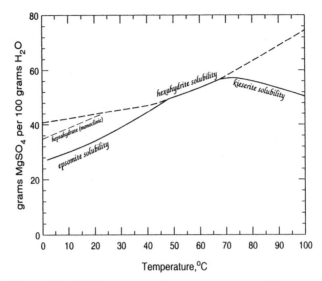

Figure 19. Solubility of the hydrates of magnesium sulfate; data from Gmelins Handbuch (1939).

the heptahydrate transforms to the hexahydrate mineral (hexahydrite), and then to the monohydrate (kieserite). No stability fields are present at 100% relative humidity for the pentahydrate (pentahydrite), the tetrahydrate (starkeyite), or the dihydrate (sanderite), and no metastable solubility data are reported. It is unusual that hexahydrite solubility can be measured far into its temperature range of metastability on both the high-temperature end and the low-temperature end. Isopiestic measurements and a thermodynamic evaluation of the $MgSO_4$–H_2O system by Archer and Rard (1998) includes a measurement of epsomite solubility at 25°C.

Aluminum sulfate hydrates. The dissolution of aluminum sulfate in water causes considerable hydrolysis to occur. Solubilities of aluminum sulfate hydrates can only be measured in sulfuric acid solutions. Hence, we have reproduced in Figure 20 the more

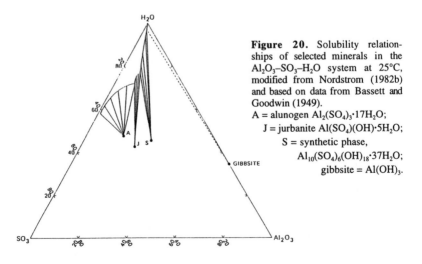

Figure 20. Solubility relationships of selected minerals in the Al_2O_3–SO_3–H_2O system at 25°C, modified from Nordstrom (1982b) and based on data from Bassett and Goodwin (1949).

A = alunogen $Al_2(SO_4)_3 \cdot 17H_2O$;

J = jurbanite $Al(SO_4)(OH) \cdot 5H_2O$;

S = synthetic phase, $Al_{10}(SO_4)_6(OH)_{18} \cdot 37H_2O$;

gibbsite = $Al(OH)_3$.

soluble portion of the ternary system Al_2O_3–SO_3–H_2O (or Al_2SO_4–H_2SO_4–H_2O) from Nordstrom (1982b), which is based on the work of Bassett and Goodwin (1949). On this diagram there are three phases, A (alunogen), J (jurbanite), and S (synthetic compound $Al_{10}(SO_4)_6(OH)_{18} \cdot 37H_2O$) that appear at 25°C. The phase at S, which has not been identified as a mineral, has a composition similar to that of alunite, but with a much higher water content. Several other relatively insoluble minerals of aluminum sulfate are known and are discussed by Bigham and Nordstrom (this volume). Reardon (1988) applied the Pitzer approach to the aqueous aluminum-sulfate system and established mixing parameters for sulfate solutions of Cu, Ni, Fe(II), and Mg (see Ptacek and Blowes, this volume). The work included refinements of the solubility for alunogen, halotrichite, and pickeringite. The solubility of potassium alum was evaluated with the Pitzer model by Reardon and Stevens (1991).

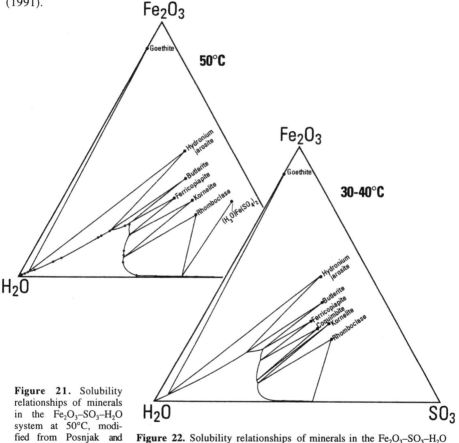

Figure 21. Solubility relationships of minerals in the Fe_2O_3–SO_3–H_2O system at 50°C, modified from Posnjak and Merwin (1922).

Figure 22. Solubility relationships of minerals in the Fe_2O_3–SO_3–H_2O system at 30 to 40°C, modified from Merwin and Posnjak (1937).

Ferric sulfate hydrates. The most extensive work on the ferric sulfate solubilities was from Posnjak and Merwin (1922); their ternary system solubilities for Fe_2O_3–SO_3–H_2O for 50°C is shown in Figure 21, and a similar diagram estimated for 30-40°C is shown in Figure 22 (Merwin and Posnjak 1937). The only significant difference between Figures 21 and 22 is the appearance of a stability field for coquimbite, replacing part of the kornelite field. The sequence of mineral formation from hydronium jarosite → butlerite → ferricopiapite → coquimbite → kornelite → rhomboclase describes

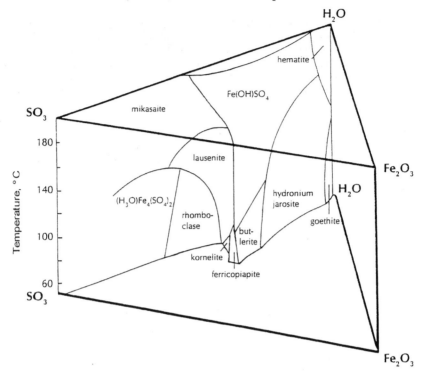

Figure 23. Phase diagram for the system Fe_2O_3–SO_3–H_2O as a function of temperature (redrawn from Atencio and Hypolito 1992, 1992/1993). Sulfate-mineral equivalents are
mikasaite = $Fe_2^{3+}(SO_4)_3$, lausenite = $Fe_2^{3+}(SO_4)_3 \cdot 6H_2O$,
rhomboclase = $(H_3O)Fe^{3+}(SO_4)_2 \cdot 3H_2O$, kornelite = $Fe_2^{3+}(SO_4)_3 \cdot 7H_2O$,
ferricopiapite = $Fe_{2/3}^{3+}Fe_4^{3+}(SO_4)_6(OH)_2 \cdot 20H_2O$, butlerite = $Fe^{3+}(SO_4)(OH) \cdot 2H_2O$,
hydronium jarosite = $(H_3O) Fe_3^{3+}(SO_4)_2(OH)_6$, goethite = FeOOH, and hematite = Fe_2O_3.

solution conditions in which the pH continually decreases and the solution SO_4/Fe ratio increases. This observation is consistent with field pH measurements of mine waters that have been found in equilibrium with jarosite (pH = 1 to 2.35), copiapite (pH = -1), and rhomboclase (pH = -2.5 to -3.6) from Iron Mountain, California (Alpers et al. 1989, Nordstrom and Alpers 1999a,b; Robinson 1999, Robinson et al. 2000a,b). A phase diagram for the ferric sulfate hydrate solubilities as a function of temperature from 50 to 200°C is shown in Figure 23.

   Sideronatrite and ferrinatrite. Sideronatrite $Na_2Fe^{3+}(OH)(SO_4)_2 \cdot 3H_2O$ and ferri-natrite $Na_3Fe^{3+}(SO_4)_3 \cdot 3H_2O$ are minerals of moderately high solubility. Ferrinatrite is trigonal and transforms to sideronatrite, which is orthorhombic, in moist air. Solubilities of these two minerals were determined by Linke and Seidell (1958) and by Flynn and Eisele (1987).

   Ferrous-ferric sulfate hydrates. There are only limited solubility data for mixed divalent–trivalent Fe sulfate salts, including römerite and bilinite (Gmelins Handbuch 1932). There has been no systematic effort to describe phase stability relations in the $FeSO_4$–$Fe_2(SO_4)_3$–H_2O system. The lack of an aqueous model that can account for mineral precipitation from Fe^{3+}-sulfate or Fe^{2+}–Fe^{3+}-sulfate solutions at high concentrations makes it difficult to obtain a complete picture of phase relations in acid mine waters and other settings where the solubility of Fe and other metals may be controlled by the formation of sulfate minerals. Ptacek and Blowes (this volume) discuss the available aqueous models for predicting sulfate-mineral solubilities in concentrated solutions.

PARAGENESIS

Paragenesis in this context refers to the sequence of mineral formation and alteration that occurs as metal-sulfate minerals form and continue to evolve with time. It was mentioned previously that the general sequence observed, both at the mineral-deposit scale and on a microscopic scale, is from initial $Fe^{2+}SO_4 \cdot nH_2O$ salts to those of Fe^{3+}, and thence to jarosite and ultimately to Fe oxyhydroxides.

Within the $Fe^{2+}SO_4 \cdot nH_2O$ minerals, the sequence melanterite ($n = 7$) → siderotil ($n = 5$) → rozenite ($n = 4$) → szomolnokite ($n = 1$) can reflect decreasing local moisture contents or increasing temperature; reversibility of hydration states among the tetrahydrate–pentahydrate–heptahydrate is easy, but the monohydrate is relatively more stable. The sequence from the heptahydrate to the monohydrate is also known to be favored as liquor acidity increases.

The sequences melanterite → fibroferrite → aluminocopiapite, or melanterite → fibroferrite → jarosite → 'limonite' (goethite), or melanterite → halotrichite → botryogen have been observed in various oxidized deposits (Cherkasov 1975, Zodrow et al. 1979, Kravtsov 1984, Minakawa and Noto 1994). Bolshakov and Ptushko (1967) noted a sequence in which melanterite altered to siderotil, szomolnokite, and copiapite, and subsequent decomposition of these products led to the formation of rhomboclase, römerite, voltaite, and bilinite. A similar sequence (Fig. 24) at another oxidized deposit was observed by Bolshakov and Ptushko (1971). The preservation of these types of sequences seems to be best in cryogenic or arid regions, wherein the sulfates can evolve with minimal climatic perturbation. In the cold, dry climate of northern Greenland, Jacobsen (1989) observed that in fault zones, römerite, melanterite, and rozenite occurred at depth, followed upward by fibroferrite, and then by surface crusts of copiapite.

Figure 24. Natural dehydration and decomposition products of melanterite, as observed by Bolshakov and Ptushko (1971).

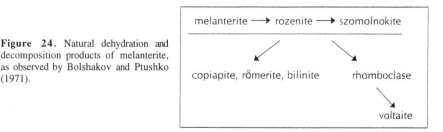

In his classical study of the sulfate deposits of Chile, Bandy (1938) observed the following (abridged) sequence at Alcaparrosa: pyrite → szomolnokite → römerite → quenstedtite → coquimbite → pickeringite → copiapite → parabutlerite → jarosite. Thus the trend is from the divalent to the mixed divalent–trivalent, and thence from the trivalent to the basic sulfates. In oxidized deposits in Kazakhstan, under arid conditions similar to those of northern Chile, Chukhrov (1953) observed the following trend downward from surface: 'limonite' (goethite) → jarosite → parabutlerite → Fe^{3+} sulfate hydrates (copiapite, metavoltine, ferrinatrite, quenstedtite, coquimbite, römerite) → divalent salts (halotrichite, szomolnokite). This general pattern was also observed to be present in oxidized ore deposits in the arid climate of northwestern China, where Tu and Li (1963) noted that the zonation was from 'limonite'-hematite at the surface, downward to Fe^{3+} sulfates (copiapite, sideronatrite, fibroferrite, römerite), and thence to gypsum, sulphur, melanterite, and

Table 5. Paragenetic sequences of Fe-sukfate minerals from pyrite oxidation.

	Alcaparrosa, Chile (Bandy 1938)	Laboratory (Buurman 1975)	Iron Mountain, California (Nordstrom & Alpers 1999a)	Formulae
early	pyrite	pyrite	pyrite	FeS_2
		melanterite	melanterite	$FeSO_4 \cdot 7H_2O$
		siderotil		$FeSO_4 \cdot 5H_2O$
		rozenite	rozenite	$FeSO_4 \cdot 4H_2O$
	szomolnokite	szomolnokite	szomolnokite	$FeSO_4 \cdot H_2O$
	römerite	rhomboclase	copiapite	$Fe^{2+} Fe^{3+}_4 (SO_4)_6(OH)_2 \cdot 20H_2O$
	quenstedtite		römerite	$Fe^{2+} Fe^{3+}_2 (SO_4)_4 \cdot 14H_2O$
	coquimbite	coquimbite	coquimbite	$Fe^{3+}_2 (SO_4)_3 \cdot 9H_2O$
	pickeringite	römerite	kornelite	$Fe^{3+}_2 (SO_4)_3 \cdot 7H_2O$
	copiapite		rhomboclase	$(H_3O)Fe^{3+}(SO_4)_2 \cdot 3H_2O$
	parabutlerite	voltaite	voltaite	$K_2 Fe^{2+}_5 Fe^{3+}_4 (SO_4)_{12} \cdot 18H_2O$
late	jarosite		halotrichite-bilinite	$Fe^{2+}(Al,Fe^{3+})_2(SO_4)_4 \cdot 22H_2O$

halotrichite at depth. These weathering profiles are consistent with the paragenetic sequences noted above, in that the deepest minerals correspond to the most recently formed, as the oxidation front descends in response to erosion and landscape evolution.

Buurman (1975) immersed pyritiferous concretions in distilled water at 25°C for one week, filtered the resulting solution, and determined the sequence of crystallization that evolved as the solution slowly evaporated. The results, summarized in Table 5, show a remarkable correlation with the approximate sequence of deposition that has been established to date for secondary minerals in the underground workings at the Iron Mountain massive sulfide deposits in California (Nordstrom and Alpers 1999a). Although there are some variations in the order of precipitation, Table 5 illustrates well the general sequential crystallization of the soluble metal salts.

ACKNOWLEDGMENTS

The assistance of L.A. Groat of the Department of Earth and Ocean Sciences, University of British Columbia, and A.C. Roberts of the Geological Survey of Canada, Ottawa, in acquiring some of the papers referred to, is much appreciated. The authors also are grateful for assistance and access to facilities at the libraries of the U.S. Geoological Survey, the University of Colorado, and the University of California at Davis. Financial support by the National Research Program of the U.S. Geoological Survey and by the U.S. Environmental Protection Agency is greatly appreciated.

REFERENCES

Africano F, Bernard A (2000) Acid alteration in the fumarolic environment of Usu volcano, Hokkaido, Japan. *In* Varekamp JC, Rowe GL Jr (eds) Crater Lakes. J Volcanol Geotherm Res 97:475-495

Agnew M (1998) The formation of hardpans within tailings as possible inhibitors of acid mine drainage, contaminant release and dusting. PhD thesis, University of Adelaide, Adelaide, Australia

Agricola G (1546) De Natura Fossilium (Textbook of Mineralogy). Translated by Bandy MC, Bandy JA (1955) Geol Soc Am Spec Paper 63

Agricola, G (1556) De Re Metallica (On Metallic Substances). Translated by Hoover HC, Hoover LH (1950) Dover Publications, New York

Alpers CN, Brimhall GH (1989) Paleohydrologic evolution and geochemical dynamics of cumulative supergene metal enrichment at La Escondida, Atacama Desert, northern Chile. Econ Geol 84:229-255

Alpers CN, Nordstrom DK, Ball JW (1989) Solubility of jarosite solid solutions precipitated from acid mine waters, Iron Mountain, California, U.S.A. Sci Géol Bull 42:281-298

Alpers CN, Nordstrom DK, Burchard JM (1992) Compilation and interpretation of water-quality and discharge data for acidic mine waters at Iron Mountain, Shasta County, California, 1940-91. US Geol Survey Water-Resources Invest Rept 91-4160

Alpers CN, Nordstrom DK, Thompson JM (1994) Seasonal variations of Zn/Cn ratios in acid mine water from Iron Mountain, California. *In* Alpers CN, Blowes DW (eds) Environmental Geochemistry of Sulfide Oxidation. Am Chem Soc Symp Series 550:324-344

Angel RJ, Finger LW (1988) Polymorphism of nickel sulfate hexahydrate. Acta Crystallogr C44:1869-1873

Archer DG, Rard JA (1998) Isopiestic investigation of the osmotic and activity coefficients of aqueous $MgSO_4$ and the solubility of $MgSO_4 \cdot 7H_2O(cr)$ at 298.15 K: Thermodynamic properties of the $MgSO_4$ + H_2O system to 440 K. J Chem Eng Data 43:791-806

Aslanian S, Balarew C (1977) On the crystal-structure differences of the $MeSO_4 \cdot nH_2O$ salt-types. Kristall Technik 12:435-446 (in German)

Aslanian S, Balarew C, Oikova T (1972) Isomorphic relations between $ZnSO_4 \cdot 7H_2O$, $NiSO_4 \cdot 7H_2O$ and $CoSO_4 \cdot 7H_2O$. Kristall Technik 7:525-531 (in German)

Atencio D, Hypolito R (1992) The system Fe_2O_3–SO_3–H_2O: Mineralogical aspects. Anais Assoc Brasil Quím 41-42(1-4):32-38 (in Portugese)

Atencio D, Hypolito R (1992/1993) Minerals and synthetic equivalents in the system Fe_2O_3–SO_3–H_2O. Bol IG-USP Sér Cient 24:67-75

Atencio D, Carvalho FMS, Hypolito R (1996) Synthesis and X-ray ray powder diffraction data for Mg-, Al- and Ni- end-members of the copiapite group. Anais Assoc Brasil Quím 45:66-72

Avdonin VN (1978) Poitevinite from the Gai deposit as the first discovery in the USSR. Mineral Petrogr Urala 1:71-76 (Chem Abstr 92:61932b)

Avdonin VN (1984) Gunningite from the industrially induced oxidation zone of the Uchaly deposit. Chem Abstr 102:188215a

Avdonin VN, Maloshag VP, Federova TV (1988) Zinc sulfates (goslarite, boyleite, and gunningite) in the technology generated oxidation zone of the Degtyarsk deposit. Chem Abstr 110:216425e

Bacon GE, Titterton DH (1975) Neutron-diffraction studies of $CuSO_4 \cdot 5H_2O$ and $CuSO_4 \cdot 5D_2O$. Z Kristallogr 141:330-341

Badak J (1959) Occurrence of secondary minerals in the oxidized zone of the menilite beds in the Carpathians. Bull Acad Polon Sci Ser Sci Chim Geol Geogr 7:759-763

Baes CF Jr, Reardon EJ, Moyer BA (1993) Ion interaction model applied to the $CuSO_4$–H_2SO_4–H_2O system at 25°C. J Phys Chem 97:12343-112348

Balarew C, Karaivanova V (1975) Change in the crystal structure of zinc sulphate heptahydrate and magnesium (II) sulphate heptahydrate due to isodimorphous substitution by copper (II), iron (II), and cobalt (II) ions. Kristall Technik 10:1101-1110

Balarew C, Karaivanova V (1976a) Preparation of pure crystal hydrates of the type $M^{II}SO_4 \cdot nH_2O$ free of isodimorphous admixtures. Z anorg allg Chem 422:173-178 (in German)

Balarew C, Karaivanova V (1976b) Isodimorphous cocrystallization in sulfate systems as a possibility of predicting the existence of crystal hydrates of the type $MSO_4 \cdot nH_2O$ (M = Mn^{2+}, Fe^{2+}, Co^{2+}, Ni^{2+}, Cu^{2+}, Zn^{2+}, Cd^{2+}). Z anorg allg Chem 422:283-288 (in German)

Balarew C, Karaivanova V (1976c) Effect of isodimorphously included Co(II), Fe(II) and Cu(II) ions on the crystal structures of $ZnSO_4 \cdot 7H_2O$ and $MgSO_4 \cdot 7H_2O$. *In* Mullin JW (ed) Industrial Crystallization 6th Proc 1975. Plenum, New York, p 239-242

Balarew C, Karaivanova V, Aslanian S (1973) Isomorphic relations among the heptahydrate sulfates of single divalent metals (Mg^{2+}, Zn^{2+}, Ni^{2+}, Fe^{2+}, Co^{2+}). Kristall Technik 8:115-125 (in German)

Balarew C, Karaivanova V, Stefanov I (1984) Effect of isodimorphic guest component on the host crystal structures and habits. *In* Jancic SJ, de Jong EJ (eds) Industrial Crystallization '84. Elsevier Science, Amsterdam, The Netherlands, p 335-338

Baltatzis E, Stamatakis MG, Kyriakopoulos KG (1986) Rozenite and melanterite in lignitic layers from the Voras mountain, western Macedonia, Greece. Mineral Mag 50:737-739

Bandy MC (1938) Mineralogy of three sulphate deposits of northern Chile. Am Mineral 23:669-760

Bannister FA (1937) The preservation of minerals and meteorites. Museums J 36:465-476

Bari H, Catti M, Ferraris G, Ivaldi G, Permingeat F (1982) Phaunouxite, $Ca_3(AsO_4)_2 \cdot 11H_2O$, a new mineral strictly associated with rauenthalite. Bull Soc franc Minéral Cristallogr 105:327-332 [as cited in Am Mineral 68:850 (1983)]

Bassett H, Goodwin TH (1949) The basic aluminum sulphates. J Chem Soc 2239-2279

Bastidas JM, Lopez-Delgado A, Lopez FA, Alonso MP (1997) Characterization of artificially patinated layers on artistic bronze exposed to laboratory SO_2 contamination. J Mater Sci 32(1):129-133

Baur WH (1962) On the crystal chemistry of salt hydrates: Crystal structure of $MgSO_4 \cdot 4H_2O$ (leonhardtite) and $FeSO_4 \cdot 4H_2O$ (rozenite). Acta Crystallogr 15:815-826 (in German)

Baur WH (1964a) On the crystal chemistry of salt hydrates: III. The determination of the crystal structure of FeSO$_4$·7H$_2$O (melanterite). Acta Crystallogr 17:1167-1174

Baur WH (1964b) On the crystal chemistry of salt hydrates: IV. The refinement of the crystal structure of MgSO$_4$·7H$_2$O (epsomite). Acta Crystallogr 17:1361-1369

Baur WH (1964c) On the crystal chemistry of salt hydrates: II. A neutron diffraction study of MgSO$_4$·4H$_2$O. Acta Crystallogr 17:863-869

Baur WH, Rolin JL (1972) Salt hydrates: IX. The comparison of the crystal structure of magnesium sulfate pentahydrate with copper sulfate pentahydrate and magnesium chromate pentahydrate. Acta Crystallogr B28:1448-1455

Bayless ER, Olyphant GA (1993) Acid-generating salts and their relationship to the chemistry of groundwater and storm runoff at an abandoned mine site in southwestern Indiana, U.S.A. J Contam Hydrol 12:313-328

Bayliss P, Atencio D (1985) X-ray powder-diffraction data and cell parameters for copiapite-group minerals. Can Mineral 23:53-56

Beevers CA, Lipson H (1932) The crystal structure of nickel sulphate hexahydrate, NiSO$_4$·6H$_2$O. Z Kristallogr 83:123-135

Beevers CA, Schwartz CM (1935) The crystal structure of nickel sulfate heptahydrate, NiSO$_4$·7H$_2$O. Z Kristallogr 91:157-169

Berdesinki W (1952) Sanderite, leonhardtite, allenite and hexahydrite, new minerals of marine salt deposits. Neues Jahrb Mineral Monatsh 28-29 (in German)

Berry LG (1947) Composition and optics of copiapite. Univ Toronto Studies Contrib Can Mineral 51: 21-34

Blain CF, Andrew RL (1977) Sulfide weathering and the evaluation of gossans in mineral exploration. Minerals Sci Eng 9:119-149

Blanchard R (1967) Interpretation of leached outcrops. Nevada Bur Mines Bull 66:1-196

Blount AM (1993) Nature of the alterations which form on pyrite and marcasite during collection storage. Collection Forum 9:1-16

Blowes DW, Reardon EJ, Jambor JL, Cherry JA (1991) The formation and potential importance of cemented layers in inactive sulfide mine tailings. Geochim Cosmochim Acta 55:965-978

Bolshakov AP, Ptushko LI (1967) New supergene iron sulfates from the Mikitivka mercury deposit. Akad Nauk Ukr RSR Dopov Seriya B 29:696-699 (in Ukrainian)

Bolshakov AP, Ptushko LI (1971) Alteration products of melanterite from the Nikitov mercury ore deposits. Internat Geol Rev 13:849-854

Bonnell DFR, Burridge LW (1935) The dissociation pressures of some salt hydrates. Farad Soc Trans 31:473-478

Boscardin M, Colmelet G (1977) Morenosite of the Val d'Ayas, Aosta. Riv Mineral Ital 8:31-32 (in Italian)

Brigatti MF, Poppi L (1976/1977) Crystallochemical study and thermal behavior of natural heptahydrate sulfates: I. Research on some terms of the melanterite series. Mineral Petrogr Acta 21:157-164 (in Italian)

Brousse R, Gasse-Fournier R, Lebouteiller F (1966) Rozenite and melanterite crystals at the La Bade (Cantal) diatomite mine. Bull Soc franc Minéral Cristallogr 89:348-352 (in French)

Bullough W, Canning TA, Strawbridge MI (1952) The solubility of ferrous sulphate in aqueous solutions of sulphuric acid. J Appl Chem 2:703-707

Bursa S, Stanisz-Lewicka (1982) Solubility isotherms of the iron (II) sulfate–sulfuric acid–water system at 318, 328, and 333 K. Pol Pr Nauk Politech Szczecin 183:27-37

Buurman P (1975) In vitro weathering products of pyrite. Geol Mijnbouw 54:101-105

Byerly DW (1996) Handling acid-producing material during construction. Environ Eng Geosci II:49-57

Carpenter CE, Jette ER (1923) The vapor pressure of certain hydrated metal sulfates. J Am Chem Soc 45: 578-590

Caven RM, Ferguson J (1922) The dissociation pressures of hydrated double sulphates: Part I. Hydrated cupric alkalki sulfates. J Chem Soc London 121:1406-1414

Caven RM, Ferguson J (1924) The dissociation pressures of hydrated double sulphates: Part II. Various double sulphates of the type M^{++}SO$_4$, M$^+_2$SO$_4$·6H$_2$O. J Chem Soc London 125:1307-1312

Cerny P (1956) Contribution to the mineralogy of the sulfates of Smolnik. Spisy přírodovdecké fakulty Masarykovy Univ 376:1-26 (Mineral Abstr 14:226)

Charles RW, Vidale Buden RJ, Goff F (1986) An interpretation of the alteration assemblages at Sulphur Spring, Valles Caldera, New Mexico. J Geophys Res 91:1887-1898

Cherkasov GN (1975) Sulfuric acid weathering crusts of western Yakutia as a possible source of aluminum oxide. Trudy Sibirsky Nauchno-Issled Inst Geol Geofiz Mineral Syr'ya 189:33-43 (Chem Abstr 86:58164y)

Chernikov AA, Dorfman MD, Dvurechenskaya SS (1994) Genesis of mineral associations of cryogenic and deep supergene zones. Dokl Akad Nauk 335:485-488 (in Russian)

Chou I-M, Seal RR II, Hemingway BS (2000) Low-temperature thermodynamic properties of hydrated ferrous sulfates: Experimental results and evaluation of published data. Abstracts with programs, Geol Soc Am 32(7):A-108

Christov C (2000) Thermodynamic study of the Na–Cu–Cl–SO$_4$–H$_2$O system at the temperature 298.15 K. J Chem Thermodyn 32:285-295

Chukhrov FV (1953) Sequence of the formations in oxidation zones in ore deposits of the Kazakhstan steppe. Voprosy Petrog Mineral Akad Nauk SSSR 2:93-99 (in Russian)

Cody AD, Grammer TR (1979) Magnesian halotrichite from White Island. NZ Geol Geophys 22:495-498

Collins EM, Menzies AWC (1936) A comparative method for measuring aqueous vapor and dissociation pressures, with some of its applications. J Phys Chem 40:379-397

Collins HF (1923) On some crystallized sulphates from the province of Huelva, Spain. Mineral Mag 20:32-38

Cravotta CA III (1994) Secondary iron-sulfate minerals as sources of sulfate and acidity: Geochemical evaluation of acidic groundwater at a reclaimed surface coal mine in Pennsylvania. *In* Alpers CN, Blowes DW (eds) Environmental Geochemistry of Sulfide Oxidation. Am Chem Soc Symp Series 550:345-364

Cripps JC, Edwards RL (1997) Some geotechnical problems associated with pyrite bearing mudrocks. *In* Hawkins AB (ed) Ground Chemistry: Implications for Construction. Proc internat conf implications ground chem microbiol construction, Bristol, UK, 1992. AA Balkema, Rotterdam, The Netherlands, p 77-88

Crockford HD, Brawley DJ (1932) The system CuSO$_4$–CoSO$_4$–H$_2$O. J Phys Chem 36:1594-1596

Dagenhart TV Jr (1980) The acid mine drainage of Contrary Creek, Louisa County, Virginia: Factors causing variations in stream water chemistry. MSc thesis, Univ Virginia, Charlottesville, Virginia

Delines M (1975) Volcanic sublimates of Rugarama, Kiva region, Republic of Zaire. Bull Surv Geol (Rwanda) 8:1-11 (Chem Abstr 86:58182)

Doner HE, Lynn WC (1977) Carbonate, halide, sulfate, and sulfide minerals. *In* Dixon JB, Weed SB (eds) Minerals in Soil Environments. Soil Sci Soc Am Spec Publ 75-98

Dristas JA (1979) Zincian pisanite in the Las Picazas mine, San Rafael department, Mendoza province, Argentina. Assoc Geol Argent Rev 34:108-112 (Chem Abstr 93:10826h)

Ducloux J, Guero Y, Fallavier P, Valet S (1994) Mineralogy of salt efflorescences in paddy field soils of Kollo, southern Niger. Geoderma 64:57-71

Eckel EB (1933) Stability relations of a Colorado pisanite (cuprian melanterite). Am Mineral 18:449-454

Elerman Y (1988) Refinement of the crystal structure of CoSO$_4$·6H$_2$O. Acta Crystallogr C44:599-601

Eliseev EN, Smirnova SI (1958) Iron- and magnesium-bearing retgersite. Zap Vses Mineral Obshch 87:3-13 (in Russian)

Escobar C, Gifford G (1961) Some new powder X-ray diffraction diagrams of copper ores from Chile. Bol Acad Nacional Ciencias 42:245-256

Fanfani L, Nunzi A, Zanazzi PF (1970) The crystal structure of roemerite. Am Mineral 55:78-89

Fanfani L, Nunzi A, Zanazzi PF, Anzari AR (1973) The copiapite problem: The crystal structure of a ferrian copiapite. Am Mineral 58:314-322

Fang JH, Robinson PD (1970) Crystal structures and mineral chemistry of hydrated ferric sulphates: I. The crystal structure of coquimbite. Am Mineral 55:1534-1540

Fang JH, Robinson PD (1974) Polytypism in coquimbite and paracoquimbite. Neues Jahrb Mineral Monatsh 89-91

Fang JH, Robinson PD (1976) Alunogen, Al$_2$(H$_2$O)$_{12}$(SO$_4$)$_3$·5H$_2$O: Its atomic arrangement and water content. Am Mineral 61:311-317

Fanning DS, Rabenhorst MC, Bigham JM (1993) Colors of acid sulfate soils. *In* Bigham JM, Ciolkosz EJ (eds) Soil Color. Soil Sci Soc Am Spec Publ 31:91-108

Fedotova MG (1967) Retgersite of the Allarechensk deposit. Mater Mineral Kol'sk Poluostrova Akad Nauk SSSR Kol'sk Filial 5:70-73 (Chem Abstr 69:53489x)

Ferraris G, Jones DW, Yerkess J (1973) Refinement of the crystal structure of magnesium sulphate heptahydrate (epsomite) by neutron diffraction. J Chem Soc (Dalton Trans) 1973:816-821

Fleischer M (1962) New mineral names. Am Mineral 47:805-810

Fleischer M (1964) New mineral names. Am Mineral 49:1774-1778

Flynn CM Jr, Eisele JA (1987) Solubilities of sideronatrite and ferrinatrite in the system Na$_2$SO$_4$–Fe$_2$O$_3$–H$_2$SO$_4$–H$_2$O. US Bur Mines Report Invest RI-9100:1-17

Foote HW, Scholes SR (1911) The vapour pressure of hydrates, determined from their equilibria with aqueous alcohol. J Am Chem Soc 33:1309-1326

Foscolos AE, Goodarzi F, Koukouzas CN, Hatziyannis G (1989) Reconnaissance study of mineral matter and trace elements in Greek lignites. Chem Geol 76:107-130

Frondel C, Palache C (1949) Retgersite NiSO$_4$·6H$_2$O, a new mineral. Am Mineral 34:188-194

Gaines RV, Skinner HCW, Foord EE, Mason B, Rosenzweig A (1997) Dana's New Mineralogy. John Wiley & Sons, New York

Garavelli C (1955) Some sulfates of the iron deposit of Terranera (Elba Island, Italy). Rend Soc Mineral Ital 11:100-146 (in Italian)

Garavelli CL (1957) Bonattite: A new mineral in the Elban deposit at Cape Calamita. Lincei–Rend Sc Fis Mat Nat 22:318-327 (in Italian)

García García G (1996) The Rio Tinto mines, Huelva, Spain. Mineral Record 27:275-285

Gary GL (1939) Sulphate minerals at the Leviathan sulphur mine, Alpine County, California. Calif J Mines Geol 35:488-489

Gerkin RE, Reppart WJ (1988) Structure of monoclinic nickel(II) sulfate hexahydrate. Acta Crystallogr C44:1486-1488

Gibson RI (1974) Descriptive human pathological mineralogy. Am Mineral 59:1177-1182

Giester G (1988) The crystal structures of copper(II) sulfate monohydrate and copper(II) selenate monohydrate, and their relationships to kieserite. Mineral Petrol 38:277-284

Giester G, Wildner M (1992) The crystal structures of kieserite-type compounds: II. Crystal structures of Me(II)SeO$_4$·H$_2$O (Me = Mg,Mn,Co,Ni,Zn). Neues Jahrb Mineral Monatsh 135-144

Giester G, Lengauer CL, Redhammer G (1994) Characterization of the FeSO$_4$·H$_2$O–CuSO$_4$·H$_2$O solid-solution series, and the nature of poitevinite, (Cu,Fe)SO$_4$·H$_2$O. Can Mineral 32:873-884

Gmelins Handbuch der Anorganischen Chemie (1932) Eisen: System-Nummer 59. Verlag Chemie, Berlin

Gmelins Handbuch der Anorganischen Chemie (1939) Magnesium: System-Nummer 27. Verlag Chemie, Berlin

Gmelins Handbuch der Anorganischen Chemie (1956) Zink: System-Nummer 32. Verlag Chemie, Weinheim

Gmelins Handbuch der Anorganischen Chemie (1958) Kupfer: System-Nummer 60. Verlag Chemie, Weinheim

Gmelins Handbuch der Anorganischen Chemie (1961) Kobalt: System-Nummer 58. Verlag Chemie, Weinheim

Gmelins Handbuch der Anorganischen Chemie (1966) Nickel: System-Nummer 57. Verlag Chemie, Weinheim

Grønvold F, Meisingset KK (1982) Thermodynamic properties and phase relations of salt hydrates between 270 and 400 K. I. Ammonium aluminum sulfate, potassium aluminum sulfate, aluminum sulfate, zinc sulfate, sodium sulfate, and sodium thiosulfate hydrates. J Chem Thermodyn 14:1083-1098

Gruner D, Hood WC (1971) Three iron sulfate minerals from coal mine refuse dumps in Perry County, Illinois. Trans Ill State Acad Sci 64:156-158

Grybeck D (1976) Some additions to the ore mineralogy of Colorado. Mineral Record 7:274-276

Guenot J, Manoli J-M (1969) Thermogravimetry of solid–gas systems: III. Thermogravimetry of the hydrate salts: Influence of low pressure. Bull Soc Chim France (8):2663-2665 (in French)

Hammel F (1939) Contribution to the study of the sulfates of the magnesium series. Ann Chim 11:247-358 (in French)

Hawkins AB, Higgins MD (1997) The generation of sulphates in the proximity of cast in situ piles. In Hawkins AB (ed) Ground Chemistry: Implications for Construction. Proc internat conf implications ground chem microbiol construction, Bristol, UK, 1992. AA Balkema, Rotterdam, The Netherlands, p 101-110

Hawkins AB, Pinches GM (1997) Understanding sulphate generated heave resulting from pyrite degradation. In Hawkins AB (ed) Ground Chemistry: Implications for Construction. Proc internat conf implications ground chem microbiol construction, Bristol, UK, 1992. AA Balkema, Rotterdam, The Netherlands, p 51-76

Hawthorne FC, Groat LA, Raudsepp M (1987) Kieserite, Mg(SO$_4$)(H$_2$O), a titanite-group mineral. Neues Jahrb Mineral Abh 157:121-132

Hay RF (1960) The geology of Mangakahia subdivision. NZ Geol Survey Bull 61:1-109

Hayes CW (1906) The Gila River alum deposits. US Geol Survey Bull 315:215-223

Heitmankova J, Cerny C (1974) Solid–gas equilibrium in the binary system sodium hydrogen suphate – water. Coll Czech Chem Comm 39:1787-1793

Hepburn JRL, Phillips RF (1952) The alums: Part I. A study of the alums by measurement of their aqueous dissociation pressures. J Chem Soc London 2569-78 (as cited by Waller 1992)

Ievlev AA (1988) Mineralogy and formation of weathering zones in phosphorite-bearing shales of Pai Khoi. Sov Geol (10):81-93 (Chem Abstr 110:157951y)

International Critical Tables (1926-30) Vol I to VI. McGraw-Hill, New York

Jacobsen UH (1989) Hydrated iron sulfate occurrences at Navarana Fjord, central North Greenland. Bull Geol Soc Denmark 37:175-180

Jambor JL (1962) Second occurrence of bonattite. Can Mineral 7:245-252

Jambor JL (1981) Mineralogy of the Caribou massive sulphide deposit, Bathurst area, New Brunswick. CANMET Nat Res Can Rept 81-8E

Jambor JL (1994) Mineralogy of sulfide-rich tailings and their oxidation products. *In* Jambor JL, Blowes DW (eds) Environmental Geochemistry of Sulfide Mine-Wastes. Mineral Assoc Can Short Course 22:59-102

Jambor JL, Blowes DW (1998) Theory and applications of mineralogy in environmental studies of sulfide-bearing mine wastes. *In* Cabri LJ, Vaughan DJ (eds) Modern Approaches to Ore and Environmental Mineralogy. Mineral Assoc Can Short Course 27:367-401

Jambor JL, Boyle RW (1962) Gunningite, a new zinc sulphate from the Keno Hill – Galena Hill area, Yukon. Can Mineral 7:209-218

Jambor JL, Boyle RW (1965) Moorhouseite and aplowite, new cobalt minerals from Walton, Nova Scotia. Can Mineral 8:166-171

Jambor JL, Traill RJ (1963) On rozenite and siderotil. Can Mineral 7:751-763

Jambor JL, Lachance GR, Courville S (1964) Poitevinite, a new mineral. Can Mineral 8:109-11

Jambor JL, Blowes DW, Ptacek CJ (2000) Mineralogy of mine wastes and strategies for remediation. *In* Vaughan DJ, Wogelius RA (eds) Environmental Mineralogy. EMU Notes Mineral 2:255-290

Jamieson HE, Przybylowicz WJ (1997) The incorporation of toxic elements in iron sulfates precipitated from acid mine waters. Geol Assoc Can – Mineral Assoc Can Program Abstr 22:A73

Jamieson HE, Alpers CN, Nordstrom DK, Peterson RC (1999) Substitution of zinc and other metals in iron-sulfate minerals at Iron Mountain, California. *In* Goldsack D, Belzile N, Yearwood P, Hall G (eds) Conference Proc, Sudbury '99: Mining and the Environment II, September 12-16, 1999, Sudbury, Ontario, Canada. 1:231-241

Jangg G, Gregori H (1967) Crystallization equilibria in the system $CuSO_4$–$NiSO_4$–H_2O, $CuSO_4$–$ZnSO_4$–H_2O, $NiSO_4$–$ZnSO_4$–H_2O. Z anorg allg Chem 351:81-99 (in German)

Janjic S, Gakovic M, Dordevic D, Bugarski P (1980) Nickeloan hexahydrite from Droskovac near Vares, Bosnia, Yugoslavia. Glas Prir Muz Beogradu Ser A, 34:5-11 (Chem Abstr 95:46247a)

Jenkins RE II, Misiur SC (1994) A complex base-metal assemblage from the Sterling mine, New Jersey. Mineral Record 25:95-104

Jolly JH, Foster HL (1967) X-ray diffraction data of aluminocopiapite. Am Mineral 52:1220-1223

Karup-Møller S (1973) Nickelhexahydrite from Finland. Bull Geol Soc Finl 45(2):155-158

Keating LF, Berry LG (1953) Pisanite from Flin Flon, Manitoba. Am Mineral 38:501-505

Keith DC, Runnells DD (1998) Chemistry, mineralogy, and effects of efflorescent sulfate salts in acid mine drainage areas. Abstracts with programs, Geol Soc Am 30:254

Keith DC, Runnells DD, Esposito KJ, Chermak JA, Hannula SR (1999) Efflorescent sulfate salts: Chemistry, mineralogy, and effects on ARD streams. *In* Tailings Mine Waste '99, Proc 6th internat conference. Balkema, Rotterdam, The Netherlands, p 573-579

Keith TEC, Casadevall TJ, Johnston DA (1981) Fumarole encrustations: Occurrence, mineralogy, and chemistry. US Geol Survey Prof Paper 1250:239-250

Keller LP, McCarthy GJ, Richardson JL (1986a) Mineralogy and stability of soil evaporites in North Dakota. Soil Sci Soc Am J 50:1069-1071

Keller LP, McCarthy GJ, Richardson JL (1986b) Laboratory modeling of northern Great Plains salt efflorescence mineralogy. Soil Sci Soc Am J 50:1363-1367

Kellersohn T (1992) Structure of cobalt sulfate tetrahydrate. Acta Crystallogr C48:776-779

Kellersohn T, Delaplane RG, Olovsson I (1991) Disorder of a trigonally planar coordinated water molecule in cobalt sulfate heptahydrate, $CoSO_4 \cdot 7D_2O$ (bieberite). Z Naturforsch 46b:1635-1640

Keys JR, Williams K (1981) Origin of crystalline, cold desert salts in the McMurdo region, Antarctica. Geochim Cosmochim Acta 45:2299-2309

King RJ, Evans AM (1964) An occurrence of morenosite in Ireland. Mineral Mag 33:1110-1113

Kossenberg M, Cook AC (1961) Weathering of sulphide minerals in coal; production of ferrous sulphate heptahydrate. Mineral Mag 32:829-830

Kravtsov ED (1971) Minerals from the permafrost oxidation zone of the D'yakhtardakh deposit. Zap Vses Mineral Obshch 100:282-290 (in Russian)

Kravtsov ED (1974) Cryogenic oxidation zone of cassiterite – sulfide deposits in the northeastern Yakutsk ASSR. Probl. Kriolitologii 4:165-174 (Chem Abstr 83:196339y)

Kravtsov ED (1984) Characteristics of sulfate metasomatism of sulfide ores in a cryogenic zone. Metasomatizm Rudobraz (5th Mater–Vses Konf 1982) 189-198 (Chem Abstr 101:94744z)

Kubisz J (1960a) Rozenite, $FeSO_4 \cdot 4H_2O$, a new mineral. Bull Acad Polon Sci, Ser Sci Geol Geogr 8:107-113

Kubisz J (1960b) Magnesium szomolnokite, (Fe,Mg)SO$_4$·H$_2$O. Bull Acad Polon Sci, Ser Sci Geol Geogr 8:101-105

Kubisz J, Michalek Z (1959) Minerals of the oxidized zone of the menilite beds in the Carpathians. Bull Acad Polon Sci, Ser Sci Chim Geol Geogr 7:765-771

Kyriakopoulos KG, Kanaris-Sotiriou R, Stamatakis MG (1990) The authigenic minerals formed from volcanic emanations at Soussaki, west Attica Peninsula, Greece. Can Mineral 28:363-368

Lazarenko EK, Orlov OM, Panov BS (1973) Recent mineral formation in the Donets Basin. Mineral Sbornik (Lvov) 27:254-257 (in Russian)

Le Fur Y, Coing-Boyat J, Bassi G (1966) Structure of the monoclinic monohydrate sulfates of the transition metals, MSO$_4$, H$_2$O (M = Mn, Fe, Co, Ni and Zn). C R Acad Sci Paris Ser C 262:632-635 (in French)

Lengauer CL, Giester G (1995) Rietveld refinement of the solid-solution series: (Cu, Mg)SO$_4$·H$_2$O. Powder Diffraction 10:189-194

Li Wanmao, Chen Guoying (1990) Lishizhenite: A new zinc sulfate mineral. Acta Mineral Sinica 10:299-305 (in Chinese; Am Mineral 76:2022)

Li Wanmao, Chen Guoying, Sun Shurong (1987) Zincovoltaite: A new sulfate mineral. Acta Mineral Sinica 7:309-312 (in Chinese; Am Mineral 75:244-245)

Linke WF (1965) Solubilities of Inorganic and Metal Organic Compounds, Vol. II. Am Chem Soc, Washington, DC

Linke WF, Seidell A (1958) Solubility of Inorganic and Organic Compounds, Vol. I. Am Chem Soc, Washington, DC

Lizalek NA, Ivlev NF, Madaras A, Speshilova MA (1989) The Namana occurrence of secondary sulfates in Yakutia. Sov Geol (9):46-52 (Chem Abstr 113:155920a)

Lovas GA (1986) Structural study of halotrichite from Recsk (Matra Mts., N-Hungary). Acta Geol Hung 29:389-398

Lowson RT (1982) Aqueous oxidation of pyrite by molecular oxygen. Chem Rev 82:461-497

Luzgin BN (1990) Mineralogy of cryogenic oxidation zones in ore deposits of southeastern Gorny Altai. Zap Vses Mineral Obshch 119(6):100-106 (in Russian)

Magnus A (1205-1280) Book of Minerals. Translated by Wyckoff D (1967). Clarenden Press, Oxford, UK

MalstrIIm M, Banwart S (1997) Biotite dissolution at 25°C: The pH dependence of dissolution rate and stoichiometry. Geochim Cosmochim Acta 61:2779-2799

Mandarino JA (1999) Fleischer's Glossary of Mineral Species. Mineralogical Record, Tucson, Arizona

Martin R, Rodgers KA, Browne PRL (1999) The nature and significance of sulphate-rich aluminous efflorescences from the Te Kopia geothermal field, Taupo Volcanic Zone, New Zealand. Mineral Mag 63:413-419.

Matsubara S, Kato A, Bunno M (1973) Occurrence of szmikite from the Toyoho mine, Hokkaido, Japan. Bull Nat Sci Museum Tokyo 16:561-570

McCaughey WJ (1918) Copiapite in coal. Am Mineral 3:162-163

McCord TB, Hansen GB, Fanale FP, Carlson RW, Matson DL, Johnson TV, Smyth WD, Crowley JK, Martin PD, Ocampo A, Hibbits CA, Granahan JC (1998) Salts on Europa's surface detected by Galileo's Near Infrared Mapping Spectrometer. Science 280:1242-1245

Mellor JW (1923) A Comprehensive Treatise on Inorganic and Theoretical Chemistry, Vol 3. John Wiley, New York

Mellor JW (1932) A Comprehensive Treatise on Inorganic and Theoretical Chemistry, Vol 12. John Wiley, New York

Melon J (1944) Epsomite and melanterite of Vedrin. Ann Soc Gépl Belg Bull 67:B56-B59 (Mineral Abstr 10:116-117)

Menchetti S, Sabelli C (1976) The crystal structure of apjohnite. Mineral Mag 40:599-608

Mereiter K (1974) The crystal structure of rhomboclase H$_5$O$_2^+$\{Fe[SO$_4$]$_2$·2H$_2$O\}$^-$. Tschermaks Mineral Petrog Mitt 21:216-232

Merwin HE, Posnjak E (1937) Sulphate incrustations in the Copper Queen Mine, Bisbee, Arizona. Am Mineral 22:567-571

Milton C, Johnston WD Jr (1938) Sulphate minerals of the Comstock Lode, Nevada. Econ Geol 33:749-771

Milton C, Evans HT Jr, Johnson RG (1982) Dwornikite, (Ni,Fe)SO$_4$·H$_2$O, a member of the kieserite group from Minasragra, Peru. Mineral Mag 46:351-353

Minakawa T, Noto S (1994) Botryogen from the Okuki mine, Ehime Prefecture, Japan. Chigaku Kenkyu 43:175-179 (Chem Abstr 123:61439e)

Mitchell AG (1984) The preparation and characterization of ferrous sulfate hydrates. J Pharm Pharmacol 36:506-510

Miura H, Niida K, Hirama T (1994) Mikasaite, $(Fe^{3+},Al)_2(SO_4)_3$, a new ferric sulphate mineral from Mikasa city, Hokkaido, Japan. Mineral Mag 58:649-653

Montoroi J-P (1995) Supply of an acid sulfate minerals sequence in a lower Casamance valley (Senegal). C R Acad Sci Paris Ser IIa 320:395-402 (in French)

Nambu M, Tanida K, Kitamura T, (1978) Jôkokuite, $MnSO_4 \cdot 5H_2O$, a new mineral from the Jôkoku mine, Hiyama district, Hokkaido, Japan. Mineral J 9:28-38

Nambu M, Tanida K, Kitamura R, Kato E (1979) Mallardite from the Jôkoku mine, Hokkaido, Japan. Ganseki Kobutsu Kosho Gakkai-Shi [J Jap Assoc Mineral Petrol Econ Geol] 74:406-412

Naray-Szabo I (1969) Aluminum sulfate hydrates. Acta Chim 60:27-36

Nawaz R (1973) Nickel-hexahydrite from Tasmania, Australia. Mineral Mag 39:246-247

Nickel EH (1995) The definition of a mineral. Can Mineral 33:689-690

Nord AG, Mattsson E, Tronner K (1998) Mineral phases on corroded archaeological bronze artifacts excavated in Sweden. N Jahrb Mineral Monatsh 265-277

Nordstrom DK (1982a) Aqueous pyrite oxidation and the consequent formation of secondary iron minerals. *In* Kittrick JA, Fanning DS, Hossner LR (eds) Acid Sulfate Weathering. Soil Sci Soc Am Spec Publ 10:37-56

Nordstrom DK (1982b) The effect of sulfate on aluminum concentrations in natural waters: some stability relations in the system Al_2O_3-SO_2-H_2O at 298 K. Geochim Cosmochim Acta 46:681-692

Nordstrom DK, Alpers CN (1999a) Negative pH, efflorescent mineralogy, and consequences for environmental restoration at the Iron Mountain Superfund site, California. Proc Natl Acad Sci USA 96:3455-3462

Nordstrom DK, Alpers CN (1999b) Geochemistry of acid mine waters. *In* Plumlee GS, Logsdon MJ (eds) The Environmental Geochemistry of Mineral Deposits: Part A. Processes, Methods, and Health Issues. Rev Econ Geol 6A:133-160

Nordstrom DK, Southam G (1997) Geomicrobiology of sulfide mineral oxidation. *In* Banfield JA, Nealson KH (eds) Geomicrobiology: Interactions between Microbes and Minerals. Rev Mineral 35:361-390

Nordstrom DK, Alpers CN, Ptacek CJ, Blowes DW (2000) Negative pH and extremely acidic mine waters from Iron Mountain, California. Environ Sci Technol 34:254-258

Novikov VP, Suprychev VV (1986) Conditions of recent mineral formation during the subsurface burning of coals in the Fan-Yagnovka deposit. Mineral Tadzh (7):91-104 (Chem Abstr 106:70434a)

Nuhfer EB (1967) Efflorescent minerals associated with coal. MSc thesis, West Virginia University, Morgantown, West Virginia

Nuhfer EB (1972) Coal blooms: Description of mineral efflorescences associated with coal. Earth Sci. (Sept-Oct), 237-241

O'Brien FEM (1948) The control of humidity by saturated salt solutions. J Sci Instr 25:73-76

Oleinikov BV, Shvartsev SL, Mandrikova NT, Oleinikova NN (1965) Nickelhexahydrite, a new mineral. Zap Vses Mineral Obshch 94:534-547 (in Russian)

Olyphant GA, Bayless ER, and Harper D (1991) Seasonal and weather-related controls on solute concentrations and acid drainage from a pyritic coal-refuse deposit in southwestern Indiana, U.S.A. J Contam Hydrol 7:219-236

Omori K, Kerr PF (1963) Infrared studies of saline sulfate minerals. Geol Soc Am Bull 74:709-734

Ossaka J (1965) Volcanic sublimates. Kazan 10:205-213 (Chem Abstr 67:4923j)

Otto J, Schuerenberg H (1974) New mineral finds in the southern Black Forest. Aufschluss 25:205-211 (in German)

Ovtracht A, Pierrot R (1962) Contributions to the mineralogy of Norway. No. 14. On the presence of melanterites at Lökken. Norsk Geol Tidssk 42:275-276

Pabst A (1940) Cryptocrystalline pyrite from Alpine County, California. Am Mineral 25:425-431

Palache C, Berman H, Frondel C (1951) The System of Mineralogy, Seventh Edn, Vol 2. John Wiley and Sons Inc, New York

Parkinson KJ, Day W (1981) Water vapour calibration using salt hydrate transitions. J Exp Bot 32:411-418

Parsons AL (1922) The preservation of mineral specimens. Am Mineral 7:59-63

Pasava J, Breiter K, Huka M, Korecky J (1986a) Chvaleticeite, $(Mn,Mg)SO_4 \cdot 6H_2O$, a new mineral. Neues Jahrb Mineral Monatsh (3):121-125

Pasava J, Breiter K, Huka M, Korecky J (1986b) Paragenesis of secondary ferrous, magnesium, and manganese sulphates from Chvaletice. Vestnik Ústredniko ústavu Geol 61:73-82 (in Czech)

Paulis P (1991) Magnesian apjohnite from Chvalitice in the Zelezne hory Mts. Vestnik Ústredniko ústavu Geol 66:245-246 (in Czech)

Pemberton HE (1983) Minerals of California. Van Nostrand Reinhold, New York

Perroud P, Meisser N, Sarp H (1987) Presence of zincocopiapite at Valais. Schweiz mineral petrogr Mitt 67:115-117 (in French)

Petlicka J (1971) Effect of impurities on the crystallization of zinc sulfate heptahydrate. Statni Vyzk Ustav Mater Rudny 19:22-25

Plumlee GS (1999) The environmental geology of mineral deposits. *In* Plumlee GS, Logsdon MJ (eds) The Environmental Geochemistry of Mineral Deposits. Rev Econ Geol 6A:71-116

Posnjak E, Merwin HE (1922) The system $Fe_2O_3-SO_3-H_2O$. J Am Chem Soc 44:1965-1994, 2629

Posnjak E, Tunell G (1929) The system, cupric oxide–sulphur trioxide–water. Am J Sci 18:1-34

Powder Diffraction File: International Centre for Diffraction Data, Swarthmore, Pennsylvania

Querol X, Alastuey A, Lopez-Soler A, Plana F, Zeng RS, Zhao J, Zhuang XU (1999) Geological controls on the quality of coals from the West Shandong mining district, Eastern China. Internat J Coal Geol 42:63-88

Randall M (1930) Free energy of chemical substances, activity coefficients, partial molal quantities, and related constants. *In* Washburn EW (ed) International Critical Tables. Vol VII. McGraw-Hill, New York, p 224-353

Reardon EJ (1988) Ion interaction parameters for $AlSO_4$ and application to the prediction of metal sulfate solubility in binary salt systems. J Phys Chem 92:6426-6431

Reardon EJ (1989) Ion interaction model applied to equilibria in the $NiSO_4-H_2SO_4-H_2O$ system. J Phys Chem 93:4630-4636

Reardon EJ, Beckie RD (1987) Modelling chemical equilibria of acid mine-drainage: The $FeSO_4-H_2SO_4-H_2O$ system. Geochim Cosmochim Acta 51:2355-2368

Reardon EJ, Stevens R(1991) Aluminum potassium sulfate dodecahydrate solubility in mixed $K_2SO_4 + Al_2(SO_4)_3$ solutions. J Chem Eng Data 36:422-424

Robinson, C (1999) The role of jarosite and copiapite in the chemical evolution of acid drainage waters, Richmond mine, Iron Mountain, California. MSc thesis, Queen's Univ, Kingston, Ontario, Canada

Robinson C, Jamieson HE, Peterson RC, Alpers CN, Nordstrom DK (2000a) Major and trace element composition of copiapite from Richmond mine, Iron Mountain, California. Abstracts with programs, Geol Soc Am 32(7):A-110

Robinson C, Jamieson HE, Alpers CN, Nordstrom DK (2000b) The composition of co-existing jarosite and water from the Richmond mine, Iron Mountain, CA. Abstracts with programs, Geol Soc Am 32(7):A-109

Robinson PD, Fang JH (1971) Crystal structures and mineral chemistry of hydrated ferric sulphates: II. The crystal structure of paracoquimbite. Am Mineral 56:1567-1572

Robinson PD, Fang JH (1973) Crystal structures and mineral chemistry of hydrated ferric sulphates: III. The crystal structure of kornelite. Am Mineral 58:535-539

Rodgers KA, Hamlin KA, Browne PRL, Campbell KA, Martin R (2000) The steam condensate alteration mineralogy of Ruatapu cave, Orakei Korako geothermal field, Taupo Volcanic Zone, New Zealand. Mineral Mag 64:125-142

Rohmer R (1939) Contribution to the study of nickel sulfate and cobalt sulfate. Ann Chim 12:611-725 (in French)

Rose AW and Cravotta CA III (1998) Geochemistry of coal mine drainage. *In* Brady KBC, Smith MW, Schueck J (eds) Coal Mine Drainage Prediction and Pollution Prevention in Pennsylvania. Pennsylvania Dept Environ Protection 1:1-22

Rosenzweig A, Gross EB (1955) Goldichite, a new hydrous potassium ferric sulfate from the San Rafael Swell, Utah. Am Mineral 40:469-480

Ruotsala AP, Babcock LL (1977) Zaherite, a new hydrated aluminum sulfate. Am Mineral 62:1125-1128

Rutstein MS (1980) Nickeloan melanterite from Sudbury Basin. Am Mineral 65:968-969

Sabina AP (1977) New occurrences of minerals in parts of Ontario. Geol Survey Can Paper 77-1A:335-339

Sabina AP (1978) Some new mineral occurrences in Canada. Geol Survey Can Paper 78-1A:253-258

Schaller WT (1903) Minerals from Leona Heights, Alameda Co., California. Bull Dept Geol Univ California 3:191-217

Schnitzer WA (1977) An unusual efflorescence (aplowite) on Neogene rocks from the Ismuth of Corinth. Ann Geol Pays Hell 28:349-351 (Chem Abstr 91:24121v)

Schumb WC (1923) The dissociation pressures of certain salt hydrates by the gas-current saturation method. J Am Chem Soc 45:342-354

Sejkora J, Ridkosil T (1993) Retgersite from Jáchymov, Krusné hory Mountains, Czech Republic. Neues Jahrb Mineral Monatsh (9):393-400

Shcherbakova E, Korablev G (1998) Autoconservation of mining industry waste in the South Ural, Russia. Proc Latvian Acad Sci 52:286-288

Siebke W, Spiering H, Meissner E (1983) Cooperative pseudo-Jahn-Teller effect of the $Fe(H_2O)_6^{2+}$ complexes in the sulfate heptahydrates. Phys Rev B, 27:2730-2739

Skarie RL, Richardson JL, McCarthy GJ, Maianu A (1986) Evaporite mineralogy and groundwater chemistry associated with saline soils in eastern North Dakota. Soil Sci Soc Am J 51:1372-1377

Skounakis S, Economou M (1983) An occurrence of magnetite with network texture in serpentinites from Veria area, Greece. Neues Jahrb Mineral Monatsh (3):97-102

Smirnova SK (1971) Voltaite find in Uzbekistan. Dokl Akad Nauk Uzb SSR 28(2):92-93 (in Russian)

Snetsinger KG (1973) Ferroan starkeyite from Del Norte County, California. Can Mineral 12:229

Soboleva OS (1958) Equilibriums in the system MgSO₄–NiSO₄–H₂O: I. Solubility isotherms. Nauk Zap L'vivsk Derzhav Univ im I Franka 46, Ser Khim (5):91-106 (Chem Abs 55:24207)

Soboleva OS (1960) Equilibrium polytherm of the system MgSO₄–NiSO₄–H₂O system. Dokl Akad Nauk SSSR 135:91-93 (in Russian)

Sokolov PN, Korobov YuI, Kosukhina IG, Kolotova LV (1985) Mineralogy of water-soluble sulfates in the Namansk manifestation of aluminite (Yakutsk ASSR). *In* Kazanskii YuP, Korel VG (eds) Morfol genesis zakonomern razmeshchiniya mineral obraz Altae-Sayan skladchatoe obl Sib Platformy. Nauka Sib Otd, Novosibirsk, p 120-124 (Chem Abstr 104:21986g)

Srebrodol'skiy BI (1971) Zone of oxidized sulfur ores in the Vodinsk deposit. Geol Rudnyk Mestorozhd 13(3):106-114 (in Russian)

Srebrodol'skiy BI (1974a) First find of lausenite in the USSR. Dokl Akad Nauk SSSR 219:441-442 (in Russian)

Srebrodol'skiy BI (1974b) Sulfate from the halotrichite group. Akad Nauk Ukr RSR Dopov Seriya B 36:600-601 (in Ukrainian)

Srebrodol'skiy BI (1974c) An occurrence of millosevichite in the U.S.S.R. Dokl Akad Nauk SSSR 214:429-430 (in Russian)

Srebrodol'skiy BI (1977) Products of melanterite alteration. Akad Nauk Ukr RSR Dopov Seriya B 39:797-800 (in Ukrainian)

Stamatakis MG, Baltatzis EG, Skounakis SB (1987) Sulfate minerals from a mud volcano in the Katakolo area, western Peleponnesus, Greece. Am Mineral 72:839-841

Stoiber RE, Rose WI Jr (1974) Fumarole incrustations at active Central American volcanoes. Geochim Cosmochim Acta 38:495-516

Sunderman JA, Beck CW (1969) Hydrobasaluminite from Shoals, Indiana. Am Mineral 54:1363-1673

Süsse P (1972) Crystal structure and hydrogen bonding of copiapite. Z Kristallogr 135:34-55

Taylor D, Bassett H (1952) The system Al₂(SO₄)₃–H₂SO₄–H₂O. J Chem Soc 4431-4442

Taylor RK, Hardy RG (1974) Sulphate species in colliery spoil. Trans Inst Mining Metall 83A:123-126

Thomas JN, Robinson PD, Fang JH (1974) Crystal structures and mineral chemistry of hydrated ferric sulfates: IV. The crystal structure of quenstedtite. Am Mineral 59:582-586

Tiegeng Liu, Guolong Gong, Lin Ye (1995) Discovery and investigation of zinc-melanterite in nature. Acta Mineral Sinica 15:286-289 (in Chinese)

Timpson ME, Richardson JL, Keller LP, McCarthy GJ (1986) Evaporite mineralogy associated with saline seeps in southwestern North Dakota. Soil Sci Soc Am J 50:490-493

Tu Kwang-Chih, Li His-Lin (1963) Characteristic features of the oxidation zone of the sulfide deposits in arid to extremely arid regions (with special reference to observations obtained from five sulfide deposits in the Northwestern China. Ti Chih Hsueh Pao 43:361-377 (Chem Abstr 60:14263)

Van Breemen N (1982) Genesis, morphology, and classification of acid sulfate soils in coastal plains. *In* Kittrick JA, Fanning DS, Hossner LR (eds) Acid Sulfate Weathering. Soil Sci Soc Am Spec Publ 10:95-108

van Doesburg JDJ, Vergouwen L, van der Plas L (1982) Konyaite, Na₂Mg(SO₄)₂·5H₂O, a new mineral from the Great Konya Basin, Turkey. Am Mineral 67:1035-1038

Van Loan PR, Nuffield EW (1959) An X-ray study of roemerite. Can Mineral 6:348-356

Vdovichenko GM, Lazarev IS, Srebrodol'sky BI (1974) Geological-mineralogical characteristics and genesis of the oxidation zone at the Gaurdak sulfur deposit. *In* Sokolov AS (ed) Genezis mestorozhdenii samorodnoi sery perspekt. Ikh Poiskov Nauka, Moscow, p 106-119 (Chem Abstr 83:82739y)

Velizade SF, Efendieva EN, Aliev AA, Mustafazade BV, Pokidin AK (1976) Some minerals from the oxidized zone of the Katsdag deposit. Dokl Akad Nauk Az SSR 32(8):40-45 (in Russian)

Vergasova LP (1983) Fumarole incrustations of lava flows of the effusive-explosive period of the Great Tolbachik fissure eruption. Vulkanol Seismol (6):75-87 (in Russian)

Vergasova LP, Filatov SK, Serafimova EK, Starova GL (1982) Piypite K₂Cu₂O(SO₄)₂—a new mineral of volcanic sublimates. Dokl Akad Nauk SSSR 266:707-710 (in Russian)

Vlasov VV, Kuznetsov AV (1962) Melanterite and its oxidation products. Zap Vses Mineral Obshch 91:490-492 (in Russian)

Wagner DP, Fanning DS, Foss JE, Patterson MS, Snow PA (1982) Morphological and mineralogical features related to sulfide oxidation under natural and disturbed land surfaces in Maryland. *In* Kittrick JA, Fanning DS, Hossner LR (eds) Acid Sulfate Weathering. Soil Sci Soc Am Spec Publ 10:109-125

Walenta K (1978) Boyleite, a new sulfate mineral from Kropback, southern Black Forest. Chem Erde 37:73-79 (in German)

Waller R (1992) Temperature- and humidity-sensitive mineralogical and petrological specimens. *In* Howie FM (ed) The Care and Conservation of Geological Material: Minerals, Rocks, Meteorites and Lunar Finds. Butterworth-Heinemann, p 25-50

Wang Qiguang, Li Wanmao (1988) Crystal structure of a new ferric sulfate mineral. Kexue Tongbao 33:1783-1787 (in Chinese, English abstr)

Ward C (1991) Mineral matter in low-rank coals and associated strata of the Mae Moh basin, northern Thailand. Internat J Coal Geol 17:69-93

Wexler A, Hasegawa S (1954) Relative humidity–temperature relationships of some saturated salt solutions in the temperature range of 0 to 50°C. J Res Natl Bur Stds 53:19-26

Whittig LD, Deyo AE, and Tanji KK (1982) Evaporite mineral species in Mancos Shale and salt efflorescence, Upper Colorado River Basin. Soil Sci Soc Am J 46:645-651

Wiese RG Jr, Powell MA, Fyfe WS (1987) Spontaneous formation of hydrated iron sulfates on laboratory samples of pyrite- and marcasite-bearing coals. Chem Geol 63:29-38

Wildner M, Giester G (1991) The crystal structures of kieserite-type compounds: I. Crystal structures of Me(II)SO$_4$·H$_2$O (Me = Mn, Fe, Co, Ni, Zn). Neues Jahrb Mineral Monatsh (7):296-306

Williams SA, Cesbron FP (1995) Wupatkiite from the Cameron uranium district, Arizona, a new member of the halotrichite group. Mineral Mag 59:553-556

Wilson RE (1921) Some new methods for the determination of the vapor pressure of salt-hydrates. J Am Chem Soc 43:740-25 (as cited by Waller 1992).

Wood MM (1970) The crystal structure of ransomite. Am Mineral 55:729-734

Workman WF, Rader EK (1961) Comments on magnesium sulfate minerals formed in Brooks Museum on serpentine from Impruenta, Italy. Virginia J Sci 12:189

Yakhontova LK, Dvurechenskaya SS, Sandomirskaya SM, Sergeeva NE, Pal'chik NA (1988) Sulfates from the cryogenic supergene zone: First finds: Nomenclature problems. Mineral Zhurnal 10(4):3-15 (in Russian)

Young B, Nancarrow PHA (1988) Rozenite and other sulphate minerals from the Cumbrian coalfield. Mineral Mag 52:551-553

Yurgenson G (1997) Oxidized zones in permafrost rocks. Zap Vseross Mineral Obshch 126(5):15-27 (in Russian)

Zachmann DW (1999) The Brunswick Lion (Germany): Environment and corrosion. Mater Corros 50:17-26

Zahrobsky RF, Baur WH (1968) On the crystal chemistry of salt hydrates: V. The determination of the crystal structure of CuSO$_4$·3H$_2$O (bonattite). Acta Crystallogr B24:508-513

Zalkin A, Ruben H, Templeton DH (1962) The crystal structure of cobalt sulfate hexahydrate. Acta Crystallogr 15:1219-1224

Zalkin A, Ruben H, Templeton DH (1964) The crystal structure and hydrogen bonding of magnesium sulfate hexahydrate. Acta Crystallogr 17:235-240

Zavalía MFM, Galliski MA (1995) Goldichite of fumarolic origin from the Santa Bábara mine, Jujuy, northwestern Argentina. Can Mineral 33:1059-1062

Zhdanov YuYa, Solov'ev LI (1998) Geology and mineralogical composition of the oxidized zone in the Deputatsk tin ore deposit. Otechestvennaya Geol (6):77-79 (Chem Abstr 131:132401

Zhu Meixiang, Tong Wei (1987) Surface hydrothermal minerals and their distribution in the Tengchong geothermal area, China. Geothermics 16:181-195

Zidarov N (1970) Cupriferous bianchite from the Borieva mine, Madan mining region. Dokl Bolg Akad Nauk 23:1283-1286

Zodrow EL (1980) Hydrated sulfates from Sydney Coalfield, Cape Breton Island, Nova Scotia, Canada: The copiapite group. Am Mineral 65:961-967

Zodrow EL, McCandlish K (1978) Hydrated sulfates in the Sydney coalfield, Cape Breton, Nova Scotia. Can Mineral 16:17-22

Zodrow EL, Wiltshire J, McCandlish K (1979) Hydrated sulfates in the Sydney coalfield of Cape Breton, Nova Scotia: II. Pyrite and its alteration products. Can Mineral 17:63-70

7   Iron and Aluminum Hydroxysulfates from Acid Sulfate Waters

J. M. Bigham

*School of Natural Resources
2021 Coffey Road
The Ohio State University
Columbus, Ohio 43210*

D. Kirk Nordstrom

*U. S. Geological Survey
3215 Marine Street
Boulder, Colorado 80303*

Acid sulfate waters are produced mostly by the oxidation of common sulfide minerals such as pyrite, chalcopyrite, pyrrhotite, and marcasite in rocks, soils, sediments, and industrial wastes. This spontaneous process of mineral weathering plays a fundamental role in the supergene alteration of ore deposits, the formation of acid sulfate soils, and the mobilization and release of acidity and metals to surface and ground waters. The purely natural process of "acid rock drainage" is often intensified by human activities related to mining, mineral processing, construction, soil drainage, and dredging. Geochemical reaction rates are accelerated because physical disturbance gives greater exposure of mineral surfaces to air and water, and to microbes that catalyze the reaction process. Large quantities of reactive sulfides are also concentrated and exposed to air as a result of mining and mineral processing. Acid sulfate waters produce a number of fairly insoluble hydroxysulfate and oxyhydroxide minerals that precipitate during oxidation, hydrolysis, and neutralization. The objective of this chapter is to describe the formation, properties, fate, and environmental implications of the nano- to microphase hydroxy-sulfates of Fe and Al that are precipitated from acid sulfate waters. These minerals are commonly of poor crystallinity and difficult to characterize. Much remains to be learned about their occurrence, formation, and properties.

INTRODUCTION TO ACID SULFATE ENVIRONMENTS

Mine drainage

The best known examples of acid sulfate waters are those released from mines where coal and metallic sulfide ores have been exploited (Ash et al. 1951, Barton 1978, Nordstrom 1982a, Rose and Cravotta 1998, Nordstrom and Alpers 1999). There may be as many as 500,000 inactive or abandoned mine sites in the United States alone (Lyon et al. 1993). Although most of these pose no immediate water-quality problem, Kleinmann (1989) estimated that about 19,300 km of streams and more than 72,000 ha of lakes and reservoirs have been seriously damaged by mine effluents. Contamination of natural waters by mine drainage has killed enormous numbers of fish and other aquatic organisms, destroyed natural vegetation, induced massive erosion and sedimentation, caused widespread corrosion of bridge abutments, culverts, roads, and other structures, and made many streams and lakes so turbid as to be unfit for recreational activities.

Most acid mine waters have pH values in the range of 2 to 4 (Nordstrom 1991,

1529-6466/00/0040-0007$05.00

Plumlee et al. 1999). Acid mine drainage occurs when acid production exceeds the buffering capacity of the host rock or the surrounding spoil and soil. In areas where carbonate rocks are available to neutralize acidity, higher pH drainage may be common. Cravotta et al. (1999), for example, reported a bimodal frequency distribution for the pH of a large number of mine drainage waters from the eastern U.S. coal province (Fig. 1). Most samples were either distinctly acidic (pH 2-4) or near neutral (pH 6–7), with relatively few samples having pH values in the range of 4 to 6. A similar distribution is apparent in the data of Plumlee et al. (1999). Neutralization promotes the removal of Fe, Al, and other metals from solution but has a much less noticeable affect on the concentrations of SO_4.

Surface Coal-Mine Drainage

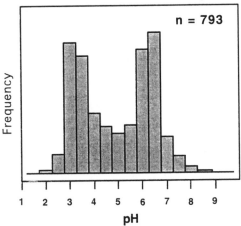

Figure 1. The pH of mine drainage from 793 sites in the eastern U.S. coal province showing a strong bimodal distribution of acid mine drainage and neutral to alkaline mine drainage. (modified from Cravotta et al. 1999).

Precipitates of Fe, Al, and sometimes Mn provide a highly visible means of identifying mine-impacted waters. Those precipitates composed primarily of Fe compounds are yellow-to-red-to-brown in color and have long been referred to as "yellow boy" by North American miners. Acid mine waters rich in Al, on the other hand, typically produce a milky-white precipitate that may be more abundant than is commonly recognized because its color can be masked by associated Fe compounds. These chemical precipitates are environmentally significant because they (1) add to the suspended sediment and bed load of receiving streams, (2) decrease the effective life of wetlands, limestone drains, and ponds constructed for mine drainage abatement, and (3) play a major role in the binding and transport of toxic elements. There have been relatively few quantitative measurements of deposition rates, and published results obviously vary with the chemistry and hydrology of the drainage system. Letterman and Mitsch (1978) observed particulate Fe accumulations of up to 3 g m^{-2} d^{-1} in a Pennsylvania stream receiving mine drainage. Typical Fe loading rates for constructed wetlands range from 0.4 to 250 g·$m^{-2}d^{-1}$ (Brodie et al. 1988, Fennessy and Mitsch 1989). Kirby et al. (1999) noted that a drainage of 10^6 L d^{-1} with an Fe concentration of 25 mg L^{-1} will yield 17 tonnes of "Fe(OH)$_3$" per year. The ability of these precipitates to scavenge metals and oxyanions from solution has been documented in numerous studies (see Singh et al. 1997 for a recent review). This process is so efficient that most contaminant species are transported in particulate form (e.g. Kimball et al. 1995) and any factor influencing the stability of the colloidal precipitates (photoreduction, pH changes, and composition changes) must also have an impact on the availability of sorbed trace elements (McKnight et al. 1988, Fuller and Davis 1989).

Residues from mineral extraction and ore processing

Mineral mining and processing usually result in the production of an extremely large volume of unwanted material. For example, an average metal mine immediately rejects 42% of the total mined material as waste rock (spoil), 52% of the ore is separated at the mill as tailings, 4% leaves the smelter as slag, and only 2% is retained as the commodity for which the ore was extracted (Godin 1991). The unwanted by-products represent a problem not just because of their volume, but because many of them contain chemically reactive sulfide minerals that have a high potential for producing acid sulfate drainage.

Tailings contain large volumes of sand- and silt-sized particles and are usually disposed of by slurry-pumping to an impoundment constructed close to the mill. As mining progresses, the impoundment is commonly increased in height by the construction of retaining dams. Elevated impoundments result in hydraulic loading and water-table mounding so that the impoundments become areas of ground-water recharge unless the impoundment areas are prepared with impermeable liners and dams are constructed from low-conductivity materials. Maintaining water-table depth is a concern because oxygen penetration can lead to accelerated sulfide oxidation and acid generation. Analyses of pore waters in the vadose zone of several Canadian tailings impoundments have shown high concentrations of Fe, Al and trace metals (Blowes and Jambor 1990, Blowes et al. 1998). Chemical discontinuities in the tailings may also give rise to cemented layers or "hardpans" composed of a variety of solid phases. Hardpans or "ferricretes" have formed naturally as well as in mining-impacted areas, commonly where an acid stream reacts with oxygenated soil or sediments. At some sites, these precipitated layers have suppressed the movement of oxygen and dissolved metals through the tailings and have thereby moderated the environmental impact of sulfide oxidation (Blowes et al. 1991).

Waste-rock dumps and tailings impoundments have properties in common; however, important differences also exist. Most waste-rock dumps usually have a much greater height-to-base ratio and contain coarser materials than associated tailings deposits (Ritchie 1994). Moreover, waste rock has not been subject to the processes of ore beneficiation that typically include crushing, sizing, and concentration of the desired ore mineral(s) through flotation, magnetic, and gravity processes. Waste-rock dumps are frequently unsaturated systems, so the chemical environment is heterogeneous and wide ranges in temperature, gas composition, and solution chemistry are common. The oxidation of iron sulfides is the primary mechanism for generating pollution in both tailings and waste rock environments, but drainage from the toe of a waste-rock dump provides only an integrated estimate of processes occurring within the pile (Ritchie 1994). Pyrite and pyrrhotite oxidize by an exothermic reaction that can generate air temperatures of 50 to 65°C in waste-rock piles (Cathles 1994). Higher temperatures of more than 220°C in underground massive sulfide mines (Wright 1906), and more than 530°C at nearly 300-m depth at the Mt. Isa sulfide deposits, Australia, have been documented (Ninteman 1978). Thermal gradients from heat generation cause convective air flow which can then become an important oxygen-supply mechanism (Ritchie 1994).

Rock weathering

Sulfide oxidation occurs in the absence of mining or land disturbance and numerous instances of naturally acidic waters containing high concentrations of sulfate and metals have been documented (Runnells et al. 1992, Posey et al. 2000, Mast et al. 2000, Yager et al. 2000). Other examples include streams (Theobald et al. 1963, Schwertmann et al. 1995) and lakes (Childs et al. 1997) receiving drainage from sulfide-bearing rocks, as well as craters (Rowe et al. 1992), fumaroles, and hot springs in active geothermal areas (Tkachenko and Zotov 1982). The H_2SO_4 associated with volcanic activity may be derived by oxidation of magmatic gases (H_2S and SO_2) or elemental sulfur oxidation (Nordstrom et al. 1997) rather than metal sulfides. Not all acid rock drainage is recent in origin. For example, ancient deposits of ferricretes in the Rocky Mountains indicate that the discharge of acidic, metal-bearing waters has occurred over thousands of years in response to the weathering and oxidation of exposed sulfide ore bodies (Furniss et al. 1999). The ferricrete is cemented by iron oxyhydroxide and hydroxysulfate minerals and occurs as aprons around acidic springs and as remnants of ancient stream terraces.

The existence of gossans and related ferricretes on Earth as aqueous weathering products has been used as evidence for similar episodes of weathering on Mars under an earlier climatic regime that was presumably both warmer and wetter (Burns 1988). Spectroscopic analyses, coupled with data from the Viking and Pathfinder missions, have provided much information about the surface materials on Mars (Morris et al. 2000). The data suggest that many of the Martian bright regions are covered by ferricretes and fine-grained "soil" rich in silicates and nanophase Fe minerals. The LANDER data (Clark et al. 1982, Rieder et al. 1997) have also revealed relatively high concentrations of S (7% SO_3) in Martian soil as compared to local rocks (<1% SO_3). Burns (1994) suggested that permafrost exists beneath the martian regolith and that iron oxides and hydroxysulfates were precipitated in equatorial regions of Mars where seepages of saline meltwaters were once exposed to the atmosphere. These meltwaters are generally thought to have been acidic (Clark and Van Hart 1981) and might therefore have been similar to acid sulfate waters on Earth. The proposed iron hydroxysulfate precipitates also have spectral characteristics comparable to those of terrestrial analogs that can be formed both by biotic and abiotic processes (Bishop and Murad 1996). The possible existence of water and biominerals on Mars has been used as a logical indicator of exobiology (Bishop 1998).

Acid sulfate soils (cat clays)

Wet, sulfidic soils that become hyperacid with drainage have been recognized for many years (Bloomfield and Coulter 1973, Van Mensvoort and Dent 1998). The famous plant taxonomist, Linnaeus (1735), described *argilla vitriolacea* (clay with sulfuric acid) in European swamps and developed a classification system for Dutch soils that included *argilla mixta fusca, vitriolica salsa* (mixed brown clays with the taste of salts from sulfuric acid). Areas of acid sulfate soils were produced by drainage of the great Haarlemmermeer polder in 1852, and such areas came to be known as kattaklei (cat clay) (Pons 1973, Bloomfield 1972). "Cat" was used in the Dutch vernacular to indicate harmful, mysterious qualities (Fanning and Burch 1997), and "cat clay" was associated with the excrement of cats (Westerveld and van Holst 1973). Dent and Pons (1995) succinctly described acid sulfate soils as "the nastiest soils in the world" and noted that their evil reputation is derived from an unusual combination of properties, including odd colors, foul odors, stunted vegetation, and rapid chemical deterioration following drainage (Fig. 2). Modern drainage systems designed to reclaim swamps and wetlands to exploit their anticipated fertility for agricultural production have, in some cases, laid waste large areas underlain by sulfidic materials. Not only have local farmers suffered from the development of acid sulfate conditions but water pollution and damage to fisheries have also occurred over wide areas in some coastal regions. Ochre from the oxidation products of acid sulfate weathering often reduces the capacity of drain lines (Bloomfield 1972, Trafford et al. 1973, Fanning et al. 1993) and may inadvertently slow the acidification process by halting efficient drainage.

Most acid sulfate soils have developed from sulfidic deposits accumulated under mangroves and reedswamps in tidal areas where abundant organic matter and the

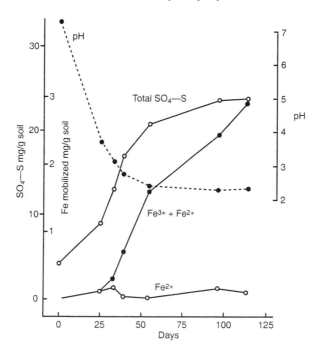

Figure 2. Changes in pH and the concentrations of SO_4^{2-} and Fe in the pore-waters of a green-colored soil containing pyrite after exposure to oxidation (modified from Trafford et al. 1973).

constant accretion of sulfate from seawater stimulate the activity of sulfate-reducing bacteria to produce FeS and FeS_2. Concentrations of up to 15 wt % FeS_2 have been reported. Extensive areas of such sediments are found in Indonesia, Indochina, the Guyanas, the Orinoco Delta, west Africa, and northern Australia (Dent and Pons 1995). It has been estimated that there are some 12-14 million ha of soils in Holocene coastal plains and tidal swamps where the topsoil is severely acid or will become so if drained (van Breeman 1980). There may be twice this area of sulfidic materials covered by thin veneers of alluvium or peat that could easily be removed through erosion or oxidation. Some Holocene sediments have been drained naturally by isostatic rebound, tectonic uplift, or through changes in delta distribution systems; however, most acid sulfate soils are clearly the result of artificial drainage.

Unusual but significant areas of saline-sulfidic soils have formed in the rolling uplands of South and West Australia in response to rising saline ground waters from the extensive removal of native *Eucalyptus* forests over the past century (Fitzpatrick et al. 1992, 1996; Fitzpatrick and Self 1997). Waterlogging and the formation of sulfidic materials occur during the wet winter months and are followed by oxidation and acidification during the dry summer months. The result is a cancerous growth of barren, red scalds across the landscape.

Much has been written about soil formation in mine spoils, rock tailings, and dredged materials around the globe. Many of these constructed soils may qualify as acid sulfate depending on the character of the waste material. Numerous studies have been dedicated to halting the acidification process and modifying root-zone chemistry to enable the establishment of higher plants for erosion control and agronomic production. Efforts include re-soiling with borrowed soil material, treatment with bactericides, and the addition of alkaline amendments (agricultural limestone, alkaline coal-combustion by-products, cement-kiln dusts) or organic wastes (yard waste compost, sewage sludge, paper-mill sludge, etc.). The reader is referred to a comprehensive review by Barnhisel et al. (2000) of current technology for managing disturbed lands.

Because of their ability to create extreme agricultural and environmental problems, acid sulfate soils are given special recognition in many soil classification systems. Soils containing sulfide minerals that can oxidize to form sulfuric acid under appropriate conditions are commonly referred to, in the U.S., as potential acid sulfate soils (Fanning and Burch 1997) or in Europe as "unripe" sulfidic soils (Dent and Pons 1995). Active or "raw" acid sulfate soils are recognized by acid drainage waters and by the appearance of straw-yellow accumulations of jarosite in the soil profile (Schwertmann 1961, van Breemen and Harmsen 1975, Öborn and Berggren 1995), both of which indicate active oxidation of sulfides above the local ground water. Materials in which the sulfides have been completely oxidized may be described as post-active or "ripe" acid sulfate soils. Most of these soils have conspicuous red mottles and may still be very acid with high concentrations of soluble Al.

FORMATION OF ACID SULFATE WATERS AND ASSOCIATED WEATHERING PRODUCTS OF FE AND AL

Pyrite (cubic FeS_2) and pyrrhotite or other ferrous sulfide phases such as chalcopyrite, arsenopyrite, greigite, and mackinawite are the most abundant Fe sulfide minerals in nature, and their decomposition is essential to obtain the conditions and products needed for the formation of iron and aluminum weathering compounds, including hydroxysulfate minerals. Pyrrhotite and ferrous sulfide phases are more reactive than pyrite (Bhatti et al. 1993, 1994; Jambor 1994). Reactivity, however, largely depends on grain size. Field and laboratory studies have largely focused on pyrite oxidation to understand the mechanism by which acid sulfate waters are generated. The weathering of pyrite is commonly described by the single, incongruent reaction given in Equation (1); however, this reaction is a gross oversimplification. As has been noted by Nordstrom (1982a, 2000) and Nordstrom and Alpers (1999), pyrite decomposition is a complex biogeochemical process involving hydration, hydrolysis, and oxidation reactions as well as microbial catalysis. Oxidation rates are dependent on temperature, pH, Eh, relative humidity, and the surface area of reactant pyrite.

$$FeS_{2(aq)} + 3.75\ O_{2(g)} + 3.5\ H_2O_{(l)} \leftrightarrow Fe(OH)_{3(s)} + 2\ H_2SO_{4(aq)} \tag{1}$$
$$\text{(pyrite)}$$

The Fe system

Reaction (1) demonstrates that oxygen and water provide the ultimate driving force for pyrite oxidation, that both the sulfur and the ferrous iron in pyrite are subject to oxidation, and that final products include an insoluble form of oxidized Fe and sulfuric acid. However, the complexity of possible solid-phase weathering products represented generically by $Fe(OH)_3$ is not depicted. Nor does the reaction indicate the important role of chemolithotrophic Fe- and S-oxidizing bacteria, the effect of the semi-conducting pyrite surface on intermediate reactions, or the role of intermediate sulfoxyanions.

The best known of the chemoautotrophic bacteria catalyzing pyrite oxidation, *Thiobacillus ferrooxidans*, oxidizes both Fe and S over the pH range 1.0 to 3.5. *Leptospirillum ferrooxidans*, an Fe-oxidizer, is now thought to play an important role in the oxidative dissolution of iron sulfide minerals at very low pH (Sand et al. 1992,

Schrenk et al. 1998, Rawlings et al. 1999). A third species, *Thiobacillus thiooxidans*, oxidizes S only. These acidophilic bacteria require only O_2, CO_2, a reduced form of Fe and/or S, and minor N and P for their metabolism. They produce enzymes that catalyze the oxidation reactions and use the energy released from these processes to transform inorganic carbon into cellular material (Gould et al. 1994). A newly described Fe-oxidizing Archaeon, *Ferroplasma acidarmanus*, has been found at Iron Mountain at elevated temperatures (40°C) and low pH (0.7, Edwards et al. 2000). Microbial diversity probably enhances the oxidation of pyrite relative to that accomplished by a single species.

The importance of bacterial oxidation becomes apparent when Reaction (1) is considered as the composite of a sequence of reactions (Fig. 3). It is thought that pyrite weathering is initiated by oxygen and water (Eqn. 1.1 of Fig. 3) because circumneutral pH values are not conducive to the activity of acidophilic microorganisms; however, the literature also indicates controversy regarding the role of bacteria in propagating this reaction. In complex natural systems, it seems logical that an interplay of chemical and biological processes may occur on a microscale (Nordstrom and Alpers 1999). Thus, it is entirely possible that extremely acid conditions might develop adjacent to pyrite grain surfaces before changes in pore water chemistry are detected. Under such conditions, measurements of bulk-water chemistry are highly unlikely to reflect the chemistry at the surface of the pyrite. It is also possible that neutrophilic *Thiobacilli* may catalyze the initial stage of pyrite oxidation (Blowes et al. 1995, 1998).

Nordstrom and Alpers (1999) summarized available data from McKibben and

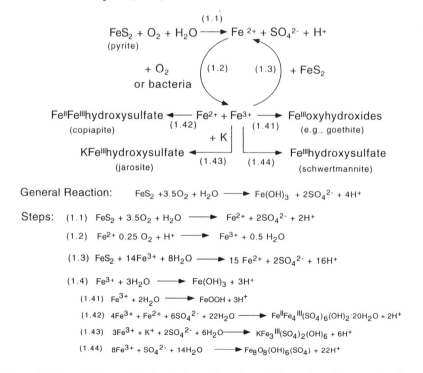

Figure 3. Schematic diagram showing the "steps" in pyrite oxidation and possible secondary Fe minerals that may form as weathering products (modified from Rose and Cravotta 1998).

Barnes (1986) and Olson (1991) for abiotic vs. microbial rates of pyrite oxidation under comparable conditions and concluded that the rates are similar at about 10^{-8} mol m^{-2} s^{-1}. As suggested by these units, oxidation rates are dependent on the surface area of pyrite exposed to solution as well as the crystallinity and structure of the pyrite surface (McKibben and Barnes 1986). The high surface area of the raspberry-like clusters of pyrite grains (framboidal pyrite) found in many sedimentary rocks at least partly accounts for their observed reactivity (Carrucio 1975). Hammack et al. (1988) observed that sedimentary pyrite was more reactive than could be explained by surface area alone, and suggested that the cause was a difference in crystal structure or the frequency of surface defects.

The S moiety in pyrite oxidizes more quickly than Fe giving rise to aqueous solutions enriched with SO_4^{2-}, H^+, and Fe^{2+}. Intermediate sulfoxyanions of lower oxidation state, such as sulfite (SO_3^{2-}), thiosulfate ($S_2O_3^-$), and polythionates ($S_nO_6^{2-}$), may also form but are subject to rapid oxidation by pyrite (Xu and Schoonen 1995), ferric iron (Williamson and Rimstidt 1993), or biodegradation (Luther 1990, Gould et al. 1994). A summary of the current research on the catalytic effect of the pyrite surface and of Fe^{3+} on intermediate sulfoxyanions (thiosulfate, polythionates, sulfite) formed during pyrite oxidation can be found in Nordstrom (2000). Several studies have confirmed that these highly efficient catalysts rapidly degrade intermediate sulfoxyanions to sulfate and that these intermdediate sulfoxyanions are not likely to be detected in acid mine waters.

Ferrous iron released in the initial stage of pyrite decomposition is also subject to oxidation (Eqn. 1.2 of Fig. 3). It is well known that the abiotic oxidation of Fe^{2+} proceeds rapidly above pH 4 and that the rate is pH dependent (Fig. 4). At lower pH values, the rate becomes very slow and is independent of pH. Singer and Stumm (1968), for example, calculated a rate of 2.7×10^{-12} mol L^{-1} s^{-1} at pH values below 4. They found that acidophilic, Fe-oxidizing bacteria increased the oxidation rate by 10^5 to about 3×10^{-7} mol L^{-1} sec^{-1} (Singer and Stumm 1970). The importance of this difference in reaction rates at low pH is related to the fact that dissolved Fe^{3+} rapidly oxidizes pyrite according to the stoichiometry shown in Equation 1.3 of Figure 3. The reaction is considerably faster than when oxygen is the oxidant (Fig. 4); however, significant concentrations of Fe^{3+} only occur at low pH due to the low solubility of hydrolyzed ferric iron at pH > 4. Without active bacterial populations to ensure the production of Fe^{3+}, the process of pyrite oxidation would proceed very slowly because the abiotic rate of Fe^{2+} oxidation at low pH is substantially slower than the rate of pyrite oxidation by Fe^{3+} (Fig. 4). For this

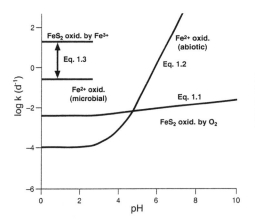

Figure 4. Comparison of rate constants for pyrite oxidation by O_2 (Eqn. 1.1 in Fig. 3) and Fe^{3+} (Eqn. 1.3 in Fig. 3) and the abiotic (Eqn. 1.2 in Fig. 3) and biotic oxidation of Fe^{2+} (modified from Nordstrom 1982a).

reason, Equation (1.2) of Figure 3 has frequently been described as the rate-limiting step in pyrite oxidation (Singer and Stumm 1970). When pyrite oxidation reaches an advanced stage, Fe is aggressively cycled between reduced and oxidized forms as a result of pyrite oxidation by Fe^{3+} and bacterial oxidation of Fe^{2+} (Fig. 3) (Kleinmann et al. 1981). The system approaches a steady state unless pyrite is consumed or the cycle of Fe oxidation-reduction is somehow broken.

Natural systems are "leaky" and Fe is eventually released to the environment either in reduced or oxidized form depending on the local pH and Eh. Equation (1.4) of Figure 3 suggests that the final repository for Fe is as an insoluble form of iron oxide or oxyhydroxide. Equation (1.4) does not reflect the multitude of possible intermediates that may be produced (Table 1; Alpers et al 1994). For example, soluble iron-sulfate minerals may precipitate directly from acid sulfate waters in virtually all climates during dry periods, especially at the interface between saturated and unsaturated zones, at which evaporation leads to an accumulation of dissolved species (Eqn. 1.1 of Fig. 3). As these waters become progressively more concentrated, salts accumulate in surface efflorescences reminiscent of those formed in closed drainage basins in arid environments. Melanterite ($FeSO_4 \cdot 7H_2O$), rozenite ($FeSO_4 \cdot 4H_2O$), and szomolnokite ($FeSO_4 \cdot H_2O$) are probably the most common of the efflorescent Fe sulfate minerals. With partial oxidation, they may be converted to soluble hydroxysulfates such as copiapite $[Fe^{II}(Fe^{III})_4(SO_4)_6(OH)_2 \cdot 20H_2O]$. Copiapite may also precipitate directly from acid sulfate

Table 1. Selected sulfates and hydroxysulfates of Fe and Al

Fe Mineral	Formula	Al Mineral	Formula
		Soluble	
melanterite	$Fe^{II}SO_4 \cdot 7H_2O$		
siderotil	$Fe^{II}SO_4 \cdot 5H_2O$		
rozenite	$Fe^{II}SO_4 \cdot 4H_2O$		
szomolnokite	$Fe_{II}SO_4 \cdot H_2O$		
copiapite	$Fe^{II}Fe_4^{III}(SO_4)_6(OH)_2 \cdot 20H_2O$	aluminocopiapite	$Al_{2/3}Fe_4^{III}(SO_4)_6(OH)_2 \cdot 20H_2O$
römerite	$Fe^{II}Fe_2^{III}(SO_4)_4 \cdot 14H_2O$	halotrichite	$Fe^{II}Al_2(SO_4)_4 \cdot 22H_2O$
coquimbite	$Fe_2^{III}(SO_4)_3 \cdot 9H_2O$	pickeringite	$MgAl_2(SO_4)_4 \cdot 22H_2O$
rhomboclase	$(H_3O)Fe^{III}(SO_4)_2 \cdot 3H_2O$	alunogen	$Al_2(SO_4)_3 \cdot 17H_2O$
fibroferrite	$Fe^{III}(SO_4)(OH) \cdot 5H_2O$	jurbanite	$Al(SO_4)(OH) \cdot 5H_2O$
amarantite	$Fe^{III}(SO_4)(OH) \cdot 3H_2O$		
butlerite	$Fe^{III}(SO_4)(OH) \cdot 2H_2O$		
		Less Soluble	
jarosite	$KFe_3^{III}(SO_4)_2(OH)_6$	alunite	$KAl_3(SO_4)_2(OH)_6$
natrojarosite	$NaFe_3^{III}(SO_4)_2(OH)_6$	natroalunite	$NaAl_3(SO_4)_2(OH)_6$
hydronium jarosite	$(H_3O)Fe_3^{III}(SO_4)_2(OH)_6$		
schwertmannite	$Fe_8^{III}O_8(SO_4)(OH)_6$	basaluminite	$Al_4(SO_4)(OH)_{10} \cdot 4H_2O$
		aluminite	$Al_2(SO_4)(OH)_4 \cdot 7H_2O$

waters as suggested by Equation (1.42) of Figure 3; however, there appears to be little field evidence for this mechanism of formation (Jambor et al., this volume). Rapid dissolution of accumulated soluble salts during subsequent rainfall events may release stored acidity and produce major "pulses" of contaminants, often with severe consequences for downstream ecosystems.

As the Fe in acid sulfate waters becomes fully oxidized, it eventually reaches saturation with respect to a variety of less soluble oxides, oxyhydroxides, and hydroxysulfate minerals that comprise the ochreous precipitates found in many streams and lakes impacted by mine drainage. The hydrolysis reactions giving rise to these minerals generate additional acidity (Eqns. 1.41–1.44 of Fig. 3) and are often responsible for suppressing the pH even after acid sulfate solutions are mixed with waters of higher alkalinity. The most common oxides and oxyhydroxides include goethite (α-FeOOH), lepidocrocite (γ-FeOOH) and ferrihydrite (nominally $Fe_5HO_8 \cdot H_2O$), each of which seems to occupy a specific geochemical niche (Bigham 1994). A comprehensive treatment of the structure, properties, and occurrence of these minerals was reported by Cornell and Schwertmann (1996) and Jambor and Dutrizac (1998).

Jarosite, $[K,H_3O,Na]Fe_3(SO_4)_2(OH)_6$ (Eqn. 1.43), is a straw-yellow mineral that is common in the weathered zone of sulfide ore deposits and in acid sulfate soils. Speciation-saturation analyses of low-pH surface waters high in sulfate commonly indicate supersaturation with respect to this mineral (e.g. Chapman et al. 1983, Filipek et al. 1987). Supersaturation and the infrequent detection of jarosite suggest that kinetic barriers may prevent its rapid precipitation from most surface waters. More details concerning the properties and geochemistry of the jarosite-group minerals are presented by Dutrizac and Jambor (this volume) and by Stoffregen et al. (this volume). Schwertmannite $[Fe_8O_8(OH)_6SO_4]$ is a less soluble hydroxysulfate mineral that is a much more common phase than jarosite in ochre deposits from acid sulfate solutions. Later sections of this chapter include descriptions of schwertmannite and its Al analog(s).

The Al system

The primary source of soluble Al in acid sulfate waters is the same as in most natural weathering environments, namely, various aluminosilicates that are subject to proton attack and dissolution as typified by Reaction (2) for K-feldspar:

$$KAlSi_3O_{8(s)} + 4\ H^+_{(aq)} + 4\ H_2O_{(l)} \leftrightarrow Al^{3+}_{(aq)} + 3\ Si(OH)_4{}^0_{(aq)} + K^+_{(aq)} \qquad (2)$$
(K-feldspar)

Such minerals occur in spoil and tailing deposits as gangue material and also comprise the matrix of most soils and sediments. The release of Al and Si from host aluminosilicates is not directly mediated by bacterial processes; however, leaching studies of ore samples have shown that rates are clearly enhanced in inoculated specimens due to the accelerated generation of sulfuric acid by microbially-mediated pyrite oxidation (Bhatti et al. 1994) (Fig. 5).

Because monosilicic acid (Reaction 2) is a very weak acid, it can be removed from the system by simple leaching. The Al component, on the other hand, is usually conserved as a solid phase through the formation of gibbsite or kaolinite.

$$KAlSi_3O_{8(s)} + H^+_{(aq)} + 7\ H_2O_{(l)} \leftrightarrow Al(OH)_{3(s)} + 3\ Si(OH)_4{}^0_{(aq)} + K^+_{(aq)} \qquad (3)$$
(gibbsite)

$$KAlSi_3O_{8(s)} + H^+_{(aq)} + 4.5\ H_2O_{(l)} \leftrightarrow 0.5\ Al_2Si_2O_5(OH)_{4(s)} + 2\ Si(OH)_4{}^0_{(aq)} + K^+_{(aq)} \qquad (4)$$
(kaolinite)

Due to Reactions (3) and (4), the concentration of dissolved Al in most natural

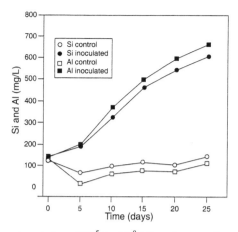

Figure 5. Changes in soluble Al and Si concentrations during the oxidative dissolution of a black schist ore by a mixed culture of Fe- and S-oxidizing thiobacilli as compared to non-inoculated controls (modified from Bhatti et al. 1994).

waters is low (10^{-5} to 10^{-8} M) so long as the pH is near neutral and the concentration of strong complexing agents, such as organic acids, is not significant (Nordstrom 1982b). In most acid systems, such as highly weathered soils, the activity of soluble Al is controlled by the solubilities of gibbsite and kaolinite. In acid sulfate waters, however, the geochemistry of Al is significantly modified by sulfate so that gibbsite and kaolinite are probably not the most stable phases. Instead, a variety of soluble sulfate and insoluble hydoxysulfate minerals may form. Studies of efflorescences from coal deposits and dumps have identified halotrichite [$Fe^{II}Al_2(SO_4)_4 \cdot 22H_2O$], pickeringtite [$MgAl_2(SO_4)_4 \cdot 22H_2O$], aluminocopiapite [$Al_{2/3}Fe^{III}_4(SO_4)_6(OH)_2 \cdot 20H_2O$], and alunogen [$Al_2(SO_4)_3 \cdot 17H_2O$] as common alteration products from pyrite oxidation (Zodrow and McCandlish 1978) (Table 1). When acid sulfate solutions containing dissolved Al are mixed with waters of higher pH, or are buffered to higher pH by carbonate minerals, insoluble hydroxysulfates form by reactions analogous to those in the Fe^{III} system (Reactions 5 and 6).

$$3\ Al^{3+}_{(aq)} + K^+_{(aq)} + 2\ SO_4^{2-}_{(aq)} + 6\ H_2O_{(l)} \leftrightarrow KAl_3(SO_4)_2(OH)_{6(s)} + 6\ H^+_{(aq)} \tag{5}$$
$$\text{(alunite)}$$

$$4\ Al^{3+}_{(aq)} + SO_4^{2-}_{(aq)} + 14\ H_2O_{(l)} \leftrightarrow Al_4(SO_4)(OH)_{10} \cdot 4H_2O_{(s)} + 10\ H^+_{(aq)} \tag{6}$$
$$\text{(basaluminite)}$$

Alunite has been observed in naturally acidic, hypersaline lakes (Alpers et al. 1992, Long et al. 1992), and it is commonly found in volcanic regions where hydrothermal alteration has occurred (Hemley et al. 1969, Raymahashay 1968). Alunite does not appear to be a major component of the white precipitates that are formed at some localities by mixing and dilution of acid sulfate waters at low temperature (e.g. Theobald et al. 1963) but it has been found in a mixing zone at Doughty Springs, Colorado (Nordstrom et al. 1984). Most aluminous precipitates have the composition of basaluminite (Nordstrom et al. 1984, Alpers et al. 1994) but are very poorly crystalline so that structural details are lacking. The precipitates, however, do not form unless the pH is nearly 5.0 or higher because the pK_1 for Al hydrolysis is 5.0 (Nordstrom and May 1996). For pH values much less than 5.0, dissolved Al behaves as a conservative constituent, whereas above a pH of 5.0 Al becomes highly insoluble and maintains an apparent solubility corresponding to that of microcrystalline to amorphous gibbsite in surface water (Nordstrom and Ball 1986, Nordstrom and Alpers 1999) even though the precipitate is an apparently amorphous basaluminite.

FE AND AL HYDROXYSULFATES OF LOW CRYSTALLINITY

Schwertmannite [$Fe_8O_8(OH)_6SO_4 \cdot nH_2O$]

Schwertmannite, first described from an occurrence at Pyhäsalmi, Finland (Bigham et al. 1994), is probably the most common direct precipitate of Fe from acid sulfate waters in the pH range of 2 to 4 (Bigham et al. 1992); however, its existence has been viewed with caution because it is poorly crystalline, is metastable, and is commonly admixed with other nanophase Fe minerals. The mineralogical history of schwertmannite is interesting and reflects the difficulties encountered when attempting to define the properties of materials with short-range structural order (Murad et al. 1994). A brief summary follows.

Mineralogical history. E.F. Glocker (1853) published a report characterizing a "new ferric sinter" that was rich in sulfate and derived from dripstones in the medieval Alt-Hackelsberg silver and gold mine near Zlaté Hory, The Czech Republic. The morphology of the dripstones, some reaching over 1 m in length, was described in a succeeding paper (Glocker 1858). Glocker (1853) suggested the material was a new hydrous ferric sulfate, and it was subsequently named glockerite by Naumann (1855). Cornu (1909) expressed the view that glockerite was not a new mineral but rather limonite with adsorbed sulfate. Glockerite was given the approximate formula $Fe_4^{III}(SO_4)(OH)_{10} \cdot 1\text{-}3H_2O$ by Palache et al. (1951). Fojt (1975) revisited the type location identified by Glocker, photographed the dripstones, and obtained new samples for chemical and mineralogical analysis. He noted that the dripstones always originated from portions of the ore that were rich in pyrite and that the infiltrating water was strongly acid (pH 1.5-4.5). Chemical analyses of five samples yielded Fe/S ratios in the range of 5.4 to 8.2 and infrared spectra showed clear evidence of SO_4 absorption features. X-ray diffraction patterns gave no measurable peaks, but thermal analyses produced an exothermic effect thought to indicate recrystallization of lepidocrocite to maghemite. Fojt (1975) concluded that glockerite was cryptocrystalline lepidocrocite with variable amounts of water and "unstable" sulfate content, and the name glockerite was discredited by the IMA. At about the same time, Margulis et al. (1975, 1976) prepared and characterized an "amorphous" basic sulfate by the hydrolytic precipitation of Fe^{III} from the $Fe_2(SO_4)_3$–KOH–H_2O system at various temperatures. The amorphous precipitate had the approximate composition $2Fe_2O_3 \cdot SO_3 \cdot nH_2O$ and, depending on the synthesis conditions, was associated with jarosite or goethite. Its thermal and infrared properties were similar to those reported by Fojt (1975). Flynn (1990) prepared a "dense amorphous basic ferric sulfate," or DABS, by neutralizing a ferric sulfate solution with a sodium bicarbonate solution. He found that the material, when carefully washed, gave a consistent formula of $Fe_4(OH)_{10}SO_4 \cdot 2\text{-}3H_2O$ and converted to goethite in 8 to 12 days at 25°C. He observed, like Bigham et al. (1996b) when dissolving schwertmannite, that Fe(III) concentrations first increased then decreased during the conversion to goethite. Spectroscopic data, however, were either lacking or inadequate to further characterize this material.

Lazaroff et al. (1982, 1985) conducted infrared (IR) and scanning electron microscope (SEM) studies of ferruginous sediments that had formed through the oxidation of acid, ferrous sulfate solutions by resting suspensions of *T. ferrooxidans* in the laboratory. Jarosites were produced when the $FeSO_4$ solutions contained Na, K, or NH_4 and an excess of sulfate at pH 2.5. Without these cations, an "amorphous" ferric hydroxysulfate was produced that was indistinguishable from precipitates collected from acid mine waters. Mössbauer analyses of similar synthetic precipitates led Murad (1988a) to conclude that the "amorphous" material was a well-crystallized ferrihydrite. Brady et al. (1986) studied the mineralogy of an ochreous sediment isolated from a stream receiving acid mine drainage and determined that it consisted of nanophase goethite and another poorly crystalline phase that yielded an X-ray diffraction profile with characteristics of both ferrihydrite and feroxyhite (δ'-FeOOH). Precipitates from other localities were collected and characterized (Bigham et al. 1990), and data from the purest specimens of this "mine drainage mineral" were used to support the recognition of schwertmannite as a new mineral. The proposal was submitted without knowledge of the prior work by Glocker (1853, 1858) or Fojt (1975) which had been responsible for the recognition and discrediting, respectively, of glockerite as a mineral. Eventually, samples of "glockerite" were obtained, matched with schwertmannite, and the historical record was clarified (Schwertmann and Fojt 1996). Schwertmannite now stands as the accepted name for this poorly crystalline ferric hydroxysulfate.

Geochemical history. A multitude of laboratory studies have made some form of ferric hydroxide or hydrous ferric oxide (HFO) by numerous pathways (see Fox 1988a). A few of these are worth mentioning because they have a bearing on the formation of minerals such as schwertmannite. The first definitive chemical study on the solubility and stability relations of Fe hydroxysulfate minerals, for a range of sulfuric acid concentrations, was the classic paper of Posnjak and Merwin (1922). The solubility data in this study were for more soluble salts than schwertmannite and temperatures of 60°C and higher, but a supplemental report (Merwin and Posnjak 1937) made some extrapolations to 30°C. One of the first studies to clarify the solubility of freshly precipitated HFO was reported by Biedermann and Schindler (1957). They showed that a goethite equilibrium solubility behavior is obeyed ($Fe^{3+}/H^+ = 1/3$) if ferric perchlorate solutions are used in a constant ionic medium of 3 M $NaClO_4$ and the precipitates are allowed to age for about 200 hours. When a NaCl medium of 0.5 M ionic strength is used, the goethite dissolution stoichiometry is not obeyed and a precipitate of composition $Fe(OH)_{2.7}Cl_{0.3}$ is formed (Biedermann and Chow 1966). Matijevic and Janauer (1966) noted the complexities of Fe colloid formation and were able to produce very uniform colloids of hydronium and other jarosites at 98°C (Matijevic et al. 1975). Several titration studies involving the hydrolysis of soluble Fe(III) salts indicated the incorporation of chloride into Fe(III) colloids (Feitknecht et al. 1973, Dousma and de Bruyn 1976, Dousma et al. 1978, 1979). Dousma and his colleagues titrated acid Fe(III) solutions with base and noted incorporation of anions into the precipitating phase when chloride, nitrate, or sulfate medium was used. Fox (1988a) recognized the change in the stoichiometry of precipitated HFO when chloride and nitrate media were used and derived a more appropriate stoichiometry of $Fe(OH)_{2.35}$ with the remainder of the charge balanced by chloride or nitrate or phosphate. He used this hypothesis to examine iron colloids in a river system and demonstrated its applicability (Fox 1988b, 1989).

Byrne and Luo (2000) have re-evaluated this issue by using a sensitive potentiometric technique similar to that employed by Biedermann and Chow (1966) to measure the stoichiometry for the HFO precipitation reaction over a wide pH range (3-7.5). It was concluded that, for the reaction:

$$Fe(OH)_n^{3-n} + nH^+ \leftrightarrow Fe^{3+} + nH_2O \tag{7}$$

the solubility product constant should be

$$K^*_{SO} = [Fe^{3+}][H^+]^{-2.86} = 10^{4.28} \tag{8}$$

or in log form

$$\log [Fe^{3+}] = 4.28(\pm0.05) - 2.86(\pm0.009) \, pH \tag{9}$$

Biedermann and Schindler (1957) reported

$$\log [Fe^{3+}] = 3.96(\pm 0.10) - 3.0 \text{ pH} \qquad (10)$$

The range of log K_{sp} values reported by Nordstrom et al. (1990) for ferrihydrite solubility range from 3.0 to 5.0 based on several other reported literature studies. These results and their implications for natural systems are discussed further in the section of this chapter that deals with geochemical controls on mineral formation.

Properties. Schwertmannite usually occurs in mixtures with other minerals that range from poorly crystalline (ferrihydrite) to moderately crystalline (goethite, lepidocrocite) to well crystalline (jarosite). Its presence in such mixtures complicates the processes of identification and characterization. The objective of this discussion is to define the properties of schwertmannite and to compare them with those of coexisting phases (Table 2).

Schwertmannite has the ideal formula $Fe_8O_8(OH)_6SO_4 \cdot nH_2O$, which implies an Fe/S molar ratio of 8 and a composition that is intermediate between those of jarosite (Fe/S = 1.5) and the common iron oxides (no S) with which it is generally associated. In fact, the sulfate content is somewhat variable, and natural samples are best described by the formula $Fe_8O_8(OH)_{8-2x}(SO_4)_x \cdot nH_2O$ where $1 \le x \le 1.75$ and Fe/S ranges from 8 to 4.6 (Bigham et al. 1996b). Slightly higher sulfate contents were recently reported for Korean samples (Yu et al. 1999). Available chemical data for both natural and synthetic schwertmannite are summarized in Table 3. The sulfate in schwertmannite may be partly or fully substituted by anions such as selenate, arsenate, and nitrate when co-precipitated (Waychunas et al. 1995). Therefore, sulfate deficient specimens and sulfate-free analogs are possible. Ferrihydrite can sorb enough sulfate to approximate the composition found for schwertmannite (Smith 1991), making it difficult to distinguish between HFO with sorbed sulfate and schwertmannite. Barham (1997) has suggested that the sulfate in schwertmannite can also be replaced through exchange reactions by equilibrating specimens with solutions containing other anions such as carbonate, oxalate, and chromate. These studies confirm that the exact character of sulfate in schwertmannite needs further investigation.

Synthesis studies have indicated that schwertmannite may have a structure similar to that of akaganéite, β-FeO(OH)$_{1-x}$Cl$_x$ (nominally β-FeOOH), which is composed of double chains of FeO$_3$(OH)$_3$ octahedra sharing corners to produce square tunnels extending parallel to the c axis (Bigham et al. 1990). The tunnels are actually a series of adjoining cavities created by framework oxygens/hydroxyls. In akaganéite, the structure is stabilized by Cl$^-$ or F$^-$ or OH$^-$ occupying every second cavity (Childs et al. 1980), and it has been suggested that sulfate could play a similar role in schwertmannite. Because of size restrictions, sulfate ions could not occupy structural cavities without sharing the oxygen atoms with surrounding Fe atoms and without severe distortion of the structure. Although distortion is consistent with the poor crystallinity of schwertmannite, this model deserves further testing.

Perhaps the best physical evidence for structural sulfate in schwertmannite has been obtained from [57]Fe Mössbauer analyses (Murad et al. 1990). These data show the Fe in schwertmannite to be exclusively trivalent and in octahedral coordination. The mineral has a Neél temperature (temperature of transition from paramagnetic to ferrimagnetic behavior) of 75±5 K and a saturation magnetic hyperfine field of about 45.5 T. The former is intermediate to the Neél temperatures of commonly associated iron oxides and jarosite, and the magnetic hyperfine field is lower by about 1 T than those of even the most poorly crystalline ferrihydrite (Table 2). Presumably, the sulfate in schwertmannite

Table 2. Properties of Minerals[†] Encountered in Mine Drainage Ochres.

Mineral Name:	Goethite	Lepidocrocite	Ferrihydrite	Schwertmannite	Jarosite[‡]
Ideal Formula:	α-FeOOH	γ-FeOOH	~ $Fe_5HO_8 \cdot 4H_2O$	$Fe_8O_8(SO_4)(OH)_6$	$KFe_3(OH)_6(SO_4)_2$
Crystal system	Orthorhombic	Orthorhombic	Trigonal	Tetragonal	Hexagonal
Cell dimensions (Å)	a = 4.608 b = 9.956 c = 3.022	a = 3.88 b = 12.54 c = 3.07	a = 5.08 c = 9.4	a = 10.66 c = 6.04	a = 7.29 c = 17.16
Color (Hue)	Yellowish-brown (7.5YR - 10YR)	Orange (5YR - 7.5YR)	Reddish-brown (5YR - 7.5YR)	Yellow (10YR - 2.5Y)	Straw yellow (2.5Y - 5Y)
Crystal shape Crystallinity	Short rods Moderate	Laths Moderate	Spherical Poor	Pin-cushion Poor	Pseudocubic Good
Most intense XRD spacings (Å)	4.18, 2.45 2.69	6.26, 3.29 2.47, 1.937	2.54, 2.24, 1.97, 1.73, 1.47	4.86, 3.39, 2.55, 2.28, 1.66, 1.51	5.09, 3.11, 3.08
Major IR bands (cm^{-1})	890, 797	1161, 1026 753	Nil	1175, 1125, 1055, 975, 680, 615	1181, 1080, 1003, 628, 493, 472
Néel temp. (K)	400	77	28-115[¶]	75	55-60
DTA events (°C) En. = endotherm Ex. = exotherm	En. 280 - 400	En. 300 - 350 Ex. 370 - 500	En. 150	En. 100 - 300 Ex. 540 - 580 En. 680	En. 475, 740 - 800
Magnetic hyperfine field (T) at:					
295 K	38.2	---	---	---	---
77 K	50.0	---	≤ 45.1	---	---
4.2 K	50.6	46.0	46.5 - 50.0	45.4	47.0
† Data taken from: Bigham et al. (1990) Bigham et al. (1992) Doner and Lynn (1989)	Fanning et al., (1993) Murad (1988) Murad et al. (1988)	Murad et al. (1994) Powers et al. (1975) Schwertmann (1993)		Schwertmann and Fitzpatrick (1992) Schwertmann and Taylor (1989) Takano et al. (1968)	

† Properties listed are specific to jarosite, but natrojarosite, hydronium jarosite or solid solutions may occur. ¶ Blocking temperature

Table 3. Chemical composition of natural and synthetic schwertmannites.

Element	(1)	(2)	(3)	(4)	(5)	(6)	(7)	(8)
Fe_2O_3 (%)	62.6	60.4–61.5	61.3–64.7	61.2–66.7	67.3	58.3–66.5	67.3	65.6
SO_3(%)	12.7	7.4–9.0	11.5–12.9	8.2–11.5	14.7	10.1–11.3	14.7	10.2
$H_2O\pm$ (%)	23.1	17.2	18.1–20.3	21.8–27.4	20.7	?	20.7	23.8
Fe/S	4.9	6.7–8.3	4.7–5.4	5.4–8.2	4.6	5.2–6.6	4.6	6.4

1. Type specimen, sample Py-4 from Finnish mine drainage (Bigham et al. 1994).
2. Samples Y1 and Y3b from Finnish mine drainage (Bigham et al. 1990).
3. Samples Bt-4, La-1, and Nb-1 from Ohio mine drainage (Bigham et al. 1990).
4. Five samples from Czech dripstone (Fojt 1975).
5. Sample PI3 from Korean mine drainage (Yu et al. 1999).
6. Samples MG1118 and DP4226 from Japanese lake sediments (Childs et al. 1998).
7. Synthetic sample Z510b formed by bacterial oxidation of ferrous sulfate (Bigham et al. 1990).
8. Synthetic sample B-2000s formed by hydrolysis of ferric nitrate in the presence of sulfate (Bigham et al. 1990).

inhibits magnetic exchange interactions between neighboring Fe atoms and is therefore responsible for the low ordering temperature and magnetic field. Spectra from both paramagnetic and magnetically ordered schwertmannite are asymmetric (Fig. 6), indicating multiple environments for Fe^{III} that could also reflect incorporation of SO_4 into the structure.

Schwertmannite is readily soluble in acidic (pH 3.0) solutions of ammonium oxalate (Brady et al. 1986) through a reaction that is catalyzed both by light (De Endredy 1963) and the presence of Fe^{2+} (Fischer 1972). This reagent has been widely employed for the selective dissolution of poorly crystalline Fe oxides (e.g. ferrihydrite) in soils by a 2- to 4-h extraction in the dark (Schwertmann 1964), and recent studies (Dold 1999) suggest that it may be equally efficient for partitioning schwertmannite from associated minerals of higher crystallinity in mine spoils and sediments. Studies of schwertmannite

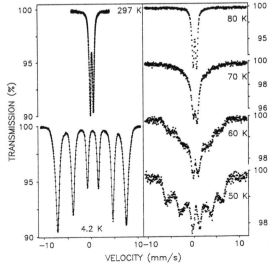

Figure 6. Mössbauer spectra of synthetic schwertmannite at 297 and 4.2 K and in the vicinity of the Néel temperature (modified from Bigham et al. 1990).

dissolution kinetics in 0.1 M HCl (Bigham et al. 1990) showed that about 15% of the total SO_4 was immediately released and was probably derived from surface sites. Half-reaction times for the remaining Fe and SO_4 were similar and indicated congruent dissolution of the bulk sample.

The thermal decomposition of schwertmannite yields a low-temperature endotherm between 100 and 300°C (Table 2) that is associated with a weight loss of 15 to 20% and is probably caused by the vaporization of sorbed water as well as structural OH/H_2O. This event is followed by an exotherm at 540-580°C that immediately gives way to a second endotherm at about 680°C. The exothermic event coincides with the formation of $Fe_2(SO_4)_3$ and may or may not be apparent in natural samples. The final endotherm can be attributed to the decomposition of $Fe_2(SO_4)_3$ to form hematite (α-Fe_2O_3) with release of SO_3. Volatilization of the latter produces a weight loss of 6 to 12%. The thermal characteristics of jarosite are similar except that the vaporization of structural hydroxyls occurs at a higher temperature (400-500°C) and the weight loss due to SO_3 evolution is greater (see Dutrizac and Jambor, this volume).

The infrared profile of schwertmannite has been examined in considerable detail (Lazaroff et al. 1985, Bigham et al. 1990, Bishop and Murad 1996) and is characterized by absorption features reflective of its composition. Four sulfate vibrational modes are possible. These modes and their approximate frequencies are υ_1 (symmetric stretch) at 983 cm^{-1}, υ_2 (symmetric bend) at 450 cm^{-1}, υ_3 (asymmetric stretch) at 1105 cm^{-1}, and υ_4 (asymmetric bend) at 611 cm^{-1} (Ross 1974). When the sulfate group exhibits symmetry lower than Td (isolated, tetrahedral SO_4^{2-}), the υ_2, υ_3, and υ_4 sulfate bending vibrations may be split into multiple vibrations. A typical schwertmannite spectrum (Table 2) exhibits a strong but degenerate υ_3 vibration near 1100 cm^{-1} with shoulders at about 1050 and 1170 cm^{-1}, a medium strength υ_4 vibration at about 610 cm^{-1}, a medium strength υ_1 vibration at 980 cm^{-1} and a probable, weak υ_2 vibration at 465 cm^{-1}. Additional features in the schwertmannite spectrum at about 850 and 700 cm^{-1} have been assigned to OH bending vibrations (Bishop and Murad 1996).

The color of schwertmannite is intermediate to those of goethite and jarosite (Table 2), and a recent study by Swayze et al. (2000) has suggested that the reflectance spectra of iron-bearing secondary minerals may provide a useful tool for mapping acidic mine wastes. Diffuse reflectance spectroscopy in the visible (400-700 nm) and extended visible range (400-1200 nm) has also been employed to identify secondary iron minerals in soils and sediments. Because of overlapping crystal-field band positions, it seems unlikely that this technique can be used to discriminate goethite, lepidocrocite, and schwertmannite (Scheinost et al. 1998). Visible color identification is inadequate to determine schwertmannite in the field because of high color variability and similar average colors to other hydrous Fe oxide minerals (Scheinost and Schwertmann 1999).

The crystal morphology of most minerals is variable and is not a reliable tool for identification purposes. This conclusion is particularly true of nanophase materials. Schwertmannite particles, by contrast, usually occur as fine needles that coalesce to form rounded aggregates that are 200 to 500 nm in diameter with electron-dense interiors (Fig. 7). This unique "pin-cushion" morphology is commonly observed in loose precipitates of both synthetic and natural samples but may not be apparent when materials are taken from consolidated sediments or surface crusts. The morphology of schwertmannite particles typically contrasts with those of associated jarosite or Fe-oxyhydroxides. Moreover, the morphology is responsible for high specific surface areas in the range of 100-200 m^2/g and should enhance the reactivity of schwertmannite in the environment.

The poor crystallinity of schwertmannite places limitations on the utility of

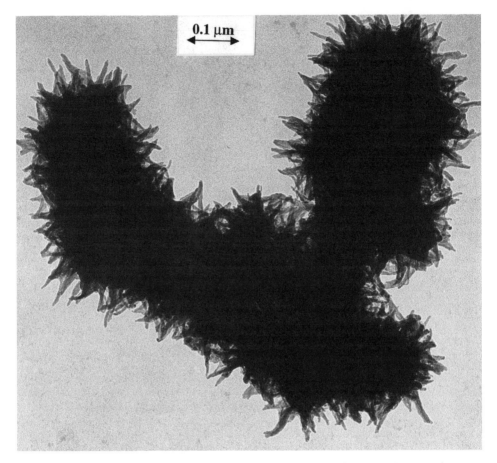

Figure 7. Transmission electron microscopy (TEM) photomicrograph of synthetic schwertmannite.

conventional X-ray diffraction (XRD) analysis, but the mineral has a unique XRD profile that can be readily distinguished from those of associated minerals if specimens are reasonably pure (Fig. 8). The powder diffraction pattern of schwertmannite consists of eight broad peaks for d > 1.4 Å and is perhaps most easily confused with that of well crystallized (6-line) ferrihydrite. The strongest peak for both minerals occurs at about 2.54 Å, but that of schwertmannite is characteristically more symmetrical. Both minerals exhibit reflections at 1.51 and 1.46 Å, but the intensity ratios are reversed. The 1.64 Å peak of schwertmannite is significantly displaced from that at 1.72 Å for ferrihydrite. Also, schwertmannite exhibits two additional reflections at about 3.31 and 4.95 Å. The detection of schwertmannite in mixed assemblage with other minerals may be difficult because of its poor crystallinity. When admixed with well crystalline minerals, such as jarosite, schwertmannite may be overlooked because its diffraction peaks become "lost" as background noise. Even when coarse-grained minerals are excluded, the diffraction patterns from mixtures may be difficult to evaluate. In such cases, the technique of differential X-ray diffraction, DXRD (Schulze 1981), may enable schwertmannite to be detected. For example, Figure 9 shows XRD patterns from a mine-drainage ochre before

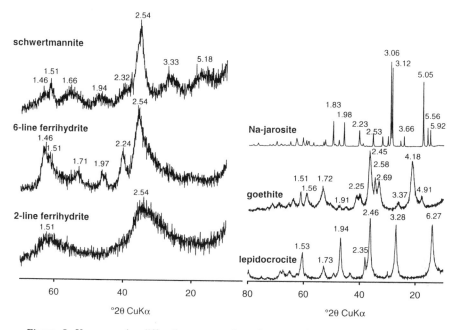

Figure 8. X-ray powder diffraction patterns for schwertmannite and commonly associated minerals; d values are in Å.

and after a 15-min extraction with acid ammonium oxalate that left a residue of goethite and silicate minerals. Subtraction of the residue diffraction pattern from that of the original sample yields the diffraction profile of the component dissolved by acid ammonium oxalate. In this case, the pattern of schwertmannite is clearly recognizable.

Hydroxysulfates of Al

Several Al hydroxysulfates of poor crystallinity and low solubility are known to form in acid sulfate environments. A summary of these minerals and their properties is provided in Table 4a. These minerals occur as fine-grained materials with such small particle size and poor crystallinity that only a few had had crystal-structure studies and space group assignments. There are four distinct compositional sets of aluminum hydroxysulfate hydrates, characterized by their Al:SO$_4$ mole ratios. Hydrobasaluminite and basaluminite have the highest ratio (4:1), followed by zaherite (2.4:1), then aluminite and meta-aluminite (2:1), and finally jurbanite and rostite (1:1), which are well-crystallized and kept separate in Table 4b. These compositional changes must reflect the relative proportion of the dissolved Al to dissolved SO$_4$ (or sulfuric acid) in the aqueous solution, and, hence, the relative pH of the solution from which they precipitate. In other words, the basaluminite set (actually only hydrobasaluminite because basaluminite forms by dehydration of hydrobasaluminite) should precipitate at higher pH values than zaherite, and with decreasing pH the sequence should be zaherite to the aluminite set to the jurbanite set to alunogen [Al$_2$(SO$_4$)$_3$·17H$_2$O, see Jambor et al., this volume]. Jurbanite and rostite (the jurbanite set) are not poorly ordered, insoluble precipitates. Rather, they crystallize as soluble salts with good crystallinity, and they will receive special mention in a later section.

Geochemical history. The chemistry of basic Al sulfates has been confusing because

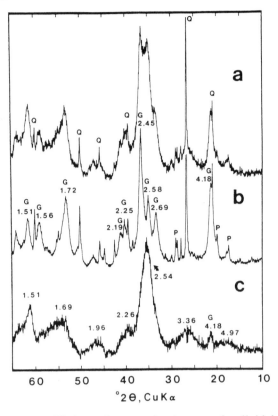

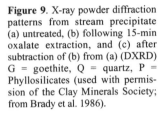

Figure 9. X-ray powder diffraction patterns from stream precipitate (a) untreated, (b) following 15-min oxalate extraction, and (c) after subtraction of (b) from (a) (DXRD) G = goethite, Q = quartz, P = Phyllosilicates (used with permission of the Clay Minerals Society; from Brady et al. 1986).

of the rapid formation of colloidal disequilibrium phases and mixtures of colloidal phases, similar to the formation of hydrous ferric oxides with substituted anions in the iron system. After the initial precipitation, slow transformations continue as the solution ages. For example, Singh (1982) showed that with sufficient aging, a 0.5 molar $Al_2(SO_4)_3$ solution will precipitate simultaneously gibbsite, böhmite, alumina, and alunite. Adams and Hajek (1978) formed both single-mineral precipitates and mixtures of gibbsite, basaluminite, and alunite when acid solutions of Al sulfate were aged for 18 weeks at 50°C. The existence of basic flocculants of Al sulfate was conjectural for many years; whereas some titration curves and chemical compositions supported the existence of such phases, the ease of displacement of the sulfate by washing seemed to suggest that sulfate was adsorbed (Weiser et al. 1941). Studies that combined XRD data with chemical compositions were most fruitful, and the identification of minerals with Al-hydroxysulfate composition provided inescapable evidence that such compounds exist (see next section).

The definitive study of Bassett and Goodwin (1949) combined phase-equilibria solubility data with careful XRD analysis and chemical composition of all precipitates that formed over a wide range of Al sulfate and sulfuric acid concentrations. Several striking results came from their 13-year study. First, they clearly defined the solubility range for alunogen and jurbanite before jurbanite was recognized as a mineral (Anthony and McLean 1976). Second, Bassett and Goodwin (1949) were unable to synthesize aluminite, basaluminite, hydrobasaluminite, zaherite, or rostite. Third, they showed a stable solubility range for a compound of composition $Al_{10}(SO_4)_6(OH)_{18} \cdot 37H_2O$ that has

Table 4a. Properties of Al hydroxysulfate minerals

Mineral Name Formula	Hydrobasaluminite $Al_4(OH)_{10}SO_4 \cdot 15H_2O$	Basaluminite $Al_4(OH)_{10}SO_4 \cdot 4H_2O$	Zaherite $Al_{12}(OH)_{26}(SO_4)_5 \cdot 20H_2O$	Aluminite $Al_2(OH)_4SO_4 \cdot 7H_2O$	Meta-Aluminite $Al_2(OH)_4SO_4 \cdot 5H_2O$
Crystal System	Monoclinic	Monoclinic	Triclinic	Monoclinic	Monoclinic
Space Group	?	$P2_1$	$P1$ or $P\bar{1}$	$P2_1/c$	?
Cell Dimension	a = 14.911 b = 9.993 c = 13.640 β = 112.24°	a = 12.954 b = 10.004 c = 11.064 β = 104.1°	a = 18.475 α = 95.24° b = 19.454 β = 91.48° c = 3.771 γ = 80.24°	a = 7.440 b = 15.583 c = 11.700 β = 110.18°	a = 7.930 b = 16.879 c = 7.353 β = 106.73°
Color	White to light yellow-brown	White	Chalk-white, to light bluish green when contains Cu	White	Silky white
Texture and Crystallinity	Clay-like Usually moist and plastic	Clay-like Conchoidal fracture	Densely-packed aggregates of micro- to cryptocrystalline	Clay-like, friable, nodular masses of minute fibers	Nodular microcrystalline aggregates and concretions
Most intense XRD spacings (Å)	12.6, 6.18, 5.29, 4.70	9.39, 4.73, 3.69, 1.438	18.1	8.98, 7.79, 4.70	8.46, 4.52, 4.39, 3.54
Stability	Dehydrates to basaluminite	Formed from hydro-basaluminite	Dehydrates under ambient conditions		Dehydrates to aluminite at 55°C

Table 4b. Properties of jurbanite and rostite.

Mineral Name	Jurbanite	Rostite
Formula	$Al(OH)SO_4 \cdot 5H_2O$	$Al(OH)SO_4 \cdot 5H_2O$
Crystal System	Monoclinic	Orthorhombic
Space Group	$P2_1/n$ or $P2_1/c$	$Pcab$
Cell dimensions	a = 8.396	a = 11.175
	b = 12.479	b = 13.043
	c = 8.155	c = 10.878
	$\beta = 101.92°$	
Color and texture	Colorless	Chalky mass
Most intense XRD spacings (Å)	6.75, 5.73, 4.48, 4.00, 3.99, 3.90	4.26, 4.19, 3.91

not yet been identified as a mineral. Johansson (1962) made relatively large crystals of jurbanite from which he was able to determine the crystal structure.

Another study by Johansson (1963) demonstrated that $Al_{13}O_{40}$ units form the building blocks for several, if not all, Al hydroxysulfates. Although the occurrence and proportion of mononuclear versus polynuclear species in titrations and other lab studies has been debated considerably in the literature, these Al_{13} polynuclear units have been recognized as rapidly forming, metastable colloids, that do not decompose easily (Bertsch and Parker 1996). Furthermore, anions such as sulfate usually decrease the pH of maximum precipitation and increase the rate of precipitation and coagulation (Bertsch and Parker 1996). The $13Al_2O_3 \cdot 6SO_3 \cdot xH_2O$ compound synthesized by Bassett and Goodwin (1949) and for which the crystal structure was determined by Johansson (1963), has compositional similarities to hydrobasaluminite/basaluminite but the XRD data do not bear a resemblance. This observation would suggest that there are other minerals, not yet identified, that might form in these systems.

Mineralogy and genesis. The most common Al hydroxysulfate minerals are hydrobasaluminite and basaluminite. Hydrobasaluminite seems to be the precipitate that forms when acid rock drainage is neutralized by mixing with a neutral, buffered water or by reaction with carbonate minerals such that the resultant pH is nearly 5.0 or higher. Examples of such precipitates and their compositions are given in Table 5. Theoretical compositions of hydrobasaluminite along with that for hydronium alunite and the analyzed compositions of two basaluminite samples (one of which is well-characterized, Clayton 1980) are provided for comparison. Samples C and E of Table 5 precipitated in cold, mountain-stream waters upon mixing of acid mine water with neutral-pH water. A typical TEM photograph of sample C (but also similar to sample E) in Figure 10 shows the lack of any crystal morphology and the appearance of aggregation of spherical colloids. Sample D of Table 5 is unusual in that it precipitated by mixing of an acid mine water with a highly carbonated, warm (20°C), mineral spring with elevated concentrations of chloride (700-800 mg/L). These conditions probably enhanced the

Table 5. Composition (wt %) of aluminous precipitates relative to theoretical and analyzed compositions of Al hydroxysulfate minerals.

	Hydronium alunite	Basaluminite	Basaluminite Sussex, UK	Basaluminite Dorset, UK	Hydro-basaluminite	C			D	E	F
	(Theoretical)	(Theoretical)	A	B	(Theoretical)	(1)	(2)	(3)	(4)	(5)	(6)
Al_2O_3	38.80	45.8	46.4	44.75	31.7	36.2	42.8	47.0	39.0	40.1	44.1
CaO				0.30		0.4	0.77	1.5	1.2		
Na_2O						0.0	0.05	0.18	0.5	0.1	
K_2O						0.0	0.04	0.00		0.0	0.0
H_2O	20.57	36.3	33.4	35.60	55.9	46.9	40.8	39.5	46.0	45.6	35.0
SO_3	40.63	17.9	17.4	18.10	12.4	10.7	12.4	8.7	17.4	11.6	22.3
Sum	100.00	100.00	97.2	98.75	100.00	94.2	96.9	96.9	104.2	97.4	101.4
X-ray diffr.	----	----	Basaluminite	Basaluminite	----	Amorphous	Amorphous	Amorphous	Alunite	Amorphous	Amorphous
Electron diffr.	----	----	Basaluminite	Basaluminite	----	Amorphous	Amorphous	Amorphous	crystalline	Amorphous	Basaluminite

A: Bannister and Hollingworth (1948a,b)
B: Clayton (1980)
C: Precipitates 1, 2, and 3 from Paradise portal drainage, San Juan Mountains, CO (Nordstrom et al., 1984)
D: Precipitate from Doughty Springs, Delta County, CO (Headden, 1905; Cunningham et al., 1996)
E: Precipitate from Leviathan mine drainage, Alpine County, CA (Ball and Nordstrom, 1989)
F: Precipitate synthesized in laboratory from aluminum sulfate solution, pH = 4.1 (Charles Roberson and John Hem, USGS, unpublished data)

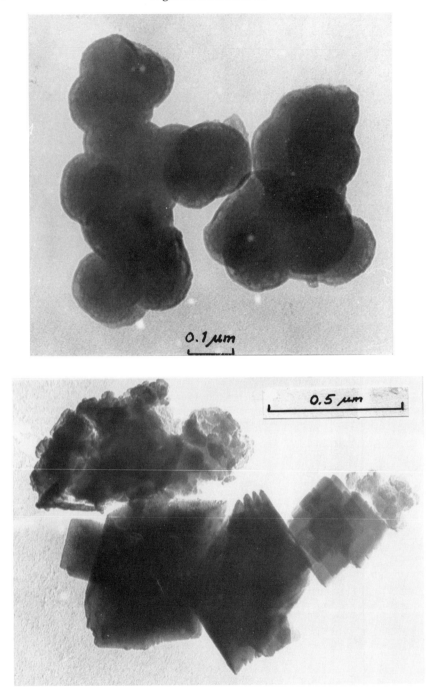

Figure 10 (top). TEM photograph of Al-hydroxysulfate precipitate C (hydrobasaluminite/basaluminite) in Table 5.

Figure 11 (bottom). TEM photograph of Al-hydroxysulfate precipitate D (alunite) in Table 5.

formation of alunite at this site. A TEM photograph of sample D (Fig. 11) clearly shows crystal faces and both electron diffraction and XRD show the presence of alunite. The elevated water content of some of the "amorphous" precipitates may reflect the hydrated character of hydrobasaluminite relative to basaluminite. All of the analyzed precipitates fall within the range of composition between the theoretical values for hydrobasaluminite and basaluminite. The analyses that are short of 100% probably reflect admixed silica. Frequently, quartz fragments can be seen mixed with the basaluminite when observed with a microscope.

Bannister and Hollingworth (1948a,b) first identified basaluminite and hydro-basaluminite in the Northampton Ironstone, near Wellingsborough, UK. One or both of these minerals have since been found in the caves of the Guadalupe Mountains, New Mexico (Polyak and Provencio 1998), in the Flysch formations of the Carpathian Mountains (Wieser 1974), near Brno, Czech Republic (Batik and Hruskova 1971), associated with a Middle Pennsylvanian coal bed in southeastern Kansas, USA (Tien 1968), in several prefectures in Japan (Matsubara et al. 1990, Minakawa et al. 1996), in Jurassic strata, Namana, Yakutia, Russia (Lizalek et al. 1989), in the bauxite deposit of Csordut, Hungary (Toth et al. 1984), near Newhaven, Sussex, UK (Wilmot and Young 1985), and in the black schist of Karatau, Kazakhistan (Zazubina and Ankinovich 1982). Formation of the minerals occurs by the reaction of acid sulfate solutions (from pyrite oxidation) with Al-rich clays, or by mixing of acid, Al-rich sulfate waters with dilute neutral waters, or through the neutralization of acid sulfate solutions, rich in Al, by carbonate minerals. The most detailed study was reported by Clayton (1980) on hydrobasalumunite and basaluminite from Chickerell, Dorset, UK and includes transmission electron microscopy with electron-diffraction data, XRD data, chemical analysis, thermogravimetric analysis (TGA), and differential thermal analysis (DTA). Crystal structure determinations of felsöbanyaite (thought to be a dimorph of basaluminite) and basaluminite have demonstrated that these two minerals are identical (Farkas and Pertlik 1997, Jambor et al. 1998).

Zaherite was first described by Ruotsala and Babcock (1977) as a white, massive, fine-grained mineral from the Salt Range, Pakistan. A second occurrence has been found in South Africa in a biotite schist near Pofadder (Beukes et al. 1984) and new XRD, IR, DTA, and compositional data was obtained. However, the XRD data gave incorrect cell dimensions that were corrected later (de Bruiyn et al. 1985).

Aluminite and meta-aluminite have been found in similar types of environments as those with basaluminite and hydrobasaluminite. Aluminite has been found in the mineral deposits of the southern Donbass area of the Ukraine (Chernitsyna et al. 1999), in the oxidized zone of the Deputatsk tin ore deposit in Yakutia, Russia (Zhdanov and Solov'ev 1998), in the oxidized zone of a silver ore deposit in the Magadan area, Russia (Oycollonov et al. 1994), in Paleozoic and Mesozoic carbonates of Siberia (Lizalek and Filatov 1986), in shales and sandstones at the contact between Cambrian and lower Jurassic rocks at Namansk, near Yakutsk, Siberia (Sokolov et al. 1985), with alunite in clay deposits near Buenos Aires, Argentina (Zalba 1982), with basaluminite and gibbsite in Carboniferous and Permian limestones and dolostones (Van'shin and Gutsaki 1982), in pyritized Proterozoic sericitic schists near Lake Baikal, Siberia (Dombrovskaya et al. 1976), in the bauxite deposits of Montenegro, Yugoslavia (Lukovic 1970), and in the Bshlybel copper deposits, Kelbadzhar area, Azerbaijan (Mamedov et al. 1969) and at several other localities (France, Germany, Italy, England, and USA). Meta-aluminite was first described by Frondel (1968) in association with basaluminite and gypsum at the Fuemrole mine in Emery County, Utah. It appears to have formed by reaction of acidic Al-rich waters with clays and carbonates. Meta-aluminite also occurs in the upper

Proterozoic schists of the Greater Goloustan area, Siberia (Mazilov et al. 1975) in a similar setting to the Utah occurrence. Sizia (1966) reported on the IR spectra of 9 sulfate minerals including aluminite.

Although numerous geochemical modeling studies have reported acidified waters in equilibrium with jurbanite (van Breeman 1973 1976; Kram et al. 1995; Monterroso et al. 1994; Alvarez et al. 1993), the only two known occurrences of the mineral are at the San Manuel copper mine, Pinal County, Arizona, USA (Anthony and McLean (1976) and the Cetine mine in Tuscany, Italy (Sabelli 1984, Brizzi et al. 1986). Sabelli (1985) refined the crystal structure for jurbanite. Rostite has had an eventful history because of the difficulty in characterizing it. Rost (1937) originally found it on burning coal heaps in Czechoslovakia, but reliable crystallographic data were not available until Cech (1979). However, there was found to be considerable F-OH substitution and it is now recognized that rostite refers to the compositional range OH>F, whereas khademite refers to F > OH.

The reason for apparent solubility equilibrium with respect to jurbanite for acid sulfate waters with pH generally less than 4.5 originated with the publications of van Breemen (1973) and Nordstrom (1982b). Van Breemen (1973) found good agreement between ion-activity products calculated from speciation computations of acid sulfate waters and an unidentified mineral that had the equilibrium stoichiometry of $Al(OH)SO_4$. Nordstrom (1982b), noting that there was a mineral with the same stoichiometry, whose solubility had been measured by Bassett and Goodwin (1949), reduced the solubility data to a solubility-product constant. Many investigators have used this information for chemical speciation modeling and the calculation of saturation indices. The agreement between the ion-activity product (IAP) for jurbanite and the solubility-product constant (K_{sp}), however, should be considered fortuitous and not indicative of solubility equilibrium for the following reasons. Jurbanite is a rare and very soluble efflorescent mineral. Bassett and Goodwin (1949) found that it was at equilibrium with a solution containing 15-20 wt % sulfate which would be equivalent to pH values of 0.0 or less, not pH values of 1 to 4.5. There must be some error in the K_{sp} value reported by Nordstrom (1982b) that shows its stability in pH values of 4 or less depending on sulfate concentrations. The solubility data of Bassett and Goodwin (1949) needs to be re-evaluated with the Pitzer. The fortuitous coincidence between the IAP and the K_{sp} most likely reflects the well-behaved correlation of increasing sulfate with increasing Al and with decreasing pH for pH < 4.5, the pH below which Al behaves conservatively (Nordstrom and Ball 1986). In other words, for pH values below the first hydrolysis constant for Al (pK = 5.0), Al remains predominately dissolved and any changes in concentration are caused by dilution in an identical manner as changes in sulfate concentration (see later section on Geochemical controls on formation and transformation: The Al system).

FORMATION AND DECOMPOSITION OF FE- AND AL-HYDROXYSULFATES OF LOW CRYSTALLINITY

Biological influences on mineral formation

The important role played by acidophilic bacteria in the aqueous oxidation of pyrite and related sulfide minerals was discussed previously. The $Fe^{2+}_{(aq)}$ released during this microbially catalyzed process is ultimately oxidized, hydrolyzed and precipitated as ferric oxyhydroxides or hydroxysulfates. At pH values <3.5, there is no doubt that the enzymatic oxidation of $Fe^{2+}_{(aq)}$ as an energy source for acidophilic organisms, such as *T. ferrooxidans* and *L. ferrooxidans*, is an essential requirement for the formation of secondary Fe^{III} minerals. Because little energy is produced by the oxidation reaction

(ΔG = 6.5 kcal/mol with O_2 as the electron acceptor at pH 2.5), these bacteria must oxidize large quantities of $Fe^{2+}_{(aq)}$ to sustain metabolic processes (Lees et al. 1969). Ehrlich (1996) has estimated that a "consumption" of 90.1 mol of $Fe^{2+}_{(aq)}$ is required to assimilate 1.0 mol of C. As a result, even a modest bacterial population is capable of oxidizing large quantities of $Fe^{2+}_{(aq)}$.

The extended role of microorganisms in guiding the actual "mineralization" of Fe, Al, and other dissolved species has been a topic of considerable debate. Mineral precipitates of Fe have been observed in direct contact with bacterial cells that grew in acid mine drainage (Ferris et al. 1989a,b; Clarke et al. 1997). Electron microscope images commonly show cells partly to completely encapsulated with Fe-rich epicellular material. This material has usually been identified as ferrihydrite (Ferris et al. 1989a) or goethite (Konhauser and Ferris 1996) but, in at least one case (Clarke et al. 1997), energy dispersive X-ray analyses have indicated the presence of ferric hydroxysulfates with Fe:S ratios ranging from 3.5:1 to 1.9:1. Bacterial surfaces have also been shown to enhance the immobilzation of other cations, especially Al (Urrutia and Beveridge 1995). Bacterial slimes may thus facilitate the undesirable accumulation of Al-hydroxysulfate precipitates and clogging of anoxic limestone drains designed for treating acid mine drainage (Robbins et al. 1999).

The formation of epicellular precipitates is related to the character of bacterial surfaces. The cell walls of both Gram-positive and Gram-negative bacteria contain functional groups, such as carboxylates and phosphates, that impart a negative surface charge and provide reactive sites for sorption of metal cations from the surrounding solution (Schultze-Lam et al. 1996). In natural environments, bacteria produce an extracellular sheath or capsule composed of polysaccharides with molecular components that are also capable of accumulating metals from around the cell. Once a metal is complexed, it effectively reduces the activation energy for nucleation of solid phases. In this way, the bacterial surface may function as a template for heterogenous nucleation (Konhauser 1998, Warren and Ferris 1998). Precipitates grow rapidly and incorporate available counter-ions from the surrounding solution to form mineral aggregates that may approach the mass of the cell. These initial precipitates are usually poorly crystalline because such phases have lower interfacial free energies than more crystalline products (Steefel and van Cappellen 1990).

It is tempting to conclude that surface-mediated mineralization is responsible for the formation of all poorly crystalline hydroxysulfates and oxyhydroxides of Fe and Al associated with acid sulfate waters and that mineral speciation is somehow biologically controlled, however, evidence suggests otherwise. There is probably little, if any, active genetic control on the process because mineralization can develop on the remains of dead cells (Ferris et al. 1989b) and occurs independently of cell morphology or physiological condition. Moreover, mineral phases can form spontaneously under conditions of rapid flow where acid sulfate waters are mixed with solutions of higher pH in the field. Geochemical parameters such as pH, SO_4, and HCO_3 must direct the ultimate mineralogical fate of Fe and Al compounds precipitated from acid sulfate waters, even when mineralization is bacterially induced. For example, Bigham et al. (1996a) identified mineral precipitates obtained by oxidizing 0.1 M $FeSO_4 \cdot 7H_2O$ solutions at room temperature and pH values of 2.3, 2.6, 3.0, 3.3 and 3.6 using a stirred bioreactor and a strain of *T. ferrooxidans* (Fig. 12). Measured oxidation rates were similar across the pH range studied. Schwertmannite formed at pH 3.0. Jarosite increased in abundance with decreasing pH, whereas goethite appeared at pH 3.3 and 3.6. These results indicate that pH is a master variable influencing the speciation of secondary Fe minerals under conditions typical of acid sulfate waters.

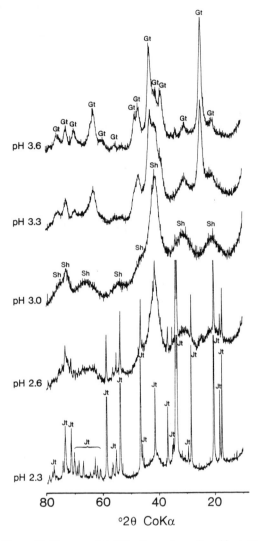

Figure 12. X-ray powder diffraction patterns from bioreactor
precipitates at various pH values.
Gt = goethite; Sh = schwertmannite; Jt = jarosite
Modified from Bigham et al. (1996a).

GEOCHEMICAL CONTROLS ON MINERAL FORMATION
AND TRANSFORMATION

If complete and reliable solution analyses have been obtained, geochemical
speciation computations can be performed to determine the state of saturation of an
aqueous system with respect to any particular mineral (provided that thermodynamic data
for mineral solubility reactions are available, see Nordstrom et al. 1990, Nordstrom and
May 1996). The usual goal of such computations is to understand the control of dissolved

major elements in terms of mineral solubilities. Such computations can be used to predict the formation of various mineral species on the basis of saturation indices (SI):

$$SI = \log (IAP/K_{sp}) \tag{11}$$

where IAP is the ion-activity product determined from observed solution concentrations after appropriate activity and speciation calculations are performed, and K_{sp} is the solubility-product constant (Nordstrom and Munoz 1994). Positive SI values indicate supersaturation and negative SI values indicate undersaturation. SI values plotted as a function of a relevant compositional variable, such as pH, should show a linear, horizontal trend close to zero when equilibrium saturation is reached with respect to the stoichiometry of a given mineral. SI calculations (and geochemical modeling in general) require reliable thermodynamic data and certain assumptions regarding chemical equilibrium. Both factors may be of concern when attempting to understand the formation of poorly crystalline Fe and Al hydroxysulfates from acid sulfate waters.

The Fe System

Numerous solutions from mines, mine tailings, and cat clays have been subjected to speciation calculations (van Breeman 1973, Nordstrom 1982a, Chapman et al. 1983, Nordstrom and Ball 1986, Karathanasis et al. 1988, Sullivan et al. 1988a,b; Blowes and Jambor 1990, Winland et al. 1991, Blowes et al. 1998, Nordstrom and Alpers 1999). In most instances, jarosite and ferrihydrite have been assumed to be the phases controlling Fe^{III} activities over relevant pH ranges. Saturation-index plots typically indicate supersaturation with respect to ferrihydrite at pH values above 4 for both surface and ground waters. An example of ferrihydrite supersaturation for about 200 samples collected from Wightman Fork and the Alamosa River system near Summitville, Colorado (Nordstrom et al. 2000) is shown in Figure 13. Supersaturation may be explained either by colloidal Fe particles that were not removed by filtering prior to conducting solution analyses, by incorrect identification of the solid phase, or by sampling artifacts.

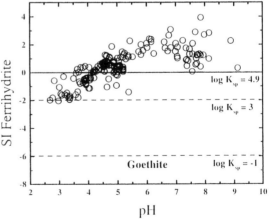

Figure 13. About 200 saturation indices for ferrihydrite plotted against pH for samples from the Summitville-Wightman Fork-Alamosa River system, Colorado.

The stoichiometry of the phase controlling the solubility of Fe can be predicted from an appropriate ion-activity plot. For example, if ferrihydrite is the phase controlling the solubility of Fe^{3+}, the reaction

$$Fe(OH)_{3(s)} + 3H^+_{(aq)} \leftrightarrow Fe^{3+}_{(aq)} + 3H_2O_{(l)} \tag{12}$$

and its log equilibrium constant expression

$$\log K = \log a_{Fe^{3+}} - 3\log a_{H^+} + 3\log a_{H2O} \tag{13}$$

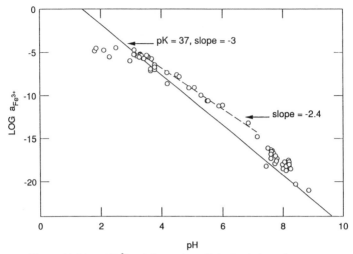

Figure 14. Plot of Fe^{3+} activity against pH for Leviathan mine waters (from Nordstrom 1991, Nordstrom and Alpers 1999). The solid line is equilibrium solubility for freshly precipitated ferrihydrite (log K_{sp} = 5.0) and the dashed line shows best fit of data between pH of 3.5 and 7.0 with a slope of -2.4.

would indicate that a plot of Fe^{3+} activity versus pH should have a slope of -3.0. Nordstrom (1991) reported a slope of -2.4 for Fe^{3+} data from mine waters at the Leviathan mine, CA, (Fig. 14). Similar results for mine drainage from St. Kevin Gulch, CO (Kimball et al. 1994), Korea (Yu et al. 1999) and numerous streams in Ohio (Bigham et al. 1996b) suggest that the relationship is fairly universal. The observed slope indicates that another anion, such as SO_4^{2-}, is replacing OH⁻ in the suspended mineral phase. This would be the case if schwertmannite is controlling the aqueous solubility of Fe^{3+} (Nordstrom 1991); however, the proposed compositional range for schwertmannite [$Fe_8O_8(OH)_{8-2x}(SO_4)_x \cdot nH_2O$ where $1 \leq x \leq 1.75$] should yield greater slopes, in the range of -2.75 to -2.56. A more likely scenario to explain solubility discrepancies is that mixtures of different Fe minerals, including schwertmannite, precipitate over the pH range represented by acid sulfate waters, and the slope is not clearly resolvable into a particular reaction.

On the basis of mineralogical and chemical analyses of ochreous precipitates and associated waters from Ohio mine drainage, Bigham et al. (1996b) proposed a solubility "window" of $\log IAP_{sh}$ = 18.0±2.5 for schwertmannite dissolution according to the reaction:

$$Fe_8O_8(OH)_x(SO_4)_{y(s)} + (24 - 2y) H^+_{(aq)} \leftrightarrow 8 Fe^{3+}_{(aq)} + y SO_4^{2-}_{(aq)} + (24 - 2y + x)/2 H_2O_{(l)} \quad (14)$$

and the log equilibrium expression

$$\log K_{sh} = 8 \log a_{Fe^{3+}} + y \log a_{SO_4^{2-}} + (24 - 2y) pH \quad (15)$$

Bigham et al. (1996b) also suggested that the discrepancy between observed and calculated slopes in pH vs. log $a_{Fe^{3+}}$ plots for acid sulfate waters was because the data define at least three slopes corresponding to the precipitation of different minerals over the full pH range examined. Mineralogical analyses of the precipitates indicated that those data in the pH range of 1.5 to 2.5 were influenced by the formation of jarosite, those above pH 5.5 by ferrihydrite, and those at intermediate pH values by

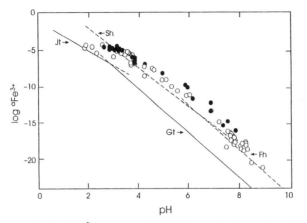

Figure 15. Plot of Fe^{3+} activity against pH for combined data from Ohio (closed circles) and Leviathan mine drainage waters (open circles) with solubility lines for Gt = goethite (log $a_{Fe^{3+}}$ = 1.40 – 3pH), Jt = K-jarosite (log $a_{Fe^{3+}}$ = -0.30 – 2pH), Fh = ferrihydrite (log $a_{Fe^{3+}}$ = 5.0 – 3pH), and Sh = schwert-mannite (log $a_{Fe^{3+}}$ = 2.67 – 2.60pH). Solubility lines calculated using an average log $a_{SO_4^{2-}}$ = -2.32, an average log a_{K^+} = -3.78 (from Nordstrom and Alpers 1999 and used with permission of Elsevier Science from Bigham et al. 1996b).schwertmannite (Fig. 15). Both schwertmannite and ferrihydrite were assumed to be metastable with respect to goethite.

The instability of schwertmannite with respect to goethite was confirmed in a long-term (1740 d) aqueous equilibrium study wherein a pure, synthetic specimen of schwertmannite was completely transformed to goethite over a period of 543 days via the overall reaction:

$$Fe_8O_8(OH)_{5.5}(SO_4)_{1.25(s)} + 2.5\ H_2O_{(l)} \leftrightarrow 8\ FeOOH_{(s)} + 2.5\ H^+_{(aq)} + 1.25\ SO_4^{2-}_{(aq)} \quad (16)$$

Solution measurements corroborated a decrease in pH and an increase in soluble SO_4^{2-} (Fig. 16). Levels of dissolved $Fe^{3+}_{(aq)}$ temporarily peaked at concentrations that were 0.5 to 1.0 orders of magnitude higher than those represented by the proposed equilibrium situation associated with goethite formation. This phenomenon may also contribute to the frequently reported supersaturation of acid sulfate waters with respect to Fe minerals.

Because the precipitation of schwertmannite may be controlled by not only the activities of H^+ and Fe^{3+}, but also of SO_4^{2-}, Yu et al. (1999) rewrote the dissolution reaction for schwertmannite as:

$$Fe_8O_8(OH)_{8-2x}(SO_4)_{x(s)} + (8 + 2x)\ H_2O_{(l)} \leftrightarrow 8\ Fe^{3+}_{(aq)} + 24\ OH^-_{(aq)} + x\ SO_4^{2-}_{(aq)} + 2x\ H^+_{(aq)}$$
$$(17)$$

so that the solubility relation becomes:

$$(pFe + 3pOH) = -x/8\ (pSO_4 + 2pH) + 1/8\ (pK_s + 24pK_w) \quad (18)$$

where K_w is the dissociation constant for water and K_s is the equilibrium constant for Reaction (17). Equation (18) then generates a line on a plot of (pFe + 3pOH) against (pSO$_4$ + 2pH) (Fig. 17).

The aqueous geochemistry of Fe is further complicated by the fact that Fe is a reduction–oxidation (redox) active species. The traditional method for delineating the

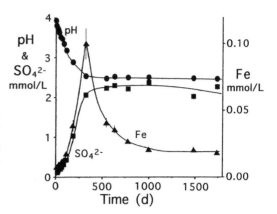

Figure 16. Changes in pH, SO$_4$ and Fe with time for a beginning suspension of synthetic schwertmannite in water. Goethite with log K$_{Gt}$ = 1.40±0.01 was the final mineral product (modified from Bigham et al. 1996b).

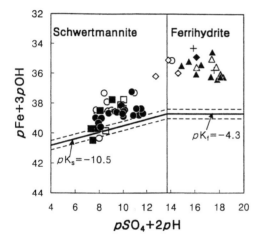

Figure 17. Plot of (pFe + 3pOH) vs. (pSO$_4$ + 2pH) for mine drainage waters from Korea (filled symbols) and Ohio (open symbols); pK$_s$ and pK$_f$ indicate proposed solubility products for schwertmannite and ferrihydrite, respectively. Dashed lines indicate proposed error term (±2.5) in the solubility product for schwertmannite (modified from Yu et al. 1999).

stabilities of such species in geochemical systems has been through the use of Eh–pH (or pe–pH) diagrams. Although such diagrams can be very useful for showing general stability relationships among minerals containing redox-sensitive ions, they are also subject to many limitations, as described by Nordstrom and Alpers (1999).

Iron speciation in acid sulfate waters is typically described using diagrams for the Fe–S–K–O$_2$–H$_2$O system in which fields for jarosite, goethite, and water are stable phases over relevant pe–pH ranges. It is important to note that stability boundaries between minerals can vary significantly with crystallinity, particle size, and composition (as is reflected in the range of the solubility products reported for both jarosite and goethite). In addition, metastable phases such as schwertmannite and ferrihydrite may form more readily than stable phases, thereby imparting kinetic controls to the system. Using reported log K$_{sp}$ values for jarosite and ferrihydrite, a measured log K$_{sp}$ for goethite, and an apparent log K$_{sp}$ for schwertmannite, Bigham et al. (1996b) proposed a pe–pH diagram for acid sulfate waters with fields of metastability for schwertmannite and ferrihydrite (Fig. 18). As is often the case, this diagram places constraints on the total log activities of Fe^{2+}, Fe^{3+}, K^+, and SO_4^{2-}. According to Figure 18, jarosite and goethite ultimately control the solubility of Fe in high-pe, acid sulfate waters, and schwertmannite is unstable relative to goethite over the pH range 2 to 6. At higher pH values, ferrihydrite precipitation is kinetically favored, but ferrihydrite is also unstable with respect to goethite.

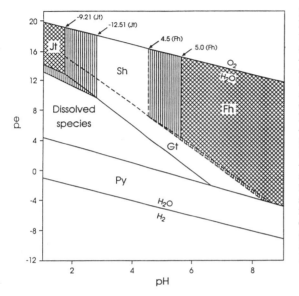

Figure 18. pe-pH diagram for Fe–S–K–O–H system at 25°C where pe = Eh(mV)/59.2, total log activities of Fe^{2+} = –3.47, Fe^{3+} = –3.36 or –2.27, SO_4^{2-} = –2.32, K^+ = –3.78. log K_{sp} values are given in Figure 13. Jt = K-jarosite, Sh = schwertmannite, Fh = ferrihydrite, Gt = goethite, and Py = pyrite. Line equations are: Gt (pe = 17.9 – 3pH), Jt (pe = 16.21 – 2pH), Fh (pe = 21.50 – 3pH), Sh (pe = 19.22 – 2.6pH), and Py (pe = 5.39 – 1.14pH). Fields of metastability shown by dashed lines. Single-hatched areas demonstrate expansion of K-jarosite and ferrihydrite fields if lower log K values are chosen from available literature. (used with permission of Elsevier Science from Bigham et al. 1996b).

Field data supporting a paragenetic relationship between jarosite, schwertmannite, ferrihydrite, and goethite have been obtained from a naturally acid alpine stream originating as a spring in talus from a pyritic schist (Schwertmann et al. 1995). Although the spring water (pH 2.3–2.8) was visibly clear, the saturated talus was laden with jarosite, indicating that its precipitation was limited by the release of K^+ from micas occurring in the schist. Downslope, the spring water merged with multiple, non-acid tributaries and produced a strong geochemical gradient. At the first confluence (pH 3.0), schwertmannite precipitated with some goethite. Further dilution down the watercourse eventually raised the solution pH to 7.0 and produced a gradual increase in the proportions of goethite and ferrihydrite in the sediments.

In addition to the Fe^{III} minerals represented in Figure 18, lepidocrocite has been reported as a product of acid sulfate weathering (e.g. Blowes et al. 1991, Milnes et al. 1992). Whereas most Fe^{III}oxides can be synthesized from either Fe^{II} or Fe^{III} salts, lepidocrocite appears to form almost entirely by oxidation of Fe^{II}. It has been identified as a direct pseudomorph after pyrite (Romberg 1969) and pyrrhotite (Jambor 1986), but its formation from Fe^{2+} solutions seems normally to pass through the blue-green (Fe^{II}, Fe^{III})-hydroxy compounds known as green rusts (Schwertmann and Fechter 1994, Lewis 1997). The green rusts have a pyroaurite structure consisting of sheets of $Fe^{II}(OH)_6$ octahedra in which some of the Fe^{II} is replaced by Fe^{III} thereby creating a positive layer charge that is balanced by anions (primarily SO_4^{2-}, Cl^-, and CO_3^{2-}) located between the octahedral sheets. Green rusts are common corrosion products of steel, but they have rarely been identified as occurring in soils (Trolard et al. 1997) or sediments (Bender Koch and Mørup 1991) because they rapidly oxidize to form either goethite or lepidocrocite through competing reactions that are strongly influenced by solution parameters. Factors that appear to suppress lepidocrocite in favor of goethite formation under laboratory conditions are low pH (<5) and the presence of excess dissolved SO_4^{2-}, HCO_3^-, or Al^{3+} (Schwertmann and Thalmann 1976, Taylor and Schwertmann 1978, Carlson and Schwertmann 1990). If similar factors affect natural systems, the precipitation of lepidocrocite from acid sulfate waters seems unlikely under most conditions.

The Al system

An approach similar to that for Fe has been used for interpreting the behavior of Al in acid sulfate systems, but with somewhat different results. Several plots have been made for log $a_{Al^{3+}}$ against pH with results indicating a solubility control by microcrystalline gibbsite (e.g. Driscoll et al. 1984, Hendershot et al. 1996) for waters having a pH greater than about 4.5. At lower pH values, a clear deviation from this trend is observed (e.g. Losno et al. 1993). Using water samples over a wide range of pH, Al concentrations, and sulfate concentrations from a single watershed in speciation calculations Nordstrom and Ball (1986) showed that two trends are apparent, with a change in slope close to the first hydrolysis constant for Al ($pK_1 = 5.0$) as shown in Figure 19. When the pH is significantly less than pK_1 then Al hydrolysis is insignificant; the Al is little affected by geochemical reactions, and it travels in surface-water systems as if it were a conservative constituent. Once the pH of the water reaches the pK_1, hydrolysis causes the Al to become insoluble and precipitate. Although Neal (1988) has argued that plots of log $a_{Al^{3+}}$ against pH are autocorrelated because pH is used to calculate the activity of free Al so that the plot looks better than it should, this argument is flawed. It is clear from a plot of dissolved Al against dissolved sulfate that the correlation (indicative of conservative solute transport) is excellent for pH values <4.5 and that for higher pH values a sudden decrease in Al concentrations is observed (Nordstrom and Ball 1986). Plotting the saturation indices against specific conductance

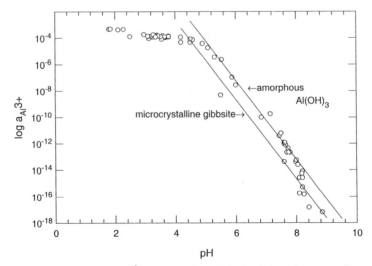

Figure 19. Plot of the log Al^{3+} activity against pH for Leviathan mine waters (from Nordstrom and Alpers 1999). Solid lines show equilibrium solubilities of microcrystalline and amorphous $Al(OH)_3$ to which the data fits well, suggesting an equilibrium solubility control.

or sulfate concentrations or viewing a histogram plot (Nordstrom and Ball 1989) results in the same conclusion. A similar trend can be seen in dilute acid sulfate waters from other watersheds; a distinct change is seen in the behavior of Al from conservative to non-conservative at the same pH value and a similar slope but with a different intercept (May and Nordstrom 1991). It is well known that the first hydrolysis constant greatly affects the chemical behavior and separation of metals (Hem and Roberson 1990, Baes and Mesmer 1976, Rubin 1974), which also explains the tendency for the separation of Al ($pK_1 = 5.0$) and Fe^{III} ($pK_1 = 2.2$).

Solubility data and solubility-product constants for alunite (log K_{sp} = -85.4), crystalline basaluminite (log K_{sp} = 117.7), and amorphous basaluminite (log K_{sp} = -116.0) at 25°C have been determined by Adams and Rawajfih (1977) and seem reasonably consistent with other data (Nordstrom 1982b).

ENVIRONMENTAL IMPLICATIONS OF TRACE ELEMENT SORPTION

Mining activities usually involve ore or coal deposits that have accumulations of such elements as Ag, As, Au, Cd, Co, Cu, Hg, Mo, Ni, Pb, Se, Sb, Tl, and Zn in either economic or sub-economic quantities. These elements occur as mineral sulfides or sulfosalts (sometimes as solid solution impurities) or in gangue minerals. When acid drainage waters are produced by oxidation of mining residues, these elements may be mobilized and transported offsite at concentrations that are sufficiently elevated to be of environmental concern. In such cases, the precipitation of hydroxysulfates and oxyhydroxides of Fe and Al may attenuate the problem by scavenging associated trace elements through processes of adsorption and/or coprecipitation (hereafter referred to collectively as sorption). Although cleansing of attendant drainage waters is a desirable effect, the sediments have a potential for pollution through changing redox conditions and their long-term stability in downstream environments may be a concern.

A substantial body of information has accumulated on the chemical and microbial composition of waters and sediments affected by acid mine drainage, and the reader is referred to recent reviews by Singh et al. (1997), Nordstrom and Southam (1997), Plumlee et al. (1999), Nordstrom and Alpers (1999), and Nordstrom (2000) for a summary of this literature. Smith (1999) has reviewed sorption, especially as it pertains to metal attenuation in mine waste environments. It can generally be concluded that most trace elements, including both cations and oxyanions, tend to accumulate in mine drainage sediments at the expense of associated solutions. Concentration factors vary widely depending on local geochemical and hydrologic conditions. Fuge et al. (1994) examined water and ochre deposits from several metal-sulfide mines in Wales and concluded that As, Mo, Ag, Sn, Sb, Hg, Ba, Tl, Pb, and Bi were strongly concentrated into the solid phase, whereas, other common trace metals (e.g. Mn, Co, Ni, Cu, Zn, Cd) showed less segregation (Fig. 20). In the case of Ba and Pb, metal solubilities may have been controlled by the precipitation of barite ($BaSO_4$) and anglesite ($PbSO_4$), respectively. Other elements were thought to have accumulated primarily by sorption to Fe compounds comprising the ochreous precipitates.

Compared to drainage waters from ore bodies like those in Wales, most drainage waters from coal mines are substantially lower in ionic strength and have lower concentrations of trace contaminants. Nevertheless, significant partitioning of trace elements into Fe and Al precipitates from these waters also occurs. For example, Winland et al. (1991) examined trace-element distributions between solid and aqueous phases at 28 sites in the Ohio coalfield by the use of distribution coefficients (K_d) calculated as:

$$K_d = (\text{mol } X \text{ kg}^{-1} \text{ solid})/(\text{mol } X \text{ L}^{-1} \text{ stream water}) \tag{19}$$

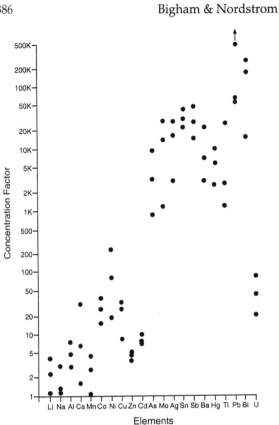

Figure 20. Relationship between ochre and water chemistry in acid mine drainage from Wales; concentration factor = [element conc. in ochre (mg/kg) / element conc. in water (mg/L)] (modified from Fuge et al. 1994).

Table 6. Minimum, maximum and mean distribution coefficients $(K_d)^†$ for trace elements from 28 coal mine drainage sites in Ohio.[‡]

Element	Min K_d	Max K_d	Mean K_d	SD^*	N^{**}
As	5.8	6.7	6.3	6.1	9
B	1.8	4.0	3.5	3.3	27
Ba	3.3	5.9	4.9	5.2	27
Co	0.2	3.1	2.3	2.5	21
Cu	2.2	3.4	2.9	2.8	15
Cr	2.5	3.3	3.1	2.8	10
Mn	0.1	4.1	2.9	3.4	28
P	3.0	5.2	4.6	4.6	28
Zn	1.2	3.7	2.9	3.1	27

[†] modified from Winland et al. (1991)
[‡] $K_d = (mol\ X\ kg^{-1}\ solid)/(mol\ X\ L^{-1}\ drainage\ water)$
*SD = standard deviation
**n = number of samples included in the analysis

where X is the trace element of interest. As in Fuge et al. (1994), the mean distribution coefficients were determined to be highest for As and Ba and lowest for Co, Cu, Mn and Zn (Table 6).

Bulk chemical analyses of mine drainage waters and sediments are valuable, but they provide little insight into either geochemical factors that control trace-element sorption or the behavior of specific mineral sorbents that may be involved in this process. Relatively few attempts have been made to sort out such interactions in mine drainage environments because of the heterogeneous character of both the mine effluents and the associated chemical precipitates of Fe and Al. An understanding of relevant processes must be based primarily on studies of well-defined systems through experiments completed under controlled conditions. To that end, the adsorption of foreign species at metal-oxide surfaces, especially the Fe and Al oxides, is a subject of widespread interest because of industrial applications and the environmental implications of this process. Consequently, a huge literature has accumulated on this topic over the past few decades. The reader is referred to a number of comprehensive texts and reviews for general information on this subject (e.g. Sposito 1984, Hochella and White 1990, Dzombak and Morel 1990, Stumm 1992, Brown et al. 1999).

Interest in the interfacial chemistry of Fe and Al oxides stems largely from the fact that their surface charge is pH-dependent. Thus, H^+ and OH^- are potential-determining ions whose exchange at the surface is responsible for establishing the net surface charge. Each surface also has a characteristic point of zero charge (pzc), the pH value at which the net surface charge is zero. The net surface charge is positive at pH values below the pzc but becomes negative at pH values higher than the pzc. Experimentally determined pzc values of the Fe and Al oxides, hydroxides, and oxyhydroxides typically range between 5.0 and 9.0 (Parks 1965, 1967; Stumm 1992).

Sorption of metal cations

The development of surface complexes between foreign metal cations and the surface of a hydrous Fe or Al oxide involves coordination of the foreign ion with oxygen donor atoms to form either monodentate or bidentate complexes, with the release of protons as per:

$$—Fe–OH + Pb^{2+} \leftrightarrow —Fe–OPb^+ + H^+ \tag{20}$$

or

$$\begin{array}{cc} —Fe–OH & —Fe–O \\ | \quad + Pb^{2+} \leftrightarrow & | \quad \diagdown Pb + 2\,H^+ \\ —Fe–OH & —Fe–O \diagup \end{array} \tag{21}$$

The complexes are considered to be inner-sphere if a mostly covalent bond is formed between the metal and the electron-donating oxygen ions. Otherwise, the complex is outer-sphere, with solute molecules providing separation between the metal and oxide surface. As might be expected, the attraction and binding of a foreign metal ion to an oxide surface is strongly pH dependent and is favored by pH values that produce a net negative surface charge (above the pzc). The formation of inner-sphere complexes with foreign metal ions effectively shifts the pzc of the oxide surface to higher pH values. As shown by Dzombak and Morel (1990), there is a narrow interval of 1-2 pH units where the extent of sorption for each metal on a hydrous iron oxide (ferrihydrite) surface rises from zero to almost 100%, thereby defining a sorption "edge" (Fig. 21a).

Evidence for the pH-dependence of metal sorption to mine-drainage precipitates and for the existence of metal-specific sorption edges is shown in Figure 22 and Table 7. The former (Karlsson et al. 1987) presents the distribution of Pb, Cu, Cd, and Zn between suspended particles and the solution phase in a stream receiving leachates from a deposit

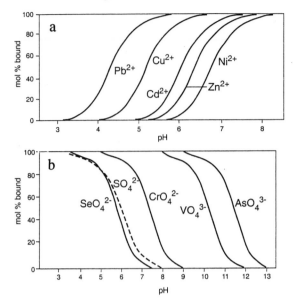

Figure 21. Binding of (a) metal ions and (b) oxyanions to ferrihydrite as a function of pH (modified from Dzombak and Morel 1990).

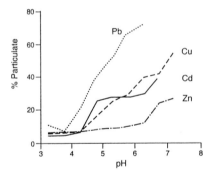

Figure 22. Distribution of metals between suspended precipitates and the solution phase as a function of pH in a stream affected by mine drainage (modified from Karlsson et al. 1987).

Table 7. Concentration factors[†] for selected metals in Welsh ochre deposits as a function of solution pH (modified from Fuge et al. 1994).

Locality[‡]	1	2	3	4	5	6
pH	5.0	3.5	2.9	2.5	2.4	2.3
	----------------------------Concentration factor----------------------------					
Mn	200	51	275	4.3	1.1	2.7
Cu	307,000	19,500	7,380	384	169	111
Zn	382	160	14	3.8	5.2	4.6
Cd	381	333	34	7.6	10	7.6

[†] Concentration factor = [conc. of element in ochre (mg/Kg) / conc. in water (mg/L)].
[‡] Localities: 1 = Y Fan mine; 2 = Gwynfynedd mine; 3 = Cwmrheidol mine; 4,5,6 = Parys Mountain.

of mine tailings. The data show a significant transfer of contaminant species from the solution phase to suspended particulates with increasing downstream pH in the order Pb > Cu ~ Cd > Zn. The data in Table 7, from Fuge et al. (1994), give concentration factors for selected metals (Mn, Cu, Zn, Cd) in ochre and water samples collected at different pH values from a number of mining localities in Wales. Note that the concentration factors for all metals increase with increasing pH and that the concentration factor for Cu at pH 5 is significantly higher than for Cd and Zn, as would be predicted from the sorption edges in Figure 20a. A number of additional studies (e.g. Filipek et al. 1987, Rampe and Runnels 1989, Webster et al. 1994, Schemel et al. 2000) have demonstrated that Mn and Zn behave as conservative ions at pH < 6 in surface waters affected by mine drainage.

At this point, no information is available concerning the surface-charge properties of schwertmannite or a theoretically corresponding Al hydroxysulfate precipitate from acid sulfate waters. Recently, however, Webster et al. (1998) compared the adsorption of Cu, Pb, Zn, and Cd on synthetic specimens of schwertmannite and 2-line ferrihydrite, as well as a natural mine drainage precipitate composed mostly of goethite. They found that the adsorption edges of all four metals occurred at a lower pH with the natural precipitate; that edges for Pb and Cd were similar for the two synthetic materials; and that the edges for Zn and Cu occurred at a lower pH for schwertmannite as compared to ferrihydrite. Removal of adsorbed sulfate from schwertmannite and the AMD precipitate by titration with $Ba(NO_3)_2$ before conducting the sorption experiments generally decreased the sorption of Pb, Cu, and Zn indicating that adsorbed sulfate may alter electrostatic conditions at the mineral surface or lead to the formation of ternary surface complexes. In another study, it was found that Al precipitates had little affect on metal removal, attenuating Cu slightly and Zn and Ni not at all (Rothenhofer et al. 1999). The only known significant uptake on Al colloids seems to be rare-earth elements (Auque et al. 1993, Gimeno et al. 1994, Nordstrom et al. 1995, Landa et al. 2000).

Sorption of oxyanions

The process of oxyanion accumulation at oxide surfaces can also be described in terms of sorption edges (Dzombak and Morel 1990). Unlike metal cations, however, the sorption of oxyanions decreases with increasing pH because of competition with OH$^-$ and electrostatic repulsion by the negatively-charged oxide surface at pH values above the pzc (Fig. 21b). As with metal cations, the surface complexes formed by oxyanion sorption may be inner-sphere or outer-sphere in character. Inner-sphere sorption may result in both monodentate and bidentate surface complexes, and the latter may be bidentate mononuclear or bidentate binuclear as per:

$$—FeOH^+ + H_2AsO_4^- \leftrightarrow —Fe\begin{matrix} O \\ \diagdown \diagup \\ O \end{matrix} As\begin{matrix} OH \\ \diagup \\ \diagdown \\ O \end{matrix} + H_2O \tag{22}$$

$$\begin{matrix} —FeOH \\ | \\ —FeOH \end{matrix} + H_2AsO_4^- \leftrightarrow \begin{matrix} —Fe—O \\ | \\ —Fe—O \end{matrix} As\begin{matrix} O^- \\ \diagup \\ \diagdown \\ O \end{matrix} + 2H_2O \tag{23}$$

Trivalent oxyanions, such as PO_4^{3-}, AsO_4^{3-}, and VO_4^{3-}, are much more aggressive in forming inner-sphere complexes as compared to their divalent counterparts (e.g. SO_4^{2-} and SeO_4^{2-}). As a result, the trivalent species yield sorption edges that are non-reversible.

Sulfate is clearly the dominant anion in most acid sulfate waters; therefore, its behavior at the mineral–water interface is important to an understanding of surface-mediated reactions in these systems. There has been considerable controversy regarding the inner-sphere vs. outer-sphere character of sorbed SO_4 (see Myneni, this volume). Macroscopic experiments involving SO_4 sorption on reference minerals and soils have generally been consistent with the formation of outer-sphere complexes because little retention occurs above the mineral pzc, and the ionic strength of the aqueous phase has a large effect on the amount of SO_4 accumulated (e.g. Charlet et al. 1993). Studies involving spectroscopic techniques have generally been less conclusive. For example, early transmission IR studies of SO_4 sorption on iron oxides (e.g. Parfitt and Smart 1978, Harrison and Berkheiser 1982) supported the formation of binuclear bridging surface complexes. More recently, Persson and Lövgren (1996) concluded that the adsorption of SO_4 on goethite was outer-sphere on the basis of results from diffuse reflectance infrared (DRIFT) spectroscopy. In contrast, Hug (1997) identified mainly monodentate, inner-sphere complexes of SO_4 on hematite surfaces between pH 3 and 5 by using *in situ* attenuated total reflection (ATR)-FTIR methods. Peak et al. (1999), employing similar techniques, concluded that SO_4 formed both outer-sphere and inner-sphere surface complexes on goethite at pH < 6. At higher pH values, SO_4 adsorption was entirely outer-sphere.

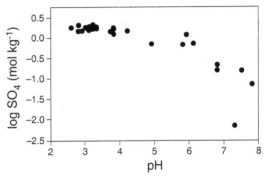

Figure 23. Relationship between solution pH and the concentration of SO_4 in mine drainage precipitates from Ohio (modified from Winland et al. 1991).

Ochreous precipitates from coal-mine drainage in Ohio contain 0.1 to 20 wt % SO_4, and a general decrease in solid-phase SO_4 content was noted with increasing pH of the source waters (Fig. 23) (Winland et al. 1991). Molar Fe/S ratios for mine drainage ochres formed at pH < 5 in the Appalachian region have been reported to range between 3.5 and 10 (Karathanasis and Thompson 1995, Bigham et al. 1996b). Rose and Ghazi (1997) examined the release of SO_4 from such materials in batch experiments involving equilibration of sediments with solutions of various pH values and containing a variety of exchange ligands (nitrate, chloride, bicarbonate, oxalate, phosphate). A significant portion of the total SO_4 load was determined to be labile; at neutral pH (with OH^- as the only exchange ligand), 33-50% of the total was released to solution. Similar results were obtained at neutral pH in the presence of nitrate and chloride, but higher quantities (50-68%) were mobilized when bicarbonate and phosphate were present. The overall results indicated that about 70% of the total sulfate was associated with specific binding sites; that is, as inner-sphere complexes on mineral surfaces or as a structural component in minerals such as schwertmannite and jarosite. Because of similarities in size and valence, the binding characteristics of SeO_4^{2-} should be similar to those of SO_4^{2-} (Waychunas et al. 1995).

Next to SO_4, the most common oxyanions in acid sulfate waters are probably arsenate (AsO_4^{3-}) and arsenite (AsO_3^{3-}) because of the common occurrence of As as an impurity in pyrite and as a stoichiometric component of such minerals as arsenopyrite (FeAsS) and cobaltite (CoAsS). Accumulations of As have been noted in numerous acid sulfate soils (Dudas 1987, Bowell 1994, Bowell et al. 1994), mine tailings (Moore and Luoma 1990), and mine drainage precipitates (Chapman et al. 1983, Johnson and Thornton 1987, Webster et al. 1994, Bowell and Bruce 1995, Kimball et al. 1995, Hudson-Edwards et al. 1999). Arsenic is extremely toxic to humans, it is a known carcinogen, and it may cause severe health effects even at low concentrations (Morton and Dunnette 1994). Consequently, its bioavailability in soils, sediments and acid leach waters is a major environmental concern.

At pH values <2.0 and AsO_4 concentrations >0.01 molal, the aqueous solubility of AsO_4 may be controlled by the precipitation of scorodite [$FeAsO_4 \cdot 2H_2O$] (Robbins 1990) or its "amorphous" counterpart ($FeAsO_4 \cdot xH_2O$) (Krause and Ettel 1988, Tuovinen et al. 1994). At higher pH values and in solutions with lower AsO_4 activity, the sorption of AsO_4^{3-} at oxide surfaces is thought to be a more important mechanism controlling dissolved As levels. Consequently, As sorption has been widely studied in relation to pH, background electrolyte, and oxidation state (arsenate vs. arsenite) as major experimental variables (e.g. Anderson et al. 1976, Pierce and Moore 1982, Goldberg 1986, Raven et al. 1998). Co-precipitation phenomena involving As and Fe have also been examined by hydrometallurgists as a means of removing As from process wastewaters (e.g. Robins et al. 1988, Papassiopi et al. 1988). Coprecipitation of AsO_4 and Fe^{III} to form ferrihydrite apparently yields smaller particles and higher As site densities than when AsO_4 is adsorbed on pre-formed ferrihydrite (Waychunas et al. 1993, Fuller et al. 1993). By contrast, coprecipitation of AsO_4 destabilizes the structure of schwertmannite and poisons crystal growth (Waychunas et al. 1995).

Mineral instability and possible affects on sorbed species

Laboratory studies of the aging of schwertmannite have shown it to be unstable with respect to goethite (Bigham et al. 1996b), and similar instability has recently been demonstrated under field conditions (Brill 1999, Peine et al. 2000). The transformation gives rise to increased proton and sulfate activities in associated solutions and would presumably have a dramatic affect on sorbed contaminants. However, data to this effect are currently lacking. Variations in dissolved arsenate in a contaminated stream have been coupled to diurnal fluctuations in pH arising from changes in photosynthetic activity as evidenced by the coincidence of pH, arsenate concentrations, and light-intensity maxima (Fuller and Davis 1989). Increases of up to 0.5 pH units during daylight hours apparently caused desorption of arsenate from iron oxyhydroxides in the streambed. The photoreduction of dissolved Fe^{3+} and colloidal Fe minerals may also have a significant impact on the concentrations of Fe^{2+} in oxygenated surface waters. McKnight et al. (1988) found that dissolved Fe^{2+} in a stream impacted by mine drainage reached a maximum during midday. Daytime production of Fe^{2+} occurred at rates that were nearly four times faster than the night–time oxidation of Fe^{2+}. An opposite effect has been observed in mine drainage wetlands, where Fe^{2+} concentrations reached a minimum during daylight hours, apparently due to daytime oxygenation by algae (Wieder 1994). In either case, exposure of acid sulfate waters to the sun could promote cycling of Fe between dissolved and particulate phases with a release of sorbed contaminant species. Alternatively, the generation of fresh mineral surfaces might provide an increased opportunity for sorption of trace elements (McKnight and Bencala 1989).

SUMMARY

Acid sulfate waters provide a complex biogeochemical stage for numerous reactions involving the weathering, formation, and transformation of minerals. Important secondary precipitates from such waters include hydroxysulfates of Fe (schwertmannite) and Al (primarily hydrobasaluminite). The environmental significance of these compounds stems from the fact that they (1) form spontaneously under pH regimes that are common to most acid sulfate waters; (2) have very short-range structural order which translates to high specific surface area and chemical reactivity with respect to dissolved contaminants; and (3) are metastable with respect to more common oxyhydroxides of Fe and Al. Because of their poor crystallinity, much remains to be learned concerning both their properties and occurrence. When admixed with other minerals commonly found in secondary precipitates from acid sulfate waters, the hydroxysulfates of Fe and Al can be difficult to identify. Considering pH as a master variable, schwertmannite seems to form primarily over the pH range of 2.5-4.5, whereas Al hydroxysulfate is strongly linked to pH values in the range of 4.5-5.5 (Fig. 24). For this reason, aluminous precipitates form as colloids separate from the Fe precipitates and there is little, if any, mutual substitution of Fe in the Al hydroxysulfates and vice versa. Minerals in the jarosite group generally form at lower pH (less than 2.5), and ferrihydrite (perhaps with microcrystalline gibbsite) is a common component of ochreous precipitates formed in the pH range of 6-8. Both schwertmannite and ferrihydrite are metastable with respect to goethite. Minerals such as hydrobasaluminite and aluminite are metastable with respect to alunite and gibbsite.

Eastern U.S. Coal Mine Drainage

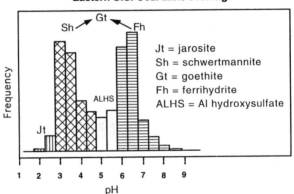

Figure 24. Schematic showing the distribution of common minerals in mine drainage precipitates as a function of pH (modified from Cravotta et al. 1999).

More research is needed to define/refine the chemical and mineralogical properties of schwertmannite and Al hydroxysulfates. The formation of both consumes substantial amounts of sulfate, but the structural role of sulfate has not been rigorously defined in either. Almost nothing is known about the surface characteristics of these minerals or how surface reactivity may vary with changes in environmental parameters. Relevant thermodynamic data have been developed to only rudimentary levels, and information regarding the possiblity of structural solid solution by other metals or oxyanions is lacking. More research is needed to understand better the processes of biomineralization as they relate to both the formation and transformation of schwertmannite and Al hydroxysulfate.

REFERENCES

Adams F, Hajek BF (1978) Effects of solution sulfate, hydroxide, and potassium concentrations on the crystallization of alunite, basaluminite, and gibbsite from dilute aluminum solutions. Soil Sci 126:169-173

Adams F, Rawajfih Z (1977) Basaluminite and alunite: a possible cause of sulfate retention by acid soils. Soil Sci Soc Am J 41:686-692

Alpers CN, Rye RO, Nordstrom DK, White LD, King BS (1992) Chemical, crystallographic, and isotopic properties of alunite and jarosite from acid hypersaline Australian lakes. Chem Geol 96:203-226

Alpers CN, Blowes DW, Nordstrom DK, Jambor JL (1994) Secondary minerals and acid mine-water chemistry. *In* The Environmental Geochemistry of Sulfide Mine-Wastes. Jambor JL, Blowes DW (Eds) Mineral Assoc Canada Short Course 22:249-270

Alvarez E, Perez A, Calvo R (1993) Aluminum speciation in surface waters and soil solutions in areas of sulfide mineralization in Galicia. Sci Total Environ 133:17-37

Anderson MA, Ferguson JF, Gavis J (1976) Arsenate adsorption on amorphous aluminum hydroxide. J Colloid Interface Sci 54:391-399

Anthony JW, McLean WJ (1976) Jurbanite, a new post-mine aluminum sulfate mineral from San Manuel, Arizona. Am Mineral 61:1-4

Ash SH, Felegy EW, Kennedy DO, Miller PS (1951) Acid Mine Drainage Problems—Anthracite Region of Pennsylvania. US Bur Mines Bull 508

Auqué LF, Tena JM, Gimeno MJ, Mandado J, Zamora A, López-Julian P (1993) Distribución de tierras raras en soluciones y coloides de un sistema natural de aguas ácidas (Arroyo del Val, Zaragoza). Est Geol 49:41-48

Baes CF Jr, Mesmer RE (1976) The Hydrolysis of Cations. Wiley-Interscience, New York

Ball JW, Nordstrom DK (1989) Final revised anaylyses of major and trace elements from acid mine waters in the Leviathan mine drainage basin, California and Nevada—October 1981 to October 1982. US Geol Survey Water-Res Invest Peport 89-4138, 49 p

Bannister FA, Hollingworth SE (1948a) Two new British minerals. Nature 162:565

Bannister FA, Hollingworth SE (1948b) Basaluminite hydrobasaluminite. Am Mineral 33:787

Barham RJ (1997) Schwertmannite: A unique mineral, contains a replaceable ligand, transforms to jarosites, hematite, and/or basic iron sulfate. J Mater Res 12:2751-2758

Barnhisel RI, Daniels WL, Darmondy RG (2000) Reclamation of Drastically Disturbed Lands. Am Soc Agron Monograph No 41. Am Soc Agron, Madison, Wisconsin

Barton P (1978) The acid mine drainage. *In* Sulfur in the Environment, Part II. Ecological Impacts. Nriagu JO (Ed) John Wiley and Sons, New York, p 313-358

Batik P, Hruskova J (1971) Hydrobasaluminite and basaluminite from Nikolcice near Brno. Czech Sb Nar Muz Praze Rada B 27:9-16

Bassett H, Goodwin TH (1949) The basic aluminum sulphates. J Chem Soc:2239-2279

Bender Koch C, Mørup S (1991) Identification of green rust in an ochre sludge. Clay Minerals 26:577-582

Bertsch PM, Parker BR (1996) Aqueous polynuclear aluminum species. *In* The Environmental Chemistry of Aluminum. Sposito G (Ed) CRC/Lewis Publishers, Boca Raton, Florida, p 117-168

Beukes GJ, Schoch AE, De Bruiyn H, Van der Westhuizen WA, Bok LDC (1984) A new occurrence of the hydrated aluminum sulfate zaherite, from Pofadder, South Africa. Mineral Mag 48:131-135

Bhatti TM, Bigham JM, Carlson L, Tuovinen OH (1993) Mineral products of pyrrhotite oxidation by *Thiobacillus ferrooxidans*. Appl Environ Microbiol 59:1984-1990

Bhatti TM, Bigham JM, Vuorinen A, Tuovinen OH (1994) Alteration of mica and feldspar associated with the microbial oxidation of pyrrhotite and pyrite. *In* The Environmental Geochemistry of Sulfide Oxidation. Alpers CN, Blowes DW (Eds) Am Chem Soc Symp Series 550: 90-105

Biedermann G, Chow TJ (1966) Studies on the hydrolysis of metal ions 57. The hydrolysis of the iron (III) ion and the solubility product of $Fe(OH)_{2.7}Cl_{0.3}$ in 0.5 M $(Na^+)Cl^-$ medium. Acta Chem Scand 20:1376-1388

Biedermann G, Schindler P (1957) On the solubility product of precipitated iron (III) hydroxide. Acta Chem Scand 11:731-740

Bigham JM (1994) Mineralogy of ochre deposits formed by sulfide oxidation. *In* The Environmental Geochemistry Of Sulfide Mine-Wastes. Jambor JL, Blowes DW (Eds) Mineral Assoc Canada Short Course 22:103-132

Bigham JM, Schwertmann U, Carlson L, Murad E (1990) A poorly crystallized oxyhydroxysulfate of iron formed by bacterial oxidation of Fe(II) in acid mine waters. Geochim Cosmochim Acta 54:2743-2758

Bigham JM, Schwertmann U, Carlson L (1992) Mineralogy of precipitates formed by the biogeochemical oxidation of Fe(II) in mine drainage. *In* Biomineralization Processes of Iron And Manganese—Modern and Ancient Environments. Skinner HCW, Fitzpatrick RW (Eds) Catena Supplement 21:219-232, Cremlingen-Destedt

Bigham JM, Carlson L, Murad E (1994) Schwertmannite, a new iron oxyhydroxysulfate from Pyhäsalmi, Finland, and other localities. Mineral Mag 58:641-64

Bigham JM, Schwertmann U, Pfab G (1996a) Influence of pH on mineral speciation in a bioreactor simulating acid mine drainage. Appl Geochem 11:845-849

Bigham JM, Schwertmann U, Traina SJ, Winland RL, Wolf M (1996b) Schwertmannite and the chemical modeling of iron in acid sulfate waters. Geochim Cosmochim Acta 60:2111-2121

Bishop JL (1998) Biogenic catalysis of soil formation on Mars. Origins Life Evol Biosphere 28:449-459

Bishop JL, Murad E (1996) Schwertmannite on Mars? Spectroscopic analyses of schwertmannite, its relationship to other ferric minerals, and its possible presence in the surface material on Mars. *In* Mineral Spectroscopy: A Tribute to Roger G Burns. Dyar MD, McCammon C, Schaefer MW (Eds) Geochemical Society Spec Publ No 5:337-358

Bloomfield C (1972) The oxidation of iron sulphides in soils in relation to the formation of acid sulphate soils, and of ochre deposits in field drains. J Soil Sci 23:1-16

Bloomfield C, Coulter JK (1973) Genesis and management of acid sulfate soils. Adv Agron 25:265-326

Blowes DW, Jambor JL (1990) The pore-water geochemistry and the mineralogy of the vadose zone of sulfide tailings, Waite Amulet, Quebec, Canada. Appl Geochem 5:327-346

Blowes DW, Reardon EJ, Jambor J, Cherry JA (1991) The formation and potential importance of cemented layers in inactive sulfide mine tailings. Geochim Cosmochim Acta 55:965-978

Blowes DW, Al T, Lortie L, Gould WD, Jambor JL (1995) Microbiological, chemical, and mineralogical characterization of the Kidd Creek mine tailings impoundment, Timmins area, Ontario. Geomicrobiol J 13:13-21

Blowes DW, Jambor JL, Hanton-Fong CJ, Lortie L, Gould WD (1998) Geochemical, mineralogical and microbiological characterization of a sulphide-bearing carbonate-rich gold-mine tailings impoundment, Joutel, Québec. Appl Geochem 13:687-705

Bowell RJ (1994) Sorption of arsenic by iron oxides and oxyhydroxides in soils. Appl Geochem 9:279-286

Bowell RJ, Bruce I (1995) Geochemistry of iron ochres and mine waters from Levant Mine, Cornwall. Appl Geochem 10:237-250

Bowell RJ, Morley NH, Din VK (1994) Arsenic speciation in soil porewaters from the Ashanti Mine, Ghana. Appl Geochem 9:15-22

Brizzi G, Ciselli I, Santucci A (1986) Geology of the Cetine mine, Siena, Italy. Riv Mineral Ital:145-155

Brady KS, Bigham JM, Jaynes WF, Logan TJ (1986) Influence of sulfate on Fe-oxide formation: Comparisons with a stream receiving acid mine drainage. Clays Clay Minerals 34:266-274

Brill MR (1999) Sediment mineralogy and geochemistry in a constructed mine drainage wetland. MS Thesis, The Ohio State University, Columbus, Ohio

Brodie GA, Hammer DA, Tomljanovich DA (1988) Constructed wetlands for acid drainage control in the Tennessee Valley. *In* Mine Drainage and Surface Mine Reclamation. US Bur Mines Inf Circ 9183: 325-331

Brown GE Jr, Henrich VE, Casey WH, Clark DL, Eggleston C, Felmy A, Goodman DW, Grätzel M, Maciel G, McCarthy MI, Nealson KH, Sverjensky DA, Toney MF, Zachara JM (1999) Metal oxide surfaces and their interactions with aqueous solutions and microbial organisms. Chem Rev 99: 77-174

Burns RG (1988) Gossans on Mars. Proc 18th Lunar Planet Sci Conf LPI, Houston, p 713-721

Burns RG (1994) Schwertmannite on Mars: Deposition of this ferric oxyhydroxysulfate mineral in acidic saline meltwaters Lunar Planet Sci 25:203-204 (abstr)

Byrne RH, Luo Y-R (2000) Direct observations of nonintegral hydrous ferric oxide solubility products. Geochim Cosmochim Acta 64:1873-1877

Carlson L, Schwertmann U (1990) The effect of CO_2 and oxidation rate on the formation of goethite versus lepidocrocite from an Fe(II) system at pH 6 and 7. Clay Minerals 25:65-71

Caruccio FT (1975) Estimating the acid potential of coal mine refuse. *In* The Ecology of Resource Degradation and Renewal, Chadwick MJ, Goodman GT (Eds) Blackwell Science, London:197-203

Cathles LM (1994) Attempts to model the industrial-scale leaching of copper-bearing mine waste. *In* The Environmental Geochemistry of Sulfide Oxidation. Alpers CN, Blowes DW (Eds) Am Chem Soc Symp Ser 550:123-131

Cech F (1979) A new name for orthorhombic $Al(SO_4)(OH) \cdot 5H_2O$. N Jahrb Mineral Monatsch:193-196

Cesbron FP, Bayliss P (1988) Mineral nomenclature: Khademite. Mineral Mag 52:133-134

Chapman BM, Jones DR, Jung RF (1983) Processes controlling metal ion attenuation in acid mine drainage streams. Geochim Cosmochim Acta 47:1957-1973

Charlet L, Dise N, Stumm W (1993) Sulfate adsorption on a variable charge soil and on reference minerals. Agric Ecosys Environ 47:87-102

Chernitsyna OM, Artemenko VM, Artemenko OV, Ponomarev VE (1999) Aluminite from gold-mercury ores in terrigenous-carbonate complex of southern Donbass. Mineral Zh 21: 29-32

Childs CW, Goodman BA, Paterson E, Woodhams FWD (1980) The nature of iron in akaganéite (β-FeOOH). Aust J Chem 33:15-26

Childs CW, Inoue K, Mizota C (1997) Natural and anthropogenic schwertmannites from Towada-Hachimantai National Park, Honshu, Japan. Chem Geol 144:81-86

Clark BC, Baird AK, Weldon RJ, Tsusaki DM, Schnabel L, Candelaria MP (1982) Chemical composition of martian fines. J Geophys Res 87:10059-10067

Clark BC, Van Hart D (1981) The salts of Mars. Icarus 45:370-378

Clarke W, Konhauser KO, Thomas J, Bottrell SH (1997) Ferric hydroxide and ferric hydroxysulfate precipitation by bacteria in an acid mine drainage lagoon. FEMS Microbiol Rev 20:351-361

Clayton T (1980) Hydrobasaluminite and basaluminite from Chickerell, Dorset. Mineral Mag 43:931-937

Cornell R, Schwertmann U (1996) The Iron Oxides. VCH Publishers, Weinheim

Cornu F (1909) Über die Verbreitung gelartiger Körper im Mineralbereich, ihrre chemisch-geologische Bedeutung und ihre systematische Stellung. Zbl Mineral Geol 12:324-326

Cravotta CA III, Brady KBC, Rose AW, Douds JB (1999) Frequency distribution of the pH of coal-mine drainage in Pennsylvania. *In* U.S. Geological Survey Toxic Substances Hydrology Program—Proc Technical Meeting. Morganwalp DW, Buxton H (Eds) US Geol Surv Water-Res Invest Rep 99-4018A:313-324

Cunningham KM, Wright WG, Nordstrom DK, Ball JW (1996) Water-quality data for Doughty Springs, Delta County, Colorado, 1903-1994, with emphasis on sulfur redox species. US Geol Survey Open-File Rep 96-619, 69 p

De Bruiyn H, Schoch AE, Beukes GJ, Bok LDC, Van Der Westhuizen WA (1985) Note on cell parameters of zaherite. Mineral Mag 49:145-146

De Endredy A S (1963) Estimation of free iron oxides in soils and clays by a photolytic method. Clay Minerals 5:209-217

Dent DL, Pons LJ (1995) A world perspective on acid sulphate soils. Geoderma 67:263-276

Dold B (1999) Mineralogical and geochemical changes of copper flotation tailings in relation to their original composition and climatic setting: Implications for acid mine drainage and element mobility. PhD Thesis #3125, Université de Gèneve, Switzerland

Dombrovskaya ZV, Yashina RS, Piloyan GO (1976) Aluminite from the zone of oxidation of pyritized schists in the Lake Baikal region. Kora Vyvetrivaniya 15:162-167

Doner HE, Lynn WC (1989) Carbonate, halide, sulfate and sulfide minerals. *In* Minerals in Soil Environments. 2nd edition. Dixon JB, Weed SB (Eds) Soil Sci Soc Am, Madison, Wisconsin, p 279-330

Dousma J, de Bruyn PL (1976) Hydrolysis-precipitation studies of iron solutions I. Model for hydrolysis and precipitation from Fe(III) nitrate solutions. J Colloid Interface Sci 56:527-539

Dousma J, van den Hoven TJ, de Bruyn PL (1978) The influence of chloride ions on the formation of iron(III) oxyhydroxide. J Inorg Nucl Chem 40:1089-1093

Dousma J, den Ottelander D, de Bruyn PL (1979) The influence of sulfate ions on the formation of iron(III) oxides. J Inorg Nucl Chem 41:1565-1568

Driscoll CT, Baker JP, Bisogni JJ, Schofield CL (1984) Aluminum speciation and equilibria in dilute acidic surface waters of the Adirondack region of New York State. *In* Geological Aspects of Acid Deposition, 4:55-75. Bricker OP (Ed) Ann Arbor Science, Boston

Dudas MJ (1987) Accumulation of native arsenic in acid sulphate soils in Alberta. Can J Soil Sci 67: 317-331

Dzombak DA, Morel FMM (1990) Surface Complexation Modeling: Hydrous Ferric Oxide. John Wiley, New York

Edwards KJ, Bond PL, Gihring TM, Banfield F (2000) An archaeal iron-oxidizing extreme acidophile important in acid mine drainage. Science 287:1796-1799

Ehrlich HL (1996) Geomicrobiology (3rd Edition) Marcel Dekker, New York

Fanning DS, Burch SN (1997) Acid sulphate soils and some associated environmental problems. *In* Soils and Environment: Soil Processes from Mineral to Landscape Scale. Auerswald K, Stanjek H, Bigham JM (Eds) Adv. Geoecol 30:145-158, Reiskirchen, Catena Verlag

Fanning DS, Rabenhorst MC, Bigham JM (1993) Colors of acid sulfate soils. *In* Soil color. Bigham JM, Ciolkosz EJ (Eds) Soil Sci Soc Am Spec Publ 31:91-108 Soil Sci Soc Am, Madison, WI.

Farkas L, Pertlik F (1997) Crystal structure determinations of felsöbányaite and basaluminite, $Al_4(SO_4)(OH)_{10} \cdot 4H_2O$. Acta Mineral-Petrogr 38:5-15

Feitknecht W, Giovanoli R, Michaelis W, Müller M (1973) Über die Hydrolyse von Eisen (III) Salzlösungen. I. Die Hydrolyse der Lösungen von Eisen (III) Chlorid. Helv Chim Acta 56:2847-2856

Fennessy MS, Mitsch WJ (1989) Treating coal mine drainage with an artificial wetland. Res J Water Pollut Cont Fed 61:1691-1701

Ferris FG, Tazaki K, Fyfe WS (1989a) Iron oxides in acid mine drainage environments and their association with bacteria. Chem Geol 74:321-330

Ferris FG, Schultze S, Witten TC, Fyfe WS, Beveridge TJ (1989b) Metal interactions with microbial biofilms in acidic and neutral pH environments. Appl Environ Microbiol 55:1249-1257

Filipek LH, Nordstrom DK, Ficklin WH (1987) Interaction of acid mine drainage with waters and sediments of West Squaw Creek in the West Shasta Mining District, California. Environ Sci Technol 21:388-396

Fischer WR (1972) Die Wirkung von zweiwertigem Eisen auf Lösung un Umwandlung von Eisen(III)-hydroxiden. In Pseudogley und Gley. Schlichting E, Schwertmann U (Eds) Weinheim, Verlag Chemie, p 37-44

Fitzpatrick RW, Self PG (1997) Iron oxyhydroxides, sulfides and oxyhydroxysulfates as indicators of acid sulfate weathering environments. In Soils and Environment: Soil Processes from Mineral to Landscape Scale. Auerswald K, Stanjek H, Bigham JM (Eds) Adv. Geoecol 30:227-240, Reiskirchen, Catena Verlag

Fitzpatrick R W, Naidu R, Self PG (1992). Iron deposits and microorganisms in saline sulfidic soils with altered soil water regimes in South Australia. In Biomineralization Processes of Iron and Manganese—Modern and Ancient Environments. Skinner HCW, Fitzpatrick RW (Eds) Catena Supplement 21: 263-286, Cremlingen-Destedt

Fitzpatrick R. W, Fritsch E, Self PG (1996) Interpretation of soil features produced by ancient and modern processes in degraded landscapes: V. Development of saline sulfidic features in non-tidal seepage areas. Geoderma 69:1-29

Flynn CM Jr (1990) Dense hydrolysis products from iron (III) nitrate and sulfate solutions. Hydrometall 25:257-270

Fojt B (1975) On the problem of glockerite as a secondary mineral of ore deposits. Scripta Fac Sci Nat UJEP Brunensis, Geologia I. 5:5-20

Fox LE (1988a) Solubility of colloidal ferric hydroxide. Nature 333:442-444

Fox LE (1988b) The solubility of colloidal ferric hydroxide and its relevance to iron concentrations in river water. Geochim Cosmochim Acta 52:771-777

Fox LE (1989) A model for inorganic control of phosphate concentrations in river waters. Geochim Cosmochim Acta 53:417-428

Frondel C (1968) Meta-aluminite, a new mineral from Temple Mountain, Utah. Am Mineral 53:717-721

Fuge R, Pearce FM, Pearce NJ, Perkins WT (1994) Acid mine drainage in Wales and influence of ochre precipitation on water chemistry. In The Environmental Geochemistry of Sulfide Oxidation. Alpers CN, Blowes DW (Eds) Am Chem Soc Symp Series 550:261-274

Fuller CC, Davis JA (1989) Influence of coupling of sorption and photosynthetic processes on trace element cycles in natural waters. Nature 340:52-54

Fuller CC, Davis JA, Waychunas GA (1993) Surface chemistry of ferrihydrite: Part 2. Kinetics of arsenate adsorption and coprecipitation. Geochim Cosmochim Acta 57:2271-2282

Furniss G, Hinman NW, Doyle GA, Runnells DD (1999) Radiocarbon-dated ferricrete provides a record of natural acid rock drainage and paleoclimatic changes. Environ Geol 37:102-106

Gimeno MJ, Tena JM, Auqué LF, Mandado J (1994) Caracterización geoquímica del sistema de agues ácidas del Arroyo del Val (Zaragoza). Bol R Soc Esp Hist Nat (Sección Geología) 89:5-17

Glocker EF (1853) Über einen neuen Eisensinter von Obergrund bie Zuckmantel. Poggendorff's Annalen Physik Chemie 89:482-488

Glocker EF (1858) Über den sulphatischen Eisensinter von Obergrund bei Zuckmantel. Acta Nova Leopold Carol 26:189-220

Godin E (Ed) (1991) Canadian Minerals Yearbook—Review and Outlook. Energy, Mines and Resources, Ottawa, Canada

Goldberg S (1986) Chemical modeling of arsenate adsorption on aluminum and iron oxide minerals. Soil Sci Soc Amer J 50:1154-1160

Gould WD, Béchard G, Lortie L (1994) The nature and role of microorganisms in the tailings environment. In The Environmental Geochemistry of Sulfide Mine-Wastes. Jambor JL, Blowes DW (Eds) Mineral Assoc Canada Short Course 22:185-200

Hammack RW, Lau RW, Diehl JR (1988) Methods for determining fundamental chemical differences between iron disulfides from different geologic provenances. US Bur Mines Inf Circ IC-9183:136-146

Harrison JB, Berkheiser VE (1982) Anion interactions with freshly prepared hydrous iron oxides. Clays Clay Minerals 30:97-102

Headden WP (1905) Mineralogical notes, no. II. Proc Colorado Sci Soc 8:62-66

Hem JD, Roberson CE (1990) Aluminum hydrolysis reactions and products in mildly acidic aqueous systems. *In* Chemical modeling in aqueous systems II. Melchior DC, Bassett RL (Eds) Am Chem Soc Symp Series 416:429-446

Hemley JJ, Hostetler PB, Gude AJ, Mountjoy WT (1969) Some stability relations of alunite. Econ Geol 64:599-612

Hendershot WH, Courchesne F, Jeffries DS (1996) Aluminum geochemistry at the catchment scale in watersheds influenced by acidic precipitation. *In* The Environmental Chemistry of Aluminum. 2nd Edition. Sposito G (Ed) CRC Press/Lewis Publishers, Boca Raton, Florida, p 419-449

Hochella MF Jr, White AF (Eds) (1990) Mineral-Water Interface Geochemistry. Rev Mineral 23, Mineral Soc Am, Washington, DC, 603 p

Hudson-Edwards KA, Schell C, Macklin MG (1999) Mineralogy and geochemistry of alluvium contaminated by metal mining in the Rio Tinto area, southwest Spain. Appl Geochem 14:1015-1030

Hug SJ (1997) *In situ* Fourier transform infrared measurements of sulfate adsorption on hematite in aqueous solutions. J Colloid Interface Sci 188:415-422

Jambor JL (1986) Detailed mineralogical examination of alteration products in core WA-20 from Waite Amulet tailings. CANMET Div Rep MSL 86-137(IR). Dep Energy Mines Res Canada.

Jambor JL (1994) Mineralogy of sulfide-rich tailings and their oxidation products. *In* The Environmental Geochemistry of Sulfide Mine-Wastes. Jambor JL, Blowes DW (Eds) Mineral Assoc Canada Short Course Vol. 22:59-102

Jambor JL Dutrizac JE (1998) Occurrence and consititution of natural and synthetic ferrihydrite, a widespread iron oxyhydroxide. Chem Rev 98:2549-2585

Jambor JL, Grew ES, Roberts AC (1998) New mineral names. Am Mineral 83:1347-1352

Johansson G (1962) The crystal structure of $[Al_2(OH)_2(H_2O)_8](SO_4)_2 \cdot 2H_2O$ and $[Al_2(OH)_2(H_2O)_8](SeO_4)_2 \cdot 2H_2O$. Acta Chem Scand 16:403-420

Johansson G (1963) On the crystal structure of the basic aluminum sulfate $13Al_2O_3 \cdot SO_3 \cdot xH_2O$. Arkiv Kemi 20:321-342

Johnson CA, Thornton I (1987) Hydrological and chemical factors controlling the concentrations of Fe, Cu, Zn and As in a river system contaminated by acid mine drainage. Water Res 21:359-365

Karathanasis AD, Evangelou VP, Thompson, YL (1988) Aluminum and iron equilibria in soil solutions and surface waters of acid mine watersheds. J Environ Qual 17:534-543

Karathanasis AD, Thompson YL (1995) Mineralogy of iron precipitates in a constructed acid mine drainage wetland. Soil Sci Soc Am J 59:1773-1781

Karlsson S, Sandén P, Allard B (1987) Environmental impacts of an old mine tailings deposit—metal adsorption by particulate matter. Nordic Hydrol 18:313-324

Kimball BA, Broshears RA, McKnight DM, Bencala KE (1994) Effects of instream pH modification on transport of sulfide-oxidation products. *In* The Environmental Geochemistry of Sulfide Oxidation. Am Chem Soc Symp Series 550:224-243, Am Chem Soc, Washington, DC

Kimball BA, Callender E, Axtmann EV (1995) Effects of colloids on metal transport in a river receiving acid mine drainage, upper Arkansas River, Colorado, U.S.A. Appl Geochem 10:285-306

Kirby CS, Decker SM, Macander NK (1999) Comparison of color, chemical and mineralogical compositions of mine drainage sediments to pigment. Environ Geol 37:243-254

Kleinmann RLP (1989) Acid mine drainage in the United States: Controlling the impact on streams and rivers. *In* 4th World Congress on the Conservation of Built and Natural Environments, Univ Toronto, p 1-10

Kleinmann RLP, Crerar DA, Pacelli RR (1981) Biogeochemistry of acid mine drainage and a method to control acid formation. Mining Eng 33:300-303

Konhauser KO (1998) Diversity of bacterial iron mineralization. Earth Sci Rev 43:91-121

Konhauser KO, Ferris FG (1996) Diversity of iron and silica precipitation by microbial mats in hydrothermal waters, Iceland: Implications for Precambrian iron formations. Geology 24:323-326

Kram P, Hruska J, Driscoll CT, Johnson CE (1995) Biogeochemistry of aluminum in a forest catchment in the Czech Republic impacted by atmospheric inputs of strong acids. Water Air Soil Pollut 85:1831-1836

Krause E, Ettel VA (1988) Solubility and stability of scorodite, $FeAsO_4 H_2O$: New data and further discussion. Am Mineral 73:850-854

Landa ER, Cravotta CA III, Naftz DL, Verplanck PL, Nordstrom DK, and Zielinski RA (2000) Geochemical investigations by the U.S. Geological Survey on uranium mining, milling, and environmental restoration. Tech 7:381-396

Lazaroff N, Sigal W, Wasserman, A (1982) Iron oxidation and precipitation of ferric hydroxysulfates by resting *Thiobacillus ferrooxidans* cells. Appl Environ Microbiol 43:924-938

Lazaroff N, Melanson L, Lewis E, Santoro N, Pueschel C (1985) Scanning electron microscopy and infrared spectroscopy of iron sediments formed by *Thiobacillus ferrooxidans*. Geomicrobiol J 4:231-268

Lees H, Kwok SC, Suzuki I (1969) The thermodynamics of iron oxidation by ferrobacilli. Can J Microbiol 15:43-46

Letterman RD, Mitsch WJ (1978) Impact of mine drainage on a mountain stream in Pennsylvania. Environ Pollut 17:53-73

Lewis DG (1997). Factors influencing the stability and properties of green rusts. *In* Soils and Environment: Soil Processes from Mineral to Landscape Scale. Auerswald K, Stanjek H, Bigham JM (Eds) Adv Geoecol 30:345-372, Reiskirchen, Catena Verlag

Linnaeus C (1735) Systema Natural XII, vol. 111, Gen 52:11

Lizalek NA, Filatov VF (1986) Geology and origin of aluminites of Siberia. Sov Geol, p 41-49

Lizalek NA, Ivlev NF, Madaras A, and Speshilova MA (1989) The Namana occurrence of secondary sulfates in Yakutia [USSR]. Sov Geol, p 46-52.

Long DT, Fegan NE, McKee JD, Lyons WB, Hines ME and Macumber PG (1992) Formation of alunite, jarosite and hydrous iron oxides in a hypersaline lake system: Lake Tyrrell, Victoria, Australia. Chem Geol 96:183-202

Losno R, Colin JL, Le Bris N, Bergametti G, Jickells T, Lim B (1993) Aluminum solubility in rainwater and molten snow. J Atmosph Chem 17:29-43

Lukovic SM (1970) Weathering processes and formation of secondary aluminite and gibbsite in boehmite-kaolin bauxite of Montenegro, Yugoslavia. Zb Rad Rud-Geol Fak, Univ Beogradu 11-12:55-63

Luther GW II (1990) The frontier-molecular-orbital theory approach in geotechnical processes. *In* Aquatic chemical kinetics. W Stumm (Ed) John Wiley and Sons, Inc., New York, p 173-198

Lyon JS, Hilliard TJ, Bethell TN (1993) Burden of Guilt. Mineral Policy Center, Washington, DC, 68 p

Mamedov AI, Makhmudov SA, Babaev IA (1969) Aluminite from the Kelbadzhar area (Azerbaidzhan SSR). Uch Zap Aerb Gos Univ, Ser Geol-Geogr Nauk, p 136-142

Mazilov VN, Kashik SA, Kashaeva GM (1975) First finding of meta-aluminite in the USSR. Zap Vses Mineral Obshch 104:202-203

Margulis EV, Savchenko LA, Shokarev MM, Beisekeeva LI, Vershinina FI (1975) The amorphous basic sulphate $2Fe_2O_3 \cdot SO_3 \cdot mH_2O$. Russ J Inorg Chem 20:1045-1048

Margulis EV, Getskin LS, Zapuskalova NA, Beisekeeva LI (1976) Hydrolytic precipitation of iron in the $Fe_2(SO_4)_3$--KOH--H_2O system. Russ J Inorg Chem 21:996-999

Mast MA, Verplanck PL, Yager DB, Wright WG, Bove DJ. (2000) Natural sources of metals to surface waters in the upper Animas River watershed, Colorado. ICARD 2000, Proc. 5th Internat Conf Acid Rock Drainage, Vol. 1. Soc Mining Metall Explor, Littleton, CO, p 513-522

Matijevic E, Janauer GE (1966) Coagulation and reversal of charge of lyophobic colloids by hydrolyzed metal ions. II. Ferric nitrate. J Colloid Interface Sci 21:197-223

Matijevic E, Sapieszko RS, Melville JR (1975) Ferric hydrous oxide sols I. Monodispersed basic iron (III) sulfate particles. J Colloid Interface Sci 50:567-581

Matsubara S, Kato A, Hashimoto E (1990) Basaluminite in fissure of the Takanuki metamorphic rocks, [Fukushima prefecture], northeast Japan. Chigaku Kenkyu 39:165-169

May HM, Nordstrom DK (1991) Assessing the solubilities and reaction kinetics of aluminous minerals in soils. *In* Soil Acidity. Ulrich B, Sumner ME (Eds) Springer-Verlag, Berlin, p 125-148

McKibben MA, Barnes HL (1986) Oxidation of pyrite in low temperature acidic solutions: Rate laws and surface textures. Geochim Cosmochim Acta 50:1509-1520

McKnight DM, Bencala KE (1989) Reactive iron transport in an acidic mountain stream in Summit County, Colorado: A hydrologic perspective. Geochim Cosmochim Acta 53:2225-2234

McKnight DM, Kimball BA, Bencala KE (1988) Iron photoreduction and oxidation in an acidic mountain stream. Science 240:637-640

Merwin HE, Posnjak E (1937) Sulphate encrustations in the Copper Queen mine, Bisbee, Arizona. Am Mineral 22:567-571

Milnes AR, Fitzpatrick RW, Self PG, Fordham AW, McClure SG (1992) Natural iron precipitates in a mine retention pond near Jabiru, Northern Territory, Australia. *In* Biomineralization Processes of Iron and Manganese—Modern and Ancient Environments. Skinner HCW, Fitzpatrick RW (Eds) Catena Supplement 21:233-261, Cremlingen-Destedt

Minikawa T, Inaba S, Tamura Y (1996) A few occurrences of basaluminite in Japan. Chigaku Kenkyu 44:193-197

Monterroso C, Alvarez E, Macias F (1994) Speciation and solubility control of Al and Fe in minesoil solutions. Sci Total Environ 158:31-43

Moore JN and Luoma SN (1990) Hazardous wastes from large-scale metal extraction: A case study. Environ Sci Technol 24:1278-1285

Morris RV, Golden DC, Bell JF III, Shelfer TD, Scheinost AC, Hinman NW, Furniss G, Mertzman SA, Bishop J, Ming DW, Allen CC, Britt DT (2000) Mineralogy, composition, and alteration of Mars Pathfinder rocks and soils: Evidence from multispectral, elemental, and magnetic data on terrestrial analogue, SNC meteorite, and Pathfinder samples. J Geophys Res 105:1757-1817

Morton WE, Dunnette DA (1994) Health effects of environmental arsenic. *In* Arsenic in the Environment. Part II: Human Health and Ecosystem Effects. Nriagu JO (Ed) John Wiley, New York, p 17-34

Murad E (1988a) The Mössbauer spectrum of "well" crystallized ferrihydrie. J Magnet Magnetic Mat 74: 155-157.

Murad E (1988b) Properties and behavior of iron oxides as determined by Mössbauer spectroscopy. *In* Iron in Soils and Clay Minerals. Stucki JW, Goodman BA, Schwertmann U (Eds) NATO Am Soils Inst Series C 217:309-350, D. Reidel, Dordrecht

Murad E, Bowen LH, Long GJ, Quin TG (1988) The influence of crystallinity on magnetic ordering in natural ferrihydrites. Clay Minerals 23:161-173

Murad E, Bigham JM, Bowen LH, Schwertmann U (1990) Magnetic properties of iron oxides produced by bacterial oxidation of Fe^{2+} under acid conditions. Hyperfine Interactions 58:2373-2376

Murad E, Schwertmann U, Bigham JM, Carlson L (1994) The mineralogical characteristics of poorly crystalline precipitates formed by oxidation of Fe^{2+} in acid sulfate waters. *In* The Environmental Geochemistry of Sulfide Oxidation. Alpers CN, Blowes DW (Eds) Am Chem Soc Symp Series 550: 190-200

Naumann CF (1855) Elemente der Mineralogie, 4 Aufl. V.W. Engelmann, Leipzig.

Neal C (1988) Aluminum solubility relationships in acid waters—A practical example of the need for a radical reappraisal. J Hydrol 194:141-159

Ninteman DJ (1978) Spontaneous Oxidation and Combustion of Sulfide Ores in Underground Mines, a Literature Survey. US Bur Mines Inf Circ 8775

Nordstrom D K (1982a) Aqueous pyrite oxidation and the consequent formation of secondary iron minerals. *In* Acid Sulfate Weathering. Kittrick JA, Fanning DS, Hossner LR (Eds) Soil Sci Soc Am Spec Publ 10:37-56

Nordstrom DK (1982b) The effect of sulfate on aluminum concentrations in natural waters: some stability relations in the system Al_2O_3-SO_3-H_2O at 298 K. Geochim Cosmochim Acta 46:681-692

Nordstrom DK (1991) Chemical modeling of acid mine waters in the western United States. US Geol Survey Water Res Invest Rep No 91-4034:534-538

Nordstrom DK (2000) Advances in the hydrogeochemistry and microbiology of acid mine waters. Internat Geol Rev 42:499-515

Nordstrom DK, Alpers CN (1999). Geochemistry of acid mine waters. *In* The Environmental Geochemistry of Mineral Deposits. Part A. Processes, methods and health issues. Plumlee GS, Logsdon MJ (Eds) Rev Econ Geol 6A:133-160

Nordstrom DK, Ball JW (1986) The geochemical behavior of aluminum in acidified surface waters. Science 232:54-56

Nordstrom DK, Ball JW (1989) Mineral saturation states in natural waters and their sensitivity to thermodynamic and analytic errors. Sci Géol Bull 42:269-280

Nordstrom DK, May HM (1996) Aqueous equilibrium data for mononuclear aluminum species. *In* The Environmental Chemistry of Aluminum. Sposito G (Ed) 2nd edition. CRC Press/Lewis Publishers, Boca Raton, Florida, p 39-80

Nordstrom DK, Munoz JL (1994) Geochemical Thermodynamics, 2nd edition. Blackwell Science, Boston

Nordstrom DK, Southam G (1997) Geomicrobiology of sulfide mineral oxidation. *In* Geomicrobiology: Interactions between Microbes andMinerals. Banfield JF, Nealson KH (Eds) Rev Mineral 35:361-390

Nordstrom DK, Ball JW, Roberson, CE, Hanshaw BB (1984) The effect of sulfate on aluminum concentrations in natural waters. II. Field occurrences and identification of aluminum hydroxysulfate precipitates. Proc Geol Soc Am Ann Mtg 16:611

Nordstrom DK, Plummer LN, Langmuir D, Busenberg E, May HM, Jones BF, Parkhurst DL (1990) Revised chemical equilibrium data for major water-mineral reactions and their limitations. *In* Chemical Modeling of Aqueous Systems II. Melchior DC, Bassett RL (Eds) Am Chem Soc Symp Series 416:398-413

Nordstrom DK, Carson-Foscz V, Oreskes N (1995) Rare earth element (REE) fractionation during acidic weathering of San Juan tuff, Colorado. Proc Geol Soc Am Ann Mtg, New Orleans, p A-199 (abstr)

Nordstrom DK, Ball JW, Southam G, Donald R (1997) Biogeochemistry of natural elemental sulur oxidation and derivative acidic waters at Brimstone Basin, Yellowstone National Park, Wyoming: I. Chemical and isotopic results.

Nordstrom DK, Ball JW, McClesky RB (2000) On the interpretation of saturation indices for iron colloids in acid mine waters. Abstracts with Programs Geol Soc Am Ann Mtg, p A-78

Olson GJ (1991) Rate of pyrite bioleaching by *Thiobacillus ferrooxidans*: Results of an interlaboratory comparison. Appl Environ Microbiol 57:642-644

Öborn I, Berggren D (1995). Characterization of jarosite-natrojarosite in two northern Scandinavian soils. Geoderma 66:213-225

Oycollonov VN, Dolinina YV, Ogovodova LP, Sokolov VN (1994) Aluminite from the oxidation zone of sulfide-poor silver ore deposit. Vestn Mosk Univ Ser 4 Geol, p 58-61

Palache C, Berman H, Frondel C (1951) The System of Mineralogy, Vol. 2. John Wiley and Sons, New York

Papassiopi N, Stefanakis M, Kontopoulos A (1988) Removal of arsenic from solutions by precipitation as ferric arsenates. *In* Arsenic Metallurgy Fundamentals and Applications. Reddy RG, Hendrix JL, Queneau PB (Eds) Symp Proc TMS-AIME Ann Mtg, Phoenix, AZ, p 321-334

Parfitt RL, Smart R StC (1978) The mechanism of sulfate adsorption on iron oxides. Soil Sci Soc Am J 42:48-50

Parks GA (1965) The isoelectric points of solid oxides, solid hydroxides, and aqueous hydroxo complex systems. Chem Rev 65:177-198

Parks GA (1967) Aqueous surface chemistry of oxides and complex oxide minerals. *In* Equilibrium Concepts in Natural Water Systems. Stumm W (Ed) Am Chem Soc. Adv Chem Series 67:121-160

Peak D, Ford RG, Sparks DL (1999) An *in situ* ATR-FTIR investigation of sulfate bonding mechanisms on goethite. J Colloid Inter Sci 218:289-299

Persson P, Lövgren L (1996) Potentiometric and spectroscopic studies of sulfate complexation at the goethite–water interface. Geochim Cosmochim Acta 60:2789-2799

Peine A, Tritschler A, Küsel K, Peiffer S (2000) Electron flow in an iron-rich acidic sediment—evidence for an acidity-driven iron cycle. Limnol Oceanogr 45:1077-1087

Pierce ML, Moore CB (1982) Adsorption of arsenite and aresenate on amorphous iron hydroxide. Water Res 16:1247-1253

Plumlee GS, Smith KS, Montour MR, Ficklin WS, Mosier E (1999) Geologic controls on the composition of natural waters and mine waters draining diverse mineral-deposit types. *In* The Environmental Geochemistry of Mineral Deposits. Part B. Case studies and research topics. Filipek LH, Plumlee GS (Eds) Rev Econ Geol 6B:373-432

Polyak VJ, Provencio P (1998) Hydrobasaluminite and aluminite in caves of the Guadalupe Mountains, New Mexico. J Cave Karst Stud 60:51-57

Pons LJ (1973) Outline of the genesis, characteristics, classification and improvement of acid sulphate soils. *In* Acid Sulphate Soils. Dost H (Ed) Proc 1st Internat Symp Land Reclamation Institute Publ 18, 1:3-27, Wageningen, The Netherlands

Posey HH, Renkin ML, Woodling J (2000) Natural acidic drainage in the upper Alamosa River watershed, Colorado. ICARD 2000 Proc 5th Internat Conf Acid Rock Drainage 1:485-498 Soc Mining Metall Explor, Littleton, CO

Posnjak E, Merwin HE (1922) The system Fe_2O_3-SO_3-H_2O. J Am Chem Soc 44:1965-1994

Powers DA, Rossman GR, Schugar HJ, Gray HB (1975) Magnetic behavior and infrared spectra of jarosite, basic iron sulfate, and their chromate analogs. J Solid State Chem 13:1-13

Rampe JJ, Runnells DD (1989) Contamination of water and sediment in a desert stream by metals from an abandoned gold mine and mill, Eureka District, Arizona, U.S.A. Appl Geochem 4:445-454

Raven KP, Jain A, Loeppert RH (1998) Arsenite and arsenate adsorption on ferrihydrite: kinetics, equilibrium, and adsorption envelopes. Environ Sci. Technol 32:344-349

Rawlings EE, Tributsch H, Hansford GS (1999) Reasons why *'Leptospirillum'*-like species rather than Thiobacillus ferrooxidans are the dominant iron-oxidizing bacteria in many commercial processes for the biooxidation of pyrite and related ores. Microbiol 145:5-13

Raymahashay BC (1968) A geochemical study of rock alteration by hot springs in the Paint Pot Hill area, Yellowstone Park. Geochim Cosmochim Acta 32:499-522

Rieder RT, Ecomomou T, Wanke H, Turkevich A, Crisp J, Bruckner J, Drebus G, McSween HY Jr (1997) The chemical composition of Martian soil and rock returned by the mobile alpha proton X-ray spectrometer: Preliminary results from the x-ray mode. Science 278:1771-1774

Ritchie AIM (1994) The waste-rock environment. *In* The Environmental Geochemistry of Sulfide Mine-Wastes. Jambor JL, Blowes DW (Eds) Mineral Assoc Canada Short Course 22:131-161

Robbins EI, Cravotta CA III, Savela CE, Nord GL Jr (1999) Hydrobiogeochemical interactions in 'anoxic' limestone drains for neutralization of acidic mine drainage. Fuel 78:259-270

Robins, RG (1990) The stability and solubility of ferric arsenate: An update. *In* EPD Congress, Gaskell DR (Ed) TMS Ann Mtg, Feb 18-22, 1990, Anaheim, California, p 93

Robins RG, Huang JCY, Nishimura T, Khoe GH (1988) The adsorption of arsenate ion by ferric hydroxide. *In* Arsenic Metallurgy: Fundamentals and Applications. Reddy RG, Hendrix JL, Queneau PB (Eds) Symp Proc TMS-AIME Ann Mtg, Phoenix, AZ, p 99-112

Romberg IB (1969) Lepidocrocite at Rossvatn, north-Norway, an example of pseudomorphism after pyrite cubes. Norsk Geol Tidsskr 49:251-256

Rose AW, Cravotta CA III (1998) Geochemistry of coal mine drainage. p 1-1 to 1-22. *In* Coal Mine Drainage Prediction and Prevention in Pennsylvania. Brady KBC, Smith MW, Schueck J (Eds) Pennsylvania Dept Environ Protection, Harrisburg, PA

Rose S, Ghazi AM (1997) Release of sorbed sulfate from iron oxyhydroxides precipitated from acid mine drainage associated with coal mining. Environ Sci Technol 31:2136-2140

Rose S, Elliott WC (2000) The effects of pH regulation upon the release of sulfate from ferric precipitates formed in acid mine drainage. Appl Geochem 15: 27-34

Ross (1972) Inorganic Infrared and Raman Spectra. McGraw-Hill, New York

Rost R (1937) The minerals forming on burning coal heaps in the coal basin of Kladno. Bull Internat Acad Sci Bohême 1937:1-7

Rothenhofer P, Sahin H, Peiffer S (1999) Attenuation of heavy metals and sulfate by aluminum precipitates in acid mine drainage. Acta Hydrochim Hydrobiol 28:136-144

Rowe GL Jr, Ohsawa S, Takano B, Brantley SL, Fernandez JF, Barquero J (1992) Using crater lake chemistry to predict volcanic activity at Poas Volcano, Costa Rica. Bull Volcanol 54:494-503

Rubin AJ (Ed) (1974) Aqueous-Environmental Chemistry of Metals. Ann Arbor Science, Ann Arbor, Michigan.

Runnells DD, Shepherd TA, Angino EE (1992) Estimation of natural background concentrations of metals in water in mineralized regions after disturbance by mining. Environ Sci Technol 26:2316-2323

Ruotsala AP, Babcock LL (1977) Zaherite, a new hydrated aluminum sulfate. Am Mineral 62:1125-1128

Sabelli C (1984) On the mineralogy of the Cetine mine of Cotorniano: The sulfate dimorphs jurbanite and rostite. Per Mineral (Roma), p 53-65

Sabelli C (1985) Refinement of the crystal structure of jurbanite, $Al(SO_4)(OH)\cdot5H_2O$. Z Kristallogr 173:33-39

Sand W, Rhode K, Zenneck C (1992) Evaluation of *Leptospirillum ferrooxidans* for leaching. Appl Environ Microbiol 58:85-92

Scheinost AC, Schwertmann U (1999) Color identification of iron oxide and hydroxysulfates: Use and limitations. Soil Sci Soc Am J 63:1463-1471

Scheinost AC, Chavernas A, Barrón V, Torrent J (1998) Use and limitations of second-derivative diffuse reflectance spectroscopy in the visible to near-infrared range to identify and quantify Fe oxide minerals in soils. Clays Clay Minerals 46:528-536

Schemel LE, Kimball BA, Bencala KE (2000) Colloid formation and metal transport through two mixing zones affected by acid mine drainage near Silverton, Colorado. Appl Geochem 15:1003-1018

Schrenk MO, Edwards KJ, Goodman RM, Hamers RJ, Banfield JF (1998) Distribution of *Thiobacillus ferrooxidans* and *Leptospirillum ferrooxidans*: Implications for generation of acid mine drainage. Science 279:1519-1522

Schulze DG (1981) Identification of soil iron oxide minerals by differential x-ray diffraction. Soil Sci Soc Am J 45:437-440

Schultze-Lam S, Fortin D, Davis BS, Beveridge TJ (1996) Mineralization of bacterial surfaces. Chem Geol 132:171-181

Schwertmann U (1961) Über das Vorkommen und die Enstehung von Jarosit in Marschboden (Maibolt). Naturwiss 45:159-160

Schwertmann U (1964) Differenzierung der Eisenoxide des Bodens durch Extraktion mit Ammoniumoxalat-Lösung. Z Pflanzenern. Bodenkunde 105:194-202

Schwertmann U (1993) Relations between iron oxides, soil color, and soil formation. *In* Soil color. Bigham JM, Ciokosz EJ (Eds) Soil Sci Soc Am Spec Publ No 31:51-69, Soil Sci Soc Am, Madison, Wisconsin

Schwertmann U, Taylor RM (1989) Iron oxides. *In* Minerals in Soil Environments. 2nd edition. Dixon JB, Weed SB (Eds) Soil Sci Soc Am, Madison, Wisconsin

Schwertmann U, Thalmann H (1976) The influence of [Fe (II)], [Si], and pH on the formation of lepidocrocite and ferrihydrite during oxidation of aqueous $FeCl_2$ solutions. Clay Minerals 11:189-200

Schwertmann U, Fechter H (1994) The formation of green rust and its transformation to lepidocrocite. Clay Minerals 29:87-92

Schwertmann U, Fojt B (1996) Schwertmannit—ein neues Mineral und seine Geschichte. Lapis, p 33-34

Schwertmann U, Bigham JM, Murad E (1995) The first occurrence of schwertmannite in a natural stream environment. Eur J Mineral 7:547-552

Singer PC, Stumm W (1968) Kinetics of the oxidation of ferrous iron. 2nd Symp Coal Mine Drainage Res National Coal Assoc Bitum Coal Res, p 12-34

Singer PC, Stumm W (1970) Acid mine drainage: The rate-determining step. Science 167:1121-1123

Singh B, Harris PJ, Wilson MJ (1997). Geochemistry of acid mine waters and the role of micro-organisms in such environments: a review. *In* Soils and Environment: Soil Processes from Mineral to Landscape

Scale. Auerswald K, Stanjek H, Bigham JM (Eds) Adv Geoecol 30:159-192, Reiskirchen, Catena Verlag

Singh SS (1982) The formation and coexistence of gibbsite, boehmite, alumina and alunite at room temperature. Can J Soil Sci 62:327-332

Sizia R (1966) Infrared spectra of some sulfates. Rend Semin Fac Sci Univ Cagliari 36:82-91

Smith KS (1991) Factors Influencing Metal Sorption onto Iron-Rich Sediment in Acid-Mine Drainage. PhD thesis T-3925, Colorado School of Mines, Golden, Colorado

Smith KS (1999) Metal sorption on mineral surfaces: An overview with examples relating to mineral deposits. In The Environmental Geochemistry of Mineral Deposits. Part A. Processes, methods and health issues. Plumlee GS, Logsdon MJ (Eds) Rev Econ Geol 6A:161-182

Sokolov PN, Korobov YI, Kosukhina IG, Kolotova LV (1985) Mineralogy of water-soluble sulfates in the Namansk manifestation of aluminite (Yakutsk ASSR). Morfol Genezis Zakonomern Razmeshcheniya Mineral Obraz Altae-Sayan Skladchatoi Obl Sib Platformy, p 120-124

Sposito G (1984) The Surface Chemistry of Soils. Oxford Univ Press, New York

Steefel CI, van Cappellen P (1990) A new kinetic approach to modeling water-rock interaction: the role of nucleation, precursors, and Ostwald ripening. Geochim Cosmochim Acta 54:2657-2677

Stumm W (1992) Chemistry of the Solid-Water Interface: Processes at the Mineral-Water and Particle-Water Interface in Natural Systems. John Wiley, New York

Sullivan PJ, Yelton JL, Reddy KJ (1988a). Solubility relationships of aluminum and iron minerals associated with acid mine drainage. Environ Geol Water Sci 11:283-287

Sullivan PJ, Yelton JL, Reddy KJ (1988b). Iron sulfide oxidation and the chemistry of acid generation. Environ Geol Water Sci 11:289-295

Swayze GA, Smith KS, Clark RN, Sutley SJ, Pearson RM, Vance JS, Hageman PL, Briggs PH, Meier AL, Singleton MJ, Roth S (2000) Using imaging spectroscopy to map acidic mine waste. Environ Sci Techno 34:47-54

Takano M, Shinjo T, Kiyama M, Takada T (1968) Magnetic properties of jarosites, $RFe_3(OH)_6(SO_4)_2$ (R = NH_4, Na or K). J Phys Soc Japan 25:902

Taylor RM and Schwertmann U (1978) The influence of aluminum on iron oxides. Part I. The influence of Al on Fe oxide formation from the Fe(II) system. Clays Clay Minerals 26:373-383

Theobald PK, Lakin HW, Hawkins DB (1963) The precipitation of aluminum, iron, and manganese at the junction of Deer Creek with the Snake River in Summit County, Colorado. Geochim Cosmochim Acta 27:121-132

Tien P-L (1968) Hydrobasaluminite and basaluminite in Cabaniss Formation (Middle Pennsylvanian), southeastern Kansas. Am Mineral 53:722-732

Tkachenko RI, Zotov AV (1982) Ultra-acidic therms of volcanic origin as mineralizing solutions. In Hydrothermal Mineral-Forming Solutions in the Areas of Active Volcanism. Naboko SI (Ed) Nauka Publishers, Novosibirsk, p 126-131

Toth A, Gecse E, Popity J (1984) Aluminite and basaluminite in the bauxite of Csordakut [Hungary]. Magy All Foldt Intez. Evi Jel 1982:423-430

Trafford BD, Bloomfield C, Kelso WI, Pruden G (1973) Ochre formation in field drains in pyritic soils. J Soil Sci 24:453-460

Trolard F, Génin J-MR, Abdelmoula M, Bourrié Humbert B, Herbillon A (1997) Identification of a green rust mineral in a reductomorphic soil by Mössbauer and Raman spectroscopies. Geochim Cosmochim Acta 61:1107-1111

Tuovinen OH, Bhatti TM, Bigham JM, Hallberg KB, Garcia O Jr, Lindström EB (1994) Oxidative dissolution of arsenopyrite by mesophilic and moderately thermophilic acidophiles. Appl Environ Microbiol 60:3268-3274

Urrutia MM, Beveridge TJ (1995) Formation of short-range ordered aluminosilicates in the presence of a bacterial surface (Bacillus subtilis) and organic ligands. Geoderma 65:149-165

Van Breemen N (1973) Dissolved aluminum in acid sulfate soils and in acid mine waters. Soil Sci Soc Am Proc 37:694-697

Van Breemen N (1976) Genesis and solution chemistry of acid sulfate soils in Thailand. Agric Res Rep 848, PhD thesis, Wageningen Univ, The Netherlands

Van Breemen N (1980) Acid sulphate soils. In Land Reclamation and Water Management. ILRI Publ 27, Wageningen, The Netherlands, p 53-57

Van Breemen N, Harmsen K (1975) Translocation of iron in acid sulfate soils: I. Soil morphology, and the chemistry and mineralogy of iron in a chronosequence of acid sulfate soils. Soil Sci Soc Am Proc 39:1140-1148

Van Mensvoort MEF, Dent DL (1998) Acid Sulfate Soils. In Lal R, Blum WH, Valentine C, Stewart BA (Eds) Methods for assessment of soil degradation. Advances in Soil Science, CRC Press, Boca Raton, p 301-335

Van'shin YV, Gutsaki VA (1982) Occurrence of aluminum minerals in the lower Volga River region. Dokl Akad Nauk SSSR 262:160-162

Warren LA, Ferris FG (1998) Continuum between sorption and precipitation of Fe(III) on microbial surfaces. Environ Sci Technol 32:2331-2337

Waychunas GA, Rea BA, Fuller CC, Davis JA (1993) Surface chemistry of ferrihydrite: Part I. EXAFS studies of the geometry of coprecipitated and adsorbed arsenate. Geochim Cosmochim Acta 57:251-2269

Waychunas GA, Xu N, Fuller CC, Davis, JA, Bigham JM (1995) XAS study of AsO_4^{3-} and SeO_4^{2-} substituted schwertmannites. Physica B 208/209:481-483

Webster JG, Nordstrom DK, Smith KS (1994) Transport and natural attenuation of Cu, Zn, As, and Fe in the acid mine drainage of Leviathan and Bryant Creeks. *In* The environmental geochemistry of sulfide oxidation, Alpers CN, Blowes DW (Eds) Am Chem Soc Symp Series 550:244-260

Webster JG, Swedlund PJ, Webster KS (1998). Trace metal adsorption onto an acid mine drainage(III) oxy-hydroxy sulfate. Environ Sci Technol 32:1361-1368

Westerveld GJW, van Holst AF (1973) Detailed survey and its application in areas with actual and potential acid sulphate soils in the Netherlands. p 243-262. *In* Acid Sulphate Soils. Dost H (Ed) Proc 1st Internat Symp Land Reclamation Institute Publ 18, 1:243-262, Wageningen, The Netherlands

Wieder RK (1994) Diel changes in iron(III)/iron(II) in effluent from constructed acid mine drainage treatment wetlands. J Environ Qual 23:730-738

Weiser HB, Milligan WO, Purcell WR (1941) Composition of floc formed at pH values below 5.5. Ind Eng Chem 33:669-672

Wieser T (1974) Basaluminite in the weathering zone of Carpathian Flysch deposits. Mineral Pol 5:55-66

Williamson MA, Rimstidt JD (1993) The rate of decomposition of ferric-thiosulfate complex in acidic aqueous solutions. Geochim Cosmochim. Acta 57:3555-3561

Wilmot RD, Young B (1985) Alunite and other aluminum minerals from Newhaven, Sussex. Proc Geol Assoc 96:47-52

Winland RL, Traina SJ, Bigham JM (1991) Chemical composition of ochreous precipitates from Ohio coal mine drainage. J Environ Qual 20:452-460

Wright LT (1906) Controlling and extinguishing fires in pyritous mines. Eng Mining J 81:171-172

Xu Y, Schoonen MAA (1995) The stability of thiosulfate in the presence of pyrite in low-temperature aqueous solutions. Geochim Cosmochim Acta 59:4605-4622

Yager DB, Mast MA, Verplanck PL, Bove DJ, Wright WG, Hageman PL (2000) Natural versus mining-related water quality degradation to tributaries draining Mount Moly, Silverton, Colorado. ICARD 2000 Proc 5th Internat Conf Acid Rock Drainage 1:535-550. Soc Mining Metall Explor, Littleton, CO

Yu J-Y, Heo B, Cho I-K, Chang H-W (1999) Apparent solubilities of schwertmannite and ferrihydrite in natural stream waters polluted by mine drainage. Geochim Cosmochim Acta 63:3407-3416

Zalba PE (1982) Scanning electron micrographs of clay deposits of Buenos Aires Province, Argentina. Dev Sedimentol 35th Internat Clay Conf 1981:513-528

Zazubina IS, Ankinovich EA(1982) Basaluminite from black-schist strata of northwestern Karatau Vopr Metall Strukt Osob Veshchestv Sostava Mestorozhd Kaz, p 3-8

Zhdanov Y, Solov'ev LI (1998) Geology and mineralogical composition of the oxidized zone in the Deputatsk tin ore deposit. Otechestvennaya Geol, p 77-79

Zodrow FL and McCandlish K (1978) Hydrated sulfates in the Sydney coalfield, Cape Breton, Nova Scotia. Can Mineral 16:17-22

8   Jarosites and Their Application in Hydrometallurgy

John E. Dutrizac

Mining and Mineral Sciences Laboratories
CANMET, Natural Resources Canada
555 Booth Street
Ottawa, Ontario, Canada K1A OG1

John L. Jambor

Leslie Research and Consulting
316 Rosehill Wynd
Tsawwassen, British Columbia, Canada V6T 1Z4

INTRODUCTION

The alunite supergroup consists of more than 40 minerals with the general formula $DG_3(TO_4)_2(OH,H_2O)_6$, wherein D represents cations with a coordination number greater or equal to 9, and G and T represent sites with octahedral and tetrahedral coordination, respectively (Smith et al. 1998). The supergroup is commonly subdivided into various groups, but the simplest primary subdivision is on the basis of the G cations. For all of the minerals in the supergroup, the dominant G cation is trivalent; most of the minerals have G represented by Fe^{3+} or Al^{3+}, but exceptions are the rare minerals gallobeudantite, in which G is Ga^{3+}, and springcreekite, in which G is V^{3+} (Table 1). Thus, the primary grouping adopted here is on whether formula Fe^{3+} exceeds or is subordinate to Al^{3+}. The hierarchical sequence in mineralogy seems to be variable, but here the decreasing sequence is given as supergroup, family, group, and subgroup. Minerals with $Fe^{3+} > Al^{3+}$ are referred to as belonging to the jarosite family, and those with $Al^{3+} > Fe^{3+}$ are allocated to the alunite family.

Subdivision of the alunite and jarosite families has also been variable; Scott (1987), for example, used seven groups, Novák et al. (1994) used six, Gaines et al. (1997) used four, and Mandarino (1999) used three. The arbitrary decision here is to use three groups, which differ from those of Mandarino (1999) but which, in general, indicate whether sulfate, phosphate, or arsenate predominates in the TO_4 tetrahedra. The three groups are the alunite group, in which TO_4 is dominated by SO_4, the crandallite group, in which (PO_4) is predominant for most of the minerals, and the arsenocrandallite group, wherein (AsO_4) is predominant for most minerals. Note that in Table 1 several of the minerals do not specify which of the two TO_4 anions predominates. This ambiguity is a consequence of the unconventional nomenclature system, which has been approved by the International Mineralogical Association, wherein the boundaries for TO_4 were set at 0.5 formula $(SO_4)_2$ and 1.5 formula $(SO_4)_2$ for the (SO_4)–(PO_4) minerals. Although the limit for $(AsO_4):(PO_4)$ substitution was not defined (Scott 1987, Birch et al. 1992), it seems to have been accepted that the boundary should lie at $(PO_4):(AsO_4) = 1:1$ (Jambor 1999, 2000, Scott 2000). The compositional compartments of the (SO_4)–(PO_4)–(AsO_4) ternary system thereby take the form shown in Figure 1.

The alunite and jarosite subgroups are characterized by their predominance (≥ 75 mol %) of SO_4 in TO_4 (Fig. 1). For end-member formulas, a monovalent cation on the alkali site provides a charge-balanced formula. Thus, partial substitution of the monovalent cation by a divalent or trivalent cation requires compensation, which is typically achieved by having some of the alkali sites remaining unoccupied. Compensation, however, can also be achieved by an appropriate amount of substitution

1529-6466/00/0040-0008$05.00

Table 1. Minerals of the alunite supergroup.

Alunite Family ($Al^{3+} > Fe^{3+}$)		Jarosite Family ($Fe^{3+} > Al^{3+}$)	
Alunite Group			
Alunite Subgroup		*Jarosite Subgroup*	
alunite	$KAl_3(SO_4)_2(OH)_6$	jarosite	$KFe_3(SO_4)_2(OH)_6$
natroalunite	$NaAl_3(SO_4)_2(OH)_6$	natrojarosite	$NaFe_3(SO_4)_2(OH)_6$
ammonioalunite	$(NH_4)Al_3(SO_4)_2(OH)_6$	ammoniojarosite	$(NH_4)Fe_3(SO_4)_2(OH)_6$
schlossmacherite	$(H_3O,Ca)Al_3(SO_4)_2(OH)_6$	hydronium jarosite	$(H_3O)Fe_3(SO_4)_2(OH)_6$
—		argentojarosite	$AgFe_3(SO_4)_2(OH)_6$
—		dorallcharite	$TlFe_3(SO_4)_2(OH)_6$
osarizawaite	$Pb(Al,Cu)_3(SO_4)_2(OH,H_2O)_6$	beaverite	$Pb(Fe,Cu)_3(SO_4)_2(OH)_6$
—		plumbojarosite	$PbFe_6(SO_4)_4(OH)_{12}$
minamiite	$(Na,Ca)_2Al_6(SO_4)_4(OH)_{12}$	—	
huangite	$CaAl_6(SO_4)_4(OH)_{12}$	—	
walthierite	$BaAl_6(SO_4)_4(OH)_{12}$	—	
Crandallite Group			
crandallite	$CaAl_3(PO_4)_2(OH,H_2O)_6$	—	
woodhouseite*	$CaAl_3[(P,S)O_4]_2(OH,H_2O)_6$	—	
goyazite	$SrAl_3(PO_4)_2(OH,H_2O)_6$	benauite	$SrFe_3(PO_4)_2(OH,H_2O)_6$
svanbergite*	$SrAl_3[(P,S)O_4]_2(OH,H_2O)_6$	—	
gorceixite	$BaAl_3(PO_4)_2(OH,H_2O)_6$	—	
plumbogummite	$PbAl_3(PO_4)_2(OH,H_2O)_6$	kintoreite	$PbFe_3(PO_4)_2(OH,H_2O)_6$
hinsdalite*	$PbAl_3[(P,S)O_4]_2(OH,H_2O)_6$	corkite*	$PbFe_3[(P, S)O_4]_2(OH,H_2O)_6$
florencite-(Ce)	$CeAl_3(PO_4)_2(OH)_6$	—	
florencite-(La)	$LaAl_3(PO_4)_2(OH)_6$	—	
florencite-(Nd)	$NdAl_3(PO_4)_2(OH)_6$	—	
waylandite	$(Bi,Ca)Al_3(PO_4,SiO_4)_2(OH)_6$	zairite	$BiFe_3(PO_4)_2(OH)_6$
eylettersite	$(Th,Pb)Al_3(PO_4,SiO_4)_2(OH,H_2O)_6$	—	
Arsenocrandallite Group			
arsenocrandallite	$CaAl_3(AsO_4)_2(OH,H_2O)_6$	—	
arsenogoyazite	$SrAl_3(AsO_4)_2(OH,H_2O)_6$	—	
kemmlitzite*	$(Sr,Ce)Al_3[(As,S)O_4]_2(OH,H_2O)_6$	—	
arsenogorceixite	$BaAl_3(AsO_4)_2(OH,H_2O)_6$	dussertite	$BaFe_3(AsO_4)_2(OH,H_2O)_6$
philipsbornite	$PbAl_3(AsO_4)_2(OH,H_2O)_6$	segnitite	$PbFe_3(AsO_4)_2(OH,H_2O)_6$
hidalgoite*	$PbAl_3[(As,S)O_4]_2(OH,H_2O)_6$	beudantite*	$PbFe_3[(As,S)O_4]_2(OH,H_2O)_6$
arsenoflorencite-(Ce)	$CeAl_3(AsO_4)(OH)_6$	—	
Others			
gallobeudantite*	$PbGa_3[(As,S)O_4]_2(OH,H_2O)_6$		
springcreekite	$BaV_3(PO_4)_2(OH,H_2O)_6$		

*Either of the indicated TO_4 anions may exceed the other. Only the alunite group has been subdivided, here into the alunite subgroup (Al > Fe) and jarosite subgroup (Fe > Al). The crandallite and arsenocrandallite groups can be subdivided similarly, but subgroup names are not assigned here.

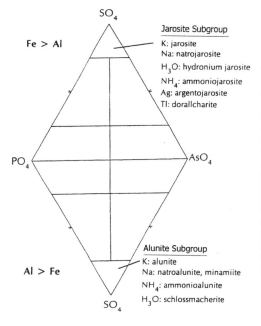

Fe > Al

SO₄

Al > Fe

SO₄

Jarosite Subgroup
K: jarosite
Na: natrojarosite
H₃O: hydronium jarosite
NH₄: ammoniojarosite
Ag: argentojarosite
Tl: dorallcharite

PO₄

AsO₄

Alunite Subgroup
K: alunite
Na: natroalunite, minamiite
NH₄: ammonioalunite
H₃O: schlossmacherite

Figure 1. Compositional limits for the minerals of the alunite and jarosite subgroups in which the alkali site is occupied by a monovalent cation. Substitution of >25 mol % AsO₄ or PO₄ for SO₄ would place such a mineral out of the fields of the alunite or jarosite subgroups and into the unnamed fields illustrated. The vertical bar separating the P-dominant and As-dominant fields, and the non-extension of the bar to the SO₄ apices, are boundaries that have not been specifically approved by the International Mineralogical Association; nevertheless, the subdivisions are assumed for functional purposes. For compositional diagrams in which the minerals have the alkali site occupied by divalent or trivalent ions, see Jambor (1999).

of SO_4, which is a divalent anion, by PO_4 or AsO_4, which are trivalent anions.

Various minerals in the alunite supergroup are of special interest to metallurgists. For example, plumbojarosite and argentojarosite at some localities have been, or are, sufficiently abundant to form ore minerals of Pb and Ag (e.g. Schempp 1923, Simons and Mapes 1956). The oldest documented exploitation of Ag-bearing jarosite was at Rio Tinto, Spain, where the pyritiferous massive sulfide deposits have been oxidized to gossan that has a jarositic accumulation near its base. It has been estimated that about 2 million tonnes of jarositic ore, probably most of it Ag-bearing, were mined in pre-Roman or early Roman times (Dutrizac et al. 1983), and the ensuing wealth may have had a profound influence on the development of the ancient world. Recent studies by Roca et al. (1999) have confirmed that solid-solution variants of jarosite and beudantite (Table 1) are the most important Ag carriers in the Rio Tinto gossan. As in Europe, exploitation of jarosites likely had a significant impact on the wealth of South America. For example, at the Matagente mine, Peru, argentiferous plumbojarosite in pockets up to 40 m thick was mined and processed by the Incas prior to the arrival of the Spaniards (Allan 1973).

Most modern-day metallurgical interest in jarositic compounds, however, is in the field of hydrometallurgy. Over the past 30 years or so, precipitation of these compounds has been widely adopted as a means of controlling iron, sulfate, alkalis, and other impurities in hydrometallurgical circuits, especially in the zinc industry. As a practical technology, jarosite precipitation was discovered simultaneously by Norzink AS (Steintveit 1965), Asturiana de Zinc (Sitges and Arregui 1964), and the Electrolytic Zinc Company of Australasia (Haigh and Pickering 1965). Although the three companies developed their processes independently, they quickly formed a consortium to market collectively the jarosite precipitation technology. The chief advantage of jarosite precipitation is that the Fe precipitate settles and filters well, and gives low losses of divalent metals, such as Zn, Cu, and Ni, in the washed residue.

SYNTHESIZED MEMBERS OF THE JAROSITE SUBGROUP

Various members of the alunite supergroup have been synthesized. Cell dimensions and sources of the data for most of the minerals of the supergroup were given by Lengauer et al. (1994), who also provided results for the synthetic Cr^{3+}, V^{3+} (Tudo et al. 1973), and various Ga^{3+}-sulfate members (Tananaev et al. 1967a,b). The In^{3+} sulfate has also been synthesized (Dutrizac 1984, Dutrizac and Mingmin 1993). Phosphate analogs containing REE, and the arsenate analogs with Ca, Sr, Ba, Pb, and of several of the REE have been prepared by Schwab et al. (1990, 1991). Although in all minerals of the supergroup the TO_4 tetrahedra are dominated by S, P, or As, the chromate (Powers et al. 1976, Cudennec et al. 1980, Baron and Palmer 1996a) and selenate (Dutrizac et al. 1981, Breitinger et al. 1997) analogs have been synthesized.

Table 2. Minerals of the jarosite subgroup, and synthetic analogs.

Formula	Mineral name	Synthetic equivalent
$KFe_3(SO_4)_2(OH)_6$	jarosite	potassium jarosite
$NaFe_3(SO_4)_2(OH)_6$	natrojarosite	sodium jarosite
$RbFe_3(SO_4)_2(OH)_6$	no mineral equivalent	rubidium jarosite
$AgFe_3(SO_4)_2(OH)_6$	argentojarosite	silver jarosite
$(NH_4)Fe_3(SO_4)_2(OH)_6$	ammoniojarosite	ammonium jarosite
$TlFe_3(SO_4)_2(OH)_6$	dorallcharite	thallium jarosite
$PbFe_6(SO_4)_4(OH)_{12}$	plumbojarosite	lead jarosite
$HgFe_6(SO_4)_4(OH)_{12}$	no mineral equivalent	mercury jarosite
$Pb(Fe,Cu)_3(SO_4)_2(OH,H_2O)_6$	beaverite	lead-copper jarosite
$(H_3O)Fe_3(SO_4)_2(OH)_6$	hydronium jarosite	hydronium jarosite (synth)

The jarosite subgroup consists of only eight minerals (Table 1). These and their synthetic counterparts are listed in Table 2 to illustrate the nomenclature used here to distinguish the two origins if they are not obvious from the context. All of the compounds correspond to the general formula $DG_3(SO_4)_2(OH)_6$, but in plumbojarosite and mercury jarosite the D cation is divalent. Consequently, only half of the D sites are filled, and ordering of the cation and vacancies results in a superstructure mineral in which the c axis of the typical unit cell is doubled to 34 Å.

OCCURRENCES OF THE JAROSITE SUBGROUP

Minerals of the jarosite subgroup are fairly common, and they form in diverse environments. Summaries of notable localities for the subgroup minerals are given by Palache et al. (1951), Gaines et al. (1997), and other textbooks; numerous European localities are discussed by Kubisz (1964). The following is only a brief discussion to provide an indication of the variety of occurrences and genesis of the minerals in the subgroup. The four principal modes of occurrence are (1) in the oxidized parts of sulfide-bearing ore deposits or barren pyritiferous rocks, including coal, and their effluents; (2) as nodules and disseminations in clays; (3) as constituents of acid soils; and (4) as hypogene minerals.

Oxidized sulfide deposits and pyritiferous rocks

Alunite has been mined in Italy for the production of alum since the middle of the fifteenth century, but jarosite was not defined until some 300 years later. Unlike alunite, the recognition of jarosite as a hypogene mineral is relatively recent. Palache et al. (1951), for example, report jarosite only as "A secondary mineral widespread as crusts and coatings on ferruginous ores and in cracks of adjoining rocks."

Argentojarosite was first observed (Schempp 1925) as a secondary product from the weathering of sulfide-rich ore veins at the Tintic Standard mine, Dividend, Utah. The mineral was deposited in a post-ore fault, and the required Ag, Fe, and sulfate were derived by weathering of the associated sulfide ore veins. End-member argentojarosite is rare, likely because of the scarcity of Ag-rich pyritiferous ores that are poor in base metals, and because Ag commonly precipitates as native silver or as a halide. Argentiferous plumbojarosite, however, is relatively common and, where abundant, is typically formed by the weathering of Ag-bearing galena-rich veins.

Plumbojarosite was first described in 1902 from an occurrence at Cook's Peak, New Mexico, and the Pb-Cu mineral beaverite (Table 1) was first found at the Horn Silver mine, Utah (Butler and Schaller 1911, Butler 1913). At the Horn Silver mine, both minerals occurred in the upper, oxidized portions of veins containing galena, sphalerite, chalcopyrite, and pyrite. The jarositic minerals had precipitated adjacent to the ore sulfides as a result of oxidation and reaction under moderately acidic conditions. In Poland, Kubisz and Zabinski (1958) reported the occurrence of plumbojarosite and jarosite on the walls of mine tunnels in Pb-Zn deposits in which the mine water was at pH 1-2. The Fe, Pb, and sulfate of the jarositic minerals were derived from the oxidation of the sulfide minerals, whereas the alkali ions likely originated from the acidic attack on the gangue minerals. Extensive substitution of Pb by alkalis, and vice versa, has been documented for the jarosite-subgroup minerals from several localities (Matvienko 1964, Enikeev 1954, Oliveira et al. 1996, Roca et al. 1999).

The mobility of Pb in sulfate media is limited by the low solubility of $PbSO_4$. There is a consequent tendency for plumbojarosite to form, during weathering, in close proximity to the source Pb minerals, especially galena; plumbojarosite is commonly observed as a capping on weathered Pb sulfide deposits, as at Chihuahua, Mexico, and at Keno Hill, Yukon, Canada, and is common in the oxidized portions of Pb-sulfide veins and other Pb-rich deposits (e.g. Enikeev 1954, Dubinina and Kornilovich 1959, Golovanov 1960, Vasilevskaya 1970). In ore deposits, plumbojarosite has been observed to have formed fine-grained nodules that resulted from the low-temperature attack on galena by cold acidic sulfate waters, as at the polymetallic ores of the Tallas Range, northern Kyrgzstan (Vakhrushev 1954). Plumbojarosite is commonly associated with other members of the subgroup, as well as with cerussite, $PbCO_3$ (Kasymov 1957) and anglesite, $PbSO_4$ (Kakhadze 1971). The association of plumbojarosite with sparingly soluble Pb minerals may indicate that Pb commonly precipitates as a carbonate or sulfate before it has the opportunity to form plumbojarosite. For example, Simons and Mapes (1956) stated that plumbojarosite is commonly the last mineral to form in the oxidation of galena, and the usual sequence observed by Simons and Mapes in the ore deposits at Hidalgo, Mexico, was galena → anglesite → cerussite → plumbojarosite. In rare cases, however, plumbojarosite has been observed as a coating on crystals of galena.

Beaverite, $Pb(Fe,Cu)_3(SO_4)_2(OH,H_2O)_6$, is similar to plumbojarosite in that it typically forms in close proximity to Fe-Pb-Cu sulfides. At the original locality, the Horn Silver mine, Utah, beaverite was associated with anglesite in oxidized portions of the ore veins. Later, beaverite was noted to occur similarly at other localities in Utah and Nevada. At copper mines in Cornwall, UK (Kingsbury 1952), beaverite occurred as partial pseudomorphs of galena, and in the Lake District, Cumbria, UK, beaverite was observed to have altered from beudantite, $PbFe_3[(As,S)O_4]_2(OH,H_2O)_6$, and thence to plumbojarosite (Kingsbury and Hartley 1957). Beaverite at the Prince Leopold mine at

Kipushi, and at Shinkolowbwe, Zaire, was observed by Van Tassel (1958a) to occur as a yellow impregnation associated with jarosite, natrojarosite, pyromorphite, and pseudomalachite in the oxidized portions of the ore veins. Beaverite has also been reported in the oxidized portions of Fe-Pb-Cu sulfide deposits in Japan (Taguchi et al. 1972, Ito 1969), at Sierra Gorda, Chile (Paar et al. 1980), at several localities in the former Soviet Union (Bolgov 1956, Kasymov 1958, Vitovskaya 1960, Moiseeva 1970, Yakhontova et al. 1988), and in several other countries. At Altyn-Topkan, Kazakhstan, beaverite replaced cerussite and did not generally precipitate directly from solution. At this locality, as in the Lake district, UK, the beaverite may well have formed at some distance from the site of weathering of the sulfides.

Alkali sulfates are highly mobile and, hence, alkali-dominant jarosite can form at considerable distance from the oxidizing sulfides. Nevertheless, alkali-jarosite minerals commonly occur in the oxidized portions of sulfide deposits (Chuhkrov 1950, 1959; Saksela 1952, Kazakhashvili 1968, Raade 1971, Kunov and Mandova 1997), in many instances as coatings on, or pseudomorphs of, Fe sulfide minerals. The occurrence of jarosite pseudomorphs after pyrite (Furbish 1963, Gorbach et al. 1967) indicates that the groundwater must already have contained sufficient K^+ to produce jarosite, and that the jarosite formed immediately at the site of pyrite oxidation. In most cases, however, direct contact between precursor sulfides and jarosite is less intimate even though a genetic relationship may be apparent. The alkali ions may be available either from distally sourced groundwater, or from local aluminosilicates that are susceptible to partial dissolution by the low-pH solutions developed by oxidation of the sulfides (Jambor et al. 2000). The relative proportion of K:Na determines which of the Na-K subgroup members is precipitated; because of the greater thermodynamic stability of the K-rich member, only solutions with a fairly high Na/K ratio will give natrojarosite (Stoffregen et al., this volume; Glynn, this volume).

Because of the stability of the alkali-rich jarosites relative to that of hydronium jarosite, the latter can form only from solutions poor in ions such as Na^+ and K^+. Kubisz (1961) concluded that such conditions exist in nature only where pyrite or other iron sulfides undergo extremely rapid oxidation. Favorable conditions may be more common in arid environments, where the water supply is sufficient to promote the local, initial oxidation to Fe sulfates, but is insufficient to allow extensive interaction with alkali-laden groundwater. Kubisz (1960) identified several occurrences of hydronium jarosite, and although numerous other occurrences have been reported (Roberts et al. 1974, Gaines et al. 1997), the mineral overall seems to be relatively rare and quantitatively sparse. All of the source materials for the analyses of hydronium jarosite ('carphosiderite') listed in Palache et al. (1951) have proved to be natrojarosite and jarosite (Moss 1956, Van Tassel 1958b). The identification of hydronium jarosite is complicated by the similarity of the X-ray pattern to those of the alkali jarosites, and by the insensitivity of Na to analysis by energy-dispersion SEM (scanning electron microscopy). Many jarosite minerals "identified" as hydronium jarosite on the basis of X-ray diffraction and SEM–energy-dispersion analyses have been shown to be natrojarosite by more sensitive wavelength-based electron microprobe analyses.

The oxidation of iron sulfides and subsequent precipitation of jarosite associated with mineral deposits is a common, widespread phenomenon (Blanchard 1968, Anderson 1981), and hundreds of occurrences have been reported in the literature. Equally numerous are occurrences not directly associated with mineral deposits. For example, jarosite is a common alteration product of pyritiferous shales (Dudek and Fediuk 1955, Kubisz and Michalek 1959, Khusanbaev and Nurullaev 1969, Penner et al. 1970, Zakirov 1978, Ievlev 1988) and other sedimentary rocks (Kubsiz 1962, Savellev and Trostyanskii

1970, Kashkai et al. 1974, Horakova and Novák 1989). Other occurrences have resulted from the oxidization of pyrite in phyllites, schists, various igneous rocks, and coal (e.g. Lofvendahl 1980, Bigham and Nordstrom, this volume). At some localities the alkali required to form the jarosite in sedimentary rocks was derived by acid attack of glauconite (Tyler 1936, Mitchell 1962), and jarosite pseudomorphs after glauconite have been observed (Biggs 1951).

The formation of ammoniojarosite is rare in nature because of the scarcity of concentrated ammonium-bearing solutions. For this reason, NH_4-rich jarosite and ammoniojarosite are typically associated with carbonaceous beds (Erickson 1922, Shannon 1927, Odum et al. 1982). The ammonium originates from the decomposition of organic substances whereas the Fe and sulfate are derived from the oxidation of associated pyrite. Aside from the scarcity of NH_4-rich solutions, the $K^+:NH_4^+$ partition coefficient strongly favors the formation of jarosite rather than ammoniojarosite.

Nodules and disseminations in clays

Jarosite commonly occurs intimately mixed with clays in seams and in thick clay beds. The general assumption is that the Fe and sulfate originate from the oxidation of pyrite whereas acid leaching of the clay minerals provides the alkalis (Warshaw 1956). The pyrite may occur within the clay (Krazewski 1972, Goldbery 1978), thereby generating the ferric sulfate solutions *in situ*. Alternatively, the jarosite could have precipitated directly, such as from ponded solutions (Gryaznov 1957, Alpers et al. 1992, Long et al. 1992), or the solutions could have been transported into the sedimentary unit from distal sources (Khlybov 1976).

Jarosite and natrojarosite have also been observed as large concretions in clay (Prokhorov 1970), as microconcretions and 1-4 mm seams in clay (Gorbach et al. 1967), and as small aggregates (Van Tassel 1965). The constituent elements in all of these occurrences were interpreted to have been derived through the *in situ* oxidation of pyrite and the alteration of the associated aluminosilicates.

Acid soils

Although the oxidized portions of sulfide ore deposits and pyritized rocks provide localized concentrations of jarosite and an obvious link between jarosite formation and oxidizing sulfides, probably the most widespread occurrence of jarosite is in acid soils. Throughout the world there are extensive regions of acid (pH 3-4) soils which have developed in marine sediments containing pyrite. These soils present serious problems for agricultural utilization; mine tailings are an equivalent anthropogenic 'soil', and most present the same difficulties concerning their utilization. Bacteria in soils have been shown (Ivarson 1973, Ivarson et al. 1976, 1979; Banfield and Nealson 1997, Banfield and Welch 2000) to play major roles both in mineral weathering and in the oxidation of pyrite to solutions rich in ferric sulfate and sulfuric acid. It is known that the alkali ions needed for jarosite precipitation can be liberated readily from local sources, particularly the trioctahedral micas (Jambor 1998, Jambor and Blowes 2000).

Jarosite- and natrojarosite-bearing soils are widespread, and among the many occurrences that have been described are those in Canada (Clark et al. 1961, Dudas 1984, 1987, Mermut and Arshad 1987), Germany (Schwertmann 1961), Scandanavia (Oeborn and Berggren 1995), Senegal (Le Brusq et al. 1986, Montoroi 1995), Korea (Shin and Jang 1993, Jung and Hoo 1994), Japan (Hyashi 1994), and the United States (Fleming and Alexander 1961). Jarosite is a probable constituent of many yellow soils, especially those formed from marine sediments containing small concentrations of pyrite. Research has established jarosite to be a common constituent of acid soils throughout the world

(Van Breemen and Harmson 1975, Kittrick et al. 1982, Fanning et al. 1993, Van Breemen and Buurman 1998).

Hypogene jarosite

Although jarosite was long known to be a supergene mineral, its recognition specifically as a hypogene mineral is relatively recent. Allen and Day (1935) reported the presence of jarosite in hot-spring deposits at Yellowstone National Park, and similar deposits were also noted decades ago in Japan (Kinoshita et al. 1955, Saito 1962), in the Kuril Islands (Naboko 1959, Kulugin 1967), and in Java, Indonesia (Zelenov and Tkachenko 1970). Zotov (1970) showed that deposition of jarosite at one of the Kuril Island sites (Mendeleev Volcano, Kunashir Island) occurred from relatively dilute sulfate-chloride (2-4.5 g/L) waters at pH 1.5-2.5.

Many of the early published occurrences of jarosite were summarized by Hutton and Bowen (1950) who, in their study of altered, tourmaline-bearing volcanic rock at Stoddard Mountain, California, concluded that the associated, disseminated jarosite was of hydrothermal origin. Low-temperature, chloride- and sulfate-rich hydrothermal solutions that reacted with sulfides were invoked by Seeliger (1950) to account for an extensive suite of sulfate, carbonate, and chloride minerals, among which were natrojarosite and plumbojarosite. Bonorino (1959) included woodhouseite and jarosite in his hydrothermal alteration assemblages associated with ore veins of the Front Range mineral belt, Colorado, and Nicholas and de Rosen (1963) included natrojarosite and gorceixite as part of their low-temperature hydrothermal alteration assemblages in the Colettes massif, France. Berzina et al. (1966) observed jarosite, with molar K:Na = 6.0, in an albitized, magnetite-free granite in which the pyrite showed no sign of oxidation. The jarosite was present predominantly as fine-grained aggregates, but as some occurred in small crystals encapsulated in quartz, either the two minerals were coeval or the quartz was younger. The formation of the jarosite was attributed to hypogene crystallization during albitization. Hypogene jarosite was subsequently reported to occur in "secondary quartzites" [high-sulfidation deposits] that formed in massifs in northern Balkhash, Kazakhstan (Kalinichenko 1973). Johnson (1977) observed the presence of non-supergene encrustations of jarosite, also reported as natrojarosite by Slansky (1975), that were deposited chiefly on andesitic boulders and rock fragments at gently steaming areas and fumarole vents at White Island, New Zealand. Jarosite was described by Lombardi and Sonno (1979) as a hydrothermal alteration product in alunitized volcanic rocks at Casale de Mezzano, Italy, and Nekrasov and Nikishova (1987) reported the occurrence of argentojarosite as a primary mineral in polymetallic ore deposits in northeastern Siberia, Russia. Jarosite and natrojarosite have since been recognized to be common, although generally in small amounts and rare relative to alunite, in high-sulfidation (acid sulfate) deposits (Cunningham et al. 1984, Stoffregen 1987, Ebert and Rye 1997, Rye and Alpers 1997).

Erd et al. (1964) reported the possible existence of ammoniojarosite in hydro-thermally altered rock at the Sulphur Bank hot-spring mercury deposit in California. The ammoniated minerals were formed at temperatures below 120°C by reaction of the host rock and ascending waters containing 460-540 mg/L NH_4. Significantly, the waters were low in K so that the ammoniojarosite could form in preference to (K) jarosite. Sulfate ion was present in the circulating waters, and ferric ion likely originated with the host rocks. Ammoniojarosite has also been reported (Altaner et al. 1988) in association with ammonioalunite as a precipitate at The Geysers hot springs, Sonoma County, California.

Alteration of minerals of the jarosite subgroup

It is evident that members of the jarosite subgroup can form under a variety of geological conditions. The essential requirement seems to be an Fe^{3+}-bearing, acid environment (pH <3); once formed, however, the minerals decompose readily upon removal from their stability region. Although estimates of the rate of solubility of jarosite in the weathering environment have varied widely (Baron and Palmer 1996b), the typical scenario is that the minerals, regardless of rate, alter to goethite. Thus, in gossans, such as those of the Iberian pyrite belt, the near-surface portion of the gossan is goethite-rich and jarosite-poor, and only at depth does jarosite become abundant. This type of profile is common in mature gossans, but can develop in naturally weathered tailings after only a few years of sulfide oxidation.

Minerals in the jarosite subgroup are not stable at elevated temperatures such as those that occur during intense thermal metamorphism. The phosphate members of the supergroup, however, are more stable and several occurrences have been reported in high-grade, Al-rich rocks. Stabilities of the alunite and jarosite minerals are discussed in detail in the papers by Bigham and Nordstrom (chapter 7, this volume), and Stoffregen et al. (chapter 9, this volume).

In dynamic heating experiments, all of the compounds of the jarosite subgroup commence to lose water at about 400°C, e.g.

$$KFe_3(SO_4)_2(OH)_6 \rightarrow KFe(SO_4)_2 + Fe_2O_3 + 3 H_2O \qquad (1)$$

$$2 KFe(SO_4)_2 \rightarrow K_2SO_4 + Fe_2O_3 + 3 SO_3 \qquad (2)$$

Although the water loss occurs within a few hours at 400°C, decomposition occurs over several days at temperatures as low as 325°C. Hence, in nature even moderately low-temperature metamorphic environments would result in the eventual destruction of jarosite, thereby producing goethite-hematite and soluble sulfates. Each mineral in the subgroup behaves slightly differently with respect to the initial decomposition temperature. Also, plumbojarosite and argentojarosite apparently do not form mixed sulfates, $MFe(SO_4)_2$; rather, mixtures of ferric sulfate and Pb sulfate or Ag sulfate, respectively, are formed at about 400°C. In aqueous media, potassium jarosite, silver jarosite, and thallium jarosite are stable somewhat above 200°C, but sodium jarosite, ammonium jarosite, and lead jarosite typically decompose in the range 180-190°C if the acid concentration is ~0.3M H_2SO_4. The decomposition at high temperatures yields hematite.

$$2 NaFe_3(SO_4)_2(OH)_6 \rightarrow Na_2SO_4 + 3 Fe_2O_3 + 3 H_2SO_4 + 3 H_2O \qquad (3)$$

In solutions, the jarosite precipitation reaction is reversible:

$$K_2SO_4 + 3 Fe_2(SO_4)_3 + 12 H_2O \Leftrightarrow 2 KFe_3(SO_4)_2(OH)_6 + 6 H_2SO_4 \qquad (4)$$

In a batch precipitation reaction without the addition of a neutralizing agent, precipitation continues until the equilibrium acid concentration is reached, at which point the net precipitation of jarosite ceases. If acid is added to the system, part of the precipitated jarosite will dissolve. Accordingly, high initial acid concentrations will result in the complete alteration of the jarosite. At about 100°C, an acid concentration in excess of 0.3 M (~30 g/L H_2SO_4) is required to dissolve sodium jarosite or lead jarosite, and appreciably higher acid concentrations (<0.5 M or ~50 g/L H_2SO_4) are needed to decompose the more stable jarosite species, such as potassium jarosite or silver jarosite. Such elevated concentrations of acid are rare in nature, but are commonly encountered in metallurgical processing circuits. For example, Jackson (1976) proposed dissolving ammonium and potassium jarosites, which might be produced in a hydrometallurgical zinc operation, in concentrated sulfuric acid to yield an ammonium sulfate–potassium sulfate–iron sulfate solution. The intent of the leaching operation was to generate a solution which could serve as the basis for the manufacture of enriched fertilizers. Sahoo and Das (1986) proposed leaching a commercially produced sodium jarosite–lead jarosite residue in 50 g/L H_2SO_4 at 90°C for 20 h to decompose the jarosites completely and to convert the Pb to $PbSO_4$ for a subsequent recovery process. Their process clearly demonstrated the solubility of jarosite-subgroup species in concentrated acid media. Although most commercial leaching operations are based on the utilization of low-cost H_2SO_4, leaching with HCl offers the advantage that both Ag and Pb are directly solubilized and can be subsequently recovered from the solution. As an example, it was possible to completely dissolve a plumbojarosite ore in 20 min at 90°C using a 0.25 M HCl solution also containing 200 g/L $CaCl_2$ to enhance the solubility of the $PbCl_2$ reaction product (Viñals and Nunez 1988, Viñals et al. 1991a,b). The leaching rate increased as the 0.8 power of the HCl concentration in the absence of added $CaCl_2$, but only as the 0.5 power of the HCl concentration in the presence of 200 g/L $CaCl_2$. The equilibrium solubility of sodium jarosite in H_2SO_4 solutions was recently determined by Sasaki et al. (1998), and the results confirmed the reversibility of reaction (4).

Although acid dissolution of jarosite is feasible, alkaline decomposition is a more effective technique because the associated Fe is reprecipitated simultaneously. In neutral or alkaline media, jarosite exists in equilibrium with K, Fe^{3+}, SO_4, and OH:

$$KFe_3(SO_4)_2(OH)_6 \Leftrightarrow K^+ + 3\ Fe^{3+} + 2\ SO_4^{2-} + 6\ OH^- \tag{5}$$

If the pH is increased by the addition of a neutralizing reagent such as ammonia or lime, the low concentration of dissolved Fe is precipitated as ferrihydrite or goethite; for example:

$$Fe^{3+} + 3\ OH^- \Leftrightarrow FeO(OH) + H_2O \tag{6}$$

The removal of the dissolved Fe^{3+} displaces the jarosite dissolution equilibrium such that further dissolution of the jarosite occurs. The final result is the complete conversion of the jarosite to goethite, with the formation of soluble alkali-sulfate salts and water. This reaction occurs naturally over many years, where the neutralizing agent is near-neutral water, but the reaction can be greatly accelerated in metallurgical processes.

The use of ammonia at ~100°C results in the complete reaction of jarosite in less than 15 min, with the formation of hematite in place of goethite (Kunda and Veltman 1979). An excess of ammonia is required according to the following stoichiometry shown for a commercial ammonium jarosite:

$$NH_4Fe_3(SO_4)_2(OH)_6 + 3\ NH_3 \Leftrightarrow 1.5\ Fe_2O_3 + 2\ (NH_4)_2SO_4 + 1.5\ H_2O \tag{7}$$

If the decomposition parameters were controlled properly, a pregnant solution containing >250 g/L $(NH_4)_2SO_4$ could be produced, and it was stated that such solutions could be employed for the recovery of $(NH_4)_2SO_4$ by crystallization. The use of other jarosite species would result in mixed K_2SO_4–$(NH_4)_2SO_4$ or Na_2SO_4–$(NH_4)_2SO_4$ solutions.

Silver is commonly present in solid solution both in natural jarosites and in the jarosites formed during metallurgical processing. The recovery of the Ag requires that the jarosite crystal-structure be destroyed, and alkaline decomposition is commonly used for this purpose. Viñals et al. (1991a,b) used a 60-min pretreatment with lime at pH 11-12 to decompose plumbojarosite present in hematite tailings. A temperature of >60°C was necessary to ensure complete alteration of the plumbojarosite so that high Ag recoveries could be obtained in a subsequent cyanidation operation. Cruells et al. (2000) showed that NaOH and $Ca(OH)_2$ were similarly effective in decomposing jarosite. At 30°C, less than 1 h was required to effect the complete alteration of the jarosite. In the jarosite decomposition reaction in NaOH media, the rate varied as the 0.6 power of the NaOH concentration and had an apparent activation energy of 43 kJ/mol. The rate in $Ca(OH)_2$ solutions varied as the 0.5 power of the $Ca(OH)_2$ concentration, and the apparent activation energy was 80 kJ/mol. Microscopic studies of partly reacted jarosite particles revealed a continuous rim of poorly crystalline $Fe–O–OH–H_2O$ surrounding each jarosite core. Complementary work (Roca et al. 1999) indicated that the alkaline decomposition of an arsenical potassium jarosite is characterized by the removal of K and sulfate from the jarosite, with the accumulation of Fe, Pb, and As in the altered residue. Temperatures above 70°C were required to achieve a high degree of alteration within 5 h if a saturated $Ca(OH)_2$ solution [2 g/L $Ca(OH)_2$] was used. The apparent activation energy was 87 kJ/mol, and this value led the authors to conclude that the reaction for jarosite decomposition was chemically controlled. It was noted that the alkaline decomposition of argentian plumbojarosite was also chemically controlled, and that the decomposition rate seemed to decrease as the saturation limit of $Ca(OH)_2$ was approached. This was attributed to the formation of $CaCO_3$, by reaction with atmospheric CO_2, that tended to block the surface of the plumbojarosite (Patino et al. 1994). Patino et al. (1998) also investigated the use of lime (CaO) to decompose synthetic plumbojarosite, and although the reaction products were not exhaustively studied, it was claimed that the lead jarosite decomposed to form $Pb(OH)_2$:

$$Pb_{0.5}Fe_3(SO_4)_2(OH)_6 + 4\ OH^- \Leftrightarrow 0.5\ Pb(OH)_2 + 3\ Fe(OH)_3 + 2\ SO_4^{2-} \qquad (8)$$

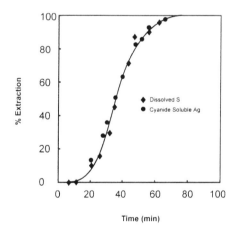

Figure 2. Decomposition curve for argentian plumbojarosite reacted at 40°C in a $Ca(OH)_2$ solution at pH 11.2. The diamonds indicate the direct dissolution of sulfate from the jarosite, and the solid circles indicate the extraction of Ag from the decomposed jarosite in a subsequent cyanidation step. (After Patino et al. 1998).

As shown in Figure 2, the synthetic plumbojarosite decomposed completely within 60 min at 40°C in a lime solution at pH 11.2. Increasing temperatures increased the rate of jarosite decomposition, and the reported activation energy was 45 kJ/mol. The decomposition rate in weakly alkaline media increased as the 0.5 power of the OH concentration, but the rate was apparently independent of OH concentration for concentrations >0.001 M.

In the above studies, a reagent base such as NH_3, NaOH, or CaO was used to effect the decomposition of the jarosite. The nature of the base, however, seems to be largely immaterial; for example, Ding et al. (1997) utilized an alkaline fly ash from coal combustion to decompose precipitates of commercial jarosite. The intent was to use the decomposition products, resulting from the layering of fly ash with deposited residue from a zinc plant, to "seal" the mass and thereby prevent negative environmental consequences that might arise from the impounded jarosite. It was shown that the alkaline fly ash began to react with the layered jarosite within one day, but that the reaction continued for several months as the alkaline constituents diffused into the massive jarosite layer. The resulting $Fe(OH)_3$–gypsum mixture "sealed" the resulting mass and significantly reduced its permeability.

CONDITIONS AFFECTING THE SYNTHESES
OF THE JAROSITE SUBGROUP

Sodium, potassium, and ammonium jarosites

Most natural occurrences of the subgroup are of minerals in the Na–K solid-solution series, commonly with additional minor H_3O^+ substitution. In commercial applications, however, most of the activity has centered on sodium jarosite or ammonium jarosite because of the lower costs of the Na and NH_4 salts; only limited work has been done with the salts of K because of their higher cost. Nevertheless, the combination of natural occurrence and commercial syntheses has led to a substantial body of literature on sodium jarosite, potassium jarosite, and ammonium jarosite, and especially on their mutual solid solutions and conditions of formation.

Fairchild (1933) was likely the first to attempt a systematic synthesis of jarosite. In his small-scale experiments, K and Fe^{3+} sulfates at a 1:3 ratio, the same as in jarosite, were combined with 0.75 N H_2SO_4. Jarosite began to form at 110°C; lower temperatures yielded other basic Fe^{3+} sulfates. It is now recognized that a stoichiometric K:Fe ratio is not necessary, and that a 0.75 N acid concentration is too high and accounts for the need for synthesis temperatures >100°C. Other formation parameters were not studied in any detail. However, the multitude of syntheses that have been done over half a century have allowed a fairly accurate definition of many of the synthesis parameters, as is summarized below.

Temperature. Sodium, potassium, or ammonium jarosite can be formed at 25°C although the rates are very slow (Brown 1971, Babcan 1971). Precipitation of synthetic jarosite becomes rapid about 80°C and is nearly complete in several hours at 100°C (Brophy et al. 1962, Steintveit 1970, Getskin et al. 1975, Dutrizac and Kaiman 1976, Margulis et al. 1976a,b; Kershaw and Pickering 1979, Pammenter and Haigh 1981). The rate increases rapidly above 100°C, but there is an upper temperature limit for jarosite stability. This limit likely varies with solution composition and the jarosite species, but it seems to be about 180-200°C (Haigh 1967, Babcan 1971). For a given set of synthesis conditions, the yield of precipitate increases with increasing temperature, and the alkali content is slightly raised (Dutrizac 1983). Commercially useful rates generally require temperatures above 90°C.

These features are illustrated in Figure 3, which shows the effect of reaction time at 97°C, and in Figure 4, which illustrates the influence of temperature on the amount and composition of the sodium jarosite made using a constant 24 h reaction time. Figure 3 shows the percentage of Fe precipitated, which increases sharply with increasing reaction periods for about 15 h, but is nearly independent of retention time thereafter. The Fe is initially present as 0.2 M Fe^{3+}, and its precipitation is proportional to the amount of sodium jarosite formed, as indicated by the content of ~32 wt % Fe in all of the precipitates. Although ~15 h are required to reach a steady-state condition at 100°C, it is known that such a state is achieved in ~1 h at 140°C. Although the amount of precipitate varies significantly with retention time, the composition of the sodium jarosite, as reflected by the Na contents shown in Figure 3, is nearly the same throughout the

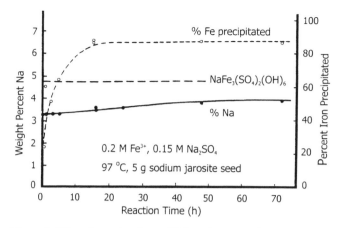

Figure 3. Effect of retention time at 97°C on the extent of Fe precipitation and the Na content of the jarosite formed. The theoretical Na content of $NaFe_3(SO_4)_2(OH)_6$ is 4.74 wt % (Dutrizac 1983).

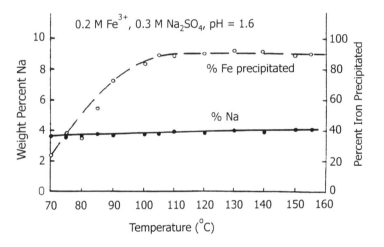

Figure 4. The percentage of the initially present Fe precipitated, and the Na content of the jarosite precipitate, as a function of the reaction temperature for a 24-h retention time (Dutrizac 1983).

reaction. Figure 4 indicates that the amount of sodium jarosite formed in a 24-h reaction period increases sharply from 70°C to ~100°C, but levels off for higher synthesis temperatures. From a metallurgical perspective, the graph illustrates the importance of operating near the boiling point of the solution to achieve maximum rejection of Fe in a reasonable time. Figure 4 also suggests that the reaction (for a 24-h retention time) is under thermodynamic control at temperatures >100°C, but is kinetically limited at the lower temperatures. Although the amount of precipitate formed is strongly temperature-dependent, the composition of the jarosite precipitate is nearly temperature-independent, as is suggested by the Na contents of the precipitates shown in Figure 4. That is, the system makes more or less of the same jarosite regardless of the temperature.

pH. The solution pH plays a major role in both the stability of jarosite (Brown 1971,

Stahl et al. 1993, Bigham et al. 1996) and the extent of its precipitation. Increasing acid concentrations retard the extent of precipitation of synthetic jarosite (e.g. Harada and Goto 1954, Brophy et al. 1962, Haigh 1967, Brown 1970, Getskin et al. 1975). The extent of precipitation increases with increasing pH until other Fe compounds are precipitated, and this occurs at pH > 2 at 100°C (Babkan 1971). A pH of 1.5-1.6 seems to be ideal for jarosite formation at 100°C (Park and Park 1978). In practice, higher temperatures can be used to offset excessive acidities.

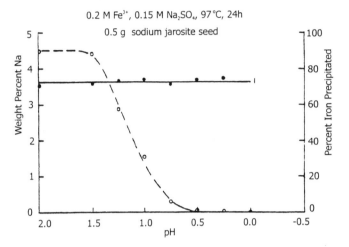

Figure 5. The influence of the initial room-temperature pH on the extent of Fe precipitation (open circles) and the Na content of the jarosite precipitate (Dutrizac 1983).

Figure 5 shows the effect of the initial solution pH on the precipitation of sodium jarosite at 97°C (Dutrizac 1982). A nearly constant amount of jarosite is precipitated for pH values <1.5, but the amount of precipitate decreases sharply at lower pH values. When the initial pH is ~0.5, no precipitate forms because the equilibrium has been displaced in favor of solution species.

$$Na_2SO_4 + 3Fe_2(SO_4)_3 + 6H_2O \Leftrightarrow 2NaFe_3(SO_4)_2(OH)_6 + 6H_2SO_4 \qquad (9)$$

Although the amount of jarosite is strongly influenced by the acidity of the solution, the composition of the precipitate, as reflected by the Na content shown in Figure 5, remains nearly constant. Once again, more or less jarosite is formed, but the composition of the jarosite is nearly constant.

Alkali concentration. The extent of Fe precipitation increases with an increasing M^+/Fe^{3+} ratio to slightly above stoichiometric and is, thereafter, nearly independent of the amount of alkali (Haigh 1967, Steintveit 1970, Getskin et al. 1975, Dutrizac and Kaiman 1976, Park and Park 1978, Kershaw and Pickering 1979). Earlier work which concluded that the stoichiometric ratio of M^+/Fe^{3+} gave the optimum product (Fairchild 1933, Brophy et al. 1962, Kubisz 1964) seems to have been in error, as higher alkali concentrations have been found to slightly improve both the yield and alkali content of the precipitate (Dutrizac 1983). Jarosites can be precipitated from solutions containing as little as 0.05 molar Na (Dutrizac 1999a) or 0.02 molar K (Brown 1970). Very high alkali-sulfate concentrations (e.g. >1 M Na_2SO_4) can result in the precipitation of alkali–Fe^{3+} sulfates instead of jarosites.

Iron concentration. Synthetic jarosite is readily precipitated from solutions containing 0.025-3.0 molar Fe^{3+} (Brophy et al. 1962, Brophy and Sheridan 1965, Brown 1970). The lower limit of Fe^{3+} for jarosite production is around 10^{-3} M (Brown 1971). If an excess of alkali is available, the fraction of Fe that is precipitated is independent of the Fe concentration of the solution; the consequence is that the total amount of jarosite formed increases linearly with increasing concentrations of Fe^{3+}, as is illustrated in Figure 6 for the precipitation of silver jarosite (Dutrizac and Jambor 1984). Despite the variations in the amount of precipitate formed, the composition of the jarosite precipitate is nearly independent of the Fe^{3+} concentration of the solution, even for relatively low Fe^{3+} concentrations. Most accurate analyses of synthetic jarosite have revealed a slight deficiency of Fe relative to sulfate (Kubisz 1970, Dutrizac and Kaiman 1976, Hartig et al. 1990).

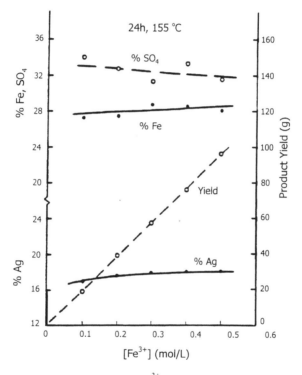

Figure 6. Effect of the initial Fe^{3+} concentration on the yield and composition of silver jarosite precipitated at 155°C at a constant ratio of dissolved Fe^{3+} to Ag. (Dutrizac and Jambor 1984).

Order of stability. The extent of Fe precipitation (and hence of jarosite stability) for the various synthetic jarosites is K > NH_4 > Na; ammonium ion is only slightly superior to Na (Haigh 1967, Steintveit 1970, Getskin et al. 1975, Margulis et al. 1977). Figure 7 provides a good indication of the greater stability of potassium jarosite relative to either sodium jarosite (Fig. 7a) or ammonium jarosite (Fig. 7b). The dashed lines in Figure 7 represent the ideal behavior which would result if K and Na were precipitated in the same molar ratio as they occurred in solution. The results indicate that K is precipitated preferentially. For example, a solution containing only 10 mole % K gives rise to a jarosite solid-solution series containing >80 mole % K. This type of dependence is useful

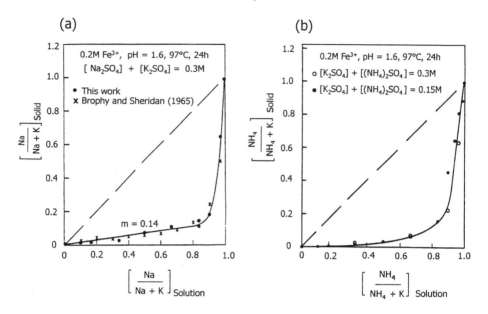

Figure 7. (a) Molar partitioning of Na and K between the solution and the jarosite precipitate which is a K-Na jarosite solid solution. Comparison with the data of Brophy and Sheridan (1965) is good. (b) Molar partitioning of NH_4 and K between the solution and the jarosite precipitate, the latter with K–NH_4 solid solution. Data are from Dutrizac (1983).

for understanding metallurgical reactions, and it also rationalizes the significantly greater abundance of K-rich jarosite relative to Na-rich jarosite in nature. Studies of synthetic systems (Dutrizac 1983, 1991, Dutrizac and Jambor 1984) have demonstrated that silver jarosite is almost as stable as potassium jarosite, and thallium jarosite is only slightly less stable. Rubidium jarosite is somewhat less stable than potassium jarosite (Dutrizac and Kaiman 1975), and lead jarosite seems to have a stability similar to that of sodium jarosite. Little is known about the relative stability of mercury jarosite, but the difficulty of its synthesis suggests that it is the least stable of the jarosite subgroup.

Solid solution. In addition to the complete solid solubility of hydronium jarosite in the alkali jarosites, there is complete solid solubility among potassium jarosite, sodium jarosite, and ammonium jarosite (Kubisz 1964, Brophy and Sheridan 1965). Rubidium jarosite forms a complete solid-solution series with potassium jarosite (Dutrizac 1983). Silver jarosite and thallium jarosite form complete solid-solution series with potassium jarosite. The implication is that solid solution likely is complete among most, if not all, of the jarosite species. Lead jarosite forms a series with the monovalent jarosites, but such series are not crystallographically perfect because order-disorder commonly leads to superstructure effects in Pb-rich members. Thus, lead jarosite commonly possesses a 34 Å *c* axis, which is double that of the monovalent jarosite members (Dutrizac and Jambor 1984, 1987a).

Ionic strength. The presence of $ZnSO_4$ has been shown to have little effect on either the yield or composition of precipitated alkali jarosites (Getskin et al. 1975, Zapuskalova and Margulis 1978). The synthesis of sodium jarosite has been shown to be independent of large quantities of Li_2SO_4 (Dutrizac and Jambor 1987b). Ionic strength by itself is not a major factor in jarosite synthesis.

effects), and purity of the reagents. Hence, it is not surprising that wide variations have been noted for the effect of seeding. Seeding has been reported by some authors to have only a slight effect (Getskin et al. 1975, Park and Park 1978, Kershaw and Pickering 1979), whereas abundant seeding has been noted by others to have a major influence (Pammenter and Haigh 1981, Dutrizac 1996, 1999a). This topic is discussed in more detail later in this chapter.

Hydronium substitution. The existence of hydronium ion (H_3O^+) in jarosite species was confirmed by Ripmeester et al. (1986) by NMR techniques. All recent workers agree that the hydronium ion substitutes for the various alkalis, typically in amounts equivalent to 15-20 mole %. The relative order of increasing hydronium substitution in synthetic jarosites is Na > NH_4 > K (Margulis et al. 1977). Brophy and Sheridan (1965) reported that increased temperature decreased the extent of hydronium substitution, but the opposite was observed by Margulis et al. (1976b, 1977).

JAROSITE PRECIPITATION IN THE ZINC INDUSTRY

Outline of the jarosite process

The jarosite process was developed to meet the need for an economic means of removing the high concentrations of Fe, typically 5 to 12 wt %, that are commonly present in Zn-sulfide (sphalerite) concentrates (Arregui et al. 1979). In practice, it is necessary to oxidize the Zn-sulfide concentrates, which in commercial operations typically contain about 50 wt % Zn and 30 wt % S, and which have averaged 7-8 wt % Fe for at least the past 15 years (Dutrizac 1999b). Roasting of the concentrate is done in air. In the roasting process, the product of which is referred to as calcine, much of the Zn that was in the concentrates is oxidized to ZnO, but part of the Zn also combines with the available Fe to form zinc ferrite, $ZnO \cdot Fe_2O_3$ (the synthetic equivalent of franklinite, $ZnFe_2O_4$), and a small amount of the Zn reacts to form Zn-bearing silicates. In the subsequent sulfuric-acid leaching of the calcine, which is referred to as the neutral-leach step (terminal pH > 4), the ZnO is readily dissolved to form a $ZnSO_4$-rich and almost Fe-free solution that is amenable to purification and electrolysis; in this neutral-leach step, however, the zinc ferrite and Zn-bearing silicates in the calcine are resistant to dissolution. Consequently, to avoid the potentially significant loss of the Zn that is present as ferrite and silicates, this residue is subsequently dissolved in hot, concentrated sulfuric acid. Although the hot acid leach effectively dissolves almost all of the Zn that is present as Zn ferrite or Zn-bearing silicate, the leach also solubilizes the associated Fe to yield a solution typically containing 20-30 g/L Fe^{3+}. The associated Zn and residual acid are readily recycled within the process, but the Fe must first be removed. This is commonly done by precipitating a jarosite compound. The resulting nearly neutral Zn-sulfate solutions are purified and then electrolyzed to yield Zn metal and sulfuric acid, and the acid (spent electrolyte) is recycled to the leaching operation (Fig. 8). Thus, from the point of view of zinc production, the roast–leach–electrolysis process is a nearly ideal technology insofar as it yields a high-purity metal directly, most impurities are rejected, and no excessive amounts of reagent other than air and electricity are consumed. Moreover, the off-gasses from the roasting process are collected and converted to H_2SO_4, which is partly utilized in the leaching process.

Over the past 15 to 20 years, the roast–leach–electrolysis process has accounted for about 80% of the world's production of Zn metal, which is currently about 8 Mt annually. An inherent part of the technology in most applications is the precipitation of jarosite as a means of eliminating the Fe that is solubilized with the Zn during the dissolution of the

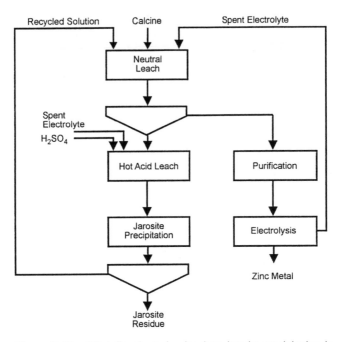

Figure 8. Simplified flowsheet showing how jarosite precipitation is incorporated into a hydrometallurgical zinc processing circuit to eliminate the Fe dissolved during hot acid leaching.

residue that contains the zinc ferrite. Mere neutralization of a Fe-pregnant solution will yield a gel-like ferric oxyhydroxide phase that does not settle and cannot be filtered easily. The jarosite process, however, provides a means of overcoming the Fe-precipitation problem. To effect the precipitation, an alkali (generally Na as Na_2SO_4 or Na_2CO_3, or NH_3 as aqueous ammonia) is added to the hot leach liquors, and the pH is adjusted by the addition of calcine. Precipitation of sodium jarosite or ammonium jarosite is spontaneous, and the precipitate settles rapidly, is easily filterable, and is readily washable to eliminate entrained solution. In addition to the excellent filtration characteristics and high degree of Fe capture, other advantages of the jarosite process are that it controls both alkalis and sulfate, it operates at atmospheric pressure rather than requiring an autoclave, it has a low capital and operating cost, the losses of divalent base metals captured by alkali jarosite are relatively low, and the process is sufficiently simple that it can be incorporated into most metallurgical circuits (Arregui et al. 1979, Dutrizac 1979, 1982).

Metallurgical problems and environmental concerns

Despite its widespread application in the zinc industry, the jarosite process is not without its problems. Chief among these is disposal of the jarosite residue, which has little potential use. For example, a plant annually producing 150,000 tonnes of metallic zinc from a concentrate containing 50 wt % Zn and 12 wt % Fe will generate about 125,000 tonnes of jarosite containing only 25-36 wt % Fe (Dutrizac 1990). These wastes are commonly stored in dedicated containment facilities, such as ponds lined with impermeable geomembranes, but in some instances the wastes are disposed by other means. For example, sodium jarosite from the Kidd metallurgical plant at Timmins, Canada, has for several years been combined with tailings slurry from the mining operation, and the two have been co-discharged to the main tailings impoundment (Al et al. 1994). Other disposal methods have been diverse, and have varied from storage in sole-purpose caverns excavated in mountains (Berg and Borve 1996), to deep-water, oceanic discharge. Whether stored in isolation or discharged, however, these practices have come under pressure because of political and environmental concerns; hence, disposal of jarositic wastes is becoming more costly and difficult (Norzinc 1986). One effect of these concerns has been the spurring of research on the conversion of the jarosite residues to hematite, thereby both reducing the volume of waste and possibly leading to utilization of at least some of the hematite by-product (Dutrizac 1990).

The success of the jarosite process as a means of rejecting Fe in the zinc industry has led to applications of the technology to the hydrometallurgy of Cu, Ni, Co, and other metals (Maschmeyer et al. 1978, Aird et al. 1980, Harris et al. 1982). As well, the technology has been extended to Zn concentrates subjected to O_2–H_2SO_4 pressure leaching at temperatures >100°C (Parker 1981, Au-Yeung and Bolton 1986). However, Zn concentrates commonly contain some Pb, and in pressure leaching much of the Pb forms lead jarosite, a product that is not metallurgically attractive. Moreover, significant amounts of other elements, such as Ag, Zn, Cd, Cu^{2+}, and In^{3+}, can be incorporated in lead jarosite by solid solution, and the losses of such metals may be significant from an economic point of view. Equally or more important, the potential release of deleterious elements from these jarositic wastes adds to the challenges associated with the environmentally safe disposal of such material. Jarosite residues containing NH_4 or K are potentially useful as fertilizer additives, but the principal impediment to such usage is the presence of toxic elements; even with the production of purer jarosites the presence of Cd entrained in the accompanying solutions will likely always be a problem.

A consequence of the failure to convert jarosite into a marketable product is that the emphasis has been shifted towards making the jarosite residue sufficiently stable so that its long-term disposal costs and environmental liabilities are minimized. Lime stabilization was the basis of early efforts, but it is not clear that the resulting "Jarochaux" product is adequately stable, especially in humid climates. The process, however, is in use at a Mexican zinc plant that is in a semi-arid region (Garcia and Valdez 1996). Mixtures with Portland cement are more stable, and cement is widely used to stabilize many other waste products. Stabilization of jarosite by addition of cement is a measure that was first used by Toho Zinc a number of years ago, and the process was recently revived by Noranda Technology with its Jarofix process (Rodier 1995, Chen and Dutrizac 1996). Like all stabilization techniques, however, the Jarofix method, which requires ~15 wt % cement, suffers philosophically in that the mass of material to be discarded is increased. The relatively long-term (50-100 a) stability of the cement-stabilized products is unknown. Mineralogical study of Jarofix products (Chen and Dutrizac 1996) indicated that the cement reacts with some of the jarosite to form gypsum and $Ca_6Fe_2(SO_4)_3(OH)_{12} \cdot nH_2O$, which is the Fe analog of ettringite, as well as minor ettringite. With such a mineralogical assemblage, spalling and dissolution would be expected to occur under normal weathering conditions over the long term.

KINETICS OF JAROSITE PRECIPITATION

The reaction for the precipitation of, e.g. sodium jarosite, involves the hydrolysis of ferric ions to form a solid compound and to produce a significant amount of sulfuric acid:

$$3 \, Fe_2(SO_4)_3 + Na_2SO_4 + 12 \, H_2O \rightarrow 2 \, NaFe_3(SO_4)_2(OH)_6 + 6 \, H_2SO_4 \qquad (10)$$

The rate of reaction would be expected to depend on seeding, because if a new solid surface is to be created from a homogeneous solution, the initiation of such reactions can be kinetically slow. Hence, induction periods prior to the start of jarosite precipitation in homogeneous systems have been observed (Margulis et al. 1976c, Hutchison and Phipps 1977, Hartig et al. 1990). The induction period can be overcome by seeding, and batch experiments by Hutchison and Phipps (1977) at 100°C showed that the precipitation rate of jarosite increased linearly with the amount of seed added. Studies by Wang et al. (1985) and Hirasawa (1993) also showed that seeding accelerated the rate of jarosite precipitation; similar effects were observed by Pammenter and Haigh (1981), Teixeira and Tavares (1986), and Elgersma et al. (1993) for ammonium jarosite, and by Won and Park (1982) for hydronium jarosite.

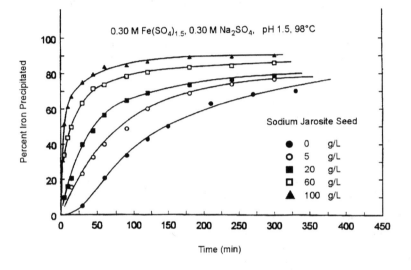

Figure 9. Iron precipitation curves for the formation of sodium jarosite in the presence of various concentrations of sodium jarosite seed. (Dutrizac 1996).

Figure 9 shows the percentage of dissolved Fe that was precipitated as a function of time when sodium jarosite was precipitated at 98°C from a 0.3 M $Fe(SO_4)_{1.5}$_0.3 M Na_2SO_4 solution in the presence of various amounts of sodium jarosite seed (Dutrizac 1996). In all instances, the amount of Fe that was precipitated increased rapidly with time, but then leveled off at a nearly stead-state value. The presence of even small amounts of seed greatly increased the rate of sodium jarosite precipitation, which is the slope of the curves for precipitation versus time. It is evident from Figure 10 that the initial rate of precipitation increases in an almost linear manner with increasing seed additions. It was also observed that seeding suppresses the induction period, and that stirring rates above those required to keep the seed suspended had no apparent benefit (Dutrizac 1996). Similar relationship were obtained for ammonium jarosite (Dutrizac 1996). Thus, a high seed recycle rate is desirable in a commercial jarosite process.

In addition to the effect of seeding, the rate of precipitation of jarosite would be expected to depend on temperature because the ferric ion must be hydrolyzed, and on the acid concentration, which affects all hydrolysis reactions. Figure 11 shows the effect of temperature on the precipitation of sodium jarosite in the presence of a constant 50 g/L of jarosite seed (Dutrizac 1996). The data show that the rate of precipitation of sodium jarosite is strongly dependent on temperature, and that relatively slow rates are obtained at <80°C even in the presence of large amounts of seed. Increasing temperatures increase

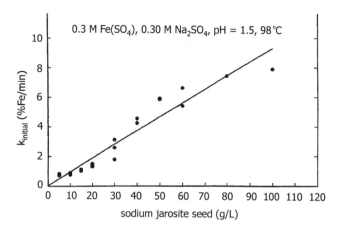

Figure 10. Effect of the amount of sodium jarosite seed on the initial rate of sodium jarosite precipitation. (Dutrizac 1996).

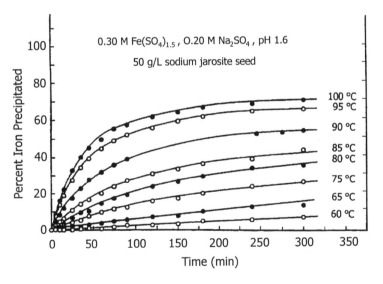

Figure 11. Curves of Fe precipitation for the formation of sodium jarosite at various temperatures in the presence of 50 g/L of sodium jarosite seed (Dutrizac 1996).

both the rate and extent of precipitation, and seeding extends the temperature rate for jarosite precipitation to lower values. This is illustrated in Figure 12, which compares the initial rate of precipitation of sodium jarosite in the presence of 50 g/L of seed (cf. Fig. 11) with the analogous rates obtained in the absence of initial seed additions. The presence of seed results in an increase in the rate of jarosite precipitation and also extends the temperature range for the rapid precipitation of jarosite. The similarity in the slopes of the two curves, however, implies that the underlying mechanism of jarosite precipitation is the same regardless of whether seed is initially present or not. Such a dependence is not unexpected as the tests done without initial seed additions produced jarosite which itself

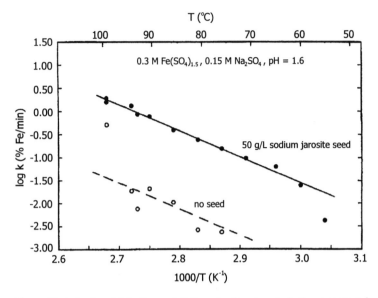

Figure 12. Arrhenius plot for the precipitation of sodium jarosite in the presence and absence of 50g/L of sodium jarosite seed (Dutrizac 1996).

became seed. There are many kinetic advantages to operating a circuit at 100°C, but generation of steam commonly constrains zinc-plant operations to somewhat lower precipitation temperatures and, as a result, significantly lower rates of Fe removal.

The effect of the solution pH on the precipitation of ammonium jarosite at 98°C in the presence of a constant 20 g/L of ammonium jarosite seed is illustrated in Figure 13. High acid concentrations effectively suppress precipitation of sodium jarosite and ammonium jarosite, and the rate of precipitation, as well as the ultimate extent of Fe removal, increases as the pH increases. Thus, operation of a jarosite precipitation circuit at as high a pH as possible is desirable, but the upper pH values are constrained by the precipitation of other Fe-bearing phases whose settling and filtration properties are less favorable.

Figure 13. Curves of Fe precipitation for the formation of ammonium jarosite at various pH values in the presence of 20 g/L of ammonium jarosite seed (Dutrizac 1996).

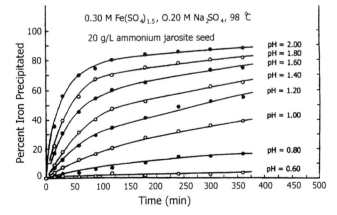

It is common practice in the zinc industry to recycle part of the jarosite residue as a means of providing seed to the jarosite precipitation stage. Hence, both the seed and the subsequent precipitates generally have approximately the same composition. It has been shown by Dutrizac (1999a) that sodium jarosite, potassium jarosite, silver jarosite, and lead jarosite are about equally effective as seed for the precipitation of sodium jarosite (Fig. 14). Thus, any combination of compositions for the seed versus the intended jarosite precipitate is likely to be effective for precipitation of all compounds in the jarosite subgroup.

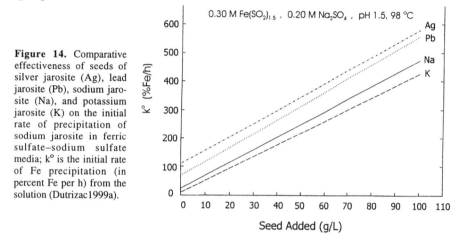

Figure 14. Comparative effectiveness of seeds of silver jarosite (Ag), lead jarosite (Pb), sodium jarosite (Na), and potassium jarosite (K) on the initial rate of precipitation of sodium jarosite in ferric sulfate–sodium sulfate media; k^o is the initial rate of Fe precipitation (in percent Fe per h) from the solution (Dutrizac 1999a).

FLOWSHEETS

Many variations of the jarosite process have been developed, and one of the simplest flowsheets is illustrated in Figure 15. In this version, the zinc plant operates a conventional circuit incorporating a neutral leach, and an Fe-free solution is generated for zinc-dust purification and subsequent electrolysis. The residue from the neutral leach is subjected to a hot acid leach to dissolve the Zn and Fe present as ferrite, but the slurry from the hot acid leach is not thickened or filtered. Instead, calcine and a jarosite-forming cation, as well as recycled jarosite seed, are added to precipitate jarosite. At this point, the pulp is thickened and filtered, and the Zn-containing solution is recycled. Part of the solids residue is recycled as seed for jarosite precipitation, and part is discarded. Although this version of the jarosite process is simple and has a low capital cost, the disadvantages are that the precious-metal values in the residue from the hot acid leach are not recovered; also lost is the zinc ferrite component in the calcine that is used to neutralize the hydrolysis acid generated in the jarosite-precipitation part of the circuit.

A partial solution to the above problem is to incorporate a liquid–solid separation step after the hot acid leach, as is shown in Figure 16. The solids from the neutral leach are treated with hot acid to dissolve the zinc ferrite that is present. The hot-acid-leach slurry is thickened, and either the solids containing the Pb, Ag, and Au can be further leached to concentrate these values, or the solids can be treated directly. The Fe-rich solution from the hot acid leaching circuit is treated with calcine, a jarosite-forming cation, and jarosite seed to precipitate a jarosite product that is filtered and discharged. This option has the advantage that a precious-metals residue is recovered; however, there is no recovery of the zinc ferrite or precious metals within the calcine that is used to neutralize the hydrolysis acid produced in the jarosite-precipitation part of the circuit.

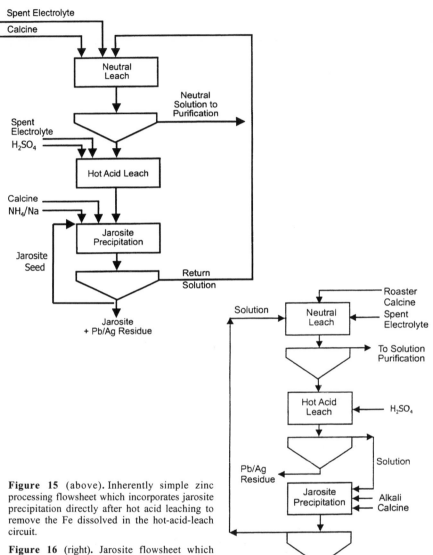

Figure 15 (above). Inherently simple zinc processing flowsheet which incorporates jarosite precipitation directly after hot acid leaching to remove the Fe dissolved in the hot-acid-leach circuit.

Figure 16 (right). Jarosite flowsheet which incorporates the recovery of a Pb/Ag residue after the hot-acid-leach circuit.

Figure 17 shows another flowsheet, this one designed to recover the zinc ferrite that is present in the calcine used for neutralization in the jarosite part of the process. A conventional neutral leach–hot acid leach is carried out; there is a liquid–solids separation step after the hot acid leach, and the residue is discarded. The Fe-rich solution from this part of the circuit is treated with calcine, a jarosite-forming cation, and jarosite seed to make a jarosite precipitate. In this instance, however, most of the jarosite residue is recycled through the leaching process. As it passes through the hot acid leach, the ferrite accompanying the jarosite is dissolved, but dissolution of the jarosite is minor if the acid concentration is properly controlled. This acid-wash version of the jarosite process gives high Zn recoveries, but the precious-metals residue is diluted with precipitated jarosite to the point where recovery of precious metals is usually not possible.

The preceding few examples illustrate that the number and location of liquid–solid separation steps can be varied considerably to recover more Zn or precious metals, or both. The choice depends largely on the value of the recovered metals versus the capital and operating costs of the additional equipment needed. In any of these options the availability of a low-Fe, low-Ag calcine for neutralization in the jarosite part of the process greatly increases the flexibility of flowsheet selection, but such calcines are relatively scarce in the industry.

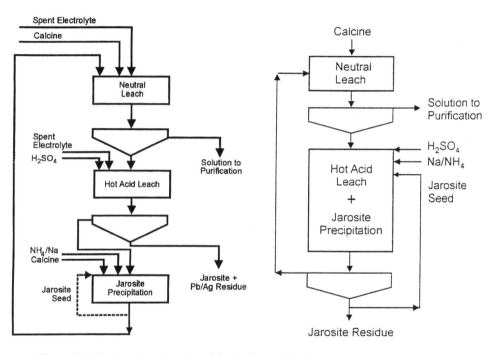

Figure 17 (left). An acid-wash version of the jarosite precipitation process that is intended to leach the zinc ferrite present in the calcine used for neutralization in the jarosite circuit.

Figure 18 (right). Schematic representation of the jarosite conversion process in which hot acid leaching and jarosite precipitation are carried out simultaneously in the same reaction vessels.

Figure 18 is a schematic flowsheet of the jarosite conversion process employed at Outokumpu's zinc plant in Finland, and at Noranda Inc., CEZinc in Canada. In this version, the residue from the neutral leach is heated with some acid, as well as with a jarosite-forming cation and recycled jarosite seed. The acid solubilizes part of the Fe present in the ferrite, and this dissolved Fe simultaneously precipitates as jarosite. The precipitation of jarosite releases sulfuric acid (Eqn. 10), and this acid is used in the same tank to leach additional ferrite. This reaction sequence continues until all of the zinc ferrite is consumed. The jarosite product is eventually filtered, the solution part is recirculated, and the solids are discharged to a residue pond. By combining the hot-acid-leach step and jarosite-precipitation operations, good extractions of Zn are achieved and capital costs are minimized. On the negative side, however, longer retention times are required and a precious-metals residue cannot be produced directly.

The solution from a typical hot-acid-leach circuit contains 20-30 g/L Fe and up to 80 g/L H_2SO_4. All of the free acid, as well as the hydrolysis acid (Eqn. 10), must be neutralized if the reaction involving jarosite precipitation is to go to completion. If the temperature is low enough, all of the acid can be neutralized without any precipitation of jarosite. This is possible because the neutralization reaction is kinetically rapid at all temperatures, whereas precipitation of jarosite at an appreciable rate requires temperatures >75°C (Fig. 11). This differential temperature dependence is the basis of the low-contaminant jarosite process illustrated in Figure 19. In this option, a conventional hot acid leach is carried out, and a subsequent liquids–solids separation step recovers a residue rich in precious metals and Pb. The clear solution from the hot-acid-leach circuit is cooled to 65°C and is then neutralized with calcine. All of the free acidity is consumed, and some of the potential hydrolysis acid can also be neutralized. The pulp is subjected to liquid–solid separation, and the solids are recycled to recover their Zn and precious-metals values. The clear, neutralized solution is then heated to >95°C, and a jarosite-forming cation and seed are added. A clean jarosite precipitates spontaneously, and this is a relatively stable residue for disposal. Although the low-contaminant jarosite process yields both high recoveries of Zn and a relatively clean jarosite residue, the process has not been adopted by the industry. Two heat-exchange steps are involved in the process, and to date it has not been possible to prevent massive scaling of the heat exchangers by jarosite and gypsum. Many different kinds of heat exchanger have been tried, but none has been sufficiently reliable to convince the industry to adopt this otherwise attractive technology.

IMPURITY INCORPORATION IN SYNTHETIC JAROSITES

Impurity incorporation is used here in a metallurgical sense insofar as the incorporation does not indicate physical inclusion; instead, the term refers to minor-

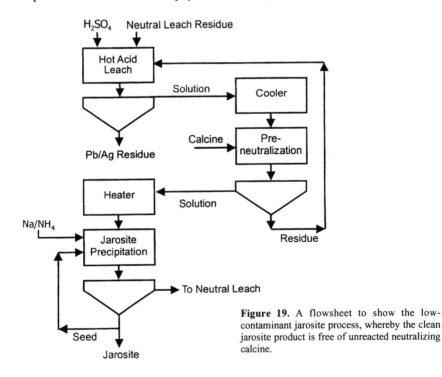

Figure 19. A flowsheet to show the low-contaminant jarosite process, whereby the clean jarosite product is free of unreacted neutralizing calcine.

element solid solution. As has been noted, various elements can be scavenged by substitution in the minerals of the jarosite subgroup. The focus here is on elements and jarosites, excluding those of K, Na, and NH_4, which for various reasons are of particular interest to the processing industry. Numerous studies have confirmed the preferential precipitation of Fe relative to Al in acidic solutions; in general, nearly Al-free jarosites are precipitated at pH values <2 if the weight ratio Al/Fe is less than 0.75.

Monovalent substitutions

Hydronium. The standard procedure in calculating formulas for the minerals of the alunite supergroup is to normalize to $TO_4 = 2$. This procedure is used because deficiencies are common for ion occupancies at both the D and G sites, especially in synthetic products, but it is structurally implausible to have vacancies for the TO_4 groups. Where a surplus of negative charges remains after calculation to $TO_4 = 2$ and (OH) = 6, it is generally assumed that this surplus is balanced by accommodation of $(H_3O)^+$ at the alkali sites if those are otherwise incompletely filled. A problem with these calculations is that $(H_3O)^+$ cannot be measured directly, and a consequence is that much effort has been expended to prove the existence of the hydronium ion in jarosite and alunite (Shishkin 1950, Kubisz 1966, 1972; Petrov et al. 1966, Kydon et al. 1968, Shokarev et al. 1972, Wilkins et al. 1974, Ripmeester et al. 1986).

The crystal structure of the isostructural Ga analog of alunite, $(H_3O)Ga_3(SO_4)_2(OH)_6$, was determined by Johansson (1963), who showed that the alkali position was occupied by an O-bearing species which he concluded was most likely to be H_3O. No crystal-structure study of hydronium jarosite or its counterpart in the alunite subgroup has been done, but candidates for such examination may be either the hydronium jarosite illustrated by Jensen et al. (1995) as brilliantly lustrous, euhedral crystals at the Gold Quarry mine, Eureka County, Nevada, or the hydronium jarosite reported to occur as 'superb' crystals at Majuba Hill, Pershing County, Nevada (Roberts et al. 1974). Schukow et al. (1999) synthesized deuterated alunite, $[K_{0.86}(D_3O)_{0.14}]Al_3(SO_4)_2(OD)_6$, and indicated that crystal-structure work on the synthesized D_3O analog is in progress.

Systematic study of the factors that affect the yield and composition of hydronium jarosite is in progress (by J.E.D). Solution pH is one of the important control parameters. At 140°C, stoichiometric hydronium jarosite was observed to form over the pH range 0.4 to 1.4; at lower pH, no product was produced, and at higher pH, discoloration of the solution was suggestive of partial ferric hydroxide precipitation. Seeding and constant agitation were essential to suppress the coprecipitation of the ferric hydroxide. At pH 0.7, the product yield was negligible for Fe concentrations <0.3 M; the yield increased with increasing Fe over the interval 0.5 to 2.0 M Fe^{3+}. Temperature also exercised a major influence on yield. At 100°C, only traces of hydronium jarosite were precipitated from 1.0 M Fe^{3+} solutions of pH = 0.8 after 24 h reaction in a well-stirred and seeded vessel. The yield rose steadily with increasing temperature to about 140°C, and then leveled off. Above 160°C, hydronium jarosite no longer formed.

The precipitation of hydronium rather than alkali jarosites in commercial operations might seem to be appealing insofar as there would be cost savings by eliminating the need to purchase chemicals for jarosite precipitation. The thermodynamics for hydronium jarosite, however, are not especially favorable: $\Delta G^\circ_{f,298}$ is about -3230 kJ mol^{-1} for hydronium jarosite (Alpers et al. 1989), and -3310 kJ mol^{-1} for potassium jarosite (Baron and Palmer 1996b), with the result that Fe concentrations are not readily reduced to low values by using hydronium jarosite. Furthermore, a pressure vessel would be required to effect efficient reaction, and this need would impose an additional burden on the Fe-removal circuit. Consequently, commercial interest has been focused on the alkali jarosites.

Rubidium. Fairchild (1933) synthesized rubidium jarosite, but the low purity of the Rb_2SO_4 available at that time resulted in precipitates containing substantial amounts of K_2O or Na_2O. Fairchild's order of stability of the jarosites was concluded to be K > Rb > Na, which was subsequently confirmed by Steintveit (1970). Rubidium jarosite devoid of K and Na, and corresponding to $Rb_{0.82}(H_3O)_{0.18}Fe_{2.7}(SO_4)_2(OH,H_2O)_6$, was synthesized by Dutrizac and Kaiman (1976).

Ivarson et al. (1979) showed that rubidium jarosite could be formed after 17 weeks reaction of a rubidium sulfate–ferrous iron solution oxidized by air in the presence of *Thiobacillus ferrooxidans* at room temperature. The rates of formation of various jarosites were observed to be K > Rb ≈ Na > NH_4.

Thallium. Thallium exists in both the monovalent and trivalent states, with Tl^{1+} by far the more stable. Nonetheless, it is theoretically possible to have two Tl-dominant jarosites: $TlFe_3(SO_4)_2(OH)_6$ and $K(Tl^{3+},Fe^{3+})_3(SO_4)_2(OH)_6$. The precipitation of both Tl^{1+} and Tl^{3+} as jarosites was reported by Yaroslavtsev et al. (1975); the nature of the precipitates was not clarified, however, and as Tl in both states was observed to behave similarly, reduction of Tl^{3+} to Tl^{1+} may have occurred prior to precipitation. Alternatively, because Tl^{3+} hydrolyzes extensively at room temperature even in fairly acid solution, it is possible that Tl_2O_3 coprecipitated with the jarosite. Synthesis of the Tl^{1+} end-member was described by Dutrizac and Kaiman (1975, 1976).

Concentration of Tl in jarosite associated with oxidized ore deposits was noted by Novokina (1957) and Mogarovskii (1961), and the latter suggested that the accumulation of Tl in jarosite could be used as a prospecting criterion for Tl. The Tl^{1+}-dominant member of the jarosite subgroup, dorallcharite, was discovered (Balic Zunic et al. 1994) in an oxidation zone at Allchar, Republic of Macedonia, a locality notable for the occurrence of numerous primary Tl-bearing sulfides and sulfosalts, several of which are known only from Allchar.

The average Tl content of Zn-sulfide concentrates submitted to plants for processing is about 20 ppm, but some concentrates have contents as high as 100 ppm (Grant 1993). Precipitation of jarosite is effective in eliminating the element, even when present at low concentrations, from Zn processing circuits. Solid solution between K and Tl^{1+} in jarosite is nearly ideal, implying that both Tl jarosite and K jarosite would have similar solubilities; Tl, however, is preferentially precipitated relative to NH_4 or Na (Dutrizac 1997). Because thallium jarosite is more stable than the ammonium or sodium analogs, the initially formed precipitates were found consistently to be the most Tl-rich.

Silver. Among the minerals of the jarosite subgroup, only argentojarosite and plumbojarosite are of economic importance as mineral species, but other members, such as jarosite, may be carriers of by-product metals, including Ag. In hydrometallurgical leaching circuits that employ jarosite precipitation, and especially in those circuits in which high temperatures and pressures are used, much of the Ag can report in the jarosite fraction.

Silver jarosite was synthesized by Fairchild (1933), May et al. (1973), and Dutrizac and Kaiman (1976). Systematic investigations of the formation of silver jarosite and various Ag–alkali solid-solution series (Dutrizac 1983, Dutrizac and Jambor 1984, 1987a) showed that, under conditions relevant to processing in zinc plants, silver jarosite is nearly as stable as potassium jarosite; during jarosite formation, Ag is selectively incorporated in preference to Na, NH_4, Pb, or (Pb+Cu). Although some of the products obtained in the Ag–(Pb+Cu) experiments were found to be three-phase (Dutrizac and Jambor 1984), Ildefonse et al. (1986) subsequently used different synthesis conditions and obtained a homogeneous Pb–Ag ($\pm H_3O$) solid-solution series up to 86 mol % Ag. For zinc-plant conditions, details of the variables that affect the yields and compositions of the silver jarosite are given by Dutrizac and Jambor (1984). The compound is extremely stable to elevated temperatures and acid concentrations.

Of particular mineralogical interest is that Dutrizac and Jambor (1987a) observed the presence of an 11-Å diffraction line in many of their synthesized Na–Ag–H_3O jarosites, thereby indicating that some ordering of the univalent, alkali-site cations had occurred. The consequence is that c of the unit cell of these products is doubled to ~34 Å, as is typical for the minerals that contain a divalent cation such as Pb or Ca at the alkali site (e.g. plumbojarosite, minamiite, huangite). However, Ildefonse et al. (1986) specifically checked their Pb–Ag jarosites for an 11-Å diffraction line and did not detect it, thus indicating complete disorder for the whole synthetic series. Similar disorder in synthesized lead jarosite had been indicated by Mumme and Scott (1966), Jambor and Dutrizac (1983), and Dutrizac and Jambor (1984). The existence of order-disorder phenomena has a potentially significant implication for the nomenclature of the minerals of the alunite supergroup (Jambor 1999).

Lithium and cesium. Although the Na, K, and Rb members of Group IA of the periodic table form jarosite compounds, it has been demonstrated that neither Li nor Cs makes an end-member jarosite species (Dutrizac and Jambor 1987b). The inability to form these end-member species is attributed to the small size of the Li$^+$ ion (r = 0.60 Å) and the large size of the Cs$^+$ ion (r = 1.69 Å) relative to the ions of the other members of the group (Na r = 0.95 Å, K r = 1.33 Å, Rb r = 1.48 Å) (Cotton and Wilkinson 1962). However, it seems that Cs-bearing potassium, sodium, and rubidium jarosite can be synthesized with Cs contents >2 wt %. The structural incorporation of Cs was greatest for potassium jarosite and was least for rubidium jarosite. Lithium does not seem to be significantly incorporated in any of the jarosite species. The highest Li content detected was 0.1 wt % Li, and such amounts are likely attributable to physical encapsulation.

Divalent substitutions

Copper and zinc. The divalent metals Cu and Zn substitute for Fe^{3+} in the jarosite structure. Dutrizac (1984) determined that potassium jarosite could incorporate about 2.1 wt % Zn, but only about half that amount was successfully substituted into sodium jarosite, and even less into ammonium jarosite. Scott (1987) reported a formula for jarosite (sensu stricto) that corresponds to a content of 2.5 wt % Zn and 6.3 wt % Pb, which seems to be about the maximum Zn content that has been described for natural occurrences of the jarosite subgroup.

Substitution of Cu^{2+} in the alkali jarosites seems to be similarly low. Dutrizac (1984) determined that sodium jarosite could incorporate only slightly more than 2 wt % Cu. In lead jarosite, however, substitution of Cu^{2+} can be substantial, and a solid-solution series extending to beaverite, Pb(Fe,Cu)$_3$(SO$_4$)$_2$(OH)$_6$ has been synthesized (Jambor and Dutrizac 1985).

Cadmium, cobalt, nickel, and manganese. Zinc concentrates typically consist largely of sphalerite, a sulfide that not only is well known as a carrier of trace amounts of Cd, but which is also the principal economic source of this metal. Thus, the deportment of Cd during alkali-jarosite precipitation is of major concern to the zinc industry. Syntheses of alkali jarosites by Dutrizac (1984) showed that up to 0.35 wt % Cu could be substituted into sodium jarosite. Under conditions applicable to an operating zinc plant, however, Cd uptake was found to be considerably lower. A subsequent study (Dutrizac et al. 1996) showed that the Cd content of the precipitated jarosites was <500 ppm, even for jarosites made from solutions containing 40 g/L Cd. Mass-balance calculations indicated that >99% of the initially present Cd remained in solution. Substitutions of Co^{2+} and Mn^{2+} in sodium jarosite were determined to be <0.5 wt % and about 2 wt %, respectively. Substitution of Co^{2+} in ammonium jarosite was marginally lower than in sodium jarosite, and was highest (~1.35 wt %) in potassium jarosite. Behavior of Ni^{2+} was similar to that of Co^{2+}, but Ni^{2+} substitution was at slightly lower levels. The order of incorporation of metals by jarosite was determined to be:

$$Cu^{2+} > Zn^{2+} > Co^{2+} \sim Ni^{2+} \sim Mn^{2+} > Cd^{2+}$$

which is the same as the ease of hydrolysis. Increasing the pH of the mother solution increases the extent of hydrolysis and promotes the structural incorporation of the divalent metals.

Lead. Little hydrometallurgical interest was shown in the synthesis of lead jarosite until the compound was found to form during the oxygen–H_2SO_4 pressure leaching of sulfides, especially Zn concentrates (Au-Yeung and Bolton 1986). Pressure leaching of Zn or Zn-Pb concentrates offers several advantages, among which is the ability to avoid the oxidation of pyrite and to produce elemental sulfur, thereby eliminating the need for the zinc plant to market roaster acid. During oxygen pressure leaching, the following reactions occur:

$$ZnS + H_2SO_4 + 1/2\ O_2 \rightarrow ZnSO_4 + S^\circ + H_2O \tag{11}$$

$$ZnS + Fe_2(SO_4)_3 \rightarrow 2\ FeSO_4 + ZnSO_4 + S^\circ \tag{12}$$

$$2\ FeSO_4 + H_2SO_4 + 1/2\ O_2 \rightarrow Fe_2(SO_4)_3 + H_2O \tag{13}$$

Pyrite remains largely unaffected. Associated Pb sulfide is known to be highly reactive, and it would be desirable to recover a Pb (and Ag) product by the following reaction:

$$PbS + H_2SO_4 + 1/2\ O_2 \rightarrow PbSO_4 + S^\circ + H_2O \tag{14}$$

The $PbSO_4$ could be recovered either by flotation or by selective leaching. However, if the terminal acid concentrations are low, lead jarosite rather than $PbSO_4$ is produced:

$$PbSO_4 + 3\ Fe_2(SO_4)_3 + 12\ H_2O \rightarrow 2\ [Pb_{0.5}Fe_3(SO_4)_2(OH)_6] + 6\ H_2SO_4 \tag{15}$$

The lead jarosite also includes the associated Ag in a largely unleachable form. The jarosite product is low in Pb (maximum 18.3 wt % Pb in the ideal formula) and is therefore not an ideal feed for a lead smelter. An additional problem associated with the formation of lead jarosite in the presence of Zn and Cu is that both elements can substitute in the compound, which at high Cu levels approaches beaverite in composition:

$$PbSO_4 + CuSO_4 + Fe_2(SO_4)_3 + 6\ H_2O \rightarrow Pb(Fe_2Cu)(SO_4)_2(OH)_6 + 3\ H_2SO_4 \tag{16}$$

Aside from the obvious loss of valuable metal, this reaction would introduce Cu and Zn impurities into any Pb-upgrading circuit designed to treat the jarosite residue. Such introduction would result in the need for a more complex purification process. The obvious solution is either to prevent the formation of lead jarosite during leaching, or to decompose the compound once it has formed. Various methods have been developed to achieve this goal (Scott 1973, Nogueira et al. 1980, Dutrizac et al. 1980, Dutrizac 1983). It is noteworthy that the inadvertent precipitation of lead jarosite can also be a problem in ferric chloride leaching, which is an alternative technology to the oxygen–H_2SO_4

pressure-leaching process (Dutrizac 1991).

In standard utilization of the jarosite process, the relative proportion of Ag undergoes selective concentration in various jarosites (Dutrizac and Jambor 1984). Substitution of Zn, which is for Fe^{3+} in lead and other jarosites, is generally limited to 1-2 wt % Zn; the maximum attainable is in lead jarosite and is slightly less than 5 wt % Zn (Dutrizac and Dinardo 1983, Jambor and Dutrizac 1983, 1985). Substitution of Cu^{2+}, like that of Zn, is for Fe^{3+}. Syntheses of cuprian lead jarosite have shown that Cu^{2+} can be incorporated in substantial amounts and that the series to synthetic beaverite (Eqn. 16) is probably complete (Jambor and Dutrizac 1985). Copper is strongly coprecipitated in preference to Zn from solutions containing both metals. From a practical point of view, it seems likely that extensive losses of Cu and Zn are to be expected when lead jarosite is precipitated from solutions rich in those metals.

Mercury. Mercury in jarosite is divalent and the formula of the compound is of the form $Hg_{0.5}Fe_3(SO_4)_2(OH)_6$, analogous to that of plumbojarosite. Mercury-dominant or even significantly mercurian jarosite has not been reported to occur naturally. Dutrizac and Chen (1981) examined jarosite minerals from mercury deposits and from other localities specifically to evaluate the Hg concentrations. None of the jarosites, including those intergrown with cinnabar or metacinnabar, contained Hg detectable by electron microprobe (detection limit <0.08 wt % Hg), and classical analyses of concentrates free from discrete Hg minerals gave <5 ppm Hg. Among the various reasons for the apparent absence of mercury jarosite are the preference to form K jarosite at the expense of the Hg analog, the narrow pH range suitable for crystallization of the Hg mineral, and the need to oxidize Hg to the divalent state.

Dutrizac and Chen (1981) synthesized mercury jarosite by a slow-addition method. Of all the synthetic jarosites, mercury jarosite was found to be the most sensitive to pH; no product was formed at pH < 1.4, and at pH > 1.8 the product was contaminated with basic Fe sulfates. Extensive $Hg–H_3O$ substitutions in the jarosite were noted, and a complete solid-solution series between K^+ and $1/2\,Hg^{2+}$ was observed. Extensive mutual substitution between Na and Hg was also observed, but it was uncertain whether complete solid solution occurred.

Trivalent substitutions

Indium and gallium. Various elements can substitute for Fe^{3+} in the minerals of the alunite supergroup. The most notable trivalent solid-solution is between Fe^{3+} and Al^{3+}. Other substitutions, such as by Sb^{3+} (Kolitsch et al. 1999a), Cr^{3+} (Walenta et al. 1982), and Nb^{3+} (Lottermoser 1990) have been reported for the minerals, and also known are the V^{3+}-dominant (Kolitsch et al. 1999b) and Ga^{3+}-dominant (Jambor et al. 1996) members (Table 1). In the hydrometallurgical processing of Zn concentrates, the behavior and deportment of In and Ga are of economic significance.

Dutrizac (1984) synthesized the indium analog of potassium jarosite and showed that, in contrast to the results of Bakaev et al. (1976), the precipitation of In relative to that of Fe increased as the solution pH increased. For the Na system, Dutrizac and Mingmin (1993) showed that In^{3+} formed a nearly ideal solid-solution series with Fe^{3+} in sodium jarosite (Fig. 20), and that low Na_2SO_4 concentrations resulted in reduced product yields and jarosite precipitates containing $H_3O^+ > Na^+$. It was concluded that In^{3+} is readily precipitated with sodium, potassium, or ammonium jarosites, and that for all three the precipitation of In from solutions is only slightly less favored than that of Fe.

Most of the world's supply of In is from the processing of sulfide concentrates (Demarthe et al. 1990, Huang 1990), but most Ga is obtained by solvent extraction from

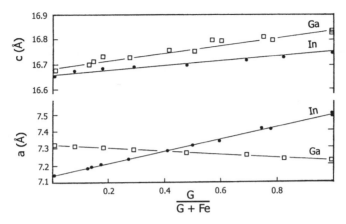

Figure 20. Variation in *a* and *c* cell parameters for In and Ga mol substitutions in sodium jarosite. The octahedral cation, which is Fe for the jarosite subgroup, is represented by G for the substitutional series. Data are from Dutrizac and Mingmin (1993) and Dutrizac and Chen (2000).

the leach liquors used in the processing of bauxite ores by the aluminum industry (Wilder and Katrak 1983). Nonetheless, some sulfide deposits and their oxidation zones, as at Tsumeb, Namibia, and at the Apex deposit in Utah, contain economically attractive amounts of Ga (Bernstein 1986, Dutrizac et al. 1986), and Ga is recovered at some zinc plants (Onozaki et al. 1986, Huang 1990).

Solid solution between Ga^{3+} and Fe^{3+} is complete for K, Na, Rb, NH_4, and H_3O synthetic jarosites (Johansson 1963, Tananaev et al. 1967a,b). In the roast–leach–electrowinning process, the Ga, like In, is solubilized with the Fe in the hot-acid leaching circuit. The ideal situation would be for Ga and In to remain in solution after jarosite precipitation so that these metals could be recovered by solvent extraction from the resulting low-Fe solution. However, both Ga^{3+} and In^{3+} form nearly ideal solid solutions with Fe^{3+} in jarosite-type compounds. Under conditions likely to be encountered in a hydrometallurgical zinc circuit, it was observed by Dutrizac and Chen (2000) that Ga^{3+} and Fe^{3+} are precipitated in approximately the same ratio as they occur in solution, and that the relative deportment of Ga is about the same regardless of whether potassium jarosite, sodium jarosite, or ammonium jarosite is precipitated. Thus, under standard commercial operating conditions, losses of both Ga and In to jarositic compounds will be appreciable.

Higher valency substitutions

Among the most notable of the substitutions in the alunite supergroup are those involving sulfate–phosphate–arsenate (Table 1). Although the end-members of many of the minerals, especially those containing Al rather than Fe, have been synthesized, little experimental work seems to have been done either for the binary joins or the ternary S–As–P solid solutions. For natural occurrences, Figure 1 shows that, for minerals with monovalent *D* cations in the alunite and jarosite subgroups, no mineral with formula S:P = <0.75 or S:As = <0.75 has yet been named. However, compositions are known to extend into the unnamed field (Jambor 2000, Stoffregen et al., chapter 9, this volume).

Dutrizac and Jambor (1987c) investigated the precipitation of arsenate (AsO_4^{3-}) in sodium jarosite and potassium jarosite at 97°C, and Dutrizac et al. (1987) carried out similar precipitation studies at 150°C. Although extensive precipitation of arsenic occurred in some of the experiments, the arsenic commonly was present as a discrete arsenate-bearing phase. At 97°C, the arsenate-bearing phase was typically amorphous to X-ray diffraction. The studies suggested that only ~2% AsO_4 could be structurally incorporated in either sodium jarosite or potassium jarosite at 97°C. At 150°C, ~2% AsO_4 could be structurally incorporated in sodium jarosite, but ~4% AsO_4 substituted for SO_4 in potassium jarosite. Preliminary investigations suggested <2% AsO_4 substituted in lead jarosite or cuprian lead jarosite. The analysis of jarosite minerals from nine localities showed the structural occurrence of AsO_4, but in amounts <4.5 wt % AsO_4. In contrast, specific compositions given by Roca et al. (1999) for jarosite from the Rio Tinto gossans extend along the S–As join (Fig. 1) to at least $(K_{0.58}Pb_{0.39})Fe_{2.98}(SO_4)_{1.65}(AsO_4)_{0.35}$ $(OH,H_2O)_6$, that is, S:As = 82.5:17.5, and even higher As contents were indicated in plotted data. On the S–P join, analyses of jarosite and natrojarosite by Horakova and Novák (1989) show S:P = ~85:15, and Ripp et al. (1998) give compositions for natroalunite that extend into the unnamed P-rich field. Thus, it seems reasonable to predict that additional compositions extending into the unnamed As-rich and P-rich fields of Figure 1 will be found. Such minerals are possible because the variety of cations that can occupy the D sites includes those that are divalent (e.g. Ca, Sr, Ba) and trivalent (REE); to balance these positive charges, a commensurate amount of PO_4 or AsO_4 may substitute for SO_4. For Pb-dominant members of the supergroup, extensive S–As–P solid solution has been documented (Fig. 21).

Silicon. Silicon is included in the formula of some minerals of the alunite supergroup, e.g. waylandite and eylettersite (Table 1), and analytically determined Si is commonly reported either as substitutional SiO_4 or as amorphous SiO_2. A Si-bearing member of the supergroup has not yet been investigated by crystal-structure study to determine whether Si is structurally incorporated, but such substitution seems crystallographically feasible especially for the phosphate members of the supergroup, which are known to exist in a variety of geological environments, including within highly metamorphosed rocks (Ripp et al. 1998, Ripp and Kanakin 1998). The existence of SiO_4^{4-} is prevalent only in media that are highly basic, whereas precipitation of members within the jarosite subgroup requires conditions that are highly acidic. In acidic conditions, Si is present mainly as large, complex anions that can lead to the formation of amorphous silica gels. It seems likely, therefore, that Si associated with minerals of the jarosite subgroup is not solid-solution Si.

Selenium. Several selenium minerals are known, but Se in sulfide ores is much more widespread as a trace solid-solution element that substitutes for S in the sulfides and sulfosalts. Selenium is a deleterious element, and its presence in small amounts in concentrates may incur penalties from the receiving smelter; larger amounts of Se may impede or negate the marketability of the concentrates, as has occurred recently for polymetallic sulfide deposits in the Finlayson Lake area, Yukon, Canada (The Northern Miner 2000).

Formation of selenate requires fairly strong oxidation conditions, but such conditions can be achieved during intensive oxidative leaching. Selenate analogs of sodium jarosite and potassium jarosite, wherein SeO_4 substitutes for SO_4, were synthesized by Dutrizac et al. (1981), who concluded that such analogs probably exist for all other members of the jarosite subgroup. During jarosite formation, selenate and sulfate are precipitated in approximately the same mol ratios that are present in the mother solution.

Other anions. Substitution of several other anions has been reported for the alunite supergroup. Small amounts of Ge are present in beudantite (Gebhard and Schlüter 1995) and in gallobeudantite and its related minerals, but substitution seems to be for Fe^{3+} rather than for the TO_4 groups (Jambor et al. 1996). Yarovslavtsev et al. (1975) reported a

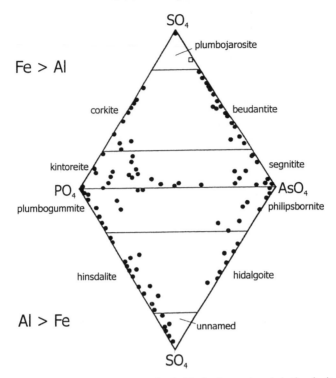

Figure 21. Solid-solution variations for the Pb-dominant minerals in the alunite supergroup. Compositions are from Sejkora et al. (1998), with additions from Rattray et al. (1996), Scott (2000), and Jambor (2000). Open square at top right is plumbian jarosite or disordered potassian plumbojarosite ($c = 17$ Å) with Pb:K = 1:1 (Roca et al. 1999). The nomenclature and compositional boundaries, including the absence of a PO_4–AsO_4 boundary, are those of the current IMA-approved system (Birch et al. 1992; Pring et al. 1995).

modest degree of Ge^{4+} incorporation in synthetic jarosite, but negligible contents were obtained in similar experiments by Dutrizac (1984).

The synthetic chromate analogue of jarosite is known (Powers et al. 1976, Cudennec et al. 1980, Baron and Palmer 1996a), and substantial CrO_4 (6.3 wt % as CrO_3) was reported by Walenta et al. (1982) to substitute for AsO_4 in philipsbornite. Other substitutions, such as molybdate in hidalgoite (Guillemin 1955) and postulated AlO_4 for AsO_4 in arsenoflorencite-(Ce) (Nickel and Temperly 1987), are rare and their existence has not been demonstrated unequivocally. Various other potential substitutions, and attempts to synthesize them, have been discussed by Dutrizac (1984).

CELL DIMENSIONS OF THE SUBGROUP

Table 3 lists the cell dimensions as taken from a few selected references, and in Figure 22 these dimensions are plotted against ionic radii for 12-fold coordination. Radii for K, Na, Pb, and Tl are from Shannon (1976). The radii of Ag and Pb are almost identical for 8-fold coordination, and for 12-fold coordination 1.48 Å is assumed for Ag. The radius for H_3O is from Okada et al. (1987), who also give 1.72 Å for NH_4, which is the same radius given by Shannon (1976) for Rb. However, Ivarson et al. (1981) have noted that Rb is larger than NH_4; thus, the latter has here been assigned a radius of

Table 3. Unit cell dimensions of the jarosite subgroup.

	a (Å)	c (Å)	Source	Comments
jarosite/	7.288	17.192	1	
potassium jarosite	7.319	17.130	2	
	7.304	17.268	3	natural
	7.315	17.119	4	92 mol % K
natrojarosite/	7.312	16.620	1	
sodium jarosite	7.329	16.703	4	
	7.335	16.747	5	
hydronium jarosite	7.355	16.980	1	
	7.347	16.994	4	
	7.356	17.010	6	
ammoniojarosite/	7.329	17.489	7	
ammonium jarosite	7.328	17.448	8	
rubidium jarosite	7.316	17.568	9	86 mol % Rb; 1.2 wt % P_2O_5
	7.32	17.36	10	
argentojarosite/	7.348	16.551	4	96 mol % Ag
silver jarosite	7.35	16.58	11	
	7.336	16.564	12	
	7.346	16.668	13	86 mol % Ag
dorallcharite/	7.329	17.554	7	
thallium jarosite	7.33	17.42	10	
	7.330	17.663	14	19 mol % K, natural
plumbojarosite*/	7.310	16.843	15	natural; 1.2 wt % P_2O_5
lead jarosite*	7.306	16.838	16	natural
mercury jarosite*	7.354	16.57	4	82 mol % Hg

Synthetic samples except where indicated otherwise. Sources of data (1) Brophy and Sheridan 1965; (2) Dutrizac and Jambor 1984; (3) Kato and Miúra 1977; (4) Dutrizac and Kaiman 1976; (5) PDF 36–425; (6) PDF 31–650; (7) Dutrizac 1997; (8) Odum et al. 1982; (9) Ivarson et al. 1981; (10) Dutrizac and Kaiman 1975; (11) May et al. 1973; (12) Dutrizac and Jambor 1987a; (13) Ildefonse et al. 1986; (14) Bali´c Zuni´c et al. 1994; (15) PDF 39–1353; (16) Szymański 1985.

*For plumbojarosite, lead jarosite, and mercury jarosite, the listed c is half true c.

1.69 Å. The radius of Hg is taken to be slightly larger than that of Ca for 12-fold coordination, and has been assigned a value of 1.36 Å.

Figure 22 shows that there is considerable scatter in the plot for a, but a good correlation is present in the data for c. As is discussed by Stoffregen et al. (this volume), a is affected mainly by Al-Fe^{3+} substitutions, whereas variations in c largely reflect substitutions at the alkali site. As all of the data points in Figure 22 are for Fe compounds with different alkali-site cations, the good correlation with c would be expected on the basis of the structural considerations.

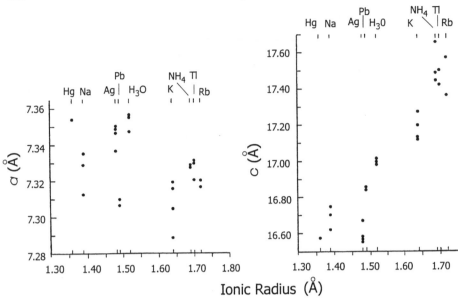

Figure 22. Variation in *a* and *c* cell parameters (Table 3) versus radii of the 12-coordinated cations in the jarosite subgroup.

CONCLUSIONS AND FUTURE TRENDS

Minerals of the jarosite subgroup occur widely in nature in a variety of oxidizing acidic environments. Eight species in the subgroup have been named, but of these only jarosite, natrojarosite, and plumbojarosite are prevalent. The most localized concentrations of jarosite are in oxidized sulfide deposits and pyritiferous rocks. These occurrences can result in massive concentrations of jarosite, some of which have been exploited for their Ag, Pb, and even for their Ga contents. In these occurrences, there is commonly a direct relationship between the oxidized sulfide minerals and the jarosite mineralization. Jarosite also occurs intimately mixed with clays or as concretions in clay. Probably the most widespread occurrence of jarosite, however, is in acid soils, wherein the required Fe is derived from oxidation of pyrite. Although most jarosite is of supergene origin, hypogene jarosite occurs in hydrothermally altered rock and as deposits directly linked to geothermal springs and fumarolic vents. Jarosite minerals alter readily, typically to goethite or hematite, and the alteration is most commonly caused by the imposition of neutral or alkaline conditions. High temperatures or strongly acidic conditions can also result in the alteration or dissolution of jarosite minerals.

Jarosite minerals are readily synthesized, and the most important synthesis parameters are temperature, pH, and the presence of jarosite seed. Both a jarosite-forming cation and a modest concentration of ferric sulfate are required, but wide variations in these concentrations are possible. The synthesis studies have demonstrated that extensive solid solution occurs in the D, G, and TO_4 sites of jarosite: $DG_3(TO_4)_2(OH,H_2O)_6$. Notable is the substitution of hydronium ion (H_3O^+) in nearly all natural and synthetic jarosites. Of industrial importance is that the common divalent base metals (Cu, Ni, Zn, Cd, Mn, Co, etc.) are not extensively incorporated into the jarosite structure, except that in lead jarosite there is substitution of Cu^{2+} for Fe^{3+}, ultimately to form beaverite $Pb(Fe,Cu)_3(SO_4)_2(OH,H_2O)_6$.

Although natural jarosites are of minor economic importance, the synthetic members have recently become important in hydrometallurgy as a means of precipitating unwanted Fe from processing solutions. Jarosite precipitation is inherently simple, and as a result, it is the lowest cost of the techniques for Fe removal, at least if the long-term costs of residue disposal are not considered. Jarosite precipitates are readily thickened, filtered, and washed; they do not usually incorporate significant amounts of the valuable divalent base metals. On the other hand, jarosites are useful for rejecting many unwanted impurities, including the alkali metals and excess sulfate. Jarosite precipitation is readily incorporated into a wide range of processing flowsheets for Zn, Ni, Co, Cu, and U.

Iron is an abundant impurity in the feeds to most hydrometallurgical operations and, regardless of the particular processing technique that is used, the elimination of the Fe impurity is an important consideration in nearly all hydrometallurgical flowsheets. Precipitation serves not only to remove Fe from the circuit, but also to reject many other impurities. This impurity-removal aspect is commonly of critical importance, and for this reason the Fe-precipitation processes will continue to play a significant role in hydrometallurgy. The disposal of Fe-bearing precipitates will continue to be a problem, as markets for such products are very limited. The ponding of large tonnages of these residues, such as jarosite precipitates, is costly and presents long-term monitoring and control obligations. Increasingly, the Fe residues are placed in ponds lined with geomembranes and fitted with underground drainage and monitoring networks. For reasons of environmental safety, however, most filled ponds must also be capped and revegetated, commonly at a considerable cost (Pophanken 1996). Some effort is being made to place the ponds in stable underground facilities, such as the cavern system operated by Norzink AS in Odda, Norway, where caverns in granitic mountains have been excavated specifically for the storage of jarosite residues. This practice greatly reduces, but does not entirely prevent, seepages from the containment areas (Berg and Borve 1996). Also evident is a growing trend to "stabilize" the Fe residues by alkaline additions prior to placing the residues in conventional open ponds. This practice simplifies the design of the pond, and thereby reduces the costs of disposal of the Fe-bearing residues. Buckle and Lorenzen (1996) investigated the use of fly ash, lime, kaolin, $CaCO_3$, and Na silicate to stabilize jarosite residues. They noted that the fly ash–Na silicate and the lime and $CaCO_3$ systems immobilized the heavy metals present in the jarosite residue, and that these stabilizers imparted a good level of mechanical strength to the product. Lime stabilization of jarosite residues, the Jarochaux process, is practiced commercially at the Torreon zinc plant of Met-Mex Penoles in Mexico. There, an ammonium jarosite precipitate is mixed with ~25% of its weight with lime, and the resulting mixture is trucked to an unlined disposal area. The resulting product seems to be stable in the semi-arid environment of the zinc plant (Garcia and Valdez 1996). Noranda Inc recently developed the Jarofix process as means of physically and chemically stabilizing jarosite residues to the point where they can be impounded on unlined clay soil adjacent to the company's CEZinc operation in Valleyfield, Canada. The jarosite residue is mixed with Portland cement and a small amount of lime, and is trucked to the adjacent disposal site. The residue cures into a hard and relatively impermeable mass. The need for a lined pond is apparently avoided (Chen and Dutrizac 1996, 2000). The recent investigation of the use of various alkaline Fe-slags to stabilize goethite residues resulted in the Graveliet process, which successfully stabilizes and hardens goethite residues (Winters et al. 2000). Unfortunately, however, such slags, which are a common waste product of the primary steel industry, do not effectively stabilize jarosite precipitates.

Although ponding of jarosite residues will continue to be practiced, the long-term objective of the industry should be to minimize the amount of Fe that needs to be discarded. Any significant reduction in the amount of jarosite sent to residue ponds implies the production of a marketable Fe product, and several such potential products can be identified: Fe-rich solutions, clean hematite, iron metal, Fe-rich slags, and stabilized residues. Iron-rich solutions presently are used for the leaching of sulfide minerals and for the stabilization of As in metallurgical operations. A potential growth market is for water purification to replace salts of Al, and another growth area could be for wastewater treatment to control phosphates and other undesirable species. Clean hematite certainly can be produced from hydrometallurgical processing solutions, but the cost of the clean hematite is relatively high. Once a clean hematite is produced, it could be used for the making of steel or for the manufacture of pigments and ferrites.

Electrolytic production of iron from hydrometallurgical processing solutions offers the great advantage that the product will be readily marketable, if only as high-grade iron scrap. The electrolytic iron process was used commercially in the first half of the twentieth century, and should be viewed as proven technology. One surprising difficulty is that copper and zinc electrodeposit with the iron, such that a preliminary Cu-Fe or Zn-Fe separation step would be required. Thus, for any of the above iron applications, it would likely be necessary to use solvent extraction or ion exchange to separate the Fe from the processing solution. A great deal of work has been done on this topic, and recent reviews of these processes have been published (Riveros et al. 1998, Riveros and Dutrizac 1999). None of the reagent systems is perfect, and all are costly. A further impediment is that the strip solutions commonly are strongly acidic, and most have an Fe concentration of less than 100 g/L Fe. However, new reagents and the development of galvanic stripping or reductive stripping techniques offer the potential to overcome these problems (Riveros and Dutrizac 1999, Barrera-Godinez et al. 2000). Another potentially marketable product is an Fe-rich slag which can be used for cement manufacture or grit blasting. Such slags are being produced in integrated lead-zinc operations (Ashman 1998), but the application of this approach is limited to the relatively small number of integrated lead-zinc producers. The smelting of goethite-type precipitates to yield marketable slags has been demonstrated using both Ausmelt technology (Lightfoot et al. 1993) and conventional pyrometallurgical methods (Kretschmer et al. 1998). If the resulting slag is fumed to volatilize Pb and Zn, it can be safely discarded or sold for various low-value applications. Although smelting of goethite-rich hydrometallurgical residues is technically and economically feasible, the treatment of jarosite precipitates is not as attractive (Rosato and Agnew 1996). The difficulties are the relatively low Fe content (25-30%) of hydrometallurgical jarosite precipitates, the high water content, and the modest sulfate content of the jarosite precipitates. These factors result in a large, capital-intensive treatment plant as well as a significant energy requirement for smelting. The sulfate level is sufficiently elevated that the process gases cannot be vented because of their SO_2 content. However, the sulfate content is too low to yield a gas stream containing enough SO_2 for acid manufacture. Hence, the off-gas treatment facility becomes both complex and costly.

In the short term, jarosite residues will continue to be ponded at most sites, and the costs of construction of residue ponds and of monitoring will increase. To minimize residue-disposal costs, the industry will seek to stabilize jarosite residues with lime or cement. The ideal solution to the residue-disposal problem, however, is to find a use for the Fe that is dissolved during the hydrometallurgical processing. The potential applications, as has been mentioned, include: Fe-rich solutions, clean hematite, Fe-rich slags, and electrolytic iron; for most of these, it will be necessary to separate the Fe from the other constituents in the processing solution, and such separations can be effected using solvent extraction or ion exchange. The development of suitable Fe-separation processes would open the door for the production of marketable iron products, and this should be the long-term objective of the hydrometallurgical industry.

REFERENCES

Aird J, Celmer RS, May AV (1980) RCM's phase 4 program. *In* Cobalt 80. Canadian Institute Mining, Metallurgy and Petroleum, Montreal, Paper 6

Al TA, Blowes DW, Jambor JL, Scott JD (1994) The geochemistry of mine-waste pore water affected by the combined disposal of natrojarosite and base-metal sulphide tailings at Kidd Creek, Timmins, Ontario. Can Geotech J 31:502-514

Allan JC (1973) The accumulations of ancient slag in the south-west of the Iberian Peninsula. Hist Metall Group Bull London 7:47-50

Allen ET, Day AL (1935) Hot Springs of the Yellowstone National Park. Carnegie Inst Wash Publ 466

Altaner SP, Fitzpatrick JJ,the Krohn MD, Bethke PM, Hayba DO, Goss JA, Brown ZA (1988) Ammonium in alunites. Am Mineral 73:145-152

Alpers CN, Nordstrom DK, Ball JW (1989) Solubility of jarosite solid solutions precipitated from acid mine waters, Iron Mountain, California, U.S.A. Sci Géol Bull 42:281-298

Alpers CN, Rye RO, Nordstrom DK, White LD, King B-S (1992) Chemical, crystallographic and stable isotopic properties of alunite and jarosite from acid–hypersaline Australian lakes. Chem Geol 96:203-226

Anderson JA (1981) Characteristics of leached capping and techniques of appraisal. *In* Titley SR (Ed) Advances in Geology of the Porphyry Ccopper Deposits, Southwestern North America. University Arizona Press, Tucson, p 245-287

Arregui, V, Gordon AR, Steintveit G (1979) The jarosite process—past, present, and future. *In* Cigan JM, Mackey TS, O'Keefe TJ (Eds) Lead–Zinc–Tin '80. TMS–AIME (The Minerals, Metals and Materials Society–American Institute of Mining, Metallurgy and Petroleum Engineers), New York, p 532-564

Ashman DW (1998) Pilot plant and commercial scale test work on the Kivcet process for Cominco's new lead smelter. *In* Dutrizac JE, Gonzalez JA, Bolton GL, Hancock P (Eds) Zinc and Lead Processing. Canadian Institute Mining, Metallurgy and Petroleum, Montreal, p 783-794

Au-Yeung SCF, Bolton GL (1986) Iron control in processes developed at Sherritt Gordon Mines. *In* Dutrizac JE, Monhemius AJ (Eds) Iron Control in Hydrometallurgy. Ellis Horwood, Chichester, UK, p 131-151

Babcan J (1971) Synthesis of jarosite, $KFe_3(SO_4)_2(OH)_6$. Geol Zb 22:299-304

Bakaev AM, Kravets MV, Margulis EV (1976) Behavior of indium during iron precipitation in the form of jarosite. Izvest Vyssh Ubechn Zaved Tsvetn Metall 76:149-150

Balic Zunic T, Moëlo Y, Loncar Z, Micheelsen H (1994) Dorallcharite, $Tl_{0.8}K_{0.2}Fe_3(SO_4)_2(OH)_6$, a new member of the jarosite-alunite family. Eur J Mineral 6:255-263

Banfield JF, Nealson KH, Eds (1997) Geomicrobiology: Interactions Between Microbes and Minerals. Reviews in Mineralogy, Vol 35

Banfield JF, Welch SA (2000) Microbial controls on the mineralogy of the environment. *In* Vaughan DJ, Wogelius RA (Eds) Environmental Mineralogy. EMU (European Mineralogical Union) Notes Mineral 2:173-196

Baron D, Palmer CD (1996a) Solubility of $KCr_3(SO_4)_2(OH)_6$ at 4–35°C. Geochim Cosmochim Acta 60:3815-3824

Baron D, Palmer CD (1996b) Solubility of jarosite at 4–35°C. Geochim Cosmochim Acta 60:185-195

Barrera-Godinez JA, Sun J, O'Keefe TJ, James SE (2000) The galvanic stripping treatment of zinc residues for marketable iron product recovery. *In* Dutrizac JE, Gonzalez JA, Henke DM, James SE, Siegmund AH-J (Eds) Lead–Zinc 2000. Minerals, Metals and Materials Society, Warrendale, Pennsylvania, p 763-778

Berg D, Borve K (1996) The disposal of iron residue at Norzink and its impact on the environment. *In* Dutrizac JE, Harris GB (Eds) Iron Control and Disposal. Canadian Institute Mining Metallurgy and Petroleum, Montreal, p 627-642

Bernstein LR (1986) Geology and mineralogy of the Apex germanium-gallium mine, Washington County, Utah. US Geol Survey Bull 1577

Berzina AP, Kuznetsova IK, Sotnikov VI (1966) Hypogene jarosite. Geol Geofiz 8:112-114

Biggs LI (1951) Jarosite from the California Tertiary. Am Mineral 36:902-906

Bigham JM, Schwertmann U, Traina SJ, Winland RL, Wolf M (1996) Schwertmannite and the chemical modeling of iron in acid sulfate waters. Geochim Cosmochim Acta 60:2111-2121

Birch WD, Pring A, Gatehouse BM (1992) Segnitite, $PbFe_3H(AsO_4)_2(OH)_6$, a new mineral in the lusungite group from Broken Hill, New South Wales, Australia. Am Mineral 77:656-659

Blanchard R (1968) Interpretation of leached outcrops. Nevada Bur Mines Bull 66

Bolgov GP (1956) Beaverite and its paragenesis in the zone of oxidation of sulfide deposits. Izvest Akad Nauk Kazakh SSR, Ser Geol 23:63-73

Bonorino FG (1959) Hydrothermal alteration in the Front Range mineral belt, Colorado. Bull Geol Am 70:53-90

Breitinger DK, Krieglstein R, Bogner A, Schwab RG, Pimpl TH, Mohr J, Schukow H (1997) Vibrational spectra of synthetic minerals of the alunite and crandallite type. J Molec Struct 408/409:287-290

Brophy GP, Sheridan MF (1965) Sulfate studies IV: The jarosite–natrojarosite–hydronium jarosite solid solution series. Am Mineral 50:1595-1607

Brophy GP, Scott ES, Snellgrove RA (1962) Solid solution between alunite and jarosite. Am Mineral 47:112-126

Brown JB (1970) A chemical study of some synthetic potassium–hydronium jarosites. Can Mineral 10: 696-703

Brown JB (1971) Jarosite–goethite stabilities at 25°C, 1 atm. Mineral Deposita 6:245-252

Buckle RL, Lorenzen L (1996) The stabilization and disposal of jarosite. In Dutrizac JE, Harris GB (Eds) Iron Control and Disposal. Canadian Institute Mining, Metallurgy and Petroleum, Montreal, p 597-611

Butler BS (1913) The occurrence of complex and little known sulphates and sulpharsenates as ore minerals in Utah. Econ Geol 8:311-322

Butler BS, Schaller WT (1911) Some minerals from Beaver County, Utah. Am J Sci 32:418-424

Chen TT, Dutrizac JE (1996) Mineralogical study of Jarofix products for the stabilization of zinc industry jarosite residues. In Ramachandran V, Nesbitt CC (Eds) Second International Symposium on Extraction and Processing for the Treatment and Minimization of Wastes. TMS (The Minerals, Metals and Materials Society), Warrendale, Pennsylvania, p 659-672

Chen TT, Dutrizac JE (2000) A mineralogical study of Jarofix products for the stabilization of jarosite residues for disposal. In Dutrizac JE, Gonzalez JA, Henke DM, James SE, Siegmund AH-J (Eds) Lead–Zinc 2000. TMS (The Minerals, Metals and Materials Society), Warrendale, Pennsylvania, p 917-934

Chukkrov FV (1950) Beudantite from the Kazakhstan Steppe. Dokl Akad Nauk SSR 72:115-117

Chukhrov FV (1959) Some peculiarities of mineralogy of the oxidation zone of mineral deposits in central districts of Kazakhstan S.S.R. Chem Erde 20:1-17

Clark JS, Gobin CA, Sprout PN (1961) Yellow mottles in some poorly drained soils of the lower Fraser Valley, British Columbia. Can J Soil Sci 41:218-227

Cotton FA, Wilkinson G (1962) Advanced inorganic chemistry. Interscience, New York, p 42-49

Cruells M, Roca A, Patino F, Salinas E, Rivera I (2000) Cyanidation kinetics of argentian jarosite in alkaline media. Hydrometall 55:153-163

Cudennec Y, Riou A, Bonnin A, Caillet P (1980) Crystallographic and infrared study of the hydroxy-chromates of iron and aluminium with the alunite structure. Rev Chimie Minérale 17:158-167

Cunningham CG, Rye RO, Steven TA, Mehnert HH (1984) Origins and exploration significance of replacement and vein type alunite deposits in the Marysvale volcanic field, west central Utah. Econ Geol 79:50-71

Demarthe JM, Rousseau AM, Fernandez FL (1990) Recovery of speciality metals, mainly germanium and indium, from zinc primary smelting. In Mackey TS, Prengaman DR (Eds) Lead–Zinc '90. TMS (The Minerals, Metals and Materials Society), Warrendale, Pennsylvania, p 151-160

Ding M, Van Der Sloot HA, Gensebroek M (1997) Interaction of Fe^{3+}, Ca^{2+} and SO_4^{2-} at the jarosite/alkaline coal fly ash interface. Proc 30th Inter Geol Cong 19:207-222

Dubinina VN, Kornilovich IA (1959) Plumbojarosite in the oxidation zone of Pb-Zn ore deposits of Eastern Transbaikal. Zap Vses Mineral Obshch 88:323-328

Dudas MJ (1984) Enriched levels of arsenic in post-active acid sulfate soils in Alberta. Soil Sci Soc Am J 48:1451-1452

Dudas MJ (1987) Accumulation of native arsenic in acid sulphate soils in Alberta. Can J Soil Sci 67:317-331

Dudek A, Fediuk F (1955) A rock wall in the Vltava Valley near Kralupy. Univ Carolina Geol 1:187-228

Dutrizac JE (1979) The physical chemistry of iron precipitation in the zinc industry. In Cigan JM, Mackey TS, O'Keefe TJ (Eds) Lead–Zinc–Tin '80. Metallurgical Society AIME (American Institute of Mining, Metallurgical and Petroleum Engineers), Warrendale, Pennsylvania, p 532-564

Dutrizac JE (1982) Jarosite-type compounds and their application in the metallurgical industry. In Osseo-Assare K, Miller JD (Eds) Hydrometallurgy Research, Development and Plant Practice. TMS–AIME (The Minerals, Metals and Materials Society–American Institute of Mining, Metallurgy and Petroleum Engineers), New York, p 531-551

Dutrizac JE (1983) Factors affecting alkali jarosite precipitation. Metall Trans 14B:531-539

Dutrizac JE (1984) The behaviour of impurities during jarosite precipitation. In Bautista RG (Ed) Hydrometallurgical Process Fundamentals. Plenum Press, New York, p 125-169

Dutrizac JE (1990) Converting jarosite residues into compact hematite products. J Metals 42:36-39

Dutrizac JE (1991) The precipitation of lead jarosite from chloride media. Hydrometall 26:327-346

Dutrizac JE (1996) The effect of seeding on the rate of precipitation of ammonium jarosite and sodium jarosite. Hydrometall 42:293-312

Dutrizac JE (1997) The behavior of thallium during jarosite precipitation. Metall Mater Trans 28B:765-776

Dutrizac JE (1999a) Effectiveness of different jarosite species as seed for the precipitation of sodium jarosite. J Metals 51(12):30-42

Dutrizac JE (1999b) Iron residues. CANMET, Natural Resources Canada, Rept MMSL 99-038 (J) Draft

Dutrizac JE, Chen TT (1981) The synthesis of mercury jarosite and the mercury concentration in jarosite-family minerals. Can Mineral 19:559-569

Dutrizac JE, Chen TT (2000) The behaviour of gallium during jarosite precipitation. Can Metall Quart 39:1-14

Dutrizac JE, Dinardo O (1983) The co-precipitation of copper and zinc with lead jarosite. Hydrometall 11:61-78

Dutrizac JE, Jambor JL (1984) Formation and characterization of argentojarosite and plumbojarosite and their relevance to metallurgical processing. *In* Park WC, Hausen DM, Hagni RD (Eds) Applied Mineralogy. Metallurgical Society AIME (American Institute of Mining, Metallurgy and Petroleum Engineers), Warrendale, Pennsylvania, p 507-530

Dutrizac JE, Jambor JL (1987a) Behaviour of silver during jarosite precipitation. Trans Inst Mining Metall 96:C206-C218

Dutrizac JE, Jambor JL (1987b) Behaviour of cesium and lithium during the precipitation of jarosite-type compounds. Hydrometall 17:251-265

Dutrizac JE, Jambor JL (1987c) The behaviour of arsenic during jarosite precipitation: arsenic precipitation at 97°C from sulphate or chloride media. Can Metall Quart 26:91-101

Dutrizac JE, Kaiman S (1975) Rubidium jarosite and thallium jarosite—new synthetic jarosite-type compounds and their structures. Hydrometall 1:51-59

Dutrizac JE, Kaiman S (1976) Synthesis and properties of jarosite-type compounds. Can Mineral 14:151-158

Dutrizac JE, Mingmin D (1993) The behaviour of indium during jarosite precipitation. *In* Matthew IG (Ed) World Zinc '93. Australasian Inst Mining Metall, p 365-372

Dutrizac JE, Dinardo O, Kaiman S (1980) Factors affecting lead jarosite formation. Hydrometall 5:305-324

Dutrizac JE, Dinardo O, Kaiman S (1981) Selenate analogues of jarosite-type compounds. Hydrometall 6:327-337

Dutrizac JE, Jambor JL, O'Reilly JB (1983) Man's first use of jarosite: the pre-Roman mining–metallurgical operations at Rio Tinto, Spain. CIM Bull 76(859):78-82

Dutrizac JE, Jambor JL, Chen TT (1986) Host minerals for the gallium-germanium ores of the Apex mine, Utah. Econ Geol 81:946-950

Dutrizac JE, Jambor JL, Chen TT (1987) The behaviour of arsenic during jarosite precipitation: reactions at 150°C and the mechanism of arsenic precipitation. Can Metall Quart 26:103-115

Dutrizac JE, Hardy DJ, Chen TT (1996) The behaviour of cadmium during jarosite precipitation. Hydrometall 41:269-285

Ebert SW, Rye RO (1997) Secondary precious metal enrichment by steam-heated fluids in the Crofoot-Lewis hot spring gold-silver deposit and relation to paleoclimate. Econ Geol 92:578-600

Elgersma F, Witkamp GJ, Van Rosmalen GM (1993) Incorporation of zinc in continuous jarosite precipitation. Hydrometall 33:313-339

Enikeev MR (1954) Plumbojarosite from Central Tian-Shan. Trudy Sredneaziat Gosudarst Univ, Geol Nauki 52(5):21-27

Enikeev MR (1964) Beaverite from Altyn-Topkan. Nauchn Trudy Tashkentsk Gosudarst Univ 249:36-39

Erd RC, White DE, Fahey JJ, Lee DE (1964) Buddingtonite, an ammonium feldspar with zeolitic water. Am Mineral 49:831-850

Erickson ET (1922) Tschermigite (ammonium alum) from Wyoming. Wash Acad Sci J 12(3):49-54

Fairchild JG (1933) Artificial jarosite—the separation of potassium from cesium. Am Mineral 18:543-545

Fanning DS, Rabenhorst MC, Bigham JM (1993) Colors of acid sulfate soils. *In* Bigham JM, Ciolkosz EJ (Eds) Soil Color. Soil Sci Soc Am Spec Publ 31:91-108

Fleming JF, Alexander LT (1961) Sulphur acidity in Southern Carolina tidal marsh soils. Soil Sci Soc Am Proc 25:94-95

Furbish WJ (1963) Geologic implications of jarosite, pseudomorphic after pyrite. Am Mineral 48:703-706

Gaines RV, Skinner HCW, Foord EE, Mason B, Rosenzweig A (1997) Dana's New Mineralogy. Wiley & Sons, New York

Garcia A, Valdez C (1996) Jarosite disposal practices at the Penoles zinc plant. *In* Dutrizac JE, Harris GB (Eds) Iron Control and Disposal. Canadian Institute of Mining, Metallurgy and Petroleum, Montreal, p 643-650

Gebhard G, Schlüter J (1995) Tsumeb, Namibia: Interesting new finds and new occurrences. Lapis 10: 24-32

Getskin LS, Margulis EV, Beisekeeva LI, Yaroslavtsev AS (1975) Study of iron hydrolytic precipitation from sulfate solutions in jarosite form. Izvest Vyssh Uchebn Zaved, Tsevtn Metall 75(6):40-44

Goldbery R (1978) Early diagenetic nonhydrothermal sodium alunite in Jurassic flint clays, Makhtesh Ramon, Israel. Geol Soc Am Bull 89:687-698.

Golovanov IM (1960) Crystals of natrojarosite from the polymetallic ore deposits of Kurgashinkan (Uzbek. S.S.R.). Zap Vses Mineral Obshch 89:704-709

Gorbach LP, Suprychev VA, Shekhotin VV (1967) Jarosite pseudomorphs of iron sulfides from Maikop formations of mountainous Crimea. Dokl Akad Nauk SSR 173:413-416

Grant RM (1993) Zinc concentrate and processing trends. *In* Matthew IG (Ed) World Zinc '93. Australasian Inst Mining Metall, Parkville, Australia, p 391-397

Gryaznov VI (1957) Minerals of the alunite-jarosite group from clays of the Kharkov series. Nauch Zap Dnepropetrovsk Gosudarst Univ, Sbornik Rabot Nauch-Issledovatel Inst Geol 58:79-85 (Chem Abs 54:24155)

Guillemin C (1955) Variety of hidalgoite from Cape Garonne. Bull Soc franc Minéral Crystallogr 78:27-32

Haigh CJ (1967) The hydrolysis of iron in acid solutions. Proc Australasian Inst Mining Metall 233:49-56

Haigh CJ, Pickering RW (1965) Treatment of zinc plant residue. Australian Patent 401,724 March 31, 1965

Harada Z, Goto M (1954) Synthesis of jarosite. Kobutsugaku Zasshi 1:344-355

Harris GB, Monette S, Stanley RW (1982) Hydrometallurgical treatment of Blackbird cobalt concentrate. *In* Osseo-Aware K, Miller JD (Eds) Hydrometallurgy Research, Development and Plant Practice. TMS–AIME (The Minerals, Metals and Materials Society–American Institute of Mining, Metallurgy and Petroleum Engineers), New York, p 139-150

Hartig C, Brand P, Bohmhammel K (1990) On the mechanism of alunite and jarosite precipitation. Neue Hutte 35:205-209

Hirasawa R (1993) Studies on the utilization of jarosite. Toyama Kogyo Koto Senmon Gakko Kiyo 20:1-4

Horakova M, Novak F (1989) Occurrence of the alunite-crandallite group minerals in the Barrandian Liten Formation, Bohemia. Cas Mineral Geol 34(2):151-163

Huang Z (1990) The recovery of silver and scarce elements at Zhuzhou Smelters. *In* Mackey TS, Prengaman DR (Eds) Lead–Zinc '90. TMS (The Minerals, Metals and Materials Society), Warrendale, Pennsylvania, p 239-250

Hutchison RFS, Phipps PJ (1977) Formation and particle size of jarosite. *In* Australasian Institute of Mining and Metallurgy Conference (Tasmania). Australasian Inst Mining Metall, Melbourne, p 319-327

Hutton CO, Bowen OE (1950) An occurrence of jarosite in altered volcanic rocks of Stoddard Mountain, San Bernardino County, California. Am Mineral 35:556-561

Hyashi H (1994) Mineralogy and chemistry of jarosite and acid sulfate soils. Nendo Kagaku 34(3):118-124 (Chem Abs 123:110807j)

Ievlev AA (1988) Mineralogy and formation of weathering zones in phosphorite-bearing shales of Pai Khoi. Soviet Geol (10):81-83

Ildefonse JP, Le Toullec C, Perrotel V (1986) About the synthetic lead-silver jarosite solid solution. *In* Experimental mineralogy and geochemistry. Abstracts Vol Internat Symp, Nancy, France, p 72-73

Ito Y (1969) Beaverite from the Kosaka mine, Akita Prefecture, Japan. Chigaku Kenkyu 20(8):219-225 (Chem Abs 72:34201t)

Ivarson KC (1973) Microbiological formation of basic ferric sulphates. Can J Soil Sci 53:315-323

Ivarson KC, Hallberg RO, Wadsten T (1976) The pyritization of basic ferric sulphates in acid sulphate soils: a microbiological interpretation. Can J Soil Sci 56:393-406

Ivarson KC, Ross GJ, Miles NM (1979) The microbiological formation of basic ferric sulphates: II crystallization in presence of potassium, ammonium and sodium salts. Soil Sci Soc Am J 43:908-912

Ivarson KC, Ross GJ, Miles NM (1981) Formation of rubidium jarosite during the microbiological oxidation of ferrous iron at room temperature. Can Mineral 19:429-434

Jackson DA (1976) Electrolytic zinc refining process including production of by-products from jarosite residues. United States Patent 3,937,658 February 10, 1976

Jambor JL (1998) Theory and applications of mineralogy in environmental studies of sulfide-bearing mine wastes. *In* Cabri LJ, Vaughan DJ (Eds) Modern Approaches to Ore and Environmental Mineralogy. Mineral Assoc Can Short Course Vol 27:367-401

Jambor JL (1999) Nomenclature of the alunite supergroup. Can Mineral 37:1323-1341

Jambor JL (2000) Nomenclature of the alunite supergroup: reply. Can Mineral 38 (in press)

Jambor JL, Blowes DW (2000) Mineralogy of mine wastes and strategies for remediation *In* Vaughan DJ, Wogelius RA (Eds) Environmental Mineralogy. EMU (European Mineralogical Union) Notes Mineral 2:255-290

Jambor JL, Dutrizac JE (1983) Beaverite–plumbojarosite solid solutions. Can Mineral 21:101-113

Jambor JL, Dutrizac JE (1985) The synthesis of beaverite. Can Mineral 23:47-51

Jambor JL, Owens DR, Grice JD, Feinglos MD (1996) Gallobeudantite, PbGa$_3$[(AsO$_4$), (SO$_4$)]$_2$(OH)$_6$, a new mineral species from Tsumeb, Namibia, and associated new gallium analogues of the alunite-jarosite family. Can Mineral 34:1305-1315

Jambor JL, Dutrizac JE, Chen TT (2000) Contribution of specific minerals to the neutralization potential in static tests. ICARD 2000, Society for Mining, Metallurgy and Exploration (SME), Littleton, Colorado, 1:551-565

Jensen MC, Rota JC, Foord EE (1995) The Gold Quarry mine, Carlin-trend, Eureka County, Nevada. Mineral Rec 26:449-469

Johansson G (1963) On the crystal structure of a basic gallium sulphate related to alunite. Arkiv Kemi 20:343-352

Johnson JH (1977) Jarosite and akaganéite from White Island volcano, New Zealand: an X-ray and Mössbauer study. Geochim Cosmochim Acta 41:539-544

Jung P-K, Yoo S-H (1994) Genesis and mineralogical characteristics of acid sulfate soils in Gimhae Plain. II. Genesis and distribution of the soil clay minerals. Han'guk T'oyang Piryo Hakhoechi 27:168-178 (Chem Abs 123:88810f)

Kakhadze EI (1971) Plumbojarosite of the Uchambo deposit of Adzharia. Soobshch Akad Nauk Gruz SSR 61:641-644

Kalinichenko LS (1973) Hypogenic jarosite in secondary quartzites. Izvest Akad Nauk Kazakh SSR, Ser Geol 30(1):60-63

Kato T, Miúra Y (1977) The crystal structures of jarosite and svanbergite. Mineral J 8:419-430

Kashkai MA, Babaev IA, Makhmudov SA, Kashkai CM (1974) Jarosite containing strontium. Zap Vses Mineral Obshch 103:481-483

Kasymov AK (1957) Plumbojarosite from Almalyk. Akad Nauk Uzbek SSR, Ser Geol 3:75-79

Kasymov AK (1958) Beaverite in Central Asia. Uzbek Geol Zhur 2:83-87

Kazakhashvili TG (1968) Jarosite from the Madneullsk copper-barite-complex ore deposit. Trudy Gruz Politekh Inst 6:58-63

Kershaw MG, Pickering RW (1979) The jarosite process—phase equilibria. *In* Cigan JM, Mackey TS, O'Keefe TJ (Eds) Lead–Zinc–Tin '80. AIME (American Institute of Mining, Metallurgy and Petroleum Engineers), New York, p 565-580

Khlybov VV (1976) Jarosite from the oxidation zone of pyritized rocks of the western Pritimanye. Trudy Inst Geol Komi Fil Akad Nauk SSSR 20:65-71

Khusanabaev DI, Nurullaev K (1969) Jarosite in the Mesozoic weathering profile in rocks of the Bukantau Mountains. Dokl Akad Nauk Uzbek SSR 26(4):41-42

Kingsbury AWG (1952) New occurrences of rare copper and other minerals. Trans R Geol Soc Cornwall 18:386-410

Kingsbury AWG, Hartley J (1957) Beaverite from the Lake District. Mineral Mag 31:700-702

Kinoshita K, Muta K, Minato H, Takano Y (1955) Jarosite and its occurrence from Tearai Hot Spring, Kirishima Volcano, Kyushu, Japan. Kobutsugaku Zasshi 2:121-129

Kittrick JA, Fanning DS, Hossner LR, Eds (1982) Acid Sulfate Weathering. Soil Sci Soc Am Spec Publ 10

Kolitsch U, Slade PG, Tieking ER, Pring A (1999a) The structure of antimonian dussertite and the role of antimony in oxysalt minerals. Mineral Mag 63:17-26

Kolitsch U, Pring A, Taylor MR, Fallon G (1999b) Springcreekite, Ba(V^{3+},Fe)$_3$(PO$_4$)$_2$(OH,H$_2$O)$_6$, a new member of the crandallite group from the Spring Creek mine, South Australia: the first natural V^{3+} member of the alunite family and its crystal structure. N Jahrb Mineral Monatsh 529-544

Krazewski SR (1972) Jarosite from Pliocene deposits in Wloclawek. Arch Mineral 30(1):5-12

Kretschmer B, Burgess G, Sanderson I (1998) Recent changes to Pasminco's lead smelter at Port Pirie. *In* Dutrizac JE, Gonzalez JA, Bolton GL, Hancock P (Eds) Zinc and Lead Processing. Canadian Institute Mining, Metallurgy and Petroleum, Montreal, p 455-469

Kubisz J (1960) Hydronium jarosite—(H$_3$O)Fe$_3$(SO$_4$)$_2$(OH)$_6$. Bull Acad Polon Sci, Ser Sci Geol Geogr 8(2):95-99

Kubisz J (1961) Natural hydronium jarosites. Bull Acad Polon Sci, Ser Sci Geol Geogr 9(4):195-200

Kubisz J (1962) Jarositization of rocks in the Upper Silesian coal basin. Bull Acad Pol Sci, Ser Sci Geol Geogr 10:1-10

Kubisz J (1964) A study on minerals of the alunite-jarosite group. Polska Akad Nauk Prace Geol 22:1-93

Kubisz J (1966) On the existence of hydronium hydrates H$_9$O$_4^+$ and H$_{15}$O$_7^+$ in minerals. Mineral Mag 35:1071-1079

Kubisz J (1970) Studies on synthetic alkali–hydronium jarosites: I. Synthesis of jarosite and natrojarosite. Mineral Polon 1:47-59

Kubisz J (1972) Studies on synthetic alkali-hydronium jarosites: III. Infrared absorption study. Mineral Polon 3:23-36

Kubisz J, Michalek Z (1959) Minerals of the oxidized zone of the menilite beds in the Carpathians. Bull Acad Polon Sci, Ser Sci Chim Geol Geogr 7:765-771

Kubisz J, Zabinski W (1958) The jarosites from the Silesia–Cracow zinc and lead ore deposits. Bull Acad Polon Sci, Ser Sci Chim Geol Geogr 6:793-797

Kulugin IA (1967) Iron deposits on volcanoes of the Kuril Islands. Geol Geofiz 4:20-27

Kunda W, Veltman H (1979) Decomposition of jarosite. Metall Trans 10B:439-446

Kunov A, Mandova E (1997) Supergene minerals in the Obichnik Au-Ag deposit (Eastern Rhodopes). Spis Bulg Geol Druzh 58:19-24 (Chem Abs 129:124950m)

Kydon DW, Pintar M, Petch HE (1968) NMR evidence of H_3O^+ ions in gallium sulphate. J Chem Phys 48:5348-5351

Le Brusq JY, Loyer JY, Mougenot B, Carn M (1987) Some new aluminum, iron, magnesium sulfate parageneses and their distribution in acid sulfate soils of Senegal. Sci Sol 25:173-184

Lengauer CL, Giester G, Irran E (1994) $KCr_3(SO_4)_2(OH)_6$: Synthesis, characterization, powder diffraction data, and structure refinement by the Rietveld technique and a compilation of alunite-type compounds. Powder Diff 9:265-271

Lightfoot BW, Floyd JM, Robillard KR (1993) Waste processing in the zinc industry using Ausmelt technology. *In* Matthew IG (Ed) World Zinc '93. Australasian Inst Mining Metall, Parkville, Australia, p 523-529

Lofvendahl R (1980) Jarosite minerals in Sweden. Geol Förenh Stockholm Förh 102:91-94

Lombardi G, Sonno M (1979) Petrographic study of the Mezzano alunite and of other altered rocks of the Latera caldera. Per Mineral 48:21-52

Long DT, Fegan NE, McKee JD, Lyons WB, Hines ME, Macumber PG (1992) Formation of alunite, jarosite and hydrous iron oxides in a hypersaline system: Lake Tyrrell, Victoria, Australia. Chem Geol 96:183-202

Lottermoser BG (1990) Rare-earth element mineralisation within the Mt. Weld carbonatite laterite, Western Australia. Lithos 24:151-167

Mandarino JA (1999) Fleischer's glossary of mineral species. Mineralogical Record, Tucson, Arizona

Margulis EV, Getskin LS, Zapuskalova NA, Beisekeeva, LI (1976a) Hydrolytic precipitation of iron in the $Fe_2(SO_4)_3$–KOH–H_2O system. Russ J Inorg Chem 21:996-999

Margulis EV, Getskin LS, Zapuskalova NA, Vershinina FI (1976b) Concentration conditions for formation of primary crystalline and amorphous phases during hydrolytic precipitation of iron (III) from sulfate solutions. Zhur Prik Khim 49:2382-2386

Margulis EV, Getskin LS, Kravets MV, Alkatsev ML, Yaroslavtsev AS (1976c) Induction period of jarosite formation in zinc sulfate solutions. Izvest Vyssh Uchebn Zaved Tsvetn Metall 4:145-147

Margulis EV, Getskin LS, Zapuskalova NA (1977) Hydrolytic precipitation of iron in the $Fe_2(SO_4)_3$–NH_3–H_2O system. Russ J Inorg Chem 22:741-744

Maschmeyer DEG, Kawulka P, Milner EFG, Swinkels GM (1978) Application of the Sherritt–Cominco copper process to Arizona copper concentrates. J Metals 30:27-31

Matvienko VN (1964) Plumbojarosite from the southern Mugodzhar. Vestnik Akak Nauk Kazakh SSR 20:67-73

May A, Sjoberg JJ, Baglin EG (1973) Synthetic argentojarosite: physical properties and thermal behavior. Am Mineral 58:936-941

Mermut AR, Arshad MA (1987) Significance of sulfide oxidation in soil salinization in southeastern Saskatchewan, Canada. Soil Sci Soc Am J 51:247-251

Mitchell RS (1962) New occurrences of jarosite in Virginia. Am Mineral 47:788-789

Mogarovskii VV (1961) The geochemistry of thallium in the zone of oxidation of the Daraiso sulfide deposit (Central Asia). Geokhim 9:771-774

Moiseeva MI (1970) Jarosites from the weathering profile in the Kuramin Ridge and their exploration significance. Zap Uzbek Otd Vses Mineral Obshch 21:38-45

Montoroi J-P (1995) Supply of an acid sulfate mineral sequence in a lower Casamance valley (Senegal). CR Acad Sci Paris Ser IIa 320:395-402

Moss AA (1956) The nature of carphosiderite and allied basic sulphates of iron. Mineral Mag 31:402-412

Mumme WG, Scott TR (1966) The relationship between basic ferric sulfate and plumbojarosite. Am Mineral 51:443-453

Naboko SI (1959) Precipitation of jarosite from the acid sulfate water of the lower Mendeleev thermal spring (Kunashir Island). Trudy Mineral Muzeya Akad Nauk SSSR No. 10:164-170

Nekrasov I.Ya., Nikishova LV (1987) First find of argentojarosite in the USSR. Mineral Zhur 9(3):90-95

Nicholas J, de Rosen A (1963) Low-temperature hydrothermal phosphates and kaolinization: gorceixite of the Colettes massif (Allier) and the associated mineral (hinsdalite). Bull Soc franc Minéral Cristallogr 86:379-385

Nickel EH, Temperly JE (1987) Arsenoflorencite-(Ce): a new arsenate mineral from Australia. Mineral Mag 51:605-609

Nogueira ED, Regife JM, Redondo AL, Nogueira GD, Zaplana M (1980) The Comprex Process: non-ferrous metals production from complex pyrite concentrates. *In* Jones MJ (Ed) Complex Sulphides Ore Conference. Inst Mining Metall, London, p 227-233

Norzinc (1986) Norzinc AS annual report. Norzink AS, Oslo, Norway

Novák F, Jansa J, Prachar I (1994) Classification and nomenclature of alunite–jarosite and related mineral groups. Vestnik Ceského geol ústavu 69:51-57

Novikova TI (1957) Mineralogy of the yellow ochers (antimony) of the Dzhizhikrut mercury-antimony deposit. Trudy Inst Geol Akad Nauk Tadzhik SSR, No 2:283-298

Odum JK, Hauff PL, Farrow RA (1982) A new occurrence of ammoniojarosite in Buffalo, Wyoming. Can Mineral 20:91-95

Oeborn I, Berggren D (1995) Characterization of jarosite-natrojarosite in two northern Scandinavian soils. Geoderma 66:213-225

Okada K, Soga H, Ossaka J, Otsuka N (1987) Syntheses of minamiite type compounds, $M_{0.5}Al_3(SO_4)_2(OH)_6$ with M = Sr^{2+}, Pb^{2+} and Ba^{2+}. N Jahrb Mineral Monatsh 64-70

Oliveira SMB de, Blot A, Imbernon RAL, Magat P (1996) Jarosite and plumbojarosite in gossans of the Canoas District, Parana State. Rev Bras Geocienc 26:3-12

Onozaki A, Sato K, Kuramochi S (1986) Effect of some impurities on iron precipitation at the Iijima Zinc Refinery. *In* Dutrizac JE, Monhemious AJ (Eds) Iron Control in Hydrometallurgy. Ellis Horwood, Chichester, UK, p 742-752

Paar WH, Burgstaller J, Chen TT (1980) Osarizawaite–beaverite intergrowths from Sierra Gorda, Chile. Mineral Rec 11:101-104

Palache C, Berman H, Frondel C (1951) The system of mineralogy. Vol 2. Wiley & Sons, New York

Pammenter RV, Haigh CJ (1981) Improved metal recovery with the low-contaminant jarosite process. *In* Extraction Metallurgy '81. Inst Mining Metall, London, p 379-392

Park YH, Park, PC (1978) A study of iron precipitation as jarosite. Taehan Kumsok Hakhoe Chi 16:495-502

Parker EG (1981) Oxidative pressure leaching of zinc concentrates. CIM Bull 74(829):145-150

Patino F, Viñals J, Roca A, Nunez C (1994) Alkaline decomposition–cyanidation kinetics of argentian plumbojarosite. Hydrometall 34:279-291

Patino F, Arenas A, Rivera I, Cordoba DA, Hernandez L, Salinas E (1998) Decomposition of argentiferous plumbojarosite in CaO media. Revista Soc Quim Mexico 42:122-128

Penner E, Gillot JE, Eden WJ (1970) Investigation of heave in Billings shale by mineralogical and biogeochemical methods. Can Geotech J 7:333-338

Petrov KI, Pervykh VG, Bol'shakova NK (1966) Infrared absorption spectra of basic gallium salts of the alunite type. Russ J Inorg Chem 11:742-745

Pophanken H (1996) Constructing, operating and capping of the jarosite pond, Galing 1. *In* Dutrizac JE, Harris GB (Eds) Iron Control and Disposal. Canadian Institute Mining, Metallurgy and Petroleum, Montreal, p 613-626

Powers DA, Rossman GR, Schugar HJ, Gray HB (1976) Magnetic behaviour and infrared spectra of jarosite, basic iron sulphate, and their chromate analogs. J Solid State Chem 13:1-13

Pring A, Birch WD, Dawe J, Taylor AM, Deliens M, Walenta K (1995) Kintoreite, $PbFe_3(PO_4)_2(OH,H_2O)_6$, a new mineral of the jarosite-alunite family, and lusungite discredited. Mineral Mag 56:143-148

Prokhorov IG (1970) Mineral composition of sulfate-containing concretions from refractory clay deposits in the Donets Basin. Voprosy Mineral Osad Obrazov 8:35-41

Raade G (1971) Natrojarosite in Norway. Nork Geol Tidsskr 51:195-197

Rattray KJ, Taylor MR, Bevan DJM, Pring A (1996) Compositional segregation and solid solution in the lead-dominant alunite-type minerals from Broken Hill, N.S.W. Mineral Mag 60:779-785

Ripmeester JA, Ratcliffe CI, Dutrizac JE, Jambor JL (1986) Hydronium ion in the alunite-jarosite group. Can Mineral 24:435-447

Ripp GS, Kanakin SV (1998) Phosphate minerals in the metamorphosed high-alumina rocks of the Ichetui occurrence, Transbaikal region. Dokl Earth Sci 359:233-235

Ripp GS, Kanakin SV, Shcherbakova MN (1998) Phosphate mineralization in metamorphosed high-alumina rocks of the Ichetoyskoye ore occurrence. Zap Vseross Mineral Obshch 127(6):98-108

Riveros PA, Dutrizac JE (1999) The recovery of iron from metallurgical effluents using ion exchange and solvent extraction. TRAWMAR annual workshop, San Sebastian, Spain. Eur Comm Indust Mater Tech Prog, Brussels, Belgium, p 52-62

Riveros PA, Dutrizac JE, Benguerel E, Houlachi G (1998) The recovery of iron from zinc sulphate-sulphuric acid processing solutions by solvent extraction or ion exchange. Mineral Proc Extr Metall Reviews 18:105-145

Roberts WL, Rapp GR Jr, Weber J (1974) Encyclopedia of Minerals. Van Nostrand Reinhold, New York

Roca A, Viñals, J, Arranz M, Calero J (1999) Characterization and alkaline decomposition/cyanidation of beudantite-jarosite materials from Rio Tinto gossan ores. Can Metall Quart 38:93-103

Rodier DD (1995) Meeting the challenges to the zinc industry. *In* Azakami T, Masuko N, Dutrizac JE, Ozberk E (Eds) Zinc & Lead '95. Mining Materials Inst Japan, Tokyo, p 3-14

Rosato LI, Agnew MJ (1996) Iron disposal options at Canadian Electrolytic Zinc. *In* Dutrizac JE, Harris GB (Eds) Iron Control and Disposal. Canadian Institute Mining, Metallurgy and Petroleum, Montreal, p 77-89

Rye RO, Alpers CN (1997) The stable isotope geochemistry of jarosite. US Geol Survey Open-File Rept 97-88

Sahoo PK, Das SC (1986) Recovery of lead from zinc plant waste. Trans Indian Inst Metals 39:604-608

Saito J (1962) Geological study on mineral springs in Hokkaido. Rept Geol Survey Hokkaido 28:1-88

Saksela M (1952) Weathering of some sulfide ores found in Finland. Bull Comm Géol Finlande 157:27-40

Sasaki K, Ootuka K, Watanabe M, Tozawa K (1998) Solubility of sodium jarosite in sulfuric acid solution at temperature range of 70 to 110°C. Shigen-to-Sozai 114:111-120

Savellev VF, Trostyanskii GD (1970) Epigenetic minerals from Mesozoic–Cenozoic sedimentary rocks of the Kyzylkums. Zap Uzbek Otd Vses Mineral Obshch 23:97-101

Schempp CA (1923) Argento-jarosite: a new silver mineral. Am J Sci 6:73-74

Schukow H, Breitinger DK, Zeiske T, Kubanek F, Mohr J, Schwab RG (1999) Localization of hydrogen and content of oxonium cations in alunite via neutron diffraction. Z anorg allg Chemie 625:1047-1050

Schwab RG, Herold H, Götz C, Pinto de Oliveira N (1990) Compounds of the crandallite type: synthesis and properties of pure rare earth element-phosphates. N Jahrb Mineral Monatsh 241-254

Schwab RG, Götz C, Herold H, Pinto de Oliveira N (1991) Compounds of the crandallite type: synthesis and properties of pure (Ca,Sr,Ba,Pb,La,Ce to Eu)–arsenocrandallites. N Jahrb Mineral Monatsh 97-112

Schwertmann U (1961) The occurrence and origin of jarosite (maibolt) in marshy soil. Naturwiss 48:159-160

Scott KM (1987) Solid solution in, and classification of, gossan-derived members of the alunite-jarosite family, northwest Queensland, Australia. Am Mineral 72:178-187

Scott KM (2000) Nomenclature of the alunite supergroup: a discussion. Can Mineral 38 (in press)

Scott TR (1973) Continuous, co-current pressure leaching of zinc–lead concentrates under acid conditions. *In* Evans DJI, Shoemaker RD (Eds) International Symposium on Hydrometallurgy (Chicago). AIME (American Institute of Mining, Metallurgy and Petroleum Engineers), New York, p 718-750

Seeliger E (1950) Pseudohydrothermal lead-zinc ore veins in the Ruhr district and near Velbert-Lintorf. A study of the effect of hot Permian salt solutions on lead-zinc ores of Christian Levin at Essen and Stein V at Hüls near Recklinghausen. Arch Lagerstättenforsch No 80:1-46 (Chem Abs 45:2374)

Sejkora J, Cejka J, Srein V, Novotná M, Ederová J (1998) Minerals of the plumbogummite–philipsbornite series from Moldava deposit, Krusné Hory Mts., Czech Republic. N Jahrb Mineral Monatsh 145-163

Shannon EV (1927) Ammoniojarosite, a new mineral of the jarosite group from Utah. Am Mineral 12:424-426

Shannon RD (1976) Revised effective ionic radii and systematic studies of interatomic distances in halides and chalcogenides. Acta Crystallogr A32:751-767

Shin JS, Jang YS (1993) Pedogenetic jarosite of acid sulfate soil of Gimhae series. I. Some chemical characteristics. Han'guk T'oyang Piryo Hakhoechi 26:278-283

Shishkin NV (1950) Oxonium ions in the structure of inorganic compounds. Zap Vses Mineral Obshch 79:94-102

Shokarev MM, Margulis EV, Vershinina FI, Beisekeeva LI, Savchenko LA (1972) Infrared spectra of iron (III) hydroxide sulphates and hydroxides. Russ J Inorg Chem 17:1293-1296

Simons FS, Mapes VE (1956) Geology and ore deposits of the Zimapán mining district, State of Hidalgo, Mexico. US Geol Survey Prof Paper 284

Sitges F, Arregui V (1964) A process for the recovery of zinc from ferrites. Spanish Patent 304,601 October 2, 1964

Slansky E (1975) Natroalunite and alunite from White Island Volcano, Bay of Plenty, New Zealand. New Zealand J Geol Geophys 18:285-293

Smith DK, Roberts AC, Bayliss P, Liebau F (1998) A systematic approach to general and structure-type formulas for minerals and other inorganic phases. Am Mineral 83:126-132

Stahl RS, Fanning DS, James BR (1993) Goethite and jarosite precipitation from ferrous sulfate solutions. Soil Sci Soc Am J 57:280-282

Steintveit G (1965) Process for the separation of iron from metal sulphate solutions and a hydrometallurgical process for production of zinc. Norwegian Patent 108,047. April 30, 1965

Steintveit G (1970) Iron precipitation as jarosite and its application in the hydrometallurgy of zinc. Erzmetall 23:532-539

Stoffregen RE (1987) Genesis of acid-sulfate alteration and Au-Cu-Ag mineralization at Summitville, Colorado. Econ Geol 82:1575-1581

Szymanski JT (1985) The crystal structure of plumbojarosite $Pb[Fe_3(SO_4)_2(OH)_6]_2$. Can Mineral 23:659-668

Taguchi Y, Kizawa Y, Okada N (1972) Beaverite from the Osarizawa mine. Kobutsugaku Zasshi 10: 313-325

Tananaev IV, Bol'shakova NK, Zazakova TI (1967a) Potassium gallium alum and its hydrolysis products. Russ J Inorg Chem 12:185-188

Tananaev IV, Kuznetsov VG, Bol'shakova NK (1967b) Basic gallium salts of the alunite type. Russ J Inorg Chem 12:28-30

Teixeira LA, Tavares LY (1986) Precipitation of jarosite from manganese sulphate solutions. *In* Dutrizac JE, Monhemius AJ (Eds) Iron Control in Hydrometallurgy. Ellis Horwood, Chichester, UK, p 431-453

The Northern Miner (2000) Expatriate inks deal to acquire Kudz Ze Kayah. Vol 86(2), p 1, 14, 22

Tudo J, Laplace G, Tachez M, Théobald F (1973) On the hydroxysulfate $VOHSO_4$. CR Acad Sci Paris Ser C 277:767-770

Tyler SA (1936) Heavy minerals of the St. Peter sandstone in Wisconsin. J Sed Petrol 6:55-84

Vakhrushev VA (1954) Plumbojarosite from N. Kirghiz. Zap Vses Mineral Obshch 83:246-249

Van Breemen N, Buurman P (1998) Soil Formation. Kluwer Academic, London, UK

Van Breemen N, Harmson K (1975) Translocation of iron in acid sulfate soils: I Soil morphology, and the chemistry and mineralogy of iron in a chromosequence of acid sulfate soils. Soil Sci Soc Am Proc 39:1140-1147

Van Tassel R (1958a) Jarosite, natrojarosite, beaverite, leonhardtite and hexahydrite from the Belgian Congo. Bull Inst R Sci Nat Belge 34:1-12

Van Tassell R (1958b) On carphosiderite. Mineral Mag 31:818-819

Van Tassel R (1965) Natrojarosite from the Warkalli beds, Varkala, Kerala, India. Geol Soc India Bull 2(3):53-56

Vasilevskaya GB (1970) Minerals in the oxidation zone of the Shamyrsai complex ore deposits. Nauch Trudy Tashkent Gosudarst Univ 358:131-134

Viñals J, Nunez C (1988) Dissolution kinetics of argentian plumbojarosite from old tailings of sulfatizing roasting pyrites by $HCl-CaCl_2$ leaching. Metall Trans 19B:365-373

Viñals J, Nunez C, Carrasco J (1991a) Leaching of gold, silver and lead from plumbojarosite-containing hematite tailings in $HCl-CaCl_2$ media. Hydrometall 26:179-199

Viñals J, Roca A, Cruells M, Nunez C (1991b) Recovery of gold and silver from plumbojarosite-containing hematite tailings by alkaline pretreatment and cyanidation. EMC '91: Non-ferrous Metallurgy—Present and Future. Elsevier Applied Science, London, p 11-17

Vitovskaya IV (1960) New data on the mineralogy of the oxidation zone of the Akchagyl deposit in Central Kazakhstan. Kora Vyvetrivaniya Akad Nauk SSSR, Inst Rudnykh Mestorozh Petrogr Mineral Geokhim No. 3:74-116

Walenta K, Zweiner M, Dunn PJ (1982) Philipsbornite, a new mineral of the crandallite group from Dundas, Tasmania. N Jahrb Mineral Monatsh 1-5

Wang Q, Ma R, Tan Z (1985) The jarosite process—kinetic study. *In* Tozawa K (Ed) Zinc '85. Mining Metall Inst Japan, Tokyo, p 675-690

Warshaw CM (1956) The occurrence of jarosite in underclays. Am Mineral 41:288-296

Wilder TC, Katrak FE (1983) Overview of current processes for the extraction of gallium. *In* Hass LA, Weir DR (Eds) Hydrometallurgy of Copper, its Byproducts and Rare Metals. SME–AIME (Minerals, Metals and Materials Society–American Institute of Mining, Metallurgy and Petroleum Engineers), New York, p 51-57

Wilkins RWT, Mafeen A, West GW (1974) The spectroscopic study of oxonium ions in minerals. Am Mineral 59:811-819

Winters J, Vos L, Canoo C (2000) Goethite: from residue to secondary building material—Union Miniere's Graveliet process. *In* Dutrizac JE, Gonzalez JA, Henke DM, James SE, Siegmund AH-J (Eds) Lead–Zinc 2000. TMS (The Minerals, Metals and Materials Society), Warrendale, Pennsylvania, p 903-916

Won CW, Park YH (1982) Precipitation kinetics of hydronium jarosite at elevated temperatures. Taehan Kumsok Hakhoe Chi 20:594-602

Yakhontova LK, Dvurechenskaya SS, Sandomirskaya NE, Sergeeva NE, Palchik NA (1988) Sulfates from the cryogenic hypergenesis zone. New findings. Nomenclature problems. Mineral Zhur 10:3-15

Yaroslavtsev AS, Getskin LS, Usenov AU, Margulis EV (1975) Behavior of impurities when precipitating iron from sulphate zinc solutions. Tsvetn Metall 16:41-42

Zakirov MZ (1978) Jarosite and its indicator properties. Uzbek Geol Zhur 5:57-62

Zapuskalova NA, Margulis EV (1978) Study of the hydrolytic precipitation of iron (III) from zinc sulfate solutions in the presence of monovalent cations. Izvest Vyssh Uchebn Zaved Tsevetn Metall (4):38-43

Zelenov KK, Tkachenko RI (1970) Chiatera thermal springs (central Java) and their deposits. Geol Geofiz 3:29-36

Zotov AV (1970) Jarosite in sediments from thermal waters of Kunashir Island. *In* Naborko SI (Ed) Mineral gidroterm sist Kamchatki Kurillskikh ostrovov. Nauka, Moscow, USSR, p 165-187

Alunite-Jarosite Crystallography, Thermodynamics, and Geochronology

R. E. Stoffregen
AWK Consulting Engineers, Inc.
1611 Monroeville Avenue,
Turtle Creek, Pennsylvania 15145

C. N. Alpers
U.S. Geological Survey
Water Resources Division, 6000 J Street
Sacramento, California 95819

J. L. Jambor
Leslie Research and Consulting
316 Rosehill Wynd
Tsawwassen, British Columbia, Canada V4M 3L9

The alunite supergroup consists of more than 40 mineral species that have in common the general formula $DG_3(TO_4)_2(OH,H_2O)_6$. The D sites are occupied by monovalent (e.g. K, Na, NH_4, Ag, Tl, H_3O), divalent (e.g. Ca, Sr, Ba, Pb), trivalent (e.g. Bi, REE) or more rarely quadrivalent (Th) ions; G is Al or Fe^{3+}, or rarely Ga or V; and T is S^{6+}, As^{5+}, or P^{5+}, and may include subordinate amounts of Cr^{6+} or Si^{4+}. Many of the minerals in this supergroup are exotic, having been described from relatively few localities worldwide, generally in association with ore deposits. Rarely are end-member compositions attained in these natural occurrences, and extensive solid solution is typical for one or more of the D, G, and T sites. In this chapter, the two solid-solution series considered in detail are alunite-natroalunite [$KAl_3(SO_4)_2(OH)_6$ – $NaAl_3(SO_4)_2(OH)_6$] and jarosite-natrojarosite [$KFe_3(SO_4)_2(OH)_6$ – $NaFe_3(SO_4)_2(OH)_6$]. These minerals are by far the most abundant naturally occurring species of the alunite supergroup.

Minerals with the generalized formula cited above can be variously grouped, but the simplest initial subdivision is on the basis of Fe > Al versus Al > Fe. Further subdivision is generally made on the basis of the predominant cation within the two TO_4 sites. Thus, within the supergroup, the alunite group consists of minerals in which both of the T sites are occupied by sulfur. This leads to a total negative charge of four on the TO_4 sites. In the ideal formulas of some members of the supergroup [e.g. woodhouseite, $CaAl_3(PO_4)(SO_4)(OH)_6$], half of the T sites are occupied by sulfur, and the other half by arsenic or phosphorus, which produces a total negative charge of five on the TO_4 sites. In still other end-members of the supergroup [e.g. crandallite, $CaAl_3(PO_4)_2(OH)_5(H_2O)$, and arsenocrandallite, $CaAl_3(AsO_4)_2(OH)_5(H_2O)$], both of the T sites are occupied solely by phosphorus or arsenic, thus producing a total negative charge of six on the TO_4 sites (see Table 1 of Dutrizac and Jambor, this volume). In this chapter, however, the primary concern is with those minerals for which TO_4 is represented by SO_4^{2-} (Table 1).

Precipitates with compositions near those of the end-members in the system alunite-natroalunite and jarosite-natrojarosite are readily prepared using sulfate salts. The products, however, almost invariably have a slight to appreciable deficiency in G^{3+}, and have an apparent non-stoichiometry for D. The latter may reflect incorporation of H_3O^+, a

1529-6466/00/0040-0009$05.00

Table 1. Minerals of the alunite group.

Alunite Subgroup		*Jarosite Subgroup*	
alunite	$KAl_3(SO_4)_2(OH)_6$	jarosite	$KFe_3(SO_4)_2(OH)_6$
natroalunite	$NaAl_3(SO_4)_2(OH)_6$	natrojarosite	$NaFe_3(SO_4)_2(OH)_6$
ammonioalunite	$(NH)_4Al_3(SO_4)_2(OH)_6$	ammoniojarosite	$(NH_4)Fe_3(SO_4)_2(OH)_6$
schlossmacherite	$(H_3O,Ca)Al_3(SO_4)_2(OH,H_2O)_6$	hydronium jarosite	$(H_3O)Fe_3(SO_4)_2(OH)_6$
—		argentojarosite	$AgFe_3(SO_4)_2(OH)_6$
—		dorallcharite	$TlFe_3(SO_4)_2(OH)_6$
osarizawaite	$Pb(Al,Cu)_3(SO_4)_2(OH,H_2O)_6$	beaverite	$Pb(Fe,Cu)_3(SO_4)_2(OH,H_2O)_6$
—		plumbojarosite*	$PbFe_6(SO_4)_4(OH)_{12}$
minamiite*	$(Na,Ca)_2Al_6(SO_4)_4(OH,H_2O)_{12}$	—	
huangite*	$CaAl_6(SO_4)_4(OH)_{12}$	—	
walthierite*	$BaAl_6(SO_4)_4(OH)_{12}$	—	

*c axis is ~34 Å, which is double that of the other members

solid solution that is difficult to prove because H_3O^+ cannot be determined directly by wet-chemistry or microprobe methods. Nevertheless, the existence of two minerals in the alunite supergroup is dependent solely on their D-site predominance of H_3O^+, namely, hydronium jarosite $[(H_3O)Fe_3(SO_4)_2(OH)_6]$ and schlossmacherite $[(H_3O,Ca)Al_3(SO_4)_2(OH,H_2O)_6]$.

This chapter is organized into four sections. In the first section, crystallographic data for alunite-natroalunite and jarosite-natrojarosite are presented and discussed. The second section describes available thermodynamic data for these two solid-solution series, in terms of properties of the end-members and mixing properties for intermediate compositions. The third section discusses the geochemistry and occurrences of alunite and jarosite, and the last section summarizes the published literature on the use of alunite and jarosite in geochronology.

CRYSTALLOGRAPHIC DATA

The crystal structures of minerals within the alunite supergroup were first determined by Hendricks (1937), who reported data for alunite, jarosite, and plumbojarosite. Structure studies of numerous additional minerals within the supergroup have since been determined; although there are several exceptions, most of the minerals have trigonal symmetry, space group $R\bar{3}m$, and unit-cell parameters of $a = ~7$, $c = ~17$ Å. Even among the exceptions, the minerals are strongly pseudotrigonal, and the basic topology of the structure remains the same regardless of chemical composition.

The basic structural motif of the supergroup consists of TO_4 tetrahedra and variably distorted octahedra of $G(O,OH)_6$. The latter corner-share hydroxyl ions, and two oxygen ions from two TO_4 tetrahedra, to form sheets parallel to the (001) plane, thereby forming a sequence that is perpendicular to the c axis. This arrangement is illustrated in Figure 1, a simplified representation of the alunite structure, and in Figure 2, a more detailed representation of the jarosite structure. Substitutions in G mainly affect the a dimension, and a increases as Fe-for-Al substitution increases. The TO_4 tetrahedra, which are aligned along [001], occur as two crystallographically independent sets within a layer; one set of TO_4 points upward along c, and this set alternates with another pointing downward. The oxygen and hydroxyl form a polyhedron amidst which is the D cation, sandwiched between the sheets of $G(O,OH)_6$ octahedra. Thus, if TO_4 sites are entirely occupied by

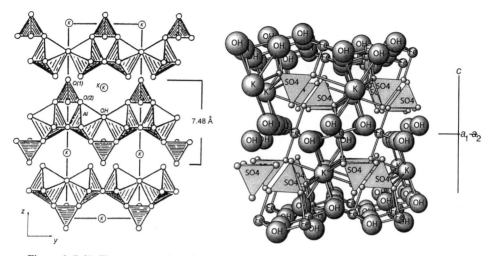

Figure 1 (left). The structure of alunite, viewed parallel to the *a* axis. (Slightly modified from Menchetti and Sabelli 1976).

Figure 2 (right). The structure of jarosite, viewed parallel to the *a* axis. (Illustration courtesy of Eric Dowty).

SO_4, variations in the length of *c* are mainly influenced by the size of the *D* cation. In the natroalunite–alunite and natrojarosite–jarosite series, therefore, as compositions approach those of alunite and jarosite the length of the *c* parameter increases as a consequence of the larger ionic radius of K versus Na. A corresponding increase in *a* does not occur because the expected expansion is instead accommodated by a closer fitting of the rough surfaces of the sheet-like layers of octahedra (Menchetti and Sabelli 1976, Okada et al. 1982).

Symmetry changes within the supergroup arise mainly because of ordering of the TO_4 tetrahedra, or because of ordering or distortion effects within *D* sites. Ordering of TO_4 reduces the symmetry, generally from $R\bar{3}m$ to $R3m$. As was discussed by Szymanski (1985), however, the space group of alunite-jarosite minerals should be assumed to be $R\bar{3}m$ unless structure data prove otherwise. Ordering of *D*-site cations can lead to a doubling of the length of *c*, an effect that is especially evident in alunite-group minerals in which *D* is occupied by a divalent element (Table 1). In plumbojarosite, for example, half of the *D* sites are vacant because of the need to maintain electroneutrality. Ordering of the Pb^{2+} and the vacancies produces a supercell with *c* = 33–34 Å.

Non-stoichiometry in both *D* and *G* is common, particularly in synthetic samples and in minerals in which *D* is occupied by a combination of monovalent and divalent elements. Calculation of formulas on the basis of $TO_4 = 2$ is the preferred approach because, whereas deficiencies in *D* and *G* occupancy may be tolerable, it is not possible to maintain a coherent structure with vacancies in TO_4.

The compositions of all of the minerals within the alunite supergroup are characterized by a predominance of SO_4, AsO_4, or PO_4. Extensive solid solution among these, well beyond that known at the time of Palache et al. (1951), is prevalent in many of the phosphate and arsenate members. Although the TO_4 groups provide well defined boundaries for a ternary system of nomenclature (Jambor 1999), the boundaries currently are set at 25 and 75 mol % (Scott 1987). For example, the composition of beudantite, which is generally assigned the formula $PbFe_3(AsO_4)(SO_4)(OH,H_2O)_6$, extends from $PbFe_3[(SO_4)_{0.75}(AsO_4)_{0.25}]_2(OH,H_2O)_6$ to $PbFe_3[(SO_4)_{0.25}(AsO_4)_{0.75}]_2(OH,H_2O)_6$, wherein OH is converted to H_2O, as necessary, to achieve charge balance. Compositions with the molar site occupancy of >0.75 total SO_4 on the T site apply to plumbojarosite, and those with AsO_4 > 0.75 are segnitite. The consequence is that, within each compositional triangle with AsO_4, PO_4, and SO_4 at the apices, five mineral names are permitted rather than the three more rigorously applicable to a ternary system.

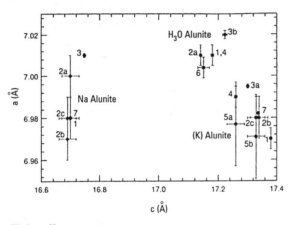

Figure 3. Variations in the unit-cell dimensions of alunite, natroalunite, and schlossmacherite (hydronium alunite). Sources of data: 1 = Kashkai (1969); 2 = Parker (1962): 2a = synthetic alunite and natroalunite (not heated, contain hydronium), 2b = synthetic alunite and natroalunite (heated, contain no hydronium), 2c = natural; 3 = Menchetti and Sabelli (1976): 3a = single crystal, 3b = powder; 4 = Kubisz (1964, 1970); 5 = Brophy and Sheridan (1965): 5a = synthetic, 5b = natural; 6 = Ripmeester et al. (1986); 7 = Stoffregen and Alpers (1992). Modified from Alpers et al. (1992).

Unit-cell parameters

Alunite. Stoffregen and Alpers (1992) provided a review of unit-cell parameters for synthetic and natural alunite. Published unit-cell dimensions for "end-member" alunite, natroalunite, and hydronium alunite are shown in Figure 3. The large range in a and c for each end-member is probably the result of non-hydroxyl water (most likely as hydronium ion) in some samples. For example, Menchetti and Sabelli (1976) reported a c value for alunite which is about 0.1 Å lower than most other published values. On the basis of their description of the synthesis method used, this lower value almost certainly reflects the presence of non-hydroxyl water in their sample. Many other studies do not provide information on the water content of the samples (e.g. Wang et al. 1965), and as such the results cannot be assumed to be representative of stoichiometric alunite or natroalunite.

Published data for unit-cell dimensions of alunite can be compared with results obtained for synthetic samples by Stoffregen and Alpers (1992; Table 2, Fig. 3). Unit-cell dimensions for the end-member alunite samples synthesized at 250 and 450°C agree within 2σ for c and a, whereas the sample obtained at 150°C has significantly higher a and lower c and cell volume (V). These results are similar to data from Parker (1962) for alunite synthesized at 100°C, for which c increased and a decreased after heating in air for one hour at 300°C. Parker attributed this shift in unit-cell dimensions to the loss of non-hydroxyl water (hydronium) during heating. This loss is confirmed by a decrease in water content from 128% of the amount predicted for stoichiometric alunite in the 150°C sample to about 105% of this amount in the other two synthetic alunite samples (Table 2).

Natroalunite. Unit-cell values for various samples of natroalunite are also shown in Figure 3. Table 2 indicates that natroalunite prepared at 150°C has a significantly larger value of a than those of the 250 and 450°C samples and also contains the most water of the three (Stoffregen and Alpers 1992).

Also given in Table 2 are unit-cell data for synthetic schlossmacherite (hydronium

Table 2. Unit-cell parameters for synthetic alunite-group minerals
(from Stoffregen and Alpers 1992).

	a, Å	c, Å	V, Å^3	Excess Water* (% of stoich.)
Alunite (150 °C)	7.000(2)	17.180(7)	729.1(4)	128.2
Alunite (250°C)	6.9831(5)	17.334(2)	732.03(9)	103.9
Alunite (450°C)	6.981(1)	17.331(4)	731.5(2)	105.5
Natroalunite (150°C)	6.9990(8)	16.690(3)	708.0(2)	119.6
Natroalunite (250°C)	6.9823(5)	16.700(2)	705.1(1)	113.0
Natroalunite (450°C)	6.9786(7)	16.696(3)	704.2(2)	99.1
Schlossmacherite	7.005(2)	17.114(7)	727.2(4)	n.d.
Jarosite (95°C)	7.310(1)	17.042(3)	788.8(2)	n.d.
Jarosite (200°C)	7.300(1)	17.216(4)	794.6(2)	n.d.
Natrojarosite (98°C)	7.336(2)	16.621(5)	774.7(3)	n.d.
Natrojarosite (200°C)	7.316(1)	16.590(3)	768.9(2)	n.d.

*Excess water computed as water content of samples divided by water content of stoichiometric end-member; n.d, not determined

alunite) prepared following the method of Ripmeester et al. (1986). For hydronium alunite, c is intermediate between those of alunite and natroalunite, but a is 7.005(2) Å. The value for a is similar to those of alunite and natroalunite samples prepared at 150°C, but is significantly larger than the 6.978–6.983 Å range obtained for the alunite and natroalunite samples prepared at 250 and 450°C. These results suggest that the presence of significant non-hydroxyl water in alunite can be recognized by an anomalously large value of a, which should be apparent regardless of the Na content of the sample.

Jarosite. Values of a and c for jarosite, natrojarosite, and hydronium jarosite are plotted in Figure 4. The c dimension for jarosite is slightly lower than that for alunite (17.24 versus 17.32 Å), and the a value for jarosite is significantly larger, at between 7.28 and 7.34 Å versus about 6.98 Å for alunite. As with alunite, the range of unit-cell values reported for jarosite seems to be due in large part to variations in the hydronium content. This relationship is particularly evident in the two samples from Stoffregen (unpublished data): the sample synthesized at 95°C had a = 7.310(1) and c = 17.042(3) Å, whereas the same sample heated to 200°C had a = 7.300(1) and c = 17.216(4) Å (Table 2, Fig. 4).

Natrojarosite. Unit-cell data for natrojarosite (Fig. 4) show values of a in the range of 7.29 to 7.33 Å, and c in the range of 16.5 to 16.7 Å. As with jarosite, the trend is for higher values of a to be associated with samples known or inferred to have an appreciable hydronium content. However, as discussed later, there also seems to be a slight increase in a that is related to the Na content.

THERMODYNAMIC DATA

Alunite and natroalunite

Overview of stability relations. The stability relations of alunite with silicate minerals in the system K–Al–Si–SO$_4$ are defined by the following reactions to form muscovite (Eqn. 1), kaolinite (Eqn. 2) (or pyrophyllite), and K-feldspar (Eqn. 3)

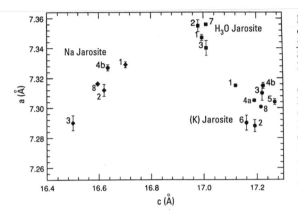

Figure 4. Variations in the unit-cell dimensions of jarosite, natrojarosite, and hydronium jarosite. Sources of data: 1 = Dutrizac and Kaiman (1976; JCPDS card 30-1203); 2 = Brophy and Sheridan (1965); 3 = Kubisz (1970); 4 = Menchetti and Sabelli (1976): 4a = powder, 4b = single crystal; 5 = Kato and Miúra (1977; JCPDS card 22-827); 6 = JCPDS card 22-827; 7 = JCPDS card 31-650; 8 = synthetic jarosite and natrojarosite (250°C), Stoffregen, unpublished data (see Fig. 12). Modified from Alpers et al. (1992).

(Hemley et al. 1969):

$$KAl_3(SO_4)_2(OH)_{6(s)} + 3\ SiO_{2(s)} \Leftrightarrow KAl_3Si_3O_{10}(OH)_{2(s)} + 2\ SO_4^{2-}{}_{(aq)} + 4\ H^+{}_{(aq)} \quad (1)$$

$$2\ KAl_3(SO_4)_2(OH)_{6(s)} + 6\ SiO_{2(s)} + 3\ H_2O \Leftrightarrow 3\ Al_2Si_2O_5(OH)_{4(s)} + 4\ SO_4^{2-}{}_{(aq)} + \\ 6\ H^+{}_{(aq)} + 2\ K^+{}_{(aq)} \quad (2)$$

and

$$KAl_3(SO_4)_2(OH)_{6(s)} + 9\ SiO_{2(s)} + 2\ K^+{}_{(aq)} \Leftrightarrow 3\ KAlSi_3O_{8(s)} + 2\ SO_4^{2-}{}_{(aq)} + 6\ H^+{}_{(aq)} \quad (3)$$

These reactions are generally represented in $\log(a^2_{H^+}a_{SO_4^{2-}}) - \log(a^2_{K^+}a_{SO_4^{2-}})$ activity space (e.g. Hemley 1969, Stoffregen 1987) as illustrated in Figure 5. The congruent reaction for alunite dissolution

$$KAl_3(SO_4)_2(OH)_{6(s)} + 6\ H^+{}_{(aq)} \Leftrightarrow 3\ Al^{3+}{}_{(aq)} + K^+{}_{(aq)} + 2\ SO_4^{2-}{}_{(aq)} + 6\ H_2O_{(l)} \quad (4)$$

can also be presented on this type of diagram for a specified value of $\log(a^2_{Al^{3+}}a^3_{SO_4^{2-}})$ (Fig. 5). This reaction limits alunite stability at high H_2SO_4 concentrations.

In the absence of silica, the stability of alunite is defined by the alunite-böhmite reaction

$$KAl_3(SO_4)_2(OH)_{6(s)} \Leftrightarrow 3\ AlO(OH)_{(böhmite)} + K^+{}_{(aq)} + 2\ SO_4^{2-}{}_{(aq)} + 3\ H^+{}_{(aq)} \quad (5)$$

or the similar alunite-to-corundum reaction (Stoffregen and Cygan 1990). These reaction boundaries parallel the kaolinite–alunite reaction in Figure 5.

Data on the stability of alunite relative to silicate minerals have been obtained by Hemley et al. (1969). Experimental results at 200, 300, and 380°C from that paper are presented in Figure 6. As is discussed later, the assemblages alunite + kaolinite and alunite + kaolinite + muscovite (or illite) are common as part of the advanced argillic alteration. The invariant point for the assemblage alunite + muscovite + kaolinite (or pyrophyllite) + quartz in terms of total molal H_2SO_4 and K_2SO_4, respectively, was reported at $0.002m$ and $0.03m$ at 200°C; $0.012m$ and $0.013m$ at 300°C; and $0.02m$ and $0.016m$ at 380°C. In contrast, the natural assemblages alunite + K-feldspar and alunite + K-feldspar + muscovite have not been reported. As noted by Hemley et al. (1969), this apparent absence suggests that the elevated K_2SO_4 concentrations required to stabilize such assemblages (Fig. 6) are rarely, if ever, obtained in nature.

The congruent reaction for alunite dissolution (Eqn. 4) has been used to evaluate the thermodynamic properties of alunite on the basis of its equilibrium relations with an aqueous phase in modern lake and hot-spring environments. Reaction (4) is also

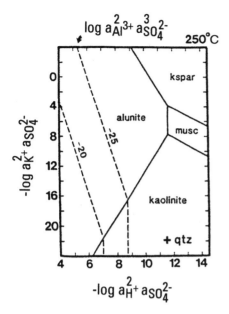

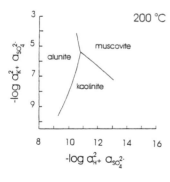

Figure 5. Alunite stability relations with silicate minerals at 250°C. Modified from Stoffregen (1987).

applicable to the interpretation of advanced argillic alteration, which is limited by the formation of an intensely leached, predominantly quartz-only assemblage. This type of alteration is typified by the so-called vuggy silica rock. The congruent reaction for alunite dissolution has not been studied experimentally at elevated temperature.

The alunite-böhmite and alunite-corundum reactions have not been studied experimentally, except indirectly in alkali-exchange experiments conducted by Stoffregen and Cygan (1990). In their experiments at 450°C, alunite reacted with $0.5m$ $(K,Na)_2SO_4 - 0.3m$ H_2SO_4 solutions to form corundum. When the initial H_2SO_4 concentration was 1.0 m, alunite reacted to form an Al-sulfate compound that is of uncertain stoichiometry and that has not been reported in nature. The silica-free system may be relevant to certain occurrences of alunite in Al-rich intermediate-grade metamorphic rocks, as discussed below.

End-member thermodynamic data.
Kelley et al. (1946) obtained heat-capacity data and an enthalpy of formation for a sample of Marysvale alunite and for a synthetic alunite. Subsequent work on the stability relations of alunite (e.g. Hemley et al. 1969, Helgeson et al. 1978) has generally been based on the data for the synthetic sample, presumably because of the impurities reported by Kelley et al. (1946) in the Marysvale

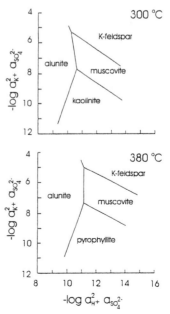

Figure 6. Experimentally determined phase relations of alunite with silicate minerals. Modified from Hemley et al. (1969).

alunite. As discussed above, many samples of synthetic alunite described in the literature seem to be non-stoichiometric because of the presence of hydronium. However, the 250°C synthesis temperature used by Kelley et al. (1946) is high enough to minimize the incorporation of hydronium, and their analytical data show the alunite to be almost stoichiometric. The Kelly et al. (1946) thermodynamic values for the synthetic alunite are also the values recommended here (Table 3).

Table 3. Recommended thermodynamic data for alunite group end-members.

	Units	Alunite	Natroalunite	Jarosite	Natrojarosite
$\Delta H_{f,298 K, 1 bar}$	kJ mol⁻¹	−5169.8[a]	−5131.97[c]	−3715.1[d]	−3673.1[d]
$S_{298 K, 1 bar}$	J mol⁻¹ (°K)⁻¹	328.0[a]	321.08[c]	388.9[d]	382.4[d]
$\Delta G_{f,298 K, 1 bar}$	kJ mol⁻¹	−4659.3[a]	−4622.40[c]	−3309.8[e]	−3256.7[f]
a	J mol⁻¹ (°K)⁻¹	642.03[b]	641.5[c]	616.89[d]	616.39[d]
b (× 10³)	J mol⁻¹ (°K)⁻²	0[b]	−7.87[c]	98.74[d]	91.21[d]
c (×10⁻⁵)	J mol⁻¹ (°K)⁻¹	229.91[b]	−234.12[c]	−199.6[d]	−203.76[d]

ᵃ Kelley et al. (1946); ᵇ Kelley (1960); ᶜ Stoffregen and Cygan (1990); ᵈ Stoffregen (1993); ᵉ Baron and Palmer (1996); ᶠ Kashkai et al. (1975)

Other published determinations of the free energy of formation of alunite have been based on mineral–water equilibrium (Kashkai et al. 1975, Raymahashay 1969, Zotov 1967). These values are considered to be less reliable than the data from Kelly et al. (1946) because the alunite that was used in the determinations was almost certainly non-stoichiometeric.

We are not aware of any published values for the entropy, enthalpy, or heat capacity of natroalunite. Stoffregen and Cygan (1990) used Na-K exchange data for alunite-natroalunite, together with a heat-capacity estimate based on the method of Helgeson et al. (1978), to obtain themodynamic data for natroalunite at 25°C (Table 3). In the absence of measured thermodynamic data on natroalunite, the values from Stoffregen and Cygan (1990) are recommended.

Jarosite and natrojarosite

Overview of stability relations. Unlike alunite, jarosite does not display equilibrium relations with silicate minerals. Instead, the principal reaction that limits its stability in nature is simply its conversion to iron oxide or oxyhydroxide minerals. At surficial conditions, conversion of jarosite to goethite occurs by the reaction

$$KFe_3(SO_4)_2(OH)_{6(s)} \Leftrightarrow 3\ FeO(OH)_{(goethite)} + K^+_{(aq)} + 2\ SO_4^{2-}{}_{(aq)} + 3\ H^+_{(aq)} \quad (6)$$

although this reaction may be complicated by the formation of metastable phases such as schwertmannite [$Fe_8O_8(OH)_{8-2x}(SO_4)_x$] and ferrihydrite (~$Fe_5HO_8 \cdot 4H_2O$) (Nordstrom and Alpers 1999, Bigham and Nordstrom, this volume). At temperatures above the goethite-to-hematite transition, which occurs at about 100°C for phases of equal surface area (Langmuir 1971), the incongruent dissolution reaction for jarosite becomes

$$KFe_3(SO_4)_2(OH)_{6(s)} \Leftrightarrow 1.5\ Fe_2O_{3(s)} + K^+_{(aq)} + 2\ SO_4^{2-}{}_{(aq)} + 3\ H^+_{(aq)} + 1.5\ H_2O_{(l)} \quad (7)$$

The compound $Fe(SO_4)(OH)$ provides a second boundary to the jarosite field by the reaction

$$KFe_3(SO_4)_2(OH)_{6(s)} + 3\ H^+_{(aq)} + SO_4^{2-}{}_{(aq)} \Leftrightarrow 3\ Fe(SO_4)(OH)_{(s)} + K^+_{(aq)} + 3\ H_2O_{(l)} \quad (8)$$

Although not reported in nature, $Fe(SO_4)(OH)$, which is the anhydrous equivalent of butlerite and related phases (see Bigham and Nordstrom, this volume), has been observed in several studies of the system $H_2O–SO_3–Fe_2O_3$ (Posnjak and Merwin 1922, Umetsu et al. 1977, Tozawa et al. 1983, Stoffregen 1993).

Reactions (7) and (8) can be used to define a field of jarosite stability in $\log(a^2_{H^+}a_{SO_4^{2-}}) – \log(a^2_{K^+}a_{SO_4^{2-}})$ space. This is illustrated in Figure 7, which indicates that jarosite should occur at $\log(a^2_{H^+}a_{SO_4^{2-}})$ values intermediate between those required to produce hematite and $Fe(SO_4)(OH)$ at a given value of $\log(a^2_{K^+}a_{SO_4^{2-}})$. Figure 7 also implies that, at a given temperature and concentration of dissolved ferric iron, there is a minimum value of $\log(a^2_{K^+}a_{SO_4^{2-}})$ necessary for jarosite stability. This minimum is the jarosite-hematite-$Fe(SO_4)(OH)$ triple point. The stability limit for jarosite is reached when this triple point occurs at $\log(a^2_{K^+}a_{SO_4^{2-}})$ equal to the saturation value for potassium sulfate.

Also shown in Figure 7 are contours for the activity product $\log(a^2_{Fe^{3+}}a^3_{SO_4^{2-}})$ in equilibrium with each of the solid phases. These contours are based on the congruent dissolution reactions

$$KFe_3(SO_4)_2(OH)_{6(s)} + 6H^+_{(aq)} \Leftrightarrow K^+_{(aq)} + 2\,SO_4^{2-}{}_{(aq)} + 3Fe^{3+}_{(aq)} + 6H_2O_{(l)} \qquad (9)$$

$$Fe_2O_{3(s)} + 6H^+_{(aq)} \Leftrightarrow 2Fe^{3+}_{(aq)} + 3H_2O_{(l)} \qquad (10)$$

and

$$Fe(SO_4)(OH)_{(s)} + H^+_{(aq)} \Leftrightarrow Fe^{3+}_{(aq)} + SO_4^{2-}{}_{(aq)} + H_2O_{(l)} \qquad (11)$$

The contours of $\log(a^2_{Fe^{3+}}a^3_{SO_4^{2-}})$ in Figure 7 imply that the amount of dissolved Fe in equilibrium with jarosite increases as $\log(a^2_{H^+}a_{SO_4^{2-}})$ increases and as $\log(a^2_{K^+}a_{SO_4^{2-}})$ decreases.

The stability relations of jarosite have been studied primarily at surficial conditions, using Reaction (9). Numerous other studies on the phase relations in the system Fe–S–H_2O have been conducted, mainly for hydrometallurgical purposes (Dutrizac and Jambor, this volume). Stoffregen (1993) conducted experiments to identify the phase boundaries given by Reactions (7) and (8) at 150 to 250°C. These experiments indicated that a H_2SO_4 molality of 0.25 to 0.63 was required to stabilize jarosite over this temperature range (Fig. 8). Natrojarosite formed from hematite in experiments with H_2SO_4 of 0.79 *m* at 200°C but could not be produced from hematite at 250°C.

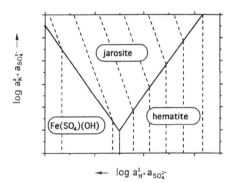

Figure 7. Schematic illustration of the stability relations of jarosite. Dashed lines represent contours of $\log(a^2_{Fe^{3+}}a^3_{SO_4^{2-}})$, increasing to the left. Used by permission of Elsevier Science Ltd., from Stoffregen (1993), *Geochim Cosmochim Acta*, Vol. 57, Fig. 1, p. 2418.

End-member themodynamic data.

The many determinations of the free energy of jarosite at surface conditions, based on assumed equilibrium between jarosite and a coexisting aqueous phase, have been summarized by Baron and Palmer (1996). The most reliable values are considered to be those of Alpers et al. (1989), Kashkai et al. (1975), Zotov et al. (1973), and Baron and Palmer (1996). The respective values for $\Delta G°_{f,298}$ are -3300.2±2.6; -3299.7±4; -3305.8±4; and -3309.8±1.7 kJ/mol. The recommended value is -3309.8±1.7 kJ/mol

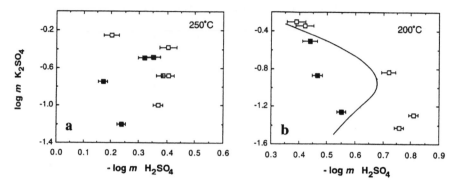

Figure 8. Experimentally determined jarosite-hematite boundary. Filled squares = hematite starting material, run product with both hematite and jarosite; open squares = jarosite starting material, run product with both hematite and jarosite. (a) 250°C; (b) 200°C. The solid line in (b) is the hematite-jarosite boundary computed with PHRQPITZ (see text). Error bars represent uncertainty in SO_4 and K measurements. Used by permission of Elsevier Science Ltd., from Stoffregen (1993), *Geochim Cosmochim Acta* Vol. 57, Fig. 2, p. 2420.

(Baron and Palmer 1996). Recommended enthalpy and entropy values for jarosite and natrojarosite are also included in Table 3. There are only a few published values for $\Delta G°_{f,298}$ of natrojarosite. The recommended value is -3256.7 kJ/mole reported by Kashkai et al. (1975).

For jarosite, Stoffregen (1993) reported a Gibbs free energy at 200°C of -3416.3 kJ/mole. On the basis of the experimental reversals shown in Figure 8, this value was computed using the program PHRQPITZ (Plummer et al. 1988) to model the aqueous phase. Because information on the temperature dependence of the H^+–HSO_4^-, K^+–HSO_4^- and Na^+–HSO_4^- interaction parameters was not available, all were assumed equal to the 25°C values in the computation. This assumption, and the high ionic strength of the coexisting aqueous phase ($I = 0.33$), introduce uncertainty into the free-energy value reported for jarosite. Nevertheless, the phase boundary predicted with PHRQPITZ is consistent with the reversals observed for Reaction (7).

No measured heat-capacity data are available for jarosite or natrojarosite. For both, Stoffregen (1993) presented values of the Meier–Kelley a, b, and c terms estimated from thermodynamic data on alunite, hematite, and corundum using the method of Helgeson et al. (1978). These values are listed in Table 3. The $\Delta G°_{f,298}$ of jarosite based on these values is -3328.4 kJ/mole, in poor agreement with the range of -3299.7 to -3310.4 kJ/mole discussed above. The discrepancy is thought to result from inaccuracies in the estimated thermodynamic parameters in Table 3, but it may also reflect errors in the determination of the aqueous activity coefficients, as noted above. Similarly, $\Delta G°_{f,298}$ for natrojarosite obtained from the data in Table 3 is 29 kJ more negative than the value reported by Kashkai et al. (1975).

Other minerals in the alunite-jarosite supergroup

The $\Delta G°_f$ values of synthetic goyazite $SrAl_3(PO_4)_2(OH,H_2O)_6$, arsenogoyazite $SrAl_3(AsO_4)_2(OH,H_2O)_6$, gorceixite $BaAl_3(PO_4)_2(OH,H_2O)_6$, arsenogorceixite $BaAl_3(AsO_4)_2(OH,H_2O)_6$, plumbogummite $PbAl_3(PO_4)_2(OH,H_2O)_6$, florencite-(Ce) to -(Eu), and As analogs of the florencite minerals were reported by Schwab et al. (1993), who determined the equilibrium constants for dissolution reactions. Because of the low solubility of these minerals, at least one year was needed for each experiment, and there

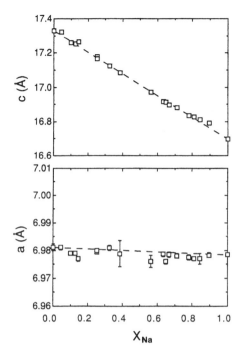

Figure 9. Unit-cell parameters of alunite as a function of Na mole fraction. From Stoffregen and Alpers (1992).

are significant uncertainties in the data because of the poor crystallinity and fine grain size of the precipitates. Nevertheless, Schwab et al. (1993) reported excellent agreement with independent estimates by Herold (1987) for the $\Delta G°_{f,298}$ of crandallite, although both studies used extrapolation from goyazite-crandallite solid solutions because pure crandallite could not be synthesized.

Mixing relations

Alunite–natroalunite. Stoffregen and Alpers (1992) presented data on unit-cell dimensions for 17 samples on the alunite–natroalunite binary. These samples were prepared in alunite–water Na-K exchange experiments at 450°C. The data are plotted as a function of composition in Figure 9, along with data for end-member alunite and natroalunite. The variation in *a* across this binary is approximately the same size as the average 2σ for a and is thus not significant. Most values of *c* and *V* plot on the ideal mixing lines connecting the 450°C alunite and natroalunite samples, which have the equations

$$X_{Na} = (17.331 - c_{meas})/0.635 \tag{12}$$

$$X_{Na} = (731.5 - V_{meas})/27.3 \tag{13}$$

If the uncertainty in the end-member values is also considered, all but four of the samples are within 2σ of ideal mixing. These samples do not define a systematic departure from ideal mixing and are believed to reflect errors in X_{Na} caused by compositional heterogeneity.

Mixing relations along the alunite-natroalunite binary were investigated in mineral–water Na-K exchange experiments by Stoffregen and Cygan (1990). The exchange reaction is

$$KAl_3(SO_4)_2(OH)_{6(s)} + Na^+_{(aq)} \Leftrightarrow NaAl_3(SO_4)_2(OH)_{6(s)} + K^+_{(aq)} \tag{14}$$

The starting fluid in these experiments contained 0.5 *m* $(Na,K)_2SO_4$, which provided the alkalis necessary to drive the exchange reaction and also helped to stabilize alunite, but did not eliminate incongruent dissolution of alunite by Reaction (5). Defining the distribution coefficient K_D for this exchange as $(X_{Na}/X_K)(m_{Na}/m_K)$, where X indicates the mole fraction in the solid and *m* the molality in the aqueous solution, gives the expression:

$$RTlnK_D = \Delta G°_{r,P,T} + (dG_{xs,alun}/dX_{Na}) - (dG_{xs,soln}/dN_{Na}) \tag{15}$$

where $\Delta G°_{r,P,T}$ is the standard-state Gibbs free-energy change for the alkali-exchange reaction at a given pressure and temperature, the G_{xs} terms are the excess Gibbs free

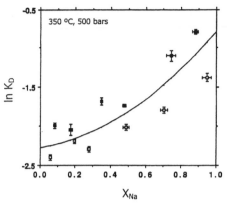

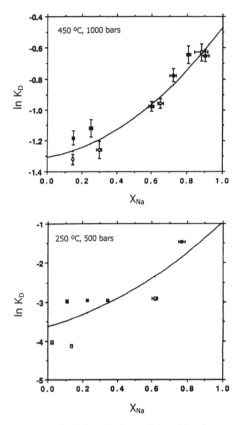

Figure 10 (upper left). Results of alunite–fluid Na-K exchange reactions at 450°C and 1000 bars. Open symbols = runs in which X_{Na} increased; filled symbols = runs in which X_{Na} decreased; curve represents subregular Margules model from the parameters listed in Table 4. From Stoffregen and Cygan (1990).

Figure 11 (above). Results of alunite–fluid Na-K exchange reactions at 350°C and 500 bars. Symbols and curve as in Figure 10. From Stoffregen and Cygan (1990).

Figure 12 (left). Results of alunite–fluid Na-K exchange reactions at 250°C and 500 bars. Symbols and curve as in Figure 10. From Stoffregen and Cygan (1990).

energy of mixing in the solid and in the aqueous solution, X_{Na} is the mole fraction of Na in the solid, and N_{Na} is the molar Na/(Na+K) ratio in the aqueous solution. Stoffregen and Cygan (1990) assumed that the excess free energy of mixing in the fluid, which is equal to the difference in the activity coefficients for Na and K, was negligible and could be ignored.

Experimentally derived values of $\ln K_D$ are plotted against X_{Na} at 450, 350, and 250°C in Figures 10, 11, and 12, respectively. Stoffregen and Cygan (1990) fit these data using a subregular Margules model which assumes that the excess free energy of the solid can be expressed as $G_{xs} = X_K X_{Na}(W_{Na} X_K + W_K X_{Na})$. This model is necessary to describe the nonlinear variation in $\ln K_D$ with X_{Na} that is evident in the figures.

Table 4 lists the parameters of fit for the excess mixing terms. The uncertainty in these terms increases with decreasing temperature, reflecting the poorer quality of the reversals at lower temperature. The data nevertheless indicate that the mixing terms increase with decreasing temperature. This behavior is typical of many solid solutions

Table 4. Alunite-natroalunite mixing parameters (from Stoffregen and Cygan 1990).

	Ln K	$W_{G,Na}$	$W_{G,K}$
450 °C, 1000 bars	−0.99(0.05)	1,837(427)	3159(435)
350 °C, 500 bars	−1.73(0.26)	2,867(1050)	4785(1229)
250 °C, 500 bars	−2.56(0.42)	4,668(2091)	6443(4836)

W terms in J/mol

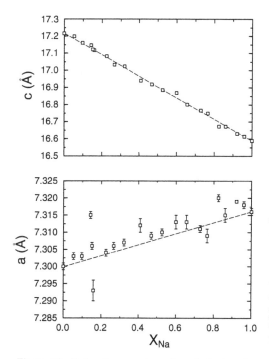

Figure 13. Unit-cell parameters of synthetic jarosite (200°C) as a function of Na mole fraction (Stoffregen, unpublished data).

(e.g. Powell 1974) and favors the formation of a solvus between alunite and natroalunite as temperature decreases. This solvus cannot occur at or above 350°C, at which the experimental results demonstrate complete miscibility, but the results do not preclude a miscibility gap at 250°C. As is discussed below, data on the chemistry of natural alunite also suggest a miscibility gap.

Jarosite-natrojarosite. Unit-cell dimensions for 19 samples of jarosite of inter-mediate composition prepared in jarosite–water Na-K exchange reactions at 200 and 250°C are plotted as a function of composition in Figure 13. Also included in Figure 13 are data for end-member jarosite and natro-jarosite. Unlike alunite, the *a* unit-cell dimension of jarosite appears to increase from the K to Na end-members, with the latter larger by bout 0.015 Å. Values of *c* and *V* lefine the following two least-quares relationships:

$$X_{Na} = (17.222 - c_{meas})/0.635 \tag{16}$$

$$X_{Na} = (795.20 - V_{meas})/25.9 \tag{17}$$

The slope on Equation (16) is identical to that observed for alunite (Eqn. 12). Most of the samples plot within two standard deviations of the ideal mixing line. Thus, the data are adequately described by an ideal mixing model.

Mixing relations along the jarosite-natrojarosite binary were investigated in mineral–water Na-K exchange experiments at 150 to 250°C (Stoffregen 1993). The exchange reaction is

$$KFe_3(SO_4)_2(OH)_{6(s)} + Na^+_{(aq)} \Leftrightarrow NaFe_3(SO_4)_2(OH)_{6(s)} + K^+_{(aq)} \tag{18}$$

As with the exchange experiments for alunite as discussed above, the initial fluid in the experiments involving jarosite contained 0.5*m* (Na,K)$_2$SO$_4$. However, to remain within the field of stability of jarosite and natrojarosite, it was also necessary to include 1.0 *m* H$_2$SO$_4$ in the starting solution.

In relation to Reaction (18), lnK$_D$ is related to the Gibbs free energy of mixing in the solid phase by an expression analogous to Equation (15) above, except with an excess-free-energy term for jarosite instead of alunite. As discussed by Stoffregen (1993), the excess-free-energy term for the fluid may be assumed constant at a given temperature and H$_2$SO$_4$ and (Na,K)$_2$SO$_4$ concentration. This allows Equation (15) to be rewritten as

$$-RTlnK_D^* = \Delta G^{\circ}_{P,T} + (dG_{xs,jar}/dX_{Na}) \tag{19}$$

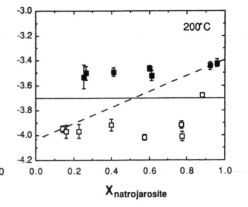

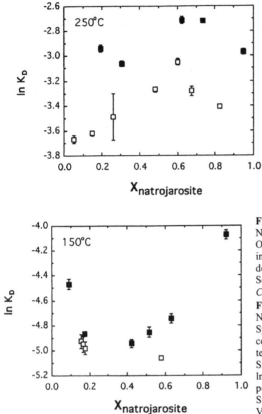

Figure 14 (upper left). Results of jarosite–fluid Na-K exchange reactions at 250°C, 100 bars. Open symbols = runs in which $X_{natrojarosite}$ increased; solid squares = runs in which $X_{natrojarosite}$ decreased. Used by permission of Elsevier Science Ltd., from Stoffregen (1993), *Geochim Cosmochim Acta* Vol. 57, Fig. 4(a), p. 2424.

Figure 15 (upper right). Results of jarosite–fluid Na-K exchange reactions at 200°C, 100 bars. Symbols as in Figure 14; dashed line was used to compute the larges value for the regular mixing term W permitted by the experimental data (see Stoffregen 1993); solid line shows position of $\ln K_D$ for an ideal mixing model. Used by permission of Elsevier Science Ltd., from Stoffregen (1993), *Geochim Cosmochim Acta* Vol. 57, Fig. 4(b), p. 2424.

Figure 16 (above). Results of jarosite–fluid Na-K exchange reactions at 150°C, vapor saturation pressure. Used by permission of Elsevier Science Ltd., from Stoffregen (1993), *Geochim Cosmochim Acta* Vol. 57, Fig. 4(c), p. 2424.

where $\ln K_D^*$ is the distribution coefficient corrected by a constant related to mixing in the fluid. For $mH_2SO_4 = 1.0$ and $m(Na,K)_2SO_4 = 0.5$, this constant is equal to 0.25 at 25°C and to 0.35 at 150°C and 200°C as calculated with PHRQPITZ (Plummer et al. 1988).

Experimental results at 250, 200, and 150°C are presented in Figures 14, 15 and 16, respectively. The best constraints on the value of $\ln K_D$ for Reaction (18) are provided by the experiments at 200°C (Fig. 15). After correction for excess mixing in the fluid phase, these experiments indicate that $\ln K_D^*$ is between -3.8 and -4.4 over the compositional range of $X_{Na} = 0.14$ to 0.96. This data set is consistent with, but does not require, ideal mixing along the jarosite-natrojarosite binary, for which $\ln K_D^*$ is a constant.

The maximum departure from ideal mixing consistent with these data can be evaluated by assuming that jarosite and natrojarosite obey a regular solution model. The excess mixing term $(dG_{xs,jar}/dX_{Na})$ in this model is equal to $W(1 - 2X_{Na})$, which implies that $\ln K_D^*$ is a linear function of X_{Na}. The maximum value of W consistent with the 200°C data is 1275 J/mole, which was obtained by connecting the values for $\ln K_D$ at X_{Na} from 0.14 to 0.96 as illustrated in Figure 15. For comparison, W must exceed 7800 J/mole for a solvus to occur at 200°C. Thus, although the experimental reversals at 200°C do not preclude some departure from ideality on the jarosite-natrojarosite binary, they demonstrate that the degree of non-ideality is small compared to that required for a solvus.

The experimental data presented above on alkali exchange in jarosite and natrojarosite demonstrate that the minerals form a nearly ideal solid solution at 200°C. This is substantially different from the alunite-natroalunite binary, which could have an asymmetric solvus at a temperature as high as 250°C, as was discussed above. The nearly ideal behavior of jarosite and natrojarosite at 200°C strongly suggests that there is complete miscibility on the jarosite-natrojarosite binary down to 25°C. Support for this suggestion is that data on natural and synthetic jarosite from a number of sources include compositions throughout the jarosite - natrojarosite - hydronium jarosite ternary (Brophy and Sheridan 1965, Alpers et al. 1989). However, these data do not provide conclusive proof of complete miscibility in this system for the following reasons: (1) amongst the samples analyzed by bulk chemical methods there may be samples that are, in reality, mixtures of two chemically distinct jarosites, and (2) the compositions may reflect non-equilibrium processes.

Alunite-jarosite. The only published studies of mixing between alunite and jarosite are those of Brophy et al. (1962) and Härtig et al. (1984). Brophy et al. (1962) synthesized a range of intermediate compositions in the series $K(Al,Fe)_3(SO_4)_2(OH)_6$ at 78, 105, and 150°C using 0.2 N sulfuric acid solutions. Recognizing the tendency for a deficiency of trivalent ions and an excess of water in alunite and jarosite synthesized at low temperature, Brophy et al. (1962) removed the excess water by heating to 300°C. Härtig et al. (1984) synthesized nine intermediate compositions in the series $K(Al,Fe)_3(SO_4)_2(OH)_6$ using mixed solutions of Fe–Al–K sulfate at 100°C, with an initial pH of 2.0. The solids produced in the experiments of Härtig et al. (1984) were non-stoichiometric with regard to (Al+Fe):S molar ratios, which ranged from 2.20:2 to 2.57:2. In both studies, for a given Al/(Al+Fe) ratio in the initial solution, the alunite-jarosite solid solution had a higher Fe content, which is expected because Fe^{3+} hydrolyzes at a lower pH than does Al^{3+}. The solids produced in both studies followed Vegard's rule with regard to the a unit-cell dimension, which varies from about 6.98 Å in alunite to about 7.30 Å in jarosite. Little to no variation in the c unit-cell dimension was observed.

Few data are available on the Al content of natural jarosite or the Fe content of natural alunite. Brophy et al. (1962, and references therein) reported the compositions of four samples of alunite, three samples of jarosite, and one sample with equimolar amounts of Al and Fe, from Kopec, Czech Republic (Jirkovsky 1927). An alunite of composition $K(Al_{0.87}Fe_{0.13})_3(SO_4)_2(OH)_6$ from the Copper Cap uranium prospect, Marysvale, Utah, was reported by Brophy et al. (1962). Unit-cell dimensions reported by Brophy et al. (1962) for the samples from Kopec and Copper Cap are consistent with those of synthetic solids of similar composition, indicating that unit-cell dimensions (especially a) can be used to make reliable estimates of Fe–Al substitution in the alunite-jarosite series. Alpers et al. (1992) used the cell dimensions of alunite and jarosite from Lake Tyrrell, Australia, together with chemical data, to estimate solid-solution compositions of approximately $[K_{0.87}Na_{0.04}(H_3O)_{0.09}](Al_{0.92}Fe_{0.08})_3(SO_4)_2(OH)_6$ for alunite and $[K_{0.89}Na_{0.07}(H_3O)_{0.04}](Fe_{0.80}Al_{0.20})_3(SO_4)_2(OH)_6$ for jarosite.

Geochemical modeling of alunite and jarosite formation (e.g. Bladh 1982) has generally considered alunite and jarosite to be distinct phases rather than members of a solid-solution series. Whether a miscibility gap exists between alunite and jarosite remains unknown, though it appears to be unlikely. It is more likely that the chemical separation of Fe and Al and the rarity of intermediate compositions in the alunite-jarosite series is caused primarily by the difference in hydrolysis constants between $Fe^{3+}_{(aq)}$ and $Al^{3+}_{(aq)}$ (Brophy et al. 1962, Nordstrom and Alpers 1999, Bigham and Nordstrom, this volume).

GEOCHEMISTRY AND OCCURRENCES

Alunite and natroalunite

Substitutions. The most significant substitution in natural alunite is that of Na^+ for K^+ on the alkali site. Hydronium (H_3O^+) substitution is difficult to prove, but has been inferred for most samples of synthetic alunite and most low-temperature natural alunite samples because of a deficiency in alkalis and an excess of water compared with the stoichiometric composition (e.g. Ripmeester et al. 1986). The non-stoichiometric water is generally attributed to the presence of hydronium (H_3O^+) but may also reflect other forms of "excess water" as discussed by Ripmeester et al. (1986). Another substitution that occurs in alunite is the coupled substitution of phosphate (or arsenate) and a divalent cation for sulfate and a monovalent cation, as described by the exchange reaction

$$SO_4^{2-} + D^+ \Leftrightarrow PO_4^{3-} + D^{2+} \tag{20}$$

The minerals woodhouseite and svanbergite consist of the alunite structure in which PO_4^{3-} substitutes for 25–75% of the total formula SO_4^{2-}, and the divalent cations Ca^{2+} and Sr^{2+}, respectively, predominate on the D site. There are a few published studies that discuss the extent of the coupled substitution represented by Reaction (20) in alunite. Significant solid solution between alunite and woodhouseite-svanbergite has been documented by Wise (1975), who observed a "strontian natroalunite" with about 12 formula % PO_4^{3-} in substitution for SO_4^{2-}, and a corresponding amount of Sr^{2+} substitution for Na^+. Stoffregen and Alpers (1987) reported that alunite from the Summitville gold deposit in Colorado shows low phosphate, Ca, and Sr contents. Alunite from La Escondida, a porphyry copper deposit in northern Chile, is somewhat richer in phosphate and Sr than are the Summitville samples.

Stoffregen and Alpers (1987) also analyzed alunite from three other localities: La Tolfa, Italy; Marysvale, Utah; and Goldfield, Nevada. La Tolfa is a hot-springs deposit, in which alunite is believed to have formed at or near the surface (Lombardi and Barbieri 1983). The alunite analyzed from Marysvale is intepreted to be of magmatic-steam origin (Cunningham et al. 1984, Rye et al. 1992). The alunites from both La Tolfa and Marysvale are not thought to be directly related to precious-metal mineralization (Lombardi and Barbieri 1983, Cunningham et al. 1984). In contrast, Goldfield, Nevada is a magmatic-hydrothermal, gold–sulfosalt deposit similar to that at Summitville, Colorado (Ashley 1974). Alunite anlyzed by Stoffregen and Alpers (1987) from La Tolfa and Marysvale has significant and fairly consistent SrO and P_2O_5 contents (samples 3 and 4 in Table 5). Alunite from Goldfield (sample 5 in Table 5) contains up to 0.20 wt % BaO but has no detectable Sr and only 0.05 wt % P_2O_5.

Allibone et al. (1995) determined that advanced argillic alteration assemblages at the Temora mine in New South Wales, Australia, contain alunite, natroalunite, woodhouseite, and svanbergite. Analyses plotted by Allibone et al. (1995) show an almost continuous increase in PO_4 that is commensurate with substitution of divalent cations (Ca, Sr, Ba) for monovalent cations (Na, K). Overall, however, the results from the various deposits suggest a limited degree of this coupled substitution in samples of magmatic-related, hypogene alunite formed at temperatures below about 300°C, although solid solution at higher temperatures is more extensive. Grains of woodhouseite, svanbergite, and related phases are found, in some cases, replacing apatite in zones of advanced argillic alteration (Stoffregen and Alpers 1987); the availability of phosphate

Table 5. Chemical composition of hydrothermal alunite (from Stoffregen and Alpers 1987).

	1	2	3	4	5
Weight %					
K_2O	7.49	9.18	10.79	10.84	9.66
Na_2O	2.41	1.44	n.d.	0.19	1.04
CaO	n.d.	n.d.	n.d.	n.d.	0.05
SrO	0.05	0.34	0.38	0.31	n.d.
BaO	n.a.	n.d.	n.d.	n.d.	0.20
Al_2O_3	37.34	36.50	36.70	36.92	37.13
SO_3	38.44	38.67	37.07	36.94	37.96
P_2O_5	0.23	0.30	0.63	0.86	0.05
H_2O^a	13.20	13.05	13.05	13.05	13.10
Total	99.16	99.78	98.62	99.11	99.19
Moles					
K	0.656	0.505	0.961	0.952	0.853
Na	0.321	0.192	n.d.	0.024	0.139
Ca	n.d.	n.d.	n.d.	n.d.	0.003
Sr	0.002	0.014	0.016	0.012	n.d.
Ba	n.a.	n.d.	n.d.	n.d.	0.005
Al	3.017	2.980	3.021	2.998	3.029
S	1.978	1.994	1.944	1.91	1.975
P	0.013	0.018	0.036	0.049	0.002
OH^b	6.048	5.981	6.080	6.003	6.034

[a] wt % H_2O calculated based on elemental composition and stoichiometry of end-member minerals in alunite supergroup (see Stoffregen and Alpers 1987)
[b] moles OH calculated based on wt % H_2O
n.d. = not detected; n.a. = not analyzed for.
Description of alunite samples: 1. Summitville, Colorado (255-629), 2. La Escondida, Chile (1067-C-298), 3. La Tolfa, Italy, 4. Marysvale, Utah, 5. Goldfield, Nevada

(and arsenate) may be the main limiting factor in determining the extent of solid solution by Reaction (20). In the high-alumina rocks east of Lake Baykal, Russia, Ripp and Kanakin (1998) and Ripp et al. (1998) have shown extensive solid solution of PO_4 in predominantly Sr-substituted natroalunite (Fig. 17). Muscovite in the prograde metamorphic assemblage was formed at 520-600°C and >5 kbar, but the associated minerals of the alunite supergroup were formed in the retrograde stage (Ripp et al. 1998).

Occurrences. Stoffregen and Alpers (1992) subdivided natural alunite into four groups on the basis of their grain size, mode of occurrence, and morphology: disseminated, coarse vein, porcelaneous vein, and fine-grained. Disseminated alunite occurs in bladed crystals 1 to 10 mm in length, which replace feldspar grains and which are accompanied by quartz and fine-grained disseminated pyrite. Coarse vein alunite occurs in >1 mm crystals in nearly monomineralic veins or pods which may be up to several meters in width. Porcelaneous vein alunite occurs as 5 to 100 μm grains with irregular surfaces within veinlets a few millimeters to a few centimeters wide and may be accompanied by minor quartz and kaolinite. Fine-grained alunite occurs as 0.5 to 5 μm pseudocubic grains within veinlets, small concretions, or bedded sediments. Fine-grained

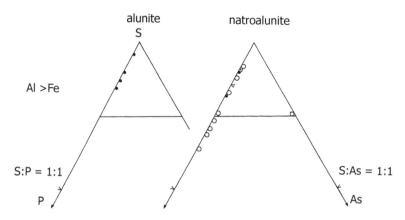

Figure 17. Compositions of PO_4-substituted alunite and natroalunite. The compositional fields for alunite and natroalunite are according to the current nomenclature system (Scott 1987); S:P = 1:1 and S:As = 1:1 indicate formula $SO_4:PO_4$ = 1:1 and $SO_4:AsO_4$ = 1:1, respectively. Data points for alunite are from Allibone et al. (1995), and sources for natroalunite are: open circles, Ripp et al. (1998) and Ripp and Kanakin (1998); triangles, Wise (1975); solid circles, Allibone et al. (1995). The open square on the far right represents the $AsO_4:SO_4$ composition of type schlossmacherite (Schmetzer et al. 1980).

alunite is generally formed at low temperature, but temperature may not be the only factor affecting grain size.

Sedimentary and low-temperature alunite. Fine-grained alunite seems to be limited to near-surface environments. This type of alunite has been reported from weathered profiles (e.g. Meyer and Peña dos Reis 1985), intertidal marine environments (e.g. Khalaf 1990), lacustrine environments (e.g. Alpers et al. 1992), and caves (e.g. Polyak et al. 1998). Porcelaneous vein alunite may form at low temperature, as illustrated by the alunite from Round Mountain, Nevada, which has been interpreted as supergene by Fifarek and Gerike (1991) and Rye et al. (1992).

Most fine-grained alunite described to date, including the three specimens investigated by Stoffregen and Alpers (1992), contain between 14 and 16 wt % H_2O. This corresponds to between 5 and 35 mol % H_3O^+ on the alkali site if all non-hydroxyl water is assumed to be present as hydronium. More hydronium-rich, fine-grained alunites, including possibly schlossmacherite (hydronium-dominant alunite) have also been reported. The common occurrence of non-hydroxyl water in fine-grained alunite suggests that at near-surface temperatures, fluid $[H_3O^+/(Na+K)]$ ratios are generally large enough to stabilize some hydronium component in alunite.

Low-temperature alunite is characterized by relatively large unit-cell values for a, which as discussed above, are interpreted to result from hydronium substitution. This substitution can be detected in XRD data, as illustrated in Figure 18. Values of a show no systematic variation with X_{Na}, but most of the fine-grained and porcelaneous vein alunite has values for a that are larger than those of the other natural specimens and the synthetic samples prepared at 450°C. These anomalously high values of a may reflect the presence of non-hydroxyl water, but they may also be an indication of Fe^{3+}-for-Al substitution, which also causes an increase in a (Brophy et al. 1962).

Figure 18 also shows, on the basis of the c unit-cell dimension, a range of Na contents from pure alunite to pure natroalunite at low temperature. However, intermediate compositions are not common.

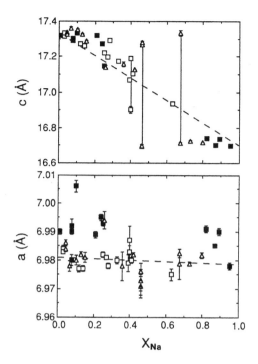

Figure 18. Unit-cell parameters of natural alunite as a function of Na mole fraction. Open squares are disseminated and coarse vein alunite, solid squares are porcelaneous vein and fine-grained alunite, and open triangles are un-classified. Vertical lines connect alunite pairs from the same specimen. Dashed lines connect values for synthetic end-members, as shown in Figure 9. Errors (1σ) are less than the symbol size except where shown. From Stoffregen and Alpers (1992).

Hydrothermal. Porcelaneous vein alunite can occur in hot-spring environments, as demonstrated by a specimen from the modern-day Waiotapu geothermal field included in the data set shown in Figure 18. Disseminated alunite forms at 200-300°C as determined by isotope geothermometry and corroborative geological evidence (Rye et al. 1992). Although formation temperatures for coarse vein alunite are not well-constrained, they are likely to be similar to those of disseminated alunite on the basis of the common spatial association of the two alunite types.

Magmatic-hydrothermal alunite characteristically occurs intimately intergrown with fine-grained disseminated pyrite in what is commonly interpreted as an equilibrium assemblage. This association, with quartz and commonly with kaolinite, is the classic "advanced argillic" assemblage of Meyer and Hemley (1967), described from many hydrothermal ore deposits. A higher temperature variant includes pyrophyllite.

The large overlap between the stability fields of alunite and pyrite is consistent with the common association of alunite and pyrite in volcanic-hydrothermal environments at temperatures in excess of 200°C (e.g. Rye et al. 1992). Log f_{O_2} in these environments is generally buffered on or near $HSO_4^- - H_2S$ equivalence by the influx of magmatic SO_2 and H_2S (e.g. Rye et al. 1992).

Some alunite from modern geothermal areas also contains hydronium (e.g. Aoki 1983), which suggests that a hydronium component may be stable to at least 100°C. Such stability is consistent with the presence of non-hydroxyl water in alunite and natroalunite synthesized at 150°C. However, as the non-hydroxyl water in these and the fine-grained natural alunite may also be a non-equilibrium effect, such water does not prove that a hydronium component is stable in alunite at any temperature. The absence of non-hydroxyl water in disseminated alunite, as indicated by H_2O determinations and by the lack of the anomalous values of unit-cell a, strongly suggests that a hydronium alunite component is not stable above 200°C. This temperature limit is consistent with the observed decrease in water content of synthetic 150°C alunite and natroalunite when heated to 250 and 450°C.

Figure 18 includes data for four natural samples that are mixtures of two alunites with different X_{Na}. The bulk X_{Na} for these specimens is between 0.40 and 0.70, but the unit-cell data indicate that three of the four are mixtures of relatively pure alunite and natroalunite. In the fourth sample, from Summitville, Colorado, alunite compositions of $X_{Na} = 0.68$ and 0.23 were determined from the XRD data. It is interesting to note that coarse vein alunite from Komatsuga, Japan, and disseminated alunite from Goldfield, Nevada, are the only specimens studied, besides the more sodic Summitville alunite, that are within the range of $X_{Na} = 0.40$ and 0.70 on the basis of Equation (12). The absence of fine-grained and porcelaneous vein alunite in this range may reflect a miscibility gap between alunite and natroalunite, a possibility also suggested by the experimental data on alunite-natroalunite mixing discussed above.

Metamorphic alunite. Alunite and natroalunite have been reported in low- to intermediate-grade, Al-rich metamorphic rocks that have no obvious hydrothermal affinities. The best documented of these occurrences are from the White Mountains of California (Wise 1975), from South Africa (Schoch et al. 1989), and from Russia (Ripp et al. 1998). At the South African locality, natroalunite occurs "as alteration veins and incrustations in massive sillimanite bodies which form part of a biotite–sillimanite schist." Quartz is not reported. The natroalunite contains only 1-6 mol % K but shows up to 18 mol % Ca and small amounts of phosphate.

No formation temperatures are reported for the alunite-supergroup minerals within these paragenetic associations. However, the association with rocks that may be generalized as of intermediate metamorphic grade suggests that alunite and natroalunite may persist in rare instances to temperatures in the range of 400 to 500°C if bulk-rock compositions are favorable. It is likely that fluid flow through such rocks was inhibited, as a significant flux of water would remove the alunite through reactions of the generalized form alunite $\rightarrow$ aluminosilicate + H_2SO_4. The alunite in these environments generally contains partial phosphate substitution for sulfate (e.g. Fig. 17), which suggests increased miscibility among the members of the group at the higher temperatures of origin inferred for this environment.

Jarosite and natrojarosite

Substitutions. Like alunite, the most common substitutions in jarosite are Na and H_3O on the alkali site. Hydronium substitution seems to be more common in jarosite than in alunite formed at similar temperature (Dutrizac and Kaiman 1976, Dutrizac and Jambor, this volume). This may be correlated with the fact that solutions that form jarosite are generally of lower pH (greater hydronium activity) compared with those that form alunite. Although plumbojarosite and argentojarosite (Table 1) occur only rarely in great abundance, these minerals and Pb- and Ag-bearing jarosite solid solutions may play a role in limiting the concentration of Pb and Ag in acidic waters such as mine drainage (e.g. Hochella et al. 1999).

Occurrences. Jarosite is common in weathering environments, where it forms as a supergene product of Fe-sulfide oxidation, but occurrences of jarosite in hydrothermal environments are uncommon (see Dutrizac and Jambor, this volume). Jarosite is generally absent as a hypogene mineral even in the sulfur-rich volcanic-hydrothermal systems that produce extensive zones of alunitic alteration. Supergene jarosite may replace alunite of either hypogene or supergene origin in situations where fluids become either more acidic or more oxidizing, or both. A mechanism for this type of evolution is long-term descent of the water table in response to erosion or climatic fluctuation. Hypogene or supergene alunite tends to form at or below the water table, and then jarosite forms in association with acidic, oxidizing conditions most commonly associated with the weathering of sulfide minerals.

The Paradise Peak Au–Ag–Hg deposit in Nevada contains perhaps the best documented example of hydrothermal jarosite. John et al. (1991) suggested that relatively coarse crystals (~2 mm) of jarosite occurring as a breccia matrix at Paradise Peak were formed at shallow depth (<200 m) over a temperature range of 150-200°C. This range is consistent with a formation temperature of 150°C that was obtained from the fractionation of ^{18}O between the sulfate and hydroxyl sites in this jarosite (Stoffregen et al. 1994). Other possible examples of hydrothermal jarosite occur in the Marysvale Volcanic Field, and are believed to have formed in a near-surface geothermal environment similar to that inferred for Paradise Peak (Cunningham et al. 1984). Polycrystalline masses of jarosite, up to 10 cm in size, at the Apex germanium-gallium mine in Utah have been described by Mahin (1990). On the basis of this coarse grain size, and on data from fluid inclusions, Mahin (1990) argued that jarosite in the Apex mine may have formed at a temperature of 200°C. Coarsely crystalline jarosite from Gilbert, Nevada, gave a formation temperature of 150°C (Rye and Alpers 1997) on the basis of ^{18}O fractionation between the sulfate and hydroxyl sites (Stoffregen et al. 1994). Such jarosite occurrences in the range of 150-200°C require fluids that were highly oxidizing and rich in sulfuric acid, and which were present roughly 20 to 100 m below the water table in these deposits. Another occurrence of coarsely crystalline jarosite, likely of hydrothermal origin, is known from the uranium deposit at Peña Blanca, Chihuahua, Mexico (P. Goodell, unpublished data).

The relatively rare occurrence of jarosite in hydrothermal environments is consistent with the experimental data described above, which demonstrate that jarosite requires more extreme acidity and more oxidizing conditions to form than does alunite. The jarosite-pyrite phase boundary lies more than two log units above the line in log f_{O_2} space representing equal activities of HSO_4^- and H_2S in the aqueous phase. As log f_{O_2} in volcanic environments is generally buffered on or near HSO_4^- –H_2S equivalence by the influx of magmatic SO_2 and H_2S (Rye et al. 1992), jarosite cannot be stabilized regardless of the pH attained in the system.

GEOCHRONOLOGY USING ALUNITE AND JAROSITE

The large concentration of K in alunite and jarosite makes these minerals potentially useful for age determinations by K-Ar and $^{40}Ar/^{39}Ar$ methods. Because both alunite and jarosite may form at either high- or low-temperature conditions, these minerals can provide information on the geochronology of both hydrothermal systems and weathering profiles. The multiple crystallographic sites available for measurement of stable isotopes in alunite and jarosite (S, H, O_{OH}, and O_{SO_4}) make weathering-related alunite and jarosite desirable candidates for paleoenvironmental studies because the direct radiometric dating of the minerals can be coupled with information about the stable isotopes of the meteoric waters from which the minerals formed (see Seal et al., this volume). This section provides an overview of the literature on K-Ar and $^{40}Ar/^{39}Ar$ methods as applied to alunite and jarosite. For additional details on geochronology using alunite and jarosite, see the recent, comprehensive reviews by Vasconcelos (1999a,b).

The first published K-Ar dates on alunite were by Shanin et al. (1968), Webb and McDonald (1968), and Chuchrov et al. (1969). Silberman and Ashley (1970) presented data on hydrothermal alunite from the epithermal gold deposit at Goldfield, Nevada, as part of a stratigraphically controlled study that provided the first geological evidence of the suitability of alunite for K-Ar dating. Argon-release experiments by Itaya et al. (1996) have since confirmed that both coarse-grained, hypogene alunite and fine-grained, supergene alunite are suitable for K-Ar dating.

Among additional studies that have used K-Ar dating of alunite to date hydrothermal events and associated precious- and base-metal mineralization are those by Mehnert et al. (1973), Ashley and Silberman (1976), Steven et al. (1979), Whalen et al. (1982), Pécksay et al. (1986), Alpers and Brimhall (1988), John et al. (1989), Sillitoe et al. (1991), Perello (1994), and Arribas et al. (1995a,b). An important factor in the interpretation of K-Ar data for hypogene alunite is the need for careful paragenetic analysis to determine the relationship between alunite deposition and sulfide mineralization. Although alunite and pyrite may be co-deposited, the acid-sulfate conditions associated with formation of hypogene alunite (and more rarely, hypogene jarosite; see Dutrizac and Jambor, this volume) are generally not simultaneous with the deposition of ore sulfides or gold. The acid-sulfate (or advanced argillic) hydrothermal alteration with which hydrothermal alunite is associated can set the stage for later mineralization by establishing permeability in the host rocks, such as the vuggy silica alteration at Summitville, Colorado (Stoffregen 1985).

The first K-Ar dating of supergene alunite from a porphyry copper deposit was reported by Gustafson and Hunt (1975) for El Salvador, Chile. They reported ages of roughly 36 Ma for two samples, in contrast with an age of about 39 Ma for hypogene alunite from advanced argillic alteration; other hypogene minerals indicated that the main hydrothermal system was active from about 42-40 Ma. Gustafson and Hunt (1975) inferred that the main period of supergene oxidation and enrichment of Cu-sulfide ores took place within about 5 Ma of the hypogene mineralization. Also reported was that five samples of supergene jarosite gave ages of less than 21 Ma, with two younger than the 10-13 Ma gravels that cap the erosional surface; thus, it was interpreted that even coarsely crystalline jarosite does not retain Ar well enough to be useful for K-Ar dating. Another interpretation of these data is that jarosite at El Salvador may have been dissolved and recrystallized in response to periodic influxes of meteoric water.

Alpers and Brimhall (1988) used K-Ar dates on supergene alunite from the porphyry copper deposit at La Escondida, Chile, to determine that supergene-Cu enrichment took place about 15-18 million years ago. K-Ar dates for hypogene alteration minerals from La Escondida reported by Alpers (1986), Ojeda (1986), and Alpers and Brimhall (1988), including two dates on hypogene alunite, indicated that the hydrothermal system was active between about 31 and 39 Ma. More recent studies by Richards et al. (1999) at La Escondida and the adjacent deposits of Zaldívar and Chimborazo, using U-Pb on zircon and the more precise $^{40}Ar/^{39}Ar$ methods on biotite, have suggested that primary mineralization and alteration was most important around 38 Ma, and that non-mineralizing systems continued to be active until 31 Ma. Alpers and Brimhall (1988) used K-Ar dates from supergene alunite to infer that regional climate change, resulting in the present hyper-arid conditions in the Atacama Desert, the lowering of long-term erosion rates, and the preservation of the supergene-enriched Cu deposits, was caused by uplift of the central Andes as well as intensification of the Humboldt Current related to build-up of the Antarctic ice cap during the middle Miocene. On the basis of results from analyses of 25 samples of supergene alunite from porphyry copper deposits in northern Chile, Sillitoe and McKee (1996) reached a similar conclusion to that of Alpers and Brimhall (1988), namely, that supergene processes were most active between 34 and 14 Ma, and that intensifying aridity related to regional uplift was the most likely reason for cessation of enrichment and preservation of the supergene deposits. K-Ar dates of supergene alunite from porphyry copper deposits in southwestern North America were provided by Cook (1994).

The K-Ar and $^{40}Ar/^{39}Ar$ methods have been used to analyze supergene alunite from several gold deposits, including breccia-hosted deposits in Queensland, Australia (Bird et al. 1990), epithermal deposits in Nevada (Ashley and Silberman 1976, Tingley and Berger 1985, Sander 1988, Arehart and O'Neil 1993, Sillitoe and Lorson 1994, Vasconcelos et al. 1994), and sediment-hosted (Carlin-type) deposits, also in Nevada (Bloomstein et al. 1990, Ilchik 1990, Sillitoe and Bohnam 1990, Arehart et al. 1992, 1993; Heitt 1992, Williams 1992, Arehart and O'Neil 1993, Teal and Branham 1997). The range for 30 published K-Ar dates of supergene alunite from the Carlin-type deposits is 3.6 ± 0.2 to 30.0 ± 1.2 Ma (Hofstra et al. 1999). Hypogene mineralization in the Carlin Trend and the Battle Mountain–Eureka Belt is thought to have taken place between about 35 to 45 Ma (Hofstra et al. 1999).

Several studies of alunite and jarosite have combined the use of geochronology and stable isotopes to provide information about paleoclimate. An important development that has allowed these studies to reconstruct the stable isotopic composition of paleo-meteoric waters was the determination of fractionation factors for O and H for the alunite–water and jarosite–water systems (see Seal et al., this volume, and references therein). Bird (1988) and Bird et al. (1989, 1990) analyzed surficial alunite from the Australian regolith and concluded that recrystallization under moist conditions might complicate the interpretation of weathering profiles. Arehart and O'Neil (1993) used the stable isotopes of supergene alunite, coupled with K-Ar dates, to reconstruct the history of paleo-meteoric waters in Nevada for the past 30 million years (see Seal et al., this volume). Vasconcelos et al. (1994) used the results of $^{40}Ar/^{39}Ar$ analysis of supergene jarosite and alunite from Goldfield, Nevada, to infer that a pervasive Middle to Late Miocene oxidation event occurred in the western USA, and they correlated this event with results from weathering profiles of similar age in western Africa, Brazil, and Chile. The epithermal Ag and base-metal deposits at Creede, Colorado, were studied using K-Ar age determinations and stable-isotope data on supergene alunite and jarosite (Rye et al. 2000). The age range of alunite was 4.8 to 3.1 Ma, whereas the range for jarosite was 2.6 to 0.9 Ma. The isotopic results and their interpretation with regard to paleoclimate and regional uplift are discussed by Seal et al. (this volume).

Vaconcelos (1999b) compiled available data for about 200 age determinations of supergene minerals (alunite, jarosite, and K-Mn oxides) by K-Ar and $^{40}Ar/^{39}Ar$ methods. Most published ages of alunite and jarosite fall within the past 20 Ma, with some as old as 92 Ma, whereas the published ages for K-Mn oxides are distributed fairly evenly over the last 60 Ma, with some ages to 100 Ma or older. It remains to be seen whether the clustering of radiometric ages of alunite and jarosite to the past 20 Ma reflects bias in the selection of study sites (predominantly Nevada and Chile), or whether there was a global increase in weathering rates during the Miocene. Regardless of the outcome, the future looks promising for alunite and jarosite to be increasingly useful on a global scale as indicators of weathering history, regional tectonic history, and paleoclimate.

REFERENCES

Allibone AH, Cordery GR, Morrison GW, Jaireth S, Lindhorst JW (1995) Synchronous advanced argillic alteration and deformation in a shear-hosted magmatic hydrothermal Au-Ag deposit at the Temora (Gidginbung) mine, New South Wales, Australia. Econ Geol 90:1570-1603

Alpers CN (1986) Geochemical and geomorphological evolution of supergene copper sulfide ore formation and preservation at La Escondida, Antofagasta, Chile. PhD thesis, University of California, Berkeley, California

Alpers CN, Brimhall GH (1988) Middle Miocene climatic change in the Atacama Desert, northern Chile: Evidence from supergene mineralization at La Escondida. Geol Soc Am Bull 100:1640-1656

Alpers CN, Nordstrom DK, Ball JW (1989) Solubility of jarosite solid solutions precipitated from acid mine waters, Iron Mountain, California, USA. Sci Géol Bull 42:281-298

Alpers CN, Rye RO, Nordstrom, DK, White LD, King BS (1992) Chemical, crystallographic and stable isotopic properties of alunite and jarosite from acid-hypersaline Australian lakes. Chem Geol 96:203-226

Aoki M (1983) Modes of occurrence and mineralogical properties of alunite solid solution in Osorezan geothermal area. Sci Rept Hirosake Univ 30:132-141 (in Japanese)

Arehart GB, O'Neil JR (1993) D/H ratios of supergene alunite as an indicator of paleoclimate in continental settings: Climate change in continental isotope records. Geophys Monograph 78:277-284

Arehart GB, Kesler SE, O'Neil JR, Foland KA (1992) Evidence for the supergene origin of alunite in sediment-hosted micron gold deposits, Nevada. Econ Geol 87:263-270

Arehart GB, Foland KA, Naeser CW, Kesler SE (1993) $^{40}Ar/^{39}Ar$, K/Ar, and fission track geochronology of sediment-hosted disseminated gold deposits at Post-Betze, Carlin trend, northeastern Nevada. Econ Geol 88:622-646

Arribas A Jr, Cunningham CG, Rytuba JJ, Rye RO, Kelly WC, Podwysocki MH, McKee EH, Tosdal RM (1995a) Geology, geochronology, fluid inclusions, and isotope geochemistry of the Rodalquilar gold alunite deposit, Spain. Econ Geol 90:795-822

Arribas A Jr, Hedenquist JW, Itaya T, Okada T, Concepción RA, Garcia JS (1995b) Contemporaneous formation of adjacent porphyry and epithermal Cu-Au deposits over 300 ka in northern Luzon, Philippines. Geology 23:337-340

Ashley RP (1974) Goldfield mining district. In Guidebook to the Geology of Four Tertiary Volcanic Centers in Central Nevada. Nevada Bur Mines Geol Rept 19:49-66

Ashley RP, Silberman ML (1976) Direct dating of mineralization at Goldfield, Nevada, by potassium-argon and fission track methods. Econ Geol 71:904-924

Babcan J (1971) Die Synthese von Jarosit $KFe_3(SO_4)_2(OH)_6$. Geol Zbornik Geol Carpath 22:299-304

Baron D, Palmer CD (1996) Solubility of jarosite at 4–35 °C. Geochim Cosmochim Acta 60:185-195

Bird MI (1988) An isotopic study of the Australian regolith. PhD thesis, Australian National University, Canberra, Australia

Bird MI, Andrew AS, Chivas, AR, Lock DE (1989) An isotopic study of surficial alunite in Austalia: 1. Hydrogen and sulfur isotopes. Geochim Cosmochim Acta 53:3223-3237

Bird MI, Chivas AR, McDougall I (1990) An isotopic study of surficial alunite in Australia: 2. Potassium-argon geochronology. Chem Geol 80:133-145

Bladh KW (1982) The formation of goethite, jarosite, and alunite during the weathering of sulfide-bearing felsic rocks. Econ Geol 77:176-184

Bloomstein EI, Massingill GL, Parratt RL, Peltonen DR (1990) Discovery, geology, and mineralization of the Rabbit Creek gold deposit, Humboldt County, Nevada. In Raines GL, Lisle, RE, Schafer, RW, Wilkinson, WH (eds) Geology and Ore Deposits of the Great Basin. Symp Proc Geol Soc Nevada (Reno) 2:821-843

Brophy GP, Scott ES, Snellgrove RA (1962) Sulfate studies: II. Solid solution between alunite and jarosite. Am Mineral 47:112-126

Brophy GP, Sheridan MF (1965) Sulfate studies: IV. The jarosite–natrojarosite–hydronium jarosite solid solution series. Am Mineral 50:1595-1607

Chukhrov FV, Yermilova LP, Shanin LL (1969) Age of alunite from certain deposits. Trans Acad Sci USSR Dokl Earth Sci Sect 185:49-51

Cook SS III (1994) The geological history of supergene enrichment in the porphyry copper deposits of southwestern North America. PhD thesis, University of Arizona, Tucson, Arizona

Cunningham CG, Rye RO, Steven TA, Mehnert HH (1984) Origins and exploration significance of replacement and vein-type alunite deposits in the Marysvale volcanic field, west central Utah. Econ Geol 79:50-71

Dutrizac JE, Kaiman S (1976) Synthesis and properties of jarosite-type compounds. Can Mineral 14:151-158

Fifarek RH, Gerike GN (1991) Oxidation of hydrothermal sulfides at Round Mountain, Nevada: Origin and relation to gold mineralization. In Raines GL, Lisle RE, Schafer RW, Wilkinson WH (eds) Geology and Ore Deposits of the Great Basin. Symp Proc, Geol Soc Nevada (Reno) 2:1111-1121

Gustafson LB, Hunt JP (1975) The porphyry copper deposit at El Salvador, Chile. Econ Geol 70:857-912

Härtig C, Brand P, Bohmhammel K (1984) Fe-Al Isomorphie und Strukturwasser in Kristallen von Jarosit-Alunit-Typ. Z anorg allg Chem 508:159-164

Heitt DG (1992) Characterization and genesis of alunite from the Gold Quarry mine, Eureka County, Nevada, and implications for Carlin-type mineralization. Soc Econ Geol Guidebook Ser 28:193-202

Helgeson HC, Delany JM, Nesbitt HW, Bird DK (1978) Summary and critique of the thermodynamic properties of rock-forming minerals. Am J Sci 278-A:1-229

Hemley JJ, Hostetler PB, Gude AJ, Mountjoy WT (1969) Some stability relations of alunite. Econ Geol 64:599-612

Hendricks SB (1937) The crystal structure of alunite and the jarosites. Am Mineral 22: 773-784

Herold H (1987) Zur Kristallchimie und Thermodynamik der Phosphate und Arsenate vom Crandallite-Typ. Thesis, University of Erlangen, Nürnberg, Germany

Hochella MF Jr, Moore JN, Golla U, Putnis AA (1999) TEM study of samples from acid mine drainage systems: Metal–mineral association with implications for transport. Geochim Cosmochim Acta 63:3395-3406

Hofstra AH, Snee LW, Rye RO, Folger HW, Phinisey JD, Loranger RJ, Dahl AR, Naeser CW, Stein HJ, Lewchuk L (1999) Age constraints on Jerritt Canyon and other Carlin-type gold deposits in the western United States: Relationship to mid-Tertiary extension and magmatism. Econ Geol 94:769-802

Ilchik RP (1990) Geology and geochemistry of the Vantage gold deposits, Alligator Ridge–Bald Mountain mining district, Nevada. Econ Geol 85:50-75

Itaya T, Arribas A Jr, Okada T (1996) Argon release systematics of hypogene and supergene alunite based on progressive heating experiments from 100 to 1000°C. Geochim Cosmochim Acta 60:4525-2535.

Jambor JL (1999) Nomenclature of the alunite supergroup. Can Mineral 37:1323-1341

Jirkovsky (1927) Casopis Narod Musea Praha 101:151 (in Palache et al. 1951)

John DA, Thomason RE, McKee EH (1989) Geology and K-Ar geochronology of the Paradise Peak mine and relationship of pre-Basin and Range extension to early Miocene precious metal mineralization in west-central Nevada. Econ Geol 84:631-649

John DA, Nash JT, Clark CW, Wulftange WH (1991) Geology, hydrothermal alteration, and mineralization at the Paradise Peak gold-silver-mercury deposit, Nye County, Nevada. *In* Raines GL, Lisle, RE, Schafer, RW, Wilkinson, WH (eds) Geology and Ore Deposits of the Great Basin. Symp Procgs, Geol Soc Nevada (Reno) 2:1020-1050

Kashkai MA (1969) Alunite group and its structural analogs. Zapiski Vses Mineral Obshch 98:150-165 (in Russian)

Kashkai MA, Borovskaya TB, Babazade MA (1975) Determination of $\Delta G°_{f,298}$ of synthetic jarosite and its sulfate analogues. Geokhim 7:771-783 (in Russian)

Kato T, Miúra Y (1977) The crystal structures of jarosite and svanbergite. Mineral J (Japan) 8:419-430

Kelley KK (1960) Contributions to the data on theoretical metallurgy: XIII. High temperature heat content, heat capacity and entropy data for the elements and inorganic compounds. US Bur Mines Bull 584

Kelley KK, Shomate CH, Young FE, Naylor BF, Salo AE, Juffman, EH (1946) Thermodynamic properties of ammonium and potassium alums and related substances with reference to extraction of alumina from clay and alunite. US Bur Mines Tech Paper 688

Khalaf FI (1990) Diagenetic alunite in clastic sequences, Kuwait, Arabian Gulf. Sedimentol 37:155-164

Kubisz J (1964) A study of minerals of the alunite-jarosite group. Polska Akad Nauk Prace Geol 22:1-90

Kubisz J (1970) Studies on synthetic alkali–hydronium jarosite: I. Synthesis of jarosite and natrojarosite. Mineral Polon 1:47-59

Langmuir D (1971) Particle size effect on the reaction goethite = hematite + water. Am J Sci 271:147-156

Lombardi G, Barbieri M (1983) The Tolfa Volcanics (Italy) alunitization process in the light of new chemical and Sr isotope data. Abstracts with Programs, Geol Soc Am 15:630

Mahin RA (1990) The mineralogy and geochemistry of the Apex germanium-gallium mine, southwestern Utah. MSc thesis, University of Utah, Salt Lake City, Utah

Menhert HH, Lipman PW, Steven TA (1973) Age of mineralization at Summitville, Colorado, as indicated by K-Ar dating of alunite. Econ Geol 68:399-412

Menchetti S, Sabelli C (1976) Crystal chemistry of the alunite series: crystal structure refinement of alunite and synthetic jarosite. N Jahrb Mineral Monatsh 406-417

Meyer C, Hemley JJ (1967) Wall rock alteration. *In* Barnes HL (ed) Geochemistry of Hydrothermal Ore Deposits. Holt, Rinehart, and Winston, New York, p 166-235

Meyer R, Peña dos Reis RB (1985) Paleosols and alunite silcretes in continental Cenozoic of western Portugal. J Sed Petrol 55:76-85

Nordstrom DK, Alpers CN (1999) Geochemistry of acid mine waters. *In* Plumlee GS, Logsdon MJ (eds) The Environmental Geochemistry of Mineral Deposits: Part A. Processes, Methods, and Health Issues. Rev Econ Geol 6A:133-160

Okada K, Hirabayashi J, Ossaka J (1982) Crystal structure of natroalunite and crystal chemistry of the alunite group. N Jahrb Mineral Monatsh 534-540

Ojeda JM (1986) Escondida porphyry copper deposit, II Region, Chile: Exploration drilling and current geological interpretation. *In* Papers, Mining Latin Am Conf, Santiago, Chile, 17-19 November 1986, Institution of Mining and Metallurgy, London UK, p 299-318

Palache C, Berman H, Frondel C (1951) The System of Mineralogy, Vol 2. John Wiley & Sons, New York

Parker RL (1962) Isomorphous substitution in natural and synthetic alunite. Am Mineral 47:127-136

Pécksay Z, Balogh K, Széky-Fux V, Gyarmati P (1986) Geochronological investigations on the Neogene volcanism of the Tokaj Mountains. Geol Carpathica 37:635-655

Perello, JA (1994) Geology, porphyry Cu-Au, and epithermal Cu-Au-Ag mineralization of the Tombulilato district, North Sulawesi, Indonesia. J Geochem Explor 50:221-256

Plummer LN, Parkhurst DL, Fleming GW, Dunkle SA (1988) A computer program incorporating Pitzer's equations for calculation of geochemical reactions in brines. US Geol Survey Water-Resources Invest Rept 88-4153

Polyak VJ, McIntosh WC, Guven N, Provencio P (1998) Age and origin of Carlsbad Cavern and related caves from $^{40}Ar/^{39}Ar$ of alunite. Science 279:1919-1922

Posnjak E, Merwin HE (1922) The system $Fe_2O_3-SO_3-H_2O$. J Am Chem Soc 44:1965-1994

Powell R (1974) A comparison of some mixing models for crystalline silicate solutions. Contrib Mineral Petrol 46:265-274

Raymahashay BC (1969) A geochemical study of rock alteration by hot springs in the Paint Pot Hill area, Yellowstone Park. Geochim Cosmochim Acta 32:499-522

Richards JP, Noble SR, Pringle MS (1999) A revised Late Eocene age for porphyry Cu magmatism in the Escondida area, northern Chile. Econ Geol 94:1231-1247

Ripmeester JA, Ratcliffe CI, Dutrizac JE, Jambor JL (1986) Hydronium ion in the alunite–jarosite group. Can Mineral 24:435-447

Ripp GS, Kanakin SV (1998) Phosphate minerals in the metamorphosed high-alumina rocks of the Ichetui occurrence, Transbaikal region. Dokl Earth Sci 359:233-235

Ripp GS, Kanakin SV, Shcherbakova MN (1998) Phosphate mineralization in metamorphosed high-alumina rocks of the Ichetuyskoye ore occurrence. Zapiski Vseross Mineral Obshch 127(6):98-108 (in Russian)

Rye RO, Alpers, CN (1997) The stable isotope geochemistry of jarosite. US Geol Survey Open-File Rept 97-88

Rye RO, Bethke PM, Wasserman MD (1992) The stable isotope geochemistry of acid-sulfate alteration. Econ Geol 87:240-262

Rye RO, Bethke PM, Lanphere MA, Steven TA (2000) Neogene geomorphic and climatic evolution of the central San Juan Mountains, Colorado: K/Ar age and stable isotope data on supergene alunite and jarosite from the Creede mineral district. *In* Bethke PM, Hay RL (eds) Ancient Lake Creede: Its volcano-tectonic setting, history of sedimentation, and relation to mineralization in the Creede Mining District, Colorado. Geol Soc Am Spec Paper 346:95-104

Sander MV (1988) Geologic setting and the relation of epithermal gold-silver mineralization to wall rock alteration at the Round Mountain mine, Nye County, Nevada. *In* Schafer RW, Cooper JJ, Vikre PG (eds) Bulk mineable precious metal deposits of the western United States. Symposium proceedings, Geol Soc Nevada, Reno, p 375-416

Schmetzer K, Ottemann J, Bank H (1980) Schlossmacherite, $(H_3O,Ca)Al_3[(OH)_6|((S,As)O_4)_2]$, ein neues Mineral der Alunit–Jarosit–Riehe. N Jahrb Mineral Monatsh 215-222

Schoch AE, Beukes GJ, van der Westhuizen WA, de Bruiyn, H (1989) Natroalunite from Koenabib, Pofadder District, South Africa. S Afr Tydskr Geol 92:20-28

Schwab RG, Götz C, Herold H, Pinto de Oliveira N (1993) Compounds of the crandallite type: Thermodynamic properties of Ca-, Sr-, Ba-, Pb-, La, Ce- to Gd- phosphates and –arsenates. N Jahrb Mineral Monatsh 551-568

Scott KM (1987) Solid solution in, and classification of, gossan-derived members of the alunite-jarosite family, northwest Queensland, Australia. Am Mineral 72:178-187

Shanin LL, Ivanov IB, Shipulin FK (1968) The possible use of alunite in K-Ar geochronometry. Geokhim 1:109-111

Silberman ML, Ashley RP (1970) Age of ore deposition at Goldfield, Nevada, from potassium-argon dating of alunite. Econ Geol 65:352-254

Sillitoe, RH, Bonham, HF Jr (1990) Sediment-hosted gold deposits: Distal products of magmatic-hydrothermal systems. Geology 18:157-161

Sillitoe, RH, Lorson RC (1994) Epithermal gold-silver-mercury deposits at Paradise Peak, Nevada: Ore controls, porphyry gold association, detachment faulting, and supergene oxidation. Econ Geol 89:1228-1248

Sillitoe RH, McKee EH (1996) Age of supergene oxidation and enrichment in the Chilean porphyry copper province. Econ Geol 91:164-179

Sillitoe RH, McKee EH, Vila T (1991) Reconnaissance K-Ar geochronology of the Maricunga gold-silver belt, northern Chile. Econ Geol 86:1261-1270

Steven TA, Cunningham CG, Naeser CW, Menhert HH (1979) Revised stratigraphy and radiometric ages of volcanic rocks and mineral deposits in the Marysvale area, west-central Utah. US Geol Survey Bull 1469

Stoffregen RE (1987) Genesis of acid-sulfate alteration and Au-Cu-Ag mineralization at Summitville, Colorado. Econ Geol 82:1575-1591

Stoffregen RE (1990) An experimental study of Na-K exchange between alunite and aqueous sulfate solutions. Am Mineral 75:209-220

Stoffregen RE (1993) Stability relations of jarosite and natrojarosite. Geochim Cosmochim Acta 57:2417-2419

Stoffregen RE, Alpers CN (1987) Woodhouseite and svanbergite in hydrothermal ore deposits: products of apatite destruction during advanced argillic alteration. Can Mineral 25:201-211

Stoffregen RE, Alpers CN (1992) Observations on the unit cell parameters, water contents and δD of natural and synthetic alunites. Am Mineral 77:1092-1098

Stoffregen RE, Cygan GL (1990) An experimental study of Na-K exchange between alunite and aqueous sulfate solutions. Am Mineral 75:209-220

Stoffregen RE, Rye RO, Wasserman MD (1994) Experimental studies of alunite: II. $^{18}O-^{16}O$ and D–H fractionation factors between alunite and water at 250–450 °C. Geochim Cosmochim Acta 58:903-916

Szymanski JT (1985) The crystal structure of plumbojarosite $Pb[Fe_3SO_4)_2(OH)_6]_2$. Can Mineral 23:659-668

Teal L, Branham A (1997) Geology of the Mike gold-copper deposit, Eureka County, Nevada. Soc Econ Geol Guidebook Series 28:257-276

Tingley JV, Berger BR (1985) Lode gold deposits of Round Mountain, Nevada. Nevada Bur Mines Geol Bull 100

Tozawa K, Sasaki K, Umetsu Y (1983) The effect of the second dissociation of sulfuric acid on hydrometallurgical processes: The electrical conductivity of sulfuric acid-containing electrolytes and the hydrolysis of ferric sulfate solutions at elevated temperatures. *In* Osseo-Asare K, Miller JD (Eds) Hydrometallurgy Research, Development and Practice. TMS–AIME, New York, p 375-389

Umetsu Y, Tozzawa K, Sasaki K (1977) The hydrolysis of ferric sulfate solutions at elevated temperatures. Can Metall Quart 16:111-117

Vasconcelos PM, Brimhall GH, Becker TA, Renne PR (1994) $^{40}Ar/^{39}Ar$ analysis of supergene jarosite and alunite: Implications to the paleoweathering history of the western USA and West Africa. Geochim Cosmochim Acta 58:401-420

Vasconcelos PM (1999a) K–Ar and $^{40}Ar/^{39}Ar$ geochronology of weathering processes. Ann Rev Earth Planet Sci 27:183-229

Vasconcelos PM (1999b) $^{40}Ar/^{39}Ar$ geochronology of supergene processes in ore deposits. *In* Lambert DD, Ruiz J (eds) Application of Radiogenic Isotopes to Ore Deposit Research and Exploration. Rev Econ Geol 12:73-113

Wang R, Bradley WF, Steinfink H (1965) The crystal structure of alunite. Acta Crystallogr 18:249-252

Webb AW, McDougall I (1968) The geochronology of the igneous rocks of eastern Queensland. J Geol Soc Aust 15:313-343

Whalen JB, Britten RM, McDougall I (1982) Geochronology and geochemistry of the Freida River prospect area, Papua New Guinea. Econ Geol 77:592-616

Williams CL (1992) Breccia bodies in the Carlin trend, Elko and Eureka counties, Nevada: Classification, interpretation, and roles in ore formation. MSc thesis, Colorado State University, Fort Collins, Colorado

Wise WS (1975) Solid solution between the alunite, woodhouseite, and crandallite mineral series. N Jahrb Mineral Monatsh 540-545

Zotov AV (1967) Recent formation of alunite in the Kipyashcheye ("boiling") crater lake, Golovnin volcano, Kunashir Island. Dokl Akad Nauk SSSR 174:124-127 (in Russian)

Zotov AV, Mironova GD, Rusinov VL (1973) Determination of $\Delta G_{f,298}$ of jarosite synthesized from a natural solution. Geokhim 5:739-745 (in Russian)

10   Solid-Solution Solubilities and Thermodynamics: Sulfates, Carbonates and Halides

Pierre Glynn

U.S. Geological Survey
432 National Center
Reston, Virginia 20192

INTRODUCTION

This review updates and expands an earlier study (Glynn 1990), that is presently out of print. The principal objectives of this chapter are (1) to review the thermodynamic theory of solid-solution aqueous-solution interactions, particularly as it pertains to low-temperature systems (between 0 and 100°C), and (2) to summarize available data on the effects of ionic substitutions on the thermodynamic properties of binary sulfate solid-solutions. Selected carbonate and halide solid-solutions are also considered.

Studies of solid-solution aqueous-solution (SSAS) systems commonly focus on measuring the partitioning of trace components between solid and aqueous phases. The effect of solid-solution formation on mineral solubilities in aqueous media is rarely investigated. Several studies, however, have examined the thermodynamics of SSAS systems, describing theoretical and experimental aspects of solid-solution dissolution and component-distribution reactions (Lippmann 1977, 1980; Thorstenson and Plummer 1977, Plummer and Busenberg 1987, Glynn and Reardon 1990, 1992; Glynn et al. 1990, 1992; Glynn 1991, 1992; Glynn and Parkhurst 1992, Königsberger and Gamsjäger 1991, 1992; Gamjäger et al. 2000). The present chapter describes and compares some of the concepts presented by the above authors, and presents methods that can be used to estimate the effect of SSAS reactions on the chemical evolution of natural waters. Sulfate minerals are particularly well suited for the investigation of the thermodynamics of SSAS systems, because their generally high solubilities facilitate the attainment of thermodynamic equilibrium states and their commonly large crystal structures tend to allow considerable ionic substitution.

Field or laboratory observations of miscibility gaps, spinodal gaps, critical mixing points, or distribution coefficients can be used to estimate solid-solution excess-free-energies, which are needed for the calculation of solid-solution solubilities and of potential component-partitioning behavior. Experimental measurements of the solubility and thermodynamic properties of solid solutions are generally not available, particularly in low-temperature aqueous solutions. A database of excess-free-energy parameters is presented here for sulfate solid solutions and also for selected carbonate and halide solid solutions, based on reported compositional ranges observed in natural environments or through experimental synthesis. When available, excess-free-energy parameters obtained from laboratory equilibration experiments are also provided.

It is hoped that this chapter will stimulate interest in obtaining thermodynamic data based on well-controlled laboratory experiments, and will encourage further research on the thermodynamics of SSAS systems.

1529-6466/00/0040-0010$05.00

DEFINITIONS AND REPRESENTATION OF
THERMODYNAMIC STATES

Several thermodynamic states are of interest in the study of SSAS systems. The following sections discuss the concepts of thermodynamic equilibrium, primary saturation, and stoichiometric saturation.

Thermodynamic equilibrium states

Thermodynamic equilibrium in a system with a binary solid solution $B_{1-x}C_xA$ can be defined by the law-of-mass-action equations:

$$[B^+][A^-] = K_{BA}a_{BA} = K_{BA}\chi_{BA}\gamma_{BA} \tag{1}$$

$$[C^+][A^-] = K_{CA}a_{CA} = K_{CA}\chi_{CA}\gamma_{CA} \tag{2}$$

where $[A^-]$, $[B^+]$ and $[C^+]$ are the activities of A^-, B^+ and C^+ in the aqueous solution; a_{BA} and a_{CA}, χ_{BA} and χ_{CA}, and γ_{BA} and γ_{CA} are the activities, mole fractions, and activity coefficients, respectively, of components BA and CA in the equilibrium solid solution. K_{BA} and K_{CA} are the solubility products of pure BA and pure CA solids.

Equations (1) and (2) can be used to construct phase diagrams that display the series of possible equilibrium states for any given binary SSAS system. By analogy to the pressure versus mole-fraction diagrams used for binary liquid-vapor systems, Lippmann (1977, 1980, 1982) defined a variable $\Sigma\Pi = [A^-]([B^+]+[C^+])$, such that adding together Equations (1) and (2) yields the following relation, known as the "solidus" equation:

$$\Sigma\Pi_{eq} = K_{BA}\chi_{BA}\gamma_{BA} + K_{CA}\chi_{CA}\gamma_{CA} \tag{3}$$

where $\Sigma\Pi_{eq}$ is the value of the $\Sigma\Pi$ variable at thermodynamic equilibrium.

For a complete description of thermodynamic equilibrium, a second relation must be derived from Equations (1) and (2) (Lippmann 1980, Glynn and Reardon 1990, 1992). The "solutus" equation expresses $\Sigma\Pi_{eq}$ as a function of aqueous solution composition:

$$\Sigma\Pi_{eq} = 1 / \left(\frac{\chi_{B,aq}}{K_{BA}\,\gamma_{BA}} + \frac{\chi_{C,aq}}{K_{CA}\,\gamma_{CA}} \right) \tag{4}$$

where the aqueous activity fractions $\chi_{B,aq}$ and $\chi_{C,aq}$ are defined as

$$\chi_{B,aq} = [B^+]/([C^+]+[B^+]) \text{ and, } \chi_{C,aq} = [C^+]/([B^+]+[C^+]).$$

Solid-solution activity coefficients can be fitted using the following equations:

$$\ln\gamma_{CA} = \chi_{BA}^2\left[a_0 - a_1(3\chi_{CA} - \chi_{BA}) + a_2(\chi_{CA} - \chi_{BA})(5\chi_{CA} - \chi_{BA}) + ...\right] \tag{5}$$

$$\ln\gamma_{BA} = \chi_{CA}^2\left[a_0 + a_1(3\chi_{BA} - \chi_{CA}) + a_2(\chi_{BA} - \chi_{CA})(5\chi_{BA} - \chi_{CA}) + ...\right] \tag{6}$$

Equations (5) and (6) are derived from Guggenheim's expansion series for the excess Gibbs free energy of mixing, G^E (Guggenheim 1937, Redlich and Kister 1948):

$$G^E = \chi_{BA}\chi_{CA}RT(a_0 + a_1(\chi_{BA} - \chi_{CA}) + a_2(\chi_{BA} - \chi_{CA})^2 + ...) \tag{7}$$

where R is the gas constant, T is the absolute temperature and the a_i parameters are dimensionless fitting parameters, occasionally referred to as excess free-energy parameters. Uppercase A_i parameters expressed in Joules/mol are also occasionally used in this paper and refer to the product: $A_i = RTa_i$.

Equation (6) differs from Equation (5) not only because of a switch between the BA and CA end members, but also because of a sign change in the second term (and other even terms) of the series. The convention adopted here follows that adopted in the MBSSAS code (Glynn 1991) for subregular solutions: a negative value for the a_1 parameter skews a possible miscibility gap towards the CA end member, that is, a negative value preferentially increases the excess free energy of solid solutions closer to the CA end member.

The first two terms of Equations (5) and (6) are generally sufficient to represent accurately the dependence of γ_{BA} and γ_{CA} composition (Glynn and Reardon 1990, 1992). Indeed, in the case of a solid solution with a small difference in the size of the substituting ions (relative to the size of the non-substituting ion), the first parameter, a_0, is usually sufficient to describe the solid solution (Urusov 1974); Equations (5) and (6) then become identical to those of the "regular" solid-solution model of Hildebrand (1936). For the case where both a_0 and a_1 parameters are needed, Equations (5) and (6) become equivalent to those of the "subregular" solid-solution model of Thompson and Waldbaum (1969), a model used extensively to describe solid-solution thermodynamics at high temperatures (Saxena 1973, Ganguly and Saxena 1987). Equations (5) and (6) can also be shown to be equivalent to the Margules activity-coefficient series (Margules 1895, Prigogine and Defay 1954). Glynn (1991) provided the relations between the "subregular" excess-free-energy parameters used in the Guggenheim series, the Thompson and Waldbaum model (Thompson 1967, Thompson and Waldbaum 1969) and the original Margules series. The Guggenheim parameters A_0 and A_1 (in dimensional form), the dimensionless Guggenheim parameters a_0 and a_1, and the Thompson and Waldbaum parameters W_{12} and W_{21} for a subregular binary solid solution (or W_{BC} and W_{CB} for components BA and CA) are related by the following equations:

$$W_{BC} = A_0 - A_1 = RT(a_0 - a_1) \tag{8}$$

$$W_{CB} = A_0 + A_1 = RT(a_0 + a_1) \tag{9}$$

According to the Thompson and Waldbaum model, the excess free energy of mixing of a binary solid solution can be expressed by the following relation:

$$G^E = \chi_{BA}\chi_{CA}(W_{BC}\,\chi_{CA} + W_{CB}\,\chi_{BA}) \tag{10}$$

Lippmann's solidus and solutus curves can be used to predict the solubility of any binary solid solution at thermodynamic equilibrium, as well as the distribution of components between solid and aqueous phases, when the solid-phase and the aqueous-phase activity coefficients are known. Figure 1 shows an example of a Lippmann phase diagram for the $(Sr,Ba)SO_4 - H_2O$ system at 25°C, modeled assuming a dimensionless a_0 value of 2.0. This value of a_0 corresponds to the maximum value that still allows complete miscibility in a regular solid-solution series. Aqueous solutions that plot below the solutus curve are undersaturated with respect to all solid phases, including the pure end-member solids, whereas solutions plotting above the solutus are supersaturated with respect to one or more solid-solution compositions.

Primary saturation states

Primary saturation is the first state reached during the congruent dissolution of a solid solution, at which the aqueous solution is saturated with respect to a secondary solid phase, (Garrels and Wollast 1978, Denis and Michard 1983, Glynn and Reardon 1990, 1992). This secondary solid will typically have a composition different from that of the dissolving solid. At primary saturation, the aqueous phase is at thermodynamic equilibrium with respect to this secondary solid but remains undersaturated with respect to the primary dissolving solid. The series of possible primary-saturation states for a given SSAS system is represented by the solutus curve on a Lippmann diagram (Fig. 1).

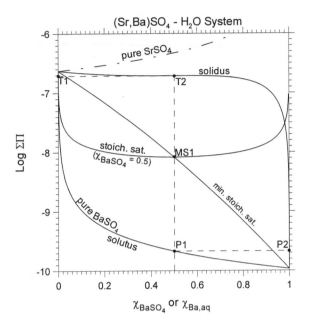

Figure 1. Lippmann diagram (with stoichiometric and pure-phase saturation curves) for the $(Sr,Ba)SO_4$ - H_2O system at 25°C. An a_0 value of 2.0 is assumed. End-member solubility products from PHREEQC (Parkhurst and Appelo 2000): $10^{-6.63}$ for celestine ($SrSO_4$), $10^{-9.97}$ for barite ($BaSO_4$). Points T1 and T2 give the aqueous and solid phase compositions, respectively, of a system at thermodynamic equilibrium with respect to a $Sr_{.5}Ba_{.5}SO_4$ solid. Points P1 and P2 describe the state of a system at primary saturation with respect to the same solid. Point MS1 gives the composition of an aqueous phase at congruent stoichiometric saturation with respect to that solid. The pure $BaSO_4$ saturation curve can not be distinguished from the solutus curve on this graph.

In the specific case of a "strictly congruent" dissolution process occurring in an aqueous phase with a $[B^+]/[C^+]$ activity ratio equal to the B^+/C^+ ratio in the solid, primary saturation can be approximately found by drawing a straight vertical line on the Lippmann diagram from the solid-phase composition to the solutus (Fig. 1). For an exact calculation, the following relations may be used to determine the primary saturation state:

$$\chi_{B,aq} = f_B \chi_{BA} \qquad (11) \qquad \chi_{C,aq} = f_C \chi_{CA} \qquad (12)$$

where f_B and f_C are factors correcting for a possible difference in the aqueous speciation and activity coefficients of B^+ and C^+.

The equation used to calculate the value of $\Sigma\Pi$ at primary saturation as a function of solid composition, for a "strictly congruent" dissolution process, may be found by combining Equation (4), the Lippmann solutus equation, with Equations (11) and (12):

$$\Sigma\Pi_{ps} = 1 / \left(\frac{\chi_{BA} f_B}{K_{BA} \, \gamma_{BA,y}} + \frac{\chi_{CA} f_C}{K_{CA} \, \gamma_{CA,y}} \right) \qquad (13)$$

where $\gamma_{BA,y}$ and $\gamma_{CA,y}$ refer to the activity coefficients of BA and CA in the secondary solid

$B_{1-y}C_yA$ with respect to which the aqueous solution (at primary saturation) is in temporary thermodynamic equilibrium. $\Sigma\Pi_{ps}$ refers to the value of the $\Sigma\Pi$ variable as specifically defined at primary saturation. The composition of the $B_{1-y}C_yA$ phase will generally not be known. By equating $\Sigma\Pi_{ps}(x)$ (Eqn. 13) to $\Sigma\Pi_{eq}(y)$ (Eqn. 3), the relation between the initial solid composition $B_{1-x}C_xA$ and the secondary solid $B_{1-y}C_yA$ is obtained:

$$1 / \left(\frac{\chi_{BA} f_B}{K_{BA}\, \gamma_{BA,y}} + \frac{\chi_{CA} f_C}{K_{CA}\, \gamma_{CA,y}} \right) = \chi_{CA,y} \gamma_{CA,y} K_{CA} + \chi_{BA,y} \gamma_{BA,y} K_{BA} \tag{14}$$

In the case of a non-ideal solid-solution series, Equation (14) must be solved graphically or by an iterative technique, because $\gamma_{BA,y}$ and $\gamma_{CA,y}$ are typically exponential functions of $\chi_{CA,y}$.

Stoichiometric saturation states

Stoichiometric saturation was formally defined by Thorstenson and Plummer (1977). These authors argued that solid-solution compositions typically remain invariant during solid-aqueous phase reactions in low-temperature geological environments, thereby preventing attainment of thermodynamic equilibrium. Thorstenson and Plummer defined stoichiometric saturation as the pseudo-equilibrium state that may occur between an aqueous phase and a multi-component solid solution, "in situations where the composition of the solid phase remains invariant, owing to kinetic restrictions, even though the solid phase may be a part of a continuous compositional series". The stoichiometric saturation concept assumes that a solid solution can, under certain circumstances, behave as if it were a pure one-component phase. In such a situation, the dissolution of a solid solution $B_{1-x}C_xA$ can be expressed as:

$$B_{1-x}C_xA \rightarrow (1-x)B^+ + xC^+ + A^- \tag{15}$$

Applying the law of mass action then gives the defining condition for stoichiometric saturation states:

$$IAP_{SS} = K_{SS} = \frac{[C^+]^x [B^+]^{1-x} [A^-]}{1} = \exp\left(\frac{-\Delta G_r^0}{RT} \right) \tag{16}$$

where x and (1-x) are equal to χ_{CA} and χ_{BA} respectively and where $-\Delta G^0_r$ is the standard Gibbs free energy change of Reaction (15). The solid has unit activity because it is assumed to behave as if it were a pure one-component phase.

According to Thorstenson and Plummer's (1977) definition of stoichiometric saturation, an aqueous solution at thermodynamic equilibrium with respect to a solid $B_{1-x}C_xA$ will always be at stoichiometric saturation with respect to that same solid. The converse statement, however, is not necessarily true: stoichiometric saturation does not necessarily imply thermodynamic equilibrium.

Stoichiometric saturation states can be represented on Lippmann phase diagrams (e.g. Fig. 1) by relating the total solubility-product variable $\Sigma\Pi_{ss}$ (defined specifically at stoichiometric saturation with respect to a solid $B_{1-x}C_xA$) to the K_{ss} constant (Eqn. 16) and to the aqueous activity fractions $\chi_{B,aq}$ and $\chi_{C,aq}$:

$$\Sigma\Pi_{ss} = \frac{K_{ss}}{\chi_{B,aq}^{1-x}\, \chi_{C,aq}^{x}} \tag{17}$$

In contrast to thermodynamic equilibrium, for which a single ($\chi_{B,aq}$, $\Sigma\Pi_{eq}$) point satisfies Equations (1) and (2), stoichiometric saturation with respect to a given solid composition is represented by a series of ($\chi_{B,aq}$, $\Sigma\Pi_{ss}$) points, all defined by Equation (17).

The minimum in each stoichiometric saturation curve for a given solid will occur at the point where $\chi_{C,aq} = x$ and $\chi_{B,aq} = 1\text{-}x$. This condition can also be generally expressed as the point at which $\chi_{B,aq}/\chi_{C,aq} = (1\text{-}x)/x$. This minimum in the stoichiometric saturation curve may be reached through a "strictly congruent" dissolution process occurring in an aqueous phase with a $[B^+]/[C^+]$ activity ratio equal to the B^+/C^+ ratio in the solid. The series of minimum stoichiometric points for a series of solid-phase compositions ranging from $x = 0$ to $x = 1$ defines the minimum stoichiometric saturation curve (Glynn and Reardon 1990), also called the equal-G curve (Königsberger and Gamsjäger 1992). The minimum saturation curve essentially defines the series of stoichiometric saturation curves that can be constructed in the system and is a function of the end-member solubility products and of the excess-free-energy of mixing, G^E, of the solid-solution series:

$$\Sigma\Pi_{ms} = K_{BA}^{1-x} K_{CA}^{x} \exp\!\left(G^E / RT\right) \tag{18}$$

In the case of an ideal solid-solution series ($G^E = 0$), the minimum stoichiometric saturation curve will be a linear function of the end-member solubility products when plotted on a log $\Sigma\Pi$ scale.

As shown in Figure 1, stoichiometric saturation states never plot below the solutus curve. Indeed, stoichiometric saturation can never be reached before primary saturation in a solid-solution dissolution experiment. The unique point at which a stoichiometric saturation curve (for a given solid $B_{1-x}C_xA$) joins the Lippmann solutus represents the composition of an aqueous solution at thermodynamic equilibrium with respect to the $B_{1-x}C_xA$ solid.

Saturation curves for the pure BA and CA end-member solids can also be drawn on Lippmann diagrams (Lippmann 1980, 1982):

$$\Sigma\Pi_{BA} = \frac{K_{BA}}{\chi_{B,aq}} \tag{19} \qquad\qquad \Sigma\Pi_{CA} = \frac{K_{CA}}{\chi_{C,aq}} \tag{20}$$

These equations define the families of (χ_{BA}, $\Sigma\Pi_{BA}$) and (χ_{CA}, $\Sigma\Pi_{CA}$) conditions for which a solution containing A^-, B^+ and C^+ ions will be saturated with respect to pure BA and pure CA solids. Thermodynamic equilibrium with respect to a mechanical mixture of the two pure BA and CA solids, in contrast to a solid solution of BA and CA, is represented on a Lippmann diagram by a single point, namely the intersection of the pure BA and pure CA saturation curves. The coordinates of this intersection are:

$$\chi_{B,aq}^{int} = \frac{K_{BA}}{K_{BA} + K_{CA}} \tag{21} \qquad\qquad \Sigma\Pi^{int} = K_{CA} + K_{BA} \tag{22}$$

COMPARISON OF SOLID-SOLUTION AND PURE-PHASE SOLUBILITIES

In predicting solid-solution solubilities, one of two possible hypotheses must be chosen. In the first, the solid solution is treated as a one-component or pure-phase solid, requiring a sufficiently short equilibration time, a sufficiently high solid-to-aqueous-solution ratio, and relatively low solubility of the solid. These requirements are needed to ensure that no significant recrystallization of the initial solid or precipitation of a secondary solid-phase occurs. For such situations, the stoichiometric saturation concept may apply.

The second hypothesis considers the solid as a multi-component solid solution, capable of adjusting its composition in response to the aqueous solution composition. This compositional adjustment requires a long equilibration period, a relatively high solubility of the solid, and a relatively low solid-to-aqueous-solution ratio. In this case, the assumption of thermodynamic equilibrium may apply. If the equilibration period is too short for thermodynamic equilibrium to have been achieved, but if an outer surface layer of the solid has been able to recrystallize (because of the high solubility of the solid), the concept of primary saturation may apply.

Currently, there are insufficient data to determine the exact conditions for which each of these assumptions may apply, especially in field situations. In many instances, neither one of these hypotheses will explain the observed solubility of a solid solution, which may lie between the "maximum" stoichiometric saturation solubility and the "minimum" primary saturation solubility. Nonetheless, these solubility limits can often be estimated.

Stoichiometric saturation solubilities

The case of stoichiometric saturation states attained after "strictly congruent" dissolution is examined here. Two hypothetical $B_{1-x}C_xA$ solid solutions are considered: (1) calcite ($pK_{sp} = 8.48$; Plummer and Busenberg 1982) with a more soluble trace $NiCO_3$ component ($pK_{sp} = 6.87$; Smith and Martell 1976) and (2) calcite with a less soluble trace $CdCO_3$ component ($pK_{sp} = 11.31$; Davis et al. 1987). Contour plots of saturation indices ($SI = \log[IAP/K_{sp}]$), with respect to major and trace end-member components, are shown in Figure 2 as a function of the a_0 value (assuming a regular solid-solution model) and of the logarithm of the mole fraction of the trace component (where $10^{-6} \leq \chi_{CA} \leq 0.5$). SI values are calculated for major (BA) and trace (CA) components using the relations:

$$SI_{BA} = -\log(K_{BA}) + \log\left[K_{SS}\left(\frac{x_{B,aq}}{x_{C,aq}} \right)^x \right] \tag{23}$$

$$SI_{CA} = -\log(K_{CA}) + \log\left[K_{SS}\left(\frac{x_{C,aq}}{x_{B,aq}} \right)^{(1-x)} \right] \tag{24}$$

K_{ss} values are evaluated from Thorstenson and Plummer's (1977) Equation (22), modified assuming a regular solid-solution model:

$$K_{SS} = K_{BA}^{(1-x)} K_{CA}^x (1-x)^{(1-x)} x^x \exp\left[a_0 x(1-x) \right] \tag{25}$$

Assuming that the dissolution to stoichiometric saturation takes place in initially pure water and that the aqueous activity ratio of the major and minor ions is equal to their concentration ratio, the relation $\chi_{B,aq} / \chi_{C,aq} = (1-x)/x$ will apply. Using this relation, applicable only at "minimum stoichiometric saturation" (Glynn and Reardon 1990, 1992), the following equations may be derived from Equations (23), (24) and (25):

$$SI_{BA} = x\log\left(\frac{K_{CA}}{K_{BA}} \right) + \log(1-x) + \frac{a_0 x(1-x)}{\ln(10)} \tag{26}$$

$$SI_{CA} = (1-x)\log\left(\frac{K_{BA}}{K_{CA}} \right) + \log x + \frac{a_0 x(1-x)}{\ln(10)} \tag{27}$$

The SI contour plots drawn using Equations (26) and (27) show the miscibility gap and spinodal gap lines separating intrinsically stable, metastable, and unstable solid solutions (Prigogine and Defay 1954, Swalin 1972). In natural environments, metastable

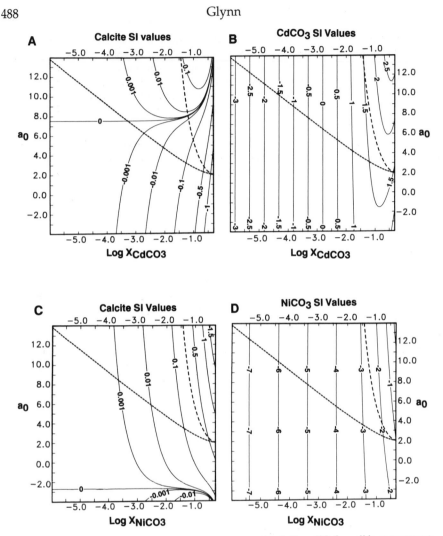

Figure 2. Values of major- and trace-component saturation indices (*SI*) for solids at congruent stoichiometric saturation (in solid lines): (a) calcite *SI* at saturation with (Ca,Cd)CO₃; (b) otavite *SI* at saturation with (Ca,Cd)CO₃; (c) calcite *SI* at saturation with (Ca,Ni)CO₃; (d) NiCO₃ *SI* at saturation with (Ca,Ni)CO₃. Miscibility-gap lines (short dashes) and spinodal gap lines (long dashes) are also shown.

solid-solution compositions may in some cases persist on a geological time scale (depending on the solubility of the solid), whereas unstable solid solutions formed in low-temperature environments are not likely to do so. Busenberg and Plummer (1989), in their study of magnesian calcite solubilities, observed that the highest known Mg contents in modern natural biogenic calcites correspond to the predicted spinodal composition. Metastable compositions will probably not persist, however, in solid solutions that have higher solubilities or which have been reacted for a longer period of time (or at higher temperatures) than the biogenic magnesian calcites. Although intrinsically unstable solid solutions, formed at low temperatures, may not be found in geological environments, they commonly can be synthesized in the laboratory (e.g.

strontian aragonites: Plummer and Busenberg 1987; barian strontianites: Glynn et al. 1988).

Figures 2a and 2b show the case of a solid-solution series, $(Ca,Cd)CO_3$, with a much less soluble trace end member. If the mole fraction of the trace component is sufficiently high ($\chi_{CdCO_3} > 10^{-2.8}$), the aqueous phase at stoichiometric saturation will be supersaturated with respect to the trace end member (except at unrealistic negative a_0 values not shown on the plot). The lower solubility of the trace component will generally cause negative SI values (undersaturation) for the major component, except at high a_0 values [higher than 7.5 in the case of $(Ca,Cd)CO_3$] for which the solid solutions will generally be metastable or unstable. Calcite SI values in Figure 2a show that the mole fraction of the trace component must be sufficiently high ($\chi_{CdCO_3} > 10^{-2.5}$) for this effect to be measurable in field environments (typical uncertainty on calcite SI values ≈ 0.01 minimum) or in laboratory experiments.

A laboratory example of the above principle is given by the "strictly-congruent" dissolution experiment of Denis and Michard (1983) on an anhydrite containing 3.5% Sr (on a mole basis). Analysis of their results shows that maximum $SI_{celestine}$ and SI_{gypsum} values of 0.37 and -0.04 respectively were attained after four days. Their last sample (after six days) gave $SI_{celestine}$ and SI_{gypsum} values of 0.35 and -0.04, respectively. Although these results show that stoichiometric saturation was not obtained with respect to the original anhydrite phase (probably because of back-precipitation of gypsum), supersaturation did occur with respect to the less soluble celestine[1] component.

In the case of solid solutions with a more soluble trace component, aqueous solutions at "minimum stoichiometric saturation" will generally be supersaturated with respect to the major component, and will be undersaturated with respect to the more soluble trace end member (Figs. 2c and 2d). Aqueous solutions at minimum stoichiometric saturation with respect to $(Ca,Ni)CO_3$ solid solutions will be supersaturated with respect to pure calcite at a_0 values greater than -2.7. $SI_{calcite}$ values greater than +0.01, however, will only be found at mole fractions of $NiCO_3$ greater than approximately $10^{-2.5}$. In contrast, SI_{NiCO_3} values will exhibit significant undersaturation even at high χ_{NiCO_3} mole fractions ($-2 < SI_{NiCO_3} < -1$ at $\chi_{NiCO_3} = 0.5$).

Primary saturation and thermodynamic equilibrium solubilities

A detailed discussion of solid-solution solubilities at primary saturation states and at thermodynamic equilibrium states was given by Glynn and Reardon (1990, 1992). Some fundamental principles governing these thermodynamic states are discussed below.

The composition of a SSAS system at primary saturation or at stoichiometric saturation will be generally independent of the initial ratio of solid to aqueous solution, but will depend on the initial aqueous-solution composition (prior to the dissolution of the solid). In contrast, the final thermodynamic equilibrium state of a SSAS system attained after a dissolution or recrystallization process will generally depend not only on the initial composition of the system but also on the initial ratio of solid to aqueous solution (Glynn et al. 1990, 1992).

An aqueous solution at primary saturation or at thermodynamic equilibrium with respect to a solid solution will be undersaturated with respect to all end-member component phases of the solid solution (Figs. 1, 3, 4 and 5). A positive excess-free-energy of mixing (that is, a positive a_0 value in the case of a regular solid solution) will

---

[1] The mineral celestine is also commonly known as celestite.

raise the position of the solutus curve relative to that of an ideal solid-solution system. A positive excess-free-energy of mixing will therefore increase the solubility of a solid solution at primary saturation or at thermodynamic equilibrium. The pure end-member saturation curves on a Lippmann diagram offer a upward limit on the position of the solutus. Conversely, a negative excess-free-energy of mixing will lower the position of the solutus relative to that of an ideal solid-solution series. Solid solutions with negative excess-free-energies of mixing may therefore show a greater degree of undersaturation relative to the pure end-member phases, although no examples of this have been found so far in ionic solids.

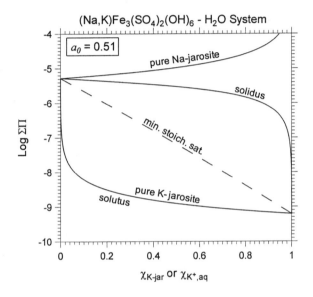

Figure 3. Lippmann phase diagram for the $(Na,K)Fe_3(SO_4)_2(OH)_6$–H_2O system at 25°C. The excess-free-energy-of-mixing parameter a_0 was extrapolated from Stoffregen's (1993) experimental determination of a_0 at 200°C. End-member solubility products from PHREEQC (Parkhurst and Appelo 2000): $10^{-5.28}$ for natrojarosite [$NaFe_3(SO_4)_2(OH)_6$], $10^{-9.21}$ for jarosite [$KFe_3(SO_4)_2(OH)_6$]. The pure jarosite saturation curve can not be distinguished from the solutus curve on this graph.

The solutus curve, in binary SSAS systems with ideal or positive solid-solution free-energies of mixing and with large differences (more than an order of magnitude) in end-member solubility products, will closely follow the pure-phase saturation curve of the less soluble end member (except at high aqueous activity fractions of the more soluble component, e.g. Figs. 1 and 3). In contrast, ideal solid solutions with very close end-member solubility products (less than an order of magnitude apart) will have a solutus curve up to 2 times lower in $\Sigma\Pi$ than the pure end-member saturation curves. The factor of 2 is obtained for the case where the two end-member solubility products are equal and can be derived from Equations (4), (21), and (22) (e.g. Fig. 4).

Figures 3, 4, and 5 show examples of pure end-member saturation curves plotted on Lippmann phase diagrams. Figure 3, for the natrojarosite-jarosite solid-solution series, is similar to the diagram for the celestine-barite series (Fig. 1) in that there is a large difference between the end-member solubility products. As a result of this large

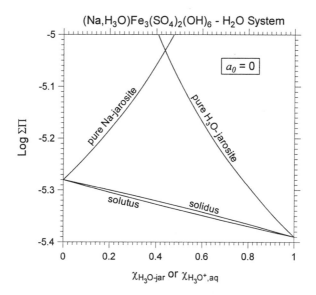

Figure 4. Lippmann phase diagram for the $(Na,H_3O)Fe_3(SO_4)_2(OH)_6$–$H_2O$ system at 25°C. Ideal mixing was assumed for the solid-solution series (a_0 = 0). End-member solubility products from PHREEQC (Parkhurst and Appelo 2000): $10^{-5.28}$ for natrojarosite [$NaFe_3(SO_4)_2(OH)_6$], $10^{-5.39}$ for hydronium-jarosite [$H_3OFe_3(SO_4)_2(OH)_6$].

difference, the saturation curve for pure jarosite, the least soluble end member, is very close to the solutus curve, only slightly above it, except in the region where the K/(K+Na) ratio in the aqueous solution (and consequently the χ_{K^+} value) goes to 0. In that region, the jarosite saturation curve climbs sharply towards an infinite $\Sigma\Pi$ value, whereas the solutus curve converges to the $\Sigma\Pi$ value of the natrojarosite end-member solubility product. In contrast, the saturation curve for the more soluble end member, natrojarosite, plots well above the solutus curve. This behavior is typical for SSAS systems with large differences in end-member solubility products. Such systems are typically controlled by the less soluble end-member component. Increasing the excess free energy of mixing in these systems, away from the 0 value of an ideal system, typically causes very little effect on the position of the solutus curve; only the solidus curve is affected. Figures 1 and 3 show the difference in the solidus curve for a nearly ideal system (Fig. 3) and for a system with a maximum a_0 value that nevertheless allows complete miscibility (Fig. 1). If the excess-free-energy-of-mixing is increased further, a miscibility gap develops as can be seen in the example for the $(Pb,Ba)SO_4$ series (Fig. 6).

Figures 4 and 5 provide an example of a SSAS system with close end-member solubility products: the solid-solution series natrojarosite–hydronium jarosite. Both pure end-member saturation curves in such a system plot well above the solutus curve, in the case of an assumed ideal system (Fig. 4). If the excess free energy of mixing is increased, however, the solutus curve will plot progressively closer to the end-member saturation curves (Fig. 5) and the solidus curve will also be significantly affected.

ESTIMATION OF THERMODYNAMIC MIXING PARAMETERS

There are two main applications for SSAS theory in the chemical modeling of aqueous systems: (1) the prediction of solid-solution solubilities, (2) the prediction of the distribution of trace components between solid and aqueous phases. Currently, a big problem with both types of predictions is the lack of low-temperature data on solid-solution excess-free-energy functions, and therefore on solid-phase activity coefficients. The two-parameter Guggenheim expansion series (the "subregular" model) has been successfully used to fit laboratory solubility data for the systems $(Sr,Ca)CO_3$-H_2O (Plummer and Busenberg 1987), $(Ba,Sr)CO_3$-H_2O (Glynn et al. 1988), $(Ca,Mg)CO_3$-H_2O (Busenberg and Plummer 1989), K(Cl,Br)-H_2O, and Na(Cl,Br)-H_2O (Glynn et al. 1990, 1992). The one-parameter Guggenheim series (the "regular" model) also has been frequently used (Lippmann 1980, Kirgintsev and Trushnikova 1966). More laboratory determinations of thermodynamic mixing parameters are needed, not only to acquire data on binary and multicomponent solid solutions, but also to test the applicability of the regular and subregular models, and to compare them with other excess-free-energy models.

In the mean time, as a better approximation than the commonly used assumption of "ideal" solid solutions, the a_0 and a_1 dimensionless parameters can be estimated for many binary systems. The MBSSAS computer code (Glynn 1991) uses either observed or estimated (1) miscibility-gap data, (2) spinodal-gap data, (3) critical temperature and

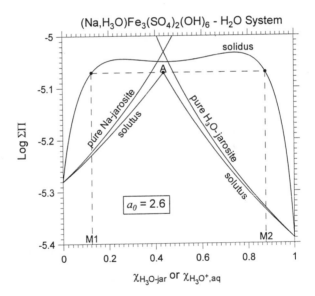

Figure 5. Lippmann phase diagram for the $(Na,H_3O)Fe_3(SO_4)_2(OH)_6$–$H_2O$ system at 25°C. An a_0 value of 2.6 was used for comparison with Figure 4. This a_0 value is probably a maximum one, based on Brophy and Sheridan's (1993) experimental observations and assumption of complete mixing for the (Na,H_3O,K)-jarosite series at 114°C. The maximum excess-free-energy-of-mixing that would still allow complete miscibility was assumed for that temperature and the corresponding value of a_0 was extrapolated down to 25°C. At that temperature, the system exhibits a miscibility gap as can be seen from the eutectic point (A) in the solutus curve. At that point the aqueous solution is in equilibrium with two solid solutions, of compositions M1 and M2 given by the intersections between the solidus curve and a horizontal line passing through the eutectic point (A).

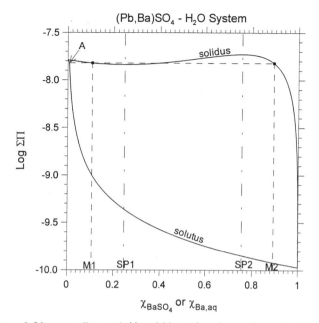

Figure 6. Lippmann diagram (with stoichiometric and pure-phase saturation curves) for the $(Pb,Ba)SO_4$– H_2O system at 25°C. An estimated a_0 value of 2.7 is used, based on observed compositions in natural minerals (Palache et al. 1951). End-member solubility products from PHREEQC (Parkhurst and Appelo 2000): $10^{-7.79}$ for anglesite ($PbSO_4$), $10^{-9.97}$ for barite ($BaSO_4$). The miscibility gap in the system is revealed by the presence of a peritectic point (A) which defines the composition of the aqueous solution at equilibrium with two miscibility-gap solids of composition M1 and M2. The spinodal gap in the system is given by the local minimum and maximum in the solidus curve (solid compositions SP1 and SP2).

critical mole-fraction of mixing data, (4) distribution-coefficient data, (5) alyotropic point data (Lippmann 1980) or (6) activity-coefficient data to calculate a_0 and a_1 parameters for binary solid solutions. Depending on whether one or two datum points are given, the program computes either a regular or subregular model.

Miscibility-gap data

Miscibility gaps, as determined from mineral compositions observed in the field, can be used to estimate thermodynamic mixing parameters in the absence of more accurate laboratory data. This approach suffers from several problems. The maximum mole fraction of trace component found may not correspond to the miscibility-gap fraction. In this case, if the solid-solution composition is stable, then the calculated excess free energy will be overestimated. On the other hand, if the analyzed "limiting" composition really represents a mechanical mixture of two end members, rather than a solid-solution mineral, the calculated excess free energy may be underestimated.

If the mineral was formed at much higher temperatures than the temperature of interest and if the mineral is fairly unreactive at the lower temperature, the maximum solid-solution mole fraction observed may well be metastable or even unstable. The temperature of formation (and of equilibration) of the solid solution may not be known. If a lower temperature is assumed, the excess free energy will be underestimated.

Additionally, the extrapolation to 25°C of excess free energies estimated at higher temperatures will introduce an additional error; typically, excess enthalpy and excess entropy data are not available for solid-solution series.

A partial solid-solution series may not be isomorphous (i.e. the end members may not have the same structure). In that case, the excess-free-energy parameters should be calculated only on a single side of the miscibility gap. On the other side of the miscibility gap, a different model will apply.

Despite the above problems, mixing parameters estimated from miscibility-gap information are still an improvement on the assumption of an ideal solid-solution model. The a_0 parameters estimated from potential miscibility-gap data provided by Palache et al. (1951), Jambor et al. (this volume), and other investigators, and also from various experimental determinations, are presented in Tables 1, 2, and 3 for several low-temperature mineral groups. Because of the large uncertainties in the data and in the estimation procedure, a subregular model is usually unwarranted. As a result, the estimated a_0 values that are presented should be used only for solid-solution compositions on a single side of the miscibility gap, i.e. only up to the given miscibility fraction. As can be seen in the tables, binary solid-solution series that have a miscibility gap but nevertheless show significant amounts of substitutional variation will have an a_0 excess-free-energy parameter slightly above a value of 2 (dimensionless). This assumes that a regular solid-solution model can be applied. This is also illustrated in Figure 7, which provides an easy way of determining the applicable a_0 value for a solid-solution series, given its miscibility-gap fraction.

Spinodal-gap data

Spinodal gaps can be used to estimate low-temperature solid-solution mixing parameters. This method is considerably less accurate than the miscibility-gap estimation technique. The estimation assumes that intrinsically unstable compositions are not likely to form during a precipitation process, although metastable compositions may do so. According to thermodynamic theory, solid-solution compositions inside a spinodal gap are intrinsically unstable, whereas compositions inside the miscibility gap, but outside the spinodal gap are metastable (Prigogine and Defay 1954). The presence of a miscibility gap is necessarily always accompanied by that of a spinodal gap. The $(Pb,Ba)SO_4$ solid-solution series provides an example of a system with hypothesized miscibility and spinodal gaps (Fig. 6).

One of the problems with the theory of spinodal gaps, at least for low-temperature SSAS systems, is that supposedly intrinsically unstable solid solutions can commonly be precipitated from aqueous solutions in the laboratory. An inability of the regular and subregular excess-free-energy models to describe adequately the relation between miscibility and spinodal gaps could offer one possible explanation. Another possible explanation might be that the thermodynamic properties of solids initially precipitated from aqueous solution are typically quite different from those of well crystallized phases. The initial precipitates are generally much finer grained, have a higher proportion of crystal defects, and may contain water, occluded solution, or other impurities. Defects and impurities in these initial solids, high surficial energy of the solids, and strong hydration bonds of the substituting ions may favor the formation of extensive solid-solution series.

Plummer and Busenberg (1987) were able to synthesize a complete suite of compositions in their investigation of the strontianite-aragonite solid-solution series, despite the presence of extensive spinodal and miscibility gaps, that were later determined on the basis of measurements of stoichiometric saturation states. Solids with compositions inside the spinodal gap, however, were much more difficult to synthesize (Busenberg, pers. commun.). Similarly, Glynn et al. (1988, and unpublished data) were able to synthesize and conduct stoichiometric saturation experiments on the entire witherite-strontianite series, despite the supposed presence of extensive spinodal and miscibility gaps (Table 2b).

Investigations of the barite-anglesite and barite-celestine series also indicate the ability to synthesize solids over the entire compositional range, despite the possible

Table 1a. Estimated mixing parameters for binary sulfate solid solutions at 25°C. References correspond to author initials and year of publication (JNA = Jambor et al., this vol.) syn. = synthetic solids; nat. = natural solids; therm. equil. = thermodynamic equilibrium experiment

System (main, trace)	Misc. Frac. (X_{trace})	Temp. (°C)	Structure	Estimated a_0 at 25°C	Reference
Melanterite group (monoclinic heptahydrates)					
$(Fe,Co)SO_4 \cdot 7H_2O$	complete (syn.)	25	mono.	< 2.00	PBF 1951
$(Fe,Mn)SO_4 \cdot 7H_2O$	up to at least 8.2% (nat.)	25	mono.	< 2.89	PBF 1951
$(Fe,Mn)SO_4 \cdot 7H_2O$	complete (syn., therm. equil.) $a_0 = 0.90$ at 0°C	0	mono.	0.83	W 1933 and present paper
$(Fe,Ni) SO_4 \cdot 7H_2O$	46% (syn.)	25	mono.-orth.	2.00	JNA 2000
$(Fe,Mg)SO_4 \cdot 7H_2O$	up to 58% (syn.)	25	mono.-orth.	subregular model needed	PBF 1951
$(Fe,Cu)SO_4 \cdot 7H_2O$	up to 66.6% (syn.)	25	mono.	subregular model needed	PBF 1951
$(Fe,Zn)SO_4 \cdot 7H_2O$	up to 30.7% (nat.)	25	mono.-orth.	2.11	PBF 1951
$(Fe,Zn)SO_4 \cdot 7H_2O$	55% (syn.)	25	mono.-orth.	subregular model needed	JNA 2000
$(Co,Mn)SO_4 \cdot 7H_2O$	18% (syn.)	20	mono.	2.37	JNA 2000
$(Co,Ni)SO_4 \cdot 7H_2O$	30% (syn.)	25?	mono.	2.12	JNA 2000
$(Co,Mg)SO_4 \cdot 7H_2O$	50% (syn.)	25?	mono.-ortho.	subregular model needed	JNA 2000
$(Co,Cu)SO_4 \cdot 7H_2O$	32% (syn.)	25	mono.	2.09	JNA 2000
$(Co,Zn)SO_4 \cdot 7H_2O$	54% (syn.)	25?	mono.-ortho.	subregular model needed	JNA 2000
$(Mn,Mg)SO_4 \cdot 7H_2O$	47% (nat.)	25?	mono.-orth.	2.00	JNA 2000
$(Cu, Fe) SO_4 \cdot 7H_2O$	possibly small misc. gap in Cu-rich region	25	mono.	subregular model needed	JNA 2000
Epsomite group (orthorhombic heptahydrates)					
$(Mg,Ni)SO_4 \cdot 7H_2O$	complete (syn.)	25	orth.	<2	PBF 1951
$(Mg,Zn)SO_4 \cdot 7H_2O$	complete (syn.)	25	orth.	<2	PBF 1951
$(Mg,Fe)SO_4 \cdot 7H_2O$	16.7% (syn.)	25	orth.-mono	2.41	PBF 1951
$(Mg,Co)SO_4 \cdot 7H_2O$	30% (syn.)	25	orth.-mono	2.12	JNA 2000
$(Mg,Mn)SO_4 \cdot 7H_2O$	28.6% (syn.)	25	orth.-mono	2.14	PBF 1951
$(Mg,Cu)SO_4 \cdot 7H_2O$	a few % (between 2 and 3% assumed) (syn.)	25	orth.-mono	4.05 - 3.70	JNA 2000
$(Zn,Fe)SO_4 \cdot 7H_2O$	37% (syn.)	25	orth.-mono	2.05	JNA 2000
$(Zn,Co)SO_4 \cdot 7H_2O$	27% (syn.)	25	orth.-mono	2.16	JNA 2000
$(Zn,Cu)SO_4 \cdot 7H_2O$	22.2% (nat.)	25	orth.-mono	2.26	PBF 1951
$(Zn,Mn)SO_4 \cdot 7H_2O$	25.6%? (nat.)	25	orth.-mono	2.22	PBF 1951
$(Ni,Fe)SO_4 \cdot 7H_2O$	19% (syn.)	25	orth.-mono	2.34	JNA 2000
$(Ni,Co)SO_4 \cdot 7H_2O$	55% (syn.)	25	orth.-mono	subregular model needed	JNA 2000
$(Ni,Cu)SO_4 \cdot 7H_2O$	18% (syn.)	25	orth.-mono	2.37	JNA 2000
$(Ni,Zn)SO_4 \cdot 7H_2O$	complete (syn.)	25	orth.	<2	PBF 1951

Table 1b. Estimated mixing parameters for binary sulfate solid solutions at 25°C. References correspond to author initials and year of publication (JNA = Jambor et al., this vol.) syn. = synthetic solids; nat. = natural solids.

System (main, trace)	Misc. Frac. (X_{trace})	Temp. (°C)	Structure	Estimated a_0 at 25°C	Reference
Hexahydrite group (monoclinic hexahydrates)					
$(Mg,Ni)SO_4 \cdot 6H_2O$	27% (nat.) but probably complete	25	mono.	<2.16	JNA 2000
$(Zn,Fe)SO_4 \cdot 6H_2O$	33% (syn.)	25	mono.	2.08	PBF 1951
$(Mn,Mg)SO_4 \cdot 6H_2O$	41% (nat.)	25	mono.	2.02	JNA 2000
$(Co,Ni)SO_4 \cdot 6H_2O$	complete above 61°C (syn.)	61	mono.	2.24	JNA 2000
$(Co,Fe)SO_4 \cdot 6H_2O$	27% (syn.)	50	mono.	2.35	JNA 2000
Chalcanthite group (triclinic pentahydrates)					
$(Cu,Mn)SO_4 \cdot 5H_2O$	complete (syn.)	21	tricl.	<1.97	JNA 2000
$(Cu,Ni)SO_4 \cdot 5H_2O$	20% (syn.)	25	tricl.	2.31	JNA 2000
$(Cu,Mg)SO_4 \cdot 5H_2O$	6% (syn.)	25	tricl.	3.13	JNA 2000
$(Cu,Zn)SO_4 \cdot 5H_2O$	6% (syn.)	25	tricl.	3.13	JNA 2000
$(Cu,Zn)SO_4 \cdot 5H_2O$	19% (syn.)	40	tricl.	2.46	JNA 2000
$(Cu,Fe)SO_4 \cdot 5H_2O$	complete? (syn. & nat.)	25	tricl.	<2.00	JNA 2000
$(Cu,Co)SO_4 \cdot 5H_2O$	5% (syn.)	25	tricl.	3.27	JNA 2000
$(Cu,Co)SO_4 \cdot 5H_2O$	16% (syn.)	60	tricl.	2.72	JNA 2000
$(Mg,Cu)SO_4 \cdot 5H_2O$	34% (nat..)	25	tricl.	2.07	PBF 1951
$(Mn,Mg)SO_4 \cdot 5H_2O$	32% (nat.)	25	tricl.	2.09	JNA 2000
Rozenite Group (monoclinic tetrahydrates)					
$(Mn,Mg)SO_4 \cdot 4H_2O$	39% (nat.)	25	mono.	2.03	JNA 2000
$(Zn,Mg)SO_4 \cdot 4H_2O$	16% (nat.)	25	mono.	2.44	JNA 2000
Kieserite group (monoclinic monohydrates)					
$(Fe,Mg)SO_4 \cdot H_2O$	41% (nat.)	25	mono.	2.02	JNA 2000
$(Ni,Fe)SO_4 \cdot H_2O$	10% (nat.)	25	mono.	2.75	JNA 2000
$(Zn,Mn)SO_4 \cdot H_2O$	39% (syn.)	25	mono.	2.03	JNA 2000
$(Fe,Zn)SO_4 \cdot H_2O$	probably complete (syn.)	25	mono.	<2.00	JNA 2000
$(Fe,Cu)SO_4 \cdot H_2O$	20% (syn.)	25	mono.-tricl.	<2.31	JNA 2000
$(Mg,Cu)SO_4 \cdot H_2O$	20% (syn.)	25	mono.-tricl.	<2.31	JNA 2000
Halotrichite group (monoclinic mixed divalent trivalent sulfate salts)					
$(Fe,Mg)Al_2(SO_4)_4 \cdot 22H_2O$	complete (nat.)	25	mono.	<2.00	PBF 1951
$(Mn,Fe)Al_2(SO_4)_4 \cdot 22H_2O$	probably complete (nat.)	25	mono.	<2.00	PBF 1951
$(Mn,Mg)Al_2(SO_4)_4 \cdot 22H_2O$	probably complete (nat.)	25	mono.	<2.00	PBF 1951
$(Co,Mg)Al_2(SO_4)_4 \cdot 22H_2O$	36% (nat.)	25	mono.	2.05	JNA 2000
$Fe(Al,Fe)_2(SO_4)_4 \cdot 22H_2O$	38% (nat.)	25	mono.	2.04	JNA 2000
Copiapite group (ideal formula: $(A)(R)_4(SO_4)_6(OH)_2 \cdot 20H_2O$)					
$(Fe,Mg)Fe_4(SO_4)_6(OH)_2 \cdot 20H_2O$	complete	25	tricl.	<2.00	PBF 1951
$(Fe,Cu)Fe_4(SO_4)_6(OH)_2 \cdot 20H_2O$	probably complete	25	tricl.	<2.00	PBF 1951
$(Fe,Zn)Fe_4(SO_4)_6(OH)_2 \cdot 20H_2O$	probably complete	25	tricl.	<2.00	PBF 1951
$(Fe,Fe_{2/3})Fe_4(SO_4)_6(OH)_2 \cdot 20H_2O$	probably complete	25	tricl.	<2.00	JNA 2000
$(Fe,Al_{2/3})Fe_4(SO_4)_6(OH)_2 \cdot 20H_2O$	probably complete	25	tricl.	<2.00	JNA 2000

All binary solid solutions formed with the major and minor ions present in the above A-site copiapite solid solutions are probably complete at 25°C. Significant substitution of Na, Ca, Mn,Ni and Co can also occur on the A site. Substantial substitution of Al for Fe on the R site is also likely.

Table 1c. Estimated mixing parameters for binary sulfate solid solutions at 25°C. References correspond to author initials and year of publication.

syn. = synthetic solids; nat. = natural solids; extrapol. = extrapolation; const. = constant; thermo. equil. = thermodynamic equilibrium experiment.

System (main, trace)	Misc. Frac. (X_{trace})	Temp. (°C)	Structure	Estimated a_0 at 25°C	Reference
Jarosite/Alunite group (ideal formula: $AR_3(SO_4)_2(OH)_6$)					
$(K,Na)Fe_3(SO_4)_2(OH)_6$	29% (nat.)	25	hexa.	2.13	PBF 1951
$(Na,K)Fe_3(SO_4)_2(OH)_6$	24% (nat.)	25	hexa.	2.22	PBF 1951
$(K,Na)Fe_3(SO_4)_2(OH)_6$	complete (syn. therm. equil.) $a_0 = 0.32$	200	hexa.	0.51	S 1993 SAJ 2000
$(K,H_3O)Fe_3(SO_4)_2(OH)_6$	complete series (syn.)	114	hexa.	<2.60	BS 1965
$(K,Na)Fe_3(SO_4)_2(OH)_6$	probably complete series (syn.)	114	hexa.	<2.60	BS 1965
$(Na,H_3O)Fe_3(SO_4)_2(OH)_6$	probably complete series (syn.)	114	hexa.	<2.60	BS 1965
$KFe_3(SO_4,CrO_4)_2$	complete series (syn. stoic. sat.)	23	hexa.	-5.0±0.9	BDP 2000
$K(Fe,Al)_3(SO_4)_2(OH)_6$	probably complete series (nat.)	25	hexa.	<2.00	PBF 1951
$K(Fe,Al)_3(SO_4)_2(OH)_6$	complete series (syn.)	100 105	hexa.	<2.50	BSS 1962 HBB 1984
$(K,Na)Al_3(SO_4)_2(OH)_6$	64% (nat.) probably complete series	25	hexa.	<2.00	PBF 1951
$(K,Na)Al_3(SO_4)_2(OH)_6$	thermo. equil. (syn.) extrapol. based on const. A_0 and A_1	250	hexa.	$a_0 = 2.24$ $a_1 = -0.358$	SC 1990 SAJ 2000
$(K,Na)Al_3(SO_4)_2(OH)_6$	thermo. equil. (syn.) linear extrapolation of A_0 and A_1 with temp.	250, 350, 450	hexa.	$a_0 = 3.65$ $a_1 = -0.424$	SC 1990 SAJ 2000
Barite group					
$(Ba,Pb)SO_4$	20% (nat.)	75	orth.	2.70	PBF 1951
$(Ba,Sr)SO_4$	complete	75	orth.	2.34	PBF 1951
$(Ba,Ca)SO_4$	7.7% (nat.)	75	orth.	3.43	PBF 1951
$(Sr,Ca)SO_4$	8.3% (nat.)	50	orth.	3.12	PBF 1951
Miscellaneous Sulfate Salts					
$(Fe,Al)_3(SO_4)_2·9H_2O$ coquimbite	42% (nat.)	25	hexa.	2.02	PBF 1951
$Fe(Fe,Al)_2(SO_4)_4·14H_2O$ römerite	16% (nat.)	25	tricl.	2.44	JNA 2000
$(Fe,Zn)Fe_2(SO_4)_4·14H_2O$ römerite	30% (nat.)	25	tricl.	2.12	JNA 2000
$Na_6(SO_4,CO_3)_2(CO_3,SO_4)$ burkeite	Sol. sol'n occurs with CO_3 between 25-50% around the 33% burkeite phase (syn.)	150	orth.	insufficient information available	SBG 1936 J 1962
$(Mg,Fe)Fe(SO_4)_2OH·7H_2O$ botryogen	39% (nat.)	25	mono.	2.03	PBF 1951
$(Mg,Zn)Fe(SO_4)_2OH·7H_2O$ botryogen	26% (nat.)	25	mono.	2.18	PBF 1951
$(K,NH_4)Al(SO_4)_2·12H_2O$ potash alum	complete (syn.)	25	isom.	<2.00	PBF 1951
$(K,Na)_2Mg(SO_4)_2·4H_2O$ leonite	25% (syn.)	25	mono.	2.20	PBF 1951
$(Na,K)_2Mg(SO_4)_2·4H_2O$ blödite	3% (syn.) not isostructural with leonite	25	mono.	3.70	PBF 1951
$K_2(Mg,Ca)_2(SO_4)_3$ langbeinite	complete (syn.)	75?	mono.	<2.34	PBF 1951

Glynn

Table 2. Estimated mixing parameters for binary carbonate solid solutions at 25°C.
References correspond to author initials and year of publication.

syn. = synthetic solids; nat. = natural solids; stoich. sat. = stoichiometric saturation experiment;
therm. equil. = thermodynamic equilibrium experiment.

System (main, trace)	Misc. Frac. (X_{trace}) (All nat., except as noted)	Temp. (°C)	Structure	Estimated a_0 at 25 °C	Reference
Calcite group					
$(Ca,Mn)CO_3$	complete?	25	rhomb.	<2?	PBF 1951
$(Ca,Fe)CO_3$	18%	50	rhomb.	2.56	PBF 1951
$(Ca,Fe)CO_3$	max 2.5% (nat. stable solids)	50	rhomb.	4.18	BP 1983
$(Ca,Zn)CO_3$	min 4.9%	50	rhomb.	3.56	PBF 1951
$(Ca,Co)CO_3$	min 2.7%	50	rhomb.	4.12	PBF 1951
$(Ca,Mg)CO_3$	2% (nat. stable solids)	25	rhomb.	4.05	BP 1983
$(Ca,Mg)CO_3$	3.2% (nat. stable solids)	50	rhomb.	3.95	BP 1983
$(Ca,Mg)CO_3$	4.5% (nat.stable solids)	150	rhomb.	4.76	BP 1983
$(Ca,Ca_{.5}Mg_{.5})CO_3$ "non-defective"	misc. gap: $2.0\% \leq \chi_{Mg} \leq 49.96\%$ (nat. & syn., stoich. sat.)	25	rhomb.	$a_0 = 5.08$ $a_1 = 1.90$	BP 1989
$(Ca,Ca_{.5}Mg_{.5})CO_3$ "defective"	$10.7\% \leq \chi_{Mg} \leq 47.5\%$ (nat. & syn., stoich. sat.)	25	rhomb.	$a_0 = 2.54$ $a_1 = 0.71$	BP 1989
$(Ca,Pb)CO_3$	min 10.5% (nat.?)	50	rhomb.-orth.	2.94	PBF 1951
$(Mg,Fe)CO_3$	complete (nat.?)	25	rhomb.	<2	PBF 1951
$(Mg,Mn)CO_3$	min 9.7%	25	rhomb.	2.77	PBF 1951
$(Mg,Ca)CO_3$	min 10.1%	25	rhomb.	2.74	PBF 1951
$(Mg,Co)CO_3$	6.2%	25	rhomb.	3.10	PBF 1951
$(Fe,Mn)CO_3$	complete (nat.?)	25	rhomb.	<2	PBF 1951
$(Fe,Ca)CO_3$	max 22.7%	25	rhomb.	2.24	PBF 1951
$(Fe,Co)CO_3$	13.0%	25	rhomb.	2.57	PBF 1951
$(Mn,Mg)CO_3$	min 34.1%	75	rhomb.	2.42	PBF 1951
$(Mn,Zn)CO_3$	min 44.6%	75	rhomb.	2.34	PBF 1951
$(Co,Ca)CO_3$	min 7%	50	rhomb.	3.26	PBF 1951
$(Co,Mn)CO_3$	complete (syn., stoich. sat.)	25	rhomb.	$a_0 = 1.46$ $a_1 = -0.25$	KG 1990
$(Zn,Ca)CO_3$	24%	25	rhomb.	2.22	PBF 1951
$(Zn,Fe)CO_3$	min 38.6% possibly 54.0%	25	rhomb.	2.04	PBF 1951
$(Zn,Co)CO_3$	15.9%	25	rhomb.	2.44	PBF 1951
$(Zn,Cu)CO_3$	9.7%	25	rhomb.	2.77	PBF 1951
$(Zn,Mn)CO_3$	min 16.1% possibly complete	25	rhomb.	2.44	PBF 1951
$(Zn,Cd)CO_3$	2.7%	25	rhomb.	3.79	PBF 1951
$(Zn,Mg)CO_3$	20.8%	25	rhomb.	2.29	PBF 1951
$(Zn,Pb)CO_3$	< 0.3%	25	rhomb.-orth.	5.84	PBF 1951
Aragonite group					
$(Ca,Cd)CO_3$	complete (syn.)	25	orth.	0.0	KHG 1991
$(Ca,Sr)CO_3$	3.8%	25	orth.	3.50	PBF 1951
$(Ca,Sr)CO_3$	misc. gap: $0.48\% \leq \chi_{Sr} \leq 85.7\%$ (syn., stoich. sat.)	25	orth.	$a_0 = 3.43$ $a_1 = -1.82$	PB 1987
$(Ca,Sr)CO_3$	misc. gap: $2.32\% \leq \chi_{Sr} \leq 79.0\%$ (syn., stoich. sat.)	76	orth.	$a_0 = 2.66$ $a_1 = -1.16$	PB 1987
$(Sr,Ca)CO_3$	18.2% (--)	75	orth.	2.76	PBF 1951
$(Ba,Sr)CO_3$	misc. gap: $0.23\% \leq \chi_{Sr} \leq 98.74\%$ (syn., stoich. sat.)	25	orth.	$a_0 = 5.26$ $a_1 = -0.82$	GPB 1988 Unpubl. data
$(Ca,Pb)CO_3$	7.7%	25	orth.	2.94	PBF 1951
$(Pb,Sr)CO_3$	8.7%	25	orth.	2.84	PBF 1951

Table 3. Estimated mixing parameters for binary halide solid solutions at 25°C. References correspond to author initials and year of publication.

syn. = synthetic solids; nat. = natural solids; stoich. sat. = stoichiometric saturation experiment; therm. equil. = thermodynamic equilibrium experiment.

System (main, trace)	Misc. Frac. (X_{trace})	Temp. (°C)	Structure	Estimated a_0 at 25°C	Reference
Ag(Cl,Br)	complete (syn.)	25	isom.	<2	PBF 1951
Ag(Cl,Br)	complete (syn.) therm. equil. (VB 1952)	25	isom.	$a_0 = 0.30$ $a_1 = -0.18$	G 1990
K(Cl,Br)	complete (syn., therm. equil.)	25	isom.	$a_0 = 1.40$ $a_1 = -0.08$	GRPB 1990
K(Br,I)	misc. gap: $0.178 < X_I < 0.870$ (syn. thermo. equil.)	25	isom.	$a_0 = 2.46$ $a_1 = 0.14$	KG 1992
(K,Rb)I	complete (syn. thermo. equil.)	25	isom.	$a_0 = 1.00$ $a_1 = -0.15$	KG 1992
Na(Cl,Br)	complete series at 28°C (syn., therm. equil.)	35	isom.	$a_0 = 1.89$ $a_1 = -0.351$	GRPB 1992

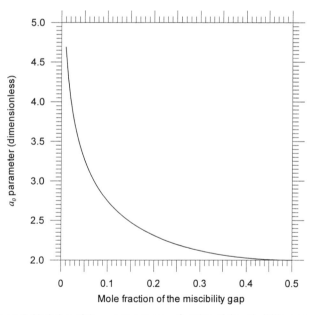

Figure 7. Variation of the a_0 parameter as a function of the miscibility gap fraction in a symmetric regular solid-solution series.

presence of spinodal gaps and miscibility gaps in the series. Takano et al. (1969) succeeded in synthesizing the complete range of compositions in the (Pb,Ba)SO$_4$ series at 20°C. Boström et al. (1967) had earlier succeeded in synthesizing complete solid-solution series in the (Ba,Pb,Sr)SO$_4$ system at higher temperatures of 100, 300, and 375°C. Takano et al. (1969) found that, despite the lack of any discontinuity in lattice parameters as a function of composition (they found that Vegard's rule applied), the results of their measurements of infrared spectra, heat of precipitation, dissolution rate, and the amounts of coprecipitation of Sr and Ca in their solid solutions indicated the presence of a "structural gap" between 25% and possibly 75% $PbSO_4$ in the $(Ba,Pb)SO_4$ solid-solution series. They hypothesized that the solid compositions synthesized in that range were metastable. The gap in the $(Ba,Pb)SO_4$ series is close to the maximum of 20% Pb found in barite (Palache et al. 1951), a limit used here to determine an a_0 value of 2.7 at 25°C (Table 1c) and to construct the Lippmann diagram for the $(Pb,Ba)SO_4$–H_2O system (Fig. 6).

The possibility of synthesizing the entire $(Ba,Sr)SO_4$ solid-solution series at low temperatures, despite the fairly large difference in the ionic radii of Ba and Sr, has been known for a long time. Initially the series was considered to be ideal at room temperature (Brower and Renault 1971). More recent investigations have concluded, however, that the series is non-ideal. Hanor (this volume) reviews some of the evidence. Malinin and Urusov (1983) argue for the existence of a miscibility gap below 100°C. This conclusion is consistent with the a_0 value of 2 at 75°C, and the consequent extrapolation of a_0 to a value of 2.34 at 25°C (Table 1c), based on compositional data reported by Palache et al. (1951). Galinier et al. (1989) estimated an a_0 value of 1.61 based on experimental data for a single solid, $Ba_{.63}Sr_{.37}SO_4$. After equilibrating a series of $(Ba,Sr)SO_4$ solids at room temperature for a period of up to 3 years, Felmy et al. (1993) concluded that the system was controlled by the equilibrium of the aqueous solution with two solids: (1) a barite which controlled the Ba concentrations, and (2) a $(Ba,Sr)SO_4$ solid solution which controlled the Sr concentrations and in which the Ba component acted as an inert component. Felmy et al. calculated an a_0 value of 0.8 based on their fit of the $SrSO_4$ solid-phase activity data. More recently, Prieto et al. (2000) calculated the excess-free-energy of mixing function for the $(Ba,Sr)SO_4$ series from a "first principles," direct calculation of lattice energies. They concluded that a five-term expansion of the Guggenheim excess-free-energy series, with null assymmetric terms, was required to fit their calculations. Their calculations then indicated the presence of a symmetric miscibility gap between 2.1 and 97.9% $SrSO_4$ at 25°C, and a local minimum in the enthalpy of mixing around 50% $SrSO_4$ due to an ordering tendency in the series. They argued that although intermediate compositions are often synthesized, these compositions probably represent metastable solids. Given the conflicting results in the literature, an intermediate a_0 value of 2.0 was used to construct the Lippmann diagram for the $(Ba,Sr)SO_4$–H_2O system shown in Figure 1.

Despite the above evidence of the ability to synthesize "unstable" solid solutions at low temperatures in SSAS system, calculated spinodal gaps have occasionally been consistent with some laboratory or field observations. Using a subregular Guggenheim model to fit the solubilities of "highly-defective" magnesian calcites, Busenberg and Plummer (1989) calculated a spinodal composition of 19.8 % Mg. They concluded that this result was in agreement with the magnesium content of modern biogenic magnesian-calcites (which range from 1 to 20% Mg, coralline algae excepted). Similarly, from stoichiometric saturation measurements on the $(Sr,Ca)CO_3$ - H_2O system, Plummer and Busenberg (1987) determined spinodal gaps at 25 and 76°C of $0.065 \leq \chi_{Sr} \leq 0.620$ and $0.103 \leq \chi_{Sr} \leq 0.585$, respectively. Synthesis of the unstable solid solutions inside these compositional ranges proved to be much more difficult than that of the stable and metastable solids (Plummer and Busenberg 1987, Busenberg, pers. commun.) outside the spinodal limits. In general it is very difficult, however, to distinguish between intrinsi-cally unstable and metastable solid-solution compositions. Therefore it is not recom-mended that a_0 values be estimated from spinodal gap information.

Critical mixing-point data

Critical mixing points can also be used to calculate the non-ideality of a solid-solution series. The critical temperature of mixing, T_c, of a solid-solution series refers to the lowest possible temperature at which a complete solid-solution series may form (Prigogine and Defay 1954). In a regular solid-solution series, a miscibility gap will occur only if a_0 is greater than 2. The following relation may be derived between a_0, T_c, and T, the current temperature of the system (Lippmann 1980):

$$a_0 = \frac{2T_c}{T} \tag{28}$$

Thus, if the critical temperature of mixing of a binary solid-solution series is known or can be estimated, a regular solid-solution model can be assumed as a first approximation, and the value of a_0 can be calculated from Equation (28). If the critical mole fraction of mixing is also known, another set of relations can be derived and both the a_0 and a_1 parameters in the subregular model can be calculated. Because of the usual lack of excess enthalpy and excess entropy data, the use of critical temperatures and critical mole fractions of mixing to estimate a_0 and a_1 data at low temperatures commonly assumes that the free energy of mixing of a solid solution decreases with increasing temperature because of the increase in $TS^{m,id}$, the temperature–ideal-entropy-of-mixing term. At the critical temperature of mixing, this term becomes sufficiently large to overcome the repulsive energy exerted between the solid-solution components and the miscibility gap disappears.

An estimate of the maximum possible a_0 value for ionic crystals can be obtained by substituting the highest temperature of melting of ionic solids for the critical temperature. The alkaline-earth sulfates and carbonates, for example, melt at temperatures as high as 1700°C. The calculated maximum a_0 value for these ionic solids is therefore approximately 13-14. The justification for this estimate depends on two postulates: (1) most ionic melts are completely miscible with each other, (2) the short-lived structures of an ionic melt often closely resemble that of the solid solution from which the melt was formed (Samoilov 1965). According to Samoilov (1965), the last postulate can be justified by comparing the lattice energy of an ionic solid with its heat of fusion. Most commonly, the heat of fusion is a small fraction of the solid's lattice energy, and consequently the lattice energies (and presumably the structures) of the solid and the melt do not differ considerably. If the above reasoning is correct (i.e. the maximum value of a_0 is 13-14), then a minimum miscibility mole-fraction of 10^{-6} can be calculated. This estimate agrees with the observation that the purest chemical reagents are at best "five nines" pure, i.e. 99.999% pure (Frye 1974).

Distribution coefficient measurements

Distribution coefficients observed in the field or in laboratory experiments have often been used to estimate excess free energies of mixing. Glynn (1990) used distribution coefficient data of Vaslow and Boyd (1952) to determine excess-free-energy-of-mixing parameters for the Ag(Cl,Br) system. Glynn and Reardon (1990) used Crocket and Winchester's (1966) $(Ca,Zn)CO_3$ distribution-coefficient data to confirm the value of a_0 estimated from solid-solution compositions given in Palache et al. (1951). There are two main problems in using this technique to calculate solid-solution mixing parameters: (1) the ratio of the component activities in the adjoining aqueous (or non-aqueous) phase must be known, and (2) the distribution coefficients measured must be representative of thermodynamic equilibrium between the two phases. The large variations in coefficients reported for SSAS systems (e.g. the magnesian calcites; Mucci 1987), are evidence of the problems in obtaining thermodynamic equilibrium coefficients (Plummer and Busenberg 1987).

Stoichiometric saturation solubilities

Stoichiometric saturation measurements in carefully controlled laboratory experiments offer perhaps the most promising technique for the estimation of thermodynamic mixing parameters at low temperatures of relatively insoluble minerals that do not undergo rapid recrystallization, such as many carbonate minerals (Thorstenson and Plummer 1977, Glynn and Reardon 1990, 1992; Königsberger and Gamsjäger 1991, 1992). Recently, stoichiometric saturatuion states have also been obtained in laboratory investigations of the jarosite–chromian-jarosite series (Baron 1996, Baron et al. 2000). Unfortunately, most of the results have not been verified by a second independent and accurate method, such as reaction calorimetry or measurement of thermodynamic equilibrium solubilities (Plummer and Busenberg 1987). The conditions necessary to obtain good stoichiometric saturation data (as opposed to thermodynamic equilibrium data) were discussed earlier.

Thermodynamic equilibrium solubilities

Measurements of solid-solution solubilities at thermodynamic equilibrium are the ideal method to obtain excess-free-energy data for solid-solution series. Unfortunately, thermodynamic equilibrium states are commonly difficult, if not impossible, to obtain in the laboratory, as solid-solution recrystallization experiments have indicated (Plummer and Busenberg 1987). SSAS systems with highly soluble solid solutions capable of rapid recrystallization, such as K(Cl,Br)–H$_2$O and Na(Cl,Br)–H$_2$O, are a significant exception to this rule (Glynn and Reardon 1990, 1992; Glynn et al. 1990, 1992).

Because of the large size of the sulfate ion, many sulfate minerals are very soluble and have relatively large crystal structures that favor the incorporation of substitutional impurities. Their high solubility should allow relatively rapid recrystallization, and consequent attainment of thermodynamic equilibrium states in laboratory experiments. Conversely, the experimental difficulties found in obtaining good thermodynamic equilibrium data for minerals in the barite group are largely explained by the relatively low solubilities of these minerals. The commonly high solubilities of sulfate minerals, however, and the consequently high ionic strengths of the equilibrated aqueous solutions, may pose the problem of significant uncertainties in the thermodynamic model for the equilibrium aqueous solutions (see Ptacek and Blowes, this volume). Generally, however, the excess-free-energy of mixing of the solid solutions will be significantly greater than that of the aqueous solution and thermodynamic parameters for the solid solution may still be estimated with relative accuracy. The assumptions that need to be made are typically that the ratio of the aqueous activity coefficients of the substituting ions and the ratio of the aqueous ion-association factors of the substituting ions can be considered constant for the entire range of equilibrated solid compositions. Generally, the ratio of the end-member solubility products must also be known. Glynn and Reardon (1992) and Glynn et al. (1992) discussed these assumptions and provided examples of such estimations for the K(Cl,Br)-H$_2$O and Na(Cl,Br)-H$_2$O systems, respectively.

White (1933) equilibrated differing amounts of pure FeSO$_4$·7H$_2$O (presumably melanterite) and MnSO$_4$·7H$_2$O (presumably mallardite) with aqueous solutions at 0 and 25°C for up to a year. Unfortunately, there are no X-ray diffraction data for the solids. The 25°C experiments resulted in heptahydrate or pentahydrate solid-solution phases at equilibrium, depending on the amount of Mn in the system. The Mn-rich systems produced pentahydrate solids whereas Fe-rich systems produced heptahydrate solids. In contrast, the 0°C experiments resulted in the presence of only heptahydrate solid-solutions. Because of a lack of knowledge of the applicable solubility products for the pentahydrate ferrous sulfate and heptahydrate manganous sulfate salts at 25°C, and an insufficient amount of experimental data by which to determine these end-member solubility products through best-fit procedures, the 0°C experiments are considered here as an example of analyzing equilibrium solubility data for a solid-solution series.

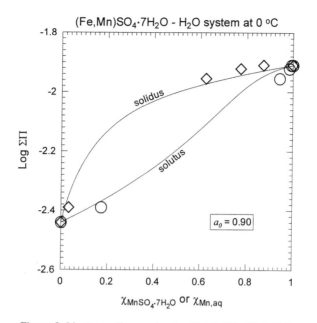

Figure 8. Lippmann diagram for the $(Fe,Mn)SO_4 \cdot 7H_2O$–H_2O system at 0°C, based on experimental solubility data from White (1933). The aqueous compositional data were speciated using PHRQPITZ (Plummer et al. 1989). Solubility products of $10^{-2.441}$ and $10^{-1.910}$ were obtained respectively for the $FeSO_4 \cdot 7H_2O$ and $MnSO_4 \cdot 7H_2O$ end members and an a_0 value of 0.90 was obtained through a best fit of the solidus data. Diamond symbols represent solid-phase compositions and circles represent the equilibrium aqueous activity fractions and log $\Sigma \Pi$ values obtained from aqueous speciation of the experimental data. The accuracy of the aqueous model is questionable because of the very high ionic strengths of the aqueous solutions (between 4.2 and 12.5) particularly for the Mn-rich systems.

Aqueous speciation of the solubility data for the end-member minerals with the PHRQPITZ computer code (Plummer et al. 1989) resulted in the following solubility products at 0°: $10^{-2.441}$ for $FeSO_4 \cdot 7H_2O$ and $10^{-1.910}$ for $MnSO_4 \cdot 7H_2O$. Aqueous speciation of the intermediate experimental systems containing both Mn and Fe was also conducted using the PHRQPITZ code. The computed aqueous activities of Fe^{2+}, Mn^{2+}, SO_4^{2-} and H_2O were used to determine the experimental aqueous Mn^{2+} activity fractions,

$$\chi_{Mn^{2+},aq} = [Mn^{2+}]/([Mn^{2+}] + [Fe^{2+}]),$$

and $\Sigma \Pi$ values, which along with solid-solution Mn/(Mn+Fe) fractions were then plotted on a Lippmann diagram (Fig. 8). Attempts were made to fit several different equations to this system assuming regular or subregular solid-solution models (distribution coefficients, the equal G function, the solutus curve). The most reasonable and justifiable results were obtained by a best fit of the solidus curve, under the assumption of a regular solid-solution model. An a_0 value of 0.90 was determined. The accuracy of the aqueous thermodynamic model is questionable because of the high ionic strengths (from 4.2 to 12.5) of the equilibrium aqueous solutions, particularly for the Mn-rich systems. The Pitzer interaction coefficients used in the PHRQPITZ code, particularly those relating to

Mn^{2+} interactions, are probably not valid at such high ionic strengths. In addition, there is a paucity of data for intermediate systems (with solid Mn mole fractions between 0.1 and 0.6). As a result, a subregular model fit is probably not warranted for this system. The results, nevertheless, provide an example of the types of experiments and interpretations that can be conducted for soluble sulfate solid-solution minerals.

Königsberger et al. (1991), in their investigation of the $(Ca,Cd)CO_3$–H_2O system, demonstrated another technique by which solid-solution excess-free-energy parameters may be determined, or at least checked, at thermodynamic equilibrium. The technique consists of (1) assuming, a priori, a set of excess-free-energy parameters for the solid-solution series, (2) determining the suite of possible thermodynamic equilibrium states for the SSAS system (for example, by means of a Lippmann diagram), (3) preparing "equilibrium" experimental SSAS systems best suited to test the validity of the hypothesized excess-free-energy model, and (4) determining whether the prepared experimental systems show any compositional deviations with time from their assumed equilibrium state. This technique may work especially well for solid-solutions that are less soluble and therefore less likely to undergo extensive recrystallization on laboratory time scales. The solids, however, must still be sufficiently reactive so that potential deviations from the assumed equilibrium states can be observed on a laboratory time scale.

Excess-free-energy data for selected sulfate, carbonate and halide solid solutions

Tables 1, 2, and 3 offer a compendium of calculated excess-free-energy parameters for binary solid-solution series at 25°C, primarily for sulfate minerals (Table 1) but also for selected carbonate (Table 2) and halide minerals (Table 3). Most of the parameters were calculated assuming a regular solution model; in a few cases a subregular model was used. As discussed earlier, most of the dimensionless parameters were obtained by assuming that observed compositional limits reported by Palache et al. (1951) or by Jambor et al. (this volume) can be equated to miscibility gap limits. In some systems, a completely miscible solid-solution series was observed (or assumed) at the indicated temperatures. That knowledge was used to place a limit of 2.0 for the maximum a_0 value at the specified temperature. In other cases, the excess-free-energy parameters were obtained by interpretation of laboratory experiments that attained either thermodynamic equilibrium or stoichiometric saturation.

Generally, a_0 and a_1 parameters determined for temperatures different from 25°C were recalculated at 25°C by assuming that the initial A_0 and A_1 values (in J/mole) did not vary as a function of temperature. In other words, it was assumed that the free energy of mixing of the solid-solution series decreased with increasing temperature only because of the greater contribution of $TS^{m,id}$ with increasing temperature (Glynn 1991). In actuality, A_0 and A_1 values are likely to increase with decreasing temperature rather than remain constant. Glynn (1991) provided an example of this phenomenon: the estimated critical temperature of mixing for the calcite-dolomite solid-solution series increased with a decrease in the experimental temperature at which the subregular excess-free-energy parameters were determined. The solid-solution series also became less assymetrical with decreasing temperature.

In another example, Stoffregen and Cygan (1990; also reported by Stoffregen et al. this volume) determined subregular parameters for the alunite-natroalunite system at three different temperatures: 250, 350, and 450°C. The Thompson and Walbaum parameters they determined (or equivalently the A_0 and A_1 parameters) increased almost linearly with decreasing temperature, instead of remaining constant. Table 1 gives the a_0 and a_1 parameters (3.65 and –0.424) estimated at 25°C based on a linear extrapolation of the decreasing A_0 and A_1 values with temperature. These are probably the most accurate values. For the sake of comparison, the table also provides the a_0 and a_1 parameters (2.24 and −0.358) calculated at 25°C, assuming that the Thompson and Waldbaum parameters (or equivalently the A_0 and A_1 parameters) determined at 250°C are also applicable at 25°C. Those values are significantly lower than those determined by linear extrapolation of the Thompson and Waldbaum parameters, but are nevertheless higher than the values that might be calculated based on the observations of natural alunite compositions by Palache et al. (1951), who suspected a complete solid-solution series, presumably at temperatures near 25°C.

In several cases, excess-free-energy parameters were estimated based on 2 or more sources of information. These are reported, with appropriate details, on adjoining lines of the tables.

For some minerals, the reported compositional data are not compatible with the symmetric miscibility gaps calculated by a regular solid-solution model, and a subregular model would be required. Generally, the subregular model could not be calculated in cases where the solid-solution series was not isomorphous. Similarly, information on compositional variations for the mineral burkeite (Schroeder et al. 1936, Jones 1962) were reported although excess-free-energy parameters could not be calculated because of the unique nature of this mineral for which solid-solution occurs between 25% and 50% CO_3 content on two different structural sites.

In summary, the excess-free-energy parameters reported in Tables 1, 2, and 3 are likely to be significantly inaccurate, particularly in the case of the values determined from observations of compositional limits. Experimental (or theoretical) determinations of excess-free-energy parameters, however, often provide conflicting results even for well researched solid-solution series, such as the $(Ba,Sr)SO_4$–H_2O system, as discussed earlier. Nevertheless, in many cases the parameters provided in Tables 1, 2, and 3 provide a better estimate of the excess-free-energy of mixing for a binary solid-solution series than the value of 0 assumed using an ideal solution model. The values provided can help place limits on the excess-free-energies of mixing that might be estimated for various solid-solution series.

Multicomponent solid-solution systems

Sulfate minerals commonly exhibit significant amounts of solid solution; therefore, many natural sulfate minerals can not be adequately described by a simple, binary, single-site, solid-solution model. Their large crystal structures tend to allow substitution by many different naturally occurring ions, in many cases on multiple crystallographic sites. Minerals in the copiapite, jarosite, alunite, and halotrichite groups and uncommon minerals such as römerite and botryogen offer excellent examples of these multicomponent, multisite, solid-solution series. The thermodynamics of multi-site solid-solutions are beyond the scope of this paper, but an introduction to the thermodynamics of multicomponent solid-solution series is given below.

The thermodynamic theory of stoichiometric saturation states and of thermodynamic equilibrium states can be extended to multicomponent solid solutions. Gamsjäger et al. (2000), for example, provided an elegant derivation, from a unifying point of view, of thermodynamic equilibrium and stoichiometric saturation states in multicomponent solid-solution systems. The greatest complication in interpreting or predicting the behavior of multicomponent SSAS systems is currently the paucity of available thermodynamic data. Another problem is the difficulty of determining the types of thermodynamic states attained during laboratory experiments, or observed through field measurements. The concept of "constrained equilibria" (Königsberger and Gamsjäger 1991) has a more general application to multicomponent systems than the concept of stoichiometric saturation, at least in its original definition by Thorstenson and Plummer (1977). The concept of "constrained equilibria" describes saturation states for multicomponent systems in which sets of two or more components are constrained to react, for a limited period of time, as a single component. Glynn and Reardon (1992) suggested that the term "stoichiometric saturation" be reserved to describe solid solutions behaving as fixed composition solids, irrespective of the number of components. Experimental evidence is still needed to document the possible existence of multicomponent SSAS systems under "constrained equilibria."

Models describing the excess free energy of mixing of a multicomponent solid solution become more complex and require a greater number of fitting parameters as the number of components is increased. For example, whereas an asymmetric binary solid solution requires at least 2 parameters (the a_o and a_1 parameters mentioned in this paper), a ternary solid solution with asymmetric binary constituent solid solutions will require 7 parameters, and a quaternary solid solution with asymmetric binary constituent solid solutions will require 16 fitting parameters. Multicomponent systems with symmetric binary constituent solid solutions may require significantly fewer fitting parameters (1 for a binary system, 3 for a ternary system, 6 for a quaternary system), but symmetric binary solid solutions are not very common except in systems in which the difference in the volume of the susbstituting ions is small relative to that of the crystallographic structure. Mukhopadhyay et al. (1993) provided a comprehensive review of Margules-type formulations for multicomponent solutions. For illustrative purposes, their equations describing the component activity coefficients and the excess-free-energy of mixing of a solid solution with 3 components (labeled 1,2,3) are given below:

$$G^E = \chi_1\chi_2(\chi_2 W_{12} + \chi_1 W_{21}) + \chi_1\chi_3(\chi_3 W_{13} + \chi_1 W_{31}) + \chi_2\chi_3(\chi_3 W_{23} + \chi_2 W_{32})$$
$$+ \chi_1\chi_2\chi_3\left[\frac{(W_{12} + W_{21} + W_{13} + W_{31} + W_{23} + W_{32})}{2} - W_{123}\right] \tag{29}$$

$$RT\ln\gamma_1 = 2(\chi_1\chi_2 W_{21} + \chi_1\chi_3 W_{31}) + \chi_2^2 W_{12} + \chi_3^2 W_{13}$$
$$+ \chi_2\chi_3\left[\frac{W_{12} + W_{21} + W_{13} + W_{31} + W_{23} + W_{32}}{2} - W_{123}\right] - 2G^E \tag{30}$$

where the W_{ij} are the various binary interaction parameters, W_{123} is the ternary interaction parameter, and the χ_i are the mole fractions of each component i. The equations for the activity coefficients of components 2 and 3 can be derived by cyclically rotating the subscripts in Equation (30). Equation (29) reduces to Equation (10), which is the Waldbaum and Thomson subregular model, in the simpler case of a binary solid-solution series.

It is hoped that this brief and incomplete introduction to the thermodynamics of ternary and multicomponent solid solutions will encourage future research in this area, particularly for low temperature SSAS systems. Sulfate minerals should make excellent candidates for such investigations.

The sophisticated models used to describe the thermodynamics of solid solutions at high temperatures (e.g. Ghiorso 1994, 1990; Sack and Ghiorso 1994), typically for solid-melt equilibria, have generally not been applied to low-temperature systems. In addition to descriptions of multicomponent mixing, these models commonly account for multi-site mixing and variable types (short range and long range) and degrees of order-disorder. They occasionally use mixing models distinctly different from the Margules approach. Much could be gained by trying to apply some of the thermodynamic concepts used in these models to low temperature systems. Nevertheless, their application to experimental data obtained under low temperature conditions may not always be warranted. Solid solutions synthesized at low temperatures generally do not show as much long-range order as solids synthesized at higher temperatures. In general, slow recrystallization kinetics commonly hinder the attainment of thermodynamic equilibrium states under low-temperature conditions. As a result, the data and the thermodynamic properties determined for the investigated systems are typically subject to significant uncertainties. Consequently, the complexity of the models used to describe the behavior of SSAS systems at low temperatures remains limited.

SUMMARY AND CONCLUSIONS

The thermodynamic theory of solid-solution aqueous-solution interactions was reviewed, particularly as it applies to low-temperature systems (between 0 and 100°C). Excess-free-energy data for binary solid-solution series were presented for sulfate, carbonate, and halide minerals. The data presented were calculated using a regular, and in a few cases a subregular, solid-solution model. More specifically, the conclusions of this paper are the following:

(1) Lippmann phase diagrams can be used to describe and compare thermodynamic equilibrium (Eqns. 3, 4), primary saturation (Eqns. 13, 14), stoichiometric saturation (Eqn. 17) and pure end-member saturation states (Eqns. 19, 20) in binary solid-solution aqueous-solution (SSAS) systems.

(2) Dissolution of a solid solution to stoichometric saturation may result in either supersaturation or undersaturation with respect to the pure end-member phases. Commonly, solid-solution formation will have only a minor to insignificant effect on the saturation indices of the major component, as long as the mole fraction of trace component in the solid solution is less than about 1% (Eqn. 26). For large differences in end-member solubility products, the limiting mole fraction may be lower. Trace-component saturation index (SI) values (Eqn. 27) generally show a greater departure from pure-phase saturation (compared with major-component SI values). If the trace component is much more insoluble than the major component, a high apparent degree of supersaturation with respect to the end-member may result. In the case of the $(Ca,Cd)CO_3$ system, supersaturation is predicted with respect to the $CdCO_3$ end member if the mole fraction of $CdCO_3$ in the solid solution is higher than about $10^{-2.75}$.

(3) Aqueous solutions at primary saturation or at thermodynamic equilibrium with respect to a solid solution will always be undersaturated with respect to the pure end-member solids. SSAS systems with more than one order of magnitude difference between end-member solubility products, and with a zero or positive solid-solution excess free energy of mixing, will be only slightly undersaturated with respect to the less soluble end-member. The $(Ba,Sr)SO_4$, $(Ba,Pb)SO_4$, and $(Na,K)Fe_3(SO_4)_2(OH)_6$ solid solutions are examples of SSAS systems that are primarily controlled by the large differences between the solubility products of the end-members.

(4) The $(Na,H_3O)Fe_3(SO_4)_2(OH)_6$ and $(Fe,Mn)SO_4 \cdot 7H_2O$ series are examples of solid-solution series that have relatively small differences between the solubility products of their end members. In such systems, the excess free energies of the solid solutions will strongly affect both the distribution of components between the solid and aqueous phases and the solubility of the solid solutions.

(5) An aqueous solution at equilibrium with an ideal solid solution will be undersaturated, with a $\Sigma\Pi$ value up to two times lower, relative to an aqueous solution at equilibrium with a mechanical mixture of the pure end-member phases. The maximum factor of 2

difference in $\Sigma\Pi$ value will occur in a system with identical end-member solubility products. The relative degree of undersaturation will decrease with a greater difference in end-member solubility products.

(6) Techniques are presented for the estimation of solid-solution excess free energies. These techniques can be used, as an alternative to an ideal mixing model, in systems for which no experimental data are available for stoichiometric saturation or thermodynamic equilibrium states. Laboratory determinations and interpretations of the thermodynamic properties of solid solutions are often difficult, however, and uncertainties abound even in the case of often investigated solid-solution series such as $(Ba,Sr)SO_4$.

(7) The thermodynamics of the melanterite ($FeSO_4 \cdot 7H_2O$)–mallardite ($MnSO_4 \cdot 7H_2O$) SSAS system are calculated at 0°C, based on equilibrium experiments conducted by White (1933) and on aqueous speciation computations with the PHRQPITZ code (Plummer et al. 1989). A regular solid-solution model was assumed and resulted in a calculated a_0 value of 0.90 for the solid-solution series. End-member solubility products of $10^{-2.441}$ and $10^{-1.910}$ were calculated for melanterite and mallardite, respectively, at 0°C.

(8) Despite the lack of thermodynamic data available for low-temperature SSAS systems, and despite the commonly large uncertainties associated with the available data, the equations and techniques presented in this paper can be used to estimate the importance of solid-solution aqueous-solution interactions on the chemical evolution of natural waters. Future research should improve the quantity and quality of thermodynamic data and thermodynamic models available for the description and prediction of SSAS interactions.

(9) Although binary solid-solution series are the primary focus of this paper, many of the thermodynamic principles presented are also applicable to ternary and other multicomponent solid-solution series. Nevertheless, much more research is needed to understand the thermodynamic properties and behavior of multicomponent solid solutions, particularly in aqueous solutions at low temperatures. As just one example, the applicability of the concepts of stoichiometric saturation and constrained equilibria to multicomponent SSAS systems should be investigated and, if possible, experimentally demonstrated.

(10) Sulfate minerals have large crystal structures that allow significant amounts of chemical substitution. Because of their generally high solubilities that may allow easier attainment of thermodynamic equilibrium states, sulfate minerals provide excellent opportunities for the study of the thermodynamics of both binary and multicomponent SSAS systems.

ACKNOWLEDGMENTS

Many thanks are due to John Jambor, Kirk Nordstrom, and Charlie Alpers, for providing a preprint of their chapter on metal sulfate salts, which is included in this volume. Reviews by Charlie Alpers, Ed Busenberg, John Jambor, and two anonymous reviewers resulted in significant improvements to this manuscript. I greatly appreciate their help. Thanks also go to Paul Ribbe for his wonderful editing skills and for his remarkable patience.

REFERENCES

Baron D (1996) Iron-chromate precipitates in Cr(VI)-contaminated soils: Identification, solubility, and solid-solution/aqueous-solution reactions. PhD Dissertation, Oregon Graduate Inst of Sci and Tech, Portland, Oregon, 129 p

Baron D, Draucker S, Palmer CD (2000) Solid solutions between jarosite and its chromate analog: Synthesis, characterization, and solubility. Geol Soc Am Ann Mtg Abstr with Program, p A-109

Boström K, Frazer J, Blankenburg J (1967) Subsolidus phase relations and lattice constants in the system BaSO₄-SrSO₄-PbSO₄. Arkiv Mineralogi Geologi 27(4):477-485

Brophy GP, Sheridan MF (1965) Sulfate studies IV: The jarosite-natrojarosite-hydronium jarosite solid solution series. Am Mineral 50:1595-1607

Brophy GP, Scott ES, Snellgrove RA (1962) Sulfate studies II: Solid solution between alunite and jarosite. Am Mineral 47:112-126

Brower E, Renault J (1971) Solubility and enthalpy of the barium–strontium sulfate solid solution series. New Mexico State Bureau Mines Mineral Res Circ 116

Busenberg E, Plummer LN (1983) Chemical and X-ray analyses of 175 calcites and marbles from the collection of the National Museum of Natural History, Smithsonian Institution. US Geol Survey Open-File Rept 83-863

Busenberg E, Plummer LN (1989) Thermodynamics of magnesian calcite solid-solutions at 25°C and 1 atm total pressure. Geochim Cosmochim Acta 53:1189-1208

Crocket JH, Winchester JW (1966) Vapour–liquid equilibria of non-ideal solutions. Geochim Cosmochim Acta 30:1093-1109

Davis JA, Fuller CC, Cook AD (1987) A model for trace metal sorption processes at the calcite surface: Adsorption of Cd^{2+} and subsequent solid solution formation. Geochim Cosmochim Acta 51:1477-1490

Denis J, Michard G (1983) Dissolution d'une solution solide: Etude theorique et experimentale. Bull Minéral 106:309-319

Felmy AR, Rai D, Moore DA (1993) The solubility of (Ba,Sr)SO₄ precipitates: Thermodynamic equilibrium and reaction path analysis. Geochim Cosmochim Acta 57:4345-4363

Frye K (1974) Modern Mineralogy. Prentice-Hall, Englewood, New Jersey

Galinier C, Dandurand JL, Souissi F, Schott J (1989) Sur le caractère non-idéal des solutions solides (Ba,Sr)SO₄: Mise en évidence et détermination des paramètres thermodynamiques par des essais de dissolution at 25°C. CR Acad Sci, Ser 2, Mécanique, Physique, Chimie, Sciences de l'Univers, Sciences de la Terre 308:1363-1368

Gamsjäger H, Königsberger E, Preis W (2000) Lippmann diagrams: Theory and application to carbonate systems. Aquatic Geochem 6:119-132

Ganguly J, Saxena SK (1987) Mixtures and Mineral Reactions. Springer-Verlag, Berlin

Garrels RM, Wollast R (1978) Discussion of "Equilibrium criteria for two-component solids reacting with fixed composition in an aqueous phase. Example: The magnesian calcites". Am J Sci 278:1469-1474

Ghiorso MS (1990) Thermodynamic properties of hematite-ilmenite-geikelite solid solutions. Contrib Mineral Petrol 104:645-667

Ghiorso MS (1994) Algorithms for the estimation of phase stability in heterogeneous thermodynamic systems. Geochim Cosmochim Acta 58:5489-5501

Glynn PD (1990) Modeling solid-solution reactions in low temperature aqueous systems. *In* Chemical Modeling of Aqueous Systems II. Melchior DC, Bassett RL (Eds) Am Chem Soc Symp Ser 416:74-86

Glynn PD (1991) MBSSAS: A code for the computation of Margules parameters and equilibrium relations in binary solid-solution aqueous-solution systems. Comp Geosci 17:907-966

Glynn PD (1992) Effect of impurities in gypsum on contaminant transport at Pinal Creek, Arizona. *In* US Geological Survey Toxic Substances Hydrology Program. Proc Technical Mtg, Monterey California, March 11-15, 1991. Mallard GE, Aronson DA (Eds) US Geol Survey Water-Resources Invest Rpt 91-4034:466-474

Glynn PD, Parkhurst DL (1992) Modeling non-ideal solid-solution aqueous-solution reactions in mass-transfer computer codes. *In* Water-Rock Interaction 7. Kharaka YK and Maest AS (Eds) Balkema Press, Rotterdam 175-179

Glynn PD, Reardon EJ (1990) Solid-solution aqueous-solution equilibria: thermodynamic theory and representation. Am J Sci 278:164-201

Glynn PD, Reardon EJ (1992) Reply to a comment by Königsberger E. and Gämsjager H. on "Solid-solution/aqueous-solution equilibria: thermodynamic theory and representation". Am J Sci 292:215-225

Glynn PD, Plummer LN, Busenberg E (1988) Thermodynamics of (Ba,Sr)CO₃ and (Sr,Ca)CO₃ solid-solutions: Results from low-temperature aqueous coprecipitation and dissolution experiments. Terra Cognita 8:178

Glynn PD, Reardon EJ, Plummer LN, Busenberg E (1990) Reaction paths and equilibrium end-points in

510   Glynn

solid-solution aqueous solution systems. Geochim Cosmochim Acta 54:267-282

Glynn PD, Reardon EJ, Plummer LN, Busenberg E (1992) Reply to a comment by R.K. Stossell on "Reaction paths and equilibrium end-points in solid-solution aqueous solution systems." Geochim Cosmochim Acta 56:2559-2572

Guggenheim EA (1937) Theoretical basis of Raoult's law. Trans Faraday Soc 33:151-159

Härtig C, Brand P, Bohmhammel K (1984) Fe-Al Isomorphie und Strukturwasser in Kristallen von Jarosit-Alunit-Typ. Z Anorg Allg Chem 508:159-164

Hildebrand JH (1936) Solubility of Non-Electrolytes. Reinhold Publ Co, New York

Jones BF (1962) Stability of burkeite and its significance in lacustrine evaporites. J Geophys Res 67:3569

Kirgintsev AN, Trushnikova LN (1966) Thermodynamics of MCl-MBr solid solutions. Russ J Inorg Chem. 11:1250-55

Königsberger E, Gamsjäger H (1990) Solid-solute equilibria in aqueous solution: III. A new application of an old chemical potentiometer. Marine Chem 30:317-327

Königsberger E, Gamsjäger H (1991) Graphical representation of solid-solute phase equilibria in aqueous solution. Ber Bunsenges Phys Chem 95:785-790

Königsberger E, Gamsjäger H (1992) Comment on "Solid-solution aqueous-solution equilibria: Thermodynamic theory and representation" by Glynn PD and Reardon EJ. Am J Sci 292:199-214

Königsberger E, Hausner R, Gamsjäger H (1991) Solid-solute equilibria in aqueous solution: V. The system CdCO$_3$-CaCO$_3$-CO$_2$-H$_2$O. Geochim Cosmochim Acta 55:3505-3514

Lippmann F (1977) The solubility product of complex minerals, mixed crystals and three-layer clay minerals. N Jahrb Mineral Abh 130:243-263

Lippmann F (1980) Phase diagrams depicting the aqueous solubility of binary mineral systems. N Jahrb Mineral Abh 139:1-25

Lippmann F (1982) Stable and metastable solubility diagrams for the system CaCO$_3$-MgCO$_3$-H$_2$O. Bull Minéral 105:273-279

Malinin SD, Urusov VS (1983) The experimental and theoretical data on isomorphism in the (Ba,Sr)SO$_4$ system in relation to barite formation. Geokhimiya 9:1324-1334

Margules M (1895) Über die Zussamensetzung der gesättigten Dämpfe von Mischungen[:1]. Akad Wiss Wien, Sitzungs, Math-Naturwiss Klasse 104(IIa):1243-1278

Mucci A (1987) Influence of temperature on the composition of magnesian calcite overgrowth precipitated from seawater. Geochim Cosmochim Acta 51:1977-1984

Mukhopadhyay B, Sabyasachi B, Holdaway MJ (1993) A discussion of Margules-type formulations for multicomponent solutions with a generalized approach. Geochim Cosmochim Acta 57:277-283

Palache C, Berman H, Frondel C (1951) The System of Mineralogy. Wiley & Sons, New York

Parkhurst DL, Appelo CAJ (2000) User's guide to PHREEQC (Version 2)—A computer program for speciation, batch-reaction, one-dimensional transport, and inverse geochemical calculations. US Geol Survey Water-Resources Invest Rept 99-4259.

Plummer LN, Busenberg E (1982) The solubilities of calcite, aragonite and vaterite in CO$_2$-H$_2$O solutions between 0 and 90°C, and an evaluation of the aqueous model for the system CaCo$_3$-CO$_2$-H$_2$O. Geochim Cosmochim Acta 46:1011-1040

Plummer LN, Busenberg E (1987) Thermodynamics of aragonite-strontianite solid solutions: Results from stoichiometric solubility at 25 and 76°C. Geochim Cosmochim Acta 51:1393-1411

Plummer LN, Parkhurst DL, Fleming GW, Dunkle SA (1989) A computer program incorporating Pitzer's equations for calculation of geochemical reactions in brines. U S Geol Survey Water-Resources Invest Rept 88-4153

Prieto M, Fernandez-Gonzalez A, Becker U, Putnis A (2000) Computing Lippmann diagrams from direct calculation of mixing properties of solid solutions: Application to the barite-celestite system. Aquatic Geochem 6:133-146

Prigogine I, Defay R (1954) Chemical Thermodynamics. Longmans, Green & Co, London

Redlich O, Kister AT (1948) Algebraic representation of thermodynamic properties and the classification of solutions. Ind Eng Chem 40:345-48

Sack RO, Ghiorso MS (1994) Thermodynamics of multicomponent pyroxenes: Formulation of a general model. Contrib Mineral Petrol 116:277-286

Samoilov OYa (1965) Structure of aqueous electrolyte solutions and the hydration of ions. (Monograph originally in Russian, 1957), Translation by Consultants Bureau, New York, 1965

Saxena SK (1973) Thermodynamics of Rock-Forming Crystalline Solutions. Springer-Verlag, Berlin

Schroeder WC, Berk AA, Gabriel A (1936) Solubility equilibria of sodium sulfate at temperatures from 150 to 350°: II. Effect of sodium hydroxide and sodium carbonate. J Am Chem Soc 58:843-849

Smith RM, Martell AE (1976) Critical Stability Constants, Vol 4: Inorganic Complexes, Plenum Press, New York

Stoffregen RE (1993) Stability relations of jarosite and natrojarosite. Geochim Cosmochim Acta 57:

2417-2419

Stoffregen RE, Cygan G (1990) An experimental study of Na-K exchange between alunite and aqueous sulfate solutions. Am Mineral 75:209-220

Swalin RA (1972) Thermodynamics of Solids. Wiley & Sons, New York

Takano B, Yanagiswa M, Watanuki K (1969) Structure gap in $BaSO_4$–$PbSO_4$ solid solution series. Mineral J 6:159-171

Thompson JB Jr (1967) Thermodynamic properties of simple solutions. *In* Researches in Geochemistry. Abelson PH (Ed) Wiley & Sons, New York, p 340-361

Thompson JB Jr, Waldbaum D (1969) Analysis of the two-phase region halite-sylvite in the system NaCl–KCl. Geochim Cosmochim Acta 33:671-690

Thorstenson DC, Plummer L N (1977) Equilibrium criteria for two-component solids reacting with fixed composition in aqueous-phase. Am J Sci 277:1203-1223

Urusov VS (1974) Energetic criteria for solid solution miscibility gap calculations. Bull Soc fr Minéral Cristallogr 97:217-222

Vaslow F, Boyd GEJ (1952) Thermodynamics of coprecipitation: Dilute solid solutions of AgBr in AgCl. Am Chem Soc 74:4691-4695

White AM (1933) The system ferrous sulfate-manganous sulfate-water at 0 and 25°C. J Am Chem Soc 55:3182-3185

11 Predicting Sulfate-Mineral Solubility in Concentrated Waters

Carol Ptacek

National Water Research Institute, Environment Canada
Burlington, Ontario, Canada L7R 4A6
Department of Earth Sciences, University of Waterloo
Waterloo, Ontario, Canada N2L 3G1

David Blowes

Department of Earth Sciences
University of Waterloo
Waterloo, Ontario, Canada N2L 3G1

INTRODUCTION

Sulfate minerals participate in a variety of mineral-water reactions with waters containing elevated concentrations of dissolved solids. Reliable prediction of sulfate-mineral solubility in concentrated waters is required for a multitude of geological, hydrogeological, industrial, and meteorological applications. Precipitation of pure sulfate minerals or sulfate-bearing solid solutions can limit concentrations of dissolved metals, radionuclides, and other cations at waste-disposal sites, in marine and other naturally occurring saline waters, in industrial waters, and in smog. Formation of these solids can control the major-ion geochemistry of water and, in turn, influence the behavior of trace constituents.

One of the most widely used and robust approaches for predicting mineral solubility in concentrated waters is the Pitzer ion-interaction approach as first introduced by Pitzer (1973), as applied to geological systems by Harvie and Weare (1980) and Harvie et al. (1980), and as summarized by Weare (1987). Clegg and Whitfield (1991) and Pitzer (1979a,b; 1991) provided further comprehensive discussions on the developments and application of the ion-interaction approach to predict mineral solubility in natural waters. This chapter focuses on the application of the ion-interaction approach, or its variations, for predicting sulfate-mineral solubility in a variety of geochemical settings. Included are a summary of the Pitzer approach as applied to predictions of mineral solubility, a summary of the available constants for modeling sulfate-mineral solubility, and examples of applications for natural and contaminated sites. The discussion focuses on recent additions to earlier models developed for application to near-Earth-surface temperature and pressure conditions. Applications outside this temperature and pressure range are described only briefly.

BACKGROUND

Compositions of concentrated waters

The majority of concentrated waters in nature form as a result of solar evaporation of marine and continentally-derived waters. The evaporation of seawater leads to predictable sequences of mineral formation, including precipitation of several common sulfate minerals (e.g. gypsum, mirabilite, thenardite, epsomite; Table 1; see chapter by Spencer, this volume). Evaporation of continentally-derived waters leads to waters of more variable composition (e.g. Jones et al. 1977; Monnin and Schott 1984; Domagalski et al. 1990). Entrapment of evaporative brines is thought to be the principal mechanism leading to the formation of subsurface brines (Landes 1960). Saline waters can form through other

1529-6466/00/0040-0011$05.00

Table 1. Formulae for selected sulfate minerals and compounds discussed in the text.

Phase	Formula	Phase	Formula
alunogen	$Al_2(SO_4)_3 \cdot 17H_2O$	kieserite	$MgSO_4 \cdot H_2O$
anglesite	$PbSO_4$	melanterite	$FeSO_4 \cdot 7H_2O$
anhydrite	$CaSO_4$	mirabilite	$Na_2SO_4 \cdot 10H_2O$
arcanite	K_2SO_4	morenosite	$NiSO_4 \cdot 7H_2O$
barite	$BaSO_4$	nickelhexahydrite	$\beta\text{-}NiSO_4 \cdot 6H_2O$
blödite	$Na_2Mg(SO_4)_2 \cdot 4H_2O$	pickeringite	$MgAl_2(SO_4)_4 \cdot 22H_2O$
celestine	$SrSO_4$	picromerite	$KMg(SO_4)_2 \cdot 4H_2O$
chalcanthite	$CuSO_4 \cdot 5H_2O$	polyhalite	$Ca_2K_2Mg(SO_4)_4 \cdot 2H_2O$
cyanochroite	$K_2SO_4 \cdot CuSO_4 \cdot 6H_2O$	potassium alum	$KAl(SO_4)_2 \cdot 12H_2O$
epsomite	$MgSO_4 \cdot 7H_2O$	portlandite	$Ca(OH)_2$
glauberite	$Na_2Ca(SO_4)_2$	retgersite	$\alpha\text{-}NiSO_4 \cdot 6H_2O$
gypsum	$CaSO_4 \cdot 2H_2O$	rubidium sulfate	$RbSO_4$
halotrichite	$FeAl_2(SO_4)_4 \cdot 22H_2O$	szomolnokite	$FeSO_4 \cdot H_2O$
hexahydrite	$MgSO_4 \cdot 6H_2O$	thenardite	Na_2SO_4
jarosite	$KFe_3(SO_4)_2(OH)_6$	tschermigite	$(NH_4)Al(SO_4)_2 \cdot 12H_2O$

mechanisms, including the dissolution of evaporite deposits, mineral alteration, and hydrothermal and magmatic processes. Different combinations of these processes can lead to brines of differing composition (e.g. Gavrieli et al. 1995). The quantitative prediction of sulfate-mineral solubility in natural brines has been a focus of research for many decades.

Among other concentrated waters are industrial waters, such as those used in metallurgical processes, and wastewaters, such as landfill leachates and waters associated with mining activities. In these waters, the formation and dissolution of sulfate minerals have the potential to influence the chemical composition of the water and its surrounding solids. Prediction of sulfate-mineral solubility in industrial waters has received widespread attention. Numerous studies have focused on quantification of sulfate-mineral solubility in industrial settings over a wide range in temperature and pressure. Quantification of sulfate-mineral solubility in wastewaters, mine waters, mine-waste leachate, and in other environmental settings has received less attention. For example, concentrated mine waters commonly develop in mine workings and in mine-waste and refinery-waste containment facilities. Very high concentrations of dissolved constituents form through a combination of reactions involving mineral alteration and evaporation. Oxidation of sulfide minerals, especially the Fe-bearing sulfides pyrite (FeS_2) and pyrrhotite ($Fe_{1-x}S$) leads to elevated concentrations of dissolved Fe(II), Fe(III), SO_4, and H^+ in mine workings and subsurface pore waters. Large volumes of water containing high concentrations of metals, sulfate, and acid, in excess of 900,000 mg/L total dissolved solids, have been reported (Nordstrom and Alpers 1999). Management of such mine sites requires prediction of the solubility of sulfate minerals. Development of a comprehensive geochemical model for the extreme concentrations observed at many mine and metallurgical sites is a relatively new area of study.

Modeling approaches

Ion-association approach. The first comprehensive ion-association model for application to natural waters was developed in the early 1960s (e.g. Garrels and Thompson 1962). Models based on the ion-association approach have been widely applied to predict sulfate-mineral solubility in a variety of settings. The original models were generally limited to applications at 25°C and waters of low ionic strength (typically <0.1 m). Enhancement of the early versions of the ion-association model to extend its applicability to broader salinity and temperature ranges has been described thoroughly (Parkhurst et al. 1985; Ball and Nordstrom 1987; Allison et al. 1990; Millero and Hawke 1992). For example, the addition of various extensions of the Debye–Hückel equation to cover a wider concentration range, and inclusion of a large number of ion-association constants have provided reliable calculations to ionic strengths of up to 1 m and a broader temperature range (e.g. Ball and Nordstrom 1987; Parkhurst 1990; Nordstrom et al. 1990; Liu and Millero 1999). Robust models based on the ion-association approach have been developed for specific applications, such as speciation calculations in seawater (see Millero and Hawke 1992). Ion-association models developed for natural systems incorporate experimental data for a large number of ion pairs, activity correction equations for single ions and ion pairs over the temperature and concentration range of interest, and mineral-solubility data as a function of temperature (Parkhurst et al. 1985; Ball and Nordstrom 1987; Allison et al. 1990). Models based on the ion-pair approach have not been developed extensively for solutions containing complex mixtures of electrolytes to very high concentrations.

Pitzer ion-interaction approach. The most widely accepted approach for prediction of ion activities and mineral solubilities in complex concentrated electrolyte mixtures is based on the Pitzer ion-interaction formalism (Pitzer 1973, 1974, 1975; Pitzer and Mayorga 1973; Pitzer and Kim 1974). Models based on the Pitzer approach have been developed for application to a variety of geochemical systems, from acidic to basic, dilute to concentrated, over a range in temperature, total pressure, and partial pressures of component gases (Pitzer 1982, 1991; Clegg and Whitfield 1991). The earliest geochemical applications of Pitzer-based models involved mostly calculations of ion-activity and mineral-solubility relations in waters containing seawater-type components. Whitfield (1975) applied Pitzer's equations to calculate ion activities in seawater. Harvie et al. (1980), Harvie and Weare (1980), and Harvie et al. (1982, 1984) applied Pitzer's equations to calculate ion activities and mineral precipitation sequences in evaporating seawater.

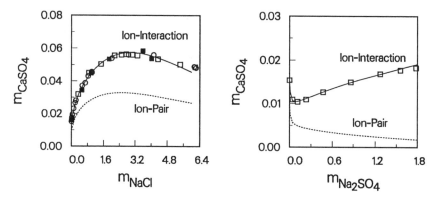

Figure 1. Comparison of results from the ion-pair model (PHREEQE) and the ion-interaction model (PHRQPITZ) for predicting gypsum solubility in NaCl and Na₂SO₄ solutions. Symbols represent experimental data.

A comparison of two widely used computer codes, one based on the ion-pair approach (PHREEQE, Parkhurst et al. 1985), and the other the ion-interaction approach (PHRQPITZ, Plummer et al. 1988), indicates better agreement between experimental solubility data and the ion-interaction model than between the experimental data and the ion-pair model (Fig. 1). Although enhancements to the ion-pair model by modifying ion-association constants and incorporating extended versions of the Deby–Hückel equation have yielded improved agreement between experimental data and ion-pair model predictions at higher concentrations (e.g. Nordstrom and Ball 1984; Parkhurst 1990), the majority of model calculations for concentrated solutions of complex composition typically rely on the Pitzer formulism for calculating ion activities.

Models based on the ion-interaction approach have been developed extensively for a variety of industrial and geochemical applications. A number of excellent reviews on the original development of the Pitzer ion-interaction approach, and database compilations dedicated to predicting the geochemical behavior of the main components present in seawater, are available (e.g. Weare 1987; Plummer et al. 1988; Clegg and Whitfield 1991; Pabalan and Pitzer 1991). For predictions of the solubility of minerals in non-seawater systems, and for trace components in seawater-type waters, the application of Pitzer-based models requires special consideration to determine whether the geochemical database incorporated in the model (interaction parameters, mineral-solubility constants, etc.) accurately encompasses the composition of the water to be modeled. This review summarizes recent additions to the earlier work on developing Pitzer-based models, with a particular emphasis on recently evaluated ion-interaction parameters and the application of ion-interaction models to field settings.

Combined or "hybrid" model. The Pitzer ion-interaction equations, in their original form, assume complete dissociation of all ions, and represent the interaction between ions solely with ion-interaction equations with parameters derived from experimental data. It became apparent early in the development of models based on the ion-interaction approach that the assumption of complete dissociation provided an inadequate description of the chemical behavior of some electrolyte solutions over a broad range of concentration, temperature, and pressure. Introduction of association constants for strongly associated species provided a large improvement to the original 'strict' Pitzer approach (e.g. Millero and Schreiber 1982; Harvie et al. 1984; Reardon and Beckie 1987; Rard and Clegg 1997). Many applications require the introduction of ion pairs or association constants to provide reliable predictions of mineral solubility in concentrated solutions. Models which include both association constants and ion-interaction equations have been dubbed "combined" or "hybrid" models, or modified Pitzer models.

An alternative to including ion-association and stability constants directly into an ion-interaction model is the approach taken by Millero and coworkers. Here, stability constants are corrected using ionic strength-dependent equations based on the ion-interaction formulism (e.g. Millero and Hawke 1992). The approach described by Millero and Hawke (1992) is particularly well-suited to the description of trace constituents in waters covering a range of concentration but with a similar matrix.

MODEL DESCRIPTION

Summary of model formulation

Development of specific models based on the Pitzer formulism to predict mineral solubility in natural waters requires ion-interaction parameters and mineral free-energy data. Binary and ternary interaction coefficients, obtained from solution properties for all binary and ternary combinations of ions in the water from which the solid is precipitating or dissolving, are required. Higher order interaction coefficients are generally not required. The interactions between binary and ternary combinations of ions are estimated through statistical regression of experimental data. For example, the equilibrium reaction for dissolution of a simple divalent metal sulfate:

$$MSO_4 \Leftrightarrow M^{2+} + SO_4^{2-} \tag{1}$$

is described by the mass-action expression:

$$K_{sp} = a_{M^{2+}} a_{SO_4^{2-}} = m_{M^{2+}} m_{SO_4^{2-}} \gamma_{M^{2+}} \gamma_{SO_4^{2-}} \tag{2}$$

where a represents the activity, m the molality (mol/kg solvent), and γ the single-ion activity coefficient. To calculate concentrations of dissolved species m in equilibrium with a phase (or solubility), estimates of γ are required. Values of γ depend strongly on the solution composition and concentration, and are defined for cations and anions as (Harvie and Weare 1980; Plummer et al. 1988):

$$\ln \gamma_M = z_M^2 F + \sum_a m_a (2B_{Ma} + ZC_{Ma}) + \sum_c m_c (2\Phi_{Mc} + \sum_a m_a \psi_{Mca})$$
$$+ \sum_{a<a'} \sum m_a m_{a'} \psi_{aa'M} + |z_M| \sum_c \sum_a m_c m_a C_{ca} \tag{3}$$

and

$$\ln \gamma_X = z_X^2 F + \sum_c m_c (2B_{cX} + ZC_{cX}) + \sum_a m_a (2\Phi_{Xa} + \sum_c m_c \psi_{Xac})$$
$$+ \sum_{c<c'} \sum m_c m_{c'} \psi_{cc'X} + |z_X| \sum_c \sum_a m_c m_a C_{ca} \tag{4}$$

where the subscripts M, c, and c' represent cations, and X, a, and a' represent anions. The summation index c represents the sum over all cations in the system, and the double summation $c < c'$ represents the sum over all distinguishable pairs of dissimilar cations. Similar definitions for a and $a < a'$ apply for anions in the system. The quantity F includes the Debye–Hückel term and other terms as follows:

$$F = -A^i \left[\frac{\sqrt{I}}{1+b\sqrt{I}} + \frac{2}{b}\ln(1+b\sqrt{I}) \right] + \sum_c \sum_a m_c m_a B'_{ca} + \sum_{c<c'} \sum m_c m_{c'} \varphi'_{cc'} + \sum_{a<a'} \sum m_a m_{a'} \varphi'_{aa'} \tag{5}$$

where $b = 1.2$ at 25°C, I is the ionic strength, and A^ϕ is defined as:

$$A^\phi = \frac{1}{3}(2\pi N_0 \rho_w / 1000)^{1/2} (e^2 / DkT)^{3/2} \tag{6}$$

In Equation (6), N_0 is Avogadro's number, ρ_w is the density of water, e is the electronic charge, k is Boltzmann's constant, and D is the dialectric constant of water.

For salts containing monovalent ions, the terms B^ϕ_{MX}, B_{MX}, and B'_{MX} are defined as:

$$B^\phi_{MX} = \beta^{(0)}_{MX} + \beta^{(1)}_{MX} e^{-\alpha\sqrt{I}} \tag{7}$$

$$B_{MX} = \beta^{(0)}_{MX} + \beta^{(1)}_{MX} g(\alpha\sqrt{I}) \tag{8}$$

$$B'_{MX} = \beta^{(1)}_{MX} g'(\alpha\sqrt{I}) / I \tag{9}$$

where $\alpha = 2$. The functions g and g' are:

$$g(x) = 2\left[1 - (1+x)e^{-x}\right] / x^2 \tag{10}$$

$$g'(x) = -2\left[1 - (1+x+\frac{1}{2}x^2)e^{-x}\right] / x^2 \tag{11}$$

where $x = \alpha\sqrt{I}$.

For solutions containing higher valence salts, an extra term, $\beta^{(2)}$, is generally included in the virial expansion. In this case, the terms B^ϕ_{MX}, B_{MX}, and B'_{MX} are:

$$B^\phi_{MX} = \beta^{(0)}_{MX} + \beta^{(1)}_{MX} e^{-\alpha_1\sqrt{I}} + \beta^{(2)}_{MX} e^{-\alpha_2\sqrt{I}} \tag{12}$$

$$B_{MX} = \beta^{(0)}_{MX} + \beta^{(1)}_{MX} g(\alpha_1\sqrt{I}) + \beta^{(2)}_{MX} g(\alpha_2\sqrt{I}) \tag{13}$$

$$B'_{MX} = \beta^{(1)}_{MX} g(\alpha_1 \sqrt{I})/I + \beta^{(2)}_X g(\alpha_2 \sqrt{I})/I \qquad (14)$$

For 2-2 salts, $\alpha_1 = 1.4$ and $\alpha_2 = 12.0$, and for 3-2 and 4-2 salts, $\alpha_1 = 2.0$ and $\alpha_2 = 50.0$.

The variable CMX is also required to define the solution chemistry for single-salt systems and is given by:

$$C_{MX} = C^\varphi_{MX} / \left(2\sqrt{|z_M z_X|} \right) \qquad (15)$$

The coefficient Z in Equations (3) and (4) is:

$$Z = \sum_i m_i |z_i| \qquad (16)$$

The parameters $\beta^{(0)}$, $\beta^{(1)}$, $\beta^{(2)}$, and C^φ that define the variables B and C in Equations (12)-(15) are determined from statistical fits of experimental data, usually osmotic coefficient data, obtained for binary solutions.

For solutions containing more complex mixtures of electrolytes, the additional parameters required are Φ, to describe cation–cation and anion–anion interactions, and Ψ, to describe cation–cation–anion, and anion–anion–cation interactions. Values of Φ are calculated using:

$$\Phi^\varphi_{ij} = \theta_{ij} + {}^E\theta_{ij}(I) + I{}^E\theta'_{ij}(I) \qquad (17)$$

$$\Phi_{ij} = \theta_{ij} + {}^E\theta_{ij}(I) \qquad (18)$$

$$\Phi'_{ij} = {}^E\theta'_{ij}(I) \qquad (19)$$

where θ_{ij} is the only adjustable parameter and is determined empirically from experimental data collected for solutions of more complex composition, and ${}^E\theta_{ij}(I)$ and ${}^E\theta'_{ij}(I)$ are defined in Pitzer (1975) and Harvie and Weare (1980). More recent modifications of the original Pitzer's molality-based equations have been developed to improve predictions in highly concentrated solutions (e.g. Rard and Clegg 1997).

Values of Φ and Ψ required to calculate mixing in ternary systems are generally obtained from the same two-salt mixtures so that the parameters are internally consistent. Binary and ternary interaction parameters are usually derived from isopiestic data at concentrations above about $0.1\ m$, and at lower concentrations from other data types, such as freezing-point depression or electrochemical cell data. Commonly, solubility products and mineral free-energy values are adjusted so that they are compatible with a given solution-chemistry model (see Pitzer 1979b, 1991, 1993; Pabalan and Pitzer 1991). As an independent check, mineral free-energy values often are compared to values obtained from independent thermodynamic methods. The resulting solubility products may or may not be consistent with values obtained using alternative solution-chemistry models, such as the conventional ion-association model. The ion-interaction parameters are also temperature dependent, allowing for reliable calculation of solubility relations versus temperature (Sylvester and Pitzer 1976, 1977; Pabalan and Pitzer 1987; see Fig. 2).

Internal consistency of model data-sets

A fundamental requirement in the development of a robust ion-interaction model is consistency or compatibility of ion-interaction parameters and other thermodynamic constants. The development of a set of ion-interaction parameters for a specific system requires binary interaction parameters for all binary combinations of ions, and ternary interaction coefficients for all ternary combinations of ions. If an extra charged species is included, as in the modified or hybrid approach, then all binary and ternary combinations parameters. Once a set of interaction parameters is developed for all binary and ternary combinations, then the model is usually further developed for predicting mineral solubility by adjusting mineral free-energy data until the difference between predicted concentrations and experimental concentrations is minimized through, for example, linear least-squares regression. Weighting coefficients are commonly utilized to accommodate different expected or actual degrees of precision in the regression analysis (see Harvie et al. 1984). The final model should be an "internally consistent" set of equations and data which can be used as a combined package to predict mineral solubility within that system. This condition of internal consistency is the same as that required in the development of an ion-association model, as described in detail by Nordstrom and Muñoz (1994).

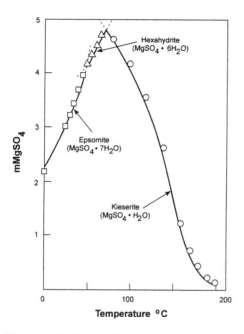

Figure 2. Prediction of metal-sulfate solubility in single-salt solutions: example solubility relations for $MgSO_4$ solutions versus temperature. (Redrawn from Pabalan and Pitzer 1987, Fig. 7, p. 2435).

A generalized ion-interaction model, intended to describe most of the constituents typically encountered in natural waters and waste sites, requires that all of the binary interaction parameters are evaluated using the same form of Pitzer's equations (e.g. original equations versus extended equations, completely dissociated form versus incorporation of association constants, etc.), that all ternary interaction coefficients are determined using a consistent set of binary interaction coefficients, and that all mineral solubilities are estimated using mineral free-energy values in a manner that is consistent with the aqueous model. Improvements in the techniques of parameter calculation, such as the ridge regression technique to avoid intercorrelation (van Gaans 1991), the use of methods for simultaneously representing different data types (e.g. Marliacy et al. 1998), and utilization of empirically extended versions of Pitzer's equations (e.g. Rard and Clegg 1997), need to be applied in a consistent fashion.

Archer and Rard (1998) provided a concise description of the complexities involved with the development of internally consistent data-sets within the framework of the ion-interaction formulism. In the example system described, Archer and Rard (1998) described a case for which the tabulated standard-state properties were adjusted to bring agreement between calculated and observed phase behavior (a common practice); it was concluded that adjustments made to the standard-state properties may have compensated for errors in the excess properties.

Constants and fitting parameters assembled for specific modeling efforts cannot simply be combined with those assembled for other applications unless a systematic evaluation of the data compatibility has been performed. This incompatibility of model databases and formulations has been an important limitation in applying the ion-interaction approach to complex field sites. The following sections focus on various models assembled to apply the Pitzer's equations to predict mineral solubility in natural waters, the compatibility of these models, and what might be needed to make the models more compatible.

Table 2. Summary of aqueous systems studied involving sulfate-mineral solubilities in waters of bulk seawater-type composition.

System	Sulfate Compounds	T Range[1]	Source
Na–K–Mg–Ca–Cl–SO$_4$	alkali-earth phases in Table 1	25°C	Harvie and Weare (1980); Harvie et al. (1982, 1984)
Na–K–Mg–Cl–SO$_4$–OH	epsomite, hexahydrite, kieserite, arcanite, mirabilite, thenardite	0 to200°C	Pitzer (1986, 1991), Pabalan and Pitzer (1987)
Na–K–Mg–Ca–Cl–SO$_4$	mirabilite, thenardite, gypsum, anhydrite, glauberite	25 to250°C	Møller (1988)
Na–Ca–Cl–SO$_4$	anhydrite, gypsum	to 200°C	Monnin (1990)
Ca–Na–Cl–SO$_4$	gypsum, anhydrite	0 to 250°C	Raju and Atkinson (1990)
Na–K–Ca–Mg–Cl–SO$_4$	mirabilite, thenardite, arcanite, epsomite, picromerite	-60 to25°C	Spencer et al. (1990)
Na–K–H–Ca–Mg–Cl–OH–SO$_4$	gypsum, anhydrite		He and Morse (1993)
Na–Cl–SO$_4$	thenardite, mirabilite	0 to100°C	Marliacy et al. (1998)
Na–K–Mg–Ca–Cl–SO$_4$–H$_2$O	mirabilite, thenardite, epsomite, arcanite, gypsum, + others	-60 to25°C	Marion and Farren (1999)

[1] $P = 1$ bar, except for Monnin (1990), which is 1 kbar

MODEL DEVELOPMENT TO PREDICT SULFATE-MINERAL SOLUBILITY

Sulfate-mineral formation in the system Na–Ca–Mg–K–Cl–SO$_4$–H$_2$O

Much effort has been focused on the prediction of sulfate-mineral solubility both in seawater-type mixtures and in waters containing the major ions observed in seawater (Table 2). In papers by Harvie et al. (1980), Harvie and Weare (1980), and Harvie et al. (1982, 1984), as summarized by Weare (1987), the Pitzer approach was applied to predict ion activities and mineral solubilities in waters containing the major constituents in seawater. These papers described the development of a comprehensive model (referred to as the Harvie-Møller-Weare or HMW model) for the formation of evaporite minerals, including sulfate minerals, from concentrated solutions. Prediction of the solubility of the common sulfate minerals (e.g. anhydrite, aphthitalite, blödite, epsomite, gypsum, glauberite, hexahydrite, kainite, kieserite, leonite, mirabilite, picromerite, polyhalite, syngenite, and thenardite) over a range in concentration in the binary, ternary, and through the full six-component Na–Ca–Mg–K–Cl–SO$_4$ system was described in detail.

The initial efforts of Harvie and coworkers were expanded to predict mineral solubilities at higher temperature in the work of Pabalan and Pitzer (1987), Møller (1988), Greenberg and Møller (1989), and Møller et al. (1998). In these papers, the ion activities and mineral solubilities were described for the mixed Na–Ca–K–Cl–SO$_4$–H$_2$O system to 250°C.

Pabalan and Pitzer (1987) and Pitzer (1991) provided sulfate-mineral solubility relations for several binary and ternary sulfate mixtures, and reported the corresponding ion-interaction parameters. Data for mirabilite and thenardite solubility between 0 and 350°C were presented. Also given were data for arcanite solubility between 0 and 225°C, and the solubility of the $MgSO_4$ minerals epsomite, hexahydrite, and kieserite between 0 and 200°C, with the corresponding derived ion-interaction parameters. Solubility isotherms for $NaCl$ and Na_2SO_4 mixtures, for KCl and K_2SO_4 mixtures, and for $MgCl_2$ and $MgSO_4$ were provided for several temperatures.

Monnin (1990) expanded the efforts of Møller (1988) to include the influence of pressure to 1 kbar, and presented solubility isotherms for gypsum and anhydrite at various pressures. Monnin used density and compressibility data to estimate P dependence in the interaction parameters. Monnin determined that the introduction of ion pairs was necessary to predict relations at higher T and P conditions. This is in contrast to the approach of Harvie and Weare (1980), who found that the assumption of complete dissociation provided a reliable representation of sulfate-mineral solubility in the simplified seawater system. The expansion of Monnin (1990) allows the ion-interaction approach to be applied to higher pressure systems, such as those that occur in deep oceanic basins and in the crustal subsurface. Others also expanded the original HMW model to include a larger temperature and pressure range for specific applications, such as was done by He and Morse (1993) to describe halite, gypsum, and anhydrite solubility in oil-field brines.

The original parameterization of the $Na-K-Mg-Cl-Ca-SO_4-H_2O$ system by Harvie and Weare (1980) was expanded by Spencer et al. (1990) to include temperatures to -60°C for prediction of mineral-solubility relations during seawater freezing. The focus of the Spencer et al. parameterization was mainly on chloride minerals, but included sulfates such as mirabilite and thenardite. Spencer et al. recommended that the sulfate portion of their model not be extended below -37°C, and be limited to waters with low SO_4 concentrations. Marion and others (Marion and Farren 1999; Morse and Marion 1999) revised and expanded the Spencer et al. model to improve the parameterization for Na and Mg sulfate phases, and to add several sulfate solids not previously included in the Spencer et al. model. This latter contribution allows for prediction of sulfate-mineral formation to temperatures as low as -37°C, and to lower temperatures if concentrations of SO_4 are also limited to low values. This limitation at very low temperatures is attributed to the lack of experimental data from which constants can be derived, and not to deficiencies in the modeling approach.

More recent studies, such as that by Archer and Rard (1998) on the $MgSO_4-H_2O$ system from 298.15 to 440 K and the solubility of $MgSO_4\cdot7H_2O$ at 25°C, have used modified versions of the original Pitzer's equations, and include a $MgSO_4^0$ ion pair to represent this system. Marliacy et al. (1998) provided a consistent approach for parameterization of the $Na_2SO_4 + NaCl$ system between 273.15 K and 373.15 K which allows simultaneous representation of molar dissolution enthalpies and solubility data for halite, thenardite, and mirabilite in binary and ternary systems. Solubility isotherms provided for $Na_2SO_4 + NaCl$ mixtures at 303.15 and 373.15 K indicate good agreement between predicted and experimental data. The crystallization enthalpies were further considered in this system for application to materials that are mirabilite-based and have latent heat storage (Marliacy et al. 2000). Other studies, such as that described by Pavicevic et al. (1999) on the $NaH_2PO_4-Na_2SO_4$ system at 298.15 K, allow calculations to be performed in other industrially important mixtures.

Geochemical interpretations and predictions involving sulfate minerals in various settings commonly integrate geochemical models with physical flow and solute-transport equations. These integrated models require estimation of various solution properties such as density. Pitzer's equations can be used to estimate solution densities (e.g. Monnin 1989; Obsil et al. 1997) and mutual diffusion coefficients (Albright et al. 1998). These models require data on binary and ternary mixtures, from which physical properties of more complex mixtures can be calculated, similar to the approach used to calculate ion activities in complex mixtures through Equations (3) and (4). Systematic measurements and compilations of partial molar density and diffusion-coefficient data for concentrated binary and ternary electrolyte mixtures are required. For calculation of physical properties involving sulfate waters, these compilations must also include data for mixtures containing sulfate (e.g. Albright et al. 1998 and references within), and evaluation of the interaction model parameters for use with Pitzer's equations. Integration of these physical properties with solute-transport equations is an important step in predicting sulfate-mineral solubility in non-static systems.

Table 3. Summary of aqueous systems studied for sulfate minerals containing trace components in seawater and continental waters.

System	Phases	T Range[1]	Source
Sr–Na–Cl	celestine		Reardon and Armstrong (1987)
Ba–Na–Cl–SO$_4$	barite	25 to 300°C	Raju and Atkinson (1988)
Ba–Ca–K–Mg–Na–Sr–Cl–SO$_4$	celestine, barite		Monnin and Galinier (1988)
Sr–Na–Cl–SO$_4$	celestine	25 to 100°C	Raju and Atkinson (1989)
Ba–Na–Sr–SO$_4$	celestine, barite	25°C	Felmy et al. (1990)
Na–Li–SO$_4$, Na–Cu–SO$_4$, Na–Ni–SO$_4$, Na–Co–SO$_4$, Na–Mn–SO$_4$, Cu–Li–SO$_4$, Ni–Li–SO$_4$, Fe(II)–Li–SO$_4$, Co–Cl–SO$_4$, Ni–Cl–SO$_4$, Mg–Cd–SO$_4$, Al–Mg–SO$_4$, and Al–Fe(II)–SO$_4$	solid phases indicated by formula	25°C	Pabalan and Pitzer (1991)
Fe(II)–Na–Cl–SO$_4$	mirabilite, melanterite, mixed Fe(II)Na phase	25°C	Ptacek (1992)
K–Rb–SO$_4$	(K,Rb)SO$_4$ series	25°C	Kalinkin and Rumyantsev (1996)
Na–K–Ca–Mg–Ba–Sr–Cl–SO$_4$	barite, celestine	to 200°C	Monnin (1999)

[1]P = 1 bar, except for Monnin (1999), which is 1 kbar.

Formation of Ba, Sr, and Rb sulfates

Reardon and Armstrong (1987) presented experimental data and ion-interaction parameters derived to describe celestine solubility in water, seawater, and NaCl solutions over a range in temperature (Table 3). Monnin and Gallinier (1988) evaluated the solubility of celestine in the system Ba–Ca–K–Mg–Na–Sr–Cl–SO$_4$–H$_2$O and observed good agreement between predicted solubility values using Pitzer's equations and celestine solubility in Cl solutions, but less satisfactory agreement for celestine solubility in Na_2SO_4 solutions. Felmy et al. (1990) reported new solubility studies for barite and celestine in Na_2SO_4 solutions, and interpreted the data using Pitzer's equations.

Phase equilibria in the system K_2SO_4–Rb_2SO_4–H_2O system at 25°C were described by Kalinkin and Rumyanstev (1996). This system involves formation of continuous solid solutions. To address this problem, Kalinkin and Rumyantsev (1996) combined solid-solution theory (see chapter by Glynn, this volume) with Pitzer's equations. Problems encountered during their analysis were described in detail. The need to quantify the formation of solid solutions in concentrated solutions is apparent (Gamsjäger 1993). Although there have been some outstanding advances in this area (e.g. Königsberger et al. 1999), general deficiency in predictive capabilities is evident when existing ion-interaction models are applied to solid solutions or systems with multiple closely related phases.

Barite and celestine solubility in the Na–K–Ca–Mg–Ba–Sr–Cl–SO_4–H_2O system to 200°C and 1 kbar was parameterized by Monnin (1999). The results were compared to measured mineral solubility in pure water to 1 kbar, and in NaCl solutions to 500 bars. Activity coefficients of Ba^{2+} and SO_4^{2-} in seawater were also calculated for temperature, pressure, and salinity values observed in the ocean, and were compared to published values. Again, inclusion of the $BaSO_4^0$ ion pair was necessary to provide agreement to experimental data. Monnin (1999) provided a thorough review of previous studies on barite and celestine solubility, including studies that had focused on concentrated solutions (see chapter by Hanor, this volume).

Boerlage et al. (1999) compared the application of a model based on Pitzer's equations with other models for calculating the saturation state for $BaSO_4$ in a reverse-osmosis system. It was observed that at higher ionic strengths, the Pitzer approach was more reliable than the models typically used in water-treatment applications.

Formation of metal sulfates

The majority of the early effort on metal-sulfate systems has been dedicated to predicting the solution properties of metal-sulfate binary mixtures (e.g. see Pitzer 1974), with less attention focused on the metal-sulfate ternary mixtures and the formation of metal-sulfate solids. Binary interaction parameters are available for most common metal-sulfate mixtures, with the exception of questionable parameters for Fe(II)–SO_4 mixing, and no parameters for Fe(III)–SO_4 mixing. Osmotic coefficient data for Fe(II)–SO_4 are inconsistent with data for other metal-sulfate mixtures (see Fig. 3). The trends in osmotic coefficient data for $FeCl_2$ solutions are consistent with those for other metal chlorides, but the experimental data for Fe(II) by Oykova and Balarew (1974) deviate from the trends of other metal-sulfate solutions, and from more recent data by Nikolaev et al. (1989). Fe(III) hydrolyzes strongly, and parameterization for this system is complex.

In a series of papers by Reardon and others (Reardon and Beckie 1987; Reardon 1988, 1989; Baes et al. 1993), interaction parameters were evaluated for various metal-sulfate mixtures (Table 4). Reardon and Beckie (1987) provided solubility products for melanterite ($FeSO_4 \cdot 7H_2O$) and szomolnokite ($FeSO_4 \cdot H_2O$) over the temperature range 0 to 100°C. Here the binary interaction parameters for $FeSO_4$ mixing were derived from the osmotic coefficient data of Oykova and Balarew (1974). Reardon (1988) expanded the concentration range for the binary parameters for Al–SO_4 mixing, from which a solubility product for $Al_2(SO_4)_3 \cdot 17H_2O$ (alunogen) was obtained. Reardon (1988) then provided solubility isotherms for mixing in the ternary systems between $Al_2(SO_4)_3$ and $CuSO_4$, $FeSO_4$, $MgSO_4$, $NiSO_4$, or $CaSO_4$ (e.g. Fig. 4). In most cases, the experimental solubilities for these systems were closely predicted using ternary interaction parameters set

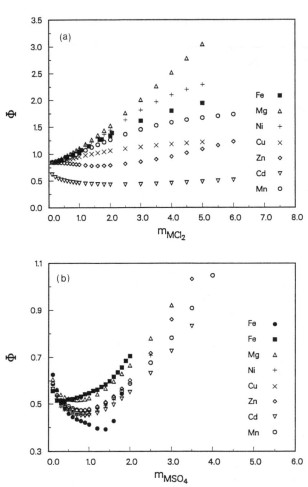

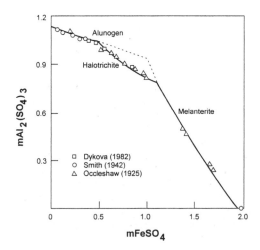

Figure 3. (a) Osmotic co-efficients for MgCl$_2$ and transition-metal chloride solutions as a function of concentration. (b) Osmotic coefficients for MgSO$_4$ and transition-metal sulfate solutions as a function of concentration. Experimental data for FeCl$_2$ are those reported by Stokes and Robinson (1941) and Kangro and Groeneveld (1962), data for FeSO$_4$ are those reported by Oykova and Balarew (1974) (squares) and Nikolaev et al. (1989) (solid circles), and for the remaining elements, by Robinson and Stokes (1970).

Figure 4. Prediction of metal-sulfate solubilities for mixed-salt solutions: example solubility isotherms for FeSO$_4$–Al$_2$(SO$_4$)$_3$–H$_2$O at 25°C. (Redrawn from Reardon 1988, Fig. 5, p. 6430).

Table 4. Summary of aqueous systems studied involving Mg-sulfate and
metal-sulfate solubilities in acidic solutions.

System	Phases	P,T Range	Source
Fe–HSO$_4$–SO$_4$	melanterite, szomolnokite	0-90°C	Reardon and Beckie (1987)
Al–SO$_4$, Al–K–SO$_4$, Al–Cu–SO$_4$, Al–NH$_4$–SO$_4$, Al–Ni–SO$_4$, Al–Fe–SO$_4$, Al–Mg–SO$_4$, Al–Ca–SO$_4$	potassium alum, tschermigite, melanterite, chalcanthite, alunogen, morenosite, halotrichite, epsomite, pickeringite, gypsum	25°C	Reardon (1988)
Ni–HSO$_4$–SO$_4$	morenosite, retgersite nickelhexahydrite	25°C	Reardon (1989)
Cu–HSO$_4$–SO$_4$	chalcanthite		Baes et al. (1993)
Pb–HSO$_4$–SO$_4$, Pb–Na–SO$_4$	anglesite		Paige et al. (1992)
Mg–HSO$_4$–SO$_4$	MgSO$_4$·7H$_2$O, MgSO$_4$·5H$_2$O, MgSO$_4$·H$_2$O	25°C	Rard (1997); Rard and Clegg (1999)
Ba–HSO$_4$–SO$_4$, Ra–HSO$_4$–SO$_4$	barite	25°C, 60°C	Paige et al. (1998)
K–Cu–Ni–SO$_4$	K$_2$SO$_4$·NiSO$_4$·6H$_2$O cyanochroite K$_2$SO$_4$(Ni,Cu)SO$_4$·6H$_2$O	25°C, 1 bar	Christov (1999)

to 0.0. The temperature dependence of three hydrates of NiSO$_4$ were provided by Reardon (1989), together with newly derived ion-interaction parameters. Baes et al. (1993) obtained binary interaction coefficients for the CuSO$_4$ mixture and a solubility product for CuSO$_4$·5H$_2$O at 25°C.

Pabalan and Pitzer (1991) summarized application of the ion-interaction approach to calculate solubilities for a series of metal-sulfate phases. The results included sulfate solubility in the aqueous ternary mixtures Na–Li–SO$_4$, Na–Cu–SO$_4$, Na–Ni–SO$_4$, Na–Co–SO$_4$, Na–Mn–SO$_4$, Cu–Li–SO$_4$, Ni–Li–SO$_4$, Fe(II)–Li–SO$_4$, Co–Cl–SO$_4$, Ni–Cl–SO$_4$, Mg–Cd–SO$_4$, Al–Mg–SO$_4$, and Al–Fe(II)–SO$_4$ at 25°C. Pabalan and Pitzer (1991) also provided solubility relations in the quaternary mixtures Na–Li–Cu–SO$_4$ and Na–Co–Cl–SO$_4$ at 25°C, and enthalpy and other thermal data for the crystalline phases.

Ion-interaction parameters and solubility products for mixing in the ternary systems Fe(II)–Na–SO$_4$ and Fe(II)–Cl–SO$_4$ were evaluated by Ptacek (1992) (see Fig. 5). Also evaluated were the solubility products for melanterite, mirabilite, and a mixed Fe,Na hydrated sulfate phase at other temperatures. These studies were conducted to provide interaction parameters for Fe(II) mixing in solutions containing Na, SO$_4$, and Cl, and to predict the solubility of siderite in brines and mine drainage waters.

Christov (1999) summarized solubility data for a study which included measurements in the system K–Ni–SO$_4$ and K–Cu–SO$_4$. The study described previously determined Pitzer parameters for the binary systems NiSO$_{4(aq)}$ and CuSO$_{4(aq)}$, and solubility products

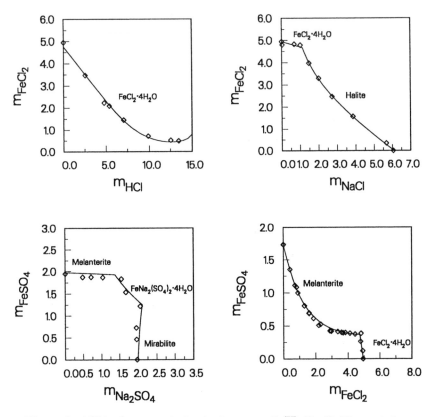

Figure 5. Additional parameterization in the system Fe(II)–Na–Cl–SO$_4$ required to predict sulfate-mineral solubility in waters containing Fe(II) (after Ptacek 1992).

were calculated for K$_2$SO$_4$, NiSO$_4$·7H$_2$O, CuSO$_4$·5H$_2$O, and the double salts K$_2$SO$_4$·NiSO$_4$·6H$_2$O and K$_2$SO$_4$·CuSO$_4$·6H$_2$O. Christov (1999) attributed the small difference between K^0_{sp} and the values that had been obtained in previous studies to small differences in measured saturation concentrations used in the calculations, not to errors in the ion-interaction parameters. For the mixing in these ternary systems, K$_2$SO$_4$·NiSO$_4$·6H$_2$O is stable for mK$_2$SO$_4 > \sim0.8m$, and K$_2$SO$_4$·CuSO$_4$·6H$_2$O is stable for mK$_2$SO$_4 > \sim0.2m$ (Fig. 6). At lower concentrations of K$_2$SO$_4$, the single hydrates of Ni and Cu are stable. In addition to the double sulfates, Christov determined the free energy of mixing for K$_2$SO$_4$(Ni,Cu)SO$_4$·6H$_2$O, which is expected to form in solutions containing both Ni and Cu. The solubility of this latter phase is described well with the Pitzer approach assuming regular mixing, which is consistent with the similarity in unit-cell parameters of the co-crystallizing isostructural double salts. The efforts of Herbert and Mönig (1997) to include Cd and Zn in the HMW model were described by van Gaans (1998); these authors investigated the interaction of Cd and Zn with Cl and SO$_4$.

Formation of sulfates in acid systems

Prediction of equilibria reactions in systems containing elevated concentrations of acid is important for a variety of applications (Tables 4 and 5). These include industrial and metallurgical applications in which acid is added for manufacturing, ore extraction, and other purposes (e.g. Paige et al. 1992, 1998; Schuiling and van Gaans 1997), the

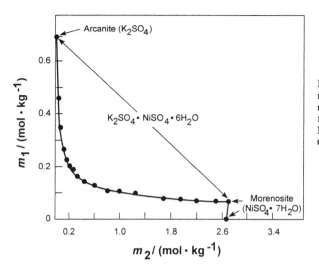

Figure 6. Prediction of metal-sulfate solubilities for mixed-salt solutions with the formation at 25°C of K_2SO_4–$NiSO_4$–H_2O. (Redrawn from Christov 1999, Fig. 3, p. 78).

Table 5. Applications of Pitzer model for predicting sulfate-mineral solubilities in natural waters.

System	Phases	Source
Seawater components	evaporation sequence	Harvie et al. (1980, 1982, 1984)
Natural brines	gypsum	Weare (1987)
Natural brines	gypsum	Krumgalz and Millero (1982, 1983)
Marine evaporite	evaporation sequence gypsum	Brantley et al. (1984)
Deep basin brines	Ca, Sr, Ba, Ra sulfates	Langmuir and Melchior (1985)
Mine drainage	gypsum, melanterite, jarosite, anglesite	Blowes et al. (1991); Ptacek and Blowes (1994)
Seawater	evaporation sequence gypsum	De Lange et al. (1990)
Deep basin brines	gypsum, anhydrite	He and Morse (1993)
Deep basin brine	Ca, Mg sulfates	Williams-Stroud (1994)
Evaporite lake	gyspum, mirabilite, thenardite, polyhalite	Donovan and Rose (1994)
Acid waste pond	gypsum, natrojarosite	van Gaans and Schuiling (1997)
Marine brines	anhydrite	Monnin and Ramboz (1996)
Geothermal water	Ca, Ba, Sr sulfates	Pátzay et al. (1998)
Seawater evaporation and freezing	Na, Ca sulfates	Marion and Farren (1999); Morse and Marion (1999)
Mine drainage	pH, melanterite	Nordstrom et al. (2000)

environmental geochemistry of mine-waste sites where elevated concentrations of acid and metals are common (e.g. Blowes and Jambor 1990; Blowes et al. 1991; Ptacek and Blowes 1994; Ridley et al. 1997, 1999; Nordstrom and Alpers 1999; Nordstrom et al. 2000), and the formation of acid-sulfate aerosols where highly evaporative conditions lead to extreme concentrations of dissolved acid and the formation of acid-sulfate solids (e.g. Clegg et al. 1998). Although ion-interaction models have been developed for specific applications, much of the information is directly transferable to other systems.

Prediction of sulfate-mineral solubility in acidic solutions requires, foremost, interaction parameters for describing the pure acid solutions prior to the addition of other acids or dissolved salts. The aqueous properties of H_2SO_4 using Pitzer's equations were first represented by assuming complete dissociation of H_2SO_4. Later, the inclusion of the bisulfate species HSO_4^- improved the predictions for this system (e.g. Reardon and Beckie 1987). These contributions provide reliable predictions up to 6 m H_2SO_4. Hovey et al. (1993) and Clegg et al. (1994) introduced a mole fraction-based version of the ion-interaction equations and introduced an ionic-strength-dependent third viral coefficient which further improved predictions in the H_2SO_4 system. The Clegg et al. (1994) approach seems to be widely used to predict the chemistry of formation of atmospheric aerosols, wherein very high concentrations of acid are common. Hachimi et al. (1996) derived parameters for the H_2SO_4 system between 0 and 27 mol kg^{-1}.

The previously described studies by Reardon and Beckie (1987), Reardon (1989), and Baes et al. (1993) also included parameterization and mineral solubilities for solutions containing H_2SO_4. Parameterization of the system $FeSO_4$–H_2SO_4–H_2O between 0 and 90°C approaches 6 m H_2SO_4. Solubility isotherms were presented for several temperature increments. Reardon (1989) provided an analysis of mixing in the system $NiSO_4$–H_2SO_4–H_2O, mostly at 25°C, and concentrations of $NiSO_4$ were closely predicted to 6 m H_2SO_4. In the parameterization of the $CuSO_4$–H_2SO_4–H_2O at 25°C, Baes et al. (1993) established solubility relations for $CuSO_4\cdot5H_2O$ versus H_2SO_4 up to a concentration of 6 m H_2SO_4. Baes et al. (1993) also gave a useful discussion of the role of parameter covariance on the predicted solution concentrations.

In a series of papers focusing on application to the processing of uranium ore, Paige et al. (1992, 1998) evaluated solution equilibria in the systems $PbSO_4$–H_2SO_4–H_2O, $PbSO_4$–Na_2SO_4–H_2O, $BaSO_4$–H_2SO_4–H_2O, and $RaSO_4$–H_2SO_4–H_2O. Concentrations of H_2SO_4 ranged between 0 and 6.19 mol kg^{-1}. The parameterization by Paige et al. relied on that of the H_2SO_4 system by Reardon and Beckie (1987). In this case, the HSO_4^- species was incorporated into the data analysis, but ion pairs were not.

Hovey et al. (1993) described the thermodynamics of Na_2SO_4(aq) between 0 and 100°C for mixtures containing H_2SO_4 and Na_2SO_4. Hovey et al. (1993) included new experimental data for this system, and modeled data presented by Rard (1992). The analysis by Hovey et al. (1993) explicitly considered the formation of the HSO_4^- species, and provided a molarity-based analysis to 6 m and a molality-based analysis to 15 m. Rard (1997) and Rard and Clegg (1999) provided experimental data and analysis with the Pitzer model for $MgSO_4$ mixtures with H_2SO_4 at 298.15 K. Their analysis included the formation of a $MgSO_4^0$ ion pair, and the HSO_4^- species. Solubility curves for $MgSO_4\cdot7H_2O$, $MgSO_4\cdot5H_2O$, and $MgSO_4\cdot H_2O$ as a function of H_2SO_4 concentration were provided.

Formation of sulfate minerals in solution and temperature ranges representing conditions in atmospheric aerosols was described by Potukuchi and Wexler (1995), Massucci et al. (1996), Clegg et al. (1996, 1998), and Pierrot et al. (1997), among others. The formation of numerous sulfate phases containing H^+, NH_4^+, Na^+, NO_3^-, Cl^-, and others, was quantified. Complex solubility diagrams were provided over a range of solution-composition and temperature conditions, similar to the original diagrams for mineral stability as had been depicted by Harvie et al. (1984). Information related to these studies can likely be used to predict mineral solubility in other acidic systems.

Formation of sulfates in basic systems

The application of the Pitzer approach to predict sulfate precipitation from basic waters has received less attention than in other systems. Harvie et al. (1984) gave solubility predictions for mirabilite, thenardite, arcanite, and other mixed sulfate and carbonate compounds in the system N–K–Mg–Ca–Cl–SO$_4$–H$_2$O at 25°C. Solubility curves for SO$_4$ and CaOH, KOH, NaHCO$_3$, and Na$_2$CO$_3$ solutions were provided. Pitzer (1991) summarized the solubility relations of thenardite in NaOH, NaCl, and Na$_2$SO$_4$ mixed solutions at various temperatures. Duchesne and Reardon (1995) derived interaction coefficients for predicting the solubility of portlandite in concentrated solutions, and included solubility isotherms for mixtures containing Ca(OH)$_2$ with CaSO$_4$, K$_2$SO$_4$, or Na$_2$SO$_4$. The solubility of gypsum was closely described in these mixtures, including the transitions to Ca(OH)$_2$ at lower SO$_4$ concentrations.

APPLICATION OF THE ION-INTERACTION APPROACH TO FIELD SETTINGS

Various model formulations based on the ion-interaction approach have been developed and applied to predict mineral saturation indices, equilibrium concentrations, and sequences of mineral precipitation and dissolution in field systems. These applications include comparison of precipitation sequences for evaporating seawater (Harvie et al. 1984; Brantley et al. 1984), evaporation of continental lake waters (Monnin and Schott 1984; Felmy and Weare 1986), mineral stability in subsurface brines (Langmuir and Melchior 1985), and mineral stability in stratified ocean brines (e.g. Monnin and Ramboz 1996; De Lange et al. 1990; van Cappellen et al. 1998). An ion-interaction model has been used to predict the aqueous geochemistry and mineral solubility controls in acid mine waters (Blowes et al. 1991; Ptacek and Blowes 1994; Nordstrom et. al. 2000) and acid waste ponds (van Gaans and Schuiling 1997). Most of the above applications involve either the direct formation of sulfate minerals, or address the indirect influence of elevated concentrations of dissolved sulfate ions on ion activities.

In a geological context, the application of the Pitzer model has relied mainly on the original parameterization of the Na–K–Mg–Ca–SO$_4$–Cl–H$_2$O system as developed by Harvie and Weare (1980) and Harvie et al. (1984), which is referred to as the HMW model. This development was incorporated into the geochemical computer code PHRQPITZ (Plummer et al. 1988), which is based on the earlier computer code PHREEQE (Parkhurst et al. 1985). The introduction of PHRQPITZ made calculations involving the HMW model easily accessible, and the result has been widespread use of the HMW model by the geochemical community. A variety of ion-interaction models, includeing the HMW model for 25°C, the TEQUIL geothermal model by Møller et al. (1998), the freezing model by Spencer et al. (1990), the aerosol models by Clegg and others (e.g. Clegg et al. 1998), are also available for use online or can be downloaded from the internet (e.g. http://geotherm.ucsd.edu; and http://www.hpc1.uea.ac.uk/~e770/aim.html). Over the past decade, the original HMW model and variations have been applied to numerous case studies of field settings.

In the applications of geochemical models, mineral solubility relations are typically expressed in terms of the saturation state of a water with respect to a specific mineral. The degree of saturation is expressed as a saturation index (see Plummer et al. 1988), or a normalized saturation index (e.g. Felmy and Weare 1986). In PHRQPITZ, the saturation index, SI, is defined as $SI = \log IAP - \log K$, where $SI = 0$ represents equilibrium conditions, $SI < 0$ undersaturated conditions, and $SI > 0$ supersaturated conditions. The solubility constant K is usually corrected for deviation in temperature from standard conditions, and the ion-activity product (IAP) is calculated for the appropriate temperature and pressure conditions; therefore, the degree of saturation is corrected accordingly.

Results of the applications

Prediction of sulfate-mineral solubilities in waters containing the major seawater components. Harvie et al. (1980) applied Pitzer's equations to calculate the sequence of mineral formation from evaporating seawater. The calculations were consistent with observed precipitation sequences in seawater.

Another early application of Pitzer's equations to calculate solubility relations for sulfate minerals was that by Krumgalz and Millero (1982, 1983). In this application, the Harvie and Weare (1980) formulation of the Pitzer model was used to calculate saturation indices for Dead Sea water, and mixtures of Dead Sea water and Mediterranean water, to assess the degree of gypsum saturation and the likelihood of gypsum precipitation under mixing conditions. It was concluded that the Dead Sea waters are supersaturated with respect to gypsum, that gypsum precipitation is limited by kinetic factors, and that supersaturation of the Dead Sea water with respect to gypsum is consistent with other experimental data. Later applications of this model included calculations to determine the maximum extent that the Dead Sea could evaporate before the activity of water in the overlying atmosphere would exceed the activity of water in the evaporated Dead Sea water (Krumgalz et al. 2000).

The ion-interaction model was applied to a modern marine evaporite at Bocana de Virrilá, Peru (Brantley et al. 1984). The estuary has a pronounced horizontal salinity gradient that has formed in response to evaporative losses, with salinity values ranging from marine composition at its mouth, to brines of approximately 330,000 ppm total dissolved solids (TDS) in the innermost waters. Application of a revised version of the Harvie and Weare (1980) model, assuming a fractional crystallization pathway, yielded close agreement between predicted and observed concentrations of the major ions. Predicted locations of calcite, gypsum, and halite precipitation also agreed closely with field observations.

The geochemistry of alkaline earth sulfates in the slightly acidic, Ca-rich brines of the Palo Duro Basin of Texas was investigated by Langmuir and Melchior (1985) to evaluate mineral stabilities at a potential repository for highly radioactive wastes. An ion-interaction model was applied to calculate saturation indices for gypsum, anhydrite, celestine, barite, and $RaSO_4$. The calculations indicated that the brines in this basin are at saturation with respect to gypsum (SI between -0.11 and +0.07), celestine (SI between -0.19 and +0.22), and anhydrite (SI between -0.18 and -0.02), and are near saturation with respect to barite at several locations (SI between -0.65 and +0.34) except one. The brines are strongly undersaturated with respect to pure $RaSO_4$, consistent with the expectations of a Ra-Ba solid-solution control on Ra concentrations.

De Lange et al. (1990) described sulfate-related equilibria in hypersaline anoxic brines of the Tyro and Bannock basins of the eastern Mediterranean. The Bannock brines are probably formed from recent dissolution of underlying late-stage evaporites into seawater. Pitzer's equations were used to calculate gypsum, dolomite, and barite SI values. Near-saturated conditions were observed in the Tyro and Bannock basins, with the saturation state being exceeded as a result of mixing of two brine types. The predicted precipitation of gypsum is consistent with observations of freshly precipitated gypsum in the Bannock Basin. There is no obvious evidence for substantial amounts of gypsum precipitation in the

Tyro brines. Supersaturation was predicted for barite at the interface between brine types where stagnation of waters occurs. At the time of the calculations, however, there was only limited information for predicting barite saturation indices in complex brines using Pitzer's equations.

Marion and Farren (1999) applied the extended version of the Spencer et al. (1990) model (the FREZCHEM model) to predict mineral precipitation sequences during evaporation and freezing of seawater. Marion and Farren (1999) specifically expanded the Spencer et al. model to improve sulfate mineral stability relations and to incorporate additional sulfate solids. The predicted sequences of mineral formation, and the predicted concentrations of dissolved ions agree remarkably well with experimental data (Fig. 7). This example illustrates the benefit of continuously improving the model parameterization and evaluation process.

Saturation indices for anhydrite were calculated by Monnin and Ramboz (1996) for the ponded brines and sediment pore waters of the Red Sea deeps. Changes in pressure were observed to have the same magnitude of effect as temperature on the calculated saturation indices for anhydrite.

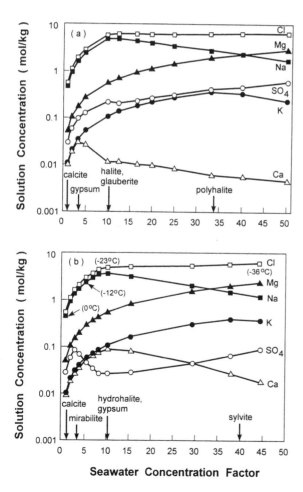

Figure 7. Application of ion-interaction model (FREZ-CHEM) for predicting the composition of seawater and the minerals precipitated as a function of concentration factor for evaporation at 25°C (a) and freezing (b). (Redrawn from Morse and Marion 1999, Fig. 10, p. 754).

Yechieli and Ronen (1997) described sequential precipitation of gypsum, halite, carnallite, and possibly sylvite and bischofite in newly exposed sediments of the Dead Sea. The presence of these minerals was verified by SEM, and their depth of occurrence was inferred from saturation indices, calculated using PHRQPITZ.

From the above examples, it can be concluded that the ion-interaction approach has been well-developed for many geochemical applications involving the formation of sulfate minerals formed with the major ions Na and Ca, and to a certain extent, the trace components Ba and Sr.

Prediction of sulfate-mineral solubilities in deep subsurface brines. As part of a study on the geochemical evolution in the Culebra Dolomite Formation, New Mexico, Siegel and Anderholm (1994) calculated saturation indices for common evaporite minerals using PHRQPITZ and compared the results with mineralogical observations. Calculated saturation indices for gypsum were 0.00±0.09 for 30 water samples, consistent with the ubiquitous presence of gypsum and the rapid approach to equilibrium typically observed for this mineral. With the exception of one sample, the waters were undersaturated with respect to anhydrite.

He and Morse (1993) gave a summary of ion-interaction parameters that were derived to describe the solubility of halite, gypsum, and anhydrite in Na–K–H–Ca–Mg–Cl–OH–SO_4–H_2O solutions over a range of temperature and pressure. Comparisons were provided for predicted solubilities and experimental data for the temperature range 25-200°C and the pressure range 1-1000 bar, and example calculations were performed to show scale formation in oil-field settings.

The potential for scale formation and corrosion from geothermal waters in geothermal plants was addressed by Pátzay et al. (1998), who gave experimental data for $CaCO_3$ and $CaSO_4$ mixtures, $CaSO_4$ solubility in NaCl solutions at elevated temperature, and temperature- and pressure-dependent Pitzer parameters. Specific applications to a geothermal plant included data such as the bubble point and partial pressures of various gases.

Prediction of sulfate-mineral solubilities in continental lakes and shallow groundwaters. Donovan and Rose (1994) applied PHRQPITZ to calculate saturation indices for lake water and groundwater collected from an area of evaporative groundwater-dominated lakes in the North American Plains. These lakes are basic, ranging between pH 8.5 to 10.3, and exhibit a large range in salinity, from 1000 to 264,000 ppm TDS. The precipitation of carbonate minerals limits Ca^{2+} activity, preventing the formation of gypsum. As a result, high concentrations of SO_4 and other dissolved constituents develop. Many of the lakes approach saturation with respect to mirabilite, occasionally blödite, and non-sulfate phases. Calculated reaction sequences using the ion-interaction model indicate that evaporation of water with a composition consistent with shallow groundwater, and to a lesser extent groundwater derived from intermediate depths, is the most probable source for both the final composition of the lakes in this region and for the observed massive accumulations of mirabilite and carbonates.

Prediction of sulfate-mineral solubilities in acid waters. Blowes et al. (1991) applied a modified version of the HMW model to a mine drainage site in eastern Canada. Oxidation of sulfide-rich tailings at this site has led to the development of high concentrations of sulfate (up to 100 g/L) and ferrous iron (up to 50 g/L), and to pH values as low as 0.5 in the pore waters of the tailings. Blowes et al. (1991) added the temperature-dependent interaction parameters for the Fe(II)–H_2SO_4–H_2O system presented by Reardon and Beckie (1987) to PHRQPITZ and provided saturation indices for a number of minerals, including gypsum and melanterite (Fig. 8). The saturation indices for gypsum

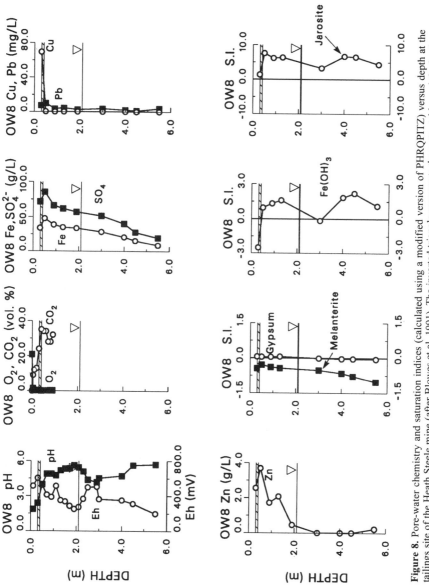

Figure 8. Pore-water chemistry and saturation indices (calculated using a modified version of PHRQPITZ) versus depth at the tailings site of the Heath Steele mine (after Blowes et al. 1991). The inverted triangle represents the water table, and the hardpan layer is shown by the cross-hatched layer.

varied between –0.0487 and 0.0601. Gypsum was observed to be present throughout the depth of the tailings. Saturation indices for melanterite ranged from –1.0228 at the base of the impoundment, to –0.2916 at 50 cm below the impoundment surface. Saturation-index values for melanterite reached a maximum at the depth of a cemented hardpan layer, 30 to 50 cm below the impoundment surface. Mineralogical study indicated that the hardpan layer was composed of tailings cemented by secondary precipitates, principally melanterite and gypsum. The melanterite was crystalline and ranged in color from clear to pale green. Energy-dispersion analyses indicated that samples that had crystallized *in situ* in the tailings were almost pure $FeSO_4$. As interaction parameters were not available for the Fe(III)–H_2SO_4–H_2O system, Blowes et al. (1991) estimated Fe(III) activities through field-measured Pt-electrode Eh values and the Nernst equation. Calculations indicated the pore water to be undersaturated with respect to Fe(OH)$_3$ near the impoundment surface. As the pH increased toward the hardpan layer, the water became supersaturated with respect to Fe(OH)$_3 \cdot 7H_2O$. Saturation indices calculated for jarosite indicated supersaturated conditions throughout the impoundment. Jarosite and goethite were detected in samples of the tailings materials, but accumulations of Fe(III)-bearing minerals were small. Blowes et al. (1991) also applied the ion-association model MINTEQA2 (Allison et al. 1990), for which the database was updated with the association constants from WATEQ4F (Ball and Nordstrom 1987) to include constants for acidic waters. The saturation indices calculated for gypsum using MINTEQA2 indicated consistently highly supersaturated conditions, suggesting that this model does not reliably describe mineral solubility relations in mine drainage waters at high concentrations of dissolved solids.

Schuiling and van Gaans (1997) described geochemical reactions that occur in an acidic discharge pond at a TiO_2 plant at Armyansk, Crimea, Ukraine. Waste acid and fine-grained "phosphogypsum" have been discharged into an isolated bay since 1969. The pH in the pond has slowly decreased to 0.85 over the course of the pond's operation. The acid was initially neutralized by contact with underlying carbonate-rich muds, but the formation of a hardpan has prevented further neutralization of the acid, and concentrations of acid have slowly increased due to evaporation. Precipitation of natrojarosite occurs as the water is concentrated from evaporation. Calcium from the underlying carbonate sediments reacts to form gypsum. As occurs at many base-metal processing sites that utilize jarosite to control Fe (Dutrizac and Jambor, this volume), the jarosite at Armyansk incorporates many trace metals, including V, Cr, and Ni. PHRQPITZ was applied to describe the geochemical evolution of this acid pond through a series of titration calculations (van Gaans and Schuiling 1997). The predicted changes in geochemistry agreed closely with the observed chemistry for various periods in the history of the pond.

One of the issues surrounding application of the Pitzer approach to concentrated waters relates to the thermodynamic treatment of acid. It would be convenient for field scientists to have a method to relate measured conventional hydrogen-ion activity (determined as pH) to the hydrogen-ion activity used in the various Pitzer-type models (see Harvie et al. 1984). Nordstrom et al. (2000) described geochemical reactions at the Iron Mountain mine site in California, where the combination of sulfide-mineral oxidation, the resulting elevated temperatures (35-48°C), and evaporation has led to hundreds of grams per liter of sulfate and Fe, and high concentrations of acid (pH < -2), to remain dissolved in the mine waters. Nordstrom et al. (2000) developed pH-electrode calibration curves using PHRQPITZ and standard H_2SO_4 solutions (Fig. 9) to provide an internally consistent approach for defining pH and to separate HSO_4^- and SO_4^{2-} species at the Iron Mountain site. Among other efforts directed at the measurement of pH in concentrated waters are those by Harvie et al. (1984), Knauss et al. (1990, 1991), Mesmer (1991), Ptacek (1992), and Dorta-Rodriguez et al. (1997). These efforts assist in the speciation of pH-sensitive constituents often required to model sulfate-mineral solubility in natural waters.

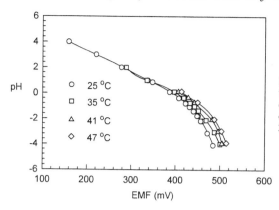

Figure 9. Application of a modified version of PHRQPITZ to construct pH calibration curves for an acid-mine-drainage site (after Nordstrom et al. 2000).

The preceding three examples highlight the need for the evaluation of interaction parameters for predicting ion activities in systems containing both Fe(II) and Fe(III). The ability to predict Fe(II) interaction in concentrated waters is improving (e.g. Ptacek 1992; Ptacek and Reardon 1992; Ptacek and Blowes 1994), but there remains extremely limited information to model Fe(III) interaction in concentrated water. In addition to the approaches taken by Blowes et al. (1991) and van Gaans and Schuiling (1997) to model Fe geochemistry in complex mixtures, other approaches have been taken which mainly involve the representation of the Fe species using other elements as surrogates. For example, King (1998) modelled oxidation rates of Fe(II) in dilute and concentrated waters by representing Fe(II)–SO_4 interaction with Mg–SO_4 parameters. The interaction coefficients for the Fe(III) system using La as a surrogate were estimated by Millero et al. (1995), who also described studies in which Fe(III) interactions were represented using Ga as a surrogate (Zhu et al. 1992).

CONCLUSIONS

The development of ion-interaction models for application to sulfate-bearing natural systems has progressed substantially over the past three decades. Since the original applications of the Pitzer approach for predicting mineral solubilities in natural waters by Harvie and Weare (1980), Harvie et al. (1982, 1984) and others, numerous advances have been made. These advances have been in multiple areas. The increased availability of user-friendly versions of the ion-interaction model has increased its applications to evaluate sulfate-mineral solubility relations in natural and anthropogenically disturbed systems. The equations in the original ion-interaction model have been modified and expanded to provide improved fits between experimental data and model predictions. More ion-association constants for species showing strong association have been included, which has improved predictions at higher concentrations and more extreme temperature and pressure conditions. Probably most notable, the systematic improvement of the databases, including ion-interaction coefficients, data on mineral solubility, and the inclusion of temperature and pressure effects, has improved prediction of the sulfate–mineral solubilities in more complex solutions. The improvements in the databases can be attributed to the addition of experimental data for major and trace constituents, and to new evaluations of ion-interaction parameters and solubility constants in an internally consistent format.

Several limitations remain, preventing the widespread application of the ion-interaction model for predicting sulfate-mineral solubility in complex concentrated solutions. These limitations include incompatibilities among the various databases, and an incomplete understanding and ability to model the mixed-sulfate phases that form from complex concentrated solutions. Inadequacies in our ability to model the formation of non-sulfate phases, such as carbonates and (oxy)hydroxides, and mixed-sulfate phases, will also contribute to error in predictions of aqueous concentrations of dissolved species, which in turn will lead to errors in determining the saturation state of sulfate phases for which data are available. Improvements in these areas, among others, will lead to the development of more robust models for predicting the important role of sulfate-mineral formation in a variety of natural settings.

REFERENCES

Albright JG, Gillespie SM, Rard JA, Miller DG (1998) Ternary solution mutual diffusion coefficients and densities of aqueous mixtures of NaCl and Na_2SO_4 at 298.15 K for six different solute fractions at a total molarity of 1.000 mol·dm^{-3}. J Chem Eng Data 43:668-675

Allison JD, Brown DS, Novo-Gradac KJ (1990) MINTEQA2/PRODEFA2, a geochemical assessment model for environmental systems: Version 3.0 user's manual. US Environmental Protection Agency, Athens, Georgia

Archer DG, Rard JA (1998) Isopiestic investigation of the osmotic and activity coefficients of aqueous $MgSO_4$ and the solubility of $MgSO_4$·$7H_2O$(cr) at 298.15 K: Thermodynamic properties of the $MgSO_4$ + H_2O system to 440 K. J Chem Eng Data 43:791-806

Baes CF Jr, Reardon EJ, Moyer BA (1993) Ion interaction model applied to the $CuSO_4$–H_2SO_4–H_2O system at 25°C. J Phys Chem 97:12343-12348

Ball JW, Nordstrom DK (1987) WATEQ4F—A personal computer Fortran translation of the geochemical model WATEQ2 with revised data base. US Geol Survey Open-File Rept 87-50

Blowes DW, Jambor JL (1990) The pore-water geochemistry and the mineralogy of the vadose zone of sulfide tailings, Waite Amulet, Quebec, Canada. Appl Geochem 5:327-346

Blowes DW, Reardon EJ, Jambor JL, Cherry JA (1991) The formation and potential importance of cemented layers in inactive sulfide mine tailings. Geochim Cosmochim Acta 55:965-978

Boerlage SFE, Kennedy MD, Witkamp GJ, van der Hoek JP, Schippers JC (1999) $BaSO_4$ solubility prediction in reverse osmosis membrane systems. J Membrane Sci 159:47-59

Brantley SL, Crerar DA, Møller NE, Weare JH (1984) Geochemistry of a modern marine evaporite: Bocana de Virrilá, Peru. J Sed Petrol 54:447-462

Christov C (1999) Study of (m_1KCl = m_2MeCl$_2$) (aq) and ($m_1K_2SO_4$) (aq) where m denotes molality and Me denotes Cu, or Ni at the temperature 298.15 K. J Chem Thermodynamics 31:71-83

Clegg SL, Whitfield M (1991) Activity coefficients in natural waters. In Pitzer KS (ed) Activity Coefficients in Electrolyte Solutions. 2nd Edn. CRC Press, Boca Raton, Florida, p 279-434

Clegg SL, Rard JA, Pitzer KS (1994) Thermodynamics properties of 0-6 mol kg^{-1} aqueous sulfuric acid from 273.15 to 328.15 K. J Chem Soc Faraday Trans 90:1875-1894

Clegg SL, Milioto S, Palmer DA (1996) Osmotic and activity coefficients of aqueous $(NH_4)_2SO_4$ as a function of temperature, and aqueous $(NH_4)_2SO_4$–H_2SO_4 mixtures at 298.15 K and 323.15 K. J Chem Eng Data 41:455-467

Clegg SL, Brimblecombe P, Wexler AS (1998) Thermodynamic model of the system H^+–NH_4^+–Na^+–SO_4^{2-}–NO_3^-–Cl^-–H_2O at 298.15 K. J Phys Chem A 102:2155-2171.

De Lange GJ, Boelrijk NAIM, Catalano G, Corselli C, Klinkhammer GP, Middelburg JJ, Müller DW, Ullman WJ, Van Gaans P, Woittiez JRW (1990) Sulphate-related equilibria in the hypersaline brines of the Tyro and Bannock basins, eastern Mediterranean. Marine Chem 31:89-112

Domagalski JL, Eugster HP, Jones BF (1990) Trace metal geochemistry of Walker, Mono and Great Salt lakes. In Spencer RJ, I-Ming Chou (eds) Fluid Mineral Interactions: A Tribute to H.P. Eugster. Geochem Soc Spec Publ 2:315-353

Donovan JJ, Rose AW (1994) Geochemical evolution of lacustrine brines from variable-scale groundwater circulation. J Hydrol 154:35-62

Dorta-Rodríguez R, Barrera-Niebla M, González S, Hernández-Luis F (1997) Calculation of liquid junction potentials. J Electroanalytic Chem 436:173-188

Duchesne J, Reardon EJ (1995) Measurement and prediction of portlandite solubility in alkali solutions. Cement Concrete Res 25:1043-1053

Felmy AR, Weare JH (1986) The prediction of borate mineral equilibria in natural waters: Application to Searles Lake, California. Geochim Cosmochim Acta 50:2271-2783

Felmy AR, Rai D, Amonette JE (1990) The solubility of barite and celestite in sodium sulfate: Evaluation of thermodynamic data. J Soln Chem 19(2):175-185

Gamsjäger H (1993) Solid–solute phase equilibria: From thermodynamic basis information to multicomponent systems. Aquatic Sci 55:314-323

Garrels RM, Thompson ME (1962) A chemical model for seawater at 25°C and one atmosphere total pressure. Am J Sci 260:57-66

Gavrieli I, Starinsky A, Spiro B, Aizenshtat Z, Nielsen H (1995) Mechanisms of sulfate removal from subsurface calcium chloride brines: Heletz-Kokhav oilfields, Israel. Geochim Cosmochim Acta 59:3525-3533

Greenberg JP, Møller N (1989) The prediction of mineral solubilities in natural waters: A chemical equilibrium model for the Na–K–Ca–Cl–SO$_4$–H$_2$O system to high concentration from 0 to 250°C. Geochim Cosmochim Acta 53:2503-2518

Hachimi S, Cote G, Bauer D, Belcadi S (1996) Modélisation des propriétés d'excès et de la distribution des espèces H$^+$, HSO$_4^-$ et SO$_4^{2-}$ à l'aide d'une version tronquée de modèle de Pitzer, dans le cas des mélanges-àerosols H$_2$O–H$_2$SO$_4$ (0 à 27 mol·kg^{-1}). J Chim Phys 93:373-385

Harvie CE, Weare JH (1980) The prediction of mineral solubilities in natural waters: the Na–K–Mg–Ca–Cl–SO$_4$–H$_2$O system from zero to high concentration at 25°C. Geochim Cosmochim Acta 44:981-997

Harvie CE, Weare JH, Hardie LA, Eugster HP (1980) Evaporation of seawater: Calculated mineral sequences. Science 208:498-500

Harvie CE, Eugster HP, Weare JH (1982) Mineral equilibria in the six-component seawater system, Na–K–Mg–Ca–SO$_4$–Cl–H$_2$O: II. Compositions of the saturated solutions. Geochim Cosmochim Acta 46:1603-1618

Harvie CE, Møller N, Weare JH (1984) The prediction of mineral solubilities in natural waters: The Na–K–Mg–Ca–H–Cl–SO$_4$–OH–HCO$_3$–CO$_3$–CO$_2$–H$_2$O system to high ionic strengths at 25°C. Geochim Cosmochim Acta 48:723-751

He S, Morse JW (1993) Prediction of halite, gypsum, and anhydrite solubility in natural brines under subsurface conditions. Comput Geosci 19:1-22

Herbert HJ, Mönig J (1997) Wechselwirkungsreaktionen UTD-relevanter chemisch-toxischer Abfäller mit hochsalinaren Lösungen. Geol Rundsch (in van Gaans 1998)

Hovey, JK, Pitzer, KS, Rard, JA (1993) Thermodynamics of Na$_2$SO$_4$(aq) at temperatures T from 273 K to 373 K and of {(1-y)H$_2$SO$_4$ + yNa$_2$SO$_4$}(aq) at T = 298.15 K. J Chem Thermodyn 25:173-192

Jones BF, Eugster HP, Rettig SL (1977) Hydrochemistry of Lake Magadi basin, Kenya. Geochim Cosmochim Acta 41:53-72

Kalinkin AM, Rumyantsev AV (1996) Thermodynamics of phase equilibria of the K$_2$SO$_4$ + Rb$_2$SO$_4$ + H$_2$O system at 25°C. J Soln Chem 25(7):695-709

Kangro W, Groeneveld A (1962) Konzentrierte wässrige Lösungen, I. Z Phys Chem NF 32:110-126

King DW (1998) Role of carbonate speciation on the oxidation rate of Fe(II) in aquatic systems (1998) Environ Sci Technol 32:2997-3003

Knauss KG, Wolery TJ, Jackson KJ (1990) A new approach to measuring pH in brines and other concentrated electrolytes. Geochim Cosmochim Acta 54:1519-1523

Knauss KG, Wolery TJ, Jackson KJ (1991) Reply to comment by R.E. Mesmer on "A new approach to measuring pH in brines and other concentrated electrolytes". Geochim Cosmochim Acta 55:1177-1179

Königsberger E, Königsberger L, Gamsjäger H (1999) Low-temperature thermodynamic model for the system Na$_2$CO$_3$–MgCO$_3$–CaCO$_3$–H$_2$O. Geochim Cosmochim Acta 63:3105-3119

Krumgalz BS, Millero FJ (1982) Physico-chemical study of the Dead Sea waters. I. Activity coefficients of major ions in Dead Sea water. Marine Chem 11:209-222

Krumgalz BS, Millero FJ (1983) Physico-chemical study of Dead Sea waters. III. On gypsum saturation in Dead Sea waters and their mixtures with Mediterranean Sea water. Marine Chem 13:127-139

Krumgalz BS, Hecht A, Starinsky A, Katz A (2000) Thermodynamic constraints on Dead Sea evaporation: Can the Dead Sea dry up? Chem Geol 165:1-11

Landes KK (1960) The geology of salt. *In* Kaufmann DW (ed) Sodium Chloride: The Production and Properties of Salt and Brine. Am Chem Soc Monogr Ser, Reinhold Publishing, New York, p 28-69

Langmuir D, Melchior D (1985) The geochemistry of Ca, Sr, Ba and Ra sulfates in some deep brines from the Palo Duro Basin, Texas. Geochim Cosmochim Acta 49:2423-2432

Liu X, Millero FJ (1999) The solubility of iron hydroxide in sodium chloride solutions. Geochim Cosmochim Acta 63:3487-3497

Marion GM, Farren RE (1999) Mineral solubilities in the Na–K–Mg–Ca–Cl–SO$_4$–H$_2$O system: A re-evaluation of the sulfate chemistry in the Spencer–Møller–Weare model. Geochim Cosmochim Acta 63:1305-1318

Marliacy P, Hubert N, Schuffenecker L, Solimando R (1998) Use of Pitzer's model to calculate thermodynamic properties of aqueous electrolyte solutions of Na$_2$SO$_4$ + NaCl between 273.15 and 373.15 K. Fluid Phase Equilibria 148:95-106

Marliacy P, Solimando R, Bouroukba M, Schuffenecker L (2000) Thermodynamics of crystallization of sodium sulfate decahydrate in H_2O–NaCl–Na_2SO_4: Application to $Na_2SO_4 \cdot 10H_2O$-based latent heat storage material. Thermochim Acta 344:85-94

Massucci M, Clegg SL, Brimblecombe P (1996) Equilibrium vapor pressure of H_2O above aqueous H_2SO_4 at low temperature. J Chem Eng Data 41:765-778

Mesmer RE (1991) Comments on "A new approach to measuring pH in brines and other concentrated electrolytes" by KG Knauss, TJ Wolery, and KJ Jackson. Geochim Cosmochim Acta 55:1175-1176

Millero FJ, Schreiber DR (1982) Use of the ion pairing model to estimate activity coefficients of the ionic components of natural waters. Am J Sci 282:1508-1540

Millero FJ, Hawke DJ (1992) Ionic interactions of divalent metals in natural waters. Marine Chem 40:19-48

Millero FJ, Yao W, Aicher J (1995) The speciation of Fe(II) and Fe(III) in natural waters. Marine Chem 50:21-39

Møller N (1988) The prediction of mineral solubilities in natural waters: A chemical equilibrium model for the Na–Ca–Cl–SO_4–H_2O system, to high temperature and concentration. Geochim Cosmochim Acta 52:821-837

Møller N, Greenberg JP, Weare JH (1998) Computer modeling for geothermal systems: Predicting carbonate and silica scale formation, CO_2 breakout and H_2S exchange. Transport Porous Media 33:173-204

Monnin C (1989) An ion interaction model for the volumetric properties of natural waters: Density of the solution and partial molal volumes of electrolytes to high concentrations at 25°C. Geochim Cosmochim Acta 53:1177-1188

Monnin C (1990) The influence of pressure on the activity coefficients of the solutes and on the solubility of minerals in the system Na–Ca–Cl–SO_4–H_2O to 200°C and 1 kbar, and to high NaCl concentration. Geochim Cosmochim Acta 54:3265-3282

Monnin C (1999) A thermodynamic model for the solubility of barite and celestite in electrolyte solutions and seawater to 200°C and to 1 kbar. Chem Geol 153:187-209

Monnin C, Schott J (1984) Determination of the solubility products of sodium carbonate minerals and an application to trona deposition in Lake Magadi (Kenya). Geochim Cosmochim Acta 48:571-581

Monnin C, Gallinari (1988) The solubility of celestite and barite in electrolyte solutions and natural waters at 25°C: a thermodynamic study. Chem Geol 71:283-296

Monnin C, Ramboz C (1996) The anhydrite saturation index of the ponded brines and sediment pore waters of the Red Sea deeps. Chem Geol 127:141-159

Morse JW, Marion GM (1999) The role of carbonates in the evolution of early Martian oceans. Am J Sci 299:738-761

Nikolaev VP, Dikaya NN, Stanish-Levitska M, Vorobév AF (1989) Activity coefficients of iron(II) sulfate in the systems $FeSO_4$–H_2O and $FeSO_4$–$FeCl_2$–H_2O at 298 K. Translated from Zhur Obschei Khimii 59:241-244

Nordstrom DK, Alpers CN (1999) Negative pH, efflorescent mineralogy, and the challenge of environmental restoration at the Iron Mountain Superfund site, California. Proc Natl Acad Sci USA 96:3455-3462

Nordstrom DK, Ball JW (1984) Chemical models, computer programs and metal complexation in natural waters. In Kramer CJM, Duinker JC (eds) Complexation of trace metals in natural waters. Martinus Nijhoff/Dr W Junk Publishers, Hague, The Netherlands, p 149-164

Nordstrom DK, Munoz JL (1994) Geochemical Thermodynamics. Blackwell Scientific, London, UK

Nordstrom DK, Alpers CN, Ptacek CJ, Blowes DW (2000) Negative pH and extremely acidic mine waters from Iron Mountain, California. Environ Sci Technol 34:254-258

Nordstrom DK, Plummer LN, Langmuir D, Busenberg E, May HM, Jones BF, Parkhurst DL (1990) Revised chemical equilibrium data for major water–mineral reactions and their limitations. In Melchior DC, Bassett RL (eds) Chemical Modeling of Aqueous Systems II. Am Chem Soc Symp Ser 416:398-413

Ob"il M, Majer V, Hefter GT, Hynek V (1997) Densities of apparent molar volumes of Na_2SO_4(aq) and K_2SO_4(aq) at temperatures from 298 K to 573 K and at pressures up to 30 Mpa. J Chem Eng Data 42:137-142

Oykova TG, Balarew C (1974) Thermodynamic study of magnesium sulfate–ferrosulfate–water system at 25°C. CR Acad Bulg Sci 27:1211-1214

Pabalan RT, Pitzer KS (1987) Thermodynamics of concentrated electrolyte mixtures and the prediction of mineral solubilities to high temperatures for mixtures in the system Na–K–Mg–Cl–SO_4–OH–H_2O. Geochim Cosmochim Acta 51:2429-2443

Pabalan RT, Pitzer KS (1991) Mineral solubilities in electrolyte solutions. In Pitzer KS (ed) Activity Coefficients in Electrolyte Solutions. CRC Press, Boca Raton, Florida, p 435-489

Paige CR, Kornicker WA, Hileman, OE Jr, Snodgrass WJ (1992) Modelling solution equilibria for uranium ore processing: The $PbSO_4-H_2SO_4-H_2O$ and $PbSO_4-Na_2SO_4-H_2O$ systems. Geochim Cosmochim Acta 56:1165-1173

Paige CR, Kornicker WA, Hileman OE Jr, Snodgrass WJ (1998) Solution equilibria for uranium ore processing: The $BaSO_4-H_2SO_4-H_2O$ system and the $RaSO_4-H_2SO_4-H_2O$ system. Geochim Cosmochim Acta 62:15-23

Parkhurst DL (1990) Ion-association models and mean activity coefficients of various salts. *In* Melchior DC, Bassett RL (eds) Chemical Modeling of Aqueous Systems II. Am Chem Soc Sym Ser 416:30-43

Parkhurst DL, Thorstenson DC, Plummer LN (1985) PHREEQE—A computer program for geochemical calculations. US Geol Survey Water-Resources Invest Rept 80-96

Pátzay G, Stáhl G, Kármán FH, Kálmán E (1998) Modeling of scale formation and corrosion from geothermal water. Electrochim Acta 43:137-147

Pavicevic V, Ninkovic R, Todorovic Miladinovic J (1999) Osmotic and activity coefficients of $\{yNaH_2PO_4 + (1-y)Na_2SO_4\}$(aq) at the temperature 298.15 K. Fluid Phase Equil 164:275-284

Pierrot D, Millero FJ, Roy LN, Roy RN, Doneski A, Niederschmidt J (1997) The activity coefficients of $HCl-Na_2SO_4$ solutions from 0 to 50°C and ionic strengths up to 6 molal. J Soln Chem 26:31-45

Pitzer KS (1973) Thermodynamics of electrolytes: I. Theoretical basis and general equations. J Phys Chem 77:2300-2308

Pitzer KS (1974) Thermodynamics of electrolytes: III. Activity and osmotic coefficients for 2-2 electrolytes. J Soln Chem 3:539-546

Pitzer KS (1975) Thermodynamics of electrolytes: V. Effects of higher-order electrostatic terms. J Soln Chem 4:249-265

Pitzer KS (1979a) Thermodynamics of aqueous electrolytes at various temperatures, pressures, and compositions. *In* Newman SA, Barner HE, Klein M, Sandler SI (eds) Thermodynamics of Aqueous Systems with Industrial Applications. Am Chem Soc Sym Ser 133:451-466

Pitzer KS (1979b) Theory: Ion interaction approach. *In* Pytkowicz RM (ed) Activity Coefficients in Electrolyte Solutions. CRC Press, Boca Raton, Florida, p 158-208

Pitzer KS (1982) Thermodynamics of unsymmetrical electrolyte mixtures. Enthalpy and heat capacity. J Phys Chem 87:2360-2364

Pitzer KS (1986) Theoretical considerations of solubility with emphasis on mixed aqueous electrolytes. Pure Appl Chem 58:1599-1610

Pitzer KS (1991) Ion interaction approach: Theory and data correlation. *In* Pitzer KS (ed) Activity Coefficients in Electrolyte Solutions. CRC Press, Boca Raton, Florida, p 75-153

Pitzer KS (1993) Thermodynamics of natural and industrial waters. J Chem Thermodynamics 25:7-26

Pitzer KS, Kim JJ (1974) Thermodynamics of electrolytes: IV. Activity and osmotic coefficients for mixed electrolytes. J Am Chem Soc 96:5701-5707

Pitzer KS, Mayorga G (1973) Thermodynamics of electrolytes: II. Activity and osmotic coefficients for strong electrolytes with one or both ions univalent. Theoretical basis and general equations. J Phys Chem 77:268-277

Plummer LN, Parkhurst DL, Fleming GW, Dunkle SA (1988) A computer program incorporating Pitzer's equations for calculation of geochemical reactions in brines. US Geol Survey Water-Resources Invest Rept 88-4153

Potukuchi S, Wexler AS (1995) Identifying solid–aqueous-phase transitions in atmospheric aerosols: II. Acidic solutions. Atmosph Environ 29:3357-3364

Ptacek CJ (1992) Experimental determination of siderite solubility in high ionic-strength aqueous solutions. PhD Thesis, University of Waterloo, Waterloo, Ontario

Ptacek CJ, Blowes DW (1994) Influence of siderite on the pore-water chemistry of inactive mine-tailings impoundments. *In* Alpers CN, Blowes, DW (eds) Environmental Geochemistry of Sulfide Oxidation. Am Chem Soc Symp Ser 550:172-189

Ptacek CJ, Reardon EJ (1992) Solubility of siderite ($FeCO_3$) in concentrated NaCl and Na_2SO_4 solutions at 25°C. *In* Karaka YK, Maest AS (eds) Water–rock Interaction 7. AA Balkema, Rotterdam, The Netherlands, p 181-184

Raju K, Atkinson G (1988) Thermodynamics of "scale" mineral solubilities: 1. $BaSO_4$(s) in H_2O and aqueous NaCl. J Chem Eng Data 33:490-495

Raju K, Atkinson G (1989) Thermodynamics of "scale" mineral solubilities: 2. $SrSO_4$(s) in aqueous NaCl. J Chem Eng Data 34:361-364

Raju K, Atkinson G (1990) The thermodynamics of "scale" mineral solubilities: 3. Calcium sulfate in aqueous NaCl. J Chem Eng Data 35:361-367

Rard JA (1992) Isopiestic determination of the osmotic and activity coefficients of $\{(1-y)H_2SO_4 + yNa_2SO_4\}$(aq) at the temperature 298.15 K: 11. Results for y = (0.12471, 0.24962, and 0.37439). J Chem Thermodyn 24:45-66

Rard JA (1997) Isopiestic determination of the osmotic and activity coefficients of $\{z H_2SO_4 + (1-z)MgSO_4\}$(aq) at the temperature T = 298.15 K: I. Results for z = (0.85811, 0.71539, and 0.57353). J Chem Thermodyn 29:533-555

Rard JA, Clegg SL (1997) Critical evaluation of the thermodynamic properties of aqueous calcium chloride. 1. Osmotic and activity coefficients of 0-10.77 mol·kg^{-1} aqueous calcium chloride solutions at 298.15 K and correlation with extended Pitzer ion-interaction models. J Chem Eng Data 42:819-849

Rard JA, Clegg SL (1999) Isopiestic determination of the osmotic and activity coefficients of $\{z H_2SO_4 + (1-z)MgSO_4\}$(aq) at T = 298.15 K: II. Results for z = (0.43040, 0.28758, and 0.14399), and analysis with Pitzer's model. J Chem Thermodyn 31:399-429

Reardon EJ (1988) Ion interaction parameters for $AlSO_4$ and application to the prediction of metal sulfate solubility in binary salt systems. J Phys Chem 92:6426-6431

Reardon (1989) Ion interaction model applied to equilibria in the $NiSO_4$–H_2SO_4–H_2O system. J Phys Chem 93:4630-4636

Reardon EJ, Armstrong DK (1987) Celestite ($SrSO_{4(s)}$) solubility in water, seawater and NaCl solution. Geochim Cosmochim Acta 51:63-72

Reardon EJ, Beckie RD (1987) Modelling chemical equilibria of acid-mine drainage: The $FeSO_4$–H_2SO_4–H_2O system. Geochim Cosmochim Acta 51:2355-2368

Ridley MK, Wesolowski DJ, Palmer DA, Bénézeth P, Kettler RM (1997) Effect of sulfate on the release of Al^{3+} from gibbsite in low-temperature acidic waters. Environ Sci Technol 31:1922-1925

Ridley MK, Wesolowski DJ, Palmer DA, Kettler RM (1999) Association quotients of aluminum sulphate complexes in NaCl media from 50 to 125°C: Results of a potentiometric and solubility study. Geochim Cosmochim Acta 63:459-472

Robinson RA, Stokes RH (1970) Electrolyte solutions. Butterworth and Co, London, UK

Schuiling RD, van Gaans PFM (1997) The waste sulfuric acid lake of the TiO_2-plant at Armyansk, Crimea, Ukraine. Part I. Self-sealing as an environmental protection mechanism. Appl Geochem 12:181-186

Siegel MD, Anderholm S (1994) Geochemical evolution of groundwater in the Culebra Dolomite near the Waste Isolation Pilot Plant, southeastern New Mexico, USA. Geochim Cosmochim Acta 58: 2299-2323

Spencer RJ, Møller N, Weare JH (1990) The prediction of mineral solubilities in natural waters: A chemical equilibrium model for the Na–K–Ca–Mg–Cl–SO_4–H_2O system at temperatures below 25°C. Geochim Cosmochim Acta 54:575-590

Stokes RH, Robinson RA (1941) A thermodynamic study of bivalent metal halides in aqueous solutions. Part VIII. The activity coefficient of ferrous chloride. Trans Faraday Soc 37:419-421

Sylvester LF, Pitzer KS (1976) Thermodynamics of electrolytes: X. Enthalpy and the effect of temperature on the activity coefficients. J Soln Chem 7:327-337

Sylvester LF, Pitzer KS (1977) Thermodynamics of electrolytes: 8. High-temperature properties, including enthalpy and heat capacity, with application to sodium chloride. J Phys Chem 81:1822-1828

van Cappellen P, Viollier E, Roychoudhury A, Clark L, Ingal E, Lowe K, DiChristina T, (1998) Biogeochemical cycles of manganese and iron at the oxic–anoxic transition of a stratified marine basin (Orca Basin, Gulf of Mexico). Environ Sci Technol 32:2931-2939

van Gaans PFM (1991) Extracting robust and physically meaningful Pitzer parameters for thermodynamic properties of electrolytes from experimental data: Ridge regression as an aid to the problem of intercorrelation. J Soln Chem 20:703-730

van Gaans PFM (1998) The role of modelling in geochemical engineering: A (re)view. J Geochem Explor 62:41-55

van Gaans PFM, Schuiling RD (1997) The waste sulfuric acid lake of the TiO_2-plant at Armyansk, Crimea, Ukraine. Part II. Modelling the chemical evolution with PHRQPITZ. Appl Geochem 12:187-201

Weare (1987) Models of mineral solubility in concentrated brines with application to field observations. *In* Carmichael ISE, Eugster HP (eds) Thermodynamic Modeling of Geological Materials: Minerals, Fluids and Melts. Rev Mineral 17:143-176

Whitfield (1975) The extension of chemical models for sea water to include trace components at 25°C and 1 atm pressure. Geochim Cosmochim Acta 39:1545-1557

Williams-Stroud SC (1994) Solution to the paradox? Results of some chemical equilibrium and mass balance calculations applied to the Paradox Basin evaporite deposit. Am J Sci 294:1189-1228

Yechieli Y, Ronen D (1997) Early diagenesis of highly saline lake sediments after exposure. Chem Geol 138:93-106

Zhu X, Prospero JM, Millero FJ, Savoie DL, Brass GW (1992) The solubility of ferric iron in marine mineral aerosol solutions at ambient relative humidities. Marine Chem 38:91-107 (in Millero et al. 1995)

12 Stable Isotope Systematics of Sulfate Minerals

Robert R. Seal, II

U.S. Geological Survey
954 National Center
Reston, Virginia 20192

Charles N. Alpers

U.S. Geological Survey
6000 J Street
Sacramento, California 95819

Robert O. Rye

U.S. Geological Survey
963 Denver Federal Center
Denver, Colorado 80225

INTRODUCTION

Stable isotope studies of sulfate minerals are especially useful for unraveling the geochemical history of geological systems. All sulfate minerals can yield sulfur and oxygen isotope data. Hydrous sulfate minerals, such as gypsum, also yield oxygen and hydrogen isotope data for the water of hydration, and more complex sulfate minerals, such as alunite and jarosite also yield oxygen and hydrogen isotope data from hydroxyl sites. Applications of stable isotope data can be divided into two broad categories: geothermometry and tracer studies. The equilibrium partitioning of stable isotopes between two substances, such as the isotopes of sulfur between barite and pyrite, is a function of temperature. Studies can also use stable isotopes as a tracer to fingerprint various sources of hydrogen, oxygen, and sulfur, and to identify physical and chemical processes such as evaporation of water, mixing of waters, and reduction of sulfate to sulfide.

Studies of sulfate minerals range from low-temperature surficial processes associated with the evaporation of seawater to form evaporite deposits to high-temperature magmatic-hydrothermal processes associated with the formation of base- and precious-metal deposits. Studies have been conducted on scales from submicroscopic chemical processes associated with the weathering of pyrite to global processes affecting the sulfur budget of the oceans. Sulfate isotope studies provide important information to investigations of energy and mineral resources, environmental geochemistry, paleoclimates, oceanography (past and present), sedimentary, igneous, and metamorphic processes, Earth systems, geomicro-biology, and hydrology.

One of the most important aspects of understanding and interpreting the stable isotope characteristics of sulfate minerals is the complex interplay between equilibrium and kinetic chemical and isotopic processes. With few exceptions, sulfate minerals are precipitated from water or have extensively interacted with water at some time in their history. Because of this nearly ubiquitous association with water, the kinetics of isotopic exchange reactions among dissolved species and solids are fundamental in dictating the isotopic composition of sulfate minerals. In general, the heavier isotope of sulfur is enriched in the higher oxidation state, such that under equilibrium conditions, sulfate minerals (e.g. barite, anhydrite) are expected to be enriched in the heavy isotope relative to disulfide minerals (e.g. pyrite, marcasite), which in turn are expected to be enriched relative to monosulfide minerals (e.g.

1529-6466/00/0040-0012$05.00

pyrrhotite, sphalerite, galena) (Sakai 1968, Bachinski 1969). The kinetics of isotopic exchange among minerals with sulfur at the same oxidation state, such as sphalerite, and galena, are such that equilibrium is commonly observed. In contrast, isotopic equilibrium for exchange reactions between minerals of different oxidation states depends on factors such as the pH, time and temperature of reaction, the direction of reaction, fluid composition, and the presence or absence of catalysts (Ohmoto and Lasaga 1982). The kinetics of oxygen isotope exchange between dissolved sulfate and water are extremely sluggish. Extrapolation of the high-temperature (100 to 300°C) isotopic exchange kinetic data of Chiba and Sakai (1985) to ambient temperatures suggests that it would take several billions of years for dissolved sulfate and seawater to reach oxygen isotopic equilibrium. In contrast, the residence time of sulfate in the oceans is only 7.9 million years (Holland 1978). However, at higher temperatures (>200°C), oxygen isotopic exchange is sufficiently rapid to permit application of sulfate isotope geothermometry to geothermal systems and hydrothermal mineral deposits. In general, equilibrium prevails at low pH and high temperatures, whereas kinetic factors preclude equilibrium at low temperatures even at low pH. Thus, the sluggish kinetics of sulfur and oxygen isotope exchange reaction at low temperatures impair the use of these isotopes to understand the conditions of formation of sulfate minerals in these environments. However, because of these slow kinetics, the oxygen and sulfur isotopic compositions of sulfate minerals may preserve a record of the sources and processes that initially produced the dissolved sulfate, because the isotope ratios may not re-equilibrate during fluid transport and mineral precipitation.

The first part of this chapter is designed to provide the reader with a basic understanding of the principles that form the foundations of stable isotope geochemistry. Next, an overview of analytical methods used to determine the stable isotope composition of sulfate minerals is presented. This overview is followed by a discussion of geochemical processes that determine the stable isotope characteristics of sulfate minerals and related compounds. The chapter then concludes with an examination of the stable isotope systematics of sulfate minerals in a variety of geochemical environments.

FUNDAMENTAL ASPECTS OF STABLE ISOTOPE GEOCHEMISTRY

An isotope of an element is defined by the total number of protons (Z) and neutrons (N) present, which sum together to give the atomic mass (A). For example, the element oxygen, abbreviated as "O", is defined by the presence of 8 protons, but can have either 8, 9, or 10 neutrons. These various combinations of 8 protons and either 8, 9, or 10 neutrons are the isotopes of oxygen and are denoted as oxygen-16 (8Z + 8N = 16A), oxygen-17 (8Z + 9N = 17A), and oxygen-18 (8Z + 10N = 18A). The isotopes of oxygen are abbreviated as ^{16}O, ^{17}O, and ^{18}O, respectively. For oxygen, all of these isotopes are stable, meaning that they do not undergo radioactive decay at significant rates. In contrast, carbon (Z = 6) has two stable isotopes, ^{12}C and ^{13}C, and one unstable isotope, ^{14}C. Unstable, or radiogenic isotopes undergo radioactive decay. For example, ^{14}C decays to ^{14}N, a stable isotope. The decay of radiogenic isotopes makes many of them useful for dating purposes. This chapter is concerned only with naturally occurring stable isotopes.

Sulfur (Z = 16) has four stable isotopes: ^{32}S, ^{33}S, ^{34}S, and ^{36}S, with approximate terrestrial abundances of 95.02, 0.75, 4.21, and 0.02 percent, respectively (Macnamara and Thode 1950). Oxygen (Z = 8) has three stable isotopes: ^{16}O, ^{17}O, and ^{18}O, with approximate terrestrial abundances of 99.763, 0.0375, and 0.1995 percent, respectively (Garlick 1969). Hydrogen (Z = 1) has two stable isotopes: ^{1}H, and ^{2}H, with approximate terrestrial abundances of 99.9844, and 0.0156 percent, respectively (Way et al. 1950). Deuterium, ^{2}H, is commonly abbreviated as "D." In this context, ^{1}H is further abbreviated merely as "H."

Stable isotope geochemistry is concerned primarily with the relative partitioning of stable isotopes among substances (i.e. changes in the ratios of isotopes), rather than their absolute abundances. Differences in the partitioning of stable isotopes, otherwise known as fractionation, are due to equilibrium and kinetic effects. In general, heavier isotopes form more stable bonds; molecules of different masses react at different rates (O'Neil 1986). Stable isotope geochemists investigate the variations of a minor isotope of an element relative to a major isotope of the element. For sulfate minerals, the principal ratios of concern are D/H, $^{18}O/^{16}O$, and $^{34}S/^{32}S$. For oxygen, the ratio $^{18}O/^{16}O$ is approximately 0.00204. Fractionation processes will typically cause variations in this ratio in the fifth or sixth decimal places. Because we are concerned with variations in isotopic ratios that are relatively small, the isotopic composition of substances is expressed in the "delta" (δ) notation as permil (or parts per thousand) variation relative to a reference material. The δ-notation for the $^{18}O/^{16}O$ composition of a substance is defined as:

$$\delta^{18}O = \left(\frac{\left(^{18}O/^{16}O\right)_{sample} - \left(^{18}O/^{16}O\right)_{reference}}{\left(^{18}O/^{16}O\right)_{reference}} \right) \times 1000 \tag{1}$$

expressed in values of parts per thousand or permil (‰), which can also be found in the literature spelled "per mil", "per mill", and "per mille". For most applications, the agreed upon reference for oxygen isotopes is Vienna Standard Mean Ocean Water (VSMOW), for which $\delta^{18}O = 0.0$ ‰ by definition. VSMOW is a hypothetical water with oxygen and hydrogen isotopic compositions similar to those of average ocean water. A sample with a positive $\delta^{18}O_{VSMOW}$ value, such as +5.0 ‰, is enriched in ^{18}O relative to VSMOW. Conversely, a sample with a negative $\delta^{18}O_{VSMOW}$ value, such as -5.0 ‰, is depleted in ^{18}O relative to VSMOW. The absolute $^{18}O/^{16}O$ ratio for VSMOW is $2005.2 \pm 0.45 \times 10^{-6}$ (Baertschi 1976). The Vienna Peedee Belemnite (VPDB) reference is only used for oxygen isotope values in carbonates, with application to studies of oceanic paleotemperature and sedimentary carbonate petrology. The VPDB reference was originally defined by a belemnite sample from the Peedee Formation in North Carolina.

Hydrogen isotopes are also defined relative to VSMOW, using the D/H ratio, such that the δD value for VSMOW is 0.0 ‰ by definition. The absolute D/H ratio for VSMOW is $155.76 \pm 0.05 \times 10^{-6}$ (Hageman et al. 1970). The selection of seawater (VSMOW) as the reference for hydrogen isotope and most oxygen isotope applications in geochemistry is useful because the oceans are the foundation of the hydrological cycle.

For sulfur isotopes, the $\delta^{34}S$ value for the $^{34}S/^{32}S$ ratio is defined relative to Vienna Cañon Diablo Troilite (VCDT) with $\delta^{34}S = 0.0$ ‰ by definition. The reference was originally defined by the isotopic composition of troilite (FeS) from the Cañon Diablo iron meteorite. The absolute $^{34}S/^{32}S$ ratio for Cañon Diablo troilite is 4500.45×10^{-6} (Ault and Jensen 1963). The selection of a meteoritic sulfide mineral as the reference for sulfur is useful because meteoritic sulfide is thought to represent the primordial sulfur isotopic composition of Earth (Nielsen et al. 1991). Thus, any variations in the isotopic composition of terrestrial sulfur relative to VCDT reflect differentiation since the formation of Earth.

For oxygen and sulfur, which have more than two stable isotopes, $^{18}O/^{16}O$ and $^{34}S/^{32}S$ are the ratios that are almost exclusively measured in studies of terrestrial systems. These ratios are chosen for two main reasons. First, they represent the most abundant isotopes of these elements, which facilitates their ease of analysis. Second, isotopic fractionation is governed by mass balance such that different isotopic ratios tend to vary systematically with one another in proportions that can be approximated by the mass differences between the isotopic ratios for closed, terrestrial systems being affected by chemical and physical processes. In other words, the variations in the $^{17}O/^{16}O$ ratio of a sample will be approximately half that of the $^{18}O/^{16}O$ ratio because of the relative differences in masses. This linear fractionation trend due to physical and chemical processes is called a "terrestrial fractionation line" (Clayton 1986). Except as recently shown by Thiemens (1999a,b) and mentioned below, isotopic data from all samples of terrestrial origins conform to terrestrial fractionation systematics. In contrast, most meteorites plot away from the terrestrial fractionation line in terms of $\delta^{17}O$ and $\delta^{18}O$; this deviation has been attributed to the different pathways of nucleosynthesis of the different oxygen isotopes and their subsequent and heterogeneous contributions to the solar nebula (Clayton 1986). All sulfide minerals from terrestrial samples and meteorites fall on a single mass-fractionation line suggesting that any heterogeneities that may have existed were homogenized early in solar history (Hulston and Thode 1965a,b). The metallic phase of iron meteorites shows anomalies in its ^{33}S and ^{36}S compositions, which have been attributed to cosmic-ray spallation (Hulston and Thode 1965b). Recent work by Thiemens (1999a,b) has shown that atmospheric sulfate aerosols have isotopic compositions that reflect mass-independent oxygen and sulfur isotopic fractionation. These anomalies have been recognized in the sediment record.

The partitioning of stable isotopes between two substances, A and B, is quantitatively described by a fractionation factor which is defined as

$$\alpha_{A-B} = \frac{R_A}{R_B} \tag{2}$$

where R is either D/H, $^{18}O/^{16}O$, or $^{34}S/^{32}S$. This equation can be recast in terms of δ values as

$$\alpha_{A-B} = \frac{1+\dfrac{\delta_A}{1000}}{1+\dfrac{\delta_B}{1000}} = \frac{1000+\delta_A}{1000+\delta_B} \tag{3}$$

Values of α are typically near unity, with variations normally in the third decimal place (1.00x). For example, the $^{18}O/^{16}O$ fractionation between anhydrite and water at 300°C yields an $\alpha_{anhydrite-H_2O}$ value of 1.0090. Thus, anhydrite is enriched in ^{18}O relative to water by 9.0 permil (i.e. the fractionation equals + 9.0 ‰). For an α value less than unity, such as $\alpha_{H_2O-anhydrite}$ which equals 0.991, the water is depleted in ^{18}O relative to anhydrite by 9.0 permil (i.e. the fractionation equals -9.0 ‰). In the literature, fractionation factors may be expressed in a variety of ways including α, 1000 lnα, and Δ, among others. The value Δ_{A-B} is defined as

$$\Delta_{A-B} = \delta_A - \delta_B \tag{4}$$

A convenient mathematical relationship is that 1000 ln(1.00X) is approximately equal to X, so that

$$\Delta_{A-B} \approx 1000 \ln\alpha_{A-B} \tag{5}$$

Isotopic fractionations may also be defined in terms of an enrichment factor (ε), where

$$\varepsilon_{A-B} = (\alpha_{A-B} - 1) \times 1000 \tag{6}$$

ANALYTICAL METHODS

Several procedures are available to determine the oxygen, sulfur, and hydrogen isotopic compositions of sulfate minerals. Conventional analyses typically involve mineral-separation procedures that may include hand-picking or gravimetric techniques (heavy liquids, panning, etc.) and (or) wet-chemical techniques. Once a suitable concentration of the desired compound is obtained, the element of interest (O, S, or H) is extracted and converted to a gaseous form that is amenable to mass-spectrometric analysis. For oxygen isotopes, the traditionally preferred gas for analysis is CO_2; for sulfur it is SO_2; and for hydrogen it is H_2. The amount of sample required varies among laboratories but typically ranges from 5 to 20 mg of pure mineral separate for $\delta^{18}O$ and $\delta^{34}S$, and up to several hundred mg for δD, using conventional techniques. Typical analytical uncertainties (1σ) for conventional techniques are ± 0.1 ‰ for $\delta^{18}O$ and $\delta^{34}S$, and ± 1 ‰ for δD.

Sulfate for oxygen and sulfur isotope analysis from simple sulfate minerals, such as barite, anhydrite, gypsum, and many of the transition metal salts (melanterite, chalcanthite, etc.) can be purified by reacting with a 5 mass % Na_2CO_3 solution to leach sulfate, followed by filtration of the solution, acidification (pH $\approx$ 4), and reprecipitation of the sulfate as barite by addition of a 10 mass % $BaCl_2 \cdot 2H_2O$ solution. Similarly, aqueous sulfate can be extracted by filtration, acidification, and precipitation by the addition of $BaCl_2 \cdot 2H_2O$ solution (Taylor et al. 1984). Details of the total isotopic analysis (δD, $\delta^{18}O_{SO_4}$, $\delta^{18}O_{OH}$, $\delta^{34}S$) of alunite-group minerals, including mineral separation techniques, have been described by Wasserman et al. (1992). Field and laboratory techniques for the extraction of dissolved sulfate and sulfide from waters for isotopic analysis have been summarized by Carmody et al. (1998).

For conventional analysis of $\delta^{18}O_{SO_4}$, the CO_2 is generally prepared by one of two methods. In the first method, $BaSO_4$ is mixed with graphite in a platinum-foil boat through which a current is passed under vacuum; CO_2 is produced directly and any CO produced is converted to CO_2 by plating out excess carbon onto electroplates (Rafter 1967, Sakai and Krouse 1971). In the second method, sulfate minerals are reacted with BrF_5 or ClF_3 at elevated temperatures (550 to 650°C) to produce O_2, which is converted to CO_2 for isotopic analysis by reaction with a hot graphite rod under vacuum (Clayton and Mayeda 1963, Borthwick and Harmon 1982). Because oxygen extraction by fluorination techniques is not quantitative, presumably due to the production of S-O-F gases, a kinetic fractionation correction must be applied to CO_2 data from this method (Pickthorn and O'Neil 1985, Wasserman et al. 1992). In hydroxy-sulfate minerals that do not have water of hydration (e.g. alunite and jarosite), the $\delta^{18}O_{OH}$ is calculated by difference from the bulk $\delta^{18}O$ of the complex sulfate mineral, where the bulk $\delta^{18}O$ is determined by fluorination techniques and the $\delta^{18}O_{SO_4}$ is determined on sulfate extracted from the mineral (Pickthorn and O'Neil 1985, Wasserman et al. 1992).

For conventional $\delta^{34}S$ analysis of sulfate minerals, SO_2 is produced for analysis by reacting the sulfate mineral with an oxidant (CuO, Cu_2O, or V_2O_5) and silica glass at elevated temperatures (1000 to 1200°C) under vacuum (Holt and Engelkemeier 1970, Haur et al. 1973, Coleman and Moore 1978). Other conventional techniques for the $\delta^{34}S$ and $\delta^{18}O$ analysis of sulfate minerals have been summarized by Rees and Holt (1991). For δD analysis, minerals are dehydrated under vacuum and the water is converted to H_2 for isotopic analysis either by reaction with uranium (Bigeleisen et al. 1952) or zinc (Coleman et al. 1982, Vennemann and O'Neil 1993).

Isotopic analysis for all of these isotopes is conducted on a gas-source, sector-type, isotope ratio mass spectrometer. In gas-source mass spectrometers, gas molecules (CO_2, SO_2, or H_2) are ionized to positively charged particles, such as SO_2^+, which are accelerated through a voltage gradient. The ion beam passes through a magnetic field, which causes separation of various masses such as 64 ($^{32}S^{16}O_2$) and 66 ($^{34}S^{16}O_2$, $^{34}S^{18}O^{16}O$). In conventional dual-inlet mass spectrometers, a sample gas is measured alternately with a reference gas. The beam currents are measured in faraday cups and can be related to the isotopic ratio when the sample and standard gases are compared.

Recent technological advances over the past decade have opened up two new frontiers in stable isotope analysis of sulfate minerals. One area is the *in situ* microanalysis of minerals, and the other is the mass production of data from small samples. For *in situ* analysis, a limited amount of sulfur isotope data has been generated from samples of anhydrite and barite using the ion microprobe or secondary ion mass spectrometer (SIMS)(McKibben and Riciputi 1998). The ion microprobe bombards a sample with a beam of charged Cs or O. The ion beam causes the sample to be ablated as secondary ionic species, which are measured by a mass spectrometer. Spatial resolution is on the order of 30 to 60 μm with an analytical uncertainty of ±1 ‰ for sulfur isotope analyses using a SHRIMP, Sensitive High Mass Resolution Ion Microprobe (McKibben et al. 1996).

The other recent advance is the development of continuous-flow techniques that use a combination of an elemental analyzer and gas chromatograph for online combustion and purification of gases that are carried in a He stream directly into the ion source of a modern mass spectrometer. Continuous-flow systems can measure the sulfur and oxygen isotopic ratios of sulfate samples in the microgram range, compared to the milligram range for conventional techniques (Giesemann et al. 1994, Kornexl et al. 1999). Sample gases are prepared by on-line peripheral devices such as elemental analyzers that are capable of processing 50 to 100 samples in a highly automated fashion. Furthermore, most sulfur isotope measurements can be made without mineral purification, if bulk sulfur data are all that is desired.

REFERENCE RESERVOIRS

The isotopic compositions of substances (minerals, waters, gases, biological material) are typically discussed in terms of geochemical reservoirs which may have served as sources for these constituents. For waters, the main reservoirs are seawater, meteoric water, and magmatic water. Modern seawater has a fairly uniform isotopic composition, with δD and $\delta^{18}O$ ranging from -7 to +5 ‰, and -1.0 to +0.5 ‰, respectively, and with mean values close to the composition of VSMOW (δD = 0.0 ‰; $\delta^{18}O$ = 0.0 ‰) (Epstein and Mayeda 1953, Friedman 1953, Craig and Gordon 1965). Isotopic studies of ancient marine carbonates suggest that the $\delta^{18}O$ of seawater in the early Phanerozoic may have been as low as -9.0 ‰ (Veizer et al. 1997).

The δD and $\delta^{18}O$ of meteoric waters, those that originate as precipitation, vary systematically and predictably (Epstein and Mayeda 1953, Friedman 1953, Craig 1961, Dansgaard 1964). The linear variation shown by meteoric waters is known as the "global meteoric water line" (Fig. 1; Craig 1961) and is described by the equation:

$$\delta D = 8\, \delta^{18}O + 10 \tag{7}$$

Subsequent work, published more than 30 years later (Rozanski et al. 1993) has slightly refined this empirical relation:

$$\delta D = 8.13\, \delta^{18}O + 10.8 \tag{8}$$

The linear relationship is the result of the kinetically controlled evaporation of moisture dominantly from the oceans, and its subsequent isotopic distillation during condensation and precipitation. The linear relationship is well-described as a Rayleigh distillation process, discussed in the next section. Numerous factors determine the position of the isotopic composition of precipitation along the meteoric water line; chief among these are temperature, which is a function of latitude, altitude, and season, followed by the "continent effect" and the "rainout effect". The "continent effect" refers to systematic decrease of the isotopic composition of precipitation as one moves toward the interior of continents. The "rainout effect" refers to the systematic decrease of the isotopic

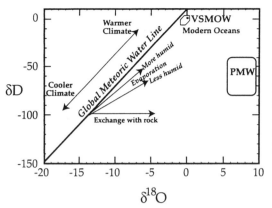

Figure 1. Plot of δD and δ^{18}O for various water reservoirs and trends for physical and chemical processes that can alter the isotopic composition of water. VSMOW indicates Vienna Standard Mean Ocean Water; PMW indicates primary magmatic water. All isotopic values in permil (VSMOW).

composition of precipitation with increased amounts of precipitation. Both effects are the result of fractionation associated with distillation. The temperature effect results in a continuous increase in δD and δ^{18}O of precipitation from cold to warm climates (Fig. 1).

The temperature dependence of the isotopic composition of meteoric waters has been used as the basis for numerous paleoclimate studies including studies of ice-sheet stratigraphy (Dansgaard et al. 1969), ancient soils (Lawrence and Taylor 1971), supergene clays (Sheppard et al. 1969), cave deposits (Schwarcz et al. 1976), supergene alunites (Rye et al. 1992, Arehart and O'Neil 1993), and fluid inclusions in hydrothermal mineral deposits (O'Neil and Silberman 1974, Seal and Rye 1993), to name a few. However, because of the variety of factors that control the isotopic composition of meteoric water, inferred compositions of meteoric water cannot be used for quantitative determination of paleotemperatures. At best, changes in composition reflect relative changes in the temperature (Dansgaard et al. 1969), and similar compositions of meteoric waters of different ages reflect grossly similar mean annual temperatures of precipitation (Seal and Rye 1993). In addition to the factors discussed above, several other factors affect the position of the meteoric water line. The relative humidity accompanying evaporation from the oceans can change the "y-intercept", known as the "deuterium excess factor" (d) of the meteoric water line resulting in regional meteoric water lines that are parallel to the global meteoric water line. In arid environments such as the eastern Mediterranean, d values can reach 23 ‰, whereas in humid environments such as the South Atlantic, d values can drop to less than 6 ‰ (Yurtsever and Gat 1981). In terms of ancient conditions, the isotopic composition of seawater directly affects d. Secondary processes such as evaporation and high-temperature reactions with rocks can alter the isotopic composition of meteoric waters.

Magmatic water as used here is water that has equilibrated with a magma, regardless of origin. Compositions of magmatic water are typically calculated rather than measured from natural samples because no unambiguous natural samples of magmatic water, other than those found in fluid inclusions have been found (e.g. Rye 1966, Deen et al. 1994). Accessible, shallow level magmatic-hydrothermal systems are invariably contaminated by meteoric waters. The isotopic composition of hypothetical magmatic water is calculated on the basis of the isotopic composition of unaltered magmatic silicate minerals (quartz, feldspar, muscovite, biotite, and hornblende), experimentally determined mineral-water fractionation factors, and estimates of magmatic temperatures. Calculated compositions of primary magmatic waters span the range δD = -40 to -80 ‰, and δ^{18}O = +5.5 to +9.5 ‰ (Sheppard et al. 1969; Fig. 1). Magmatic waters calculated by such means represent an oversimplification of the complexity of natural processes that can effect stable isotopic fractionation such as crystallization history and speciation of water in magmas (Taylor

1986, Taylor 1997). Nevertheless, calculated compositions of magmatic water provide a convenient reference from which to assess the relative importance of various geochemical processes in determining the isotopic compositions of waters and minerals precipitated from these processes.

Several common chemical and physical processes can act upon waters to alter their isotopic composition; chief among these processes are evaporation, water-rock interaction, and mixing. Evaporation causes an enrichment in D and ^{18}O in the residual water, thereby producing a positive trend from the starting composition typically with a slope between 4 and 6 (Fig. 1). Under very saline conditions, however, a different slope and a hook in the δD-$\delta^{18}O$ trajectory may be produced (Sofer and Gat 1975). For dilute water, the slope is dominantly a function of the relative humidity accompanying evaporation, but for saline waters, the hydration effect on cations becomes an important factor. Trends with lower slopes on a δD versus $\delta^{18}O$ plot reflect drier conditions and steeper slopes reflect wetter conditions. Water-rock interactions can alter both the oxygen and hydrogen isotopic composition of waters. At high water/rock ratios, the effect is detectable only in the oxygen isotopic composition of the water because of the relatively small amount of hydrogen relative to oxygen in rocks. The mass ratio of oxygen in water to oxygen in siliciclastic sedimentary and felsic igneous rocks typically varies around 1.8, whereas that ratio for hydrogen is around 60. The shape of the interaction paths for the water in terms of δD and $\delta^{18}O$ will vary depending on the starting composition of the water and rock (Ohmoto and Rye 1974, Criss and Taylor 1986, Seal and Rye 1992). Geothermal waters commonly have hydrogen isotope compositions that are indistinguishable from local meteoric waters, except where modified by mixing and boiling (Truesdell et al. 1977), whereas the oxygen isotope compositions are commonly enriched in ^{18}O by up to 20 ‰ (Truesdell and Hulston 1980).

An additional reservoir of oxygen contributes to secondary sulfate minerals that result from the weathering of sulfide minerals. The sulfate oxygen typically comprises mixtures of oxygen derived from water and oxygen derived from the atmosphere. Atmospheric oxygen is globally homogeneous, averaging $\delta^{18}O = 23.5\pm0.3$ ‰ (Dole et al. 1954, Kroopnick and Craig 1972).

For sulfur, the most common reference reservoirs are meteoritic sulfur and seawater. Meteoritic sulfur, such as Cañon Diablo troilite, provides a convenient reference because it approximates the Earth's bulk composition. The iron meteorites have an average sulfur isotope composition of $\delta^{34}S = 0.2\pm0.2$ ‰ (Kaplan and Hulston 1966) which is indistinguishable from that of pristine mid-ocean ridge basalts ($\delta^{34}S = 0.3\pm0.5$ ‰; Sakai et al. 1984). Geochemical processes, the most notable of which are oxidation and reduction, profoundly fractionate sulfur isotopes away from bulk-Earth values in geological systems (Fig. 2). Oxidation processes produce species that are enriched in ^{34}S relative to the starting material, whereas reduction produces species that are depleted in ^{34}S.

Oxidation-reduction reactions involving reduced sulfur from the interior of the Earth throughout its history have resulted in a $\delta^{34}S$ of +21.0±0.2 ‰ for dissolved sulfate in modern oceans (Rees et al. 1978). Because of the volume and importance of the ocean in the global sulfur cycle, this composition is another important reference reservoir from which to evaluate sulfur isotope variations in geological systems. The $\delta^{34}S$ of sulfate in ancient oceans as recorded by marine evaporite sequences (Claypool et al. 1980) has varied from a low of approximately 10 ‰ during Permian and Triassic time to a high of 35 ‰ during Cambrian time. The causes and implications of the secular variations in the sulfur isotope composition of seawater are discussed in a later section.

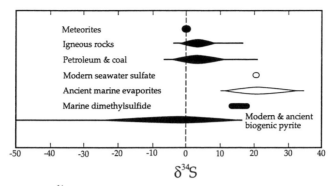

Figure 2. δ^{34}S of various geologic reservoirs. Modified from Krouse (1980). Dimethylsulfide data from Calhoun et al. (1991); isotopic values in permil (VCDT).

FACTORS THAT CONTROL STABLE ISOTOPE FRACTIONATION

Isotopic fractionation is a result of variations in thermodynamic properties of molecules that are dependent on mass. Details of the thermodynamic basis for understanding isotope fractionation have been presented by Urey (1947), Bigeleisen and Mayer (1947), and Bigeleisen (1952). Isotope fractionation results from equilibrium and kinetically controlled chemical and physical processes. Equilibrium processes include isotopic exchange reactions, which redistribute isotopes among molecules of different substances. Kinetic processes include irreversible chemical reactions and physical processes such as evaporation and diffusion (O'Neil 1986). Equilibrium isotope effects result from the effect of atomic mass on bonding; molecules containing a heavier isotope are more stable than those containing a lighter isotope. Kinetic isotope effects are related to greater translational and vibrational velocities associated with lighter isotopes. It is easier to break bonds with lighter isotopes, for example the ^{32}S-O bond, compared with those with heavier isotopes, such as the ^{34}S-O bond, during the bacterially mediated reduction of dissolved sulfate to sulfide.

Among the several factors that influence the magnitude of equilibrium stable isotope fractionations are temperature, chemical composition, crystal structure, and pressure (O'Neil 1986). For the present discussion, temperature and chemical composition are the most important. Polymorphism is rare in the sulfate minerals discussed here and pressure effects are minimal at upper crustal conditions. The temperature dependence of fractionation factors results from the relative effect of temperature on the vibrational energies of two substances. Theoretical considerations indicate that the stable isotope fractionation between two substances should approach zero at infinite temperature (Bigeleisen and Mayer 1947). These fractionations are generally described well by equations of the form:

$$1000 \ln\alpha = \frac{A}{T^2} + \frac{B}{T} + C \tag{9}$$

where A, B, and C are empirically determined constants.

The dependence of isotopic fractionation on chemical composition includes such compositional variables as oxidation state, ionic charge, atomic mass, and the electronic configuration of the isotopic elements and the elements to which they are bound (O'Neil 1986). For sulfur-bearing systems, the effect of the oxidation state of sulfur is especially important. The higher oxidation states of sulfur are enriched in the heavier isotopes relative to lower oxidation states such that ^{34}S enrichment follows the general trend $SO_4^{2-} > SO_3^{2-} >$

$S_x^0 > S^{2-}$ (Sakai 1968, Bachinski 1969). In the geological record, this trend is reflected by the fact that sulfate minerals, such as barite or anhydrite, typically have higher $\delta^{34}S$ values than cogenetic sulfide minerals in a variety of geochemical settings. Cationic substitutions also play an important role in stable isotope fractionations. O'Neil et al. (1969) documented a cation-mass dependence of ^{18}O enrichment in divalent metal-carbonate minerals with ^{18}O enrichment following the order $CaCO_3 > SrCO_3 > BaCO_3$. Likewise, the ^{18}O enrichment in divalent sulfate minerals is such that $CaSO_4 > BaSO_4$ (Lloyd 1968, Kusakabe and Robinson 1977, Chiba et al. 1981), and the ^{34}S enrichment follows the order $CaSO_4 \cdot 2H_2O > BaSO_4$ (Thode and Monster 1965, Kusakabe and Robinson 1977). Similarly, for Al^{3+}-Fe^{3+} substitution in the grossular-andradite garnets, Taylor and O'Neil (1977) found that ^{18}O was enriched in natural Al-rich garnets relative to Fe-rich garnets. In the alunite-jarosite minerals, ^{18}O is enriched in the sulfate site in alunite relative to jarosite by approximately 2.4 ‰ at 250°C (Stoffregen et al. 1994, Rye and Stoffregen 1995).

EQUILIBRIUM FRACTIONATION FACTORS

Equilibrium isotopic fractionation factors are typically derived by one of three methods: (1) experimental determination, (2) theoretical estimation using calculated bond strengths or statistical mechanical calculations based on data on vibrational frequencies of compounds, and (3) analysis of natural samples for which independent estimates of temperature are available. Each method has advantages and disadvantages. Experimental determination provides a direct measurement of the fractionation, but such efforts are commonly hampered by experimental kinetic limitations and the fact that solutions used in experiments often do not approximate natural solutions. Theoretical estimation avoids the kinetic hurdles of experimental studies but is limited by the availability and accuracy of data required for the estimation. Fractionation factors derived from the analysis of natural materials provides a means of investigating isotopic fractionations when data from neither of the other methods are available. However, this method is subject to retrograde isotopic exchange and uncertainties related to the contemporaneity of the mineral pairs and to the independent temperature estimate.

Available sulfur, oxygen, and hydrogen isotopic fractionation factors for sulfate minerals are limited to a few mineral species, despite the geological importance of numerous sulfates. Isotopic fractionation studies at temperatures ranging from approximately 0 to 600°C have been conducted on dissolved and gaseous sulfate and sulfite species, and for the minerals anhydrite, gypsum, barite, chalcanthite, mirabilite, alunite, and jarosite. Oxygen and hydrogen isotope studies of sulfates have investigated fractionation between crystallographically distinct sites as well. This chapter provides a critical evaluation of the available fractionation data with the goal of assembling an internally consistent set of sulfur, oxygen, and hydrogen isotope fractionation factors for sulfate minerals.

Sulfur

Data for the sulfur isotope fractionation for sulfate minerals are limited. The gypsum-dissolved sulfate system has been investigated by Thode and Monster (1965) and the barite-dissolved sulfate system has been investigated by Kusakabe and Robinson (1977). Both studies found limited sulfur isotope fractionation between the mineral and the solution. At ambient temperature, Thode and Monster (1965) determined a sulfur isotope fractionation between gypsum and dissolved sulfate of $\Delta_{gypsum-sulfate} = \delta^{34}S_{gypsum} - \delta^{34}S_{sulfate} = +1.65$ ‰. At temperatures between 110 and 350°C, Kusakabe and Robinson (1977) found no sulfur isotope fractionation between barite and dissolved sulfate (i.e., $\Delta_{barite-sulfate} = 0$ ‰). These results are consistent with the predictions of Sakai (1968) based on statistical mechanics. Sakai (1968) concluded that because of the minimal modification of the vibrational energy of the sulfur atom in the SO_4^{2-} radical by the crystalline field, the sulfur isotope fractionation between a sulfate mineral and aqueous sulfate should be small. He also concluded that a small, but measurable, enrichment in ^{34}S in sulfate minerals should follow $CaSO_4 > CaSO_4 \cdot 2H_2O > SrSO_4 > BaSO_4 > PbSO_4$, which is in agreement with the experimental data for gypsum by Thode and Monster (1965), and for barite by Kusakabe and Robinson (1977). A major benefit of the minimal fractionation between sulfate minerals and aqueous sulfate is that the measured isotopic composition of the sulfate mineral approximates the isotopic composition of the parent fluid. This feature is important in interpreting secular variations in the global oceanic sulfur cycle by using the sulfur isotopic composition of marine evaporites as a record of ancient seawater sulfate compositions.

From a process-oriented perspective, it is useful to be able to relate the isotopic composition of sulfates to sulfides. Ohmoto and Lasaga (1982) critically evaluated the available data on the fractionation of sulfur isotopes between SO_4^{2-} and H_2S in aqueous systems (Robinson 1973, Bahr 1976, Sakai and Dickson 1978, Igumnov et al. 1977) to eliminate suspect experimental data that may have experienced re-equilibration during quenching. Ohmoto and Lasaga (1982) derived a temperature-dependent equation to describe this sulfur isotope fractionation. Following the convention of Ohmoto and Rye (1979), zero fractionation has been assumed between all sulfate minerals and aqueous sulfate for the purpose of evaluating fractionation among sulfate minerals and H_2S. Sulfur isotope fractionations for aqueous sulfate and sulfate minerals relative to H_2S are presented in Table 1 and Figure 3. For reference, sulfur isotope fractionations for pyrite-H_2S and SO_2-H_2S from Ohmoto and Rye (1979) are also presented (Table 1; Fig. 3).

Oxygen

Data are available for oxygen isotope fractionation between the sulfate oxygen and water for the compounds aqueous sulfate (SO_4^{2-} and HSO_4^-), anhydrite ($CaSO_4$), barite ($BaSO_4$), alunite ($KAl_3(SO_4)_2(OH)_6$), and jarosite ($KFe_3(SO_4)_2(OH)_6$). Fractionation factors between hydroxyl oxygen and water are available for alunite and jarosite, as are the intramineral fractionation factors for SO_4^{2-}-OH^-. Oxygen isotope fractionation factors between crystallographic H_2O and water are also available for gypsum ($CaSO_4 \cdot 2H_2O$) and mirabilite ($Na_2SO_4 \cdot 10H_2O$). To improve internal consistency among the oxygen isotope fractionation factors presented in Table 1, published data for fractionations relative to H_2O were recalculated using a value of $\alpha_{CO_2-H_2O} = 1.04115$ at 25°C (Brenninkmeijer et al. 1983) where appropriate. In addition, data from studies that employed concentrated salt solutions

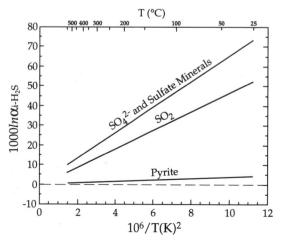

Figure 3. Temperature dependence of equilibrium sulfur isotope fractionation factors relative to H_2S for dissolved sulfate, sulfate minerals, SO_2, and pyrite. The dashed line indicates a 0.0 ‰ 1000 lnα value. Data from Table 1.

Table 1. Equilibrium isotopic fractionation factors for sulfate minerals
and related compounds.

Compound or component $i(-j)$	A	B	C	T (°C) range*	Data sources
Sulfur					
$1000\ln\alpha_{i-H_2S} = \dfrac{A\times10^6}{T^2} + \dfrac{B\times10^3}{T} + C$ (T in K)					
Sulfate minerals and aqueous sulfate	6.463		0.56	200 - 400	2
SO_2	4.70		-0.5	350 - 1050	1
FeS_2	0.40			200 - 700	1
Oxygen					
$1000\ln\alpha_{i-H_2O} = \dfrac{A\times10^6}{T^2} + \dfrac{B\times10^3}{T} + C$ (T in K)					
SO_4^{2-} (HSO_4^-)	3.26		-5.81	70 - 350	5, 6, 7
Anhydrite	3.21		-4.72	100 - 550	8
Barite	2.65		-4.97	110 - 350	5, 9
Alunite SO_4	3.09		-2.94	250 - 450	3
Jarosite SO_4	3.53		-6.91	100 - 250	4
Alunite OH	2.28		-3.9	250 - 450	3
Jarosite OH	2.1		-8.77	100 - 250	4
Oxygen					
$1000\ln\alpha_{i-j} = \dfrac{A\times10^6}{T^2} + \dfrac{B\times10^3}{T} + C$ (T in K)					
Alunite SO_4-OH	0.8		0.96	250 - 450	3
Jarosite SO_4-OH	1.43		1.86	100 - 250	4

* Temperature range refers to the experimental temperature range; note that fractionation factors may extrapolate significantly beyond these ranges (see text)

Data sources: 1. Ohmoto and Rye (1979); 2. Ohmoto and Lasaga (1982); 3. Stoffregen et al. (1994); 4. Rye and Stoffregen (1995); 5. This study; 6. Lloyd (1968); 7. Mizutani and Rafter (1969a); 8. Chiba et al. (1981); 9. Kusakabe and Robinson (1977).

were also corrected for isotopic effects of salt hydration using data from Horita et al. (1993, 1995).

Oxygen isotope exchange between aqueous sulfate, both as SO_4^{2-} and HSO_4^-, and water has been investigated by Lloyd (1968) and Mizutani and Rafter (1969a) at temperatures from 70 to 350°C (Fig. 4). Both studies determined similar fractionations between aqueous sulfate and water. Thus, the expression presented in Table 1 and Figures 4 , 5, and 6 represents a linear regression of both data sets. For anhydrite, mineral-water oxygen isotope fractionations have been studied by Lloyd (1968) and Chiba et al. (1981) with quite different results. The 1000 lnα factors for anhydrite-water of Lloyd (1968) are between 2.5 and 7.0 ‰ greater than those of Chiba et al. (1981) for the temperature range from 100 to 550°C (Fig. 5). Chiba et al. (1981) cited data from natural samples, which

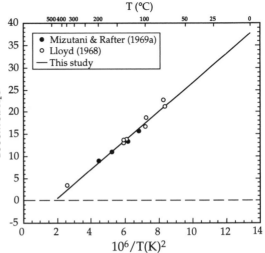

Figure 4. Comparison of the temperature dependence of oxygen isotope fractionation factors between dissolved sulfate and water from Mizutani and Rafter (1969a) and Lloyd (1968). The solid line is the linear regression of both data sets. The dashed line indicates a 0.0 ‰ 1000 lnα value.

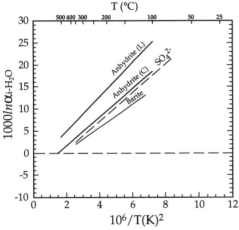

Figure 5. Comparison of the temperature dependence of oxygen isotope fractionation factors between simple sulfate minerals and water. Anhydrite (L) is from Lloyd (1968); Anhydrite (C) is from Chiba et al. (1981). Barite is modified from Kusakabe and Robinson (1977). SO₄ curve is the best-fit curve from Figure 4 and Table 1. The dashed line indicates a 0.0 ‰ 1000 lnα value.

support their experimental fractionation factors rather than those of Lloyd (1968). In addition, the anhydrite-water fractionations of Chiba et al. (1981) are more consistent with the theoretical fractionations of anhydrite relative to SO_4 and barite (Sakai 1968) than those of Lloyd (1968). Therefore, the fractionation factors of Chiba et al. (1981) are preferred over those of Lloyd (1968). Kusakabe and Robinson (1977) presented experimentally determined fractionations factors between barite and water. Their mineral-water fractionations for barite relative to published values for anhydrite are consistent with what would be expected on the basis of atomic mass and ionic radius differences between Ba^{2+} and Ca^{2+} (cf. O'Neil et al. 1969). The fractionation expression for the barite-water system in Table 1 is derived from the experimental work of Kusakabe and Robinson (1977), with a correction for the revised 25°C $\alpha_{CO_2-H_2O}$ value and hydration effects.

Stoffregen et al. (1994) presented oxygen isotope fractionation factors for the systems alunite SO_4^{2-}-water, alunite OH^--water, and intramineral alunite SO_4^{2-}-alunite OH^- (Table 1; Fig. 6). Likewise, Rye and Stoffregen (1995) presented similar fractionation data for jarosite (Table 1; Fig. 6). In both cases, explicit corrections for hydration effects were not incorporated in the expressions for isotopic fractionations because the data are lacking to make these corrections. However, preliminary experiments by Stoffregen et al. (1994) indicated that deviations from pure water due to salt effects in the alunite SO_4^{2-}-water fractionation may reach 4 ‰ at 400°C in 1.0 m KCl + 0.5 m H_2SO_4 solutions. It was also determined that K/Na substitutions in alunite have negligible effect on SO_4^{2-} oxygen isotope fractionation in the temperature range 350 to 450°C.

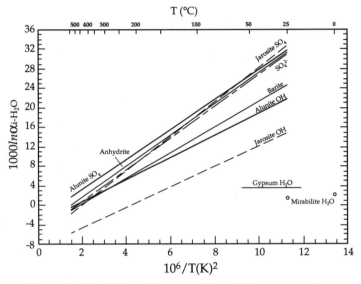

Figure 6. Compilation of recommended equilibrium oxygen isotope fractionation factors relative to H_2O for various compounds, mineralogical sites, and dissolved sulfate. See text for data sources.

Gonfiantini and Fontes (1963) determined that the oxygen isotope fractionation between the hydrate water in gypsum and water is independent of temperature from 17 to 57°C and 1000 $\ln\alpha_{gypsum-H_2O} \approx \Delta_{gypsum-H_2O} = 3.7$ ‰ (Fig. 6). Stewart (1974) determined that the oxygen isotope fractionation between hydrate water in mirabilite and solution (1000 $\ln\alpha_{mirabilite-H_2O} \approx \Delta_{mirabilite-H_2O}$) is 2.0 ‰ at 0°C and 1.4 ‰ at 25°C (Fig. 6). The experimental solutions were saturated with respect to mirabilite. On the basis of the data presented by Horita et al. (1993), the effect of the hydration of Na_2SO_4 on the oxygen isotope fractionation between mirabilite and water should be less than the analytical uncertainty; therefore, no correction for hydration effects is necessary to relate the fractionation factors to pure H_2O.

Hydrogen

Mineral-water hydrogen isotope fractionation factors have been experimentally determined for alunite, jarosite, gypsum, mirabilite, and chalcanthite. Hydrogen isotope fractionation factors generally tend not to be described by simple expressions following the form of Equation (9). Therefore, hydrogen isotope fractionation factors are presented only in Figure 7 without accompanying equations. Stoffregen et al. (1994) experimentally measured hydrogen isotope fractionation factors between alunite and aqueous solutions and found 1000 $\ln\alpha_{alunite-H_2O}$ values that ranged from -19 at 450°C to -6 at 250°C. Bird et al. (1989) proposed an alunite-water fractionation of 4 ‰ at ambient temperatures on the basis of data from natural samples. Rye and Stoffregen (1995) experimentally measured hydrogen isotope fractionation factors between jarosite and aqueous solutions and found 1000 $\ln\alpha_{jarosite-H_2O}$ ($\approx \Delta_{jarosite-H_2O}$) values that were generally independent of temperature

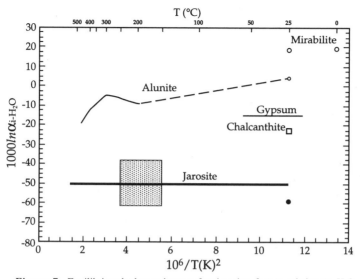

Figure 7. Equilibrium hydrogen isotope fractionation factors relative to H_2O for various minerals. The 25°C jarosite fractionation factor is from Alpers et al. (1988). Other data sources cited in text.

from 150 to 250°C and averaged -50 ‰. Alpers et al. (1988) proposed a jarosite-water fractionation of -59 ‰ at ambient temperature on the basis of data from a jarosite that precipitated from mine water in the laboratory. Stoffregen et al. (1994) determined that any variations in alunite-water fractionations associated with hydration effects of dissolved salts were within the analytical uncertainty.

Fontes and Gonfiantini (1967) investigated hydrogen isotope fractionation between gypsum and water in solutions saturated with gypsum; the 1000 $\ln\alpha_{gypsum-H_2O}$ values were independent of temperature from 17 to 57°C and averaged -15 ‰ (Fig. 7). Stewart (1974) determined hydrogen isotope fractionations (1000 $\ln\alpha_{mirabilite-H_2O}$) between mirabilite and water at 0 and 25°C and found them to be 18.8 and 16.9 ‰, respectively (Fig. 7). For both the gypsum and mirabilite studies, the compositions of the experimental solutions were not sufficiently dilute that they would not be expected to affect mineral-water fractionations due to hydration effects (Horita et al. 1993). Heinzinger (1969) measured hydrogen isotope fractionations of 1000 $\ln\alpha_{chalcanthite-H_2O}$ ($\approx \Delta_{chalcanthite-H_2O}$) = -22.7 ‰ between chalcanthite and a saturated aqueous solution at ambient temperature. Data are lacking to evaluate the effect of copper salt hydration on hydrogen isotope fractionation. In addition, Heinzinger (1969) determined that there is a 57 ‰ intramineral fractionation in chalcanthite between one hydrogen-bonded water molecule and the four water molecules bonded to Cu^{2+}.

Geothermometry

Stable isotope geothermometry is typically based on the partitioning of stable isotopes between two substances, such as sulfur between barite and pyrite. However, the partitioning of an isotope between two crystallographically distinct sites within a single mineral, such as oxygen between the sulfate and hydroxyl sites in alunite, can be used as a single mineral geothermometer (Pickthorn and O'Neil 1985, Rye et al. 1992, Rye and Alpers 1997). The use of stable isotopes for geothermometry is based on several requirements or assumptions. First, the minerals must have formed contemporaneously in equilibrium with one another. Second, subsequent re-equilibration or alteration of one or both minerals must not have occurred. Third, pure minerals must be separated for isotopic analysis. Fourth, the temperature dependence of the fractionation factors must be known. In addition, greater precision in the temperature estimate will be achieved from the use of mineral pairs that have the greatest temperature dependence in their fractionations. Kinetic considerations offer both advantages and disadvantages to geothermometry. Rapid kinetics of isotope exchange promote mineral formation under equilibrium conditions. Unfortunately, rapid exchange kinetics also make mineral pairs prone to re-equilibration during cooling. In contrast, sluggish kinetics hamper isotopic equilibration between minerals. However, once equilibrated, mineral pairs with sluggish exchange kinetics, such as sulfate-sulfide mineral pairs will tend to record peak formation conditions without subsequent re-equilibration at lower temperatures.

Coexisting pairs of sulfate minerals are uncommon and the temperature dependence of sulfur and oxygen isotope fractionation are similar among sulfate minerals (Figs. 3 and 6). Thus, thermometry based on coexisting sulfate minerals is not practical. Geothermometry based on isotopic data from sulfate minerals commonly employs sulfur isotope fractionation between sulfate minerals (most commonly barite or anhydrite) and sulfide minerals (most commonly pyrite), or oxygen isotope fractionation between sulfate minerals and other oxygen-bearing minerals, such as silicates or carbonates. Intramineral fractionation of oxygen isotopes between SO_4 and OH in alunite allows single mineral geothermometry. Oxygen isotope fractionation between dissolved SO_4^{2-} and H_2O, and sulfur isotope fractionation between dissolved SO_4^{2-} and H_2S have been used to assess reservoir temperatures in geothermal systems.

Ohmoto and Lasaga (1982) evaluated the kinetics of sulfur isotope exchange between SO_4^{2-} and H_2S and found that exchange rates were dependent on pH, temperature, and total concentrations of dissolved sulfur. Rates generally increase with increasing temperature and sulfur concentration, and with decreasing pH. It was found that for "typical" hydrothermal systems of near-neutral to slightly acidic conditions (pH ≈ 4 to 7) and $\Sigma S = 10^{-2}$ m, isotopic equilibrium was unlikely to reached at temperatures below 200°C in geologically reasonable time periods (Fig. 8). Therefore, sulfate-sulfide sulfur isotope thermometry should be most useful in the study of acidic (acid-sulfate) and high-temperature magmatic hydrothermal systems, and the study of the metamorphism of sulfate-bearing rocks. An excellent example of acidic settings is in the case of alunite from

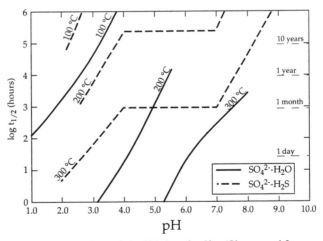

Figure 8. Comparison of the kinetics of sulfur (Ohmoto and Lasaga 1982) and oxygen (Chiba and Sakai 1985) isotope exchange in terms of pH vs $\log t_{1/2}$. Modified from Cole and Ohmoto (1986).

acid-sulfate hydrothermal systems, for which the pH is sufficiently low to permit isotopic equilibration at temperatures below 250°C, and alunite-pyrite pairs can be used for exceptionally good geothermometry (Rye et al. 1992, Deen et al. 1994).

Because of the relationships expressed in Equations (4) and (5), mineral-mineral fractionation equations can be derived from the equations in Table 1. An equation to calculate the temperature recorded by a coexisting pair of barite (Ba) and pyrite (Py) can derived as follows:

$$1000 \ln\alpha_{Ba-Py} \approx \Delta_{Ba-Py} = \delta^{34}S_{ba} - \delta^{34}S_{Py} \tag{10}$$

Thus,

$$\Delta_{Ba-Py} = \Delta_{Ba-H_2S} - \Delta_{Py-H_2S} = \delta^{34}S_{Ba} - \delta^{34}S_{H_2S} - (\delta^{34}S_{py} - \delta^{34}S_{H_2S}) \tag{11}$$

or

$$\Delta_{Ba-Py} \approx 1000 \ln\alpha_{Ba-H_2S} - 1000 \ln\alpha_{Py-H_2S} \tag{12}$$

Substituting from Table 1 yields

$$\Delta_{Ba-Py} = \frac{6.463 \times 10^6}{T^2} + 0.56 - \left(\frac{0.40 \times 10^6}{T^2}\right) \tag{13}$$

with T in K, or,

$$\Delta_{Ba-Py} = \frac{6.063 \times 10^6}{T^2} + 0.56 \tag{14}$$

Solving for T, and converting to°C yields:

$$T\ (°C) = \sqrt{\frac{6.063 \times 10^6}{\Delta_{Ba-Py} - 0.56}} - 273.15 \tag{15}$$

For example, for a sample with $\delta^{34}S_{Ba} = 21.0$ ‰ and $\delta^{34}S_{Py} = 5.1$ ‰, a temperature of 356°C is calculated using Equation (15). Temperatures calculated using Equation (15) are significantly lower (~70°C) than those calculated using the equations presented by Ohmoto and Rye (1979). The difference is the result of the revised equation for SO_4^{2-}-H_2S sulfur isotope fractionation presented by Ohmoto and Lasaga (1982). Thus, data from geothermometry studies published using the older fractionation factors should be recalculated using the updated fractionation factors before reinterpretation. Figure 9 compares temperatures calculated on the basis of the revised SO_4-H_2S fractionation factors of Ohmoto and Lasaga (1982) with those calculated using the earlier fractionation factors summarized by Ohmoto and Rye (1979).

PROCESSES THAT CAUSE STABLE ISOTOPIC VARIATIONS

Variations in the stable isotopic composition of natural systems can result from a variety of equilibrium and kinetically controlled processes, which span a continuum. The most important steps for producing sulfur and oxygen isotopic variations in sulfate minerals are the geochemical processes that initially produce (or transform) the sulfate from (to) other sulfur species, rather than precipitation of the sulfate mineral from dissolved sulfate. The more important processes generally involve the oxidation of sulfide to sulfate, or the reduction of sulfate to sulfide. In addition, the low-temperature kinetic rates of many of the oxidation and reduction processes are enhanced by bacterial mediation, which can impart distinct isotopic fractionations to these processes. Thus, the complex aqueous geochemistry of sulfur species is a key aspect of understanding the stable isotope

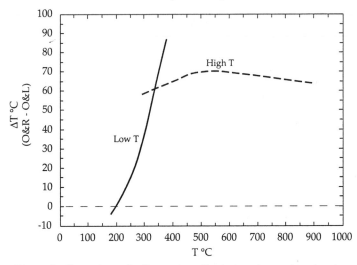

Figure 9. Comparison of sulfate-pyrite sulfur isotope temperatures based on SO_4–H_2S fractionation factors of Ohmoto and Lasaga (1982) and those summarized by Ohmoto and Rye (1979). X-axis is the temperature calculated using the SO_4–H_2S fractionations of Ohmoto and Lasaga (1980) (O&L). ΔT is the difference between the temperatures calculated using the fractionation equations of Ohmoto and Lasaga (1980) and the low-temperature (solid) and high-temperature (dashed) fractionation equations summarized in Ohmoto and Rye (1979)(O&R) from the literature.

geochemistry of sulfate minerals. Ohmoto (1972) developed the principles for application of sulfur isotope systematics to hydrothermal ore deposits. Van Stempvoort and Krouse (1994) provided a comprehensive review of controls on oxygen isotope compositions in aqueous sulfate. Taylor and Wheeler (1994) reviewed controls of the oxygen and sulfur isotope systematics of acid mine-drainage systems. Comprehensive reviews of the controls on the oxygen and sulfur isotope systematics of sulfates in ore deposits have been given by Ohmoto and Rye (1979), Ohmoto (1986), and Ohmoto and Goldhaber (1997).

If the starting reservoir is limited in size, many equilibrium and kinetic processes can be described as Rayleigh distillation processes. Rayleigh processes are described by the equation

$$R = R_o f^{(\alpha-1)} \tag{16}$$

where R_o is the initial isotopic ratio, and R is the isotopic ratio when a fraction (f) of the starting amount remains, and α is the fractionation factor, either equilibrium or kinetic. This equation can be recast in the δ notation for sulfur isotopes as

$$\delta^{34}S = (\delta^{34}S_o + 1000)f^{(\alpha-1)} - 1000 \tag{17}$$

Rayleigh models accurately describe isotopic variations associated with processes such as the precipitation of minerals from solutions, the condensation of precipitation (rain, snow) from atmospheric moisture, and the bacterial reduction of seawater sulfate to sulfide, among others. Bacterial reduction of seawater sulfate can be modeled using Equation (17). If $\alpha = 1.0408$ and $\delta^{34}S_o = 21.0$ ‰, then precipitation of pyrite from H_2S produced from bacterial reduction of sulfate will preferentially remove ^{32}S and the first pyrite formed will have $\delta^{34}S \approx -20$ ‰. The preferential removal of ^{32}S will cause the $\delta^{34}S$ of the residual aqueous sulfate to increase, which in turn will lead to an increase in the $\delta^{34}S$ of

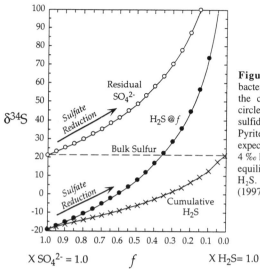

Figure 10. Rayleigh distillation curves for bacterial reduction of seawater sulfate showing the change in $\delta^{34}S$ of resultant H_2S (filled circles), residual sulfate (open circles), and bulk sulfide (X) as a function of reaction progress. Pyrite precipitated from the H_2S would be expected to have a $\delta^{34}S$ that is approximately 4 ‰ higher than that shown for H_2S assuming equilibrium fractionation between pyrite and H_2S. Modified from Ohmoto and Goldhaber (1997). All isotopic values in permil (VCDT).

subsequently formed pyrite (Fig. 10). Under closed-system behavior, after all of the sulfate has been reduced, the bulk isotopic $\delta^{34}S$ of the precipitated pyrite will equal the $\delta^{34}S$ of the initial sulfate. However, the $\delta^{34}S$ of individual pyrite grains or growth zones can be both lower and higher than the bulk composition, depending on when they formed.

Mixing is another important process that can cause isotopic variations and it can be modeled on the basis of simple mass-balance equations such as

$$\delta_{mixture} = X_A\delta_A + X_B\delta_B \qquad (18)$$

where $\delta_{mixture}$ is the resulting isotopic composition of the mixture, δ_A and δ_B are the isotopic compositions of components A and B, and X_A and X_B are the mole fractions of components A and B.

Kinetics of isotope exchange reactions

Ohmoto and Lasaga (1982) found that exchange rates between dissolve SO_4^{2-} and H_2S decreased proportionally with increasing pH at pH < 3; from pH $\approx$ 4 to 7, the rates remain fairly constant; and at pH > 7, the rate also decreases proportionally with increasing pH. Ohmoto and Lasaga (1982) proposed that the overall rate of exchange is limited by exchange reactions involving intermediate valence thiosulfate species ($S_2O_3^{2-}$), the abundance of which is dependent on pH. The rate-limiting step was postulated to be an intramolecular exchange between non-equivalent sulfur sites in thiosulfate. They calculated the most rapid equilibration rates at high temperature (T = 350°C) and low pH (pH $\approx$ 2) of approximately 4 hours for 90 percent equilibrium between aqueous sulfate and sulfide; however, at low temperature (T = 25°C) and near neutral pH (pH = 4-7), the time to attain 90 percent equilibrium reached 9×10^9 years. Disequilibrium between sulfate and sulfide minerals is prevalent in hydrothermal and geothermal systems below 350°C (Figs. 8, 11).

The rate of oxygen isotope exchange between dissolved sulfate and water has been investigated by Hoering and Kennedy (1957) from 10 to 100°C, by Lloyd (1968) from 25 to 448°C, and by Chiba and Sakai (1985) from 100 to 300°C. Both Lloyd (1968) and Chiba and Sakai (1985) used Na_2SO_4 solutions in which pH was adjusted using either H_2SO_4 or $Na(OH)$. The pH values of the experimental solutions of Lloyd (1968) ranged from 3.8 to 9.0, and those of Chiba and Sakai (1985) ranged from 2.3 to 7.3. In contrast,

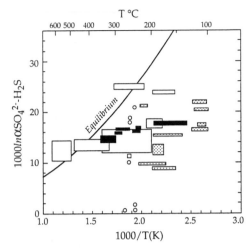

Figure 11. Comparison between observed Δ and equilibrium Δ between SO₄ and H₂S as a function of T for mineral deposits and geothermal systems. White rectangles are from porphyry hydrothermal systems, black rectangles are from seafloor massive sulfide hydrothermal systems, gray rectangles are from Mississippi Valley-type hydrothermal systems, and white circles are from modern geothermal systems. Equilibrium curve is from Ohmoto and Lasaga (1980). Diagram is modified from Ohmoto and Lasaga (1980). All isotopic values in permil (VCDT).

the experimental solutions of Hoering and Kennedy (1957) consisted of mixtures of H_2SO_4 and H_2O that ranged from 1.04 to 16.0 M H_2SO_4; although the resulting pH was not specified, it can be modeled using the computer program PHRQPITZ for electrolyte solutions of high ionic strength (Plummer et al. 1988). Calculated pH values range from 0.0 to less than -25. Pitzer interaction parameters for H_2SO_4 solutions have been calibrated up to 6 M and appear to be valid up to approximately 9 M (Ptacek and Blowes, this volume). Thus, the present study is limited to a discussion of data from Hoering and Kennedy (1957) for concentrations less than 9 M (pH > -5) even though the higher concentration data follow the same general trends as the lower concentration data in term of rate of exchange versus pH. Comparison of the three experimental studies in terms of rate of reaction versus pH at various temperatures shows excellent agreement between the studies of Hoering and Kennedy (1957) and Chiba and Sakai (1985) at 100°C, the only temperature of overlap between the two studies (Fig. 12). In contrast, the study of Lloyd (1968) covers the entire temperature range of both of the other studies, but the rates of exchange derived from Lloyd (1968) are approximately two orders of magnitude faster than the rates of Hoering and Kennedy (1957) and Chiba and Sakai (1985). Insufficient details are provided in the three studies to determine which studies are most likely to be correct; however, the agreement between the results of Hoering and Kennedy (1957) and Chiba and Sakai (1985) is noteworthy.

Chiba and Sakai (1985) concluded that exchange is limited by interactions between $H_2SO_4°$ and H_2O at low pH (pH = 2.3 to 2.7 at T = 100°C and pH = 4.0 to 5.5 at T = 200°C), and between HSO_4^- and H_2O at higher pH (pH = 6.1 to 7.3 at T = 300°C). It was also concluded that the rate of sulfate-water oxygen isotope exchange is sufficient to permit this pair to be used as geothermometer for geothermal systems. In contrast, sulfate-water isotopic exchange in ambient seawater (pH ≈ 8; T ≈ 4°C) was suggested to require on the order of 10^9 years to reach equilibrium. At very low pH, corresponding to extreme acid mine-drainage conditions, tentative extrapolation of the data of Chiba and Sakai (1985) suggested that a 50 % approach to equilibrium could be achieved in a period of several months (Taylor and Wheeler 1994).

A comparison of exchange rates of sulfur isotopes between sulfate and sulfide, and oxygen isotopes between sulfate and water indicates that the half-life of exchange for oxygen is several orders of magnitude faster than for sulfur at acidic to neutral conditions

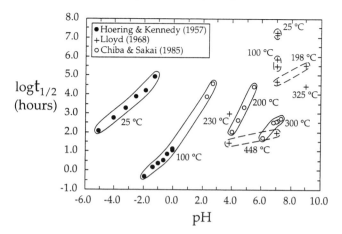

Figure 12. Comparison of the kinetics of oxygen isotope exchange data between dissolved sulfate and water in terms of pH and $\log t_{1/2}$ (hours). Data from Hoering and Kennedy (1957), Chiba and Sakai (1985), and Lloyd (1968). Values of pH for solutions of Hoering and Kennedy (1957) were computed using PHRQPITZ (Plummer et al. 1988) with MacInnes convention for scaling individual ion activity coefficients. Note agreement between Hoering and Kennedy (1957) and Chiba and Sakai (1985) studies at 100°C.

(Fig. 8). This difference is the result of differences in rate-limiting steps. In sulfate, sulfur atoms are surrounded by tetrahedrally coordinated oxygen atoms. Thus, multiple O-S bonds must be broken before sulfur isotope exchange can occur, whereas only a single O-S bond must be broken for oxygen isotope exchange. In addition, Ohmoto and Lasaga (1982) concluded that the rate-limiting step for sulfur isotope exchange is the intramolecular exchange between non-equivalent sites in thiosulfate.

Sulfate reduction, sulfide oxidation, and associated processes

The processes of sulfate reduction and sulfide oxidation in natural systems tend to have associated characteristic, kinetically controlled, non-equilibrium oxygen and sulfur isotope fractionations. Oxidation and reduction processes can occur in both biotic and abiotic environments. Isotopic variations associated with the biogenic reduction of sulfate have been studied by numerous researchers, most of whom have concentrated on the role of the dissimilatory sulfate-reducing bacteria such as *Desulfovibrio desulfuricans*. The activity of sulfate-reducing bacteria in marine sediments throughout most of geological time had a profound effect of the sulfur isotope composition of seawater sulfate, which is discussed in a later section.

Sulfate-reducing bacteria are active only in anoxic environments such as below the sediment-water interface, and in anoxic water bodies. Various species of sulfate-reducing bacteria can survive over a range of temperature (0 to 110°C) and pH (5 to 9.5) conditions, but prefer near-neutral conditions and can withstand a range of salinities from dilute up to halite saturation (Postgate 1984). The metabolism of sulfate-reducing bacteria can be described by the general reaction:

$$2\ CH_2O + SO_4^{2-} = H_2S + 2\ HCO_3^- \tag{19}$$

where CH_2O represents generic organic matter (Berner 1985). The H_2S can be lost to the water column, reoxidized, fixed as pyrite, or other sulfide minerals if reactive metals are present, or fixed as organic-bound sulfur. In near surface sediments deposited under normal (oxygenated) marine settings, the activity of sulfate-reducing bacteria is limited by the supply and reactivity of organic matter; in freshwater and euxinic basins, the activity is limited by sulfate availability (Berner 1985).

The fractionation of sulfur isotopes between sulfate and sulfide during bacterial sulfate reduction is a kinetically controlled process in which ^{34}S is enriched in the sulfate relative to the sulfide, in the same sense as equilibrium fractionation between sulfate and sulfide (Chambers and Trudinger 1979). The sulfate-reducing bacteria more readily metabolize ^{32}S and ^{16}O relative to ^{34}S and ^{18}O, respectively. Thus, the $\delta^{34}S$ and $\delta^{18}O$ of the residual aqueous sulfate increase during the reaction progress. The fractionation associated with bacterial sulfate reduction (1000 $\ln\alpha_{SO_4-H_2S}$) typically ranges from 15 to 60 ‰ (Goldhaber and Kaplan 1975) in marine settings, compared to an equilibrium, abiotic fractionation of approximately 73 ‰ at 25°C. The magnitude of the fractionation has been shown to be a function of the rate of sulfate reduction, which can be related to sedimentation rates. The smaller fractionations (~15 ‰) correspond to faster rates of sulfate reduction and sedimentation, whereas the larger fractionations (~60 ‰) correspond to slower rates of sulfate reduction and sedimentation (Goldhaber and Kaplan 1975; Fig. 13).

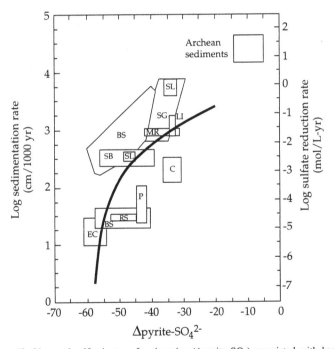

Figure 13. Observed sulfur isotope fractionation (Δpyrite-SO₄) associated with bacterial sulfate reduction as a function of rate of reduction and rate of sedimentation. Modified from Ohmoto and Goldhaber (1997). Abbreviations: BS, Black Sea; C, Carmen Basin; EC, East Cortez Basin; LI, Long Island Sound; MR, Marina Del Rey Harbor; P, Pascadero Basin; RS, Red Sea; SB, Santa Barbara Basin; SL, Solar Lake; SG, Spencer Gulf. Curved line represents best fit to the data. All isotopic values in permil.

The fractionation of oxygen isotopes during bacterial reduction of sulfate is due to a complex combination of processes among which are preferential reduction of isotopically light enzyme (adenosine 5'-phosphosulfate; APS)-bound sulfate to sulfite (SO_3^{2-}), reoxidation of sulfite to APS and sulfate, sulfite-water oxygen isotope exchange, and reoxidation of sulfide (Van Stempvoort and Krouse 1994). Experimental studies to quantify the oxygen isotope fractionation associated with bacterial sulfate reduction have been conducted by Mizutani and Rafter (1969b, 1973), and Fritz et al. (1989). Whereas the sulfur isotope systematics associated with bacterial sulfate reduction follow a Rayleigh pathway, the oxygen isotope systematics do not (Fritz et al. 1989). The kinetics of direct oxygen isotopic exchange, as discussed above, are sufficiently slow at ambient, near-neutral conditions to exclude direct sulfate-water exchange as a significant process during bacterial sulfate reduction. However, the exchange of oxygen isotopes between sulfite and water at ambient conditions (pH < 9) is extremely rapid, on the order of minutes (Betts and Voss 1970, Holt et al. 1981). Oxygen isotope exchange between sulfite and water, and the subsequent reoxidation of the sulfite to sulfate, seem to be important processes in determining the $\delta^{18}O$ of residual sulfate during bacterial reduction (Mizutani and Rafter 1969b, Fritz et al. 1989). Experimental studies using ^{35}S-labeled H_2S demonstrate the viability of concurrent reduction of SO_4^{2-} and oxidation of H_2S in anoxic environments (Trudinger and Chambers 1973, Elsgaard and Jørgensen 1992). The net result is that even though sulfate-reducing bacteria preferentially metabolize ^{16}O relative to ^{18}O to produce a kinetic fractionation ranging between 25 and 29 ‰ (depending on the rate of reduction), the resulting $\delta^{18}O$ of the residual sulfate will reflect the $\delta^{18}O$ of the water if there is sufficient time for oxygen isotope exchange to occur between the aqueous sulfite and water (Mizutani and Rafter 1969b, 1973; Fritz et al. 1989).

In a study of the abiotic (thermochemical) reduction of aqueous sulfate through high-temperature (200 to 350°C) interactions with Fe^{2+} (fayalite and magnetite), Shanks et al. (1981) found that the sulfur isotope data could be modeled as an equilibrium Rayleigh distillation process. The $\delta^{34}S$ of residual aqueous sulfate increased during reduction in accordance with published equilibrium fractionation factors (Ohmoto and Lasaga 1982; Table 1). Similarly, Sakai et al. (1980) found that sulfur isotope fractionations associated with thermochemical reduction of dissolved sulfate through reaction with olivine (X_{Fo} = 0.90) at 400°C produced results consistent with equilibrium exchange between sulfate and sulfide. Orr (1974) and Kiyosu (1980) documented sulfur isotopic effects associated with thermochemical reduction of sulfate due to interaction with organic matter and found that sulfate-sulfide kinetic fractionation was less than 10 ‰. No studies have been conducted to examine the oxygen isotope systematics associated with abiotic reduction of sulfate. However, on the basis of the rate of oxygen isotopic exchange between SO_4^{2-} and H_2O at elevated temperature, sulfate-water oxygen isotopic equilibrium would be expected at temperatures above 250°C (Chiba and Sakai 1985).

The supergene oxidation of sulfide minerals to form sulfate minerals or aqueous sulfate produces distinct effects in terms of sulfur and oxygen isotope signatures. With respect to sulfur isotopes, low-temperature oxidative alteration of sulfide minerals to sulfate minerals is a quantitative, unidirectional process that seems to produce negligible sulfur isotope fractionation; the $\delta^{34}S$ of resulting sulfate minerals is indistinguishable from that of the parent sulfide mineral (Gavelin et al. 1960, Nakai and Jensen 1964, Field 1966, Rye et al. 1992). Gavelin et al. (1960) and Field (1966) documented similar sulfur isotope composition among hypogene ore-sulfide minerals and various associated supergene sulfate minerals, including gypsum, alunite, jarosite, anglesite, and brochantite. Field (1966) proposed that the similarity of $\delta^{34}S$ of sulfate minerals formed in the supergene environment with that of the hypogene sulfide minerals from which they formed could be used to distinguish supergene sulfates from high-temperature, hypogene sulfates, which tend to form in isotopic equilibrium with the associated sulfides. A similar conclusion can be reached regarding the relationship of aqueous sulfate with sulfide minerals in acid mine-drainage settings. Taylor and Wheeler (1994) and Seal and Wandless (1997) found no discernible difference between $\delta^{34}S$ in the parent sulfides and the associated dissolved sulfate. However, some experimental and field studies, particularly those involving sulfide oxidation in neutral to alkaline conditions have documented negative sulfate-sulfide fractionations up to approximately 14 ‰ (Toran and Harris 1989), indicative of a kinetic isotope fractionation. Another notable exception occurs when oxygenated mine waters subsequently become exposed to anoxic conditions, and bacterial sulfate reduction occurs. In these settings, the $\delta^{34}S$ of dissolved sulfate is higher than that of the parent sulfide minerals (Hamlin and Alpers 1996, Seal and Wandless 1997).

In contrast, the $\delta^{18}O$ of mineral or aqueous sulfate formed during the biotic or abiotic oxidative weathering of sulfide minerals depends on the source of oxygen. The greatest attention has been focussed on the weathering of pyrite because of its common occurrence in numerous rock types. The weathering of pyrite is generally described by two unidirectional reactions that differ in their oxidizing agent (O_2 versus Fe^{3+}) and their source of oxygen (O_2 and H_2O). The two reactions are (Garrels and Thompson 1960, Singer and Stumm 1970, Nordstrom and Alpers 1999a):

$$FeS_2 + 7/2\ O_{2(aq)} + H_2O \rightarrow Fe^{2+} + 2\ SO_4^{2-} + 2\ H^+ \tag{20}$$

and

$$FeS_2 + 14\ Fe^{3+} + 8\ H_2O \rightarrow 15\ Fe^{2+} + 2\ SO_4^{2-} + 16\ H^+ \tag{21}$$

The reaction rate of the second reaction is limited by the rate of the reaction

$$Fe^{2+} + 1/4\ O_{2(aq)} + H^+ \rightarrow Fe^{3+} + 1/2\ H_2O \tag{22}$$

which is greatly enhanced by the presence of the iron-oxidizing bacteria, such as *Thiobacillus ferrooxidans* (Singer and Stumm 1970). Other bacteria, such as *Thiobacillus thiooxidans*, which oxidize sulfur, can also accelerate the rate of pyrite oxidation. Even though Equations (20) to (22) effectively describe the mass balance of pyrite oxidation, they do not define the pathways of electron transfer (Toran and Harris 1989). Equations (20) and (21) only describe the overall of pyrite weathering; actual rate-limiting steps may involve a variety of intermediate species including native sulfur ($S°$), sulfite (SO_3^{2-}), thiosulfate ($S_2O_3^{2-}$), and polythionates ($S_nO_6^{2-}$; Moses et al. 1987). The stable isotope systematics accompanying these reactions have been most recently reviewed by Taylor and Wheeler (1994) and Van Stempvoort and Krouse (1994).

Because of the sluggish kinetics of sulfate-water oxygen isotope exchange, aqueous sulfate should preserve, except under the most acidic conditions, a record of the sources of oxygen from which the oxygen originated. For the first reaction (Eqn. 20), 87.5 % of the oxygen is derived from O_2 and 12.5 % is derived from H_2O on a molar basis, whereas in the case of the second reaction (Eqn. 21), 100 % of the oxygen is derived from water. The $\delta^{18}O$ of atmospheric oxygen is globally homogeneous and averages 23.5±0.3 ‰ (Kroopnick and Craig 1972), whereas the $\delta^{18}O$ of water varies considerably depending on whether it is meteoric water or seawater, or if it has been modified by processes such as evaporation. In addition, kinetic fractionations are associated with the oxidation of pyrite and the incorporation of oxygen from O_2 or H_2O into sulfate. Taylor et al. (1984) conducted an experimental study of oxygen isotope fractionation associated with pyrite oxidation under a variety of conditions and proposed that the relationship between the $\delta^{18}O$ of dissolved sulfate and water could be used to constrain contributions of atmospheric-derived and water-derived oxygen to the weathering of pyrite. Taylor and Wheeler (1994) presented the following equation to describe the $\delta^{18}O$ of sulfate produced from the oxidation of pyrite in terms of relative contributions of the reactions portrayed in Equations (20) and (21):

$$\delta^{18}O_{SO_4} = X(\delta^{18}O_W + \varepsilon_W) + (1-X)\ [0.875\ (\delta^{18}O_{O_2} + \varepsilon_{O_2}) + 0.125\ (\delta^{18}O_W + \varepsilon_W)] \tag{23}$$

where X is the proportion of oxidation accomplished through Reaction (21), 1-X is the proportion of oxidation accomplished through Reaction (20), and εw and εo_2 are the isotopic enrichment factors for the incorporation of oxygen from water and atmospheric oxygen, respectively. Because of the presence of H_2O in Reaction (20), the permissible range of water-derived oxygen is between 12.5 and 100 mole % (Fig. 14). Taylor and Wheeler (1994) emphasized that the use of Equation (23) assumes that the oxygen isotopic enrichment factors are constant and that isotope exchange between aqueous sulfate and water does not occur after the oxidation of the pyrite. This model also assumes that the participation of O_2 and H_2O is wholly by molecular pathways. However, oxygen may also be involved in an abiotic or biotic electrochemical pathway involving only electron transfer (cf. Toran and Harris 1989). Nevertheless, interpretation of the oxygen isotope systematics of sulfate derived from the oxidation of pyrite using Equation (23) can provide useful insights into the environment of pyrite weathering as has been attempted for supergene oxidation in some ore deposits (Rye et al. 1992, Rye and Alpers 1997).

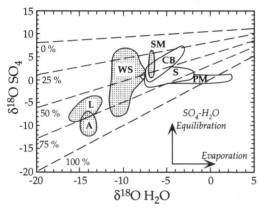

Figure 14. Oxygen isotope variations between sulfate and water associated with the oxidation of pyrite. Modified from Taylor and Wheeler (1994). Dashed isopleths mark mixtures between Reactions (20) (0%) and (21) (100%) as described by Equation (23). Fields for mine drainage sites are indicated by the abbreviations:

L	Leviathan;
A	Argo Tunnel;
WS	West Shasta district;
SM	mines in Great Smoky Mountains National Park;
CB	Cabin Branch mine;
S	Sulfur mine; and
PM	Penn mine.

Data from Taylor and Wheeler (1994), Seal and Wandless (1997), and Hamlin and Alpers (1996). Unshaded fields indicate sites where evaporation was significant, whereas shaded fields indicate sites where evaporation was insignificant. Vectors show the general trends expected for evaporation and oxygen isotopic exchange between SO_4 and H_2O. All isotopic values in permil.

Taylor et al. (1994) determined oxygen isotopic enrichment factors associated with the oxidation of pyrite to form sulfate under a variety of conditions. The origins of these fractionations are complex and involve a variety of processes including mediation by microbes, isotopic exchange between water and intermediate sulfur oxyanions, and isotopic effects associated with the microenvironments of pyrite oxidation, among others. Lloyd (1967) experimentally investigated the oxidation of aqueous Na_2S and H_2S. Under acidic conditions, Taylor et al. (1984) found that εo_2 varied between experiments conducted under sterile conditions (-4.3 ‰) and those conducted in the presence of *Thiobacillus ferrooxidans* (-11.4 ‰). The latter enrichment factor is more relevant to natural systems. Experimental studies by Taylor et al. (1984) found εw to be 4.0 ‰. Comparison with data from natural systems led Taylor and Wheeler (1994) to propose that $\varepsilon w = 0$ ‰ may be more appropriate. For neutral to alkaline environments, $\varepsilon o_2 = -8.7$ ‰ seems to be more appropriate (Lloyd 1967, Taylor and Wheeler 1994). Greater complexity to the stable isotope systematics associated with sulfide oxidation occurs at higher pH values, at which exchange reactions with intermediate sulfur oxyanions can be important (Goldhaber 1983). Under acidic conditions, the rate of pyrite oxidation is relatively rapid and the reaction tends to be complete at the pyrite surface (Wiersma and Rimstidt 1984, McKibben and Barnes 1986).

Because of the considerable overlap in the magnitude of sulfate-water fractionations found in experimental studies, $\Delta_{SO_4\text{-}H_2O}$ values cannot be used to quantify the relative importance of biotic and abiotic pyrite oxidation (Toran and Harris 1989, Taylor and Wheeler 1994). However, higher $\Delta_{SO_4\text{-}H_2O}$ values correlate qualitatively with biotic, aerobic, environments that cycle between wet and dry conditions, whereas lower $\Delta_{SO_4\text{-}H_2O}$ values correlate with sterile, anaerobic, submersed conditions (Taylor and Wheeler 1994).

Mechanisms of precipitation and dissolution of sulfate minerals

The precipitation mechanism for sulfate minerals can have important implications for the interpretation of stable isotope data for these minerals (Ohmoto and Goldhaber 1997). At temperatures above 250°C, dissolved sulfate and sulfide can be expected to be in sulfur isotope equilibrium under most circumstances. Rapid cooling, followed by precipitation of sulfate minerals may not be recorded in sulfate isotope systematics because the dissolved species may fail to re-equilibrate, thereby preserving disequilibrium $\Delta_{SO_4\text{-}H_2S}$ values that are too low for the precipitation conditions (Ohmoto and Lasaga 1982, Rye 1993). Simple cooling of hydrothermal fluids is not a likely mechanism for the precipitation of sulfate minerals because of their retrograde solubility. However, below approximately 75°C, barite, anglesite, and celestite show prograde solubility (Rimstidt 1997). Coprecipitation of sulfide and sulfate minerals occurs typically when decreasing temperature is accompanied by an increase in sulfate concentration at the site of deposition. Local increases in sulfate can result from the oxidation of H_2S, sulfide minerals, or native sulfur, the hydrolysis of SO_2, or mixing with a sulfate-rich fluid (Ohmoto and Goldhaber 1997). The rapid rates of these processes may preclude the re-establishment of isotopic equilibrium. Thus, isotopic disequilibrium can be the result of processes at the immediate site of sulfate-mineral deposition.

The hydrolysis, or disproportionation of SO_2 is a related process that can be important for coprecipitation of sulfide and sulfate minerals. Above 400°C, SO_2 is the dominant oxidized species of sulfur in hydrothermal fluids (Holland 1965, Burnham and Ohmoto 1980). During cooling, the hydrolysis of SO_2 produces both H_2S and HSO_4^- through the reaction:

$$4\,H_2O + 4\,SO_2 \leftrightarrow H_2S + 3\,H^+ + 3\,HSO_4^- \tag{24}$$

The relationships of these processes to specific geochemical settings are discussed below.

The dissolution of sulfate minerals generally proceeds as a quantitative disequilibrium process. Thus, dissolution produces no fractionation of neither sulfur or oxygen isotopes. The retrograde solubility of many sulfate minerals tends to cause dissolution during cooling when associated sulfide or silicate minerals are actively precipitating (Rimstidt 1997).

GEOCHEMICAL ENVIRONMENTS

Secular variations in seawater sulfate isotopic compositions

The geochemistry of sulfur has been intimately tied to global-scale factors such the oxidation state of the atmosphere, anoxia state of the oceans, and the rate of erosion on the continents, which are ultimately influenced by tectonic processes. The sulfur and oxygen isotope signatures of marine evaporitic sulfate minerals provide a record of secular variations in the oceans. Claypool et al. (1980) summarized published sulfur and oxygen isotopic data from marine evaporite deposits, and contributed new data from other key parts of the geological record. Since 1980, several studies have concentrated on Proterozoic and Archean isotopic variations (Hayes et al. 1992, Ohmoto 1992, Strauss 1993, Ross et al. 1995). Paytan et al. (1998) recently reported a high-resolution record for Cenozoic seawater sulfate from deep-sea sediments. Mineralogically, these data are derived dominantly from gypsum, anhydrite, and barite. Collectively, these studies and the studies referenced therein have found that the mean $\delta^{34}S$ of marine evaporitic sulfate has varied from around 4 ‰ at 3.4 Ga to a high of 33 ‰ during Cambrian time to a low of about 10 ‰ during Permian and Triassic time, ultimately to a modern value around 23 ‰ (Fig. 15). The oxygen isotopic data reported in the literature are more limited in scope. The mean $\delta^{18}O$ of marine evaporitic sulfate has varied from around 17 ‰ at 900 Ma to a low of 10 ‰ during Permian time, to a modern value of 13 ‰ (Fig. 16). On the basis of known low-temperature fractionation factors between gypsum and water, the $\delta^{34}S$ of the associated seawater sulfate should be 1.65 ‰ less than that of the mineral (Thode and Monster 1965), and the $\delta^{18}O$ of the associated dissolved sulfate should 3.7 ‰ less than that of the mineral (Gonfiantini and Fontes 1963). The $\delta^{34}S$ of modern seawater is homogeneous on a global scale, with $\delta^{34}S = 20.99 \pm 0.08$ ‰ (Rees et al. 1978). The homogeneity results from the rapid mixing time of seawater (~1000 y) relative to the residence time of sulfate in seawater (8 Ma; Holland 1978). The $\delta^{18}O$ of modern seawater sulfate is also homogeneous

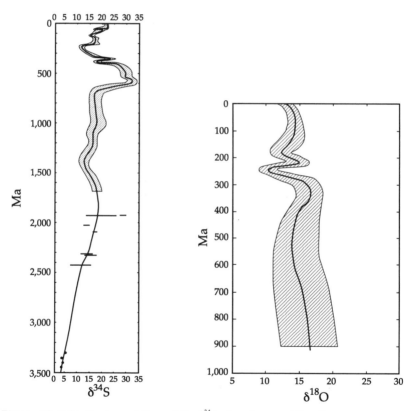

Figure 15 (left). Secular variations of the $\delta^{34}S$ of marine evaporites. Curves were compiled from Claypool et al. (1980), Strauss (1993), and Paytan et al. (1998). Curve shows best fit to the data. Shaded field shows range of most likely values for the Phanerozoic and part of the Proterozoic. Filled circles from 3.3 to 3.5 Ga are from marine evaporites; horizontal bars from 1.9 to 2.5 Ga are from trace sulfate minerals in marine carbonates and cherts (Strauss 1993). All isotopic values in permil (VCDT).

Figure 16 (right). Secular variations of the $\delta^{18}O$ of marine evaporites. Modified from Claypool et al. (1980). Black curve shows best fit to the data; shaded field shows range of values. All isotopic values in permil (VSMOW).

throughout the oceans with a value of 9.5 ‰ (Longinelli and Craig 1967, Nriagu et al. 1991).

Claypool et al. (1980) and Holser et al. (1979) interpreted qualitatively the $\delta^{34}S$ and $\delta^{18}O$ of marine sulfate in terms of a dynamic balance between (1) the removal of sulfate from seawater either by its reduction to sulfide by sulfate-reducing bacteria, or its precipitation as sulfate minerals in marine evaporite deposits and (2) the addition to seawater of sulfate derived either from the oxidative weathering of sulfide minerals on the continents, or from the weathering of evaporite minerals on the continents (Fig. 17a). Oxygen isotopic exchange between dissolved sulfate and water is not considered to be a complicating factor because of the slow rate of exchange at neutral to alkaline conditions relative to the residence time of sulfate in the oceans (Hoering and Kennedy 1957, Chiba and Sakai 1985). Holser et al. (1979) also concluded that the rapid oxidation-reduction cycling mechanism proposed by Lloyd (1968), and the hydrothermal circulation of seawater at mid-ocean ridges were not significant processes affecting seawater-sulfate stable isotope systematics.

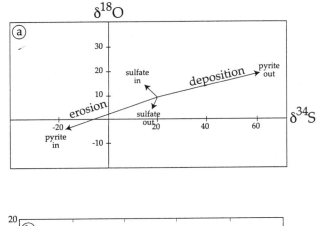

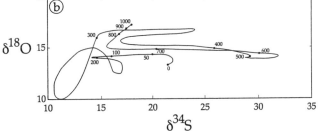

Figure 17. Covariation of oxygen and sulfur isotopes in global oceans. (a) Vector model for processes contributing to secular variations in stable isotopes for marine evaporites. Modified from Claypool et al. (1980); (b) Secular covariations of marine evaporitic $\delta^{34}S$ and $\delta^{18}O$. Modified from Claypool et al. (1980). Age along the curve is indicated in Ma. All isotopic values in permil (VSMOW or VCDT).

The oceanic sulfur-budget analysis of Claypool et al. (1980) made several assumptions about sulfur and oxygen isotope fractionations associated with the various input and output processes, some of which may require modification in light of other studies. It was assumed that the sulfate-sulfide sulfur isotope fractionation associated with bacterial sulfate reduction was 40 ‰. The sulfate-sulfide fractionation in modern sediments averages 40 ‰, but has been documented to range between 15 and 60 ‰ depending on the rate of sulfate reduction, which can be related to sedimentation rate (Goldhaber and Kaplan 1975). It was also assumed that the weathering of pyrite and other sulfide minerals generally occurs without significant sulfur isotope fractionation, which seems to be supported by more recent studies (e.g. Taylor and Wheeler 1994). Claypool et al. (1980) estimated that the modern mean $\delta^{34}S$ of sulfate input derived from the weathering of pyrite is -17±3 ‰. Likewise, they assumed that the removal of sulfate as sulfate minerals causes a decrease of 1.65 ‰ in the $\delta^{34}S$ of the residual aqueous sulfate and that the generation of dissolved sulfate derived from the weathering sulfate minerals produces no significant fractionation of sulfur isotopes. It was estimated that the modern mean $\delta^{34}S$ of sulfate input derived from the weathering of evaporites is 16±2 ‰.

Holser et al. (1979) assumed on the basis of the experimental study of Mizutani and Rafter (1969b) that the effect on the residual dissolved sulfate of oxygen isotopic fractionation associated with bacterial sulfate reduction was approximately 25 % of that for sulfur (i.e. 10 ‰). Mizutani and Rafter (1969b) found that the ratio of $\delta^{18}O$ variations to $\delta^{34}S$ variations with the progress of sulfate reduction was $(\delta^{18}O-\delta^{18}O_i)/(\delta^{34}S-\delta^{34}S_i)$ = 1/3.8, where $\delta^{18}O_i$ and $\delta^{34}S_i$ are the starting oxygen and sulfur isotopic compositions, respectively, of the dissolved sulfate, and $\delta^{18}O$ and $\delta^{34}S$ are the compositions at the termination of the sulfate reduction experiments. In contrast, data from the experimental studies of Mizutani and Rafter (1973) and Fritz et al. (1989) yielded $(\delta^{18}O-\delta^{18}O_i)/(\delta^{34}S-\delta^{34}S_i)$ ratios of 1/2.5 and 1/19.4, respectively; all three studies combined yield a ratio of 1/6.1 (Fig. 18). The significant differences among these three studies in terms of

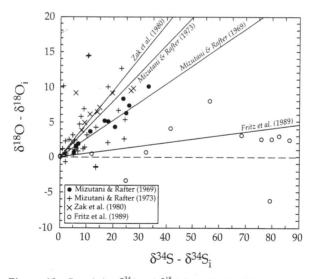

Figure 18. Covariation $\delta^{34}S$ and $\delta^{18}O$ in residual sulfate associated with bacterial sulfate reduction in laboratory (Mizutani and Rafter 1969a, Mizutani and Rafter 1973, Fritz et al. 1989) and field (Zak et al. 1980) studies. Data sets are normalized to their starting sulfur ($\delta^{34}S_i$) and oxygen ($\delta^{18}O_i$) compositions. Lines are least-squared fits to the data. Dashed line indicates $\delta^{18}O-\delta^{18}O_i$=0.0 ‰. All isotopic values in permil (VSMOW or VCDT).

temperature, culture media, bacterial cultures, and experiment duration make it difficult to determine which data might be most applicable to natural systems. However, Zak et al. (1980) reported $\delta^{18}O$ and $\delta^{34}S$ data for pore waters from modern marine sediments in the Pacific Ocean which yield a ratio of 1/2.1 associated with sulfate reduction in natural settings. This ratio suggests that a fractionation of 19 ‰ may be more appropriate to describe oxygen isotope variations associated with bacterial sulfate reduction in the oceans.

The hydrothermal barite of the Creede district, Colorado, indicates that aqueous sulfate in the saline waters of ancient Lake Creede underwent an O/S isotope enrichment during reduction of about 1/2 prior to recharge into the hydrothermal system, but the $\delta^{18}O$ values may have been affected by exchange with highly evaporated (^{18}O-enriched) waters (Rye et al. 1988, Barton et al. 2000). The covariation of sulfur and oxygen isotopes during bacterial sulfate reduction warrants further research.

For inputs, Holser et al. (1979) assumed that the oxidative weathering of pyrite introduces sulfate to the oceans with an average $\delta^{18}O$ of 2.0 ‰ on the basis of experimental studies on the oxidation of aqueous sulfides. Subsequent studies on the oxidation of pyrite, summarized by Taylor and Wheeler (1994), indicate that most weathering of pyrite occurs through processes dominated by 50 to 75 % of Reaction (21), which derives all of its oxygen from water. This range corresponds to a resultant sulfate $\delta^{18}O$ between -1 and 3 ‰, assuming global mean precipitation $\delta^{18}O$ of -4.0 ‰ (Craig and Gordon 1965). Claypool et al. (1980) assumed that the weathering of evaporites results in no fractionation of oxygen isotopes in sulfates, and assumed that the global mean $\delta^{18}O$ of evaporitic sulfate was 12 ‰. They proposed that the combined input of sulfate to the oceans from the weathering of pyrite and evaporites has a $\delta^{18}O$ of 5 ‰.

Claypool et al. (1980) concluded that the secular variations for the marine sulfate curves for $\delta^{34}S$ and $\delta^{18}O$ share some similarities, but also differ significantly. With the exception of Permian time, the secular variations are most dominant in the $\delta^{34}S$ record with a range of over 20 ‰ (excluding the Permian), whereas the $\delta^{18}O$ record has a range of 5 ‰ or less (excluding the Permian) (Fig. 17b). These coupled variations were interpreted in terms of variable inputs and output to the oceans. The large excursions in $\delta^{34}S$, with subordinate variations in $\delta^{18}O$, require that episodes of erosion and oxidation of pyrite be accompanied by erosion and dissolution of evaporites, and episodes dominated by sulfate reduction be accompanied by precipitation of evaporitic sulfate (Fig. 17a,b). To date, the role of hydrothermal circulation through ridge crests on the sulfur and oxygen isotopic systematics of seawater has yet to be addressed comprehensively.

The secular variations of marine sulfate and sedimentary sulfide minerals have been central to discussions of the evolution of the early atmosphere. The bulk $\delta^{34}S$ of Earth is assumed to be similar to meteoritic $\delta^{34}S$ (i.e. ~ 0 ‰). The $\delta^{34}S$ of the earliest marine sulfide minerals clusters around 0 ‰ (prior to ~ 2.7 Ga) and is generally interpreted to reflect the absence of significant concentrations of dissolved sulfate due to the absence of an oxygenated atmosphere (Lambert and Donnelly 1990, Hayes et al. 1992). Both factors would have caused significant scatter in the isotopic composition of marine sulfides because of the large fractionations between oxidized and reduced sulfur species. On the basis of sulfur isotope data from marine sulfate and sulfide minerals, Ohmoto and Felder (1987) proposed a sulfate-sulfide kinetic fractionation of approximately 10 ‰ for bacterial sulfate reduction in the Archean oceans. Ohmoto (1992) suggested that the smaller kinetic fractionation was due to a higher rate of sulfate reduction, which can be related to inferred higher ocean temperatures (30 to 50°C) and greater availability of oxidizable organic matter. A smaller kinetic fractionation would necessitate a greater degree of sulfate reduction to produce larger variations in the $\delta^{34}S$ of marine sulfide minerals. However, the presence of sulfate in Archean oceans does not necessarily require the presence of free oxygen in the atmosphere.

Seafloor hydrothermal systems

Modern systems. The stable isotope characteristics of modern seafloor hydrothermal systems from mid-ocean ridges have been summarized recently by Shanks et al. (1995). Studies which present stable isotope data for sulfate minerals for individual seafloor hydrothermal systems include those of Shanks and Seyfried (1987), Zierenberg and Shanks (1988, 1994), Alt et al. (1989), Peter and Shanks (1992), Goodfellow and Franklin (1993), and Zierenberg et al. (1993). Among the minerals studied were anhydrite, gypsum, and barite. Most studies have concentrated on the $\delta^{34}S$ of seafloor sulfate minerals with limited attention given to $\delta^{18}O$. All of these studies are from mid-ocean ridge hydrothermal systems, with the exception of that of Zierenberg and Shanks (1988), which described the Red Sea hydrothermal system, for which Miocene evaporite sequences may have been an important component of the sulfate budget.

In mid-ocean ridge hydrothermal systems, sulfate is stripped from seawater during heating associated with downwelling due to the retrograde solubility of anhydrite and other sulfate minerals (Bischoff and Seyfried 1978, Seyfried and Bischoff 1981, Shanks et al. 1995). Thus, sulfate minerals at the vent site form from the interaction of H_2S-bearing hydrothermal fluids with SO_4-bearing ambient seawater. The clustering of the $\delta^{34}S$ of anhydrite from off-axis basalts in the Pacific at around 21 ‰, the composition of modern seawater (Fig. 19), is consistent with the nearly quantitative stripping of sulfate during recharge of hydrothermal convection systems (Alt et al. 1989). In general, the $\delta^{34}S$ values of modern mid-ocean ridge seafloor vent-site sulfate minerals cluster around the $\delta^{34}S$ of modern seawater, but locally range from a low of 3.7 ‰ to a high of 29.3 ‰ (Fig. 19). Janecky and Shanks (1988) have modeled chemical and sulfur isotopic processes in seafloor hydrothermal systems. It was found that models based on sulfate-sulfide disequilibrium best describe the sulfur isotope systematics of modern seafloor hydrothermal systems, and that barite and anhydrite form under distinctly different thermal regimes. Both minerals form from the mixing of vent fluids with ambient

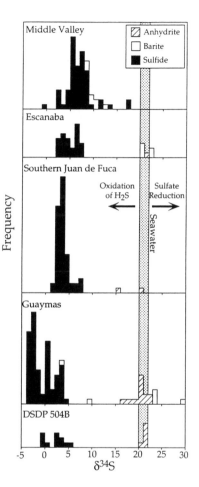

Figure 19. $\delta^{34}S$ histogram for sulfide and sulfates minerals from modern seafloor hydrothermal systems. Shaded vertical field shows the isotopic composition of modern seawater. Vectors show the effect of sulfate reduction and oxidation of H_2S on the $\delta^{34}S$ of sulfate exposed to these processes. Data are from Shanks and Seyfried (1987), Zierenberg and Shanks (1988), Peter and Shanks (1992), and Goodfellow and Franklin (1993). Data for DSDP 504B are for trace sulfide and sulfate minerals in altered oceanic crust (Alt et al. 1989). All isotopic values in permil (VCDT).

seawater. Barite forms during conductive cooling with mixing, whereas anhydrite forms either during adiabatic mixing or during conductive heating with mixing. Scatter from seawater values for both minerals fails to correlate with these inferred differences in depositional environments. Thus, sulfate minerals with a $\delta^{34}S$ less than seawater are generally interpreted to reflect the oxidation of hydrothermal H_2S (typical range $\delta^{34}S = -3$ to 8 ‰; Shanks et al. 1995) in the vent complex. Sulfate minerals with a $\delta^{34}S$ greater than that of seawater are generally interpreted to reflect the residual sulfate from thermochemical reduction through interactions with ferrous silicates or organic matter in the shallower portions of the hydrothermal system (Fig. 19).

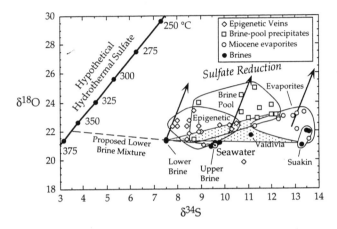

Figure 20. Sulfur and oxygen isotope data for sulfate minerals and dissolved sulfate from the Red Sea. Red Sea brine and Miocene evaporite data are from Zierenberg and Shanks (1986); epigenetic vein and brine-pool precipitate data are from Zierenberg and Shanks (1988). Dashed line represents proposed mixing of sulfate between the upper brine and a hypothetical hydrothermal fluid to produce the lower brine. The line for the hypothetical hydrothermal sulfate was calculated using the fractionation factors in Table 1 and assuming equilibrium with seawater ($\delta^{18}O = 1.2$ ‰) and H_2S ($\delta^{34}S = 5.4$ ‰) at various temperatures. The gray field represents hypothetical mixing between the lower brine and dissolved sulfate derived from Miocene evaporites. Note that many of the epigenetic veins and brine-pool precipitates fall outside of the mixing field. The bold arrows represent the trend expected for sulfate reduction. All isotopic values in permil (VSMOW and VCDT).

Studies from seafloor hydrothermal systems in the Red Sea illustrate the complex interplay of geochemical processes that control the stable isotope compositions of sulfate minerals. Zierenberg and Shanks (1986) observed distinct differences in the $\delta^{34}S$ and $\delta^{18}O$ of dissolved sulfate among the upper and lower brines of the Atlantis II deep, which hosts impressive modern massive sulfide deposits, and the brines of the nearby Valdivia and Suakin deeps. The variations were interpreted in terms of a combination of high-temperature oxygen isotope exchange between seawater and dissolved sulfate, and mixing of seawater sulfate with sulfate derived locally from Miocene evaporite deposits; the dissolved sulfate of the brines from the Valdivia and Suakin deeps are dominated by simple mixtures of seawater sulfate and Miocene evaporite sulfate, without high-temperature isotope exchange. In contrast, the brines of the Atlantis II deep show significant stratification with the lower brine having a lower $\delta^{18}O$ and a slightly higher $\delta^{34}S$ compared to those of the upper brine. The $\delta^{18}O$ and $\delta^{34}S$ of dissolved sulfate from the upper brine are indistinguishable from modern seawater (Fig. 20). Therefore, the sulfate from the lower brine probably comprises a mixture of Red Sea deep water and hydrothermal sulfate. The isotopic composition of the hydrothermal end-member sulfate can be constrained by calculating the isotopic composition of aqueous sulfate in equilibrium with local seawater ($\delta^{18}O$ = 1.2 ‰; Zierenberg and Shanks 1986) and hydrothermal H_2S ($\delta^{34}S$ = 5.4 ‰; Zierenberg and Shanks 1988) at various temperatures (Fig. 20). A mixing line drawn from the sulfate of the upper brine ($\delta^{34}S$ = 21.0 ‰, $\delta^{18}O$ = 9.5 ‰) through that of the lower brine ($\delta^{34}S$ = 21.4 ‰, $\delta^{18}O$ = 7.5 ‰) intersects the hypothetical end-member sulfate curve at $\delta^{34}S$ = 22.2 ‰, $\delta^{18}O$ = 3.6 ‰, and T = 360°C. Thus, on the basis of mass-balance calculations (Eqn. 18), approximately 35 % of the sulfate in the lower brine may have been derived from the hydrothermal fluid. The inferred temperature is consistent with temperatures commonly found in modern vent systems (Shanks et al. 1995) and with mineralogical and fluid inclusion evidence for a high temperature history for the Atlantis II deep hydrothermal fluids (Pottorf and Barnes 1983, Oudin et al. 1984). The assumption of equilibrium is reasonable with respect to the kinetics of isotopic exchange of SO_4 with H_2O and H_2S at these temperatures. However, less certain is the ability of the inferred hydrothermal fluids to transport significant amounts of dissolved sulfate because of the retrograde solubility of anhydrite.

Regardless of the ultimate origin of the sulfate in the lower brine, the oxygen and sulfur isotopic composition of anhydrite and barite in epigenetic vein and brine-pool precipitates in the Atlantis II deep are only partly explained by mixing of sulfate from the lower brine pool and sulfate from Miocene evaporites. The $\delta^{18}O$ of the lower brine and the evaporites encloses the range of the sulfate minerals from the epigenetic veins and brine-pool precipitates, but the $\delta^{34}S$ of the minerals ranges to higher $\delta^{34}S$ values than for a simple mixture. One possible explanation for the higher $\delta^{34}S$ values is that the sulfate had undergone partial thermochemical reduction prior to deposition as anhydrite or barite. Thermochemical reduction of sulfate produces an increase in the $\delta^{34}S$ of the residual sulfate (Shanks et al. 1981). The corresponding change in $\delta^{18}O$ associated with thermochemical reduction has not been investigated in natural or experimental systems, as discussed above, but can be expected to result in an increase in the $\delta^{18}O$ in the residual sulfate, consistent with bacterial sulfate reduction. If a $\delta^{34}S/\delta^{18}O$ slope of 4/1 is chosen to represent the trend produced by sulfate reduction, then the partial reduction of various admixtures of lower brine sulfate and evaporite sulfate can adequately describe the isotopic variations of the anhydrite and barite in the epigenetic veins and brine-pool precipitates. No bacterial reduction of sulfate is known to occur at present in the Atlantis II deep (Zierenberg and Shanks 1988), but low $\delta^{34}S$ compositions in oxide-rich facies of metalliferous sediments indicate such reduction in the past (Shanks and Bischoff 1980). The average $\delta^{34}S$ of sulfide minerals from the Atlantis II deep (~5.4 ‰; Shanks et al. 1980) is significantly higher than the $\delta^{34}S$ of basalts such as the ones underlying the Red Sea. Thus, the elevated $\delta^{34}S$ of sulfide minerals indicates a component of thermochemical sulfate reduction.

Ancient systems. Ancient seafloor hydrothermal systems produce two general classes of deposits containing sulfate minerals: volcanic-associated (volcanogenic) massive sulfide deposits, and sedimentary-exhalative (sedex) massive sulfide deposits, which Maynard and Okita (1991) refer to as "cratonic rift-type barite deposits". Volcanic-associated deposits form at active mid-ocean ridge spreading centers, and in arc-volcanic rocks, continental rifts, and Archean greenstone belts, whereas sedex deposits form in failed continental rift settings. Maynard and Okita (1991) further distinguish "continental margin type barite deposits", which form in convergent margins and typically are lacking associated Zn and Pb. Volcanic-associated and sedex deposits can have associated hydrothermal sulfate minerals, usually anhydrite, barite, or gypsum. Gypsum is typically a weathering product of primary anhydrite. In general, barite and anhydrite occur as veins, disseminations in massive sulfide ore, or as massive stratiform zones. Sulfur isotope data are reasonably common, but oxygen isotope results are more limited.

The stable isotope systematics of volcanic-associated massive sulfide deposits and their significance have been reviewed by Ohmoto (1986) and Huston (1999). The sulfur-isotope data for sulfate minerals plotted in Figure 21 are from published studies of deposits that range in age from the Recent back to approximately 1.9 Ga. This figure includes data from the Kuroko district, Japan (Watanabe and Sakai 1983, Kusakabe and Chiba 1983), the Eastern Black Sea district, Turkey (Çagatay and Eastoe 1995), the Buchans camp, Newfoundland (Kowalik et al. 1981), various districts in California (Eastoe and Gustin 1996), the Mount Read volcanic belt, Tasmania (Solomon et al. 1969, 1988; McGoldrick and Large 1992, Gemmell and Large 1992), Barite Hill, South Carolina (Seal et al., in press), the Skellefte district, Sweden (Rickard et al. (1979), and Buttle Lake, British Columbia (Seccombe et al. 1990). For the most part, the $\delta^{34}S$ of sulfate minerals from these deposits is similar to the $\delta^{34}S$ of coeval seawater, although deviations to higher and lower values can be found (Fig. 21). Some of the deviations from the coeval seawater composition may be attributed to uncertainties in the age of the deposits, particularly for those deposits that formed near abrupt transitions in the secular curve for marine sulfate isotopic compositions (Fig. 15). The similarity indicates that sulfate minerals associated with seafloor hydrothermal systems are generally produced by the mixing of sulfate-rich ambient seawater with Ba- or Ca-bearing hydrothermal fluids. As with modern hydrothermal systems, values lower than that of ambient seawater are

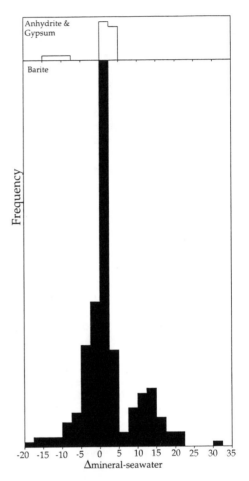

Figure 21. Histograms of the difference between the sulfur isotope composition of sulfate mineral and that of coeval seawater from ancient seafloor hydrothermal deposits. The $\delta^{34}S$ of seawater was taken from the average values of Claypool et al. (1980) and adjusted assuming a mineral-dissolved sulfate fractionation of 1.65 ‰. See text for sources of data. Huston (1999) produced a similar diagram to compare sulfide and sulfate mineral compositions relative to seawater. All isotopic values in permil (VCT).

typically interpreted to represent the oxidation of hydrothermal H_2S to produce sulfate for later precipitation, such as at Buttle Lake, British Columbia (Seccombe et al. 1990).

Values higher than that of ambient seawater are typically interpreted to reflect partial reduction of seawater sulfate and associated isotopic enrichment of the residual sulfate, either by bacterial or abiotic processes, such as in the Mount Read volcanic belt, Tasmania (Solomon et al. 1969, 1988; McGoldrick and Large 1992, Gemmell and Large 1992). The $\delta^{34}S$ of sulfate minerals from the deposits of the Mount Read volcanic belt is up to 22 ‰

greater than that of coeval seawater, with a distinct mode around 14 ‰ greater than coeval seawater (Fig. 21). Such pervasive reduction of seawater sulfate is unlikely to occur through abiotic, thermochemical interactions with ferrous iron during hydrothermal convection as suggested by Green et al. (1981) and Solomon et al. (1988), because of the limited ability of hydrothermal fluids to transport seawater sulfate due to the retrograde solubility of anhydrite and other sulfate minerals (Bischoff and Seyfried 1978, Seyfried and Bischoff 1981). The common appearance of isotopic equilibrium between sulfate and sulfide minerals likely reflects the dynamic, disequilibrium mixing environment between hydrothermal fluids and ambient seawater typically invoked to explain the formation of seafloor exhalative barite bodies. Bacterial reduction of seawater sulfate in anoxic bottom waters is an alternative to explain the high $\delta^{34}S$ of the Mount Read sulfates relative to coeval seawater. Goodfellow and Jonasson (1984) proposed a similar mechanism to explain enriched isotopic compositions of barite in the Selwyn Basin, Yukon relative to coeval seawater. The deposits of the Mount Read volcanic belt formed at a time of oceanic bottom-water anoxia in the Iapetus Ocean, which may have been of global extent (Leggett 1980). The presence of black slates associated with the Rosebery deposit in the Mount Read belt supports this hypothesis, but the presence of minor hematite in some of the facies of mineralization suggests that further evaluation may be warranted (Green et al. 1981).

The Selwyn Basin, Yukon, Canada hosts several syngenetic sedimentary exhalative Zn-Pb-barite deposits and bedded barite deposits, which range in age from Cambrian to Mississippian and formed from seafloor hydrothermal processes (Goodfellow and Jonasson 1984). Goodfellow and Jonasson (1984) documented secular trends in the $\delta^{34}S$ of pyrite and barite within the basin. The Selwyn Basin barite compositions followed the secular trends of global marine evaporites (Claypool et al. 1980), but were consistently 10 to 15 ‰ higher in $\delta^{34}S$. Goodfellow and Jonasson (1984) interpreted the higher $\delta^{34}S$ of the Selwyn barite relative to global evaporites to reflect formation in a restricted basin, with anoxic bottom waters and limited access to the open ocean. The bacterial reduction of sulfate caused the residual sulfate, now preserved as hydrothermal and sedimentary barites, to become enriched in ^{34}S relative to global averages.

"Continental margin type" bedded barite deposits, such as those in Nevada (Rye et al. 1978), in the Qinling district, China (Maynard and Okita 1991, Wang and Li 1991), and Arkansas (Maynard and Okita 1991) tend to have $\delta^{34}S$ values similar to that of coeval seawater. In Nevada, in addition to bedded barite, barite also occurs in diagenetic rosette and concretion morphologies. The $\delta^{34}S$ of the rosettes is significantly higher than that of coeval seawater (up to 30 ‰ greater) and the $\delta^{18}O$ of sulfate shows a 1/3 enrichment slope relative to $\delta^{34}S$ (R.O. Rye unpubl. data). The high $\delta^{34}S$ of these non-bedded barites was interpreted to reflect the influence of closed-system bacterial sulfate reduction on residual sulfate during diagenesis (Rye et al. 1978). Similarly, Nuelle and Shelton (1986) determined that nodular barites from Devonian shales in the Appalachian Basin, with $\delta^{34}S$ up to 28 ‰ and $\delta^{18}O$ up to 15 ‰, were higher than coeval marine evaporites.

Magmatic systems

In the spring of 1982, a series of three major pyroclastic eruptions occurred at El Chichón Volcano in eastern Mexico, involving an unusually sulfur-rich trachyandesitic magma (Luhr et al. 1984, Rye et al. 1984). The sulfur-rich character was reflected in both in the presence of microphenocrysts of anhydrite in fresh pumices and in the large size of the stratospheric cloud of sulfur acid aerosols produced by the eruptions. Anhydrite has since been recognized as a characteristic and important mineralogical component of fresh tephra from Mount Pinatubo and other arc volcanoes that erupt oxidized, hydrous sulfur-rich magma (Fournelle 1990, Bernard et al. 1991, McKibben et al. 1996). Anhydrite is not well-preserved in older pyroclastic rocks because it weathers out of tephra soon after eruption. Experimental studies have demonstrated the relatively high solubility of anhydrite and the stability of dissolved SO_4^{2-} species in oxidized melts (Carroll and Rutherford 1987, 1988, Luhr 1990). It is therefore now commonly assumed that the anhydrite in these recent pyroclastic materials is a juvenile magmatic phase. This was first demonstrated by a combination of textural evidence and highly concordant temperatures based on sulfur and oxygen isotopic compositions of anhydrite, pyrrhotite, magnetite and plagioclase crystals from fresh pumices of the 1982 eruptions of El Chichón volcano which indicated a pre-eruption temperature of $810\pm40°C$. The isotopic composition of sulfate leached from fresh ash-fall samples showed it to have been derived from a mixture of anhydrite micro-phenocrysts and adsorbed sulfate, the latter derived from the oxidation of SO_2 in the eruption plume. The isotopic data also indicate the lack of early oxidation of significant amounts of H_2S, which should have dominated the eruption cloud based on H_2S/SO_2 in equilibrium with the melt. Apparently, this oxidation occurs slowly as some sulfate aerosol clouds have been observed to grow as they circled the Earth in the stratosphere (Bluth et al. 1997). The sulfur-rich magma may have developed from the assimilation of a volcanogenic massive sulfide or sedimentary sulfate body. Finally, sulfur isotope data on sulfides and anhydrite in lithic fragments support the development of a porphyry-type system overlying the magma at El Chichón.

SHRIMP sulfur isotope studies of anhydrite in pumice from the early vent-clearing June 1991 eruption of Mount Pinatubo, Philippines, show a bimodal sulfur isotope distribution that may represent primary phenocrysts and xenocrysts derived from hydrothermal anhydrite deposits occurring within the flanks of the volcano. To explain the sulfur isotope systematics, McKibben et al. (1996) favored a mechanism involving long-term passive degassing of sulfur from underlying basalt into the much more oxidized dacite magma, and steady-state degassing of vapor enriched in ^{34}S.

Igneous apatites can contain up to 1,500 mg/kg sulfur probably as SO_4^{2-} substituting for PO_4^{2-}, and coexisting minerals such as biotite and magnetite may contain up to 200 mg/kg sulfur, probably as SH substituting for OH (Banks 1982). Various sulfate and sulfide mineral inclusions may occur in minerals such as plagioclase and magnetite. Ueda and Sakai (1984) used the different reactivity of minerals to various acids to recover separately sulfate and coexisting sulfide in I-type igneous rocks. Their data seldom show sulfur isotopic equilibrium. Lack of isotopic equilibrium between sulfate in apatite and sulfide in igneous rocks has also been shown for a series of igneous rocks that culminated with the intrusion of an anhydrite-bearing dike that interrupted ore deposition at Julcani, Peru (Deen 1990). This lack of equilibrium can be explained by a combination of assimilation of country-rock coupled with fractionations due to degassing of magma during its ascent (See also Ohmoto 1986).

Continental hydrothermal systems

Modern systems. The stable isotope systematics of active geothermal systems are dominated by water, dissolved sulfate, and dissolved sulfide, depending on redox conditions, and the interactions of these species with the reservoir rocks. The stable isotope geochemistry of geothermal systems has been reviewed by Truesdell and Hulston (1980) and Giggenbach et al. (1983a). Numerous studies have investigated the stable isotope composition of dissolved sulfate in geothermal systems, particularly for oxygen relative to H_2O, and to a lesser extent, for sulfur relative to H_2S. In addition, Rye et al. (1992) presented data from alunite from active geothermal systems. Information recorded by dissolved sulfate and dissolved sulfide depends on the temperature, pH, and on the residence time of fluids in the system. Oxygen isotopes exchange more rapidly between sulfate and water than do sulfur isotopes between sulfate and sulfide (Fig. 8). Kusakabe (1974) and Robinson (1978) compared SO_4-H_2O oxygen isotope and SO_4-H_2S sulfur isotope geothermometers for geothermal systems in New Zealand and found that the systematics of the oxygen isotopes were easier to interpret than those of the sulfur isotopes.

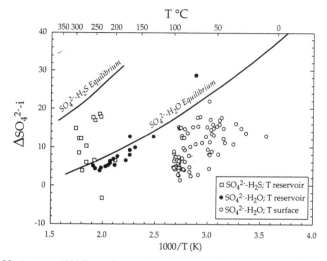

Figure 22. Δ versus 1000/T (K) for geothermal systems for oxygen isotopes between sulfate and water, and for sulfur isotopes between sulfate and sulfide. Equilibrium curves were calculated from the expressions in Table 1. "T reservoir" corresponds to temperatures estimated from geochemical thermometers; "T surface" corresponds to surface temperatures of hot springs or well heads. See text for sources of data. All isotopic values in permil (VSMOW).

Figure 22 shows the relationship of SO_4-H_2O oxygen isotope fractionation and SO_4-H_2S sulfur isotope fractionation to observed temperatures for geothermal systems in Japan (Mizutani 1972, Sakai 1977, Sakai and Matsubaya 1977, Matsuo et al. 1985), Italy (Cortecci 1974), New Zealand (Kusakabe 1974, Simmons et al. 1994), India (Giggenbach et al. 1983b), United States (McKenzie and Truesdell 1977), and Mexico (Martinez et al. 1996). Temperatures are either surface temperatures of hot springs (<100°C), or reservoir temperatures, determined by chemical thermometers (>100°C). Collectively, the SO_4-H_2O oxygen isotope data appear to reflect reservoir temperatures that average ~ 200±75°C. The good agreement among chemical thermometry (reservoir) temperatures and SO_4-H_2O oxygen isotopic temperatures suggests the attainment of isotopic equilibrium, which forms the basis of SO_4-H_2O oxygen isotope thermometry in geothermal systems (Cortecci 1974, Sakai 1977, McKenzie and Truesdell 1977). Thus, the $\delta^{18}O$ of dissolved sulfate in geothermal systems is dominated by the $\delta^{18}O$ of the water. The surface temperatures plot below the equilibrium SO_4-H_2O oxygen isotope fractionation curve, and suggest that the SO_4-H_2O pairs preserve a record of deeper and hotter reservoir conditions.

The $\delta^{34}S$ values of SO_4-H_2S pairs are less consistent, with none of the data approaching equilibrium for reservoir temperatures determined by various chemical geothermometers (Fig. 22). For the Wairakei geothermal system, New Zealand, the temperatures may reflect deep reservoir temperatures around 350°C (Kusakabe 1974). In other systems, the inferred temperatures are unrealistically high (>450°C) for deep reservoir temperatures. The interpretation of SO_4-H_2S sulfur isotopic data is also complicated by shallow-level, disequilibrium oxidation of H_2S, or reduction of sulfate, both occurring with insufficient time to re-establish equilibrium (Kusakabe 1974, Robinson 1978, Sakai 1983).

Ancient systems. The stable isotope geochemistry of ancient continental hydrothermal systems in igneous terranes is similar to that of modern geothermal systems, with the added, complicating factor that ancient systems typically represent the deeper, unexposed portions of modern geothermal systems. Thus, in associated high-temperature regimes, SO_2 (SO_3^{2-}) dominates over SO_3 (SO_4^{2-}), and direct magmatic contributions of sulfur species and water to the hydrothermal system are important. Low-temperature disequilibrium processes dominate the stable isotope geochemistry of ancient continental hydrothermal systems in sedimentary environments.

Many of the important geochemical processes that affect the stable isotope characteristics of sulfate minerals in igneous hydrothermal systems are illustrated by the minerals alunite [$KAl_3(SO_4)_2(OH)_6$] and jarosite [$KFe_3(SO_4)_2(OH)_6$], which occur in a variety of mineral-deposit types. Alunite and jarosite are especially informative because insights can be gained from four stable isotope sites by the analysis of $\delta^{34}S$ from the SO_4 site, $\delta^{18}O$ from the SO_4 and OH sites, and δD from the OH site (Rye et al. 1992, Rye and Alpers 1997). In addition, because alunite and jarosite are K-bearing minerals, they are useful as geochronometers (K/Ar, $^{40}Ar/^{39}Ar$; see Stoffregen et al., this volume). The same processes are also recorded in simple sulfate minerals, such as barite and anhydrite, in porphyry and polymetallic vein deposits. However, a key difference between acid-sulfate environments that form alunite, and porphyry and polymetallic vein environments, is that the highly acidic conditions associated with the formation of alunite are more conducive to isotopic equilibrium among sulfur and oxygen species than the less acidic conditions associated with the other deposit types.

Alunite can be found in four distinct settings known as the supergene, steam-heated, magmatic hydrothermal, and magmatic steam environments (Rye et al. 1992). In contrast, jarosite has only been found in steam-heated and supergene environments (Rye and Alpers 1997); jarosite also occurs in the "sour gas" environment, in which sulfate originates by oxidation of H_2S, for example, in the Rio Grande rift in New Mexico and Chihuahua, Mexico (Lueth et al. 2000). Other occurrences of hydrothermal jarosite are summarized by Dutrizac and Jambor (this volume). Each environment displays unique stable isotope characteristics and jarosite formation requires unique pH and fO_2 conditions for formation (Stoffregen 1993).

For both supergene alunite and jarosite, the sulfur and oxygen isotope characteristics of the sulfate are determined by the geochemical processes responsible for the weathering of sulfide minerals, such as pyrite, as described in Reactions (20) and (21), and typically reflect disequilibrium fractionations. For hydroxyl, the oxygen and hydrogen isotope characteristics are dictated by the isotopic composition of the associated water, with the principal difference between the two minerals being the -55±5 ‰ hydrogen isotope fractionation between jarosite and water. Intramineral fractionation between the SO_4 and OH also typically reflects disequilibrium. Rye et al. (1992) and Rye and Alpers (1997) defined a "supergene alunite SO_4 field" (SASF), a "supergene jarosite SO_4 field" (SJSF), a "supergene alunite OH zone" (SAOZ), and a "supergene jarosite OH zone" (SJOZ) to describe the isotopic compositions (Fig. 23) that would be expected for supergene alunites

Figure 23 (opposite page). Isotopic variations of alunite for various environments for given fluid compositions. Modified from Rye et al. (1992). SASF indicates the Supergene Alunite Sulfate Field and SAOZ indicates the Supergene Alunite OH Zone. Jarosites display similar isotopic variations (Rye and Alpers 1997), the principal difference being the large negative hydrogen isotope fractionation between jarosite and water results in a much diminished supergene jarosite sulfate field (SJSF) which is the counterpart to the SASF. The meteoric water line (MWL) and the field for hypothetical primary magmatic water (PMW) are shown for reference. (a) oxygen and sulfur isotopic variations expected for supergene alunites; (b) hydrogen and oxygen isotopic variations expected for supergene alunites; (c) oxygen and sulfur isotopic variations expected for steam-heated alunites; (d) hydrogen and oxygen isotopic variations expected for steam-heated alunites. Examples illustrate effects expected for an unmodified meteoric water ($\delta D = -80$ ‰, $\delta^{18}O = -11.3$ ‰) and a modified meteoric water ($\delta D = -110$ ‰, $\delta^{18}O = -5.0$ ‰); (e) oxygen and sulfur isotopic variations expected for magmatic-hydrothermal alunites; (f) hydrogen and oxygen isotopic variations expected for magmatic-hydrothermal alunites. The mixing trends shows the range isotopic effects expected for mixing magmatic fluids with modified meteoric waters; (g) oxygen and sulfur isotopic variations expected for magmatic-steam alunites; (h) hydrogen and oxygen isotopic variations expected for magmatic-steam alunites; all isotopic values in permil (VSMOW or VCDT).

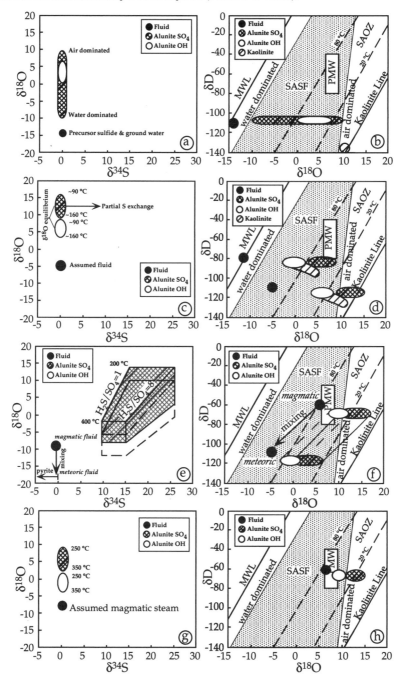

and jarosites. The limits of supergene SO_4 fields are defined by whether the oxygen used to produce the sulfate was dominantly derived from water (Reaction 21) or atmospheric oxygen (Reaction 20). The limits of the supergene OH zones are defined by temperature-dependent mineral-water fractionations between geologically reasonable temperatures of 20 to 80°C. The sulfate isotope systematics of supergene alunite and jarosite can be further complicated if the sulfate undergoes partial bacterial reduction prior to precipitation. In this case, the resulting sulfate will have a higher $\delta^{18}O$ and $\delta^{34}S$ than its starting composition as discussed above. Examples of supergene alunite and jarosite are common throughout the world, and have been studied in Australia, Spain, New Mexico, Arizona, Nevada, Colorado, Chile, and elsewhere; isotope compositions parallel the meteoric water line over a wide range of latitudes (Rye et al. 1992, Rye and Alpers 1997).

Steam-heated environments share many geochemical characteristics with supergene environments. Aqueous sulfate is derived from the oxidation, at or above the water table, of H_2S that is distilled off of hydrothermal fluids as described by the summary reaction:

$$H_2S + 2\ O_2 = H_2SO_4 \tag{25}$$

The reaction occurs at temperatures less than 100°C, but the resulting acids may descend to greater depths at higher temperatures (Rye et al. 1992). The higher temperatures and acidic conditions typical of geothermal fields promote isotopic equilibrium between SO_4^{2-} and H_2O and unlike the supergene environment, the $\Delta^{18}O_{SO_4\text{-}OH}$ values for alunite commonly give reasonable temperatures. Although sulfur isotopic equilibrium between aqueous SO_4^{2-} and H_2S is not obtained in this environment, significant partial exchange occurred before deposition of alunite at the Crofoot-Lewis deposit, Nevada (Ebert and Rye 1997). In geothermal and hydrothermal systems, the hydrothermal fluids can exchange oxygen isotopes with country rocks to a significant degree, producing fluid with $\delta^{18}O$ values that are substantially higher than those of local meteoric waters. Thus, the predicted isotopic fields for steam-heated alunite and jarosite can overlap those for supergene minerals depending on the $\delta^{18}O$ of the associated hydrothermal fluid (Fig. 23). Steam-heated alunites and jarosites can generally be distinguished from supergene minerals by their tendency for equilibrium fractionation between intramineral SO_4-OH sites, and inferred SO_4-H_2O pairs. Examples of steamed-heated alunites and jarosites can be found in Nevada, Utah, Italy, New Zealand, and elsewhere (Rye et al. 1992, Rye and Alpers 1997, Ebert and Rye 1997). Because of the exceptionally low pH and high fO_2 required for jarosite formation, steam-heated jarosites represent the last hydrothermal event in an area and can be used to date the duration of hydrothermal mineralization or to recognize a later hydrothermal event (Rye and Alpers 1997).

Magmatic hydrothermal alunite forms from sulfuric acid derived from the disproportionation of magmatic SO_2, as described by Reaction (24). At the high temperature and low pH of this environment, isotopic equilibrium is expected among all species and mineralogical sites (Rye et al. 1992). However, due to the elevated formational temperature, the OH site is prone to retrograde isotopic exchange on cooling such that intramineral SO_4-OH fractionation and the hydrogen isotopic compositions may not be representative of the conditions of formation for the alunite. Equilibrium fractionations produce alunite sulfate with $\delta^{18}O$ and $\delta^{34}S$ that are dominantly a function of temperature and the H_2S/SO_4 ratio of the hydrothermal fluid. Primary magmatic hydrothermal alunites have SO_4 $\delta^{18}O$ values that are higher than those of supergene alunites and have OH $\delta^{18}O$ values that overlap those of the supergene alunite OH zone (Fig. 23). In reality, mixing with local meteoric waters can produce SO_4 values that overlap the supergene alunite SO_4 field and OH zone, depending on the isotopic composition of meteoric water. Examples of magmatic hydrothermal alunite occur in Peru, Colorado, Chile, Spain, Nevada, and elsewhere (Rye et al. 1992).

Magmatic steam alunites exhibit some similarities with both steam-heated and magmatic hydrothermal alunites. However, magmatic steam alunite can be remarkably coarse banded with zones of P_2O_5-enrichment variation and almost constant $\delta^{34}S$ values that reflect quantitative, disequilibrium oxidation of magmatic SO_2 (and possibly H_2S) by entrainment of atmospheric oxygen or by loss of hydrogen from the hydrothermal fluid (Cunningham et al. 1997). Their oxygen and hydrogen isotope characteristics are dominated by equilibrium with magmatic water or derived steam; thus, the sulfate $\delta^{18}O$ is distinctly higher than that of the supergene alunite sulfate field (Fig. 23). The $\delta^{18}O$ of the OH is less distinctive because it lies within the supergene alunite OH zone, but the δD is consistent with magmatic waters. Examples of magmatic steam alunite can be found in Utah, California, and Colorado (Rye et al. 1992).

Stable isotope studies of simple sulfate minerals, such as anhydrite and barite, from less acidic ancient porphyry and epithermal deposits are dominated by sulfur isotope data, with fewer oxygen isotope data. Studies typically concentrate on the fractionation of sulfur isotopes between coexisting sulfate and sulfide minerals, which can be useful for constraining sources of sulfur, temperatures of mineralization, and equilibrium status of mineralizing systems. Interpretations have utilized plots of the $\delta^{34}S$ of coexisting sulfate and sulfide minerals as a function of $\Delta_{sulfate-sulfide}$ (Field and Gustafson 1976; cf. Shelton and Rye 1982, Ohmoto 1986, Krouse et al. 1990). Theoretically, "δ-Δ" diagrams can provide information about temperatures of mineralization and the SO_4/H_2S ratio of the hydrothermal fluids provided that (1) the SO_4/H_2S ratio of the fluid remained constant, (2) the bulk sulfur isotopic composition of the fluid ($\delta^{34}S\Sigma s$) remained constant, and (3) the only cause of isotopic variations in the initial $\delta^{34}S$ was exchange between SO_4 and H_2S in the fluid (Field and Gustafson 1976, Ohmoto 1986). Pairs of coexisting sulfate and sulfide minerals plot along two linear arrays that converge to a point at $\Delta = 0$ ‰, which corresponds the $\delta^{34}S\Sigma s$ of the fluid. The angle between the linear arrays for the sulfate and sulfide minerals can be related to the SO_4/H_2S ratio of the fluid (Field and Gustafson 1976). However, because δ and Δ are not totally independent variables, linear arrays can result from disequilibrium mineral pairs (Ohmoto 1986, Krouse et al. 1990). In addition, natural settings rarely satisfy all of the conditions stated above. Interpretation of natural data sets can be complicated by fluctuating fluid compositions, kinetic processes related to isotopic exchange and precipitation, and mixing of multiple sulfur reservoirs, among other processes (Shelton and Rye 1982, Ohmoto 1986, Krouse et al. 1990).

Similar δ-Δ diagrams have been interpreted in terms of equilibrium processes (Field and Gustafson 1976) and in terms of disequilibrium processes (Shelton and Rye 1982). Interpretations of equilibrium fractionation assume that the three conditions listed above are satisfied. Interpretations of disequilibrium attribute the lack of equilibrium to non-equilibrium oxidation-reduction in the fluid, mixing of fluids, dissolution and reprecipitation of pre-existing sulfate and sulfide minerals, and (or) preservation of higher temperature equilibria in the fluid (Shelton and Rye 1982). Shelton and Rye (1982) proposed disequilibrium conditions at Mines Gaspé and El Salvador because anhydrite-sulfide sulfur isotope temperatures were higher than temperatures derived from fluid inclusion measurements. However, they used the sulfate-H_2S fractionation factors from Ohmoto and Rye (1979), which can yield temperatures that are in excess of 60°C greater than those calculated using the sulfate-H_2S fractionation of Ohmoto and Lasaga (1982; Fig. 9). Thus, many of the anomalously high calculated temperatures may, in fact, be within analytical error of the fluid inclusion temperatures if the Ohmoto and Lasaga (1982) SO_4-H_2S fractionation is used to recalculate temperature. Therefore, these systems may be closer to equilibrium than previously thought.

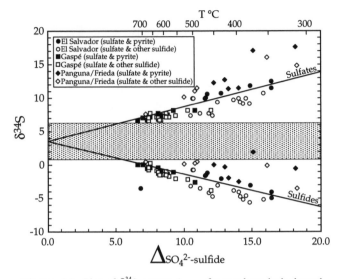

Figure 24. Plot of $\delta^{34}S$ versus $\Delta_{SO_4\text{-}Py}$ for porphyry hydrothermal systems. Values above the shaded fields are for sulfate minerals associated with the sulfide minerals indicated in the legend; values below (and within) the shaded field are for sulfide minerals. See text for sources of data. Shaded field shows the permissive range for the bulk $\delta^{34}S$ of the hydrothermal fluid for sulfide- and sulfate-bearing porphyry systems. The solid lines indicate the hypothetical $\delta^{34}S$ values for sulfates (upper line) and sulfides (lower line) in equilibrium with a hydrothermal fluid with equal proportions of SO_4 and H_2S. All isotopic values in permil (VCDT).

The isotopic characteristics of sulfate and sulfide minerals from porphyry environments suggest a general approach to equilibrium at elevated temperatures. Data from El Salvador, Chile (Field and Gustafson 1976), Gaspé, Quebec (Shelton and Rye 1982), and Papua New Guinea (Eastoe 1983) are plotted in Figure 24. The data plot in two linear trends, one for sulfate minerals, and one for sulfide minerals. The trends converge with a y-intercept of ~3.5 ‰ and the trends are symmetrically distributed above and below the intercept. The fractionations between SO_4 and pyrite ($\Delta_{SO_4\text{-}Py}$) range from 6.6 to 18.2. Assuming equilibrium, these fractionations correspond to a temperature range of 730 to 315°C, respectively. Regardless of equilibrium state, these data indicate that the bulk $\delta^{34}S_{\Sigma S}$ of the hydrothermal fluid lies between 0.5 and 6.5 ‰. Eastoe (1983) questioned equating bulk fluid compositions with the composition of magmatic sulfur, because of the complexities of the evolution of volatile phases from magmas. In high temperature porphyry environments, Shelton and Rye (1982) suggested that the discrepancy between fluid inclusion temperatures and SO_4-sulfide sulfur isotope temperatures resulted from the short time span between the disproportionation of SO_2 to SO_4^{2-} and H_2S, and the subsequent precipitation of sulfate as anhydrite.

Krouse et al. (1990) modeled the isotopic effects that would be expected from a hydrothermal system undergoing active bacterial sulfate reduction. It was found that converging trends for sulfates and sulfides, both with positive slopes, would be produced, and it was suggested that this was a criterion indicating the possible involvement of kinetic isotope effects. Krouse et al. (1990) also urged caution in the use δ vs. Δ diagrams, and encouraged their interpretation in the context of other supporting studies such as petrographic observations of mineral assemblages, and independent estimates of temperatures from fluid inclusions (cf. Shelton and Rye 1982, Eastoe 1983).

Sulfur isotope disequilibrium among aqueous sulfur species in ore-forming fluids was recognized in large banded sulfide-barite veins in the Ag and base-metal ores of the Creede mining district, Colorado (Rye and Ohmoto 1974). The barite shows a very large range of both $\delta^{34}S$ and $\delta^{18}O$ values whereas the intergrown sulfides are characterized by a narrow range of $\delta^{34}S$ values (-2±2‰) typical of magmatic sulfur in a Tertiary volcanic environment. The barite data fall in a three component mixing triangle (Fig. 25) with a trend to higher $\delta^{34}S$ and $\delta^{18}O$ from north to south in the district (Rye et al. 1988, Barton et al. 2000). The lower apex ($\delta^{18}O$ = 2 ‰; $\delta^{34}S$ = 1‰) in Figure 25 is interpreted as deep sulfate from the hydrothermal system formed by the oxidation of H_2S. The right apex ($\delta^{18}O$ = 20‰; $\delta^{34}S$ = 35‰) requires that precursor aqueous sulfate went through a bacteriogenic reduction cycle in a sedimentary environment, interpreted to have been the sediments that formed the Creede Formation. The top apex ($\delta^{18}O$ = 8‰; $\delta^{34}S$ = 48‰) represents sulfate that evolved by bacteriogenic reduction but underwent further isotopic evolution by oxygen isotope exchange with ore fluids and thermochemical reduction of sulfate in the presence of organic matter. The isotope data on the barites at Creede are so unusual for a hydrothermal deposit in a Tertiary volcanic environment that they indicated some factor was present at Creede that does not normally exist in similar hydrothermal systems. This unusual isotope data linked the origin of the ore fluids in the district to the stratified saline lake (Rye et al. 2000a) that developed in the moat of the Creede caldera and was a major factor leading to the choice of the Creede Moat by the Continental Scientific Drilling Program (Bethke and Hay 2000).

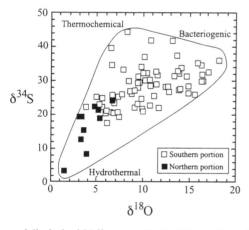

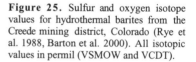

Figure 25. Sulfur and oxygen isotope values for hydrothermal barites from the Creede mining district, Colorado (Rye et al. 1988, Barton et al. 2000). All isotopic values in permil (VSMOW and VCDT).

Mississippi Valley-type Pb-Zn-F deposits typically form in continental settings in low-temperature (<200°C), near-neutral environments in which sulfur, and probably oxygen, isotope disequilibrium would be expected to dominate (Ohmoto and Lasaga 1982, Chiba and Sakai 1985). Thus, stable isotope data from sulfate minerals from Mississippi Valley-type deposits should provide information about the source of sulfate and its geochemical history. Stable isotope studies of Mississippi Valley-type deposits are dominated by sulfur isotope data, with a limited amount of oxygen isotope data (Ault and Kulp 1960, Ohmoto 1986, Kaiser et al. 1987, Richardson et al. 1988, Kesler et al. 1994, Appold et al. 1995, Kesler 1996, Misra et al. 1996, Jones et al. 1996). Most data are from barite, but some sulfur isotope values have been reported for anhydrite, gypsum, and celestine ($SrSO_4$).

Both sulfate sulfur and sulfide sulfur in Mississippi Valley-type environments can be derived from a variety of sources including organically bound sulfur, H_2S reservoir gas, evaporites, connate seawater, and diagenetic sulfides. In all cases, these sources represent seawater sulfate that has followed various geochemical pathways which impart different isotopic fractionations. The reduction of sulfate occurs either through bacterially mediated processes or abiotic thermochemical processes. Bacterial sulfate reduction, as discussed above, can produce sulfate-sulfide fractionations that typically range from 15 to 60 ‰ (Goldhaber and Kaplan 1975), whereas those associated with abiotic thermochemical reactions with organic compounds range from nil to up 10 ‰ (Orr 1974, Kiyosu 1980). Bacterial sulfate reduction has been documented at temperatures up to 110°C (Jørgensen et al. 1992), but the optimum temperature range is between 30 and 40°C. Ohmoto and Goldhaber (1997) argued that at the site of ore deposition, thermochemical reduction is not effective at T <125°C because of slow reaction kinetics. For thermochemical reduction to be an important process in Mississippi Valley-type environments, reduction must occur away from the site of ore deposition, in the deeper, hotter parts of the basin. It should be noted that although the kinetic fractionations associated with both reduction processes are distinct, they can produce H_2S of similar compositions if bacterial sulfate reduction occurs quantitatively (or nearly so) in an environment with little or no Fe to sequester the sulfide.

Stable isotope and fluid-inclusion studies by Richardson et al. (1988) of samples from the Deardorff mine from the Cave-in-Rock fluorspar district, Illinois, indicate mineralization was dominated by two formation fluids recharged by meteoric waters, one of which circulated into the basement rocks under low water/rock conditions. Liquid hydrocarbons are present in fluid inclusion in most minerals. The low $\delta^{34}S$ of sulfides (4.0 to 8.9 ‰ for sphalerite) indicates a significant contribution of H_2S from petroleum sources. These values are completely out of equilibrium with those for late stage barites which range from about 57 to 103 ‰, with some crystals showing systematic increases from the core to the edge of the crystals. Sulfate $\delta^{18}O$ values however, range from only 19.6 to 20.8 ‰ and were apparently in equilibrium with the hydrothermal fluids. These data suggest that the aqueous sulfate was derived from a small fluid reservoir in which the residual seawater sulfate underwent thermal chemical reduction with organic matter. Supporting evidence includes the decrease in the $\delta^{13}C_{CO_2}$ of the fluids during carbonate deposition.

Sulfur isotope data from Mississippi Valley-type deposits suggest two major sulfide reservoirs, one centered between -5 to 15 ‰ and one greater than 20 ‰ (Fig. 26). The higher values of sulfides typically coincide with those of the composition of associated sulfate minerals, and have been interpreted to reflect the minimal fractionation associated with abiotic thermochemical reduction (Kesler 1996). However, similar compositions of sulfide could be generated by closed-system, quantitative bacterial reduction of sulfate. A carbonate aquifer is an ideal environment for such a geochemical process due to the lack of reactive Fe to scavenge and fractionate sulfur. The lower values may reflect H_2S derived either directly or indirectly from open-system bacterial reduction of sulfate. Kesler et al. (1994) proposed that low $\delta^{34}S$ H_2S was derived from oil in the deeper parts of the basin for the Central Tennessee and Kentucky Mississippi Valley-type districts. This H_2S ultimately would have been derived from the bacterial reduction of sulfate. The H_2S from both bacterial and abiotic reduction is not in sulfur isotope equilibrium with associated sulfate minerals (Fig. 26).

Oxygen isotope data for barites from the Southeast Missouri barite district mimics the sulfur isotope data in defining hypogene ($\delta^{18}O$ = 17.0 to 19.6 ‰, $\delta^{34}S$ = 20.5 to 40.8 ‰) and supergene fields ($\delta^{18}O$ = 9.0 ‰, $\delta^{34}S$ = 15.8 ‰; Kaiser et al. 1987). Because of the extremely slow kinetics of oxygen and sulfur isotope exchange between SO_4 and H_2O and H_2S, respectively, at 100°C, the inferred temperature of mineralization, these compositions undoubtedly reflect of the sources of the sulfate. The hypogene barite is probably

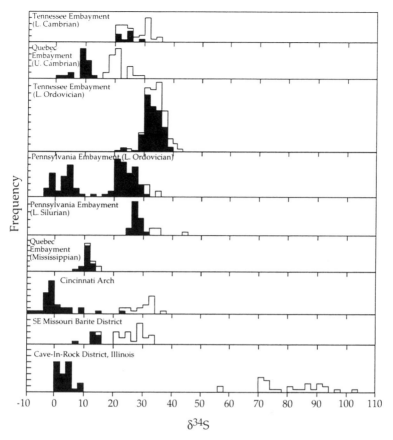

Figure 26. Sulfur isotope histograms for Mississippi Valley-type hydrothermal systems in North America. Data from sphalerite are shown in black; data from sulfate minerals are shown in white. All isotopic values in permil (VCDT).

dominated by sulfate ultimately derived from seawater, whereas the supergene barite probably represents a mixture of dissolved hypogene barite, and sulfate generated by the weathering of ore sulfide minerals.

Metamorphism of sulfate minerals

Metamorphism of assemblages containing a sulfate mineral and another sulfur-bearing mineral, such as pyrite, may result in sulfur isotopic exchange. Isotopic re-equilibration may involve both heterogeneous (mineral-mineral or mineral-fluid) exchange reactions, and homogeneous (aqueous SO_4 and H_2S) exchange reactions. Key aspects of the metamorphism of sulfate minerals are illustrated here with examples from the Barite Hill gold deposit in the Carolina slate belt, South Carolina (Seal et al., in press), the Åsen deposit, Skellefte district, Sweden (Rickard et al. 1979), and the Balmat-Edwards Zn-Pb deposits, New York (Whelan et al. 1984).

At Barite Hill, barite and pyrite occur as both separate lenses and as intergrown massive lenses hosted by metavolcanic and metasedimentary rocks and have been interpreted to have formed by seafloor exhalative processes during Late Proterozoic time (Seal et al., in press). Massive barite forms blades up to 3 cm in length and is associated

Seal, Alpers & Rye

with quartz. Grains of massive pyrite are generally less than 2 mm in diameter. In intergrown assemblages, pyrite grains, less than 1 mm, are surrounded by barite, quartz, and silicate gangue. The sulfur isotope compositions of the barite and pyrite vary systematically with sulfur-bearing mineral assemblages. The $\delta^{34}S$ values of massive barite cluster tightly between 25.0 and 28.0 ‰, and are almost identical to Late Proterozoic seawater values; the $\delta^{34}S$ of massive sulfide ranges from 1.0 to 5.3 ‰, and is consistent with seafloor hydrothermal systems. In contrast, the $\delta^{34}S$ values of finer-grain, intergrown pyrite (5.1 to 6.8 ‰) and barite (21.0 to 23.9 ‰) are higher and lower than their massive counterparts, respectively (Fig. 27). The area of sulfur isotope exchange appears to be on the scale of several millimeters, because massive barite within 1 or 2 cm of barite intergrown with pyrite has $\delta^{34}S$ values that are 3 to 5 ‰ higher than the barite intergrown with pyrite. Intergrown pairs of barite and pyrite yield temperatures from 332 to 355°C, which is consistent with the Ordovician greenschist facies regional metamorphism that affected the deposit (Fig. 28).

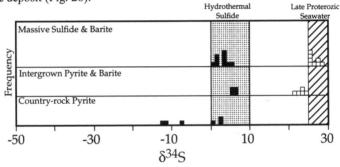

Figure 27. Sulfur isotope histogram for barite and pyrite from the Barite Hill deposit, South Carolina. Modified from Seal et al. (in press). Black symbols are for sulfide minerals; open symbols are for barite. Shaded field represents typical seafloor hydrothermal sulfide compositions; diagonally dashed fields represents Late Proterozoic seawater composition. All isotopic values in permil (VCDT).

At Balmat-Edwards, anhydrite is associated with sphalerite, pyrite, and galena in massive lenses hosted by marble (Whelan et al. 1984). The entire package was metamorphosed to amphibolite facies conditions (~ 600°C and ~ 6.5 kb) at approximately 1 Ga (Bohlen et al. 1985). The $\delta^{34}S$ of ore sulfide minerals ranges from 11.5 to 17.5 ‰, whereas that of associated anhydrite ranges from 25.6 to 27.0 ‰. The $\delta^{34}S$ of evaporitic anhydrite and associated pyrite stratigraphically above the orebodies shows a wider range from 4.2 to 30.2 ‰, and from -31.6 to 20.1 ‰, respectively (Whelan et al. 1984, 1990). Using the fractionation factors in Table 1, anhydrite-pyrite pairs from both the ores and the evaporites yield sulfur isotope temperatures that range from 504 to 595°C (excluding one temperature of 411°C), which are consistent with peak metamorphic conditions, possibly modified by minor amounts of retrograde re-equilibration (Fig. 28). In contrast, temperatures based on sulfur isotopic fractionations for sulfide-mineral pairs (pyrite-sphalerite, sphalerite-galena, pyrite-galena) are much lower, ranging from 251 to 597°C, and have been interpreted to reflect either retrograde re-equilibration to lower temperatures, or a second, lower grade metamorphic event that was not sufficiently intense to reset anhydrite-pyrite pairs (Whelan et al. 1984).

The study of Rickard et al. (1979) of the Åsen deposit in the Skellefte district of Sweden illustrates the potential pitfalls associated with the use of inaccurate fractionation factors. The Åsen deposits consist of massive stratiform and stratabound massive pyrite +

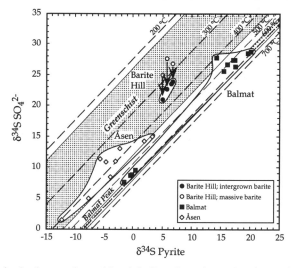

Figure 28. Plot of $\delta^{34}S$ for sulfate minerals versus $\delta^{34}S$ for pyrite from metamorphosed mineral deposits. Peak metamorphic conditions for Balmat are shown in shaded field. Greenschist conditions are shown as stippled field. The change in $\delta^{34}S$ for barite at Barite Hill is shown by the arrows going from massive barite (the presumed starting composition for the intergrown barite) to the intergrown barite. All isotopic values in permil (VCDT).

barite lenses, hosted by 1.8 Ga submarine volcanics rocks. Rickard et al. (1979) noted a correlation between $\delta^{34}S$ values for barite and pyrite and found that the barite-pyrite sulfur isotopic fractionation ($\Delta_{barite-pyrite}$) ranged from 11.7 to 17.1 ‰ with a mean of 13.8 ‰. On the basis of the older sulfate-pyrite fractionation factors (Ohmoto and Rye 1979), a mean apparent temperature of 470°C was calculated. This was concluded to be too high for both the greenschist-facies metamorphic and hydrothermal histories of the deposits; a disequilibrium model was therefore formulated to explain the observed isotopic compositions. The model included the oxidation of volcanic sulfur and bacterial sulfate reduction. However, on the basis of more recent SO_4-H_2S fractionation factors (Ohmoto and Lasaga 1982), the sulfur isotope data of Rickard et al. (1979) yield temperatures between 332 and 465°C (mean = 404°C), which are consistent with greenschist-facies metamorphism (Fig. 28). Nevertheless, many aspects of the Rickard et al. (1979) disequilibrium model for Åsen are necessary to explain the sulfur isotopic variations. However, the correlation between barite and pyrite $\delta^{34}S$ values probably reflects a metamorphic overprint.

Isotopic re-equilibration of sulfate minerals during metamorphism seems to be dependent on a variety of kinetic factors, which probably includes grain size, presence or absence of a fluid phase, presence or absence of recrystallization during metamorphism, and retrograde vs. prograde solubility. Given the appropriate circumstances, isotopic re-equilibration of sulfate minerals seems to occur at metamorphic grades at least as low as greenschist facies.

Surficial environments

Continental evaporites, saline lakes, and ground waters. Stable isotope studies of saline lakes and non-marine evaporites include investigations of Late Cretaceous to Early Tertiary evaporites in Bolivia (Rouchy et al. 1993), Tertiary evaporites in Spain (Birnbaum and Coleman 1979, Utrilla et al. 1992), recent evaporitic gypsum deposits in Namibia (Eckardt and Spiro 1999), modern playa lakes in Australia (Chivas et al. 1991, Alpers et al. 1992), and Searles Lake in California (Holser and Kaplan 1966). The data are dominantly from gypsum, and anhydrite, with some data coming from alunite, jarosite, epsomite $MgSO_4 \cdot 7H_2O$, celestine $SrSO_4$, blödite $Na_2Mg(SO_4)_2 \cdot 4H_2O$, mirabilite $Na_2SO_4 \cdot 10H_2O$, thenardite Na_2SO_4, hanksite $KNa_{22}(SO_4)_9(CO_3)_2Cl$, sulphohalite $Na_6(SO_4)_2FCl$, aphthitalite $(K,Na)_3Na(SO_4)_2$, and burkeite $Na_6(CO_3)(SO_4)_2$.

Non-marine evaporites typically form in continental interiors in rain shadows of major mountain ranges, such as the Great Basin of Nevada, California, Utah, Idaho, Colorado, and northern Arizona in the rain shadow of the Cordillera. The stable isotope characteristics of sulfate from saline lakes and continental (non-marine) evaporites reflect sources and processes within the local catchments in addition to regional, climatic influences. Sulfate in these environments typically is derived from a variety of sources, which may include: (1) atmospheric deposition of marine aerosols; (2) dissolution of pre-existing evaporites in the basin; (3) marine salts from marine sedimentary rocks (connate salts); (4) "relict" sea salts left by oceanic regression; (5) evaporated river water; and (6) weathering of sulfide minerals in the basement rocks of the basin (Chivas et al. 1991). Marine aerosols consist of both sea-spray sulfate, with $\delta^{34}S$ = 21.0 ‰ for modern spray (the composition of seawater), and "non-sea-salt" sulfate (NSS). NSS sulfate is dominantly derived from the atmospheric oxidation of organic sulfur compounds such as dimethylsulfide (CH_3SCH_3; DMS), that is produced by marine organisms. Additional non-sea-salt sulfate can be derived from the oxidation of H_2S produced by bacterial sulfate reduction in marine sediments. Dimethylsulfide from the central Pacific has $\delta^{34}S = 15.6\pm3.1$ ‰ (Calhoun et al. 1991). Bulk stratospheric sulfur, which includes oxidized DMS and H_2S, among other compounds, has $\delta^{34}S = 2.6\pm0.3$ ‰ when not perturbed by volcanic inputs (Castleman et al. 1974). As discussed above, H_2S produced from bacterial reduction of modern seawater sulfate should have $\delta^{34}S = -20\pm20$ ‰. Sea-spray droplets are expected to be deposited from rainfall in coastal regions and from low altitude air masses, whereas oxidized biogenic sulfur compounds (i.e. DMS) reach higher altitudes, and consequently are expected to precipitate farther inland (Wakshal and Nielsen 1982, Chivas et al. 1991).

The relative importance of the different sulfate sources varies with study area (Fig. 29). In Australia, the sulfate budget of the non-marine playas appears to be dominated by marine aerosols. Near the coast, the $\delta^{34}S$ is identical to modern seawater values ($\delta^{34}S \approx 21$ ‰), reflecting the sea-salt-spray component, whereas inland, values systematically range down to approximately 15 ‰, which has been interpreted to reflect greater inputs from higher altitude, oxidized DMS (Chivas et al. 1991). More negative, sedimentary basement sources are only locally important. In Namibia, sulfate in gypsum is dominantly derived from oxidized atmospheric DMS, primarily produced offshore in the Benguela Current; oxidation of sulfide-rich ore bodies is only locally important to the sulfate budget, resulting in gypsum $\delta^{34}S$ values of approximately 3 ‰ (Fig. 29; Eckardt and Spiro 1999). In Spain, sulfate in non-marine evaporites was recycled from Triassic and, to a lesser extent, Cretaceous marine evaporites exposed in the basins. Minor modification of isotopic signatures has been attributed to redox cycling associated with bacterial sulfate reduction (Birnbaum and Coleman 1979, Utrilla et al. 1992). Sulfate in Bolivian evaporites was derived from a combination of recycled sulfate in Permo-Triassic marine evaporites and weathering of igneous sulfide minerals in the basins, and may have locally been modified by bacterial sulfate reduction (Rouchy et al. 1993). Sulfate in Searles Lake, California, has been interpreted to be derived from a combination of igneous and atmospheric sources, with subsequent modification by bacterial sulfate reduction (Holser and Kaplan 1966).

The $\delta^{18}O$ of sulfate from these systems in Spain (Utrilla et al. 1992), and Bolivia (Rouchy et al. 1993) ranges from 11.3 to 19.7 ‰, with one value of -0.5 ‰. These values generally reflect the same marine evaporite sources and bacterial sulfate-reduction processes as do the $\delta^{34}S$ data. The $\delta^{18}O$ of alunite and jarosite from acidic, hypersaline Lake Tyrrell in Victoria, Australia, ranges from 22.5 to 24.9 ‰, and represents some of the highest values reported for natural sulfates (Alpers et al. 1992). Alpers et al. (1992) suggested that the anomalously high $\delta^{18}O$ was the result of oxygen isotopic exchange between dissolved SO_4 and H_2O in the ground-water aquifers that feed Lake Tyrrell; they proposed that the $\delta^{18}O$ values of the SO_4 and H_2O reflect between 63 and 89 % equilibration. The exchange was

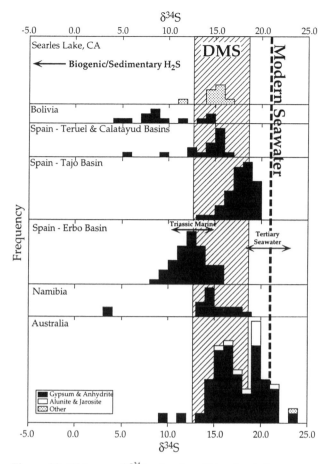

Figure 29. Histogram of $\delta^{34}S$ values for sulfate minerals from non-marine evaporites, and saline lakes. Marine DMS field is from Calhoun et al. (1991). Biogenic and sedimentary sulfide at Searles Lake has an inferred $\delta^{34}S$ below –20 ‰. The ranges of $\delta^{34}S$ for Triassic and Tertiary marine sulfate sources in the Erbo basin are shown by the arrows. All isotopic values in permil (VCDT).

facilitated by long residence times of water, and the enhanced kinetics at the acidic pH (2.9 to 4.0) of the ground waters. Using kinetic data from Lloyd (1968), Alpers et al. (1992) suggested that the equilibration state of the SO_4-H_2O oxygen isotopes indicated minimum residence times of 110 to 670 yr at 25°C for sulfate in the regional ground waters. However, extrapolation of the kinetic data of Hoering and Kennedy (1957) and Chiba and Sakai (1985) to pH = 3.0 to 4.0 at 25°C indicates significantly longer (and probably unrealistic) residence times of 17,000 to 400,000 yr to achieve 63 to 89 % equilibrium.

Acid mine drainage. The stable isotope characteristics of acid mine drainage have been reviewed by van Everdingen and Krouse (1988) and Taylor and Wheeler (1994). Dissolved sulfate concentrations can typically range up to 5 g/L, but extreme concentrations up to 760 g/L have been reported where evaporation is important (Nordstrom et al. 2000). In evaporative settings, mine drainage can precipitate a variety of secondary, efflorescent sulfate salts, such as gypsum, chalcanthite $CuSO_4 \cdot 5H_2O$, melanterite $FeSO_4 \cdot 7H_2O$, rozenite $FeSO_4 \cdot 4H_2O$, szomolnokite $FeSO_4 \cdot H_2O$, copiapite $Fe^{II}Fe^{III}_4(SO_4)_6(OH)_2 \cdot 20H_2O$, and halotrichite $Fe^{II}Al_2(SO_4)_4 \cdot 22H_2O$, to name a few (Alpers et al. 1994, Nordstrom and Alpers 1999b).

Sulfate in these systems is dominantly derived from the oxidation of pyrite and other sulfide minerals. The oxidation of pyrite is described by Reactions (20) and (21), which differ in terms of oxidizing agents and sources of oxygen for the resulting sulfate, and impart distinctly different oxygen isotopic signatures, as discussed above. The $\delta^{34}S$ of dissolved SO_4 is typically identical to that of the parent sulfide mineral, because of quantitative, disequilibrium oxidation (Field 1966, Taylor et al. 1984, Taylor and Wheeler 1994). The $\delta^{18}O$ of dissolved sulfate depends on the composition of local water and the mechanism and environment of pyrite oxidation (Toran and Harris 1989, Taylor and Wheeler 1994).

Taylor and Wheeler (1994) discussed the application of plots $\delta^{18}O_{SO_4}$ versus $\delta^{18}O_{H_2O}$ to the interpretation of the stable isotope characteristics of mine drainage. Among the several caveats and qualifiers that bear on the applicability and limitations of this approach, are the assumption that the isotopic fractionations are constant, that isotopic re-equilibration subsequent to the initial formation of sulfate has not occurred, and that oxygen sources participate in a fully molecular pathway in which atoms of oxygen are transferred. However, oxygen may only be involved in electron transfer in a strictly electrochemical pathway. In the use of $\delta^{18}O_{SO_4}$ versus $\delta^{18}O_{H_2O}$ plots, the oxidation of pyrite through Reactions (20) and (21) is represented by graphical depictions of Equation (23), where the percentages of sulfate contribution from Reaction (20) and Reaction (21) are shown (Fig. 14). In Reaction (20), pyrite is oxidized by dissolved oxygen, and 87.5 % of the oxygen is contributed to the resulting sulfate from dissolved oxygen and 12.5 % is derived from water. In Reaction (21), dissolved Fe^{3+} oxidizes pyrite and all of the oxygen in the resulting sulfate is derived from water. Samples that plot near Reaction (20) (0 % isopleth), and consequently have higher $\Delta_{SO_4-H_2O}$ values, are interpreted to represent unsaturated (or wet/dry), oxygenated environments. Samples that plot near Reaction (21) (100 % isopleth), with lower $\Delta_{SO_4-H_2O}$ values, are interpreted to represent submersed, less oxic environments (Fig. 14). The 2.7 ‰ difference in fractionation between dissolved oxygen and sulfate between acidic (ε_{O_2} = -11.4 ‰) and neutral to alkaline (ε_{O_2} = -8.7 ‰) conditions causes minor changes to the locations of isopleths for Reactions (20) and (21) as can be seen from Equation (23).

Stable isotope data are available for the West Shasta district, the Leviathan mine and the Penn mine, California, the Argo Tunnel, Colorado, the Fontana and Hazel Creek mines, North Carolina, and the Sulfur and Cabin Branch mines, Virginia (Fig. 14; Taylor and Wheeler 1994, Hamlin and Alpers 1996, Seal and Wandless 1997). All of these sites are base-metal sulfide deposits, where the primary sulfide minerals are pyrite, or pyrrhotite, with lesser chalcopyrite, sphalerite, galena, and native sulfur. Collectively, the data from these studies illustrate several important points regarding the interpretation and applicability of $\delta^{18}O_{SO_4}$ versus $\delta^{18}O_{H_2O}$ diagrams. The data from these sites spans a range of $\delta^{18}O_{SO_4}$ and $\delta^{18}O_{H_2O}$. The range of $\delta^{18}O_{H_2O}$ values reflects the temperature dependence of the $\delta^{18}O$ of meteoric waters, which can be related to variations in altitude and latitude, and the effect of evaporation, which raises the $\delta^{18}O$ of residual water (Fig. 1). The range of $\delta^{18}O_{SO_4}$ values dominantly reflects the environment of sulfide oxidation. The data for the Leviathan and Sulfur mines, and the Argo Tunnel suggest that Fe^{3+} (Reaction 21) is responsible for greater than 50 % of the sulfide oxidation from these sites, which is consistent with the water saturated nature of much of the mine workings at these sites. In the case of the Fontana and Hazel Creek mines, and parts of the West Shasta district, the data indicate that dissolved oxygen is responsible for between 50 and 75 % of the sulfide oxidation, which is consistent with the location of some parts of these deposits in the unsaturated zone.

Several processes, such as evaporation and dissolution of efflorescent salts, complicate the interpretation of data using these diagrams. Evaporation will shift the $\delta^{18}O$ of waters to more positive values without affecting the $\delta^{18}O$ of the sulfate. The importance of evaporation at a site can be assessed from the δD and $\delta^{18}O$ signatures of the waters; evaporation produces water that has higher δD and $\delta^{18}O$ values than local meteoric water. Evaporation is known to be an important process at the Penn, Sulfur, and Cabin Branch mines (Hamlin and Alpers 1996, Seal and Wandless 1997). The isotopic composition of efflorescent salts should be representative of the environment of initial sulfide oxidation. However, when the salts redissolve, the $\delta^{18}O$ of the waters may or may not be representative of the conditions of initial sulfide oxidation, because of seasonal variations in the $\delta^{18}O$ of meteoric waters due to differences in the temperature of precipitation. Thus, the $\delta^{18}O$ of water associated with dissolved efflorescent salts may be higher or lower than the $\delta^{18}O$ of the water that originally carried the sulfate, potentially leading to erroneous interpretations. In the case of the West Shasta district, the concentrations of some of the waters from Iron Mountain are the most extreme ever recorded, with pH values as low as -3.6, SO_4 concentrations up to 760 g/L, dissolved metal concentrations in excess of 160 g/L, and temperatures up to 47°C (Nordstrom et al. 2000). These conditions enhance the kinetics of oxygen isotope exchange between SO_4 and H_2O. At 50°C, SO_4 in equilibrium with a water with $\delta^{18}O = -10$ ‰ should have $\delta^{18}O = 15.4$ ‰. Alpers et al. (1996) reported $\delta^{18}O$ values in aqueous sulfate and efflorescent salts from Iron Mountain ranging from 10 to 15 ‰, indicating partial to total isotopic equilibration with the highly acidic mine waters. Extrapolation of the data from Hoering and Kennedy (1957) and Chiba and Sakai (1985) indicates that an approach to 60 % of isotopic equilibrium can be achieved in only 15 hours at pH = -4 and 1.5 years at pH = 0 (Fig. 12).

Paleoclimate studies. Stable isotope studies of sulfate minerals have received limited application to paleoclimate studies. Arehart and O'Neil (1993) investigated the continental paleoclimate record of supergene alunites from Nevada, and Khademi et al. (1997) investigated the paleoclimate associated Iranian gypsums. Alunite and jarosite are ideally suited for paleoclimate studies; K within the structure is amenable to K/Ar and $^{40}Ar/^{39}Ar$ dating and clear stable isotope criteria exist for distinguishing supergene origins alunite and jarosite from other origins (Rye et al. 1992, Arehart et al. 1992, Arehart and O'Neil 1993, Rye and Alpers 1997). Arehart and O'Neil (1993) used the δD of supergene alunites as an indicator of ancient meteoric waters over the past 30 Ma in Nevada; the results were then related to temperature variations due to the changing temperatures of precipitation. There appears to be little or no hydrogen isotopic fractionation between alunite and water at ambient conditions (Bird et al. 1989). Arehart and O'Neil (1993) were able to document changes in the δD compositions of ancient meteoric waters, and by inference, mean annual temperatures of precipitation, that ranged to higher and lower values than present-day conditions (Fig. 30).

Rye et al. (2000b) used a combination of isotopic ages and stable isotope data on supergene alunite and jarosite formed by the weathering of the Creede deposits to trace the evolution of the chemical and hydrologic processes that affected ancient oxidized acid ground waters, as well as the details of climate history and geomorphic evolution in the San Juan Mountains. Fine grained (>1μm to <10 μm) supergene alunite and jarosite occur in minor fractures in the upper, oxidized parts of the 25 Ma sulfide-bearing veins of the Creede mining district, and jarosite also occurs in adjacent oxidized Ag-bearing clastic sediments. K/Ar ages for alunite range from 4.8 to 3.1 Ma, and for jarosite from 2.6 to 0.9 Ma. The δD values for alunite and jarosite shows opposite correlations with elevation, and values for jarosite correlate with age. Calculated δD_{H_2O} values of alunite fluids approach,

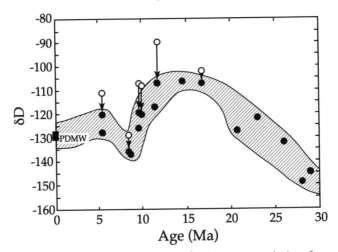

Figure 30. Secular variations in the δD of supergene alunites from Nevada. Modified from Arehart and O'Neil (1993). White values with arrows are corrected for latitude (Arehart and O'Neil 1993). PDMW indicates present-day meteoric water compositions. More negative values are inferred to represent cooler climates, and less negative values are inferred to represent warmer climates. All isotopic values in permil (VSMOW).

but are higher than, those of present-day meteoric water. Calculated δD$_{H_2O}$ values for jarosite fluids are more variable; the values of the youngest jarosite are lowest and are similar to those of present day meteoric water in the district. The narrow range for δD-δ^{18}O$_{SO_4}$ values of alunite reflects oxidation of sulfide below the water table. The greater range in these values for jarosite reflects oxidation of sulfide under vadose conditions. Ages of alunite can be used to mark the position of the paleo-water table at the end of a period of moderate erosion from 25 to 5 Ma that exposed the tops of the ore-bodies to oxidation. The younger jarosite formed in the vadose zone during or following canyon cutting related to regional uplift of the southern Rocky Mountains. The δD values suggest that climate in the area was similar to that of the present day prior to regional uplift, but went through a warm period before returning to present conditions during or after regional uplift.

Khademi et al. (1997) measured the δD and δ^{18}O of hydration water in gypsum to gain insights into the waters associated with gypsum formation. Their data demonstrated that evaporation of brines was the most important process resulting in gypsum precipitation. It was also found that an evaporation trend for the inferred "mother" waters intersected the meteoric water line at compositions lower than modern meteoric water compositions. This fact was used to suggest that climates in the Quaternary were more moist and (or) cooler than modern climates, or that Quaternary climate patterns were dramatically different from modern patterns.

SUMMARY

The stable isotope characteristics of dissolved sulfate and sulfate minerals are typically dominated by the kinetic rates of exchange reactions among sulfate, H$_2$O, and H$_2$S, although other less abundant species may also be important. The oxidation state of sulfur species plays a dominating role in causing large sulfur isotope fractionations, with oxidized species generally having a higher δ^{34}S values than reduced species. The kinetic rates of the exchange reactions are strongly dependent on temperature and pH. At high temperature and low pH, equilibrium dominates and results from abiotic processes. Due to slower kinetics, high-temperature isotopic signatures are usually preserved during cooling. Thus, stable isotope data commonly provide a record of ambient conditions, such as temperature, and fluid compositions.

In contrast, at lower temperature and less acidic pH, disequilibrium, mediated by microbial processes, prevails. Under these conditions, stable isotope data tend to record information about the source of components and the nature of (bio)geochemical processes acting on the system. Isotopic differences among the various geologically important, terrestrial sulfur reservoirs arose as a result of redox-dependent fractionation of the bulk Earth sulfur early in geological time. Under appropriate redox conditions, sulfate-reducing bacteria and iron- and sulfur-oxidizing bacteria played important roles, which are inadequately understood at present. Disequilibrium processes are responsible for the secular variations observed in the marine record. Reduced sulfur from the bulk Earth is the major reservoir, followed by seawater sulfate.

Key elements of our knowledge that appear to have the most significant impact on the interpretation of the stable isotope geochemistry of sulfate include revisions to our understanding of SO_4-H_2S sulfur isotope fractionations (Ohmoto and Lasaga 1982), and persisting uncertainties in the rates of oxygen isotope exchange between dissolved SO_4 and H_2O. Discrepancies between temperatures calculated from the $\delta^{34}S$ of sulfate-sulfide pairs using previous fractionation factors versus those of Ohmoto and Lasaga (1982) range from nil at 200°C to more than 50°C at temperatures above 300°C (Fig. 9). This discrepancy has led several researchers to formulate elaborate geochemical interpretations because calculated sulfur-isotope temperatures could not be reconciled with other geological and geochemical constraints.

The temperature and pH dependence of oxygen isotope exchange rates between dissolved sulfate and water from the experimental studies of Lloyd (1968) and Chiba and Sakai (1985) are in stark contrast to one another; the rates of Lloyd (1968) are several orders of magnitude faster than those of Chiba and Sakai (1985). Probably because of their earlier publication date, the data from Lloyd (1968) have served as a foundation for much of the interpretation of oxygen isotope data from sulfates. However, recalculation of the experimental fluid compositions of Hoering and Kennedy (1957) in terms of pH by the present study permits comparisons with the data of Lloyd (1968) and Chiba and Sakai (1985). Agreement found between the studies of Hoering and Kennedy (1957) and Chiba and Sakai (1985) suggests, although not conclusively, that the exchanges rates from these studies are more accurate than those of Lloyd (1968). Accordingly, many of the fundamental aspects of the interpretation of the oxygen isotope geochemistry of sulfates may need to be re-evaluated.

ACKNOWLEDGMENTS

The preparation of this manuscript benefited from discussions with JK Böhlke, Tyler Coplen, Wayne Goodfellow, Carol Ptacek, Pat Shanks, and Rob Zierenberg. The manuscript also benefited from reviews by Phil Bethke and Jane Hammarstrom. Adam Johnson and Greg Wandless assisted in literature searches.

REFERENCES

Alpers CN, Nordstrom DK, White LD (1988) Solid solution properties and deuterium fractionation factors for hydronium-bearing jarosites from acid mine waters. EOS, Trans Am Geophys Union 69:1480-1481
Alpers CN, Rye RO, Nordstrom DK (1996) Stable isotope systematics of S and O in aqueous and mineral sulfates from hyper-acid environments [abstr]. Chapman Conference on crater lakes, terrestrial degassing, and hyper-acid fluids in the environment. Crater Lake, Oregon, Sept. 4-9, 1996, Am Geophys Union, Washington, DC

Alpers CN, Rye RO, Nordstrom DK, White LD, King B-S (1992) Chemical, crystallographic and stable isotopic properties of alunite and jarosite from acid-hypersaline Australian lakes. Chem Geol 96:203-226

Alpers CN, Blowes DW, Nordstrom DK, Jambor JL (1994) Secondary minerals and acid mine drainage. In Environmental Geochemistry of Sulfide Mine-Wastes. Jambor JL, Blowes DW (Eds) Mineral Assoc Canada Short Course Handbook 22:247-270

Alt JC, Anderson TF, Bonnell L (1989) The geochemistry of sulfur in a 1.3-km section of hydrothermally altered oceanic crust, DSDP 504B. Geochim Cosmochim Acta 53:1011-1023

Appold MS, Kesler SE, Alt JC (1995) Sulfur isotope and fluid inclusion constraints on the genesis of Mississippi Valley-type mineralization in the Central Appalachians. Econ Geol 90:902-919

Arehart GB, O'Neil JR (1993) D/H ratios of supergene alunite as an indicator of paleoclimate in continental settings. In Climate change in continental isotopic records. Swart PK, Lohmann KC, McKenzie JA, Savin S (Eds) Geophys Monogr 78:277-284

Arehart GB, Kesler SE, O'Neil JR, Foland KA (1992) Evidence for the supergene origin of alunite in sediment-hosted micron gold deposits, Nevada. Econ Geol 87:263-270

Ault WU, Kulp JL (1960) Sulfur isotopes and ore deposits. Econ Geol 55:73-100

Ault WU, Jensen ML (1963) Summary of sulfur isotope standards. In Biogeochemistry of Sulfur Isotopes. Jensen ML (Ed) Nat Sci Foundation, Symp Proc, Yale University, New Haven, CT

Bachinski DJ (1969) Bond strength and sulfur isotopic fractionation in coexisting sulfides. Econ Geol 64:56-65

Baertschi P (1976) Absolute ^{18}O content of Standard Mean Ocean Water. Earth Planet Sci Lett 31:341-344

Banks NG (1982) Sulfur and copper in magma and rocks. In Advances in Geology of the Porphyry Copper Deposits in Southwestern North America. Titley SR (Ed) University of Arizona Press, Tucson, Arizona, p 227-258

Barton PB, Rye RO, Bethke PM (2000) Evolution of the Creede caldera and its relation to mineralization in the Creede mining district, Colorado. In Ancient Lake Creede, its volcano-tectonic setting, sedimentation history, and relation to mineralization in the Creede mining district. Bethke PM, Hay RL (Eds) Geol Soc Am Spec Paper 346:301-326

Bahr JR (1976) Sulfur isotopic fractionation between H$_2$S, S and SO$_4^{2-}$ in aqueous solutions and possible mechanisms controlling isotopic equilibrium in natural systems. MSc thesis, Pennsylvania State University

Bernard A, Demaiffe D, Mattielli N, Punongbayan RS (1991) Anhydrite-bearing pumices from Mount Pinatubo: Further evidence for the existence of sulphur-rich silicic magmas. Nature 354:139-140

Berner RA (1985) Sulphate reduction, organic matter decomposition and pyrite formation. Phil Trans R Soc London A 315:25-38

Bethke PM, Hay RL (2000) Overview: Ancient Lake Creede, its volcano-tectonic setting, sedimentation history, and relation to mineralization in the Creede mining district. In Ancient Lake Creede, its volcano-tectonic setting, sedimentation history, and relation to mineralization in the Creede mining district. Bethke PM, Hay RL (Eds) Geol Soc Am Spec Paper 346:1-8

Betts RH, Voss RH (1970) The kinetics of oxygen exchange between sulfite ion and water. Canad J Chem 48:2035-2041

Bigeleisen J (1952) The effects of isotopic substitution on the rates of chemical reactions. J Phys Chem 56:823-828

Bigeleisen J, Mayer MG (1947) Calculation of equilibrium constants for isotopic exchange reactions. J Chem Phys 15:261-267

Bigeleisen J, Perlman ML, Prosser HC (1952) Conversion of hydrogenic materials for isotopic analysis. Anal Chem 24:1356

Bird MI, Andrew AS, Chivas AR, Lock D (1989) An isotopic study of surficial alunite in Australia: 1. Hydrogen and sulfur isotopes. Geochim Cosmochim Acta 53:3223-3238

Birnbaum SJ, Coleman M (1979) Source of sulphur in the Ebro basin (northern Spain) Tertiary nonmarine evaporite deposits as evidenced by sulphur isotopes. Chem Geol 25:163-168

Bischoff JL, Seyfried WE (1978) Hydrothermal chemistry of seawater from 25° to 350°C. Am J Sci 278:838-860

Bluth GJS, Rose WI, Sprod IE, Krueger AJ (1997) Stratospheric loading of sulfur from explosive volcanic eruptions. J Geol 105:671-683

Bohlen SR, Valley JW, Essene EJ (1985) Metamorphism in the Adirondacks. I. petrology, pressure, and temperature. J Petrol 26:971-992

Borthwick J, Harmon RS (1982) A note regarding ClF$_3$ as an alternative to BrF$_5$ for oxygen isotope analysis. Geochim Cosmochim Acta 46:1665-1668

Brenninkmeijer CAM, Kraft P, Mook WG (1983) Oxygen isotope fractionation between CO$_2$ and H$_2$O. Isotope Geosci 1:181-190

Burnham CW, Ohmoto H (1980) Late-stage process of felsic magmatism. Soc Mining Geol Japan Spec Issue 8:1-11

Çagatay MN, Eastoe CJ (1995) A sulfur isotope study of volcanogenic massive sulfide deposits of the Eastern Black Sea province, Turkey. Mineral Deposita 30:55-66

Calhoun JA, Bates TS, Charlson RJ (1991) Sulfur isotope measurements of submicrometer sulfate aerosol particles over the Pacific Ocean. Geophys Res Lett 18:1877-1880

Carmody RW, Plummer LN, Busenberg E, Coplen TB (1998) Methods for collection of dissolved sulfate and sulfide and analysis of their sulfur isotopic composition. US Geol Survey Open-File Report 97-234

Carroll MR, Rutherford MJ (1987) The stability of igneous anhydrite: Experimental results and implications for sulfur behavior in the 1982 El Chichón trachyandesite and other evolved magmas. J Petrol 28:781-801

Carroll MR, Rutherford MJ (1988) Sulfur speciation in hydrous experimental glasses of varying oxidation state: results from measured wavelength shifts of sulfur X-rays. Am Mineral 73:845-849

Castleman AW Jr, Munkelwitz HR, Manowitz B (1974) Isotopic studies of the sulfur component of the stratospheric aerosol layer. Tellus 26:222-234

Chambers LA, Trudinger PA (1979) Microbiological fractionation of stable sulfur isotopes: A review and critique. Geomicrobiol J 1:249-293

Chiba H, Sakai H (1985) Oxygen isotope exchange rate between dissolved sulfate and water at hydrothermal temperatures. Geochim Cosmochim Acta 49:993-1000

Chiba H, Kusakabe M, Hirano S-I, Matsuo S, Somiya S (1981) Oxygen isotope fractionation factors between anhydrite and water from 100 to 550°C. Earth Planet Sci Lett 53:55-62

Chivas AR, Andrew AS, Lyons WB, Bird MI, Donnelly TH (1991) Isotopic constraints on the origin of salts in Australian playas: 1. Sulphur. Palaeogeog Palaeoclim Palaeoecol 84:309-332

Claypool GE, Holser WT, Kaplan IR, Sakai H, Zak I (1980) The age curves of sulfur and oxygen isotopes in marine sulfate and their mutual interpretations. Chem Geol 28:199-260

Clayton RN (1986) High temperature effects in the early solar system. Rev Mineral 16:129-140

Clayton RN, Mayeda TK (1963) The use of bromine pentafluoride in the extraction of oxygen from oxides and silicates for isotopic analysis. Geochim Cosmochim Acta 27:43-52

Cole DR, Ohmoto H (1986) Kinetics of isotopic exchange at elevated temperatures and pressures. Rev Mineral 16:41-90.

Coleman ML, Moore MP (1978) Direct reduction of sulfates to sulfur dioxide for isotopic analysis. Anal Chem 50:1594-1595

Coleman ML, Sheppard TJ, Durham JJ, Rouse JE, Moore GR (1982) Reduction of water with zinc for hydrogen isotope analysis. Anal Chem 54:993-995

Cortecci G (1974) Oxygen isotopic ratios of sulfate ions-water pairs as a possible geothermometer. Geothermics 3: 60-64.

Craig H (1961) Isotopic variations in meteoric waters. Science 133: 1702-1703.

Craig H, Gordon LI (1965) Deuterium and oxygen-18 variations in the ocean and marine atmosphere. Symposium on marine geochemistry, Narragansett Marine Laboratory, Univ Rhode Island Publ 3:277-374

Criss RE, Taylor HP Jr. (1986) Meteoric-hydrothermal systems. Rev Mineral 16:373-424

Cunningham CG, Rye RO, Bethke PM, Logan MAV (1997) Vein alunite formed by degassing of an epizonal stock. Eos, Trans 78:F746

Dansgaard W (1964) Stable isotopes in precipitation. Tellus 16:436-468

Dansgaard W, Johnsen SJ, Møller J, Langway CC Jr. (1969) One thousand centuries of climatic record from Camp Century on the Greenland ice sheet. Science 166:377-381

Deen, JA (1990) Hydrothermal ore deposition related to high-level igneous activity: a stable isotope studies of the Julcani mining district, Peru. Unpublished PhD thesis, Univ Colorado, Boulder, 246 p

Deen JA, Rye RO, Munoz JL, Drexler JW (1994) The magmatic hydrothermal system at Julcani, Peru: Evidence from fluid inclusions and hydrogen and oxygen isotopes. Econ Geol 89:1924-1938

Dole M, Lane GA, Rudd DP, Zaukelies DA (1954) Isotopic composition of atmospheric oxygen and nitrogen. Geochim Cosmochim Acta 6:65-78

Eastoe CJ (1983) Sulfur isotope data and the nature of the hydrothermal systems at the Panguna and Frieda porphyry copper deposits, Papua New Guinea. Econ Geol 78:201-213

Eastoe CJ, Gustin MM (1996) Volcanogenic massive sulfide deposits and anoxia in the Phanerozoic oceans. Ore Geol Rev 10:179-197

Ebert SW, Rye RO (1997) Secondary precious metal enrichment by steam-heated fluids in the Crofoot-Lewis hot spring gold-silver deposit and relation to paleoclimate. Econ Geol 92:578-600

Eckardt FD, Spiro B (1999) The origin of sulphur in gypsum and dissolved sulphate in the Central Namib desert, Namibia. Sed Geol 123:255-273

Elsgaard L, Jørgensen BB (1992) Anoxic transformations of radiolabeled hydrogen sulfide in marine and freshwater sediments. Geochim Cosmochim Acta 56:2425-2435

Epstein S, Mayeda T (1953) Variations in the O^{18} content of waters from natural sources. Geochim Cosmochim Acta 4:213-224

Field CW (1966) Sulfur isotopic method for discriminating between sulfates of hypogene and supergene origin. Econ Geol 61:1428-1435

Field CW, Gustafson LB (1976) Sulfur isotopes in the porphyry copper deposit at El Salvador, Chile. Econ Geol 71:1533-1548

Fontes JC, Gonfiantini R (1967) Fractionnement isotopique de l'hydrogéne dans l'eau de cristallisation du gypse. Acad Sci Comptes Rendus 265:4-6

Fournelle J (1990) Anhydrite in Nevado del Ruiz November pumice: Relevance to the sulfur problem. J Volcanol Geoth Res 42:189-201

Friedman I (1953) Deuterium content of natural water and other substances. Geochim Cosmochim Acta 4:89-103

Fritz P, Basharmal GM, Drimmie RJ, Ibsen J, Qureshi RM (1989) Oxygen isotope exchange between sulphate and water during bacterial reduction of sulphate. Chem Geol 79:99-105

Garlick GD (1969) The stable isotopes of oxygen. In Handbook of Geochemistry, 8B. Wedepohl KH (Ed) Springer, Heidelberg, New York

Garrels RM, Thompson ME (1960) Oxidation of pyrite by iron sulfate solutions. Am J Sci 258A:57-67

Gavelin S, Parwel A, Ryhage R (1960) Sulfur isotope fractionation in sulfide mineralization. Econ Geol 55:510-530

Gemmell JB, Large RR (1992) Stringer system and alteration zones underlying the Hellyer volcanic-hosted massive sulfide deposit, Tasmania, Australia. Econ Geol 87:620-649

Giesemann A, Jäger H-J, Norman AL, Krouse HR, Brand WA (1994) On-line sulfur-isotope determination using an elemental analyzer coupled to a mass spectrometer. Anal Chem 66:2816-2819

Giggenbach W, Gonfiantini R, Panichi C (1983a) Geothermal systems. Guidebook on Nuclear Techniques in Hydrology. International Atomic Energy Agency Tech Report Ser 91:359-379

Giggenbach WF, Gonfiantini R, Jangi BL, Truesdell AH (1983b) Isotopic and chemical composition of Parbati Valley geothermal discharges, north-west Himalaya, India. Geothermics 12:199-222

Goldhaber MB (1983) Experimental study of metastable sulfur oxyanion formation during pyrite oxidation at pH 6-9 and 30°C. Am J Sci 283: 193-217.

Goldhaber MB, Kaplan IR (1975) Controls and consequences of sulfate reduction rates in recent marine sediments. Soil Sci 119:42-55

Gonfiantini R, Fontes JC (1963) Oxygen isotopic fractionation in the water of crystallization of gypsum. Nature 200:644-646

Goodfellow WD, Jonasson IR (1984) Ocean stagnation and ventilation defined by $\delta^{34}S$ secular trends in pyrite and barite, Selwyn Basin, Yukon. Geology 12:583-586

Goodfellow WD, Franklin JM (1993) Geology, mineralogy, and chemistry of sediment-hosted clastic massive sulfides in shallow cores, Middle Valley, northern Juan de Fuca Ridge. Econ Geol 88:2037-2068

Green GR, Solomon M, Walshe JL (1981) The formation of the volcanic-hosted massive sulfide ore deposit at Rosebery, Tasmania. Econ Geol 76:304-338

Hageman R, Nief G, Roth E (1970) Absolute isotopic scale for deuterium analysis of natural waters. Absolute D/H ratio for SMOW. Tellus 22: 712-715

Hamlin SN, Alpers CN (1996) Hydrogeology and geochemistry of acid mine drainage in ground water in the vicinity of Penn Mine and Camanche Reservoir, Calaveras County, California: Second-year summary. US Geol Survey Water-Resources Invest Report 96-4257, 44 p

Haur A, Hladikova J, Smejkal V (1973) Procedure of direct conversion of sulfates into SO_2 for mass spectrometric analysis of sulfur. Isotopenpraxis 18: 433-436

Hayes JM, Lambert IB, Strauss H (1992) The sulfur-isotopic record. In The Proterozoic Biosphere. Schopf JW, C Klein C (Eds) Cambridge Univ Press, 129-132

Heinzinger K (1969) Ein wasserstoffisotopieeffekt im $CuSO_4 \cdot 5H_2O$. Z Naturforschung 24a:1502-1511

Hoering TC, Kennedy JW (1957) The exchange of oxygen between sulfuric acid and water. J Am Chem Soc 79: 56-60

Holland HD (1965) Some applications of thermochemical data to problems of ore deposits, II. Mineral assemblages and the composition of ore-forming fluids. Econ Geol 60:1101-1166

Holland HD (1978) The Chemistry of the atmosphere and oceans. John Wiley and Sons, New York, 351 p

Holser WT, Kaplan IR (1966) Isotope geochemistry of sedimentary sulfates. Chem Geol 1:93-135

Holser WT, Kaplan IR, Sakai H, Zak I (1979) Isotope geochemistry of oxygen in the sedimentary sulfate cycle. Chem Geol 25:1-17

Holt BD, Engelkemeier AG (1970) Thermal decomposition of barium sulfate to sulfur dioxide for mass spectrometric analysis. Anal Chem 42:1451-1453

Holt BD, Kumar R, Cunningham PT (1981) Oxygen-18 study of the aqueous-phase oxidation of sulfur dioxide. Atmos Environ 15:557-566

Horita J, Wesolowski DJ, Cole DR (1993) The activity-composition relationship of oxygen and hydrogen isotopes in aqueous salt solutions: I. Vapor-liquid water equilibration of single salt solutions from 50 to 100°C. Geochim Cosmochim Acta 57:2797-2817

Horita J, Cole DR, Wesolowski DJ (1995) The activity-composition relationship of oxygen and hydrogen isotopes in aqueous salt solutions: III. Vapor-liquid water equilibration of NaCl solutions to 350°C. Geochim Cosmochim Acta 59:1139-1151

Hulston JR, Thode HG (1965a) Variations in the S^{33}, S^{34}, and S^{36} contents of meteorites and their relation to chemical and nuclear effects. J Geophys Res 70:3475-3484

Hulston JR, Thode HG (1965b) Cosmic-ray produced S^{33} and S^{36} in the metallic phase of iron meteorites. J Geophys Res 70:4435-4442

Huston DL (1999) Stable isotopes and their significance for understanding the genesis of volcanic-associated massive sulfide deposits: A review. *In* Volcanic-associated Massive Sulfide Deposits: Processes and examples in modern and ancient settings. Barrie CT, Hannington MD (Eds) Rev Econc Geol 8:157-179

Igumnov SA, Grinenko VA, Poner NB (1977) Temperature dependence of the distribution coefficient of sulfur isotopes between H_2S and dissolved sulfates in the temperature range 260-400°C. Geokhimiya 7:1085-1087

Janecky DR, Shanks WC III (1988) Computational modeling of chemical and sulfur isotopic reaction processes in seafloor hydrothermal systems: chimneys, massive sulfides, and subjacent alteration zones. Can Mineral 26:805-825

Jones HD, Kesler SE, Furman FC, Kyle JR (1996) Sulfur isotope geochemistry of southern Appalachian Mississippi Valley-type deposits. Econ Geol 91:355-367

Jørgensen BB, Isaksen MF, Jannasch HW (1992) Bacterial sulfate reduction above 100°C in deep-sea hydrothermal vent sediments. Science 258:1756-1757

Kaiser CJ, Kelly WC, Wagner RJ, Shanks WC III (1987) Geologic and geochemical controls on mineralization in the Southeast Missouri Barite district. Econ Geol 82:719-734

Kaplan IR, Hulston JR (1966) The isotopic abundance and content of sulfur in meteorites. Geochim Cosmochim Acta 30:479-496

Kesler SE (1996) Appalachian Mississippi Valley-type deposits: Paleoaquifers and brine provinces. Soc Econ Geol Spec Pub 4:29-57

Kesler SE, Jones HD, Furman FC, Sassen R, Anderson WH, Kyle JR (1994) Role of crude oil in the genesis of Mississippi Valley-type deposits: Evidence from the Cincinnati Arch. Geology 22:609-612

Khademi H, Mermut AR, Krouse HR (1997) Isotopic composition of gypsum hydration water in selected landforms from central Iran. Chem Geol 138:245-255

Kiyosu Y (1980) Chemical reduction and sulfur-isotope effects of sulfate by organic matter under hydrothermal conditions. Chem Geol 30:47-56

Kornexl BE, Gehre M, Höfling R, Werner RA (1999) On-line $\delta^{18}O$ measurement of organic and inorganic substances. Rapid Commun Mass Spectrom 13:1685-1693

Kowalik J, Rye RO, Sawkins FJ (1981) Stable-isotope study of the Buchans, Newfoundland, polymetallic sulphide deposit. Geol Assoc Canada Spec Paper 22:229-254.

Kroopnick P, Craig H (1972) Atmospheric oxygen: isotopic composition and solubility fractionation. Science 175:54-55

Krouse HR (1980) Sulphur isotopes in our environment. *In* Handbook of Environmental Isotope Geochemistry I. The terrestrial environment. Fritz P, Fontes J-Ch (Eds.), Elsevier, Amsterdam, p 435-472.

Krouse HR, Ueda A, Campbell FA (1990) Sulfur isotope abundances in coexisting sulphate and sulphide: Kinetic isotope effects *versus* exchange phenomena. *In* Stable Isotopes and Fluid Processes in Mineralization. Herbert HK, Ho SE (Eds) Univ of Western Australia, Univ Extension Pub 23:226-243

Kusakabe M (1974) Sulfur isotopic variations in nature, 10. Oxygen and sulphur isotope study of Wairakei (New Zealand) geothermal well discharges. New Zealand J Sci 17:183-191

Kusakabe M, Robinson BW (1977) Oxygen and sulfur isotope equilibria in the $BaSO_4\text{-}HSO_4^-\text{-}H_2O$ system from 110 to 350°C with applications. Geochim Cosmochim Acta 41:1033-11040

Kusakabe M, Chiba H (1983) Oxygen and sulfur isotope composition of barite and anhydrite from the Fukazawa deposit, Japan. Econ Geol Monogr 5:292-301

Lambert IB, Donnelly TH (1990) The paleoenvironmental significance of trends in sulphur isotope compositions in the Precambrian: A critical review. *In* Stable Isotopes and Fluid Processes in Mineralization. Herbert HK, Ho SE (Eds) Univ of Western Australia, Univ Extension Pub 23:260-268

Lawrence JR, Taylor HP Jr. (1971) Deuterium and oxygen-18 correlation: clay minerals and hydroxides in Quaternary soils compared to meteoric waters. Geochim Cosmochim Acta 35:993-1004

Leggett JK (1980) British Lower Paleozoic black shales and their palaeo-oceanographic significance. J Geol Soc London 137:139-156

Leuth VW, Rye RO, Peters L (2000) Hydrothermal "sour gas" jarosite: Ancient and modern acid sulfate mineralization events in the Rio Grande rift. Geol Soc Am Abstracts with Programs (in press)

Lloyd RM (1967) Oxygen-18 composition of oceanic sulfate. Science 156:1228-1231

Lloyd RM (1968) Oxygen isotope behavior in the sulfate-water system. J Geophys Res 73:6099-6110

Longinelli A, Craig H (1967) Oxygen-18 variations in sulfate ions in sea water and saline lakes. Science 156:56-59

Luhr JF (1990) Experimental phase relations of water- and sulfur-saturated arc magmas and the 1982 eruptions of El Chichón volcano. J Petrol 31:1071-1114

Luhr JF, Carmichael ISE, Varenkamp JC (1984) The 1982 eruptions of El Chichón volcano, Chiapas, Mexico: Mineralogy and petrology of the anhydrite-bearing pumices. J Volcanol Geotherm Res 23:69-108

Macnamara J, Thode HG (1950) Comparison of the isotopic constitution of terrestrial and meteoritic sulphur. Phys Rev 78:307

Martinez-S. RG, Jacquier B, Arnold M (1996) The δ^{34}S composition of sulfates and sulfides at the Los Humeros geothermal system, Mexico and their application to physicochemical fluid evolution. J Volcanol Geotherm Res 73:99-118

Matsuo S, Kusakabe M, Niwano M, Hirano T, Oki Y (1985) Origin of thermal waters from the Hakone geothermal system, Japan. Geochem J 19:27-44

Maynard JB, Okita PM (1991) Bedded barite deposits in the United States, Canada, Germany, and China: two major types based on tectonic setting. Econ Geol 86:364-376

McGoldrick PJ, Large RR (1992) Geologic and geochemical controls on gold-rich stringer mineralization in the Que River deposit, Tasmania. Econ Geol 87:667-685

McKenzie WF, Truesdell AH (1977) Geothermal reservoir temperatures estimated from the oxygen isotope compositions of dissolved sulfate and water from hot springs and shallow drillholes. Geothermics 5:51-61

McKibben MA, Barnes HL (1986) Oxidation of pyrite in low temperature acidic solutions: rate laws and surface textures. Geochim Cosmochim Acta 50:1509-1520

McKibben MA, Eldridge CS, Reyes AG (1996) Sulfur isotopic systematics of the June 1991 Mount Pinatubo eruptions: A SHRIMP ion microprobe study. In Fire and mud—eruptions and lahars of Mount Pinatubo, Philippines. Newhall CG, Punongbayan RS (Eds) PHIVOLCS and Univ Washington Press, 825-843

McKibben MA, Riciputi LR (1998) Sulfur isotopes by ion microprobe. In Applications of microanalytical techniques to understanding mineralizing processes. McKibben MA, Shanks WC III, Ridley WI (Eds) Reviews in Economic Geology 7:121-139

Misra KC, Gratz JF, Lu C (1996) Carbonate-hosted Mississippi Valley-type mineralization in the Elmwood-Gordonsville deposits, Central Tennessee zinc district: A synthesis. Soc Econ Geol Spec Pub 4:58-73

Mizutani Y (1972) Isotopic composition and underground temperature of the Otake geothermal water, Kyushu, Japan. Geochem J 6:67-73

Mizutani Y, Rafter TA (1969a) Oxygen isotopic composition of sulphates, Part 3: Oxygen isotopic fractionation in the bisulphate ion-water system. New Zealand J Sci 12:54-59

Mizutani Y, Rafter TA (1969b) Oxygen isotopic composition of sulphates, Part 4: Bacterial fractionation of oxygen isotopes in the reduction of sulphate and in the oxidation of sulphur. New Zealand J Sci 12:60-68

Mizutani Y, Rafter TA (1973) Isotopic behavior of sulphate oxygen in the bacterial reduction of sulphate. Geochem J 6:183-191

Moses CO, Nordstrom DK, Herman JS, Mills AL (1987) aqueous pyrite oxidation by dissolved oxygen and ferric iron. Geochim Cosmochim Acta 51:1561-1571

Nakai N, Jensen ML (1964) The kinetic isotope effect in the bacterial reduction and oxidation of sulfur. Geochim Cosmochim Acta 28:1893-1912

Nielsen H, Pilot J, Grinenko LN, Grinenko VA, Lein AY, Smith JW, Pankina RG (1991) Lithospheric sources of sulphur. In Stable Isotopes in the Assessment of Natural and Anthropogenic Sulphur in the Environment. Krouse HR, Grinenko VA (Eds) SCOPE 43, J Wiley and Sons, New York, 65-132

Nordstrom DK, Alpers CN (1999a) Geochemistry of acid mine waters. In The Environmental Geochemistry of Mineral Deposits, Part A: Processes, techniques, and health issues. Plumlee GS, Logsdon MJ (Eds) Rev Econ Geol 6A:133-160

Nordstrom DK, Alpers CN (1999b) Negative pH, efflorescent mineralogy, and consequences for environmental restoration at the Iron Mountain Superfund site, California. Proc Nat Acad Sci 96:3455-3462

Nordstrom DK, Alpers CN, Ptacek CJ, Blowes DW (2000) Negative pH and extremely acidic mine waters from Iron Mountain, California. Environ Sci Techn 34:254-258

Nriagu JO, Ress CE, Mekhtiyeva VL, Lein AY, Fritz P, Drimmie RJ, Pankina RG, Robinson RW, Krouse HR (1991) Hydrosphere. *In* Stable Isotopes in the Assessment of Natural and Sulphur in the Environment. Krouse HR, Grinenko VA (Eds) SCOPE 43, John Wiley and Sons, New York, p 177-265

Nuelle LM, Shelton KL (1986) Geologic and geochemical evidence of possible bedded barite deposits in Devonian rocks of the Valley and Ridge Province, Appalachian Mountains. Econ Geol 81:1408-1430

Ohmoto, H (1972) Systematics of sulfur and carbon isotopes in hydrothermal ore deposits. Econ Geol 67:551-578

Ohmoto, H (1986) Stable isotope geochemistry of ore deposits. Rev Mineral 16:185-225

Ohmoto H (1992) Biogeochemistry of sulfur and the mechanisms of sulfide-sulfate mineralization in Archean oceans. *In* Early Organic Evolution: Implications for Mineral and Energy Resources. Schidlowski M et al. (Eds) Springer-Verlag, p 378-397

Ohmoto H, Rye RO (1974) Hydrogen and oxygen isotopic compositions of fluid inclusions in the Kuroko deposits, Japan. Econ Geol 69: 947-953

Ohmoto H, Rye RO (1979) Isotopes of sulfur and carbon. *In* Geochemistry of hydrothermal ore deposits. Barnes HL (Ed) J Wiley and Sons, New York, p 509-567

Ohmoto H, Lasaga AC (1982) Kinetics of reactions between aqueous sulfates and sulfides in hydrothermal systems. Geochim Cosmochim Acta 46:1727-1745

Ohmoto H, Felder RP (1987) Bacterial activity in the warmer, sulphate-bearing, Archaean oceans. Nature 328:244-246

Ohmoto H, Goldhaber MB (1997) Sulfur and carbon isotopes. *In* Geochemistry of Hydrothermal Ore Deposits. Barnes HL (Ed) J Wiley and Sons, New York, p 517-611

O'Neil JR (1986) Theoretical and experimental aspects of isotopic fractionation. Rev Mineral 16:1-40

O'Neil JR, Clayton RN, Mayeda TK (1969) Oxygen isotope fractionation in divalent metal carbonates. J Phys Chem 51:5547-5558

O'Neil JR, Silberman ML (1974) Stable isotope relations in epithermal Au-Ag deposits. Econ Geol 69:902-909

Orr WL (1974) Changes in sulfur content and isotopic ratios of sulfur during petroleum maturation—study of Big Horn Paleozoic oils. Am Assoc Petrol Geol Bull 58:2295-2318

Oudin E, Thisse Y, Ramboz C (1984) Fluid inclusion and mineralogical evidence for high-temperature saline hydrothermal circulation in the Red Sea metalliferous sediments: preliminary results. Mar Mining 5:3-31

Paytan A, Kastner M, Campbell D, Thiemens MH (1998) Sulfur isotopic composition of Cenozoic seawater sulfate. Science 282:1459-1462

Peter JM, Shanks WC III (1992) Sulfur, carbon, and oxygen isotope variations in submarine hydrothermal deposits of the Guaymas Basin, Gulf of California, USA. Geochim Cosmochim Acta 56:2025-2040

Pickthorn WJ, O'Neil JR (1985) [18]O relations in alunite minerals: potential single-mineral geothermometer. Geol Soc Am Abstracts with Program 17:689

Plummer LN, Parkhurst DL, Fleming GW, Dunkle SA (1988) A computer program incorporating Pitzer's equations for calculation of geochemical reactions in brines. US Geol Survey Water-Resources Invest Report 88-4153

Pottorf RJ, Barnes HL (1983) Mineralogy, geochemistry and ore genesis of hydrothermal sediments from the Atlantis II Deep, Red Sea. Econ Geol Monogr 5:198-223

Postgate JR (1984) The Sulfate-Reducing Bacteria. 2nd Ed., Cambridge University Press, Cambridge, UK

Rafter TA (1967) Oxygen isotopic compositions of sulphates, Part 1: A method for the extraction of oxygen and its quantitative conversion to carbon dioxide for isotopic radiation methods. New Zealand J Sci 10:493-515

Rees CE, Holt BD (1991) The isotopic analysis of sulphur and oxygen. *In* Stable Isotopes in the Assessment of Natural and Anthropogenic Sulphur in the Environment. Krouse HR, Grinenko VA (Eds) SCOPE 43, J Wiley and Sons, New York, p 43-64

Rees CE, Jenkins WJ, Monster J (1978) The sulphur isotope geochemistry of ocean water sulphate. Geochim Cosmochim Acta 42:377-382

Richardson CK, Rye RO, Wasserman MD (1988) The chemical and thermal evolution of the fluids in the Cave-in-Rock fluorspar district, Illinois: stable isotope systematics at the Deardorff mine. Econ Geol 83:765-783

Rickard DT, Zweifel H, Donnelly TH (1979) Sulfur isotope systematics in the Åsen pyrite-barite deposits, Skellefte district, Sweden. Econ Geol 74:1060-1068

Rimstidt JD (1997) Gangue mineral transport and deposition. *In* Geochemistry of Hydrothermal Ore Deposits. Barnes HL (Ed) J Wiley and Sons, New York, p 487-515

Robinson BW (1973) Sulfur isotope equilibrium during sulfur hydrolysis at high temperatures. Earth Planet Sci Lett 18:443-450

Robinson BW (1978) Sulphate-water and -H_2S isotopic thermometry in the New Zealand geothermal systems. US Geol Survey Open-file Report 78-701:354-356

Ross GM, Bloch JD, Krouse HR (1995) Neoproterozoic strata of the southern Canadian Cordillera and the isotopic evolution of seawater sulfate. Precambrian Res 73:71-99

Rouchy JM, Camoin G, Casanova J, Deconinck JF (1993) The central palaeo-Andean basin of Bolivia (Potosi area) during the late Cretaceous and early Tertiary: Reconstruction of ancient saline lakes using sedimentological, palaeoecological and stable isotope records. Palaeogeogr Palaeoclim Palaeoecol 105:179-198

Rozanski K, Araguás-Araguás L, Gonfiantini R (1993) Isotopic patterns in modern global precipitation. In Climate Change in Continental Isotopic Records. Swart PK, Lohmann KC, McKenzie JA, Savin S (Eds) Geophys Monogr 78:1-36

Rye RO (1966) The carbon, hydrogen, and oxygen isotopic composition of the hydrothermal fluids responsible for the lead-zinc deposits at Providencia, Zacatecas, Mexico. Econ Geol 61:1399-1427

Rye RO (1993) The evolution of magmatic fluids in the epithermal environment: the stable isotope perspective. Econ Geol 88:733-753

Rye RO, Alpers CN (1997) The stable isotope geochemistry of jarosite. US Geol Survey Open-file Report 97-88, 28 p

Rye RO, Ohmoto H (1974) Sulfur and carbon isotopes and ore genesis: A review. Econ Geol 69:826-842

Rye RO, Stoffregen RE (1995) Jarosite-water oxygen and hydrogen isotope fractionations: preliminary experimental data. Econ Geol 90:2336-2342

Rye RO, Stoffregen RE, Bethke PM (1990) Stable isotope systematics and magmatic hydrothermal processes in the Summitville, CO, gold deposit. US Geol Survey Open-File Report 90-626, 31 p

Rye RO, Shawe DR, Poole FG (1978) Stable isotope studies of bedded barite at East Northumberland Canyon in Toquima Range, central Nevada. US Geol Survey J Res 6:221-229.

Rye RO, Luhr JF, Wasserman MD (1984) Sulfur and oxygen isotope systematics of the 1982 eruptions of El Chichón volcano, Chiapas, Mexico. J Volcanol Geotherm Res 23:109-123

Rye RO, Plumlee GS, Barton PM, Barton PB (1988) Stable isotope geochemistry of the Creede, Colorado, hydrothermal system. US Geological Survey Open-File Report 88-356, 40 p

Rye RO, Bethke PM, Wasserman MD (1992) The stable isotope geochemistry of acid sulfate alteration. Econ Geol 87:225-262

Rye RO, Bethke PM, Finkelstein DB (2000a) Stable isotope evolution and paleolimnology of Ancient Lake Creede. In Ancient Lake Creede, its volcano-tectonic setting, sedimentation history, and relation to mineralization in the Creede mining district. Bethke PM, Hay RL (Eds) Geol Soc Am Spec Paper 346:233-266

Rye RO, Bethke PM, Lanphere MA, Steven TA (2000b) Neogene geomorphic and climatic evolution of the central San Juan Mountains, CO: K/Ar age and stable isotope data on supergene alunite and jarosite from the Creede mining district. In Ancient Lake Creede, its volcano-tectonic setting, sedimentation history, and relation to mineralization in the Creede mining district. Bethke PM, Hay RL (Eds) Geol Soc Am Spec Paper 346:95-104

Sakai H (1968) Isotopic properties of sulfur compounds in hydrothermal processes. Geochem J 2:29-49

Sakai H (1977) Sulfate-water isotope thermometry applied to geothermal systems. Geothermics 5:67-74

Sakai H (1983) Sulfur isotope exchange rate between sulfate and sulfide and its application. Geothermics 12:111-117

Sakai H, Krouse HR (1971) Elimination of memory effects in $^{18}O/^{16}O$ determinations in sulphates. Earth Planet Sci Lett 11:369-373

Sakai H, Matsubaya O (1977) Stable isotopic studies of Japanese geothermal systems. Geothermics 5:97-124

Sakai H, Dickson FW (1978) Experimental determination of the rate and equilibrium fractionation factors of sulfur isotope exchange between sulfate and sulfide in slightly acid solutions at 300°C and 1000 bars. Earth Planet Sci Lett 39:151-161

Sakai H, Takenaka T, Kishima N (1980) Experimental study of the rate and isotope effect in sulfate reduction by ferrous oxides and silicates under hydrothermal conditions. Proc 3rd Int'l Symp Water-Rock Interact, Edmonton, Alberta, p 75-76

Sakai H, Des Marais DJ, Ueda A, Moore JG (1984) Concentrations and isotope ratios of carbon, nitrogen, and sulfur in ocean-floor basalts. Geochim Cosmochim Acta 48:2433-2442

Schwarcz HP, Harmon RS, Thompson, P, Ford DC (1976) Stable isotope studies of fluid inclusions in speleothems and their paleoclimatic significance. Geochim Cosmochim Acta 40:657-665

Seal RR II, Rye RO (1992) Stable isotope study of water-rock interaction and ore formation, Bayhorse base and precious metal district, Idaho. Econ Geol 87:271-287

Seal RR II, Rye RO (1993) Stable isotope study of fluid inclusions in fluorite from Idaho: implications for continental climates during the Eocene. Geology 21:219-222

Seal RR II, Wandless GA (1997) Stable isotope characteristics of waters draining massive sulfide deposits in the eastern United States. *In* Fourth Int'l Symp Environ Geochem Proc. Wanty RB, Marsh SP, Gough LP (Eds) US Geol Survey Open-File Report 97-496, 82 p

Seal RR II, Ayuso RA, Foley NK, Clark SHB (2000) Sulfur and lead isotope geochemistry of hypogene mineralization at the Barite Hill gold deposit, Carolina Slate Belt, southeastern United States: a window into and through regional metamorphism. Mineral Deposita (in press).

Seccombe PK, Godwin CI, Krouse HR, Juras SJ (1990) Sulphur and lead isotopic studies of the Buttle Lake massive sulphide deposits, Vancouver Island, B.C.: Sources of ore constituents in a submarine exhalative environment. Proc Pacific Rim 90 Congress, Gold Coast, p 419-426.

Seyfried WE, Bischoff JL (1981) Experimental seawater-basalt interaction at 300°C, 500 bars, chemical exchange, secondary mineral formation and implications for the transport of heavy metals. Geochim Cosmochim Acta 45:135-149

Shanks WC III, Bischoff JL (1980) Geochemistry, sulfur isotope composition, and accumulation rates of the Red Sea geothermal deposits. Econ Geol 75:445-459

Shanks WC III, Seyfried WE Jr. (1987) Stable isotope studies of vent fluids and chimney minerals, southern Juan de Fuca Ridge: sodium metasomatism and seawater sulfate reduction. J Geophys Res 92:1387-11399

Shanks WC III, Bischoff JL, Rosenbauer RJ (1981) Seawater sulfate reduction and sulfur isotope fractionation in basaltic systems: interaction of seawater with fayalite and magnetite at 200-350°C. Geochim Cosmochim Acta 45:1977-1995

Shanks WC III, Böhlke JK, Seal RR II (1995) Stable isotopes in mid-ocean ridge hydrothermal systems: interactions between fluids, minerals, and organisms. *In* Seafloor hydrothermal systems: Physical, chemical, biological, and geological interactions. Humphris SE, Zierenberg RA, Mullineaux LS, Thomson RE (Eds) Geophys Monogr 91:194-221

Shelton KL, Rye DM (1982) Sulfur isotopic compositions of ores from Mines Gaspé, Quebec: an example of sulfate-sulfide isotopic disequilibria in ore-forming fluids with applications to other porphyry type deposits. Econ Geol 77:1688-1709.

Sheppard SMF, Nielsen RL, Taylor HP Jr. (1969) Oxygen and hydrogen isotope ratios of clays from porphyry copper deposits. Econ Geol 64: 755-777

Simmons SF, Stewart MK, Robinson BW, Glover RB (1994) The chemical and isotopic compositions of thermal waters at Waimangu, New Zealand. Geothermics 23:539-553

Singer PC, Stumm W (1970) Acid mine drainage: the rate-determining step. Science 167:1121-1123

Sofer Z, Gat JR (1975) The isotope composition of evaporating brines: effect of the isotopic activity ration in saline solutions. Earth Planet Sci Lett 26:179-186

Solomon M, Rafter TA, Jensen ML (1969) Isotope studies on the Rosebery, Mount Farrell and Mount Lyell ores, Tasmania. Mineral Deposita 4:172-199

Solomon M, Eastoe CJ, Walshe JL, Green GR (1988) Mineral deposits and sulfur isotope abundances in the Mount Read Volcanics between Que River and Mount Darwin, Tasmania. Econ Geol 83:1307-1328

Stewart MK (1974) Hydrogen and oxygen isotope fractionation during crystallization of mirabilite and ice. Geochim Cosmochim Acta 38:167-172

Stoffregen RE (1993) Stability relations of jarosite and natrojarosite at 100-250°C. Geochim Cosmochim Acta 57:2417-2429

Stoffregen RE, Rye RO, Wasserman MD (1994) Experimental studies of alunite: I. ^{18}O-^{16}O and D-H fractionation factors between alunite and water at 250-450°C. Geochim Cosmochim Acta 58:903-916

Strauss H (1993) The sulfur isotopic record of Precambrian sulfates: new data and a critical evaluation of the existing record. Precambrian Res 63:25-246

Taylor BE (1986) Magmatic volatiles: isotopic variation of C, H, and S. Rev Mineral 16:185-225

Taylor BE, Wheeler MC (1994) Sulfur- and oxygen-isotope geochemistry of acid mine drainage in the western United States. *In* Environmental Geochemistry of Sulfide Oxidation. Alpers CN, Blowes DW (Eds) Am Chem Soc Symp Ser 550:481-514

Taylor BE, Wheeler MC, Nordstrom DK (1984) Stable isotope geochemistry of acid mine drainage: experimental oxidation of pyrite. Geochim Cosmochim Acta 48:2669-2678

Taylor BE, O'Neil JR (1977) Stable isotope studies of metasomatic skarns and associated metamorphic and igneous rocks, Osgood Mountains, Nevada. Contrib Mineral Petrol 63:1-49

Taylor HP Jr. (1997) Oxygen and hydrogen isotope relationships in hydrothermal mineral deposits. *In* Geochemistry of Hydrothermal Ore Deposits. Barnes HL (Ed) J Wiley and Sons, New York, 229-302

Thiemens MH (1999a) Mass-independent isotope effects in planetary atmospherics and the early solar system. Science 283:341-345

Thiemens MH (1999b) Mass independent isotopic composition of atmospheric molecules. Late abstract added to 9th Goldschmidt Conf Abstract with Programs

Thode HG, Monster J (1965) Sulfur-isotope geochemistry of petroleum, evaporites, and ancient seas. Am Assoc Petrol Geol Mem 4:367-377

Toran L, Harris RF (1989) Interpretation of sulfur and oxygen isotopes in biological and abiological sulfide oxidation. Geochim Cosmochim Acta 53:2341-2348

Trudinger PA, Chambers LA (1973) Reversibility of bacterial sulfate reduction and its relevance to isotope fractionation. Geochim Cosmochim Acta 37:1775- 1778

Truesdell AH, Nathenson M, Rye RO (1977) The effects of subsurface boiling and dilution on the isotopic compositions of Yellowstone thermal waters. J Geophys Res 82:3694-3704

Truesdell AH, Hulston JR (1980) Isotopic evidence of environments of geothermal systems. *In* Handbook of Environmental Isotope Geochemistry 1:79-226. Fritz P, Fontes J-Ch (Eds)

Ueda A, Sakai H (1984) Sulfur isotope study of Quaternary volcanic rocks from the Japanese island arc. Geochim Cosmochim Acta 44:579-587

Urey HC (1947) The thermodynamic properties of isotopic substances. J Chem Soc 1947:562-581

Utrilla R, Pierre C, Orti F, Pueyo JJ (1992) Oxygen and sulphur isotope compositions as indicators of the origin of Mesozoic and Cenozoic evaporites from Spain. Chem Geol 102:229-244

Van Everdingen RO, Krouse HR (1988) Interpretation of isotopic compositions of dissolved sulfates in acid mine drainage. US Bur Mines Info Circ 9183:147-156

Van Stempvoort DR, Krouse HR (1994) Controls on $\delta^{18}O$ in sulfate. *In* Environmental Geochemistry of Sulfide Oxidation. Alpers CN, Blowes DW (Eds) Am Chem Soc Symp Ser 550:446-480

Veizer J, Bruckschen P, Pawellek F, Diener A, Podlaha OG, Carden GAF, Jasper T, Korte C, Strauss H, Azmy K, Ala D (1997) Oxygen isotope evolution of Phanerozoic seawater. Palaeogeogr Palaeoclim Palaeoecol 132:159-172

Vennemann TW, O'Neil JR (1993) A simple and inexpensive method of hydrogen isotope and water analyses of minerals and rocks based on zinc reagent. Chem Geol 103:227-234

Wakshal E, Nielsen H (1982) Variations of $\delta^{34}S(SO_4)$, $\delta^{18}O(H_2O)$ and Cl/SO_4 ratio in rainwater over northern Israel, from the Mediterranean coast to Jordan rift valley and Golan Heights. Earth Planet Sci Lett 61:272-282

Wang Z, Li G (1991) Barite and witherite deposits in Lower Cambrian shales of south China: stratigraphic distribution and geochemical characterization. Econ Geol 86:354-363

Wasserman MD, Rye RO, Bethke PM, Arribas A Jr (1992) Methods for separation and total stable isotope analysis of alunite. US Geol Survey Open-File Report 92-9

Watanabe M, Sakai H (1983) Stable isotope geochemistry of sulfates from the Neogene ore deposits in the Green Tuff region, Japan. Econ Geol Monogr 5: 282-291

Way K, Fano L, Scott MR, Thew K (1950) Nuclear data. A collection of experimental values of halflifes, radiation energies, relative isotopic abundances, nuclear moments and cross-sections. Nat Bur Standards US Circ 499

Whelan JF, Rye RO, deLorraine W (1984) The Balmat-Edwards zinc-lead deposits—synsedimentary ore from Mississippi Valley-type fluids. Econ Geol 79: 239-265

Whelan JF, Rye RO, deLorraine W, Ohmoto H (1990) Isotopic geochemistry of a mid-Proterozoic evaporite basin: Balmat, New York. Am J Sci 290:396-424

Wiersma CL, Rimstidt JD (1984) Rates of reaction of pyrite and marcasite with ferric iron at pH 2. Geochim Cosmochim Acta 48:85-92

Yurtsever Y, Gat JR (1981) Atmospheric waters. Int'l Atomic Energy Agency Tech Report 210:103-142

Zak I, Sakai H, Kaplan IR (1980) Factors controlling $^{18}O/^{16}O$ and $^{34}S/^{32}S$ isotope ratios of ocean sulfates, evaporites and interstitial sulfates from modern deep sea sediments. Isotope Marine Chem 339-373

Zierenberg RA, Shanks WC III (1986) Isotopic constraints on the origin of the Atlantis II, Suakin and Valdivia brines, Red Sea. Geochim Cosmochim Acta 50:2205-2214

Zierenberg RA, Shanks WC III (1988) Isotopic studies of epigenetic features in metalliferous sediment, Atlantis II Deep, Red Sea. Can Mineral 26:737-753

Zierenberg RA, Shanks WC III (1994) sediment alteration associated with massive sulfide formation in the Escanaba Trough, Gorda Ridge: the importance of seawater mixing and magnesium metasomatism. US Geol Survey Bull 2022:257-277

Zierenberg RA, Koski RA, Morton JL, Bouse RM, Shanks WC III (1993) Genesis of massive sulfide deposits on a sediment-covered spreading center, Escanaba Trough, southern Gorda Ridge. Econ Geol 88:2069-2098

Index of Mineral Names

Compiled by D. Kirk Nordstrom

Note: *Names in italics are*
non-sulfate minerals.